Practice of Research Methodology in Civil Engineering and Architecture

Amjad Almusaed · Asaad Almssad ·
Ibrahim Yitmen

Practice of Research Methodology in Civil Engineering and Architecture

A Comprehensive Guide

Springer

Amjad Almusaed
Department of Architectural Engineering
University of Basrah
Basrah, Iraq

Asaad Almssad
Karlstad University
Karlstad, Sweden

Ibrahim Yitmen
Jönköping University
Jönköping, Sweden

ISBN 978-3-031-97392-5 ISBN 978-3-031-97393-2 (eBook)
https://doi.org/10.1007/978-3-031-97393-2

This Springer imprint is published by the registered company Springer Nature Switzerland AG
The registered company address is: Gewerbestrasse 11, 6330 Cham, Switzerland

Competing Interests The authors have no competing interests to declare that are relevant to the content of this manuscript.

Contents

Research Design and Methodology

Foundations of Research Methodology

Many people view research as the foundation of scholarly development and pragmatic creativity. Society, science, and technology advancement rely on providing reliable knowledge, critically evaluating existing information, and generating creative ideas across various disciplines and industries. Research methodology, a disciplined, reflective, and systematic approach to addressing significant problems, supports this endeavor. Ensuring that research is methodologically sound, philosophically consistent, and intellectually rigorous relies on this parts foundational knowledge and conceptual orientation. This part aims to assist new researchers in navigating the initial stages of the research process. While practical training heavily emphasizes research techniques, less attention is sometimes given to the fundamental thinking that influences how and why specific approaches are chosen. This foundational reasoning distinguishes meaningful research from mere data collection, rooted in methodological clarity, logical organization, and introspective engagement. A robust conceptual framework enables researchers to conduct relevant and substantial studies. From developing research questions to interpreting results, a systematic rationale guides every decision at the heart of each research endeavor. Understanding research methodology encompasses more than simply selecting tools or techniques; it reflects a respect for the epistemological and ontological foundations of the researcher's perspective. Research design and the credibility and relevance of knowledge produced rely on this understanding. Identifying a precise and relevant field of study poses a significant challenge during the early stages of research. In academic contexts, this often necessitates systematically analyzing existing work to reveal where information is lacking, debatable, or outdated. Recognizing these research gaps is crucial rather than mechanical; it involves identifying underexplored topics, highlighting methodological shortcomings, and analyzing patterns of agreement and disagreement. The objective is to uncover what is missing and to assess the importance of addressing that deficiency. A well-justified research gap highlights the potential contribution of a study to the field and enhances its validity. Once a research gap is identified, the researcher must articulate it as a clearly defined issue. A research problem is a concise statement that reflects what the study seeks to understand, explain, or resolve, rather than merely indicating a subject or area of interest. Developing this issue should be informed by both theoretical insights and empirical observations. It should delineate the study's scope, clarify its relevance, and outline the investigative approach. In this context, the problem statement is the conceptual foundation for the entire research process. Closely tied to problem formulation is articulating the study's goal through consistent research questions, aims, and objectives. These components constitute the operational framework of the study. Research questions transform abstract issues into specific ones that can be examined theoretically or experimentally. Objectives provide a roadmap for the study's progression by breaking the overarching goal into manageable parts. The composition of these elements must demonstrate methodological viability, intellectual clarity, and alignment with the broader aim of the study. Well-structured goals and inquiries ensure a logically organized, focused investigation. The initial presentation of a study project—often found in the title and opening paragraphs—is fundamentally important. A research title must convey the essential aspects of the study, including its focus, scope, and orientation. A compelling title is engaging and informative, informing readers of what to expect and motivating them to delve into the work. The introduction offers the first thorough interaction between the researcher and the audience. It sets the stage through problem contextualization, a statement of the research gap, and a presentation of the goals. More than just a summary, the introduction justifies the study, positions it within relevant literature, and demonstrates its significance. A logical and persuasive beginning enhances the reader's understanding and appreciation of the importance of the research.

Beyond these practical elements lies a more abstract but equally important aspect of the investigation: the philosophical foundations of research. Every research effort operates, either openly or indirectly, on assumptions about the nature of reality and how knowledge about that reality can be acquired. These beliefs are not arbitrary; they are grounded in different philosophical traditions that influence the logic and validity of the research process. Understanding philosophical paradigms such as positivism, interpretivism, constructivism, critical realism, and pragmatism enables researchers to align their investigations with a coherent worldview. Each paradigm reflects specific ontological and epistemological perspectives shaping the types of questions asked, the techniques deemed appropriate, and the valid interpretations. For instance, a study informed by positivism may prioritize measurable factors and search for universal rules; an interpretivist approach would highlight the meanings individuals assign to their experiences. Recognizing these distinctions requires internal consistency within a research endeavor and critically engaging with the broader academic community. Philosophical reflexivity is, therefore, a methodological necessity rather than an academic luxury. It urges researchers to challenge their preconceptions, articulate their positionality, and clarify the reasoning that underlies their methodological choices. Such reflexivity strengthens the theoretical foundation of the study and fosters intellectual integrity. It also enriches the discourse by facilitating more significant interactions among scholars operating under various paradigms. The concepts and techniques outlined in this critical section serve as preliminary guidelines for a specific topic and as enduring competencies for academic research.

The ability to identify critical issues, formulate clear objectives, justify methodological choices, and articulate philosophical perspectives is essential throughout the research lifecycle. These skills empower academics to generate socially meaningful, methodologically robust, innovative knowledge. Moreover, these foundations are crucial when the complexity of social and scientific issues necessitates multidisciplinary and integrative approaches. Researchers must navigate the space between theoretical abstraction and practical application, between disciplinary boundaries and collaborative engagement. Addressing such complexity requires a cohesive research perspective rooted in solid techniques and clear philosophical justifications. In this light, the integrity and impact of the entire study hinge on the initial phases of the research process, which are often regarded as preliminary or preparatory. They shape the results obtained, the data analyzed, the methods used, and the questions posed. They influence the practical application, how colleagues perceive the study, and its contribution to ongoing discussions. Therefore, establishing a robust foundation in research techniques is not only an issue of academic preparation but also a commitment to excellence, responsibility, and quality. This part offers a conceptual framework for engaging with research as a deliberate, thoughtful, and rigorous endeavor rather than merely a procedural guide. It encourages readers to reflect on the nature of knowledge and the researcher's role in its construction. It promotes ethical engagement, methodological transparency, and strategic decision-making, laying the groundwork for the subsequent study design and implementation phases, which will utilize these concepts to develop specific methodological tools. The essential components of research methodology discussed in this book part cultivate the conceptual framework for effective research. They enable the researcher to transition from curiosity to clarity, questions to exploration, and ideas to evidence. In doing so, they produce methodologically rigorous, necessary, relevant, and evolving knowledge.

Introduction to Research Methodology

Abstract The method of inquiry is fundamental to genuine and significant research, converting concepts into practical information that fosters innovation and addresses intricate problems. This chapter thoroughly examines the approaches that support rigorous academic, industrial, and social research. This chapter lays out a complete framework for high-quality research by explaining research methodologies, essential tools like surveys and experiments, and the primary ideas behind how to use them. Come up with exciting research questions, specific hypotheses, and solid research frameworks. Learn how to combine qualitative, quantitative, and mixed-method approaches to properly and deeply explore complicated topics. Reveal the principles of generating morally responsible and replicable outcomes while maintaining the utmost standards of academic honesty. The chapter illustrates how a well-organized research process connects concepts with significant answers, ranging from revolutionary theoretical developments to practical advancements. Examine how research technique propels academic advancements, expedites industry development, and tackles urgent global challenges such as sustainability and social equality. This chapter is a starting point for experienced researchers and people who want to learn more about ethical, reliable, and transformative research. It uses techniques to improve knowledge, encourage new ideas, and help society progress.

1 Introduction

1.1 Essentials of Research Methodology: Rigor, Reliability, and Ethics

Research methodology constitutes a systematic and comprehensive framework that underpins the research process, ensuring the generation of credible and reliable knowledge. It encompasses the theoretical principles, methodological tools, and practical techniques necessary for investigating a research problem in a structured and rigorous manner. As noted by Adeoye (2024), a well-designed research methodology emphasizes the critical importance of a coherent research design,

A. Almusaed et al., *Practice of Research Methodology in Civil Engineering and Architecture*, https://doi.org/10.1007/978-3-031-97393-2_1

which facilitates the formulation of straightforward research questions, the selection of appropriate methods, and the accurate interpretation of findings. Mastery of research methodology is foundational for producing scholarly work of the highest standards, enabling researchers to deliver valid, replicable findings subject to rigorous peer review, thereby contributing to advancing knowledge across disciplines [1]. Research methods denote specific data collection and analysis procedures, such as surveys, experiments, or interviews [2]. In contrast, research methodology includes broader logic and philosophical foundations for selecting particular methods. Research methodology ensures that investigations are anchored in theory and practice, enhancing their rigor and reproducibility. A comprehensive grasp of research techniques is essential for executing significant research. The main objective is to offer a systematic approach to tackling research inquiries or hypotheses [3]. Establishing explicit rules aids researchers in circumventing biases, mistakes, and inconsistencies, resulting in more believable conclusions. In scientific research, following a clearly defined technique guarantees that experiments are replicable, and outcomes are similar across investigations. In the social sciences, methodology offers frameworks for comprehending intricate human behaviors and societal phenomena. According to Chekenyaa and Misraa (2024), research methods can be understood as the systematic tools and processes that graduate students and researchers use to conduct their inquiries, guided by principles of responsibility, ethics, and integrity. They are not only about data collection and analysis but also encompass the standards and practices that ensure research is reliable, credible, and impactful [4]. The scope of the research technique is broad, covering several facets of the research process. The process starts with issue formulation, a fundamental phase in which researchers discern a deficiency in current knowledge or an area for more inquiry. A clearly defined research problem ensures focus and relevance, serving as the cornerstone for all subsequent stages of the research process. Prediger et al. (2021) argue that integrating diverse research approaches is essential to building a robust research foundation, particularly in iterative cycles of topic-specific design research. This combination of methodologies enhances the adaptability and depth of the investigative process, fostering more comprehensive and reliable outcomes [5]. A vital element of research technique is the development of research questions and hypotheses. These function as the guiding framework for the inquiry, dictating the necessary data types and suitable analytical procedures. The methodology includes research design, which entails selecting from many qualitative, quantitative, or hybrid methods [6]. The characteristics of the study challenge determine this selection, the nature of the accessible data, and the desired outcomes. Ethical issues are also encompassed within the domain of research technique. Researchers must guarantee that their methodologies uphold participants' rights, privacy, and dignity [7]. This encompasses acquiring informed permission, preserving confidentiality, and mitigating any damage. In medical research, it is essential to adhere to ethical guidelines to protect participants' welfare while furthering scientific knowledge. Data collection and analysis are crucial aspects of research technique. Researchers must select appropriate methodologies for data collection, including experiments, surveys, interviews, or observational techniques. The selection of analytical tools and frameworks is equally significant. Quantitative studies often depend on statistical analysis, whereas qualitative research may utilize topic or content analysis to extract insights [8]. Moreover, research techniques include the interpretation and presentation of results. Research discoveries must be conveyed successfully through clear and cohesive reports, academic articles, or presentations. This guarantees that the information produced benefits the wider academic or professional community. Research technique is fundamental to every academic pursuit [9]. It offers a systematic research methodology, ensuring reliable, repeatable, and morally valid results. Its extensive range, including all aspects from issue formulation to data interpretation, underscores its significance in promoting rigorous and influential research across all fields.

1.2 Distinguishing Research Methodology from Research Methods

It is essential to differentiate between research techniques and research methodologies. Research methods refer to the instruments or processes utilized for data collection and analysis, such as surveys, experiments, or simulations [10]. Conversely, research methodology denotes the overarching conceptual framework and justification that inform the selection and application of these methodologies. The methodology examines the fundamental inquiries of "why" and "how" research is performed, integrating theoretical frameworks, ethical standards, and procedural strategies that influence the whole research process. In sustainable architecture design, methodologies may encompass energy simulations for structures or lifespan evaluations of materials. The approach will justify using these methodologies, elucidate their connection with research objectives such as advancing sustainability and energy efficiency, and describe their role in producing trustworthy and valid results [11].

1.2.1 The Extent and Importance of Research Methodology

The breadth of the research methodology is extensive, covering each phase of the research journey, from identifying and conceptualizing a problem to collecting, analyzing, and interpreting data, culminating in disseminating findings [12]. A robust methodology guarantees that research results are dependable, valid, and trustworthy. These attributes are crucial for enhancing scholarly understanding, fostering industrial progress, and tackling societal issues. By following systematic and well-structured methodological approaches, individuals contribute to significant advancements that connect theoretical insights with practical applications, promoting progress across various fields and effectively addressing urgent global challenges [13].

1.2.2 Dependability and Accuracy in Research

A study's reliability may be defined as the degree to which its results remain consistent when repeated under comparable settings [14]. This ensures that the findings can be replicated throughout time and in different surroundings. However, validity ensures that the study appropriately addresses the topic or phenomena that were supposed to be investigated. To demonstrate the credibility and robustness of research findings, it is essential to possess both reliability and validity [15]. Research methodology provides the theoretical foundations and practical instruments to achieve these objectives while avoiding biases, mistakes, and inconsistencies. The concept of dependability in the context of sustainable architecture may refer to the constant replication of the findings of energy performance modeling across various structure projects. Validity guarantees that such models anticipate energy consumption and environmental implications in an accurate manner, which in turn drives significant breakthroughs in environmentally responsible design [16].

1.3 Significance Across Disciplines

The significance of research techniques extends beyond academia, impacting industrial innovation and social advancement.

1.3.1 Within Academia

An effective research technique strengthens theoretical frameworks, fosters interdisciplinary collaboration, and generates innovative knowledge [17]. For instance, research strategies in sustainable architectural design facilitate investigations into passive cooling methods; renewable energy integration; and the use of low-carbon, locally sourced materials. These approaches not only address specific challenges in architecture but also have broader implications for sustainable development. By adopting meticulous methodologies, researchers ensure their findings are robust, credible, and applicable to diverse contexts. This rigor enhances academic discourse and equips practitioners with actionable insights, bridging the gap between theory and real-world application while advancing the global sustainability agenda [18].

1.3.2 Within the Industry

Research approaches are pivotal in supporting evidence-based decision-making and improving product development and design strategies [19]. For example, architectural firms often utilize post-occupancy evaluations and environmental performance assessments to refine sustainable construction practices. These methodologies enable compliance with stringent green certification systems such as LEED or

BREEAM and ensure that buildings meet energy efficiency benchmarks and align with occupant needs. Additionally, integrating these research-driven insights into design processes helps companies innovate, reduce operational costs, and enhance the functionality and sustainability of their projects, contributing to a more resilient and environmentally conscious built environment.

1.3.3 Within Society

Research approaches tackle urgent social issues, including reducing the carbon footprint in the built environment and advancing resilient urban design [20]. For example, sustainable architecture design methodologies often integrate community engagement to guarantee that projects are ecologically sustainable and socially inclusive. Involving stakeholders allows academics to harmonize technical advancements with cultural and social factors, facilitating equitable and sustainable urban development. The research technique constitutes the foundation of productive and meaningful research. A systematic framework guarantees that research results are dependable, valid, and widely applicable [21]. Implementing rigorous approaches in sustainable architecture design fosters innovation in developing energy-efficient structures, aiding global initiatives to battle climate change and save resources. A robust research process is essential for tackling intricate problems and furthering knowledge in academia, industry, and society [22]. Table 1 presents a comprehensive analysis of research methodology, including fundamental concepts, study design, tools, and their applications in academic, industrial, and social contexts. It differentiates between methods and techniques, emphasizing ethical issues, data management, and the need for systematic approaches. The table demonstrates the contribution of research approaches to innovation, dependable discoveries, and resolving intricate global concerns by integrating theoretical frameworks and practical implementations.

2 Types and Classifications of Research

Research may be organized to create new information, resolve existing issues, or improve understanding in a particular field [28]. Researchers can solve various academic and practical difficulties thanks to its categories and classifications, which are decided by the study's aims, methods, and applications. The following three fundamental categories of research are discussed in this chapter: basic research versus applied research, qualitative research versus mixed-methods research, and research classifications by purpose; examples from the field of smart, sustainable buildings are incorporated into the discussion to explain the principles being discussed.

Table 1 A tabular analysis of the whole research methodology framework

Topic	Subtopic	Details	Examples of applications
Research methodology [23]	Core principles	• Framework for systematic research • Emphasizes rigor, ethics, and reliability • Differentiates between research methods and methodology	• Methodology anchors research in theory and practice • Guides the design and ethical considerations for experiments or studies
	Importance	• Ensures credible, valid, and reproducible findings • Provides theoretical and practical tools	• Increases scholarly impact by enabling rigorous peer review • Advances knowledge across academic disciplines
Research design [24]	Elements	• Problem formulation • Selection of methods • Interpretation of findings	• Combines qualitative, quantitative, or hybrid methods based on the research objective • Focuses on specific issues like sustainability in architecture
	Ethical considerations	• Upholds participants' rights, privacy, and dignity • Includes informed consent and confidentiality	• Adhering to medical ethics in clinical research • Mitigating participant harm in social science surveys
Research techniques [3]	Data collection	• Methods: surveys, interviews, experiments, or observations • Ensures reliability and minimizes bias	• Using observational techniques to understand human behavior • Conducting energy performance modeling in sustainable architecture
	Data analysis	• Quantitative: statistical techniques • Qualitative: content or thematic analysis	• Quantitative example: regression analysis for social behavior studies • Qualitative example: content analysis of interviews on urban design
	Results interpretation	• Focuses on transparent and meaningful communication • Presents findings in academic articles, reports, or presentations	• Publishing findings in peer-reviewed journals • Presenting industry innovations at conferences

(continued)

Table 1 (continued)

Topic	Subtopic	Details	Examples of applications
Research methods [25]	Definition	• Specific tools or techniques for conducting research • Encompasses data collection and analysis	• Surveys for public opinion polling • Simulations for predicting environmental outcomes
	Scope	• Focused on "how" research is conducted • Includes systematic processes to address research inquiries	• Using experiments in scientific research to test hypotheses • Employing interviews to gather subjective insights
Significance [26]	Academic impact	• Advances theoretical frameworks • Encourages multidisciplinary collaboration	• Research into low-carbon materials in architecture • Development of frameworks for understanding human cognition in psychology
	Industrial applications	• Supports innovation in product and process development • Provides evidence-based strategies	• Post-occupancy evaluations in architecture to improve building design • Environmental performance assessments to meet certifications like LEED or BREEAM
	Societal contributions	• Addresses social and environmental challenges • Integrates community perspectives for inclusivity	• Urban design integrating stakeholder feedback • Policies for reducing the carbon footprint of the built environment
Methodology vs. methods [27]	Key differences	• Methodology: conceptual framework that guides research choices • Methods: specific procedures for data collection and analysis	• Methodology example: justification of using energy simulations in sustainable design • Methods, for example, conducting interviews for qualitative data
	Integration in research	• Methodology ensures the coherence of methods with objectives • Both works together to achieve reliable and valid research outcomes	Using simulations for lifespan evaluation in architecture, guided by a methodology focusing on sustainability objectives

2.1 *Basic and Applied Research*

2.1.1 Characteristics and Objectives of Basic Research

Regarding extending theoretical knowledge, primary research, also known as fundamental research, is primarily concerned with expanding theoretical knowledge without immediately considering practical applications [29]. For example, in the field of intelligent and environmentally friendly buildings, fundamental research can entail the investigation of the characteristics of advanced nanomaterials utilized for thermal insulation [30]. Such studies are conducted to gain a more in-depth understanding of the behavior of materials under various environmental situations, adding to the theoretical underpinning of energy-efficient construction materials. Fundamental research aims to test previously established hypotheses, discover previously unknown occurrences, and investigate hitherto unexplored domains of knowledge. Although these studies are not intended for immediate application, they frequently set the framework for future advancements. For instance, advancements in the molecular structure of phase-change materials have been utilized to develop smart windows that maximize temperature management within the building [31].

2.1.2 Comparison of Basic and Applied Research

As a means of finding solutions to issues that arise in the actual world, applied research uses concepts or facts that have previously been established [32]. In the topic of smart and sustainable building, for example, practical research may involve designing and testing a building management system that integrates devices connected to the Internet of Things with renewable energy sources to maximize energy use. Among the key objectives of practical research in this field instance, energy efficiency improvements, indoor air quality upgrades, and carbon emissions reduction are among the essential aims [33]. An example of an application of applied research is a study that evaluates the effectiveness of integrating solar panels with AI-driven predictive energy management. This is an example of an application of applied research defined by its focus on practical applications and finding answers.

2.2 *A Comparison of Research that is Both Basic and Applied*

There is a strong connection between fundamental and applied research, which together provide a potent engine for the progress and innovation of scientific knowledge. The discovery of fundamental truths through basic research pushes the frontiers of knowledge and lays the groundwork for discoveries that can completely reshape the world [34]. Applied research aims to transform these findings into solutions that can be immediately implemented, therefore tackling significant obstacles and propelling society's progress. The practical needs of applied research frequently led to new questions and directions for fundamental research, which fuels a constant cycle of discovery and innovation. This dynamic interplay does not just occur in one direction. To ensure that research continues to be both visionary and decisive in expanding human potential, they develop a robust framework in which practical demands refine theoretical discoveries [35]. For example, applied research may then use these sensors to produce intelligent heating, ventilation, and air conditioning (HVAC) systems that dynamically adapt themselves based on the occupancy levels within the building and the weather conditions outside. This interaction demonstrates how theoretical developments influence practical implementations and vice versa.

2.3 Qualitative, Quantitative, and Mixed-Methods Research

There are three primary techniques for examining complex phenomena: qualitative research, quantitative research, and mixed-methods research [6]. Each approach brings a distinct set of skills to the broader research process. In qualitative research, complex and context-dependent topics are investigated using non-numerical data. This study provides profound knowledge of the researcher's experiences, actions, and perceptions. Using this methodology, researchers can capture subtle and subjective aspects that are frequently unavailable using quantitative methods. Quantitative research, on the other hand, emphasizes systematic collection and statistical analysis of numerical data to identify patterns, correlations, and trends [36]. This method is beneficial for putting hypotheses to the test, determining the values of variables, and delivering findings that may be generalized. This approach provides empirical precision and objectivity, which helps to promote the generation of resilient and reproducible outcomes. Combining the benefits of qualitative and quantitative research approaches, the mixed-methods approach makes it possible to gain a thorough grasp of both qualitative and quantitative approaches. Combining comprehensive, contextual insights with quantitative, empirical data, mixed-methods research can bridge the gap between descriptive depth and analytical precision [37]. By using this technique, researchers can more successfully handle problems that include several facets, guaranteeing that both the subjective and objective aspects of a study problem are thoroughly investigated. The combination of these approaches constitutes a flexible framework that may be utilized to further knowledge and handle a variety of scientific and practical difficulties. An example of quantitative research would be a study examining how installing smart thermostats in one hundred different buildings may save some money on energy bills. It is common for these studies to entail collecting data on a massive scale to test hypotheses or discover patterns. A complete knowledge of phenomena may be obtained through mixed-methods research, which blends qualitative and quantitative findings. Mixed-methods research might be conducted in smart buildings, incorporating quantitative energy use data analysis and qualitative interviews with facility managers. The purpose of this study would be to evaluate operational issues and user feedback on the system [38].

2.3.1 Strengths and Limitations

When applied to the setting of smart and ecologically sustainable buildings, each study approach possesses a unique set of advantages and disadvantages. Qualitative research is particularly effective in providing comprehensive and contextual insights into user experiences [39]. This type of research also enables a more in-depth comprehension of subjective impressions, actions, and opportunities. On the other hand, its findings frequently have a restricted capacity for generalization and are susceptible to researcher bias. The opposite of qualitative research is quantitative research, which provides empirical precision and generalizability via the systematic study of numerical data. However, quantitative research may need more contextual richness to capture complex user experiences [40]. These limitations are addressed by a mixed-methods strategy, which integrates the benefits of both approaches to provide a comprehensive view of complex research challenges when applied. One example of this is the identification of reluctance to adopt smart devices owing to privacy concerns. However, it frequently suffers from limitations such as lower sample numbers and the possibility of researcher bias. Quantitative research presents data that can be replicated and generalized, such as a statistical study of energy savings obtained through AI-powered systems [41]. It is possible, however, that it does not consider more subtle social or behavioral issues, such as the unhappiness of tenants with highly automated systems. Mixed-methods research: This type of research amalgamates the advantages of both approaches, such as determining statistical patterns in energy efficiency and investigating stakeholders' experiences [42]. However, it requires qualitative and quantitative research skills and demands significant resources.

2.4 Research Classifications by Purpose

It is also possible to categorize research according to its goal, with exploratory, descriptive, and explanatory research being the primary categories. The field of exploratory research is the subject of this debate.

2.4.1 Exploratory Research

Exploratory research helps locate and comprehend topics that have not been explored extensively or are relatively new. Exploratory research might be conducted in smart, sustainable buildings to study the possibility of using blockchain technology in energy-sharing networks among smart houses [43]. To uncover scalability, security, and regulatory hurdles, researchers must interview technology experts, evaluate developing literature, and analyze early case studies. This study is essential to identify new opportunities and formulate research topics for further studies. For instance, an exploratory examination of user behavior would suggest that renters are more inclined to embrace smart technology when provided with financial incentives [44]. This would motivate more focused follow-up studies to be conducted.

2.4.2 Strengths and Limitations

Because of its adaptability, exploratory research can accommodate newly developing patterns and recognize essential characteristics within the population [45]. For example, research that investigates the application of artificial intelligence in anticipating the energy requirements of buildings could bring to light aspects previously ignored, such as fluctuations in the local climate [46]. On the other hand, their conclusions are frequently tentative and typically lack the statistical rigor required for generalization. When selecting acceptable approaches and attaining relevant results, having a solid understanding of the many types and categories of research is necessary. Basic and applied research address knowledge different ends of the andomdge spectrum, while qualitative, quantitative, and mixed-methods research offer various techniques to collect and interpret data at different points along the spectrum. In addition, categories based on aim, such as exploratory research, offer direction for tackling complex issues. In the context of smart, sustainable buildings, these frameworks make it possible for academics to investigate theoretical developments, develop practical solutions, and handle new difficulties transdisciplinary [47]. Both scholars and practitioners can successfully contribute to creating energy-efficient intelligent buildings that improve user happiness and environmental sustainability. This is accomplished by matching research methodologies with aims. Figure 1 illustrates exploratory, descriptive, and explanatory research, emphasizing the applicability of exploratory research to novel subjects. It highlights applications in intelligent, sustainable structures, including blockchain-enabled energy sharing, while examining strengths and weaknesses. These insights facilitate energy-efficient intelligent buildings, improving user happiness and advancing environmental sustainability.

3 Research Design and Framework

A research design functions as the structural framework for a research study, precisely outlining the approach to address research questions while ensuring coherence and alignment with the study's primary objectives [48]. This framework guarantees that the research produces trustworthy, reliable, and valid outcomes. A study design is an extensive framework that delineates the tactics and procedures employed to gather, quantify, and evaluate data to fulfill the research objectives. It amalgamates theoretical frameworks with practical applications, guaranteeing coherent development from problem identification to result interpretation [29].

A study design fundamentally addresses three essential questions:

- What will be examined?—The research problem and aims.

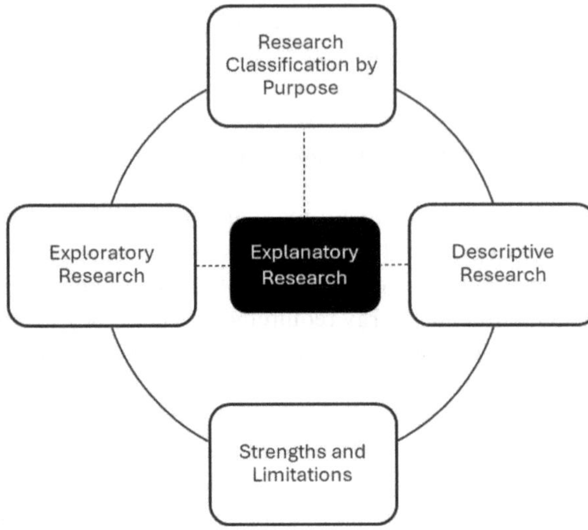

Fig. 1 Research classifications by purpose with focus on exploratory research (*source* authors)

- What methodology will be employed for the study?
- What is the rationale for employing these procedures and techniques? Rationale for the methodology.

3.1 The Function of Research Design

The function of a research design extends beyond simply organizing the research process; it plays a pivotal role in shaping the study's integrity, focus, and overall significance [49]. A well-defined research design serves as the foundation for establishing the study's objectives and scope, clearly delineating its parameters to ensure that all activities are directed toward addressing the primary research question. Providing a structured framework guides critical decisions regarding data collection methods, sample selection, and analytical techniques, maintaining coherence with the study's goals and constraints. Furthermore, a robust research design enhances the validity and reliability of findings by minimizing biases and errors, thereby improving the accuracy and consistency of results across different settings and replications [50]. It also promotes resource efficiency, enabling researchers to systematically allocate time, effort, and financial resources to maximize the study's effectiveness. Significantly, a clear and comprehensive research design facilitates reproducibility, allowing other researchers to replicate the study, verify its results, and contribute to accumulating scientific knowledge. Figure 2 demonstrates the critical function of research design in organizing a survey. Research design fundamentally delineates objectives, directs methodological selections, and guarantees coherence. It improves validity and reliability by minimizing biases, promotes effective resource utilization, and enables replication for extensive information acquisition. Each component interacts

Fig. 2 The function of research design

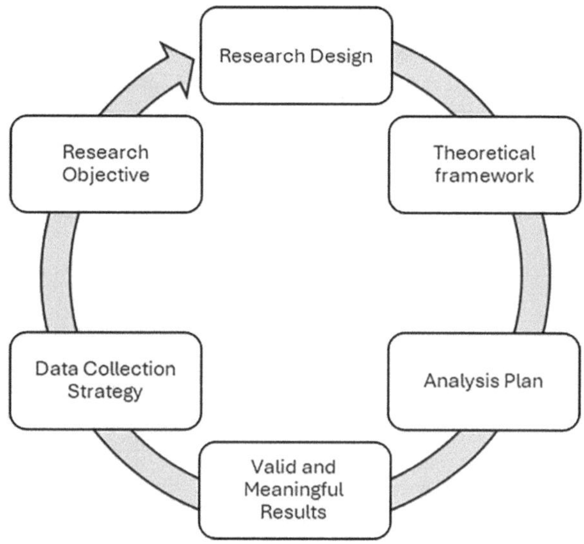

dynamically, enhancing the study's integrity and facilitating methodical, significant research consistent with its key objectives.

3.2 The Essential Elements of a Research Design

The essential components of a research design consist of several crucial components that, when taken together, guarantee the comprehensiveness and significance of the study. The research objectives serve as the primary aims, outlining the purpose of the inquiry clearly and concisely and giving direction for all future activities [51]. The theoretical framework serves as the intellectual underpinning, directing the creation of hypotheses and influencing the analysis process. A data-collecting strategy that has been well planned will specify the procedures for acquiring information. These methodologies may include surveys, interviews, experiments, or archival research, and they will be suited to the requirements of the study. Furthermore, the analysis strategy details the specific procedures utilized to understand the data, regardless of whether they were statistical, computational, or qualitative. The careful building of these components inside a research design is vital to guarantee that the study's outcomes are valid and significant [52]. Figure 3 delineates the essential elements of a study design and their interrelations. The process starts with establishing research objectives to direct the study's emphasis, underpinned by the theoretical framework that informs hypotheses and analysis. The data collection strategy outlines the techniques for acquiring information, whereas the analysis plan delineates the interpretation of the data. Collectively, these components guarantee that the study yields valid and significant outcomes.

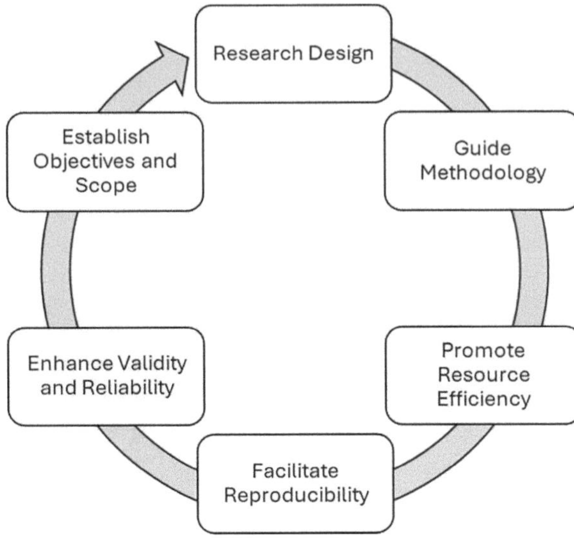

Fig. 3 Essential elements of a research design

3.3 Categories of Research Methodologies

Research designs are classified according to the nature of the research questions, the data needed, and the researcher's epistemological perspective [53]. Every design has distinct advantages and constraints, and its use is frequently customized to the research objectives.

3.3.1 Experimental Research Methodology

The primary objective of experimental research approaches is to determine the causal linkages between the variables. Among the characteristics that define these designs are the random assignment of participants, the manipulation of independent variables, and the control of extraneous influences [54]. Because of their high degree of internal validity, researchers can draw precise conclusions on the causal link. This is their main advantage. Medicine, psychology, and the natural sciences are a few fields that often use experimental techniques. For instance, researchers frequently use experimental methods to assess the thermal efficiency of materials used to construct passive houses [55]. Controlled laboratory studies examining the effects of cutting-edge insulating materials on interior temperature management under various exterior climate conditions would be a typical example of this.

3.3.2 Quasi-Experimental Design

Quasi-experimental designs provide a practical alternative to experimental methods, distinguished by the absence of random participant assignment [56]. This property renders them especially effective in applied research settings where randomization is unfeasible or unethical. These designs are essential for obtaining causal conclusions in actual contexts, providing a compromise between experimental rigor and practical practicality. The lack of random allocation presents issues in managing confounding factors, potentially compromising internal validity and the strength of causal inferences [57]. Even with these constraints, quasi-experimental designs are extensively employed, particularly in disciplines where controlled experimental conditions are challenging to establish. They are frequently utilized to evaluate the effects of educational reforms or interventions in schools where random student distribution is impractical. Quasi-experimental techniques offer a robust framework for assessing interventions by tolerating real-world constraints and preserving methodological flexibility.

3.3.3 Survey Research Methodology

In survey research methodology, the data collection process involves using organized instruments, such as questionnaires or interviews, to collect information from a broad population [3]. This strategy is very successful when acquiring information about individuals' attitudes, actions, and traits. One of its key advantages is that it can collect data from a wide variety of groups in an effective manner, which makes it an excellent choice for researching on a broad scale. On the other hand, surveys frequently have limitations, such as response biases and a lack of completeness in their insights. Social sciences, market research, and public health are just a few of the fields that use them extensively [58]. An example of this would be a nationwide poll investigating how the general population feels about using renewable energy.

3.3.4 Case Study Methodology

Case study methodology involves an in-depth examination of a single entity within its real-world context, such as an individual, organization, or event. This approach allows for a comprehensive and detailed analysis, making it particularly effective for exploring complex phenomena that require nuanced understanding. A significant strength of case studies is their ability to provide rich, context-specific insights [59]. However, their focus on specific instances often limits their generalizability to broader populations or settings. Case studies are widely used in business, academia, and ethnography. For example, a case study might analyze the leadership strategies of a rapidly successful startup, providing valuable insights into organizational dynamics and decision-making in similar contexts.

3.3.5 Ethnography Entails Intensively Observing Cultural or Social Groups to Comprehend Their Practices, Values, and Relationships

The qualitative research approach known as ethnography entails the in-depth observation and analysis of cultural or social groups to understand the behaviors, values, and interpersonal interactions prevalent within those groups [60]. When researchers immerse themselves in the natural surroundings of the group, they can get profound and contextualized insights into the cultural phenomena they are studying. This method is beneficial in examining complex social dynamics and cultural subtleties, which is why it is a popular approach in anthropology, sociology, and organizational studies. On the other hand, ethnographic research frequently demands a significant amount of time and calls for a high level of cultural sensitivity and understanding to guarantee appropriate interpretation [61]. As an illustration, ethnography may be utilized to investigate the organizational culture of a global firm. This would, in turn, shed light on how cultural norms impact the behavior and decision-making processes that occur in the workplace.

3.3.6 Longitudinal Research Design

The primary objective of longitudinal research designs is to gather data from the same individuals over time [62]. This allows researchers to monitor changes and trends throughout the study. This method is beneficial when analyzing developmental patterns and causal links since it provides vital insights into the growth of phenomena. Among its many advantages is the capacity to keep track of effects over extended periods and to build temporal sequences. It is important to note that longitudinal studies are frequently accompanied by significant expenses and the possibility of participant dropout, which can impact the continuity and validity of the research. Health sciences, developmental psychology, and economics are examples of everyday use within this field [63]. An investigation on the effects of extended exposure to air pollution on human health may, for instance, be carried out over 10 years. To guarantee the robustness of the methodology, it is necessary to fit the design with the research aims, themes, and resources.

3.4 Ethical Considerations in Research

Ethical concerns are fundamental to the research process. They guarantee that studies are executed correctly and that participants' rights are respected [64]. Research ethics preserve the integrity of the study, maintain social confidence, and safeguard vulnerable groups.

3.4.1 Fundamentals of Research Ethics

Informed Consent

Informed consent is a fundamental ethical principle requiring that participants are fully informed about a study's objectives, methods, risks, and benefits before agreeing to participate. This involves providing clear and accessible information, ensuring voluntary participation, and maintaining proper records of consent. For example, participants in a clinical study might receive a detailed consent form outlining all relevant details. Similarly, safeguarding participants' confidentiality and anonymity is crucial, achieved through data encryption, data de-identification of datasets, and limiting access to sensitive information [65]. For example, survey results can be reported in an aggregated format to protect individual responses and ensure privacy.

Mitigating Harm

It is a critical ethical obligation in research to mitigate damage, which requires researchers to foresee and limit potential hazards to participants [66]. These risks might include physical, emotional, and psychological harm. To ensure that the advantages of the study outweigh any possible risks, this procedure entails exhaustive risk–benefit analyses. In addition, researchers are obligated to make available suitable support systems, such as access to counseling or other types of aid, to mitigate potential detrimental consequences [67]. For example, it is vital to include post-interview debriefing sessions when conducting interviews on traumatic events. These sessions are designed to support participants and ensure they remain healthy throughout the many stages of the research process.

Confronting Ethical Quandaries in Research

Confronting ethical difficulties in research is vital to maintaining integrity, particularly when the researcher's intentions conflict with the participants' rights or society's standards [67]. Several different approaches are required to address these difficulties. The importance of transparency cannot be overstated, and it is imperative that all components of the study, including any conflicts of interest and sources of financing, be disclosed in their entirety. Implementing ethical oversight using Institutional Review Boards (IRBs) or ethical committees can ensure compliance with predetermined standards [68]. Equally critical is reflexivity, which requires researchers to critically evaluate their own biases, power dynamics, and ethical duties regarding their work. For instance, academics researching underrepresented groups must avoid exploiting those populations and ensure their work benefits the community. Furthermore, managing sensitive data necessitates implementing stringent security measures to safeguard participants' privacy and avoid breaches [69].

Structures and Protocols

Ethical research adheres to established guidelines that safeguard participants and uphold the integrity of the study [70]. The Belmont Report underscores three fundamental principles: respect for persons, beneficence, and justice, establishing a basis for ethical research methodologies. The Declaration of Helsinki delineates explicit ethical principles for medical research involving human subjects, emphasizing protecting their rights and welfare [71]. Additionally, many institutions and organizations have their own adapted ethical rules meant to address the unique demands and conditions of their research environments. These frameworks and processes jointly guarantee that researchers comply with stringent ethical requirements during their work.

3.5 Final Assessment

A meticulous study design and compliance with ethical norms are essential for generating valid, trustworthy, and significant research. The study design functions as the framework for the research process, guaranteeing that all elements from data collection to analysis align with the study's aims [49]. An organized design improves the study's coherence and validity, facilitating precise and significant results. Equally substantial, ethical issues maintain the integrity of the scientific process. They cultivate trust and respect among participants, researchers, and the wider academic community by protecting participants' rights and assuring ethical conduct. Collectively, these components constitute the foundation of reputable research, harmonizing scientific rigor with ethical accountability. By merging rigorous design principles with ethical practices, researchers facilitate the responsible and sustainable advancement of knowledge, fostering significant breakthroughs and considering societal and individual values [72].

4 The Research Process

A disciplined and systematic approach to gathering, evaluating, and interpreting data to answer specific questions or find solutions to complex problems defines research [73]. It is the foundation for knowledge development, enabling breakthroughs across various disciplines such as academia, industry, and policymaking. A comprehensive grasp of the research process is necessary to provide trustworthy, practical, and actionable results. This chapter delves into the essential steps of this process. These stages include identifying and articulating a research topic, comprehensively evaluating the available literature, developing an acceptable data collection methodology, and conducting a rigorous analysis. Furthermore, it highlights the significance of successfully presenting and distributing findings to guarantee that they are relevant and applicable. It is essential to master these components to ensure the study results' reliability, validity, and ethical integrity. Each step interrelates with the others. A well-structured research method not only contributes to the advancement of knowledge but also helps support effective decision-making and problem-solving in various scenarios [74]. Researchers can contribute to a more in-depth understanding of their respective professions, stimulate innovation, and address social concerns when they follow this systematic approach via their research. The purpose of this chapter is to serve as a guide for navigating the difficulties of the research process. It provides insights into the best practices and emphasizes the need for a rigorous, evidence-based approach.

4.1 Identification of a Problem Existing in Research

Determining a research problem is essential in any research method since it establishes the basis for the entire investigation [75]. Choosing a topic that is not only noteworthy but also achievable within the parameters of the research is the first step in this procedure. To identify and refine a viable research problem, it is necessary to employ both creative and analytical methodologies. Several different techniques can efficiently guide this procedure. The process of doing a brainstorming session entails coming up with a list of prospective subjects that are relevant to the area of interest and that resonate with the audience that is being targeted. Reviewing the current literature helps identify gaps or unsolved problems that may have been present in previous research, providing chances for additional investigation. Researchers can uncover practical themes by drawing on real-world difficulties or professional experiences [76]. This helps to ensure that the study targets topics that are either relevant or applicable. To obtain relevant comments and ideas, consulting peers, mentors, or experts may greatly help refine the focus. Once a topic has been selected, the next step is to refine it so that the problem is more particular, controllable, and researchable. This frequently entails limiting significant topics by concentrating on specific people, periods, or situations to understand them better. Researchers can establish an evident and practical framework by focusing on these components, paving the way for relevant and impactful investigations. A successful research process is based on the foundation of this methodical methodology [64]. A methodical strategy for determining and refining research questions is depicted in Fig. 4, which provides general information. Various approaches, including brainstorming, literature analysis, practical observation, and consultation, are applied, beginning with selecting a relevant and attainable topic. These strategies are incorporated into the refining process to ensure that the research topic is controllable and researchable. The result focuses on identifying specific study parameters, which contribute to forming a firm basis for effective investigations that have an impact.

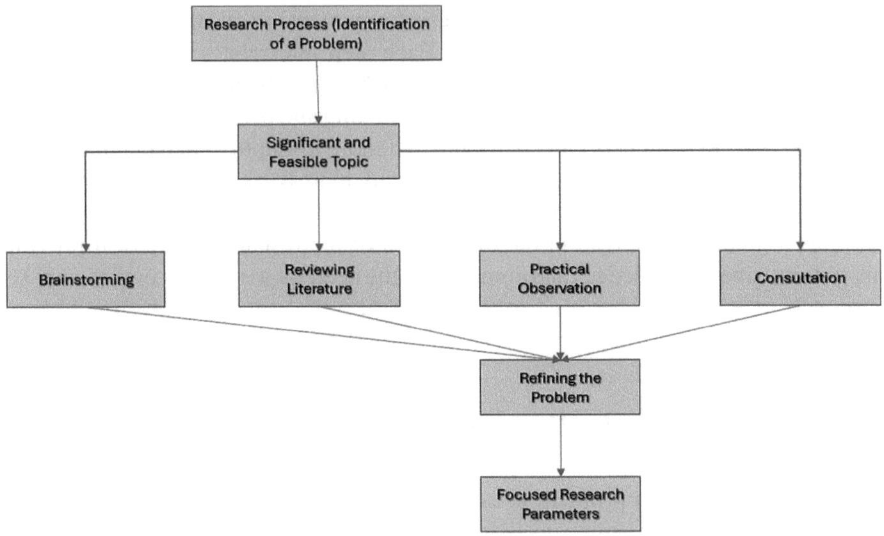

Fig. 4 Methodologies for identifying and refining research problems

4.2 The Process of Formulating Hypotheses and Research Questions

The research questions and hypotheses provide the study with its direction. A favorable research question is unambiguous, well defined, and aligned with the study's objectives [77]. Just one example: "What factors influence employee retention in small businesses?" is the subject of exploratory research. "What are the demographic characteristics of online learners?" is the question that pertains to descriptive research. Whether or not mentorship helps high school pupils enhance their academic performance comes from causal research. The assertion that "the integration of smart building technologies significantly reduces energy consumption in commercial office spaces" is an example of a hypothesis that can be considered an example of a hypothesis. The research question is the basis for formulating hypotheses and claims that can be tested and provide a foundation for empirical study and confirmation [78].

4.3 Review of Literature

The literature review provides context, underlines the study's value, and situates it within the larger academic debate. It serves as the theoretical underpinning of a research endeavor. By acting as a link between the existing body of knowledge and the addition made by the researcher, it highlights the significance of the topic under investigation.

4.3.1 Why the Literature Review is Being Conducted?

Several important goals, each of which is essential to the research process, are included in the literature review:

- Establishing a theoretical framework involves building a conceptual foundation for the study. This helps ensure that the research is based on well-established ideas and models.
- Identifying Research Deficits: This method can identify regions in which information is lacking, contradictory, or nonexistent, elucidating the necessity of conducting further research.
- Trying to Avoid Duplicating: Researchers can avoid duplicating work already done by analyzing past studies and instead concentrate on improving the area.
- Informing the Research Design: Insights gained from previous studies are direct guides while developing research topics, procedures, and analytical approaches.

4.3.2 A Systematic Approach to Conducting a Literature Review

When doing a comprehensive literature study, it is vital to use a rigorous and thorough approach. An efficient procedure is outlined in the successive phases that follow:

- The first step in defining the scope of the study is to identify the most important topics, themes, and terminologies pertinent to the investigation. These terms are the foundation for database searches.
- Choosing Sources Accredited by Scholars: Accessing publications that peers have evaluated requires using reliable databases such as PubMed, Scopus, or JSTOR. This guarantees that reliable and high-quality sources are included in the compilation.
- An Examination of Sources with Criticism: Research must be evaluated regarding its relevance, dependability, and methodological rigor. Its contributions to the field must be analyzed, and any limitations it may have must be determined.
- A Synthesis of the Conclusions: The material that has been evaluated should be organized into topic areas, methodological methods, or keys to the findings. This assists in identifying patterns, trends, and gaps that may be used to determine the direction of the research.

A well-constructed literature review will summarize the work that has already been done, synthesize findings, incorporate a variety of views, and identify areas that require future exploration [79].

4.4 Data Collection and Analysis

The "data collection" phase aims to collect information that can answer the research questions or test the hypotheses [25]. Primary or secondary data sources, each with their methodology, may be utilized in this process. Methods of information gathering that are primary and secondary:

Primary Data Collection: Involves obtaining original data directly from participants or through experimental settings:

- The use of surveys allows for the efficient collection of quantitative data from huge populations.
- In-depth qualitative insights may be obtained by investigating the participants' experiences and viewpoints through interviews.
- Observations: Record behaviors in the real world, either in natural or controlled environments.
- Establishing causal linkages through controlled testing is the purpose of doing experiments.

Secondary Data Collection: Relies on existing datasets, such as government reports, historical documents, or previously published research. This method saves time and resources while enabling longitudinal analysis.

It is common practice for researchers to blend primary and secondary methodologies to obtain comprehensive and quantitative data.

4.5 Data Analysis Techniques

Data analysis transforms raw information into relevant insights [80]. The characteristics of the data should guide the selection of the appropriate analytical methods:

- Statistical Analysis: When applied to numerical data, statistical approaches show patterns, connections, and causal links.
- The term "thematic analysis" refers to a qualitative method for recognizing recurrent themes and insights into textual material.
- Text-based information, such as interview transcripts or media content, is subjected to extensive categorization and interpretation through content analysis.

Because the selection of analytic techniques must be based on the study's objectives, the findings can be guaranteed valid and pertinent [25].

4.6 Presenting and Disseminating Research Findings

Research findings are more impactful and valuable when communicated effectively. An organized presentation is essential to ensure clarity and accessibility.

Research reporting formats that are considered standard:

- Comprehensive publications designed for policy, organizational, or practical purposes are called research reports.
- Comprehensive works submitted as part of the prerequisites for an academic degree are referred to as theses and dissertations.
- Brief contributions subjected to peer review and intended for publishing in academic journals are known as journal articles.

Each format often includes an abstract, an introduction, a methodology, findings, a discussion, and a conclusion. This standardized framework maintains coherence and makes the material easier to comprehend.

4.7 Academic Integrity and Dissemination Standards

When it comes to publishing research, it is necessary to comply to stringent academic and ethical standards:

- Peer review ensures that the work satisfies academic requirements and receives constructive criticism.
- Citations: Appropriate acknowledgment of previous work helps to avoid plagiarism and strengthens the integrity of intellectual work.
- Ethical Considerations: Transparency in how techniques, results, and reporting are carried out is necessary to sustain confidence and trustworthiness.

A systematic approach to distribution not only raises awareness of the study but also encourages academic discourse and provides information that can influence future investigations [81]. Adherence to these principles allows scholars to make significant contributions to their respective areas and maintain the integrity of the academic community.

5 Conclusion

Within the context of academic investigation, the conclusion of this chapter emphasizes the fundamental significance of research technique as a cornerstone. It offers a systematic framework that guides the whole research process, assuring that the information generated is rigorous, dependable, and morally sound. This systematic approach is not only necessary for the procedures to be followed; it is an essential method for enhancing the reliability and validity of the study's findings. The principles and practices of research methodology serve as a bridge between theoretical abstraction and empirical investigation. This bridge helps develop a firm basis for creating and distributing information across all fields of study. This chapter examines the differences between research methodologies and methodology, one of the most important contrasts discussed. Research methodology is an overall conceptual framework that justifies and combines the many research techniques for data gathering and analysis. Examples of research methods include surveys, interviews, and experiments. Research methods are the instruments and processes used for data collection and analysis. The significance of this distinction lies in the fact that it highlights the need to connect the "how" of data gathering with the "why" of research aims and theoretical foundations. Such alignment is essential to guarantee methodological consistency, advance academic discourse, and handle complicated research topics precisely and in depth. Not only does the systematic character of the research technique guarantee the internal rigor of the study, but it also guarantees that the findings may be replicated in different settings. This replicability is necessary when determining the generalizability of the findings and the possible uses of those findings in various sectors. The rigor offered by methodological frameworks enables researchers to move beyond anecdotal evidence and contribute to a cumulative body of knowledge susceptible to critical peer review and empirical verification. This is true whether the researchers work in the natural, social, or applied domains. In addition, the chapter emphasizes the relationship between basic research and applied research, illustrating how these two types of research complement and inform one another. Fundamental research strongly focuses on theoretical investigation and is the foundation upon which practical breakthroughs are ultimately built. On the other hand, the practical issues addressed by applied research frequently inspire new theoretical concerns, which helps foster a dynamic and iterative cycle of knowledge generation among researchers. This mutually beneficial partnership guarantees that theoretical breakthroughs will continue to be applicable and that practical solutions will be based on sound scientific foundations. The chapter elucidates how different methodological paradigms address different parts of research issues by examining qualitative, quantitative, and mixed-methods techniques. These research approaches are discussed in detail. Qualitative research offers a wealth of contextual and deep insights into complicated phenomena. These methods frequently capture nuances that are not accessible by quantitative assessment methods. The quantitative approaches, on the other hand, provide empirical precision and statistical generalizability, which enables researchers to test hypotheses and discover patterns in data. To provide a holistic strategy that is particularly well suited to research issues that involve several facets, the mixed-methods paradigm blends the qualities of both approaches. This pluralistic viewpoint reflects the ever-increasing complexity of current research and highlights the necessity of methodological adaptation to handle a wide range of epistemological and practical issues. When it comes to research techniques, ethics emerges as an essential component, highlighting researchers' moral responsibility toward the people they are studying, the academic community, and society in general. When it comes to the research process, ethical issues, such as informed consent, confidentiality, and damage mitigation, are not only supplementary concerns; they are essential components that guarantee the preservation of the rights of participants and the credibility of the findings. Research that adheres to ethical standards protects the participants' rights to dignity and well-being and helps maintain public confidence in the honesty of academic investigation. In addition, the chapter emphasizes the role that institutional frameworks, such as ethical review boards, play in maintaining these norms and cultivating a culture of responsibility in research practice. Much attention is also paid to the design of research projects, focusing on the structural coherence necessary to establish alignment between study aims, procedures, and outcomes. The research design serves as a plan that directs all stages of the study, beginning with developing research questions and hypotheses and continuing through data collection, analysis, and interpretation. Well-conceived designs provide methodological rigor and increase the validity and dependability of findings by methodically addressing the "what," "how," and "why" of research. This is accomplished via the use of a systematic approach. The methodological decisions taken at the beginning of a study significantly impact the study's coherence, scope, and impact. This is true regardless of whether the study is structured as an experimental, quasi-experimental, survey-based, or case study design. Furthermore, the applicability of research techniques across various disciplines, including academic, industrial, and societal institutions, is recognized as a crucial quality. Within academia, rigorous methodological frameworks advance theoretical growth, encourage partnerships across disciplinary lines, and guarantee the authenticity of scholarly contributions. Applying these frameworks in industrial settings helps foster innovation, improves the efficiency of operations, and improves evidence-based decision-making. Regarding society, research approaches address critical concerns such as sustainability, social fairness, and public health. This ensures that academic inquiry makes a meaningful contribution to the greater public good. In addition, the obstacles inherent in research are discussed. These challenges include the control of biases, the reconciliation of methodological limits, and the resolution of ethical concerns. Even though these difficulties are important, the chapter emphasizes that they are not insurmountable. Researchers must first engage in thorough preparation and methodological reflexivity and adhere to these established ethical principles to traverse these difficulties successfully. Not only does this repeated process of reflection and refinement contribute to the formation of research methodology as a dynamic and responsive discipline, but it also helps improve individual studies' robustness. In summary, this chapter aims to reaffirm the significance of research techniques as the intellectual and procedural basis of academic inquiry. Research methodology promotes the generation of credible, repeatable, and effective information. This is accomplished by blending theoretical rigor, methodological accuracy, and ethical accountability. These concepts enable investigating complicated phenomena, solving urgent problems, and bridging the gap between academic and practical understandings across various educational fields. The insights offered in this chapter provide a complete grasp of the methodological imperatives supporting scholarly research. This understanding equips researchers with the tools necessary to successfully contribute to the progress of knowledge and its application to the needs of society. This fundamental comprehension lays the groundwork for later investigations into more sophisticated methodological frameworks and the actual implementations of these frameworks in a variety of professional domains.

References

1. Adeoye MA (2024) Mastering the basics: a guide to research methodology for effective writing and publication. Chalim J Teach Learn 4(1):30–41
2. Jukkola M (2024) LIS integration and implementation process from customer perspective–survey and interview
3. Karunarathna I, Gunasena P, Hapuarachchi T, Gunathilake S (2024) The crucial role of data collection in research: techniques, challenges, and best practices. Uva Clin Res 1–24
4. Chekenyaa NS, Misraa S (2024) Responsible conduct of research for graduate students: what should they know?. Appl Econ Teach Resour 6
5. Prediger S, Barzel B, Hußmann S, Leuders T (2021) Towards a research base for textbooks as teacher support: the case of engaging students in active knowledge organization in the KOSIMA project. ZDM–Math Educ 1–16
6. Taherdoost H (2022) What are different research approaches? Comprehensive review of qualitative, quantitative, and mixed method research, their applications, types, and limitations. J Manage Sci Eng Res 5(1):53–63
7. Moriña A (2021) When people matter: the ethics of qualitative research in the health and social sciences. Health Soc Care Commun 29(5):1559–1565
8. Riffe D, Lacy S, Watson BR, Lovejoy J (2023) Analyzing media messages: using quantitative content analysis in research. Routledge
9. Bashir S, Gul S, Bashir S, Nisa NT, Ganaie SA (2022) Evolution of institutional repositories: managing institutional research output to remove the gap of academic elitism. J Librariansh Inf Sci 54(3):518–531
10. Fellows RF, Liu AMM (2021) Research methods for construction. John Wiley & Sons
11. Bibri SE, Krogstie J (2020) Environmentally data-driven smart sustainable cities: applied innovative solutions for energy efficiency, pollution reduction, and urban metabolism. Energy Inform 3(1):29
12. Arbale H, Mutisya DN (2024) Book review: "Research methods for business students" by Mark NK Saunders, Philip Lewis, and Adrian Thornhill (Pearson Education, 2019). Afri Q Soc Sci Rev 1(2):8–21
13. Kosiol J, Silvester T, Cooper H, Alford S, Fraser L (2024) Revolutionising health and social care: innovative solutions for a brighter tomorrow–a systematic review of the literature. BMC Health Serv Res 24(1):809
14. Hennink M, Kaiser BN (2022) Sample sizes for saturation in qualitative research: a systematic review of empirical tests. Soc Sci Med 292:114523
15. Vazire S, Schiavone SR, Bottesini JG (2022) Credibility beyond replicability: improving the four validities in psychological science. Curr Dir Psychol Sci 31(2):162–168
16. Ahmad T, Madonski R, Zhang D, Huang C, Mujeeb A (2022) Data-driven probabilistic machine learning in sustainable smart energy/smart energy systems: key developments, challenges, and future research opportunities in the context of smart grid paradigm. Renew Sustain Energy Rev 160:112128
17. Moirano R, Sánchez MA, Štěpánek L (2020) Creative interdisciplinary collaboration: a systematic literature review. Thinking Skills Creativity 35:100626
18. Kuran O (2024) Closing the theory-practice gap: socio-economic approach and action research in management. Syst Pract Action Res 1–28
19. Wohlin C, Papatheocharous E, Carlson J, Petersen K, Alégroth E, Axelsson J, Badampudi D, Borg M, Cicchetti A, Ciccozzi F, Olsson T (2021) Towards evidence-based decision-making for identification and usage of assets in composite software: a research roadmap. J Softw: Evol Process 33(6):e2345
20. Verma A, Singhania R (2023) Urban resilience in the face of climate change: strategies for adaptation and mitigation. J Sustain Technol Infrastruct Planning 7(1):46–66
21. Ngcobo K, Bhengu S, Mudau A, Thango B, Lerato M (2024) Enterprise data management: types, sources, and real-time applications to enhance business performance—a systematic review. Syst Rev

22. Bammer G, O'Rourke M, O'Connell D, Neuhauser L, Midgley G, Klein JT, Grigg NJ, Gadlin H, Elsum IR, Bursztyn M, Fulton EA (2020) Expertise in research integration and implementation for tackling complex problems: when is it needed, where can it be found and how can it be strengthened? Palgrave Commun 6(1):1–16
23. Schreiber F, Cramer C (2024) Towards a conceptual systematic review: proposing a methodological framework. Educ Rev 76(6):1458–1479
24. Haneef DI, Agrawal M (2024) Ethical issues in educational research. Asian Res J Arts Soc Sci 22(5):29–38
25. Taherdoost H (2021) Data collection methods and tools for research; a step-by-step guide to choose data collection technique for academic and business research projects. Int J Acad Res Manage (IJARM) 10(1):10–38
26. Bayuo BB, Chaminade C, Göransson B (2020) Unpacking the role of universities in the emergence, development and impact of social innovations–a systematic review of the literature. Technol Forecast Soc Chang 155:120030
27. Ganesha HR, Aithal PS (2022) How to choose an appropriate research data collection method and method choice among various research data collection methods and method choices during Ph.D. program in India?. Int J Manage Technol Soc Sci (IJMTS) 7(2):455–489
28. Karunarathna I, De Alvis K, Gunasena P, Jayawardana A (2024) Creating value through literature reviews: techniques for identifying research gaps
29. Corbin J, Strauss A (2014) Basics of qualitative research: techniques and procedures for developing grounded theory. Sage publications
30. Almusaed A, Almssad A, Yitmen I (2024) Sustainable built environment and its implications on real estate development: a comprehensive analysis
31. Mustafa MN, Abdah MAAM, Numan A, Moreno-Rangel A, Radwan A, Khalid M (2023) Smart window technology and its potential for net-zero buildings: a review. Renew Sustain Energy Rev 181:113355
32. Osaba E, Villar-Rodriguez E, Del Ser J, Nebro AJ, Molina D, LaTorre A, Suganthan PN, Coello CAC, Herrera F (2021) A tutorial on the design, experimentation and application of metaheuristic algorithms to real-world optimization problems. Swarm Evol Comput 64:100888
33. Almusaed A, Yitmen I, Almssad A (2023) Reviewing and integrating AEC practices into industry 6.0: strategies for smart and sustainable future-built environments. Sustainability 15(18):13464
34. Poincaré H (2022) The foundations of science: science and hypothesis, the value of science, science and method. DigiCat
35. Cooperrider D, Srivastva S (2017) The gift of new eyes: personal reflections after 30 years of appreciative inquiry in organizational life appreciative inquiry in organizational life. In: Research in organizational change and development, vol 25. Emerald Publishing Limited, pp 81–142
36. Mohajan HK (2020) Quantitative research: a successful investigation in natural and social sciences. J Econ Develop Environ People 9(4):50–79
37. Pregoner JD (2024) Research approaches in education: a comparison of quantitative, qualitative and mixed methods. IMCC J Sci 4(2):31–36
38. Zou PX, Xu X, Sanjayan J, Wang J (2018) A mixed methods design for building occupants' energy behavior research. Energy Build 166:239–249
39. Ramamoorthy N (2025) Harnessing quantitative and qualitative data for digital health experience design. In: Intelligent systems and IoT applications in clinical health. IGI Global, pp 421–442
40. Bentalha B, Alla L (2024) Revealing the subtleties: the art of qualitative studies in science and management. In: Applying qualitative research methods to management science. IGI Global, pp 1–21
41. Engelbrecht A, Yitmen I, Almusaed A (2024) Industry 4.0 transformation towards industry 5.0 paradigm-challenges, opportunities and practices: challenges, opportunities and practices
42. Papadakis N, Katsaprakakis DA (2023) A review of energy efficiency interventions in public buildings. Energies 16(17):6329

43. Baashar Y, Alkawsi G, Alkahtani AA, Hashim W, Razali RA, Tiong SK (2021) Toward blockchain technology in the energy environment. Sustainability 13(16):9008
44. Wunderlich P, Veit DJ, Sarker S (2019) Adoption of sustainable technologies: a mixed-methods study of German households. Mis Quart 43(2)
45. Kogan N, Lee KJ (2014) Exploratory research on the success factors and challenges of Smart City projects. Asia Pac J Inform Syst 24(2):141–189
46. Amiri Z, Heidari A, Navimipour NJ (2024) Comprehensive survey of artificial intelligence techniques and strategies for climate change mitigation. Energy 29:132827
47. Almusaed A, Almssad A, Yitmen I, Wallhagen M, Yang YF, Iyer-Raniga U (2024) Integrative approaches in urban sustainability: architectural design, technological innovations and social dynamics in global contexts. BoD–Books on Demand
48. Turner SF, Cardinal LB, Burton RM (2017) Research design for mixed methods: a triangulation-based framework and roadmap. Organ Res Methods 20(2):243–267
49. Andrew DP, Pedersen PM, McEvoy CD (2019) Research methods and design in sport management. Human Kinetics
50. Nosek BA, Hardwicke TE, Moshontz H, Allard A, Corker KS, Dreber A, Fidler F, Hilgard J, Kline Struhl M, Nuijten MB, Rohrer JM (2022) Replicability, robustness, and andomizability in psychological science. Annu Rev Psychol 73(1):719–748
51. Hancock DR, Algozzine B, Lim JH (2021) Doing case study research: a practical guide for beginning researchers
52. Marczyk GR, DeMatteo D, Festinger D (2010) Essentials of research design and methodology. John Wiley & Sons
53. Al-Ababneh MM (2020) Linking ontology, epistemology and research methodology. Sci Philos 8(1):75–91
54. Pirlott AG, MacKinnon DP (2016) Design approaches to experimental mediation. J Exp Soc Psychol 1(66):29–38
55. Hernández G, Cetina-Quiñones AJ, Bassam A, Carrillo JG (2024) Passive strategies towards energy efficient social housing: a parametric case study and decision-making framework in the Mexican tropical climate. J Build Eng 1(82):108282
56. Oj I (2024) Sampling and andomization in experimental and quasi-experimental CALL studies: issues and recommendations for design, reporting, review, and interpretation. ReCALL 36(1):58–71
57. Hall GJ, Putzeys S, Kratochwill TR, Levin JR (2024) Discovering internal validity threats and operational concerns in single-case experimental designs through directed acyclic graphs. Educ Psychol Rev 36(4):128
58. Khan MM (2024) Optimizing web surveys in research: methodological considerations and validity aspects. Int J Res Sci Innov 11(4):75–105
59. Ruppert S (2024) Creating context-specific critical incidents with ChatGPT for effective soft skill training. ChatGPT and global higher education: using artificial intelligence in teaching and learning, p 277
60. Denzin N, Lincoln Y (2025) Ethnography and participant observation. Qualitative Research Methods for Media Studies, p 181
61. Brooks LA, Manias E, Rasmussen B, Bloomer MJ (2025) Practice recommendations for culturally sensitive communication at the end of life in intensive care: a modified eDelphi study. Intensive Crit Care Nurs 86:103814
62. Tan MM, Barbosa MG, Pinho PJ, Assefa E, Keinert AÁ, Hanlon C, Barrett B, Dregan A, Su TT, Mohan D, Ferri C (2024) Determinants of multimorbidity in low-and middle-income countries: a systematic review of longitudinal studies and discovery of evidence gaps. Obes Rev 25(2):e13661
63. Eid NM. Al-Ofi EA, Enani SM, Mosli RH, Saqr RR, Qutah KM, Eid SM (2024) A 6-month randomised controlled trial to compare the effectiveness of telenutrition versus telenutrition supported by telemonitoring and health coaching in a weight loss program: a study protocol. Br J Nutri 1–28

64. Karunarathna I, De Alvis K, Gunasena P, Jayawardana A (2024) Designing and conducting clinical research: methodological approaches
65. Balagurunathan Y, Sethuraman RR (2024) An analysis of ethics-based foundation and regulatory issues for genomic data privacy. J Inst Eng (India): Ser B 1–11
66. Guelmami N, Ben Ezzeddine L, Hatem G, Trabelsi O, Ben Saad H, Glenn JM, El Omri A, Chalghaf N, Taheri M, Bouassida A, Ben Aissa M (2024) The ethical compass: establishing ethical guidelines for research practices in sports medicine and exercise science. Int J Sport Stud Health 7(2):31–46
67. Edwards-Fapohunda MO, Adediji MA (2024) The impact of the educational journey of special needs children on the employability, sociability, and economy of working-class parents in the UK. IRE J 8(1):421–422
68. Valencia AT (2024) Institutional review board (IRB) evaluation of policies affecting minority inclusion in clinical research. Master's thesis, The University of Arizona
69. Sargiotis D (2024) Data security and privacy: protecting sensitive information. In: Data governance: a guide. Springer Nature Switzerland, Cham, pp 217–245
70. National Academies of Sciences, Policy, Global Affairs and Committee on Responsible Science (2017). Fostering integrity in research. National Academies Press
71. Ali B, Sule MM (2024) The evolution of ethical standards in health research: an Islamic overview. Multicult Islamic Educ Rev 2(1):1–18
72. Swargiary K (2024) Research methodologies: evolution, practice, and prospects. ERA, US
73. Pandey GP (2024) Foundations and frameworks of ELT and applied linguistics research: principles, processes and practices. Int J Soc Sci Manage 11(4):126–135
74. Millalén FV (2024) Student performance in writing prompts for text-based GenAI tools in a research methodology course: text-based GenAI tools in a research methodology course. Transformar 5(2):71–90
75. Tracy SJ (2024) Qualitative research methods: collecting evidence, crafting analysis, communicating impact. John Wiley & Sons
76. Venaik A (2024) ChatGPT's role in healthcare education, research, and practice: a systematic review of promising prospects and legitimate concerns. Educ Admin: Theory Pract 29(1):337–344
77. Saharan VA, Kulhari H, Jadhav H, Pooja D, Banerjee S, Singh A (2020) Introduction to research methodology. In: Principles of research methodology and ethics in pharmaceutical sciences. CRC Press, pp 1–46
78. Graebner ME, Knott AM, Lieberman MB, Mitchell W (2023) Empirical inquiry without hypotheses: a question-driven, phenomenon-based approach to strategic management research. Strateg Manag J 44(1):3–10
79. Rimban E (2024) Building bridges of understanding: the art of constructing the literature review. Available at SSRN 4730313
80. Olaniyi O, Shah NH, Abalaka A, Olaniyi FG (2023) Harnessing predictive analytics for strategic foresight: a comprehensive review of techniques and applications in transforming raw data to actionable insights. Available at SSRN 4635189
81. Akour M, Alenezi M (2022) Higher education future in the era of digital transformation. Educ Sci 12(11):784

Identifying Research Gaps

Abstract Identifying research gaps in the current academic literature is crucial in promoting innovation and technical advancement, especially in engineering. This debate outlines many research gaps, including theoretical, methodological, technical, contextual, and practical. Each kind poses unique obstacles and opportunities for academic investigation. Readers are provided with methodical approaches to discover these gaps, including thorough literature studies, examining patents and technical trends, and collaborating with industry partners. The importance of these multidisciplinary methods and engagement with industry is to be recognized, as they are crucial in closing research gaps and promoting a more robust and influential research environment. By identifying and correcting these deficiencies, researchers may develop specific research inquiries, create ground-breaking studies, and contribute significantly to scholarly understanding and practical implementations in the business. Analyzing case studies and practical examples emphasizes the possibility of making substantial progress, illustrating the significance of accurately recognizing and resolving research deficiencies. This conversation prompts readers to acknowledge the perpetual character of the study and the need for consistent academic endeavors to broaden the limits of understanding and application. It also underscores the urgent role of research gaps in advancing engineering innovation, a call to action for all researchers.

1 Introduction

Identifying research gaps is crucial in academic investigation, especially in civil engineering and architecture, where knowledge progress is strongly linked to technical innovation and practical implementation. Your research, as a reader, has the potential to impact these fields significantly. Darwish (2018) argues that practical engineering is fundamentally multidisciplinary, which makes it challenging to identify study gaps in cognitive styles using only one technique. Hence, it is imperative to tackle expanding cognitive techniques by precisely describing and providing detailed explanations of industrial thinking while pinpointing and examining crucial areas that

A. Almusaed et al., *Practice of Research Methodology in Civil Engineering and Architecture*, https://doi.org/10.1007/978-3-031-97393-2_2

need more investigation [1]. This chapter thoroughly examines research gaps, emphasizing their importance in guiding research endeavors that contribute to academic discussion and tackling urgent engineering concerns. By identifying areas lacking comprehensive, up-to-date, or absent research, academics may produce information that advances discipline, promoting theoretical advancement and practical solutions [2]. The procedure starts by identifying uncharted subjects in the current academic domain, a crucial initial phase in formulating a research endeavor. The proposal report is crucial at this stage, providing a well-organized framework that encompasses the research issue, its significance, and the study's overall importance. To effectively communicate the purpose of the research, it is necessary to clearly explain the reasons behind it, identify the areas that the study aims to target, and outline the recommended approach to tackle these areas. This chapter comprehensively analyzes finding research gaps and choosing innovative research subjects. Before developing a research question, it is crucial to have a comprehensive grasp of the pertinent academic literature, existing views, and ongoing disputes [3]. Choosing a study area encompassing a wide range of topics is essential as it enables a comprehensive investigation of the subject. It is essential to thoroughly study influential texts, ongoing debates, and notable theories relevant to the selected subject. Although studying all the relevant literature may not be feasible, it is crucial to be acquainted with the most prominent and well-acknowledged schools of thought. Expanding upon this basis, scholars can pinpoint unsolved matters and inquiries that require more investigation, therefore adding to the developing pool of knowledge. This chapter explores the approaches used to identify research gaps, such as systematic literature reviews, research analysis, discussions with industry specialists, and analysis of publication and funding patterns after they have been identified. Every approach is meticulously scrutinized, focusing on its efficacy in detecting and resolving research deficiencies in architecture and civil engineering. In addition, the chapter explores the difficulties linked to this procedure, including constraints on resources, ethical considerations, and inherent biases within academic publication and innovation institutions. Gaining a comprehensive understanding of these issues enables researchers to formulate effective ways to overcome them, guaranteeing that their work successfully identifies and addresses research gaps with significant impact [4]. The chapter ends with a forward-thinking discourse on prospective future avenues in engineering research, emphasizing new subjects anticipated to present research deficiencies and the role of digital technologies, such as artificial intelligence and big data analytics, in remedying these deficiencies. Furthermore, it emphasizes the significance of continuous cooperation between academia and industry to close research gaps and enhance engineering knowledge and practice. Research engineers are ultimately responsible for discovering topics that have not been thoroughly examined and strategically directing their efforts toward areas with the most demand and potential for making a significant effect. This chapter thoroughly examines the many research gaps in engineering, the techniques for detecting them, and the related difficulties, improving the reader's comprehension of their importance and impact on the field's theoretical and practical advancement.

1.1 Identifying Critical Research Gaps for Further Exploration

The term "key research gaps" refers to several areas within a field of study that have not been sufficiently investigated or in which the available information is insufficient [5]. The identification of these gaps is critically important for the advancement of both understanding and innovation. These gaps may entail questions that have yet to be answered, subjects that have not been investigated, or emergent difficulties that need to be adequately addressed in the research that is currently being conducted. It is helpful to recognize these gaps so that future research efforts can be focused on them. This helps ensure that studies contribute to the creation of new ideas, technologies, or techniques, ultimately leading to the advancement of the field and the resolution of crucial challenges. In academic study, a research gap denotes a specific domain within the existing body of literature that has not been thoroughly investigated or adequately tackled, resulting in unresolved questions or unexplained occurrences [6]. Research gaps in engineering can appear in several forms, such as theoretical, methodological, technical, contextual, and practical. Every gap category signifies a distinct aspect of lacking knowledge and suggests distinct possibilities for more exploration. Theoretical gaps occur when current models, frameworks, or theories do not adequately explain specific elements of engineering phenomena, requiring the creation of new or improved ideas. Methodological gaps arise when existing research methods are insufficient to capture or evaluate specific data or occurrences, restricting the strength or relevance of study findings. Technological gaps become apparent when there is a delay in incorporating developing technology into engineering solutions, typically indicating a lack of connection between invention and its practical application. Contextual gaps arise when study findings lack generalizability across diverse settings, such as distinct geographical areas, cultures, or industrial environments [7]. Practical gaps arise when there is a noticeable difference between academic research and its implementation in business. This occurs when research fails to tackle real-world issues or is not effectively transformed into solutions that can be easily expanded and used. For example, within the field of sustainable engineering, a theoretical gap may pertain to the limited comprehension of the extended ecological consequences of innovative materials. The need for advanced simulation methods to accurately model complex systems under various conditions exposes methodological shortcomings. Technological gaps may occur from poor integration of artificial intelligence or nanotechnology in engineering processes. Context gaps are created if engineering approaches in one region are unsuccessful in another due to different environmental or socio-economic circumstances. When academic research does not coincide with industry expectations for scalability or economic viability, practical inadequacies result [4]. Similarly, "research gaps" emphasize the importance of identifying and addressing gaps in academic research, which are crucial for advancing scholarly knowledge and ensuring meaningful contributions to a field. "Less or No Information" points to underexplored areas in the literature, offering opportunities for original research to fill existing knowledge voids. It is important to

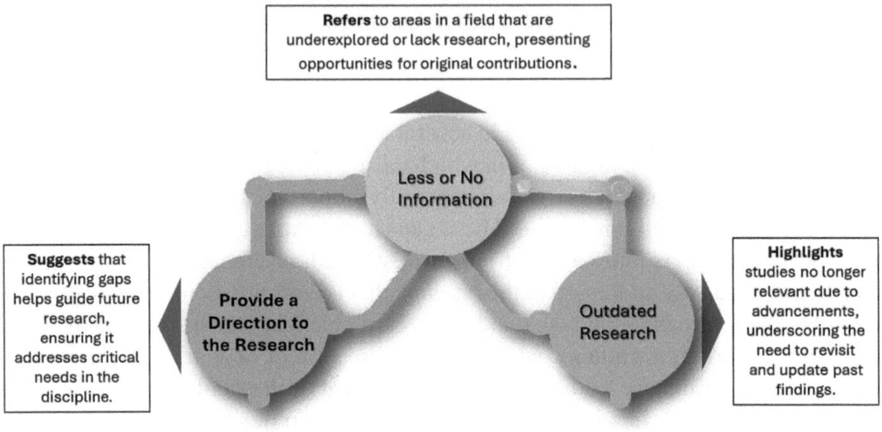

Fig. 1 Key components of identifying and addressing research gaps in academic studies (*source* authors)

note that "outdated research" refers to studies that have become obsolete due to developments in theory, technology, or data, highlighting the necessity of updating them. In conclusion, "Provide Direction to the Research" demonstrates how identifying gaps may guide future scholarly inquiry. This enables researchers to concentrate on areas requiring more examination and contributes to the advancement of the field, as seen in Fig. 1.

1.2 Significance of Identifying Research Gaps

Identifying research gaps is crucial in engineering research for several reasons. As a practical field, engineering primarily focuses on resolving practical issues and advancing technical development [8]. By identifying and correcting research gaps, researchers may focus on areas that require the most excellent attention, thereby maximizing the potential for innovation and social impact. Moreover, research gaps are a navigational tool for the academic community, guiding future research trajectories. By identifying areas lacking information, researchers can prioritize studies that contribute to the growth of theoretical frameworks, the creation of new approaches, and the application of emerging technologies [9]. This procedure improves the thoroughness and significance of academic research and guarantees that it stays in line with the changing requirements of business and society. Addressing research gaps is essential for preserving a competitive edge and promoting technological innovation from an industrial standpoint. For example, suppose there is a lack of knowledge about the long-lasting qualities of recently developed composite materials. In that case, this might result in significant progress in aircraft engineering, leading to enhanced safety measures and decreased expenses. Researchers may bridge the gap between academia and business by recognizing these gaps and developing scientifically rigorous and financially successful solutions [8]. Furthermore, identifying research gaps is crucial for advancing knowledge within the wider academic community. It facilitates the creation of novel ideas and models that may be subjected to empirical testing, improved, and included in the current knowledge base. The iterative process of constructing and validating theories is essential for the progress of any scientific field, including engineering. Scholars contribute to the ongoing growth of the intellectual environment and promote innovation and discovery by resolving research gaps [9, 10].

1.3 Categorizing and Identifying Research Gaps

The identification of research gaps is a task that is both essential and difficult for researchers. This is partly owing to the absence of formal frameworks that guide the process of identifying research gaps. Identifying research gaps is a subjective process that frequently leads to different interpretations from academics, leading to discrepancies and conflicts across the board [11]. It is proposed in this article that a systematic framework be developed to classify and identify research gaps according to their value and relevance. Researchers have the potential to systematically improve their capacity to detect and communicate research gaps by adopting this strategy, which in turn contributes to substantial breakthroughs in academic inquiry.

1.3.1 The Challenge of Subjectivity in Identifying Research Gaps

Because researchers come from a wide range of backgrounds, have a variety of experiences, and have various academic interests, the process of finding research gaps is fraught with subjectivity. Something that one researcher considers a gap may be overlooked by another researcher, which can result in a lack of consensus and slow down the pace of the study. The fact that there is no methodical strategy to identify and define research gaps is a factor that makes this problem even worse. Garg and Baliyan (2021) presented a fundamental framework that classifies research gaps into seven unique kinds to solve these issues. This framework provides researchers with a formal structure that allows them to identify and assess research gaps [12] systematically.

1.3.2 Categories of Research Gaps

Research gaps are crucial because they indicate areas that still need investigation to progress knowledge and solve open problems [13]. These gaps are significant because they show where knowledge is still lacking, where presumptions may need to be reconsidered, or where theories now fall short of explaining a phenomenon thoroughly. To ensure that research efforts are focused on the most critical and understudied subjects, academics may advance their discipline by identifying these gaps [14]. Researchers can systematically identify gaps in the literature and create studies to address them by categorizing them. This classification can also aid in prioritizing study endeavors by directing academics to concentrate on regions with the most significant possibility for noteworthy outcomes. Research gaps can be of several kinds: methodological (where current approaches are inadequate or unsuitable for addressing a research question), theoretical (where current theories fall short of fully explaining a phenomenon), and empirical (where data or research findings are lacking on a specific topic). For example, in the field of civil engineering, there may be a theoretical vacuum in our knowledge of the behavior of novel materials like ultra-high-performance concrete under harsh environments like earthquakes or extremely high temperatures. Current theories and models could not entirely predict the material's performance, requiring more study to create more precise theoretical frameworks. One methodological gap in civil engineering can be the requirement for newer or more sophisticated techniques to evaluate the long-term resilience of infrastructure under different environmental circumstances. Current evaluation procedures may be deficient because traditional methods may not effectively represent the various and harsh conditions that buildings encounter during their lifespan. The scant information on how well green infrastructure solutions—like permeable pavements—perform in lowering urban floods may reveal an empirical gap. Although there have been some studies, their breadth may be restricted. More extensive data collection across various climates and metropolitan contexts is required to give a more thorough picture. Other kinds of gaps include population gaps, which in the context of civil engineering might result from a paucity of studies on the effects of infrastructure development on various demographic groups, especially regarding equality and accessibility. There may be context gaps when results from infrastructure projects in one area cannot be applied to another because of variations in local construction customs, soil types, or climate [15]. Civil engineers may make significant progress in the area; resolve outstanding difficulties; and promote the advancement of more complete and robust theories, techniques, and empirical data through the methodical identification and closure of these diverse research gaps. In the end, this approach produces infrastructure solutions that are safer, more effective, and sustainable. A systematic framework for identifying and classifying research gaps in architectural engineering is presented in Table 1 for research convenience. This framework provides detailed descriptions of various types of gaps: conceptual, empirical, methodological, contextual, theoretical, practical, and ethical, as well as their respective definitions, levels of significance, and relevant examples from the field.

1.4 Using Research Gaps to Frame Research Questions

In academic research, utilizing research gaps to formulate research questions is an essential strategy, especially to guarantee that scholarly endeavors make a meaningful

Table 1 Framework for identifying research gaps with examples

Type of research gap	Definition	Importance	Examples
Conceptual gaps [16]	Areas where theoretical frameworks, models, or concepts are underdeveloped or not appropriately applied	Strengthens the theoretical foundation of research	– Need for robust theoretical frameworks for integrating sustainable materials into architectural design
Empirical gaps [17]	Insufficient empirical evidence on specific research topics	Requires further empirical research to validate or challenge existing theories	– Lack of comprehensive studies on the long-term energy efficiency of new sustainable building materials
Methodological gaps [18]	Inadequate research methods to address specific questions	Necessitates developing new methods or adapting existing ones	– Traditional methods may not capture subjective experiences in sustainable buildings
Contextual gaps [19]	Research findings are not generalizable due to specific contextual limitations	Requires studies in diverse settings to improve generalizability	– Preservation strategies may not apply universally across cultural contexts
Theoretical flaws or defect gaps [20]	Existing theories or models fail to explain certain phenomena adequately or lack comprehensive frameworks	Refining or developing new theories is crucial	– Current theories may not fully explain the impact of design on human behavior in indoor environments
Practical gaps [21]	Research does not provide applicable solutions to real-world problems	Requires translating research findings into actionable recommendations	– Research on energy efficiency may lack practical guidelines for architects and builders
Ethical gaps [21]	Absence of ethical considerations, especially in studies involving human subjects or sensitive topics	Essential for maintaining research integrity	– Studies on sustainable building components must consider informed consent and privacy

contribution to the progress of knowledge [8]. Research gaps exist in a field with either a lack of obsolete or inadequate knowledge, providing a chance for in-depth study. By identifying these gaps, researchers might develop inquiries that specifically tackle the most consequential and understudied problems, thus augmenting the significance and influence of their study. For instance, there may be a significant research gap in the efficient synchronization of real-time data between Building Information Modeling (BIM) and Digital Twin integration. "What methods can be developed to enhance the integration of BIM and Digital Twin technologies for improved real-time decision-making in construction management?" is one way to phrase a research topic. One concern in green building materials may be how sustainable and long lasting these materials are in various environmental settings. "What are the long-term durability and environmental impacts of emerging green building materials under varying climatic conditions?" is one such research question that might result from this. Analogously, improving AI algorithms for precise construction delay prediction may represent a need for research in artificial intelligence (AI) and machine learning in the construction industry. "How can AI and Machine Learning algorithms be refined to more precisely predict and mitigate construction delays in large-scale projects?" is one such research question. Finally, there can be a scalability issue with prefabrication and modular construction for large-scale infrastructure projects. "What challenges and strategies exist for scaling modular construction techniques in large-scale infrastructure projects?" is one possible research question that may result from this. In conclusion, a systematic technique that guarantees research is both significant and in line with the vital demands of the area is to formulate research questions around recognized gaps [22]. This strategy encourages creativity and propels the creation of new information in fields that most urgently need more investigation. Table 2 delineates the significance of identifying research gaps in engineering. These gaps are classified into theoretical, methodological, technical, contextual, and practical categories. It highlights how filling these gaps with thoughtfully crafted research questions stimulates creativity and expands the body of knowledge on the topic, both theoretically and practically.

2 Conceptual Framework of the Gap Method in Research

In the realm of research, the gap technique may be compared to the process of putting together a jigsaw puzzle, where each study represents a piece that contributes to a more thorough understanding of a specific topic [26]. However, in contrast to a conventional puzzle, the components of scientific inquiry are not fixed; instead, they can be rearranged in various ways, depending on the methodological approach the researcher chooses. This chapter examines the gap technique within the construction sector context, particularly emphasizing its use in architecture, urban design, sustainable building, and Sustainable Building Information Modeling (BIM).

Table 2 Overview of research gaps and their significance in civil engineering

Section	Key points	Examples
Identifying research gaps [23]	– Research gaps are pivotal in academic inquiry, particularly in the domains of civil engineering and architecture, where the progression of knowledge is closely linked to technical innovation and practical implementation – Practical engineering is inherently multidisciplinary, posing challenges in identifying research gaps through a singular methodological approach	
Significance of research gaps [23]	– Research gaps represent areas within a field where the existing body of knowledge is incomplete, outdated, or nonexistent – Identifying these gaps is essential for directing future research efforts toward areas of critical need, thereby advancing theoretical frameworks, fostering innovation, and contributing to practical solutions	

(continued)

Table 2 (continued)

Section	Key points	Examples
Typology of research gaps [24]	– Theoretical gaps: Arise when current models or frameworks fail to adequately explain specific engineering phenomena, necessitating the development of new or refined theories – Methodological gaps: Emerge when existing research methodologies are inadequate for capturing or analyzing specific data, thereby limiting the robustness or applicability of research findings – Technological gaps: These gaps reflect delays in integrating emerging technologies within engineering practices, indicating a disconnect between innovation and its practical application – Contextual gaps: Occur when research findings lack generalizability across different contexts, such as varying geographical regions, cultures, or industrial environments – Practical gaps: Indicate a misalignment between academic research and its practical application within industry, where research fails to address real-world challenges or is not effectively translated into scalable and usable solutions	Examples: – Theoretical gap: In sustainable engineering, a theoretical gap may involve a limited understanding of the long-term ecological impacts of innovative materials – Methodological gap: Advanced simulation techniques are needed to model complex systems accurately under diverse environmental conditions – Technological gap: Artificial intelligence or nanotechnology has not been adopted in engineering processes – Contextual gap: Engineering methodologies developed in one region may need to be more effective in another due to differences in environmental or socio-economic factors – Practical gap: Academic research needs to meet industry scalability and economic viability requirements

(continued)

Table 2 (continued)

Section	Key points	Examples
Significance of identifying research gaps [13]	– Identifying research gaps is a critical navigational tool, guiding academic research toward addressing the most pressing and underexplored issues – It facilitates the alignment of research endeavors with the evolving needs of industry and society, thereby maximizing the potential for innovation and societal impact – Moreover, addressing research gaps contributes to continuously developing theoretical constructs, methodological advancements, and practical applications	Example: – The lack of comprehensive knowledge regarding the durability of newly developed composite materials could catalyze significant advancements in aerospace engineering, leading to enhanced safety protocols and cost-reduction strategies
Framing research questions through gaps [25]	Utilizing research gaps to frame research questions is an essential methodological approach to ensuring that academic research is relevant and impactful. Research questions should be strategically formulated to address identified gaps, thereby enhancing the study's contribution to the field and ensuring that it addresses critical and underexplored issues	Examples: – BIM and digital twin integration: "What methodologies can be developed to enhance the integration of Building Information Modeling (BIM) and Digital Twin technologies for improved real-time decision-making in construction management?" – Green building materials: "What are the long-term durability and environmental impacts of emerging green building materials under varying climatic conditions?" – AI in construction: "How can Artificial Intelligence (AI) and Machine Learning algorithms be optimized to more accurately predict and mitigate construction delays in large-scale projects?" – Modular construction: "What are the challenges and strategies for scaling modular construction techniques in large-scale infrastructure projects?"

2.1 *Literature Review as a Tool for Identifying Research Gaps*

For the gap technique to be effective, conducting a thorough and coordinated analysis of the current body of literature is necessary [27]. Scholars can map the existing state of knowledge, identify gaps in understanding, and position new studies to meet these shortcomings by conducting a rigorous review of relevant earlier research [28]. For instance, a literature study in sustainable design may uncover much research on energy-efficient building materials. However, it might also indicate a dearth of studies on how these materials function in various climates. By identifying this gap, researchers can concentrate on developing materials adapted to specific climates, advancing worldwide talks on environmentally responsible building practices. Similarly, a thorough literature analysis may reveal a deficiency in incorporating lifecycle assessment (LCA) approaches through BIM software in sustainable building information modeling (BIM). The use of lifecycle assessment (LCA) to analyze environmental consequences throughout the lifespan of a structure is still underexplored despite the increased usage of building information modeling (BIM) in the construction industry. There is a possibility that addressing this gap will result in construction techniques that are more ecologically responsible and comprehensive.

2.2 *Methodological Considerations in Structuring Literature Reviews*

In order to employ the gap technique effectively, the literature review must be arranged so that the gaps can be precisely defined and the future study can be justified. Locke and Golden-Biddle (1997) propose three approaches to arranging literature reviews: synthesized, progressive, and no coherence [29].

- Connecting various aspects of literature that are not generally associated with one another is what is meant by the term "synthesized coherence." For instance, combining public health and urban planning research may indicate a gap in developing urban settings that increase mental well-being. This is the case in the field of urban design. With the help of this synthesis, academics can offer novel approaches to urban planning that put mental health first.
- Progressive coherence emphasizes the changing nature of research over time. For instance, a literature review may track the transition from early research on energy efficiency in sustainable architecture to more contemporary studies that encompass social and economic aspects of sustainability. This would be an example of a transition. Identifying gaps in this development could inspire research that more effectively addresses social justice within environmental contexts.
- Inconsistencies or contradictions throughout the literature are brought to light by No Coherence. Some studies, for example, argue for incorporating renewable

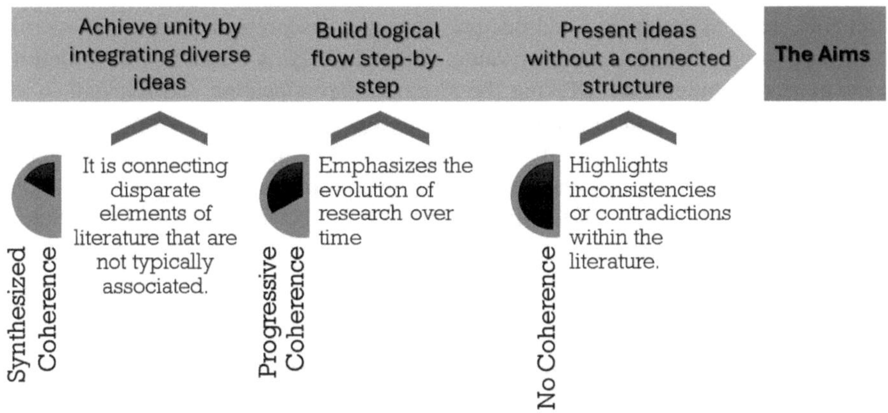

Fig. 2 Visual representation of aims in coherence—synthesized, progressive, and non-coherent structures (*source* authors)

energy sources into BIM models, while others concentrate purely on cost-saving techniques. This is an example of sustainable building information modeling (BIM). In building information modeling (BIM), researchers can conduct studies that aim to harmonize economic efficiency and environmental responsibility by highlighting these discrepancies.

The three goals of coherence in arranging thoughts are depicted in Fig. 2. These goals are to achieve unity by integrating a variety of concepts, to construct a logical flow step by step, and to offer ideas without a linked framework. The diverse degrees of coherence are reflected in the various parts of a circle that are used to illustrate these goals. There are three types of coherence: synthesized coherence, progressive coherence, and no coherence. Synthesized coherence emphasizes integration, progressive coherence emphasizes logical sequencing, and no coherence encourages disconnection.

2.3 *Challenges in the Application of the Gap Method*

Despite its usefulness, the gap approach offers several difficulties. The multidisciplinary character of the relevant content is a significant barrier that must be overcome [30]. For example, should the emphasis be placed on studying materials science, environmental psychology, or urban policy in sustainable architecture? The notion of a research gap can be somewhat different from one standpoint to another, depending on the perspective taken. Another obstacle that must be overcome is ensuring the detected gap is big and new. In the context of sustainable urban planning, for instance, a researcher can discover a gap in the existing body of literature about green roofs. On the other hand, if this problem has previously been well investigated in adjacent domains, such as landscape architecture, then the research might not necessarily contribute a significant amount of value. Consequently, it is necessary to identify gaps that contribute value. Making the research gap appealing and relevant to the larger academic conversation might be challenging. For example, incorporating LCA into sustainable BIM may initially appear technical. On the other hand, recasting this integration in the broader framework of lowering building carbon footprints can make the study more attractive to academics and professionals working in the construction sector [31].

2.4 Evaluating the Advantages and Disadvantages of the Gap Method

The gap technique provides a valuable framework for developing research in various sectors, including urban planning, sustainable building, and architecture [32]. Researchers can contribute to the resolution of critical real-world difficulties by methodically identifying and correcting gaps in the existing body of literature. These challenges include boosting urban sustainability, improving building resilience, and minimizing the environmental effects of construction projects. However, the approach has some inherent limits; detecting a gap is just the first step in the process. In addition, researchers have the responsibility of ensuring that the gap is significant, engaging, and pertinent. In order to effectively address these difficulties, it is necessary to conduct a thorough assessment of the relevant literature, interpret the findings with precision, and effectively communicate the findings to the broader academic and professional communities. After all, the gap technique continues to be an effective instrument for fostering innovation and sustainability in essential domains. Scholars can make significant contributions that advance their respective fields, provided they carefully identify and fix any knowledge gaps.

3 Framing Research Questions Through the Identification of Research Gaps

The gap approach in research introduces fresh concepts and viewpoints by locating and correcting the lack of crucial components from the corpus of current knowledge [33]. This strategy is beneficial in industries where advancement and innovation are critical. Scholars may progress in their profession, leading to significant discoveries and deep insights by identifying gaps in existing research. Finding research gaps necessitates carefully analyzing existing literature to identify topics that need more in-depth investigation or have not received enough attention [34].

For example, improving the connectivity of IoT devices with the current urban infrastructure may be necessary in the context of intelligent infrastructure and the

Internet of Things. The promise of IoT in smart cities has been extensively studied; however, the difficulties of fusing new technology with antiquated infrastructure have received less attention. Closing this disparity might result in significant progress toward more effective and flexible urban settings. Analogously, more research on human–robot interaction may be necessary in advanced robotics and automation, especially in crucial scenarios like disaster relief or healthcare. Even though robotic technologies have come a long way, further research may still be necessary to address the difficulties in establishing safe and seamless human–robot interaction in these critical contexts [35]. Concentrating on these uncharted territories may improve the creation of dependable and intuitive robotic systems that can operate well in challenging real-world scenarios. Examining the shortcomings of present infrastructure designs in the context of extreme weather events may be one way to uncover research needs in resilient and climate-adaptive infrastructure. For example, more information may be needed about how these infrastructures operate when subjected to the cumulative stress of several climate events over a lengthy period, even though building materials and structural designs that promote resilience are well understood. Closing this gap might create an infrastructure that can adjust to long-term climate changes and be resilient to current problems. There could be gaps in how densely inhabited urban areas and green infrastructure are integrated into sustainable urban planning [36]. While the advantages of green infrastructure are widely known, further study may be necessary to maximize the effectiveness of these systems in densely populated metropolitan environments with constrained space. By filling this gap, creative solutions that satisfy the demands of city dwellers and optimize the environmental advantages of green infrastructure might be created. The main cognitive obstacles that must be overcome to identify research gaps are the incapacity to develop and interact with new knowledge, the difficulty of linking complicated information, and the difficulty of integrating multiple types of expertise [37].

These cognitive barriers might be compared to interlocking gears, where each stands for a crucial cognitive function required to advance the field of inquiry. Underdeveloped fields of knowledge result from specific brain processes needing to be improved, which emphasizes the need for more research. Researchers can accelerate the growth of their profession by bridging knowledge gaps and efficiently overcoming cognitive hurdles [38]. This method encourages the exploration of uncharted ground within a given topic, which fosters creativity and makes it easier to identify areas that require more research. For instance, to solve integration issues in smart infrastructure or human–robot interaction in advanced robotics, new frameworks for comprehending complex systems and their relationships must be developed in addition to technological developments. The gap method is vital to promote advancement in sectors where ideas must constantly evolve. Researchers contribute to the continuous expansion and evolution of the body of knowledge by concentrating on topics that require more investigation or in-depth comprehension. By doing this, their area will stay dynamic and adaptable to new possibilities and obstacles. The method strongly emphasizes critical thinking, interdisciplinary teamwork, and a dedication to creativity in the search for knowledge [39]. The leading causes of research gaps are three connected cognitive obstacles, as shown in Fig. 3: incapacity to create

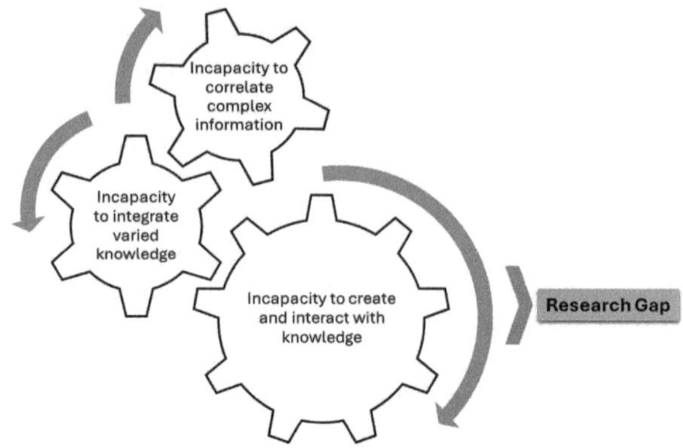

Fig. 3 Identifying research gaps through cognitive challenges (*source* authors)

and interact with knowledge, difficulty linking complicated information, and difficulty integrating various knowledge. These difficulties, which might be compared to interacting gears, impede the advancement of knowledge in a particular sector. When these cognitive processes are impaired, they reveal knowledge gaps that call for more investigation and show places where the body of information is deficient.

4 Exploring and Addressing Research Gaps: From Small Steps to Ground-Breaking Advances

Research is an ongoing process that involves questioning, investigating, and finding to ensure that information is continuously expanded [40]. This process is crucial for the progress of knowledge. During this process, researchers find gaps in the current body of knowledge. These gaps include regions in which there is a lack of comprehensive understanding, issues that have not been answered, or fresh chances for study that have not yet been taken advantage of. Research gaps can range from very modest methodological inadequacies to significant gaps in grasping key concepts within a topic [41]. A variety of factors can determine these gaps. For example, in artificial intelligence (AI) and modular construction, a research need may be discovered regarding integrating AI technologies with prefabrication procedures to improve buildings' efficient and high-quality construction. The specialized application of artificial intelligence in real-time decision-making during modular building is yet underexplored, even though AI has been the subject of much research. This presents a considerable possibility for innovation. It is essential to acknowledge that the significance of a research gap is not always equal to the size of the gap itself. Even what may appear to be relatively small gaps can have huge repercussions and make a big contribution to the development of a field [42]. A research need may, for instance, entail enhancing the interoperability of Internet-of-Things (IoT) devices inside smart building systems. This would be relevant in smart infrastructure and the Internet of Things (IoT). The resolution of this problem, even though it may appear to be of little consequence, might result in substantial breakthroughs in building automation, energy management, and comprehensive operational efficiency. For researchers to make valuable contributions, seeking new findings or building ground-breaking hypotheses is unnecessary. Instead, they might concentrate on evaluating the gaps that already exist, regardless of the magnitude of those gaps, and looking for inventive ways to remedy them [43]. As an illustration, using an artificial intelligence algorithm that has not been tested to improve the logistics of prefabricated building components might result in the acquisition of significant insights and improvements that can be implemented. Researchers can contribute to expanding knowledge and implementing novel solutions in their respective domains if they can recognize and fix the gaps in their expertise [44]. Figure 4 emphasizes the need to resolve research gaps to increase academic understanding and practical applications. It highlights that even minor contributions may significantly augment the existing knowledge base, resulting in advancements and enhancements in many domains. This strategy guarantees a consistent expansion in research and its tangible influence.

Exploring alternative ideas and theories that previous researchers have not thoroughly studied is a goal that is both vital and intellectually stimulating [45]. This is especially true within robust and climate-adaptive infrastructure and sustainable urban design. By reexamining historical data and established processes through innovative viewpoints, researchers can unearth parts of social phenomena, ideologies, or infrastructure initiatives that had been neglected. Conventional paradigms are susceptible to being challenged and may be transformed using this strategy [46]. For instance, in the field of resilient infrastructure, standard engineering techniques may be ineffective in tackling the growing consequences of climate change, such as the increased frequency and intensity of extreme weather events. The development of infrastructure systems that are resistant to these disturbances and capable of reacting dynamically to changing environmental circumstances is an essential area of study that is currently lacking. Increasing the resilience of urban settings may be accomplished by investigating adaptable materials and creative building processes, which

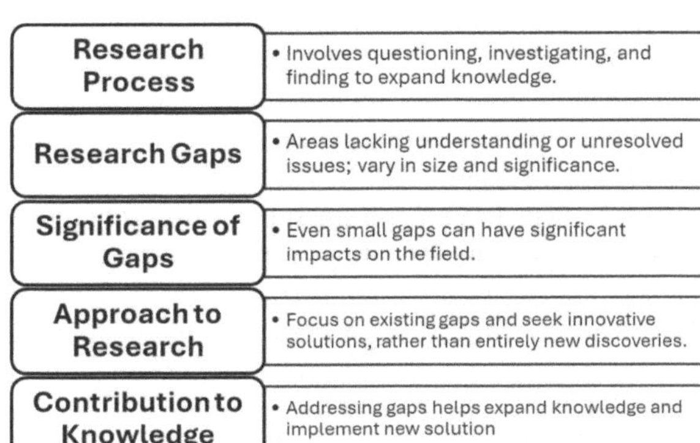

Fig. 4 Addressing research gaps to enhance knowledge contribution

can lead to designing robust and flexible infrastructure models. Expanding the boundaries of knowledge, refining existing ideas, and providing insights into practical applications depend on discovering and methodically examining research gaps [47]. Even slight gaps, such as those relevant to optimizing energy efficiency in sustainable urban design, can significantly contribute to the discipline's growth. For instance, although the advantages of green roofs and walls in urban settings, such as the reduction of the urban heat island effect and the improvement of air quality, have been widely documented, there are still gaps in our understanding of how effective they are over the long term in a variety of climatic situations. It is possible that addressing these gaps may result in considerable advances in sustainable urban planning and design, which will, in turn, increase the livability and sustainability of urban places. On the other hand, significant research gaps allow ground-breaking discoveries and paradigm changes. When it comes to climate-adaptive infrastructure, investigating the intersection of advanced technologies, ecological principles, and urban planning methodologies could create new frameworks for urban environments. These frameworks would not only reduce the negative impact on the environment, but they would also proactively improve the well-being of urban populations. In order to fill up these research gaps, it is necessary to use a multidisciplinary strategy that incorporates a wide range of academic disciplines and encourages collaborative efforts among academics, practitioners, and policymakers. This kind of collaboration makes it possible to develop creative answers to complex problems by easing the flow of ideas and skills across the borders of different fields of study. In the end, this all-encompassing technique not only fills in essential knowledge gaps but also guarantees that the research findings are transformed into meaningful and relevant social effects, contributing to an ecosystem rich in diversity and dynamic in its research. Research gaps of varying magnitude and significance present numerous prospects for improvement in civil engineering and other fields [48]. These gaps, ranging from small methodological errors to neglected historical viewpoints or underrepresented groups, provide a rich environment for improving our comprehension and fostering small-scale and big-picture discoveries. Researchers may close these gaps in understanding, improve on pre-existing ideas, and provide new opportunities for creativity. By concentrating on these gaps, the scientific community may expand our understanding and expedite development, enriching society. Table 3 thoroughly overviews significant architecture and civil engineering research needs, including sustainable materials, indoor environmental quality (IEQ), building information modeling (BIM), resilient design, human-centric design, and smart buildings. It highlights the most critical knowledge gaps, makes recommendations for future areas of inquiry to fill them, and describes the anticipated results of such efforts. This in-depth examination provides a strategic framework for improving the topic through focused academic investigation. The knowledge gained from this study might spur essential discoveries that would eventually help to create constructed environments that are more robust, sustainable, and focused on the needs of users.

Table 3 Overview of research gaps and potential impacts in civil engineering

Area of focus (example)	Specific research gap	Potential research directions	Anticipated impact
Sustainable materials [49]	Need to explore and evaluate new construction materials with minimal environmental impact	– Development of bio-based composites and self-healing materials – Assessment of material life cycles for sustainability and circularity	– Reduced environmental footprint of buildings – Enhanced energy efficiency and reduced carbon emissions
Indoor environmental quality (IEQ) [50]	Limited understanding of IEQ's impact on health, comfort, and productivity	– Investigation of long-term health effects of poor IEQ – Development of strategies to optimize air quality, lighting, acoustics, and thermal comfort	– Healthier and more productive indoor environments – Improved occupant well-being and quality of life
Building information modeling (BIM) [51]	Underutilization of BIM in enhancing construction efficiency and facility management	– Exploration of advanced BIM tools and methodologies – Integration of BIM with VR, AR, and AI – Assessment of BIM's long-term benefits	– Streamlined construction processes – Improved collaboration and project outcomes – Efficient facility management
Resilient and disaster-resistant design [52]	Insufficient strategies for resilient design to withstand extreme events	– Development of innovative structural systems and materials – Integration of resilient design into urban planning and policy frameworks	– Enhanced resilience of buildings and infrastructure – Safer communities and reduced disaster recovery time
Human-centric design [53]	Lack of research on the impact of design on human behavior and well-being	– Study of the effects of building layout and design on occupant behavior – Research on biophilic design and its impact on stress and cognition	– Improved mental and physical well-being – Design practices that enhance social connections and occupant satisfaction
Smart and adaptive buildings [54]	Need for research on intelligent systems integration in buildings	– Development of smart building automation and occupant-responsive environments – Exploration of adaptive facade systems and IoT integration	– Energy-efficient and adaptable buildings – Improved operational performance and occupant

5 Techniques for Determining Engineering Research Gaps

Identifying research gaps in the engineering industry to increase theoretical knowledge and practical solutions is essential. In addition to catalyzing innovation and technological advancement, this process is crucial in ensuring that research programs align with the ever-changing demands of society and industry [55]. The systematic discovery of these gaps helps researchers concentrate on underexplored and highly influential regions, increasing their work's relevance and significance. This is accomplished by directing their efforts into unexplored areas. The utilization of a wide variety of strategic approaches, such as systematic literature reviews, patent analysis, industry engagement, and trend analysis, makes it easier to get a thorough picture of the present research landscape. These techniques not only show areas that have not been adequately examined, but they also offer insights into prospective future directions. This helps to ensure that engineering research continues to be responsive to new challenges and possibilities as they arise. Ultimately, this method helps ensure that the field continues to develop and improve via continual evolution and advancement, which contributes to developing novel solutions that address crucial engineering challenges [56].

5.1 Engineering Systematic Literature Review

In engineering, the Systematic Literature Review (SLR) is an integral methodological framework utilized to identify research gaps within the discipline systematically. This is accomplished by undertaking an in-depth and rigorously organized review of the current corpus of academic literature. In contrast to traditional literature reviews, which may take a more narrative and selective approach, a systematic literature review (SLR) conforms to a highly severe, structured, and reproducible technique [57]. This strategy ensures that the review process is both thorough and free of bias. The SLR approach includes articulating research objectives clearly and concisely; establishing specific inclusion and exclusion criteria; and systematically identifying, selecting, and synthesizing relevant scholarly resources. When applied to the field of engineering, SLRs play a crucial role in the process of mapping the research landscape. This provides a comprehensive overview of the existing body of knowledge and facilitates identifying research topics that require more academic investigation [58]. For example, within Building Information Modeling (BIM) and Digital Twin Integration, an SLR can systematically uncover gaps in the application and integration of these technologies across various engineering subfields. These subfields include construction management, facility operations, and smart infrastructure development. Researchers can discover new trends by thoroughly analyzing the existing literature. One example of this is the integration of building information modeling (BIM) with emerging technologies such as the Internet of Things (IoT) and artificial intelligence (AI). These trends reflect unexplored areas that have significant potential for future innovation. In a similar vein, a systematic literature review (SLR) may indicate that there is a dearth of longitudinal studies that investigate the performance of green building materials under a variety of environmental situations. This would highlight the need for more concentrated and ongoing research. The quantitative evaluation of the research landscape is made possible by bibliometric analysis, which is a significant component of SLRs in engineering. This evaluation is accomplished by analyzing publication volumes, co-authorship networks, and citation patterns. The findings of this analysis shed light on the influence of research efforts within certain engineering areas and the interconnection of such activities.

Researchers can identify underexplored topics, emerging research themes, and dominant areas of focus in engineering scholarship with advanced bibliometric tools such as VOSviewer and CiteSpace [59]. These tools benefit this process because they make it easier to visualize complex bibliometric data. In the case of applying SLR approaches to the investigation of Artificial Intelligence (AI) in engineering, for instance, bibliometric analysis can shed light on the most significant publications, prominent researchers, and large collaboration networks influencing the course of AI research within the discipline. Furthermore, this study has the potential to bring to light critical research gaps, such as the requirement for additional research into the application of artificial intelligence in environmentally responsible engineering practices or the incorporation of AI with sophisticated materials for smart manufacturing processes. Researchers can systematically identify areas where the literature is insufficient, conflicting, or obsolete by utilizing a systematic literature review (SLR). This allows them to zero in on specific themes that deserve greater academic attention. In the case of the integration of Digital Twin technology with Building Information Modeling (BIM), for instance, an SLR may identify a significant gap in research that addresses the difficulties of real-time data integration and the scalability of these systems for large-scale infrastructure projects. Similarly, in green building materials, an SLR could identify a scarcity of studies that critically evaluate the environmental impacts of novel sustainable materials throughout their entire lifecycle compared to traditional construction materials [60].

This would bring to light the urgent need for comprehensive lifecycle assessments. To summarize, the SLR is an essential methodological instrument that significantly identifies research gaps within the engineering profession. Researchers can create an accurate map of the research landscape by thoroughly reviewing the current body of literature, utilizing extensive bibliometric analysis, and using advanced tools such as VOSviewer and CiteSpace. This meticulous process facilitates the identification of critical gaps. It provides strategic insights for future research directions, ensuring that scholarly inquiry in engineering remains dynamic and aligned with the discipline's evolving technological and societal challenges. The methodological framework for performing Systematic Literature Reviews (SLR) within the engineering discipline is shown in Table 4, which provides a complete overview of the approach. This table provides an overview of the goal, principal components, and tools employed in SLRs. It emphasizes SLRs' role in methodically identifying research gaps, mapping the research landscape, and giving strategic insights to guide future academic activities.

Table 4 Methodological framework for systematic literature reviews (SLR) in engineering

Aspect	Description	Key points
Purpose of SLR in engineering [61]	To methodically identify research gaps by conducting a rigorous, unbiased literature review	It facilitates a comprehensive understanding of the current state of knowledge, identifying underexplored areas that require further academic investigation
Methodological approach [62]	Adheres to a stringent, structured, and replicable process to ensure thoroughness and objectivity	It involves precisely articulating research objectives, establishing explicit inclusion and exclusion criteria, and systematically selecting and synthesizing scholarly resources
Key components of SLR [63]	– Clearly defining research objectives – Establishing rigorous inclusion/exclusion criteria – Systematic identification, selection, and synthesis of literature	Ensures the methodological rigor and comprehensiveness of the review process, minimizing bias and enhancing the reliability and validity of the findings
Bibliometric analysis [64]	Utilizes quantitative methods to evaluate publication volumes, co-authorship networks, and citation patterns	Employs advanced bibliometric tools to analyze and visualize the research landscape systematically, thereby identifying underexplored topics, emerging themes, and dominant research areas
Role in mapping the research landscape [65]	Enables a detailed and systematic mapping of the current research landscape within the engineering discipline	It provides a critical overview of existing academic knowledge, facilitates the identification of research gaps, and guides the strategic direction of future scholarly inquiries
Identification of research gaps [66]	Detects areas where the literature is sparse, contradictory, or outdated	Systematically highlights gaps in knowledge, underscoring the necessity for further research in areas that have been insufficiently explored or inadequately addressed in existing studies

(continued)

Table 4 (continued)

Aspect	Description	Key points
Strategic insights for future research [67]	Offers a structured framework for guiding future research efforts toward areas of high academic and practical significance	Ensures that future research endeavors align with the discipline's evolving needs, contributing to advancing theoretical knowledge and practical applications
Tools utilized [68]	Advanced bibliometric tools such as VOSviewer and CiteSpace	Facilitates the rigorous analysis of complex bibliometric data, enhancing the ability to identify trends and gaps in the engineering research landscape systematically

5.2 Technological Trends and Patent Analysis

Patent analysis is an essential technique for finding areas of study that need attention, especially in applied engineering fields where innovation is directly linked to technical progress [69]. Patents represent the most advanced and innovative technical advancements, providing insight into the direction of the industry and the latest breakthroughs. By doing patent analysis, researchers may spot trends in the invention, locate places with a high concentration of technical activity, and determine discrepancies between academic research and industrial accomplishments [70]. Patent landscaping is a crucial approach used in patent research to represent the distribution of patents across different technological domains graphically. This approach is precious in domains like resilient and climate-adaptive infrastructure, where incorporating new materials and design concepts is essential for creating infrastructure that can endure the effects of climate change. By looking at patent landscapes, researchers can find groups of innovations focused on materials and technologies resistant to climate change and areas with few patents, which may suggest uncharted areas for further research. For instance, a study of patents might uncover a significant focus on developing materials that can endure severe weather conditions. However, it may also indicate a scarcity of patents that specifically address how to incorporate these materials into current metropolitan infrastructures. Patent citation analysis improves our comprehension by tracking the influence of certain patents on later technical advancements [71]. This strategy is highly beneficial in sustainable urban design, where the convergence of green technology and urban planning is paramount. By analyzing patents about sustainable technologies, such as energy-efficient building systems or water management innovations, researchers can discover the technological implications and research directions that have not been extensively explored in academic literature. For example, a review of citations might show that although some patents related to green building technologies are often referenced, suggesting their significance, there may be a limited number of following patents or research articles specifically addressing incorporating these technologies into large-scale urban planning. Patent analysis is crucial for identifying current technical trends and predicting future research requirements. Researchers may anticipate prospective obstacles by surveilling technological trajectories, identifying nascent breakthroughs, and aligning their research endeavors accordingly. Monitoring patents associated with smart materials and adaptive design technologies in the resilient and climate-adaptive infrastructure field might aid in forecasting future requirements, such as the necessity for infrastructures capable of dynamically adapting to climatic fluctuations. Similarly, in sustainable urban design, the examination of patents may provide valuable guidance to academics in identifying new technologies that will play a crucial role in developing environmentally friendly cities capable of adapting to future urban difficulties. Patent analysis is a potent instrument for identifying research deficiencies and directing future academic pursuits in engineering [72]. Researchers can enhance their grasp of the technical environment, identify areas where academic research lags industrial innovation, and forecast future engineering discipline trends using techniques like patent landscaping and citation analysis. This approach is especially beneficial in practical fields such as resilient and climate-adaptive infrastructure and sustainable urban design, where keeping up with technological advancements is crucial for tackling the intricate difficulties of the future. Figure 5 thoroughly summarizes essential patent analysis methods, such as patent landscaping and citation analysis. These approaches are crucial for finding gaps in research, monitoring technological progress, and predicting future research orientations in applied engineering domains. This helps to ensure that academic research stays in line with industry trends and advances.

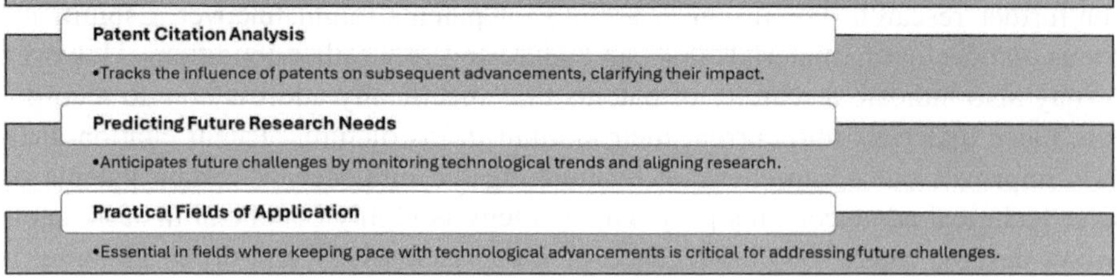

Fig. 5 Overview of patent analysis techniques (*source* authors)

5.3 Industry Cooperation and Advice

It is necessary to actively and meaningfully include industry stakeholders to identify research needs applicable in practice and significant in academia [4]. Collaboration between academia and industry is essential for directing academics in tackling real-world engineering challenges, ensuring that their work is theoretically rigorous and practically relevant [4]. This mutually beneficial partnership serves as a platform for collaboration between academia and industry, enabling the creation of knowledge advantageous to both sectors. This collaboration results in the development of practical and rigorous inventions. A significant advantage of business cooperation is the ability to utilize advanced technological tools and exclusive datasets typically unavailable in academic institutions. Industry partners usually work at the forefront of technology, employing sophisticated tools and processes to address intricate issues. Academic researchers may boost the quality and relevance of their research by working with industry leaders and utilizing their tools and data. This kind of access allows researchers to experiment with theoretical models and hypotheses in real-life situations, providing a rigorous validation process that enhances the reliability and trustworthiness of their results. Furthermore, industry collaboration enhances comprehension of professionals' difficulties in their day-to-day activities [73]. This firsthand understanding benefits academic scholars, enabling them to recognize and concentrate on the most urgent issues within the discipline. In subjects such as systems engineering, industrial processes, and materials science, professionals in the industry typically face specific obstacles that may not be immediately obvious in academic environments. Researchers can customize their studies to tackle these obstacles by engaging in frequent discussions and cooperative endeavors, guaranteeing that their work is pertinent and influential [74]. The value of industry collaboration in matching academic research with actual engineering difficulties is shown in Table 5, which describes the importance of this alignment. This highlights the advantages of having access to cutting-edge tools and exclusive data, both of which contribute to improving the quality and relevance of research. This partnership aims to bridge the gap between theory and practice and ensure that it influences the real world by concentrating on critical industrial demands. This collaboration fosters construction, prefabrication, and the Internet-of-Things innovation.

Industry–academic cooperation is pivotal in driving innovation, particularly in modular construction and prefabrication. For instance, a collaboration between a leading construction firm and a university's civil engineering department resulted in the development of innovative, lightweight prefabricated modules [75]. These modules, designed with high-strength materials, optimized shipping costs and assembly time while meeting stringent safety and energy standards. Another example is the partnership between a data science institution and an IoT startup to create an intelligent building management system for a commercial office complex. This system used real-time data from IoT sensors to optimize energy use, significantly enhancing building efficiency and occupant comfort. Additionally, a construction company worked with academic researchers to retrofit an office building with smart

Table 5 The role of industry collaboration in enhancing academic research

Key aspect	Summary
Collaboration importance	Essential for aligning academic research with real-world engineering challenges and practical relevance
Benefits of industry partnership	It provides access to advanced tools and exclusive data and enhances research quality and applicability
Understanding industry needs	Facilitates focus on critical industry challenges, ensuring research addresses relevant and pressing issues
Innovation impact	Promotes innovation in construction, prefabrication, and IoT, bridging theory, and practice

technology, resulting in adaptive energy management and improved occupant conditions. These collaborations highlight the essential role of industry partnerships in bridging the gap between theoretical research and practical application. Through such partnerships, academia gains access to advanced tools, data, and real-world challenges, ensuring that research remains relevant and impactful [76]. This synergy fosters innovation across fields like modular construction, smart buildings, and IoT, leading to practical solutions with significant real-world impacts.

5.4 Engineering Research Trends

Examining engineering research trends is crucial for comprehending the discipline's present emphasis and predicting future trajectories [77]. This study examines financial allocations, citation metrics, and publishing trends to discover areas experiencing increased interest and prospective areas where there may be a lack of research. Using databases like Web of Science and Scopus allows for a thorough analysis of academic publications, allowing researchers to monitor changes in study focus and developing trends [78]. Analyzing publication trends, which includes examining frequency, citation rates, and the appearance of new keywords, provides valuable insights into the changing engineering research field. These indications identify regions that are receiving attention and may be overlooked, guiding future study endeavors. Research in smart and adaptive buildings has grown substantially, primarily because it aligns with sustainable development goals. Building Information Modeling (BIM) is gaining more attention because of its capacity to integrate many elements of construction and lifecycle management into a unified digital framework. This demonstrates how emergent subjects may quickly become central areas of focus in the industry.

Examining research funding statistics in conjunction with publishing trends yields valuable insights into the allocation of financial resources, reflecting the goals of governmental bodies and funding agencies [79]. For instance, the increasing fascination with smart and adaptive buildings is frequently supported by funding to advance energy efficiency and sustainability. BIM has garnered significant funding due to its ability to facilitate innovation in building processes and project management.

This financial analysis assists researchers in aligning their work with established and new disciplines, guaranteeing its relevance and long-term viability. To predict future research directions, it is necessary to combine information from funding and publishing patterns with broader socio-economic, technical, and regulatory changes [78]. By employing this predictive methodology, researchers may proactively foresee forthcoming obstacles and possibilities, guaranteeing that their work stays future-oriented and adaptable to the ever-changing engineering field. The growing emphasis on energy efficiency and smart infrastructure regulations is fueling research in the smart and adaptive buildings field, suggesting that this sector will continue to be a top priority [80].

An all-encompassing method for determining research requirements entails doing systematic literature reviews, analyzing patents, engaging with industry, and analyzing trends [81]. Systematic literature evaluations identify researched topics and places where knowledge is lacking, whereas patent analysis offers valuable information on real-world applications and trends in innovation. For instance, doing a patent analysis in BIM can uncover patterns in digital twin technology, bolstering its utilization in the field of building. Engaging with industry ensures that academic research aligns with practical requirements, such as the need for increasingly inventive and adaptable construction solutions. Trend analysis utilizes this data to get insights into larger scale changes, directing study toward areas with significant effects [82].

Using these techniques, engineering research may maintain its leading position by tackling crucial obstacles and propelling the discipline forward. This strategic strategy promotes acquiring information and guarantees that research results are applicable and adaptable to technological progress and social demands. Understanding and evaluating engineering research trends is essential for guiding future research endeavors [78]. Researchers may detect developing disciplines such as smart and adaptive buildings and BIM by analyzing publishing and funding patterns and incorporating broader social trends. This allows them to ensure the continued relevance and effect of their work. This strategy maintains the vitality of engineering research, guaranteeing its ongoing ability to tackle significant obstacles and make substantial contributions to technical and social advancement.

6 Conclusion

Academic inquiry requires identifying research gaps, especially in subjects like engineering, where technical innovation and practical application drive advancement. Finding these gaps advances theoretical understanding and practical solutions. This chapter has stressed the need to methodically identify and resolve research gaps to promote innovation, influence research paths, and solve real-world problems. Research gaps show incomplete, obsolete, or nonexistent information in a field. Recognizing these gaps helps academics focus on areas where they can significantly affect. Identifying these gaps keeps engineering research current and meets social and technical requirements. Identifying research gaps requires a thorough literature study, analysis of new technologies and trends, and close engagement with industry partners. This interdisciplinary approach captures engineering difficulties' complexity and aligns research with academia and industry's changing needs. Several methods are needed to detect research gaps. Systematic literature reviews (SLRs) are excellent for mapping field knowledge. SLRs can identify understudied topics by rigorously analyzing previous research. In sustainable building information modeling (BIM), an SLR may lack research incorporating lifecycle assessment (LCA) approaches, indicating a need for additional research. Patent analysis is helpful in applied disciplines where innovation is tightly tied to technology. Patent landscapes and citation networks can reveal patterns, technical gaps, and places where academic research lags behind industrial breakthroughs. New materials and design concepts are essential for climate change mitigation in robust, climate-adaptive infrastructure. Identifying research gaps is beneficial yet challenging. The transdisciplinary character of relevant material makes defining a research gap difficult. Sustainable architecture research may focus on materials science, environmental psychology, or urban politics, each presenting a unique viewpoint on a problem. Identifying a significant and novel gap is another difficulty. If a gap in literature has been addressed in similar domains, the research may not offer anything. Thus, it is crucial to identify existing gaps that might contribute to the area. Presenting the research gap compellingly and relevant to academic discourse can be difficult. Researchers must identify and structure a gap to appeal to academics and industry practitioners. To make the gap more relevant and appealing, contextualize it within more significant aims like minimizing building carbon footprints. Identifying research gaps organizes and advances research in crucial domains, including architecture, urban design, and sustainable building. Researchers may help improve urban sustainability, build resilience, and construct environmental effects by carefully identifying and filling literature gaps. However, this method has drawbacks. After identifying a gap, researchers must ensure it is essential, compelling, and relevant. This needs a thorough literature study, precise interpretation of findings, and effective communication to a broader audience in academia and professional practice. Identifying research gaps is essential for innovation and sustainability in critical fields. Researchers can advance their disciplines by finding and filling knowledge gaps. This method requires a thorough literature analysis, knowledge of upcoming trends and technology, and constructive communication with industry stakeholders. This holistic approach keeps research active, sensitive to real-world concerns, and able to make theoretical and practical contributions.

References

1. Darwish H (2018) Expanding industrial thinking by formalizing the industrial engineering identity for the knowledge era. Doctoral dissertation, North-West University
2. Bastow S (2014) The impact of the social sciences: how academics and their research make a difference

3. May T, Perry B (2022) Social research: issues, methods and process. McGraw-Hill Education (UK)
4. Dwivedi YK, Jeyaraj A, Hughes L, Davies GH, Ahuja M, Albashrawi MA, Al-Busaidi AS, Al-Sharhan S, Al-Sulaiti KI, Altinay L, Amalaya S (2024) "Real impact": challenges and opportunities in bridging the gap between research and practice–making a difference in industry, policy, and society. Int J Inform Manage 102750
5. Kavuri AS, Milne A (2019) FinTech and the future of financial services: what are the research gaps?
6. Rawat G, Sharma D, Ashok PP, Bairathi V (2024) The fundamentals of research methodology. Academic Guru Publishing House
7. Bampoh DK, Sdunzik J, Sinfield JV, McDavid L, Burgess WD (2024) Investigating the robustness and relevance of an evidence-based sense-making construct to bridge the research-practice gap in cross-sector partnerships. Bus Strategy Dev 7(1):e301
8. Dwivedi YK, Jeyaraj A, Hughes L, Davies GH, Ahuja M, Albashrawi MA, Al-Busaidi AS, Al-Sharhan S, Al-Sulaiti KI, Altinay L, Amalaya S (2024) Real impact: challenges and opportunities in bridging the gap between research and practice–making a difference in industry, policy, and society. Int J Inform Manage 102750
9. Jacks BS, Ajala OA, Lottu OA, Okafor ES (2024) Theoretical frameworks for ICT for development: impact assessment of telecommunication infrastructure projects in Africa and the US. World J Adv Res Rev 21(3):394–400
10. Aboelmaged M, Bani-Melhem S, Al-Hawari MA, Ali I (2024) Product innovation research over the past 60 years: a bibliometric analysis of intellectual structure and emergent trends. Benchmarking: Int J 31(6):2034–2060
11. Fàbregues S, Escalante-Barrios EL, Molina-Azorin JF, Hong QN, Verd JM (2021) Taking a critical stance towards mixed methods research: a cross-disciplinary qualitative secondary analysis of researchers' views. PLoS ONE 16(7):e0252014
12. Garg S, Baliyan N (2021) Android security assessment: a review, taxonomy and research gap study. Comput Secur 100:102087
13. Hopmans JW, Qureshi AS, Kisekka I, Munns R, Grattan SR, Rengasamy P, Ben-Gal A, Assouline S, Javaux M, Minhas PS, Raats PAC (2021) Critical knowledge gaps and research priorities in global soil salinity. Adv Agron 169:1–191
14. Kriz A, Nailer C, Jansen K, Potocnjak-Oxman C (2021) Teaching-practice as a critical bridge for narrowing the research-practice gap. Ind Mark Manage 92:254–266
15. Saxe S, Guven G, Pereira L, Arrigoni A, Opher T, Roy A, Arceo A, Von Raesfeld SS, Duhamel M, McCabe B, Panesar DK (2020) Taxonomy of uncertainty in environmental life cycle assessment of infrastructure projects. Environ Res Lett 15(8):083003
16. Moody DL (2005) Theoretical and practical issues in evaluating the quality of conceptual models: current state and future directions. Data Knowl Eng 55(3):243–276
17. Williams PC, Jones M, Snelling TL, Duguid R, Moore N, Dickson B, Wu Y, Saunders J, Wijeratne P, Douangnouvong A, Ashley EA (2024) Coverage gaps in empiric antibiotic regimens used to treat serious bacterial infections in neonates and children in Southeast Asia and the Pacific. The lancet regional health-Southeast Asia, p 22
18. Meskus M, Tikka E (2024) Speculative approaches in social science and design research: methodological implications of working in 'the gap' of uncertainty. Qual Res 24(2):209–228
19. Stefaniak J (2021) Determining environmental and contextual needs. Learning design: principles, processes, and praxis. EdTech Books. https://edtechbooks.org/id/needs_analysis
20. Zheng Y, Slade TJ, Hu L, Tan XY, Luo Y, Luo ZZ, Xu J, Yan Q, Kanatzidis MG (2021) Defect engineering in thermoelectric materials: what have we learned? Chem Soc Rev 50(16):9022–9054
21. Burton S, Habli I, Lawton T, McDermid J, Morgan P, Porter Z (2020) Mind the gaps: assuring the safety of autonomous systems from an engineering, ethical, and legal perspective. Artif Intell 279:103201
22. Xames MD, Topcu TG (2024) A systematic literature review of digital twin research for healthcare systems: research trends, gaps, and realization challenges. IEEE Access

23. World Health Organization (2024) Evidence and research gaps identified during development of policy guidelines for tuberculosis. World Health Organization
24. Nishikawa-Pacher A (2023) A typology of research discovery tools. J Inf Sci 49(4):1086–1095
25. Dubois NS, Gomez A, Carlson S, Russell D (2020) Bridging the research-implementation gap requires engagement from practitioners. Conserv Sci Pract 2(1):e134
26. Kharomah S, Setiawan D, Zubaidah S (2024) Developing puzzle based case study to improve cognitive learning outcomes of high school students. Prisma Sains: Jurnal Pengkajian Ilmu dan Pembelajaran Matematika dan IPA IKIP Mataram 12(3):411–427
27. Salehi R, Masoudi-Asl I, Gorji HA, Gharaee H (2024) Gap analysis of strategies for promoting interprofessional teams in healthcare units. J Health Organ Manage
28. Govindan K, Demartini M, Formentini M, Taticchi P, Tonelli F (2024) Unravelling and mapping the theoretical foundations of sustainable supply chains: a literature review and research agenda. Transp Res Part E: Logistics Transp Rev 189:103685
29. Locke K, Golden-Biddle K (1997) Constructing opportunities for contribution: structuring intertextual coherence and "problematizing" in organizational studies. Acad Manag J 40(5):1023–1062. https://doi.org/10.2307/256926
30. Kuran O (2024) Closing the theory-practice gap: socio-economic approach and action research in management. Syst Pract Action Res 1–28
31. Verdier T (2024) A cost-optimal analysis for nearly zero-carbon timber building solutions in line with the EPBDrecast 2020 for Belgium= Multi-objective optimization, Life Cycle Cost, Life Cycle Carbon, Nearly Zero Carbon Buildings, Timber, Green Building. Doctoral dissertation, Politecnico di Torino
32. Akbari S, Sheikhkhoshkar M, Rahimian FP, El Haouzi HB, Najafi M, Talebi S (2024) Sustainability and building information modelling: integration, research gaps, and future directions. Autom Constr 163:105420
33. Laviosa S (2021) Corpus-based translation studies: theory, findings, applications, vol 17. Brill
34. Fink A (2019) Conducting research literature reviews: from the internet to paper. Sage publications
35. Makris S, Michalos G, Dimitropoulos N, Krueger J, Haninger K (2024) Seamless human–robot collaboration in industrial applications. In: CIRP novel topics in production engineering, vol 1. Springer Nature Switzerland, Cham, pp 39–73
36. Hou W, Li X (2024) Assessing urban green infrastructure: a simple and practical measure of its spatial distribution equity and a comprehensive evaluation. Ecol Ind 158:111408
37. Hew KF, Brush T (2007) Integrating technology into K-12 teaching and learning: current knowledge gaps and recommendations for future research. Educ Tech Res Dev 55:223–252
38. McCabe A, Parker R, Osegowitsch T, Cox S (2023) Overcoming barriers to knowledge co-production in academic–practitioner research collaboration. Eur Manag J 41(2):212–222
39. Moirano R, Sánchez MA, Štěpánek L (2020) Creative interdisciplinary collaboration: a systematic literature review. Thinking Skills Creativity 35:100626
40. Neuman D (2014) Qualitative research in educational communications and technology: a brief introduction to principles and procedures. J Comput High Educ 26:69–86
41. Kiley M, Wisker G (2009) Threshold concepts in research education and evidence of threshold crossing. High Educ Res Dev 28(4):431–441
42. McIntyre D (2005) Bridging the gap between research and practice. Camb J Educ 35(3):357–382
43. Stigler JW, Hiebert J (2009) The teaching gap: best ideas from the world's teachers for improving education in the classroom. Simon and Schuster
44. Grimshaw JM, Eccles MP, Lavis JN, Hill SJ, Squires JE (2012) Knowledge translation of research findings. Implement Sci 7:1–17
45. Alvesson M, Sandberg J (2024) The art of phenomena construction: a framework for coming up with research phenomena beyond 'the usual suspects.' J Manage Stud 61(5):1737–1765
46. Schuurman GW, Cole DN, Cravens AE, Covington S, Crausbay SD, Hoffman CH, Lawrence DJ, Magness DR, Morton JM, Nelson EA, O'Malley R (2022) Navigating ecological transformation: resist–accept–direct as a path to a new resource management paradigm. Bioscience 72(1):16–29

47. Hulland J (2020) Conceptual review papers: revisiting existing research to develop and refine theory. AMS Rev 10(1):27–35
48. Zar A, Hussain Z, Akbar M, Rabczuk T, Lin Z, Li S, Ahmed B (2024) Towards vibration-based damage detection of civil engineering structures: overview, challenges, and future prospects. Int J Mech Mater Des 20(3):591–662
49. Nwokediegwu ZQS, Ilojianya VI, Ibekwe KI, Adefemi A, Etukudoh EA, Umoh AA (2024) Advanced materials for sustainable construction: a review of innovations and environmental benefits. Eng Sci Technol J 5(1):201–218
50. Ackley A, Olanrewaju OI, Oyefusi ON, Enegbuma WI, Olaoye TS, Ehimatie AE, Ukpong E, Akpan-Idiok P (2024) Indoor environmental quality (IEQ) in healthcare facilities: a systematic literature review and gap analysis. J Build Eng 108787
51. Elkadi K (2024) BIM-enabled facility management framework and simulation. Doctoral dissertation, George Mason University
52. Parab RR, Jariwala HI, Karale KN, Subhan KMS, Sameen R (2022) Innovations in disaster-resistant construction and practices. Int J Res Eng Sci Manage 5(1):178–188
53. Yitmen I, Almusaed A, Alizadehsalehi S (2023) Investigating the causal relationships among enablers of the construction 5.0 paradigm: integration of operator 5.0 and society 5.0 with human-centricity, sustainability, and resilience. Sustainability 15(11):9105
54. Almusaed A, Yitmen I (2023) Architectural reply for smart building design concepts based on artificial intelligence simulation models and digital twins. Sustainability 15(6):4955
55. Aithal PS, Maiya AK (2023) Innovations in higher education industry–shaping the future. Int J Case Stud Business, IT Educ (IJCSBE) 7(4):283–311
56. Almusaed A, Yitmen I, Almssad A (2023) Reviewing and integrating AEC practices into industry 6.0: strategies for smart and sustainable future-built environments. Sustainability 15(18):13464
57. Krüger J, Lausberger C, von Nostitz-Wallwitz I, Saake G, Leich T (2020) Search. Review. Repeat? An empirical study of threats to replicating SLR searches. Empir Softw Eng 25:627–677
58. Khan MU, Sherin S, Iqbal MZ, Zahid R (2019) Landscaping systematic mapping studies in software engineering: a tertiary study. J Syst Softw 149:396–436
59. Guiling Y, Panatik SA, Sukor MSM, Rusbadrol N, Cunlin L (2022) Bibliometric analysis of global research on organizational citizenship behavior from 2000 to 2019. Sage Open 12(1):21582440221079898
60. Khan JS, Zakaria R, Shamsudin SM, Abidin NIA, Sahamir SR, Abbas DN, Aminudin E (2019) Evolution to emergence of green buildings: a review. Admin Sci 9(1):6
61. Zhang H, Babar MA (2013) Systematic reviews in software engineering: an empirical investigation. Inf Softw Technol 55(7):1341–1354
62. Aguinis H (2023) Research methodology: best practices for rigorous, credible, and impactful research. SAGE Publications
63. Thomé AMT, Scavarda LF, Scavarda AJ (2016) Conducting systematic literature review in operations management. Prod Planning Control 27(5):408–420
64. Uddin S, Hossain L, Abbasi A, Rasmussen K (2012) Trend and efficiency analysis of co-authorship network. Scientometrics 90(2):687–699
65. Godfrey E, Parker L (2010) Mapping the cultural landscape in engineering education. J Eng Educ 99(1):5–22
66. Cawley P (2018) Structural health monitoring: closing the gap between research and industrial deployment. Struct Health Monit 17(5):1225–1244
67. Grover V, Chiang RH, Liang TP, Zhang D (2018) Creating strategic business value from big data analytics: a research framework. J Manag Inf Syst 35(2):388–423
68. Meng L, Wen KH, Brewin R, Wu Q (2020) Knowledge atlas on the relationship between urban street space and residents' health—a bibliometric analysis based on VOSviewer and CiteSpace. Sustainability 12(6):2384
69. Gerken JM, Moehrle MG (2012) A new instrument for technology monitoring: novelty in patents measured by semantic patent analysis. Scientometrics 91(3):645–670

70. Knight HJ (2012) Patent strategy: for researchers and research managers. John Wiley & Sons
71. Liu P, Zhou W, Feng L, Wang J, Lin KY, Wu X, Zhang D (2024) Mapping and comparing the technology evolution paths of scientific papers and patents: an integrated approach for forecasting technology trends. Scientometrics 129(4):1975–2005
72. Brandão Neto N, Faria LDA, Melo FCLD (2024) Identifying technological trends and promoting strategies to boost innovation and technology transfer: a case study on the patent portfolio of Brazilian public research institutions. J Aerosp Technol Manag 16:e1624
73. Mavri A, Ioannou A, Loizides F (2023) A model for enhancing creativity, collaboration and pre-professional identities in technology-supported cross-organizational communities of practice. Educ Inform Technol 1–42
74. Maiya AK, Aithal PS (2023) A review based research topic identification on how to improve the quality services of higher education institutions in academic, administrative, and research areas. Int J Manage Technol Soc Sci (IJMTS) 8(3):103–153
75. Luo T, Xue X, Tan Y, Wang Y, Zhang Y (2021) Exploring a body of knowledge for promoting the sustainable transition to prefabricated construction. Eng Constr Archit Manag 28(9):2637–2666
76. Malhotra R, Massoudi M, Jindal R (2023) An alumni-based collaborative model to strengthen academia and industry partnership: the current challenges and strengths. Educ Inf Technol 28(2):2263–2289
77. Laato S, Mäntymäki M, Islam AN, Hyrynsalmi S, Birkstedt T (2023) Trends and trajectories in the software industry: implications for the future of work. Inf Syst Front 25(2):929–944
78. Dwivedi YK, Sharma A, Rana NP, Giannakis M, Goel P, Dutot V (2023) Evolution of artificial intelligence research in technological forecasting and social change: research topics, trends, and future directions. Technol Forecast Soc Chang 192:122579
79. Lepori B, Jongbloed B, Hicks D (2023) Introduction to the handbook of public funding of research: understanding vertical and horizontal complexities. In Handbook of public funding of research. Edward Elgar Publishing, pp 1–19
80. Apanavičienė R, Shahrabani MMN (2023) Key factors affecting smart building integration into smart city: technological aspects. Smart Cities 6(4):1832–1857
81. Moreira A, Navaia E, Ribau C (2024) Innovation capabilities and their dimensions: a systematic literature review. Int J Innov Stud
82. Gros M, Zilkey DR, Griffiths KT, Pham J, MacKeigan PW, Taranu ZE, Aulard C, Baud A, Garner RE, Ghanbari H, Lachapelle M (2023) Long-term environmental changes in the Canadian boreal zone: synthesizing temporal trends from lake sediment archives to inform future sustainability. Environ Rev 31(3):509–526

Formulating a Problem Statement

Abstract In academic research, the problem statement is a crucial component that forms the basis for the research process. An in-depth discussion of the significance of developing a clear issue statement is covered in this chapter, along with how it affects the study design, data gathering, and analysis. The chapter highlights how a specific and unambiguous issue description fills in the gaps between the objectives and the research question, keeping the study effective, focused, and cohesive. A well-written issue statement clearly defines the differences between the intended and present states, establishing the investigation's boundaries. It also demonstrates the research's applicability and justifies the study's significance for the scholarly community and the public. The chapter emphasizes the iterative process of creating a problem statement and the need to give the problem's context, stakeholders, and possible effects serious thought. Researchers may formulate a problem statement that effectively steers the research process and significantly contributes to the field of study by having a solid knowledge of these components. This chapter offers critical insights into the strategic development of problem statements, guaranteeing that research endeavors are grounded in solid methodology and line with more general scholarly and practical objectives.

1 Introduction to the Thematic Area

One of the most important aspects of any research project is the issue statement, which is the basis for the construction study [1]. Providing direction for both the technique and the analysis, it serves as the connecting element between the research question and the objectives. Formulating a problem statement is an essential step in the beginning stages of any research effort, commercial venture, or endeavor to solve an issue [2]. Specifically, it acts as the conceptual cornerstone that guides the future stages of the process, which include the study's design, the collecting of data, and the analysis of the outcome. Formulating a problem statement that is both specific and well-structured is essential to guarantee that the efforts made to solve the identified problem are focused, coherent, and ultimately triumphant [3]. When this information

A. Almusaed et al., *Practice of Research Methodology in Civil Engineering and Architecture*, https://doi.org/10.1007/978-3-031-97393-2_3

is available, efforts to tackle the problem may become more cohesive, coordinated, and efficient. This might result in less than optimum outcomes and the possibility for resources to be misused. A problem statement is generally characterized as a condensed summary of a problem or void that must be addressed. In addition to setting the parameters within which the study or project will be carried out, it outlines the disparity between the existing and wanted states [4]. For example, a well-crafted issue statement will answer three critical:

- *What constitutes a research problem?*
- *What makes it so significant?*
- *Who is affected by it?*

Research problems are issues that have not been addressed, questions that have not been answered satisfactorily, or subjects that need more explanation in the context of the already available information [5]—research endeavors to fill existing knowledge gaps by refining or modifying pre-existing concepts, methodologies, or processes. Different academics define the "research problem" differently, highlighting that others have varied ideas about what constitutes a research challenge. A clear issue description is essential when developing a shared understanding among stakeholders. This is especially true in projects that include many disciplines and are complicated. Because of this shared grasp, misunderstandings are reduced to a minimum, and it is ensured that all efforts are directed toward the accomplishment of a single objective [6]. Additionally, the issue statement helps to restrict the focus of the research or project, which ensures that resources and efforts are directed explicitly toward tackling the components of the problem that are the most significant. In addition, it serves as a standard against which the accomplishments of the research or project may be measured. In a well-defined problem statement, it is abundantly evident whether the solutions that have been devised have successfully addressed the problem that has been recognized. Often, numerous essential components make up a thorough issue statement [7]. These components include the problem itself, the context, the effect, and the audience. Clear articulation of the problem is required to determine the gap between the current circumstances and the results sought. It should be explicit enough to be actionable, but it should also be wide enough to allow for original and creative options. This backdrop gives the required background information, outlining the events that have led to the problem. It also includes pertinent historical facts, present trends, and a review of the already available literature. It is essential to provide this background to demonstrate the problem's relevance within its larger setting [8]. By elaborating on the implications of addressing the issue, the impact component highlights how urgent it is to find a solution to the problem. Depending on the nature of the issue, this influence may be defined in terms of the economic cost, the amount of time, the resources, or the social repercussions of the situation. Finally, the audience is determined, which includes determining who is impacted by the issue and who stands to gain from its resolution. This may include individuals, organizations, or the community [9].

Research problems are the essential guiding principle for all academic investigations. Research often centers around a specific topic or subject, and the whole research process aims to resolve this fundamental issue, including data gathering and analysis. Research questions in this context define the extent of the study, encompassing the specific population under investigation, pertinent factors, and critical topics to be examined. Research issues in architectural engineering can entail intricate complexities and necessitate a systematic approach. An example of a study subject in this field may be incorporating renewable energy technology into modern architectural designs. This project aims to investigate methods for integrating solar panels and wind turbines into urban residential buildings while ensuring that the constructions remain structurally sound, visually pleasing, and economically viable. A possible study inquiry may be: "What are the most efficient methods for seamlessly incorporating renewable energy technologies into urban residential structures while ensuring their structural robustness and aesthetic attractiveness?" This question would serve as a guiding principle throughout the study, encompassing data collection, analysis, and interpretation [10]. Architectural engineering research utilizes CAD simulations, energy consumption models, and prototype testing. The choice of methodologies is determined by their suitability and efficacy in answering the research subject. Furthermore, this research can have wider ramifications, including urban planning, mitigating building carbon emissions, and advocating for sustainable urban development [11]. Clearly defined research problems and questions are crucial for directing the research process and providing a systematic method for discovering, assessing, and implementing answers to current difficulties, especially in the built environment and sustainability fields.

- *A Review of the Definition of a Research Problem*

The research problem, as stated by the author, pertains to a situation where there is no instant remedy, suggesting that identifying a problem involves recognizing an existing gap or obstacle that has not been handled. Dwivedi et al. (2024) highlight that achieving meaningful outcomes from academic research is an intricate and diverse undertaking, marked by conflicting agendas and crucial benchmarks for success [12]. These elements jointly impact the extent and importance of the concrete and quantifiable results obtained from research endeavors. This highlights the complex nature of making a significant effect on academic research. Dehalwar (2024) takes this further; he defines a problem as a summary of information requiring a solution. He suggests that difficulties are recognized by diligent observation and study of facts or knowledge already available [13]. According to Treffinger et al. (2023), a situation is problematic if it can change. According to this viewpoint, a problem is characterized not only by the presence of a problem but also by the potential for finding and putting a solution in place. The dynamic character of problem-solving is highlighted by the emphasis on the possibility of change [14]. The problem is addressed through innovative and practical techniques. One of the most critical roles research questions play in stimulating creativity and exploration is framing a topic as a launchpad into unknown terrain. This is because research questions offer a dynamic aspect that maintains reader interest. According to Saura et al. (2023), technical developments over the past few decades have driven open innovation, resulting in substantial organizational, corporate, social, and economic shifts [15]. Biereenu-Nnabugwu (2022) contends that problem description, often handled in the beginning chapters of a research effort, is essential for defining the study context, clearly expressing the research problem, and selecting an appropriate research design [16]. Solis et al. (2023) highlight that a complete grasp of the problem must be solved at the fundamental stage in developing new designs. After the issue has been stated clearly and concisely, it is essential to determine the goals the design must accomplish to remedy it successfully. The process may start with defining the objectives, then leading to identifying the underlying problems that must be addressed to achieve these objectives [17]. This method emphasizes how iterative and dynamic the design process is. Archibald (2020) stresses that the definition of problems in program planning and evaluation should be more critically examined. The central issue is understanding why the absence of problematization is problematic—in other words, treating problems as self-evident can jeopardize the effectiveness of evaluation practices [18]. A more nuanced and reflective approach to problem definition mitigates such risks. Therefore, the research problem is a circumstance that calls for a solution, typically discovered via careful observation and study of the existing gaps or impediments. The complexity of academic research, characterized by competing agendas and essential standards, influences the impact and importance of the findings acquired from this type of study. To effectively define an issue, it is necessary to acknowledge the possibility of change and innovation. As a result, it is crucial to formulate research questions that encourage investigation. When establishing an acceptable study design and producing successful solutions, having a thorough grasp of the problem is necessary. Because of this, it is essential to take a more nuanced and analytical approach to characterizing the problem to guarantee successful evaluation and practice. In conclusion, Wacker (1998) offers a more organized definition. He considers a research problem a question about the relationship between variables, and the pursuit of the solution directs the research process. When taken as a whole, these definitions highlight the varied character of research questions and their fundamental role in developing and directing scientific investigation [19]. Figure 1 offers a picture of research questions from an academic perspective, contrasting each subject's clarity and complexity levels. According to this classification system, research questions are apparent, semi-clear, or unclear, with varied degrees of established parameters and needs. The difficulties that arise while attempting to handle complicated issues are brought to light by this picture, particularly in societal and industrial settings. The graphic emphasizes that the clarity of a research topic plays a critical part in selecting the proper research technique and methodology. This illustrates the complex nature of problem identification and resolution in academic research, a vital aspect of the field.

Formulating a problem statement in the context of academics not only defines the path for the whole research process but also ensures that the study is in line with the larger academic discourse [20]. This makes the formulation of a problem statement extremely important. It serves as a guide for establishing research questions, objectives, and hypotheses, guaranteeing that the study is pertinent and rigorous.

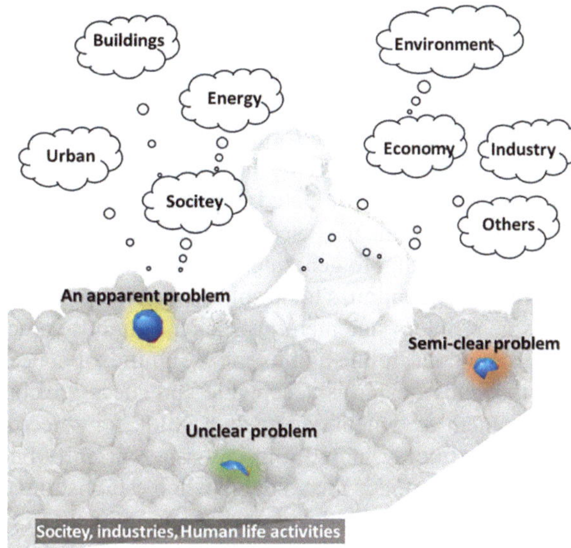

Fig. 1 Problem statement between clear, semi-clear, or unclear (*source* authors)

Justifying the study's relevance to the academic community and society is accomplished by developing a well-crafted problem statement that adds to the progress of knowledge in the subject. Several common errors should be avoided when crafting an effective issue statement. Some of these pitfalls include imprecise assumptions about the reasons or remedies and ensuring the problem is relevant to the stakeholders involved [21]. Engaging stakeholders in the formulation process is equally crucial since it guarantees that the problem is examined from many viewpoints and that all appropriate elements are considered in the formulation process. In addition, the legitimacy of the issue statement is improved by providing facts and evidence to back it up, which also serves as a solid basis for the research. Revising the issue statement iteratively is frequently required to accommodate the emergence of new information or insights during the research process. Developing a problem statement is an essential academic activity that establishes the foundation for doing productive research and finding solutions to problems. A well-formulated problem statement guarantees that research efforts are focused, methodologically sound, and can yield meaningful and practical conclusions. This is accomplished by clearly describing the issue, providing contextual information, expressing the effect, and identifying the audience affected by the problem.

2 The Significance of a Problem Statement in Academic Research

A problem statement is a formal requirement in academic research and the foundational component that drives the whole research process [22]. A problem statement that has been well formulated imparts important clarity, focus, and direction to the researcher, enabling the researcher to carry out a rigorous and pertinent investigation. It acts as an anchor to generate specific research questions and hypotheses, ensuring that the study continues to be cohesive and tailored to its intended audience. In addition, the problem statement justifies the study's value by explaining its contribution to academic discipline and its more significant consequences for society. To ensure that the study design is acceptable and successful in resolving the stated problem, the problem statement serves as a guide for selecting the technique [23]. This is done by explicitly identifying the issue currently being taken care of. As a result, the issue statement is unavoidable when establishing the foundation for an academic investigation that is both significant and influential.

2.1 Focus and a Sense of Direction

When it comes to restricting the scope of the research, having a problem statement that is well expressed is of the utmost importance. This helps prevent the study from being extremely wide or lacking clarity [24]. An example would be a researcher who, in the context of sustainable building, may first profess interest in "improving energy efficiency in buildings." On the other hand, in the absence of a specific issue statement, the study runs the danger of covering an extensive range of subjects, ranging from insulating materials to HVAC systems, which would result in the research narrowing its emphasis. By describing the problem precisely, such as "Investigating the impact of advanced glazing technologies on energy consumption and indoor thermal comfort in commercial buildings," the researcher can effectively delimit the study's scope. This makes it possible to conduct a focused investigation into how particular glazing technologies contribute to energy efficiency. Therefore, this investigation can result in insights and recommendations that can be implemented to improve sustainable building practices [25]. The accuracy that is offered by the issue statement guarantees that the study will continue to be focused and will produce results that are both applicable and relevant.

2.2 *Establishing a Foundation for Objectives and Questions of Research*

The problem statement is the fundamental part that serves as the basis for formulating research questions and objectives, which are essential to the integrity and coherence of the study [13]. In the topic of sustainable daylighting, for instance, a researcher may be interested in investigating the impact that natural light plays in lowering the amount of energy that is consumed in office buildings. It is possible to build the framework for the formulation of specific research questions by establishing a problem statement that is clearly defined. For example, "Assessing the efficacy of daylighting strategies in minimizing artificial lighting requirements and enhancing energy efficiency in office environments" is an example of such a statement. The following questions might be asked: "Which daylighting strategies are most effective in different office configurations?" In addition to "How do these strategies influence overall energy consumption?" These questions, in turn, serve as a guide for designing research objectives, ensuring that the study is geared toward addressing the fundamental difficulties mentioned in the problem statement. This systematic method is essential for preserving the academic rigor and consistency of the study.

2.3 *Significance and the Justification of It*

By providing an answer to the essential "so what?" question, a substantial problem statement also serves the purpose of justifying the necessity and value of the research [26]. As an illustration, in sustainable building research, one can come into an issue such as "the lack of standardized metrics for evaluating the long-term energy performance of green buildings." This problem characterization draws attention to a significant void in the existing body of knowledge or practice, which justifies the requirement for more research. The researcher emphasizes the significance of the study to architects, politicians, and the broader community of sustainability advocates by defining this challenge clearly and concisely. The research may lead to new standards or frameworks for evaluating energy performance. This will result in the advancement of sustainable construction and will also guide future policy and practice. Establishing the study's relevance within the context of the larger scholarly discourse depends on the academic justification provided [27].

2.4 *Providing Direction for the Methodology*

Problem description is essential in formulating the study design, methodology, and strategies for data-gathering processes [28]. An example of this would be research on sustainable daylighting that evaluates the effectiveness of various daylighting systems. In this case, the issue statement would drive the choice of a mixed-methods approach. This may comprise quantitative analysis of energy savings in buildings that use various daylighting methods, in addition to qualitative evaluations of the level of satisfaction experienced by the building's occupants by such structures. When studying energy efficiency, a problem statement focusing on the obstacles that prevent using energy-efficient technology in residential structures might result in selecting case studies, surveys, and policy analysis as the critical research methodologies. As a result, the problem statement directly impacts the methodological decisions. This helps ensure that the research design aligns with the study's objectives and can address the identified problems [29].

2.5 Analysis of the Results

The final point is that the problem statement offers a distinct criterion that may be used to evaluate the effectiveness of the investigation [30]. For example, if the issue statement directs a study on energy efficiency, such as "How can smart building technologies optimize energy consumption in high-rise residential buildings?" The research findings may be evaluated depending on how well they address this challenge. It is possible to assess the efficacy of the proposed solutions, such as automated energy management systems, by comparing them to the expectations outlined in the issue description. In the context of sustainable daylighting, the research might be assessed based on the extent to which the proposed daylighting solutions minimize dependency on artificial illumination while simultaneously retaining the quality of the interior environment. Therefore, the problem statement functions as a benchmark, enabling a systematic review to determine whether the study has successfully addressed the fundamental difficulties highlighted at the beginning of the process. When researching academic purposes, the problem statement is an essential component that is the foundation for the whole investigation process [31]. This aims to offer the methodology and direction, build the basis for research questions and objectives, justify the study's relevance, guide the selection of research methodology, and serve as a standard for assessing the findings [32]. To ensure that the research conducted in specialized sectors such as energy efficiency, sustainable construction, and sustainable daylighting is rigorous, coherent, and impactful, it is essential to have a problem statement that is well stated. The formulation of a problem statement that is both clear and precise not only improves the overall quality of the study but also makes a substantial contribution to the expansion of knowledge and the creation of solutions that are specifically tailored to address the environmental concerns of the present day. The key features of a problem statement in academic research are summarized and outlined in Table 1. The table emphasizes the function of a problem statement in defining the research subject, creating hypotheses, explaining the study's relevance, directing methodological choices, and giving criteria for assessing the study's results. Each component is demonstrated using examples from research on energy efficiency and environmentally responsible construction practices.

Table 1 The role of a problem statement in academic research

Aspect	Academic significance	Illustrative example
Articulation of research focus [33]	The problem statement precisely delineates the scope of the research, ensuring that the study remains concentrated and avoids ambiguity	Example: "Evaluating the influence of advanced glazing technologies on the energy efficiency and thermal regulation of commercial buildings"
Foundation for hypotheses and research objectives [34]	The problem statement underpins the formulation of specific hypotheses and research objectives guiding the systematic inquiry	Example: "Determining the impact of daylighting strategies on reducing artificial lighting dependency in office environments"
Rationale and scholarly justification [35]	A well-formulated problem statement provides a scholarly rationale, articulating the research's necessity and contribution to the academic field	Example: "Addressing the lack of standardized metrics for assessing the long-term energy performance of sustainable buildings"
Methodological alignment [36]	The problem statement informs the research design and methodological approach, ensuring coherence between the research questions and the methods employed	Example: Utilizing a mixed-methods approach to assess daylighting technologies, integrating quantitative energy analysis with qualitative occupant feedback
Criterion for evaluative measures [37]	The problem statement establishes the criteria against which research outcomes are evaluated, determining the effectiveness in addressing the identified issue	Example: "Assessing the efficacy of smart building technologies in optimizing energy consumption within high-rise residential structures"

3 Characteristics of a Well-Defined Problem Statement

A well-defined problem statement in academic research has numerous crucial features that guarantee its clarity, specificity, and feasibility, setting it apart from a poorly formulated one. Clarity is an essential attribute. The issue statement should be expressed using precise and unambiguous language that can be easily comprehended, especially by persons who may not have extensive knowledge of the study topic [38]. An example of a clear problem statement in sustainable building is: "How can incorporating renewable energy technologies in building design decrease total energy usage while ensuring occupant comfort?" This statement is clear because it avoids technical terms and overly complicated sentences, making it understandable to a broader audience. Precision is another crucial attribute. A practical problem statement should confine its scope to a specific issue, demographic, or occurrence rather than attempting to address a problem that is too wide or generic [39]. As an illustration, rather than discussing "energy efficiency in buildings," a more precise issue statement may be: "What is the impact of utilizing advanced insulation materials on the energy efficiency of tall residential buildings in urban settings?" This level of detail guarantees that the research is concentrated, enabling a more precise and controllable examination. In quantitative research, it is crucial to have a problem statement that can be measured. This means that the problem should be framed in a way that allows for the quantification, comparison, or testing of results. An example of a quantifiable issue statement in a sustainable Building Information Modeling (BIM) context may be: "What is the precise impact, measured in tons, of implementing sustainable BIM practices on the reduction of construction waste in large-scale commercial projects?" This attribute is essential for guaranteeing that the investigation can yield empirical, quantitative outcomes. Feasibility is an additional crucial characteristic. The issue description should accurately represent a research subject that is feasible to explore, given the limitations of time, money, and skill. An example of a viable issue statement in energy efficiency may be: "To what extent can the installation of energy-efficient windows in already constructed office buildings lead to a 20% reduction in heating expenses within five years?" This example demonstrates the significance of establishing a research issue that is both practical and attainable within the specified constraints. Relevant is equally crucial when formulating a problem statement. The challenge should focus on a subject pertinent to the field of study and wider academic and practical settings. An illustrative issue statement may be: "What are the obstacles hindering the adoption of environmentally friendly construction methods in developing nations, and how can focused policy interventions improve their implementation?" This illustrates the importance of the subject and its ability to fill a gap in the existing literature or address a pressing practical need. Furthermore, the issue statement must align with the researcher's overarching aims and objectives, establishing a strong correlation with the research questions. Ensuring this alignment is crucial for preserving the consistency and concentration of the investigation. Within sustainable BIM, the issue statement might focus on objectives like enhancing the precision of energy modeling in the design stage. Additionally, it is essential to provide a contextual history to place the problem within a wider academic and practical framework. An effective problem statement should provide enough background information to clarify the importance of the issue and the elements that contribute to its existence. In sustainable building research, the backdrop would typically examine the growing worldwide need for energy-efficient structures and the environmental consequences of traditional construction techniques. A well-structured problem statement should follow a logical structure, starting with a basic introduction to the topic, then providing a more particular articulation of the problem, and then closing with the consequences or need of the study [40]. For instance, one could begin by understanding the difficulties faced by global energy consumption. Then, the discussion could be narrowed to explore how sustainable BIM practices can help address these challenges. Finally, the conclusion could summarize the specific research question about the efficacy of these practices in decreasing energy usage. Figure 2 illustrates the key attributes of a well-defined issue statement, emphasizing the significance of clarity, specificity, measurability,

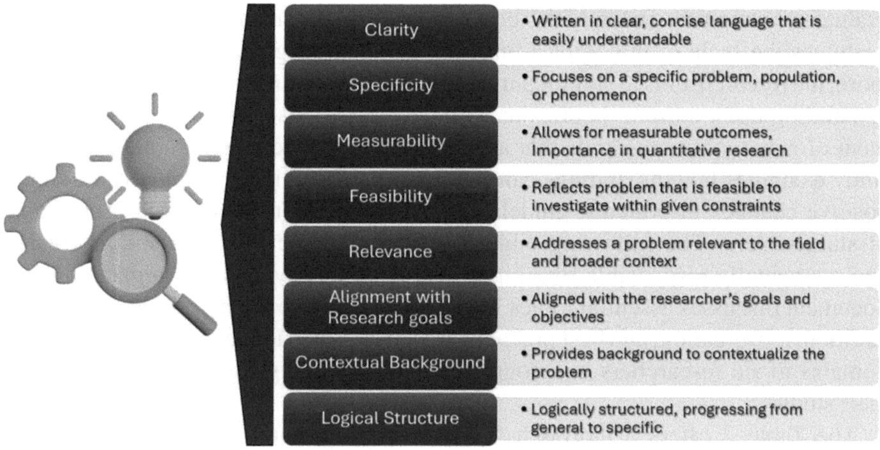

Fig. 2 Characteristics of a well-defined problem statement

feasibility, relevance, alignment with research objectives, contextual backdrop, and logical structure. Every attribute adds to the overall efficacy of the issue statement in directing scholarly study. Clarity guarantees that the issue description is comprehensible while eliminating excessive technical terminology. Specificity limits the scope of research to a particular subject, which helps to make the study more manageable. Measurability enables achieving measurable results, which is essential for empirical research. Feasibility guarantees that the problem may be practically examined within the specified limitations. The problem's relevance is connected to key concerns in the area, and its alignment with research aims assures that it is coherent with the study's objectives. Contextual context is essential for comprehending the significance of the problem, while a logical structure arranges the statement in a meaningful fashion, moving from broad to specific.

4 The Impact of Critical Research Questions on the Research Problem

The problem statement is the foundation for any research endeavor, guiding the inquiry into critical areas that must be addressed to ensure the study's significance and impact [41]. The research statement focuses on several pivotal questions, beginning with the current relevance of the issue. It explores whether the problem is still pertinent in today's rapidly evolving context and assesses its effects on contemporary society, industries, or specific groups. Next, the research considers the future relevance of the problem, questioning whether it is likely to persist or escalate in the future. It delves into how emerging trends might influence the continuation or resolution of the issue. The impact scope is another critical aspect, where the study evaluates how many individuals, communities, or organizations are affected and examines the scale of this impact on local, national, or global levels [42]. Furthermore, the research aims to significantly contribute to existing knowledge, challenging or building upon current theories and frameworks. It also focuses on the development of new tools or methods that are practical and widely applicable. Finally, the study examines how its findings can influence best practices, potentially leading to positive changes in policies, guidelines, and standards that benefit a broad range of stakeholders. The following table overviews significant research concerns for environmentally responsible structures, energy efficiency, and urban planning. The document discusses essential topics like relevance, future impact, scope, knowledge addition, tool creation, and best practices [43]. It also provides examples from various domains to aid researchers in recognizing and addressing significant difficulties in their studies.

The Table 2 offers a thorough framework for research. It illustrates how, by addressing critical issues like relevance, impact, knowledge contribution, tool creation, and best practices, research in various domains may spur notable breakthroughs and inform policies that support sustainability, equity, and innovation. In conclusion, Fig. 3, a series of crucial questions intended to assist researchers in defining and honing their issue statements, is shown in the accompanying figure. The importance, future relevance, impact, prospective additions to the body of knowledge, creation of practical tools or techniques, and effect of study findings on best practices are all evaluated by the researcher using these questions as a guide. Researchers may ensure their work has significance and influence by methodically addressing these issues.

5 Essential Attributes of an Effective Research Problem

A practical research challenge is essential to the success of an investigation as it guides the study's focus and ensures its relevance [12]. To achieve this, the problem must possess several important qualities. First, it should address critical issues or gaps in knowledge within the subject, ensuring that the study contributes valuable insights. To add credibility and significance to the research, the problem must be based on verifiable facts, reflecting the obstacles and opportunities present in the real world. Additionally, it should be controllable and relevant, with a practical and feasible scope, given the available resources and time constraints. In addition, a helpful research challenge should lead to formulating an explicit and testable hypothesis. This hypothesis guides the research process and facilitates a comprehensive analysis. This clarity ensures that the study remains focused and that its findings can be examined objectively. The final point is that the problem should not lead to inconsequential outcomes [50]. The study's aim should be to make valuable contributions that either expand the topic's understanding or improve its techniques. By incorporating these features, a research problem ensures that the study will be achievable

Table 2 Key research considerations in the research area

Key aspect	Questions to address	Considerations	Example from engineering research
Current relevance [44]	– How critical is the problem in today's socio-economic, technological, and environmental context?	– Examine the problem's alignment with current global challenges, such as sustainability, digital transformation, etc.	– Research on adopting energy-efficient technologies in existing buildings to reduce carbon footprints
	– Does the issue directly impact key sectors or vulnerable populations?	– Assess the immediate and long-term consequences on affected stakeholders	– Studies on the impact of urban heat islands on low-income communities in densely populated cities
Future relevance [45]	Given current trends, Will the problem continue to pose a challenge in the foreseeable future?	– Analyze emerging technologies, demographic shifts, or policy changes that could exacerbate or mitigate the problem	– Projections on the future energy demands of smart cities and the role of sustainable urban planning
	– How might future scenarios (e.g., economic shifts, climate change) influence the persistence or resolution of this problem?	– Consider both best-case and worst-case future scenarios	– Scenario analysis on the effects of climate change on coastal urban infrastructure and housing stability
Impact scope [46]	– What is the quantitative and qualitative extent of the problem's impact across different regions and populations?	– Utilize data-driven analysis to estimate the number of people or entities affected	– Assessment of how energy inefficiency disproportionately affects rural versus urban populations globally
	– How does the problem's impact differ across various demographic, geographic, or socio-economic groups?	– Consider disparities in impact and the potential for unequal effects	– Evaluation of green building standards' accessibility and benefits to different socio-economic groups

(continued)

Table 2 (continued)

Key aspect	Questions to address	Considerations	Example from engineering research
Contribution to knowledge [47]	– In what ways can this research fundamentally alter or refine existing theories, models, or frameworks in the field?	– Explore the potential for paradigm shifts or the filling of significant knowledge gaps	– Development of new models for integrating renewable energy sources into urban grids
	– How will the study contribute to the academic discourse, possibly challenging conventional wisdom or introducing novel perspectives?	– Evaluate the potential for interdisciplinary impacts or applications	– Research challenging conventional urban design principles by promoting biophilic urbanism
Development of tools or methods [48]	– Can the study pioneer new methodologies, analytical techniques, or technologies that offer practical solutions to the problem?	– Consider the scalability and adaptability of these tools or methods across different contexts	– Creation of a new tool for real-time energy monitoring and management in smart buildings
	– Will these innovations have a lasting impact, becoming standard practice or widely adopted within the industry?	– Assess the feasibility and potential barriers to adoption	– Introduction of modular building components that enhance sustainability and are adaptable to various climates
Influence on best practices [49]	– How can the research findings be translated into actionable strategies that improve industry standards or public policies?	– Focus on the potential for immediate and long-term implementation in real-world scenarios	– Recommendations for updating building codes to include mandatory energy efficiency standards
	– Will the study contribute to ethical practices, sustainability, or equity in its respective field?	– Consider the broader implications for societal well-being and global best practices	– Proposals for equitable urban planning policies that ensure access to green spaces for all city residents

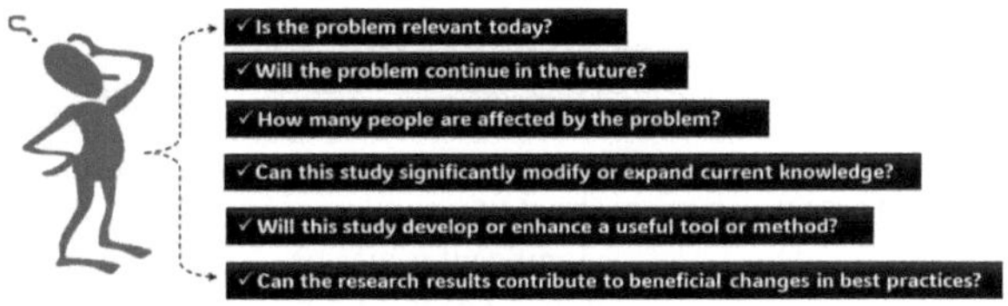

Fig. 3 Key questions for defining the research problem statement

and impactful, contributing to the broader academic debate and practical applications. Table 3 outlines the fundamental characteristics of a helpful research problem, including the importance of addressing critical issues, grounding the problem in factual evidence, maintaining manageability and relevance, proposing a clear and testable hypothesis, and the need to avoid trivial outcomes. The following table illustrates how these concepts may be used to lead relevant and meaningful examples in research.

6 Analyzing the Origins of Research Problems

When attempting to identify a research issue, it is necessary to investigate various sources that might stimulate helpful inquiry. The practical problems or knowledge gaps must be addressed which are frequently brought to light by various personal experiences [61]. Literature sources offer a complete summary of the study that has already been conducted, identifying areas that need more inquiry. Existing ideas can be put to the test or built upon, which can result in new understandings. Previous research frequently reveals issues or constraints that have not been satisfactorily answered, providing a basis for more research. In conclusion, academic experiences, such as coursework or partnerships, can generate ideas for study by exposing researchers to crucial conversations and developing trends [62]. These many sources, when taken together, lead to the formulation of a research problem that is sturdy and significant. Figure 4 shows the variety of research problem sources. It emphasizes personal experiences, literature sources, theories, research, and academic experiences. These sources inspire and influence research problems by revealing significant concerns, knowledge gaps, and innovation possibilities. These sources help researchers create meaningful, evidence-based, and impactful challenges.

7 Steps to Identify a Research Problem

Identifying a research topic is a fundamental and crucial phase in the research process. It plays a vital role in determining the focus and extent of a study [63]. The diagram illustrates a systematic approach for this work, commencing with identifying overarching topics of interest. The first phase entails thoroughly examining

Table 3 Key traits of an effective research problem

Characteristic	Description	Considerations	Examples
Addresses key issues or needs [51]	The research problem should focus on significant issues or unmet needs within the field	– Identify areas with apparent knowledge gaps or where current practices can be improved	– Investigating the barriers to the widespread adoption of energy-efficient technologies in residential buildings
It should target areas where there is a clear gap in knowledge or where current practices can be improved [52]	– Ensure the problem is relevant to stakeholders and addresses urgent issues	The research problem should focus on areas with a clear knowledge gap or where current practices need improvement, ensuring they are relevant to stakeholders and address pressing issues	BIM procedures lack sustainability, typically ignoring environmental considerations like lifecycle evaluation. The research challenge is critical for industry stakeholders since closing this gap is essential for sustainable building
Grounded in factual evidence [53]	The problem must be based on verifiable data and empirical evidence	– The research should be rooted in real-world data, reflecting challenges or opportunities	– Analyzing energy consumption patterns in existing buildings to identify inefficiencies and potential improvements
It should be rooted in the actual challenges or opportunities observed in practice [54]	– Use reliable data sources and methodologies to ground the problem in factual evidence	– A practical research problem should be rooted in real-world challenges, using reliable data and rigorous methodologies to ensure it reflects genuine issues and is grounded in factual evidence	Structures frequently don't maximize natural daylighting, which causes discomfort and increases energy use. To overcome this, research should create better tactics based on facts and difficulties from the real world
Manageable and relevant [55]	The scope of the research problem should be realistic and feasible within the given time and resource constraints	– Ensure the research problem is manageable within the constraints of time, budget, and available resources	– Developing a framework for retrofitting older buildings with energy-efficient technologies within a limited budget

(continued)

Table 3 (continued)

Characteristic	Description	Considerations	Examples
It should be directly relevant to ongoing discussions, debates, or developments in the field, ensuring the study's significance [56]	– The problem should align with current trends and debates to ensure relevance and applicability	To guarantee its relevance and importance, the research must be by the latest trends and debates in the area. It must also make a valuable contribution to ongoing conversations and advancements	Research on data privacy in smart cities should align with ongoing debates about surveillance and data security. The study will remain relevant and contribute significantly to current discussions on balancing innovation with citizens' rights by addressing these concerns
Proposes a clear and testable hypothesis [57]	The research problem should lead to the formulation of a specific, clear, and testable hypothesis	– Formulate hypotheses that are specific, measurable, and capable of being tested through empirical research	– Hypothesizing that smart building automation systems significantly reduce energy consumption compared to traditional buildings
This hypothesis should be capable of being confirmed or refuted through the research process [58]	– Data analysis should frame the hypothesis for explicit confirmation or refutation	The hypothesis should be clearly defined and explicitly confirmed or refuted through the research process. Data analysis must be designed to test the theory directly, ensuring the results provide a clear answer	Hypothesis: "Advanced insulation reduces residential energy consumption by 20%." This can be confirmed or refuted by analyzing energy usage data from buildings with and without insulation

(continued)

Table 3 (continued)

Characteristic	Description	Considerations	Examples
Avoids leading to insignificant answers [59]	The problem should be structured to avoid trivial or insignificant outcomes	– Structure the research to avoid outcomes that do not contribute meaningful insights or advancements in the field	– Avoid broad, non-specific research questions like "Do sustainable buildings help the environment?" which lack focus and depth
It must provide meaningful contributions that advance knowledge or improve practices within the field [60]	– Focus on delivering actionable, relevant findings and being capable of driving real change in practices or policies	The research should provide practical insights that enhance understanding and facilitate significant transformation in practices or policy	Research on sustainable material passports should aim to offer practical advice on standardizing and applying these passports throughout the building sector. By improving comprehension of material life cycles and encouraging reuse, the results may lead to notable modifications in sustainable construction methods and impact forthcoming policy formulation

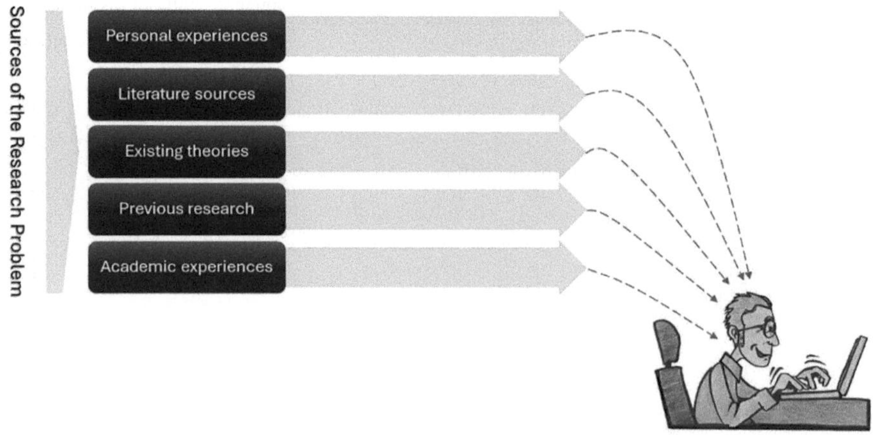

Fig. 4 Sources of inspiration for formulating research problems (*source* authors)

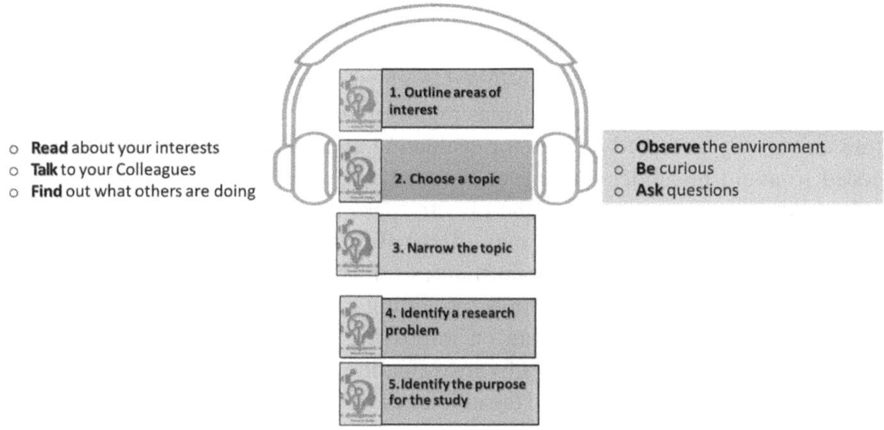

Fig. 5 A systematic approach to identifying a research problem

subjects that align with academic or professional goals, which will serve as the foundation for subsequent improvement. After selecting an area of interest, the next step is to choose a specific issue that requires more investigation. Subsequently, the chosen subject is refined to maintain a concentrated, controllable, and pertinent study scope within contemporary scholarly conversation. Afterward, the research problem is unambiguously specified within this restricted scope, identifying a specific issue or subject the study will investigate. The last phase is determining the research objective, clearly stating the study's goals and anticipated contributions to the area. This systematic methodology guarantees that the research subject is thoroughly defined and that the survey is strategically positioned to make a significant academic contribution. Figure 5 depicts a systematic method for establishing a research problem. This involves initially sketching areas of interest and then picking and narrowing down a topic. Next, the research problem is defined, and the study's purpose and objectives are identified.

7.1 Outline Areas of Interest

When selecting study topics, starting with known and recognizable issues is crucial. This familiarity establishes a solid basis for investigation, guaranteeing a fundamental comprehension of the circumstances, difficulties, and interests within the profession or sector involved [64]. Commencing research in a familiar location helps you quickly recognize captivating elements, facilitating a more thorough investigation of specific research questions. The method should start by taking a comprehensive approach to examining diverse areas within the profession or sector of interest [65]. For instance, within green buildings, individuals may contemplate the difficulties associated with incorporating environmentally friendly materials into construction or evaluate the efficacy of existing energy efficiency regulations. In bioeconomics, one can focus on the long-term viability of utilizing resources in agricultural systems or the financial consequences of shifting to bio-based products. Contemplating dissatisfying or perplexing aspects may indicate regions requiring more investigation. For example, more universally accepted criteria for evaluating energy efficiency in buildings are needed to avoid misunderstanding or dissatisfaction, suggesting a need for more study. Likewise, comprehending the effectiveness of proven green building certifications or bioeconomic models may offer valuable knowledge for enhancing or creating new ideas. Identifying opportunities for improvement reveals deficiencies or inefficiencies in current procedures, providing a promising chance to study. For example, research may concentrate on improving the economic sustainability of environmentally friendly buildings by investigating cost-efficient materials or techniques. Exploring ways to enhance the efficiency of bioenergy production is a crucial focus in bioeconomics. Creativity is an essential component of this process. Engaging in divergent thinking by exploring unorthodox viewpoints or challenging accepted conventions is vital. This technique can potentially uncover new study domains, such as investigating the convergence of energy efficiency technology with intelligent city planning or constructing bioeconomic models that incorporate social justice factors.

Brainstorming is a highly effective approach for expanding thoughts. During this stage, the primary emphasis should be on producing diverse ideas without promptly assessing their practicality [66]. For example, exploring all potentialities, ranging from incorporating renewable energy sources into urban infrastructure to formulating policies that promote the bioeconomy, even unconventional or difficult ones, might generate novel research inquiries. During brainstorming, it is crucial to record all thoughts and ideas. This guarantees the comprehensive inclusion of all forms of innovative thought, enabling further fine-tuning and narrowing of emphasis toward a specific study subject. This systematic approach to inquiry and documentation aids in identifying the most favorable areas for additional examination, ultimately resulting in a research topic that is both significant and feasible [66]. By initially identifying a general field of interest and narrowing down the scope, a firmly based, original, and influential research challenge may be formulated, with tangible applications in domains such as sustainable construction, energy conservation, and the economics of biological resources. A setup way to describe topics of research interest is shown in Table 4, which may be found here. The process is guided by it, beginning with the selection of common themes and the identification of gaps, culminating in the identification of chances for development, the utilization of creative thinking, and the generation of ideas. The use of this systematic methodology guarantees a study topic that is both well defined and significant.

7.2 Choose a Topic

The selection of a research topic is a crucial initial stage in the research process since it establishes the direction and extent of the investigation. The chosen subject

Table 4 Steps to outline areas of research interest

Step	Description
Outline areas of interest	Select familiar issues to establish a foundation for research. Explore various topics comprehensively
Identify gaps	Reflect on unsatisfactory or confusing aspects that need further investigation
Spot improvement opportunities	Identify deficiencies or inefficiencies in current practices to uncover study areas
Use creative thinking	Explore unconventional perspectives to discover new research opportunities
Brainstorm and document	Generate a wide range of ideas without judgment; document them for further refinement

should arise from a defined field of interest, guaranteeing continuous involvement and concentration throughout the study [67]. However, choosing a suitable subject requires thoughtful evaluation of several variables to ensure the research's importance, novelty, and academic relevance. Before proceeding, it is essential to thoroughly assess the significance and pertinence of the possible subject matter. Determining if the problem is relevant in the present academic and practical environment is crucial, focusing on the central issues within the discipline [68]. As an illustration, in sustainable urban design, a specific subject may center around incorporating green infrastructure into urban planning to alleviate the impacts of urban heat islands, a critical problem worsened by climate change. This issue, which is very relevant and urgent, is anticipated to make a significant contribution to current academic discussions and practical implementations. Furthermore, it is necessary to evaluate the current reservoir of knowledge and identify areas where research could be improved in the respective discipline. An exhaustive examination of the existing literature is required to determine which subjects have been adequately researched and which still need a more substantial investigation. One potential research emphasis in ecological building might be utilizing locally obtained, sustainable materials in high-density urban areas, which may have received limited study. Similarly, research in the field of energy efficiency should focus on optimizing energy-saving solutions for retrofitting older structures, which contrasts with the current emphasis on new buildings. Identifying these gaps guarantees that the selected topic will provide new and valuable insights, pushing the area forward.

Ultimately, the researcher's academic curiosity about the subject matter is crucial to consider [69]. Choosing a subject corresponding to the researcher's intellectual interest and enthusiasm, such as investigating passive design solutions in ecological construction, may cultivate long-term dedication and a more profound degree of investigation. To summarize, choosing a research topic involves carefully evaluating its importance, identifying areas where information is lacking, and ensuring it aligns with the researcher's interests. Opting for a carefully selected subject matter propels the discipline forward and guarantees that the research procedure is intellectually stimulating and professionally fulfilling. Table 5 presents a concise overview

Table 5 Important factors to consider when choosing a research topic

Stage	Description	Key considerations
Importance and relevance	Assess the significance and relevance of the potential topic in both academic and practical contexts	– Is the topic central to current academic discussions? – Does it address pressing issues within the discipline?
Review of existing knowledge	Conduct a thorough review of the existing literature to identify gaps and areas needing further investigation	– What has already been researched? – Which areas require more substantial investigation?
Gap identification	Identify areas within the field that have received limited attention and offer opportunities for new research contributions	– Are there under-researched topics? – Can the research provide novel insights that advance the field?
Alignment with researcher's interest	Consider the researcher's intellectual curiosity and enthusiasm for the subject, ensuring sustained engagement and deeper inquiry	– Does the topic align with the researcher's academic interests? – Will it sustain long-term dedication to the investigation?
Intellectual and professional fulfillment	Ensure the chosen topic is intellectually stimulating and professionally rewarding, contributing meaningfully to the discipline	– Will the research be fulfilling? – Can it propel the discipline forward and contribute to professional growth?

of the crucial stages of choosing a research topic. It highlights the significance of assessing the pertinence, examining the current body of knowledge, finding areas that require more investigation, aligning with the researcher's interests, and guaranteeing both intellectual and professional satisfaction. These factors ensure that the selected issue significantly contributes to the area and maintains ongoing involvement over a lengthy period.

This Table 5 systematically summarizes the key factors to examine when choosing a research subject. It focuses on the significance of relevance, identifying gaps, and aligning with the researcher's interests.

7.3 Narrow the Topic

Restricting the extent of a research subject is a crucial stage in honing the concentration and trajectory of a study. This process requires transitioning from broad, general topics of interest to more focused and specialized inquiries, which enable the formulation of a transparent and manageable research question. Narrowing broad themes is crucial for maintaining academic rigor while developing a research plan, as it helps create a cohesive and focused approach. A highly successful strategy for narrowing a research subject is to pose precise and targeted questions directly relevant to the area of interest. These questions aim to extract specific features of the more significant issue that may require further examination. By posing such inquiries, the investigation might be guided toward a particular matter, therefore augmenting the accuracy and practicality of the study. Polit and Beck (2008) propose a framework for refining a broad topic into a specific research question [70]. They provide question stems that guide narrowing down the issue. For example, the inquiry "What is happening with..." might be employed in Passive House research to investigate the persistent obstacles in implementing Passive House requirements in various temperature zones. Similarly, in sustainable cities, the question "What is the meaning of…," may be used to explore the importance of green infrastructure in improving urban resilience to climate change. Alternative question prompts, such as "What factors or triggers...," might be crucial in pinpointing the elements that affect the incorporation of intelligent technology in smart buildings. The phrase "What is the process by which…" may be employed to examine the mechanisms by which sustainable cities accomplish reductions in carbon emissions. In addition, the question "What factors contribute to…" might aid in identifying the crucial features that improve the energy efficiency of Passive Houses. Finally, "How effective is..." may be used to assess the efficacy of intelligent technologies in enhancing energy efficiency and occupant comfort in smart buildings. This systematic question stems from a methodical strategy for narrowing down a study topic, guaranteeing that the research emphasis is academically rigorous and pertinent. By strategically utilizing these inquiries to refine the subject matter, scholars may build a well defined and achievable research query, resulting in an influential and academically noteworthy study. Table 6 provides a concise overview of the crucial phases involved in refining a research subject. It emphasizes the significance of narrowing the topic to a specific area of interest, formulating focused questions, employing organized question prompts, and verifying the issue's relevance and practicality. This systematic technique enables the formulation of a specific and controllable research inquiry.

Narrowing down a study topic to produce a valid research question is crucial in the research process. It requires rigorous refining to guarantee that the question is empirically viable. This procedure entails methodically eliminating questions that

Table 6 Key steps in narrowing a research topic

Step	Description
Refine the topic	Focus on becoming more specific about the area of interest
Pose specific questions	Formulate targeted questions to narrow the focus of the research
Use question stems	Apply structured question stems (e.g., "What is going on with…", "What factors contribute to…") to guide refinement
Evaluate relevance and feasibility	Ensure the narrowed topic is relevant, manageable, and aligned with the research goals

cannot be empirically investigated. Brink and Wood (2001) state that a research question must meet specific criteria for success. The question should prioritize objectivity by focusing on information rather than personal opinions. This will ensure that the replies describe or explain phenomena based on objective data rather than subjective ideas [71]. Furthermore, a researchable question should be timely, focusing on current and significant concerns within the relevant field. Moreover, the research issue should be pertinent, produce practical and appropriate information, and contribute to current knowledge and practice. Ultimately, the inquiry should be focused on acting, motivating the research process by requiring specific actions, and directing the following stages of the study [72]. By ensuring that a research topic satisfies these stringent criteria, researchers may develop a question that is precise, pertinent, and capable of being tested by empirical methods. This ultimately improves the quality and influence of the research.

7.4 Identify a Research Problem

Identifying a research issue is an essential and fundamental phase in the research process. It requires thoughtful consideration and a deliberate approach to selecting the problem [73]. Once a broad research subject is identified, it is crucial to narrow it down into a specific research problem by analyzing the researchable questions that arise from it. For example, in energy efficiency, it would be beneficial to concentrate on maximizing energy utilization in already constructed structures. A relevant study issue within sustainable architecture might entail assessing the enduring effects of employing environmentally conscious materials. The study challenge in practical daylight usage might focus on optimizing natural light in urban high-rise structures to minimize energy consumption. Multiple variables should influence the choice of a study area. Personal interest is paramount since it guarantees continuous involvement and dedication throughout the study process. For instance, a researcher who strongly values environmental sustainability may be tempted to investigate novel materials for constructing sustainable structures. Conducting a comprehensive examination of the available literature is crucial to identify any deficiencies in the present knowledge base, such as the necessity for more research on the cost-effectiveness of upgrading older buildings with energy-efficient technology. Another crucial factor to examine is the practicality of tackling the issue, which includes assessing the feasibility of doing the study with available resources. Concluding the study problem necessitates a thorough assessment of several crucial factors. It is essential to analyze its relevance to guarantee that the problem is substantial within the area [50]. For example, one way to handle a big issue is by lowering carbon emissions through the improved exploitation of daylight in urban architecture. It is essential to carefully assess the feasibility of researching the topic, making sure that the issue can be examined through empirical investigation utilizing existing methodologies and instruments, such as sophisticated energy modeling software in the context of sustainable buildings. Precision is essential for adequately resolving the research issue. The study

Table 7 Key steps in identifying a research problem

Aspect	Description
Select focus [74]	Choose a specific issue within the broader topic that aligns with your interests
Review literature [75]	Identify gaps in current research to ensure the problem is underexplored
Assess importance [76]	Ensure the problem addresses a significant and relevant issue in the field
Ensure research ability [77]	Confirm that the problem can be investigated with available methods and tools
Define specificity [78]	Narrow the problem to a focused and manageable area of study
Consider ethics [79]	Evaluate ethical implications, especially if human participants are involved
Evaluate practicality [80]	Consider time, funding, expertise, and access to necessary resources and tools

becomes more manageable and focused by reducing the research topic to specifically examine the influence of daylighting schemes on energy usage under climatic circumstances. It is crucial to prioritize ethical issues while researching occupant behavior in energy-efficient buildings. This ensures that the study is performed with honesty and respect for the participants. It is essential to thoroughly assess practical factors such as the time needed, financing availability, requisite skills, participant availability, and access to appropriate facilities and resources. For instance, researching optimal daylighting may require the utilization of specialist tools and the availability of specific building types to ensure precise analysis. Through careful examination of these factors, researchers can pinpoint a research issue that is meaningful, able to be studied, ethically appropriate, and practically achievable. This sets the foundation for a thorough and influential study of energy efficiency, sustainable buildings, and daylight utilization. Table 7 concisely overviews the necessary processes to define and refine a research issue. This statement highlights the significance of choosing a specific and well-defined problem, examining previous research, evaluating the importance of the problem, ensuring that it can be studied, specifying its details, considering ethical considerations, and assessing practical limitations. These steps provide a systematic framework for developing a robust research problem.

7.5 *Identify the Purpose of the Study*

Identifying the study's purpose is vital in the research process, as it offers a concise and targeted expression of its goal, aim, focus, or objective. As stated by Burns and Grove (2014), it is essential to clearly and succinctly define the study's aim, encapsulating the research's fundamental goals. An explicit study objective directs the research methodology and guarantees conformity with the desired results, keeping concentration throughout the inquiry [81]. A well-defined study objective usually includes three essential components: "What," "Who," and "Where." The term "What" pertains to the specific variables or phenomena that are the focus of investigation in the research. For example, in energy efficiency research, the "What" may entail investigating the effects of innovative insulating materials on decreasing energy usage in residential structures. Within biophilic architecture, the emphasis lies in evaluating the mental and physical advantages of incorporating natural elements into urban workplace environments. In the context of sustainable building solutions, the "What" aspect might investigate the efficacy of green roofs in augmenting urban biodiversity and alleviating heat island phenomena. The "Who" component refers to the individuals or subjects who are part of the study. This process may entail carefully selecting a sample of residential buildings that exhibit a wide range of architectural designs and insulating materials within energy efficiency research. The participants for biophilic building research might consist of office workers in metropolitan contexts who have different degrees of exposure to biophilic design components. When researching sustainable building solutions, the individuals in the "Who" category may include urban planners or architects engaged in projects integrating green infrastructure. The term "Where" pertains to the specific location or environment in which the study will occur. Energy efficiency research often focuses on certain climatic zones with a substantial variation in insulation efficacy. When researching biophilic construction, the focus might be urban office complexes in highly populated cities. When considering sustainable building solutions, the context may involve different urban settings where green roofs are part of a more considerable sustainability effort; by explicitly expressing the "What," "Who," and "Where," the objective of the study can be accurately determined, which will aid in the creation of a targeted and significant research strategy. Ensuring clarity in the research aligns it with its aims, boosting the relevance and effect of the study's conclusions, especially in areas such as energy efficiency, biophilic design, and sustainable building solutions. Table 8 provides a comprehensive overview of the essential components for determining the objective of a study, with specific emphasis on the "What," "Who," and "Where" aspects. These elements guarantee clarity and accuracy when developing research goals.

Table 8 Key components of identifying a study purpose

Component	Description
What	Defines the variables or phenomena of interest that the study seeks to investigate, establishing the scope and focus of the research
Who	Identifies the sample or participants involved in the study, including specific characteristics or criteria for inclusion
Where	Specifies the setting or environment in which the study will be conducted, providing context and ensuring relevance to the research objectives

8 A Methodological Framework for Structuring Research Studies

Describing a systematic approach to organizing a research study is necessary, beginning with determining the research topic and creating research questions [82]. This first phase is critical because it outlines the primary issue explored in the study. It also addresses essential factors such as what will be investigated, how it will progress, and why the research is required. Not only do the research questions give direction, but they also guarantee that the study will continue to concentrate on specific problems or occurrences. The following phase, which comes after identifying the research topic, articulates the objectives and purposes of the present investigation. In contrast to the aims, which provide a more comprehensive conceptual framework that explains the study's overarching purpose, the objectives clarify the research's goals [42]. Clearly defined objectives are necessary to preserve concentration and ensure the study aligns with the anticipated outcomes. Identifying and selecting subjects, which refer to the individuals or things that will be the focus of the research, is another step in the process. When it comes to defining the extent of the research and the extent to which its findings may be generalized, this phase is quite important. To guarantee that the individuals chosen for the study correctly represent the larger population or phenomena being investigated, careful thought must be paid to each participant's selection [83]. As a result, it is essential to consider the study's constraints. Recognizing and comprehending these constraints is necessary to contextualize the research findings and provide direction for their interpretation. There may be limitations associated with the sample size, limitations linked to the methodological techniques, or external influences that might impact the results. When these limitations are acknowledged and considered, a more nuanced and thorough assessment of the study results is possible. As shown in Fig. 6, a methodological framework is presented to organize research studies. This framework highlights critical processes such as the identification of the problem, the formulation of the aim, the selection of the subjects, and the consideration of the constraints of the study, which ensures a thorough research design.

In general, this organized method of planning a research study emphasizes the significance of having well-defined objectives, precise research questions, a thorough selection of participants, and careful consideration of study restrictions. All these factors contribute to the robustness and effect of the research.

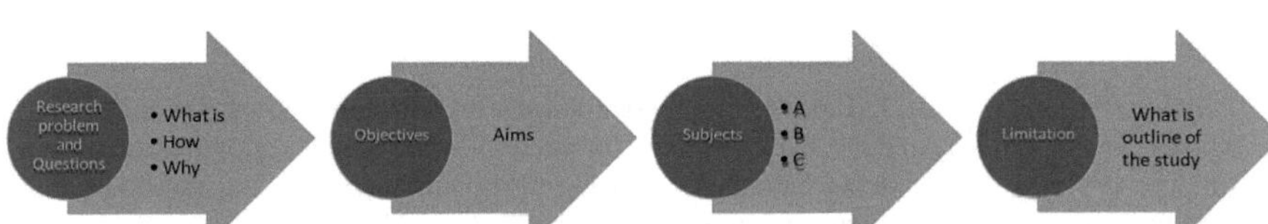

Fig. 6 Methodological framework for structuring research studies

9 Conclusion

This chapter has provided an in-depth examination of the crucial procedure of generating a problem statement in the context of academic research. This chapter emphasized the need to maintain a well-articulated problem statement as the foundation of any research project that is intended to be effective. It is not only the issue statement that guarantees that the research is aligned with wider academic and practical aims, but it also gives a clear direction and emphasis for the investigation. A problem statement that has been carefully written should serve as the basis upon which research questions, goals, and techniques are constructed. This will make it easier to conduct a cohesive and methodologically sound inquiry. Throughout the conversation, it became abundantly clear that the clarity, precision, and relevance of a problem statement cannot be overstated. To determine whether the research is founded on real-world difficulties and whether it contributes significantly to the current body of knowledge, the research must describe the gap or issue it wants to solve clearly. In addition, the chapter underlined the importance of a problem statement being specific enough to direct the study while simultaneously being wide enough to allow for unique and creative methods of issue-solving. Maintaining this equilibrium is essential to keep the research's focus intact while allowing for flexibility in investigating other facets of the topic. In addition, the chapter included an explanation of the function that the problem statement plays in determining the extent of the study. The problem statement prevents the research from being too broad or diffuse, which might dilute its influence and relevance. This is accomplished by explicitly identifying the study's boundaries to avoid confusion. In addition, it offers a structure for determining whether the research was successful, as the issue statement acts as a standard against which the study results may be evaluated. It was said that constructing a problem statement is an iterative process that needs careful consideration of various elements. These considerations include the context of the problem, the parties involved, and the possible effect of resolving the issue. As key elements in this process, it was highlighted that engaging with the existing literature and grasping the problem's historical and contemporary context was necessary. These procedures guarantee that the issue statement is pertinent, current, and informed by a profound grasp of the sector in which it is being evaluated.

References

1. Fellows RF, Liu AM (2021) Research methods for construction. John Wiley & Sons
2. Masloboev AV (2024) An index-based method for integral estimation of regional critical infrastructure resilience using fuzzy calculations (part 1. Problem statement and method generic structure). Надежность и качество сложных систем 1(45):124–141
3. Jonassen DH (2010) Learning to solve problems: a handbook for designing problem-solving learning environments). Routledge
4. Pierre J, Peters BG (2020) Governance, politics, and the state. Bloomsbury Publishing

5. Zhai C (2024) Large language models and future of information retrieval: opportunities and challenges. In: Proceedings of the 47th international ACM SIGIR conference on research and development in information retrieval, pp 481–490
6. Micklos A, Woensdregt M (2023) Cognitive and interactive mechanisms for mutual understanding in conversation. In the Oxford research encyclopedia of communication
7. Smoktunowicz E, Barak A, Andersson G, Banos RM, Berger T, Botella C, Dear BF, Donker T, Ebert DD, Hadjistavropoulos H, Hodgins DC (2020) Consensus statement on the problem of terminology in psychological interventions using the internet or digital components. Internet Interv 21:100331
8. Casino F, Dasaklis TK, Patsakis C (2019) A systematic literature review of blockchain-based applications: current status, classification, and open issues. Telematics Inform 36:55–81
9. Heath RL, Palenchar MJ (2008) Strategic issues management: organizations and public policy challenges. Sage Publications
10. Ondieki K, Njuguna M, Gariy A (2024) Aesthetic indicators for sustainable road development in Kenya. Curr Urban Stud 12(2):243–266
11. Zhang R, Xu X, Liu K, Kong L, Wang W, Wortmann T (2024) Airflow modeling for building design: a designers' review. Renew Sustain Energy Rev 197:114380
12. Dwivedi YK, Jeyaraj A, Hughes L, Davies GH, Ahuja M, Albashrawi MA, Al-Busaidi AS, Al-Sharhan S, Al-Sulaiti KI, Altinay L, Amalaya S (2024) "Real impact": challenges and opportunities in bridging the gap between research and practice–making a difference in industry, policy, and society. Int J Inform Manage 102750
13. Dehalwar K (ed) (2024) Basics of research methodology-writing and publication. EduPedia Publications Pvt Ltd
14. Treffinger DJ, Isaksen SG, Stead-Dorval KB (2023) Creative problem solving: an introduction. Routledge
15. Saura JR, Palacios-Marqués D, Ribeiro-Soriano D (2023) Exploring the boundaries of open innovation: evidence from social media mining. Technovation 119:102447
16. Biereenu-Nnabugwu M (2022) On problem definition and research designing in the logic and methods of political inquiry. Socialscientia: J Soc Sci Human 7(4)
17. Solis MM, García Alcaraz JL, Solórzano JMM, Macías EJ (2023) Definition of variables and research problem. In: Leadership and operational indexes for supply chain resilience, pp 19–30
18. Archibald T (2020) What's the problem represented to be? Problem definition critique as a tool for evaluative thinking. Am J Eval 41(1):6–19
19. Wacker JG (1998) A definition of theory: research guidelines for different theory-building research methods in operations management. J Oper Manag 16(4):361–385
20. Williamson K (2002) Research methods for students, academics and professionals: information management and systems. Elsevier
21. Brown AW, Kaiser KA, Allison DB (2018) Issues with data and analyses: errors, underlying themes, and potential solutions. Proc Natl Acad Sci 115(11):2563–2570
22. Gregor S, Hevner AR (2013) Positioning and presenting design science research for maximum impact. MIS Quart 337–355
23. Rashid Y, Rashid A, Warraich MA, Sabir SS, Waseem A (2019) Case study method: a step-by-step guide for business researchers. Int J Qual Methods 18:1609406919862424
24. Sovacool BK, Axsen J, Sorrell S (2018) Promoting novelty, rigor, and style in energy social science: towards codes of practice for appropriate methods and research design. Energy Res Soc Sci 45:12–42
25. Adeyeye GM (2024) Parental leave policy at the Nigerian civil service commission. Int J Multicult Educ 26(01):341–352
26. Yenti D, Susanti R (2024) College students' problems in writing introduction for research proposal. Jurnal Edukasi 4(1):10–19
27. Lynch J, Ramjan LM, Glew PJ, Salamonson Y (2024) How to embed a conceptual or theoretical framework into a dissertation study design. Nurse Res 32(1)
28. Rawat G, Sharma D, Ashok PP, Bairathi V (2024) The fundamentals of research methodology. Academic Guru Publishing House

29. Zou PX, Xu X (2023) Research methodology and strategy: theory and practice. John Wiley & Sons
30. Kocak B, Baessler B, Bakas S, Cuocolo R, Fedorov A, Maier-Hein L, Mercaldo N, Müller H, Orlhac F, Pinto dos Santos D, Stanzione A (2023) Checklist for EvaluAtion of Radiomics Research (CLEAR): a step-by-step reporting guideline for authors and reviewers endorsed by ESR and EuSoMII. Insights Imaging 14(1):75
31. Johnson RB, Christensen LB (2024) Educational research: quantitative, qualitative, and mixed approaches. Sage publications
32. Aguinis H (2023) Research methodology: best practices for rigorous, credible, and impactful research. SAGE Publications
33. Imran A, Ghufran F, Shah T, Ahmad B, Anwar A (2024) Optimizing research article structure and style: best practices for ensuring clarity and impact in scholarly publications. J Asian Develop Stud 13(2):1432–1444
34. Ali M (2023) Information systems research: foundations, design and theory. Springer Nature
35. Younas A, Durante A, Fàbregues S (2023) Understanding the nature of and identifying and formulating "research problems" in mixed methods research. J Mixed Methods Res 15586898231191441
36. Ramanathan S, Isaksson R (2023) Sustainability reporting as a 21st-century problem statement: using a quality lens to understand and analyze the challenges. TQM J 35(5):1310–1328
37. Dehalwar K, Sharma SN (2023) Fundamentals of research writing and uses of research methodologies. Edupedia Publications Pvt Ltd
38. Alamoodi AH, Al-Samarraay MS, Albahri OS, Deveci M, Albahri AS, Yussof S (2024) Evaluation of energy economic optimization models using multi-criteria decision-making approach. Expert Syst Appl 255:124842
39. Shilo G, Ragonis N (2023) Students' abilities to formulate exhaustive titles are key to comprehending arguments. Int J Appl Linguist 33(2):132–149
40. Callegaro N (2024) The potential of smart home for comfort and energy use optimization in residential buildings
41. Auliya PK, Amrullah QL (2024) Analyzing the flow of ideas in university students' cause and effect essay. Innov Res J 5(1):1–9
42. Tracy SJ (2024) Qualitative research methods: collecting evidence, crafting analysis, communicating impact. John Wiley & Sons
43. Di Girolami E, Kampen J, Arts B (2023) Two systematic literature reviews of scientific research on the environmental impacts of forest certifications and community forest management at a global scale. Forest Policy Econ 146:102864
44. Birkstedt T, Minkkinen M, Tandon A, Mäntymäki M (2023) AI governance: themes, knowledge gaps, and future agendas. Internet Res 33(7):133–167
45. DesJardine MR, Zhang M, Shi W (2023) How shareholders impact stakeholder interests: a review and map for future research. J Manag 49(1):400–429
46. Baer M, Gasparini M, Lancaster R, Ranger N (2023) "All scenarios are wrong, but some are useful"—toward a framework for assessing and using current climate risk scenarios within financial decisions. Front Clim 5:1146402
47. Vengroff DE (2024) Impact charts: a tool for identifying systematic bias in social systems and data. In: The 2024 ACM conference on fairness, accountability, and transparency, pp 1187–1198
48. Filippi ME, Barcena A, Trogrlić RŠ, Cremen G, Menteşe EY, Gentile R, Creed MJ, Jenkins LT, Kalaycioglu M, Poudel DP, Muthusamy M (2023) Interdisciplinarity in practice: reflections from early-career researchers developing a risk-informed decision support environment for Tomorrow's cities. Int J Disaster Risk Reduction 85:103481
49. Blok AC, Drake C, Decosimo K, Zullig LL, Hughes JM, Sperber NR, Kota S, Franzosa E, Coffman CJ, Shepherd-Banigan M, Chadduck T (2024) Adaptations and early adoption of a family caregiver intervention in the veterans affairs health care system: a multimethod pragmatic approach for national scaling. Health Serv Res

50. Savin-Baden M, Major C (2023) Qualitative research: the essential guide to theory and practice. Routledge
51. Naeem M, Ozuem W, Howell K, Ranfagni S (2024) Demystification and actualization of data saturation in qualitative research through thematic analysis. Int J Qual Methods 23:16094069241229776
52. Katzelnick LC, Coloma J, Harris E (2017) Dengue: knowledge gaps, unmet needs, and research priorities. Lancet Infect Dis 17(3):e88–e100
53. Skivington K, Matthews L, Simpson SA, Craig P, Baird J, Blazeby JM, Boyd KA, Craig N, French DP, McIntosh E, Petticrew M (2021) Framework for developing and evaluating complex interventions: gap analysis, workshop, and consultation-informed update. Health Technol Assess (Winchester, England) 25(57):1
54. Piper A (2020) Can we be wrong? The problem of textual evidence in a time of data. Cambridge University Press
55. Rose J, Johnson CW (2020) Contextualizing reliability and validity in qualitative research: toward more rigorous and trustworthy qualitative social science in leisure research. J Leis Res 51(4):432–451
56. Akanle O, Ademuson AO, Shittu OS (2020) Scope and limitation of study in social research. Contemp Issues Soc Res 105:114
57. Mohajeri K, Mesgari M, Lee AS (2020) When statistical significance is not enough: investigating relevance, practical significance, and statistical significance. MIS Quart 44(2)
58. Willis LD (2023) Formulating the research question and framing the hypothesis. Respir Care 68(8):1180–1185
59. Barker SF (2020) Induction and hypothesis: a study of the logic of confirmation. Cornell University Press
60. Jorem S (2022) The good, the bad, and the insignificant—assessing concept functions for conceptual engineering. Synthese 200(2):106
61. Maynard MT, Conroy S, Lacerenza CN, Barnes LY (2021) Teams in the wild are not extinct, but challenging to research: a guide for conducting impactful team field research with ten recommendations and best practices. Organ Psychol Rev 11(3):274–318
62. Wahlgren B, Aarkrog V (2021) Bridging the gap between research and practice: how teachers use research-based knowledge. Educ Action Res 29(1):118–132
63. Hanna-Benson C, Kroeze S, Gandhi R, Haffie T, Wahl LM (2020) Students as partners in collaborative course design and educational research. Int J Stud Partners 4(2):61–80
64. Zina O (2021) The essential guide to doing your research project. Sage
65. Fraune MR, Leite I, Karatas N, Amirova A, Legeleux A, Sandygulova A, Neerincx A, Dilip Tikas G, Gunes H, Mohan M, Abbasi NI (2022) Lessons learned about designing and conducting studies from the experts. Front Robot AI 8:772141
66. Flynn SV (ed) (2021) Research design for the behavioral sciences: an applied approach. Springer Publishing Company
67. Wilson C (2013) Brainstorming and beyond a user-centered design method. Newnes
68. Denscombe M (2017) EBOOK: The good research guide: for small-scale social research projects. McGraw-Hill Education (UK)
69. Wellington J (2015) Educational research: contemporary issues and practical approaches. Bloomsbury Publishing
70. Silvia PJ (2012) Curiosity and motivation. The Oxford Handbook of human motivation, pp 157–166
71. Polit DF (2008) Nursing research: Generating and assessing evidence for nursing practice. Lippin cott
72. Brink PJ, Wood MJ (2001) Basic steps in planning nursing research, from question to proposal, 5th edn. Jones and Bartlett, Boston, MA
73. Mertler CA (2024) Action research: improving schools and empowering educators. Sage Publications
74. Runeson P, Höst M (2009) Guidelines for conducting and reporting case study research in software engineering. Empir Softw Eng 14:131–164

75. Mosyjowski EA, Daly SR, Peters DL (2017) Drivers of research topic selection for engineering doctoral students. Int J Eng Educ 33(4):1283
76. Müller-Bloch C, Kranz J (2015) A framework for rigorously identifying research gaps in qualitative literature reviews
77. Belcher BM, Rasmussen KE, Kemshaw MR, Zornes DA (2016) Defining and assessing research quality in a transdisciplinary context. Res Eval 25(1):1–17
78. Pandey P, Pandey MM (2021) Research methodology tools and techniques. Bridge Center
79. MacLeod M (2018) What makes interdisciplinarity difficult? Some consequences of domain specificity in interdisciplinary practice. Synthese 195(2):697–720
80. Brittain S, Ibbett H, de Lange E, Dorward L, Hoyte S, Marino A, Milner-Gulland EJ, Newth J, Rakotonarivo S, Veríssimo D, Lewis J (2020) Ethical considerations when conservation research involves people. Conserv Biol 34(4):925–933
81. Chen HT (2005) Practical program evaluation: assessing and improving planning, implementation, and effectiveness. Sage
82. Grove SK, Gray JR, Burns N (2014) Understanding nursing research: building an evidence-based practice. Elsevier health sciences
83. Chua N, Miska C, Mair J, Stahl GK (2024) Purpose in management research: navigating a complex and fragmented area of study. Acad Manag Ann 18(2):755–787

Crafting Research Questions, Objectives, and Aims

Abstract The chapter strongly emphasizes how important it is for academic inquiry to formulate precise, targeted research questions, objectives, and goals. These components offer a systematic framework that guarantees consistency and pertinence during the study. The investigation's central focus and research questions drive methodological choices and data-gathering activities. Well-defined research objectives assist in keeping a clear focus and enable a systematic approach to obtaining significant findings by translating more general issues into concrete, actionable activities. Research questions and objectives must align since this establishes a rational framework that keeps the study focused and well organized. Research runs the danger of becoming dispersed, losing focus, or failing to contribute meaningfully to the field if this alignment is not achieved. By highlighting the significance of well-structured research components, the chapter draws attention to methodologically sound and impactful research that requires a deliberate and focused procedure. Formulating specific and unambiguous research questions and goals is not just a formality; it is a vital component of ensuring that research makes a meaningful and successful contribution to knowledge advancement in various fields.

1 Introduction to the Thematic Area

Establishing clear and explicit research questions, goals, and targets is crucial for establishing a solid and organized basis for any research project [1]. These factors serve as the catalysts that influence the entirety of the research process, guaranteeing a concentrated and intentional examination. A study lacking well-defined research questions, goals, and targets generally has difficulties maintaining coherence and relevance, resulting in fragmented or inconclusive findings [2]. Hence, researchers must comprehend the significance of creating these fundamental components to ensure their studies are methodologically rigorous and capable of significantly contributing to their respective fields. The fundamental components of every academic study are research questions, goals, and aims. They give a clear knowledge of what the study aims to accomplish and function as navigational aids for

A. Almusaed et al., *Practice of Research Methodology in Civil Engineering and Architecture*, https://doi.org/10.1007/978-3-031-97393-2_4

both the researcher and the reader, providing clarity of purpose. The importance of this cannot be exaggerated; without well-designed guiding factors, research risks losing its focus and deviating from its original objectives [3]. Formulating research questions, goals, and purposes is not just a routine academic task but a deliberate process that serves as the foundation for the whole research project. It guarantees that every subsequent action, such as data collecting, analysis, or interpretation, aligns with the main investigation. The research questions form the core of any research effort. These inquiries serve as the starting point for investigation, demonstrating the researcher's inquisitiveness, whether it is to tackle an unresolved issue, query established knowledge, or investigate a novel occurrence within a certain domain. Research questions are crucial in determining the course of the study and influencing choices about the research design, technique, and data collection [4]. Lacking a well-defined and concentrated research issue, a study might rapidly lose direction, resulting in vague outcomes or an incapacity to contribute substantially to the current pool of knowledge. Developing impactful research inquiries necessitates a profound comprehension of the research environment and pre-existing scholarly works. This often entails a comprehensive examination of previous research to pinpoint areas with insufficient information or subjects that want more investigation. Research inquiries should be exact, refraining from any vagueness or extensive scopes that can impede the study's capacity to provide significant insights [5]. Additionally, it is crucial to ensure that the study is viable, meaning that it can be carried out within the limitations of time, money, and data accessibility. Finally, the importance of the research topic is of utmost importance; it should tackle a pertinent matter within the discipline and can enhance knowledge or resolve practical issues [6]. After precisely defining the research questions, the subsequent stage converts them into study objectives. Research objectives are a way to divide a larger investigation into distinct, practical tasks that direct the research process. These goals outline the precise actions the researcher will take to address the research questions. Research objectives provide a framework and guidance, assisting the researcher in maintaining concentration and direction during the inquiry. Research risks becoming disorganized or deviating from its initial aim without carefully formulated objectives. Objectives are distinct from research questions in that they are more comprehensive and precise. Research questions provide a broad structure for the inquiry, while goals give a more detailed view by specifying the precise actions the researcher will do to collect and evaluate data. The study objectives are commonly structured hierarchically, starting with broad objectives and progressing to more detailed ones that tackle various areas of the study subject. It is crucial that objectives are S.M.A.RT, meaning they should be Specific, Measurable, Achievable, Relevant, and Time-bound [7]. This ensures that the research method is practical and can be easily handled within the study's limitations. Research aims, questions, and objectives offer a more comprehensive view of the study's goals. Aims are often broader in scope than aims since they articulate the overarching purpose of the research and its expected impact on the area. Objectives pertain to the specific activities required to finish the study, whereas goals encompass the wider consequences of the research, typically connecting it to greater societal, technical, or scientific progress. Having a clearly stated objective enables the study to be situated within a broader framework, emphasizing its significance and contribution to the advancement of the area. The difference between research objectives and purposes is nuanced yet significant. Objectives are focused on actions and guide the measures the researcher will undertake to accomplish the study. On the contrary, the aims are more forward-thinking, providing a glimpse of the possible influence the study may have in the future. By explicitly stating the objectives and purposes, researchers may guarantee that their investigations are rationally organized and capable of substantially contributing to academic debate or practical implementation [8]. This chapter highlights the significance of formulating research questions, goals, and targets that are closely matched with each other to establish a research framework that is logical and meaningful. These components provide a thorough manual for the researcher, guaranteeing that the study maintains its concentration, adheres to rigorous methodology, and can generate significant outcomes. By meticulously formulating research inquiries, goals, and purposes, researchers may guarantee that their work is methodically sound and pertinent, making valuable contributions to the progression of knowledge and effectively tackling substantial real-world issues. Formulating research questions, goals, and aims is crucial for success in any research undertaking. These factors provide the fundamental basis of the research process, offering clarity, concentration, and guidance. By carefully developing these components, researchers may create methodologically strong studies that can provide useful insights into their respective disciplines [9]. A research question serves as the central focus of an inquiry, guiding the direction and scope of a research project. It identifies a specific issue or problem the researcher seeks to address through systematic investigation [10]. Formulating a research question is a foundational step in the research process, establishing the framework for the study's methodology, analysis, and conclusions. Research questions are utilized across various research paradigms, including experimental, qualitative, and observational studies. Through careful data analysis and interpretation, researchers aim to comprehensively answer the question, contributing to the broader field of knowledge [11].

2 Formulating Research Questions

2.1 Identify Research Questions

Identifying research questions is a crucial feature of the research process, acting as the fundamental element of academic investigation. According to Hosseini et al. (2024), formulating research questions is essential and has a significant role in determining the course of any scientific project. The intellectual framework sets the foundation for the research, providing guidance for the methodology, defining the area of inquiry, and ensuring that the investigation remains in line with the overall objectives of the study [12]. Research questions focus on the investigation's scope, guiding the methodological approach and shaping the direction of analysis. Articulating clear and precise research questions is integral to ensuring the coherence and relevance of a study [13]. They provide a framework that enables the researcher to systematically explore, analyze, and interpret phenomena within a structured academic context [14]. The formulation of such questions reflects the depth of engagement with existing literature and the identification of gaps that warrant further exploration. The role of research questions extends beyond simply narrowing the scope of inquiry; they define the intellectual boundaries within which the research operates [15]. Well-crafted research questions facilitate a focused examination of specific aspects of a broader field, ensuring the research remains directed and methodologically sound. Doing so allows the researcher to concentrate on areas with the most significant potential for generating new insights or contributing to theoretical advancement. Moreover, research questions underpin the logical progression of the research process, from hypothesis formation to data collection and analysis [16]. The precision and clarity of these questions significantly influence the study's overall rigor, determining the relevance of the selected methodologies and the validity of the conclusions drawn. In this sense, research questions are indispensable for maintaining the research's internal coherence and scholarly integrity. Research's internal coherence and scholarly integrity in identifying research questions is a foundational task that shapes the entire trajectory of the research process. They provide the structure for a systematic and rigorous investigation, guiding the researcher through each study phase. The importance of well-formulated research questions cannot be overstated, as they are essential to both the methodological rigor and the intellectual contribution of academic work [17]. The process of creating research questions is shown in Fig. 1, which shows the iterative technique. It emphasizes three interconnected steps: beginning with the solicitation of the first questions, taking into consideration the audience to guarantee relevance and clarity, and, lastly, analyzing the question's scope, depth, and focus. This cyclical approach helps researchers modify their questions, ensuring they are well targeted, rigorous, and aligned with the intended academic or professional audience. To contribute to creating viable research questions, each stage builds upon the previous one through reinforcement.

2.2 Formulating Robust Research Questions

The formulation of a study is an essential preliminary phase in executing good research. The process entails defining the research challenge, establishing explicit objectives, and delineating the methodologies to examine the topic [18]. This phase establishes the groundwork for the whole research endeavor, guiding its trajectory and ensuring that the investigation remains concentrated on resolving inquiries or deficiencies in understanding. In this phase, researchers conduct an extensive literature analysis to comprehend the present knowledge landscape, pinpoint areas of ambiguity, and delineate the focus of their inquiry [19]. Moreover, researchers must

Fig. 1 The process of formulating research questions (*source* authors)

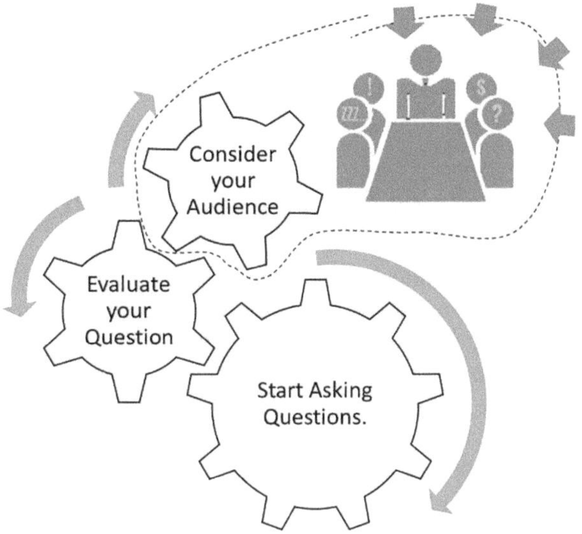

meticulously evaluate the theoretical framework that will direct the study and ascertain the suitable procedures to guarantee that the research process is rigorous, genuine, and dependable. Establishing a robust foundation at this initial stage enhances the likelihood of obtaining significant and credible results that advance the field. In material innovation, for example, researchers frequently concentrate on discovering new materials or enhancing the sustainability of existing materials [20]. A research question such as "How can biodegradable materials be integrated into large-scale construction without compromising structural integrity?" is an example of a question that may be used to influence the study by concentrating on the intersection of biodegradability and structural performance. Because of this topic, the research will be directed at measuring performance metrics under various settings, testing prototypes, and evaluating the qualities of the material. In addition, it establishes a distinct boundary for the inquiry, signaling that the investigation will give equal weight to engineering and environmental viewpoints. For example, research into the production of materials based on algae utilized for insulation shows how a well-formulated research topic may drive a study. The scope of the study can be narrowed down to comparisons between novel and conventional materials by asking questions such as "What are the thermal properties of algae-based insulation materials compared to traditional insulation methods?" The researcher can concentrate on thermal conductivity, cost–benefit analysis, and environmental advantages, which will successfully maintain the study's roundedness and capability for implementation [21]. Further, research questions assist in determining where there are gaps in the current body of literature, which is an essential component of sustainability research. The question is, "How does using recycled concrete affect the overall energy consumption in the building lifecycle?" For example, researchers are compelled to explore the energy footprint involved with the reprocessing of materials. This may entail collecting data on energy use during material manufacturing, shipping, and construction phases, ultimately providing insights into using recycled materials in environmentally responsible building practices [22].

2.3 Refining Research Focus for Effective Questions

The consolidation of the focus of a research project is an essential phase that guarantees the study will continue to concentrate on answering the primary questions and achieving its objectives until the project is completed [23]. During this stage, the research topic is refined, and the scope of the inquiry is narrowed significantly to make the investigation more manageable and meaningful. Concentrating on specific facets of the overarching subject matter makes the study more focused, increasing the likelihood that it will provide essential insights and meaningful contributions to the area [24]. Maintaining the coherence of the study and preventing deviations that are not essential are facilitated by having a clear and well-defined research topic. During this stage, the researcher must frequently review previous research, evaluate the significance of various theoretical frameworks, and determine the gaps the study intends to address. In the end, improving the clarity and accuracy of the study by consolidating the research focus ensures that all efforts are aligned with achieving the set objectives and generating dependable and actionable results [25]. Material innovation and sustainability are expansive subjects encompassing various overlapping subfields, including engineering, environmental, and material science. Research questions can narrow this broad emphasis into more achievable research aims. In the case of a material, for instance, a study topic can concentrate on a particular stage of the material's lifetime, such as manufacture or disposal, rather than investigating the entire lifecycle of the substance. Consider the following research question: "What are the environmental impacts of sourcing raw materials for sustainable concrete alternatives?" would be an example of such question. The early stages of material development are emphasized as the central focus of the investigation, enabling the researcher to directly address critical issues such as resource extraction, energy consumption during processing, and the environmental impacts associated with raw material acquisition. Concentrating on the sourcing phase makes the study more focused and manageable. This refined approach contrasts with previous research strategies that attempted to encompass every aspect of concrete production, often diluting the depth of analysis [26]. For example, in the context of material lifecycle management, a more targeted research question might be: "How does the utilization of reclaimed timber impact the embodied carbon in residential buildings?" This question focuses on a specific material (reclaimed timber) and a sustainability metric (embodied carbon), allowing researchers to focus their analysis on the carbon emissions associated with material reuse and its contribution to the overall carbon footprint of a construction project. Research questions play an equally pivotal role in building production and lifecycle management. A well-structured research framework facilitates a systematic examination of various components, encompassing design, construction processes, structural maintenance, deconstruction, and potential repurposing [27]. Within this domain, the emphasis on sustainability and resource efficiency is paramount, as the research spans the entire lifecycle of a building, from the procurement of raw materials to eventual demolition. Each phase is critically assessed to ensure that it contributes to reducing environmental impact and advancing sustainable construction methodologies. This comprehensive approach fosters the integration of environmentally responsible practices across the entire lifecycle, aligning with global objectives for resource conservation and ecological resilience.

2.4 Formulating Research Questions and Hypotheses to Bridge Knowledge Gaps

The research process begins with discovering a problem or gap within the current body of knowledge. This is an essential stage. According to Bansal et al. (2012), although the gap between research and practice is generally acknowledged, there is a limited discourse on practical techniques to bridge the gap [28]. This is the beginning of the conceptualization phase of the research process. As part of this process, relevant literature is reviewed to understand previous research and suggest areas that require future exploration. The critical phase, which comes after selecting the research subject, is formulating research questions and hypotheses that will serve as the basis for the investigation. According to Barroga and Matanguihan (2022), well-crafted research questions need to be specific, and they also require a comprehensive understanding of the issue being investigated and a comprehensive evaluation of the relevant scholarly material. There is a possibility that research questions will first be descriptive before transitioning into inferential ones [29]. These inquiries must be explicit and brief to provide a solid foundation for creating hypotheses. On the other hand, hypotheses are formal predictions that define prospective outcomes, particularly about the interactions between different groups or factors. The theoretical framework, which provides a lens for analysis utilizing known theories and models, is produced by these questions, which form the framework [30]. After this, a research methodology, which may be qualitative, quantitative, or mixed, details the tactics for data gathering and analysis. Ethical issues ensure that the study is carried out responsibly [31]. After that, the study design is established, including the population that will be selected, the sampling procedures, and the instruments used to gather data. These components constitute a systematic strategy for accomplishing the study's goals [32]. For example, a well-defined study subject might investigate the many stages of a building's lifespan, such as production, maintenance, or end of life, in building production and lifecycle management. A study can be narrowed down to a particular production method and its potential for reducing waste by asking a targeted research question: "How can modular construction techniques reduce waste during the building production phase?" According to the results of comparative research, modular construction is more efficient than traditional techniques in terms of the amount of material waste, the amount of time, and the amount of work [33]. The question "What are the benefits of integrating digital twin technology in lifecycle management for smart buildings?" is another example that directs research toward the application of digital twins. This question investigates how digital twins optimize maintenance, real-time energy monitoring, and resource allocation throughout the lifecycle of a building. Keeping the study small and focused on a specific topic is accomplished by concentrating on technology applications, such as digital twins used in intelligent buildings.

2.5 *Refining Research Focus: Crafting Targeted Questions for Effective Study Design and Impact*

The stage of any academic study that is comprised of the process of concentrating on the research topic is an essential step. During this stage, the scope of the investigation is refined and narrowed to guarantee that the study will continue to be understandable, controllable, and in line with its predetermined goals [34]. For researchers to concentrate on specific areas that contribute significantly to their subject, they must critically engage with the current literature and identify gaps. The creation of questions and hypotheses, which may then lead to the application of the methodological approach and the procedures of data collecting, is made possible by a research topic that has been established. In addition, having a distinct emphasis guarantees that the study will continue to be pertinent and firmly rooted in the ongoing conversations that are taking on within the field. This consolidation process gives the research a sense of direction, which lays a firm basis for advancing new information and contributes to ongoing academic discussions [35]. Completing this phase is necessary to create an organized and coherent study, eventually improving clarity and impact of the research findings. Several subfields fall under the umbrella of building production and lifecycle management. These subfields include energy efficiency, construction waste management, and post-occupancy performance evaluation [36]. Using research questions, these broad themes may be narrowed down into more targeted investigations. An example would be the question, "What role does predictive maintenance play in extending the lifecycle of high-rise buildings in urban areas?" This question allows the researcher to focus on a particular building type (high-rise), a particular lifecycle phase (maintenance), and a specific management strategy (predictive maintenance). By narrowing the topic, the study becomes more actionable, enabling a more in-depth analysis of the components that lead to the extension of the lifetime [37]. Similarly, a question such as "How does the utilization of lifecycle assessment (LCA) tools impact decision-making in the design of sustainable buildings?" The research is narrowed down to concentrate on decision-making processes and specific tools' function in improving building lifecycles. To answer this issue, the research will focus on gaining knowledge of how lifecycle assessment (LCA) tools assist architects, engineers, and builders in making educated decisions on the materials, energy systems, and construction processes that will result in a reduction in the environmental effect over the lifespan of the building. In "Material Innovation and Sustainability" and "Building Production and Lifecycle Management," the research method significantly emphasizes formulating research topics. They determine the course that a study will take, directing researchers toward specific goals while at the same time reducing the emphasis to ensuring that investigations are both manageable and impactful [38]. Through the provision of a defined framework, research questions make it possible to conduct in-depth investigations into significant concerns such as the environmental consequences of new materials, the energy efficiency of building manufacturing, and the implementation of innovative management systems throughout the lifespan of a building. In generating and creating research questions, an essential component of academic and scientific inquiry, the graphic shows significant phases involved in the process. The remainder of this article offers a complete study of the most important topics and a deep exploration of the essential components. The basic components are investigated, their relevance and linkages are analyzed, and the more significant ramifications for the field are brought to light. A comprehensive analysis highlights the fundamental ideas and fundamental contributions, which are as follows.

2.5.1 Conducting Preliminary Research

Conducting preliminary research before creating research questions is essential since it involves comprehensively examining the current relevant literature. This procedure fulfills several vital roles [39]. It enables a thorough comprehension of the existing knowledge landscape, facilitating the identification of widely explored regions and recognizing research gaps that require additional exploration. A comprehensive evaluation of scholarly publications and academic literature facilitates the identification of underexplored subjects or alternative methodologies that may enhance the research. In building energy efficiency, preliminary research may examine the literature on energy-efficient building materials, passive design methodologies, and smart energy management systems innovations. Recent studies on insulation methods and energy recovery systems reveal that specific geographical locations or building types must be more adequately researched. Furthermore, the analysis may indicate that although innovative energy systems have been thoroughly examined, the incorporation of artificial intelligence for real-time energy optimization in residential buildings still needs to be improved. Acquaintance with contemporary publications is crucial, as expertise in many academic disciplines evolves swiftly [40]. In building energy efficiency, remaining informed about recent achievements guarantees that research tackles current challenges, including the influence of novel materials such as phase-change materials (PCMs) on energy usage. Focusing on recent research allows for identifying developing trends, such as the transition to zero-energy buildings (ZEBs) and the growing significance of lifecycle energy assessment in construction. This foundational phase enhances comprehension of the research domain. It establishes the intellectual framework required to refine and direct the research topic, thereby ensuring its contribution to scholarly advancements, especially in sustainable building practices and innovations in energy-efficient design [41].

2.5.2 Determine the Prerequisites for Action

After completing the primary research, it is necessary to make sure that you clearly understand your research's unique requirements [42]. A thorough comprehension of them will serve as a guide for the development of your query. Some critical factors to consider are.

- The place of origin of the problem is as follows: Explain why this subject is essential. It is possible that the objective of your research is to discover a solution to a practical problem, to probe a theoretical gap, or to contribute new insights into an existing body of knowledge all at the same time. It is possible to ensure that your research question is by the broad objectives of your study by offering clarity on purpose [43].
- When the study includes examining a particular theory or notion, the subject may be intended to evaluate the validity or application of the theory across various situations or with fresh data [44]. This is referred to as testing a hypothesis or proposition. In this context, experimental or comparative methodological approaches may be utilized to test the hypothesis thoroughly and examine its applicability in various settings.

Recognizing prerequisites is crucial for sustaining attention, relevance, and alignment with academic and practical goals. Peel (2020) asserts that utilizing a case study methodology within a particular context depends on an inquiry framework to facilitate the research process efficiently [45]. By explicitly delineating these fundamental criteria, a systematic framework is created that directs the research parameters, reduces distractions, and facilitates attaining significant results. This methodology guarantees that the study tackles essential topics and significantly contributes to the relevant domain of knowledge or practice.

2.5.3 Refining the Research Focus for Targeted and Impactful Inquiry

Refining the research focus is one of the most critical steps in the research process. This helps to ensure that the study continues to be targeted and relevant [16]. During this stage, the process of breaking down broad concepts into precise, manageable questions or objectives closely connected with the study aims is carried out. By doing so, the research may maintain its scope in a manageable manner, allowing for a more in-depth investigation of the selected issue. Using a more precise focus, the study can address significant concerns, ultimately resulting in more relevant and accurate outputs [46]. An additional benefit of this procedure is that it helps explain the methodology, making it possible to pick relevant instruments and techniques to answer the research questions correctly. Furthermore, narrowing the emphasis helps contribute to a more transparent and organized study structure, which helps facilitate the logical evolution of arguments and discoveries [18]. This stage is necessary to produce clearly defined research, has a significant impact, and directly addresses the academic or practical goals that are meant to be achieved within a particular field of study. After a project's broad purpose and requirements have been established, zeroing in on a particular study issue is paramount. There is a strong correlation between the clarity of research questions and the effectiveness of academic inquiry. This is because comprehensive questions are frequently unmanageable within the confines of a single study. Simplifying the emphasis makes conducting a more thorough investigation of a specific problem possible, ultimately resulting in more significant and applicable discoveries. To promote a more specialized study, it is helpful to concentrate on a particular area or feature of the subject matter [47]. Refining the research question makes the investigation more profound into a specific topic, resulting in more pertinent and complex findings than straightforward inquiries. Restricting the scope helps to define clear boundaries for the research, preventing the study from becoming unmanageable due to excessive data or many variables. For example, rather than investigating the overarching topic of "sustainability in urban design," a more specialized inquiry such as "the impact of biophilic design on energy efficiency in office buildings in temperate climates" is an opportunity for a more in-depth and context-specific investigation. This level of detail allows the research to align with its initial aims while eliminating the possibility of diversions from concerns on the periphery. If this is done, the study will continue to be targeted, manageable, and methodologically sound. This will ensure that the research's findings will successfully contribute to the current body of knowledge.

2.5.4 Identifying and Formulating Questions for Potential Research

The procedure of determining and developing research questions is an essential component in directing a research effort toward significant and influential outcomes. After completing exploratory research, identifying requirements, and refining the study subject, a researcher is better positioned to formulate specific, well-defined questions [48]. The formulation of practical research questions is dependent on several fundamental concepts, including the following:

- *Clarity and Focus*: The research questions need to be precise and well-defined to guarantee that the study is carried out clearly and focused. Obtaining significant findings from fragmented research might be challenging due to imprecise or unduly broad questions [25].
- *Feasibility*: Accurate answers to the study questions depend on having enough data or evidence [49]. This guarantees that the study will always have a solid empirical foundation. Research risks devolve into conjecture rather than objective analysis without easy access to current, dependable data.

A research issue must close a knowledge gap or offer a practical solution to be considered significant [50]. This relevance presents fresh ideas or creative solutions and develops both the academic and professional fields. For instance, studying sustainable urban drainage systems (SUDS) in environmental design might help with the increasing problem of managing stormwater in metropolitan areas. The research may contribute to the academic knowledge of sustainable water management while providing valuable solutions that improve urban resilience to floods by creating a unique strategy for integrating green infrastructure, such as bioswales and permeable pavements.

Developing thoughtful research questions is a crucial first stage that determines the overall course of the investigation. The research techniques, data sources, and analytical approaches are chosen based on these questions. A well-crafted question acts as a road map, coordinating the research with theoretical frameworks and real-world applications [51]. To summarize, formulating research questions entails a comprehensive procedure encompassing preliminary inquiry, a precise comprehension of the study requirements, and homing in on the research emphasis. This guarantees that the study is well founded, narrowly targeted, and capable of significantly advancing academic understanding and professional practice. A well-crafted research topic may, for instance, examine how urban infrastructure might be modified to survive an increase in climate-related occurrences like heatwaves or flooding in the context of resilient and climate-adaptive infrastructure. In addition to providing valuable answers for city planners and engineers dealing with climate concerns, the research may add to theoretical conversations on resilience by examining the efficacy of various climate-adaptive materials and designs, such as cool roofs or flood-resistant foundations.

2.6 *Crafting Research Questions for Effective Academic Inquiry*

Research questions are crucial for guiding academic inquiry as they clarify and emphasize the research process. The query on the existence of a standard for research questions brings attention to a significant matter: the absence of a universally accepted criterion for research inquiries. This is because they must be tailored to the specific circumstances of each research. Research questions possess intrinsic flexibility because of the wide range of academic fields and research aims. The study's precise objectives and techniques determine its form and structure [8]. Within the context of architecture and construction, research questions need to be customized to target specific areas that need attention, such as investigating the effectiveness of material recovery or the influence of recycling procedures on the lifespan of buildings. One such question is: "What are the most efficient methods of reusing materials in modern urban construction?" Alternatively, how may recycling systems be enhanced to minimize the environmental impact of extensive undertakings? These examples demonstrate the process of formulating research questions that specifically address a specialized aspect of broader sustainability debates. The expression "From where?" highlights the difficulty of formulating research topics, frequently necessitating a thorough examination of current literature to pinpoint unsolved matters. Reuse and recycling techniques may entail examining prior research to pinpoint deficiencies in present processes that hinder the optimization of material sustainability. This research serves as the foundation for inquiries that contribute to the growth of theory and the development of practical solutions within the subject. When formulating research questions, it is essential to consider the audience and ensure they are relevant to stakeholders such as policymakers, industry experts, or scholars. An example of a research subject exploring solutions for reusing building materials might attract academic researchers interested in sustainability and urban planners looking for practical ways to minimize waste. A well-formulated research question directs the chosen methodology and guarantees consistency throughout the objectives, data gathering, and analysis [52]. When developing strategies for reusing and recycling, it is essential to formulate a research topic that is precise, achievable, and aligned with current conversations on sustainable building practices. This will guarantee that the study's findings target essential areas that need attention. Research questions must be formulated by a uniform standard, considering the field of study, the research topic, and the target audience. The formulation process requires a combination of intellectual rigor and creativity to produce questions that guide coherent and relevant research, especially in developing areas such as reuse and recycling solutions in buildings [53]. Figure 2 depicts the process of determining a research subject by following many crucial steps: fostering curiosity, discovering gaps in knowledge, acknowledging conflicting viewpoints, and completing a comprehensive literature study. Together, these factors assist researchers in identifying unsolved challenges, which contribute to developing significant and pertinent research topics that promote academic investigation.

The most important aspects of a solid research topic are emphasizing several critical criteria that guarantee the subject is resilient, practicable, and beneficial

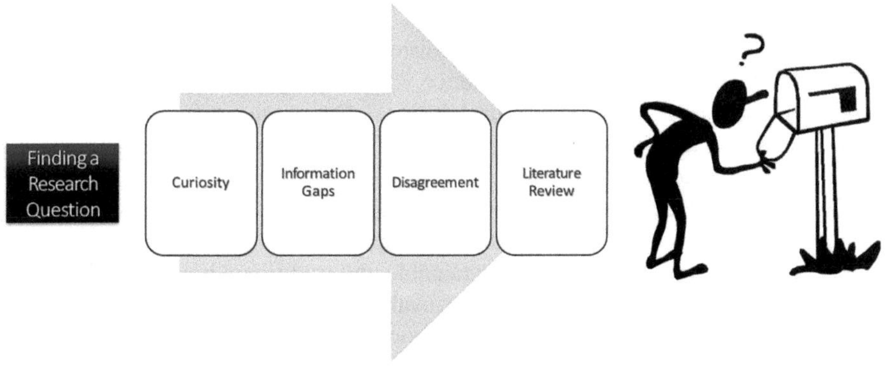

Fig. 2 Pathways to finding a research question

for academic investigation [54]. Each of these aspects contributes to the framework of producing research questions, which are among the most critical factors in determining the course of research, its breadth, and its level of involvement.

2.7 *Varieties of Research Questions*

Formulating research questions is essential in the research process since it defines the study, directs the technique, and affects the outcomes. Research inquiries are often categorized into two primary types: quantitative and qualitative. The classification is dictated by the type of data needed and the extent of the inquiry. Each inquiry fulfills various purposes and utilizes different data collection and analysis methodologies. A comprehensive overview of the many forms of research questions is presented according to the frameworks developed by Doody and Bailey [55].

2.7.1 Questions for Quantitative Research

Quantitative research inquiries concentrate on numerical data and statistical evaluation. These inquiries assess certain variables, discern linkages, and evaluate theories. Quantitative research often encompasses extensive populations, utilizing standardized data-gathering instruments like surveys or experiments to get quantifiable information. These inquiries are especially beneficial for comprehending patterns, facilitating comparisons, and deriving conclusions from statistical data. Quantitative research questions are characterized by their systematic nature and application in hypothesis testing. Berger (2015) emphasizes that quantitative inquiries typically eschew open-ended phrasing, favoring instead specific, quantifiable terminology such as "what," "how much," or "to what extent." Marshall and Rossman (2011) contend that quantitative research's objective is frequently to systematically investigate specific social, educational, or environmental issues [56]. Quantitative research inquiries can be classified into descriptive, comparative, and relational categories:

- *Descriptive Questions* [57]: These inquiries seek to delineate the attributes of a population or phenomena. For instance, "What proportion of architects utilize sustainable materials in architectural design?" Descriptive questions encapsulate the phenomena being studied, synthesizing data to yield insights into specific facets of the study subject.
- *Comparative Questions* [58]: These inquiries aim to discern distinctions or resemblances among two or more groups. An example may be, "What are the distinctions in energy consumption between sustainable and traditional buildings?" Comparative inquiries enable researchers to evaluate factors among distinct groups, uncovering discrepancies that may influence results.
- *Relational Questions* [59]: These inquiries examine the correlations between variables. For example, "What is the correlation between the utilization of recycled

materials and the decrease in construction expenses?" Relational inquiries aim to ascertain the existence of a causal or correlational connection between two or more variables and frequently use statistical techniques to validate these associations.

Quantitative research questions yield unambiguous, objective responses that may be extrapolated to broader audiences. They are frequently employed in research necessitating statistical validation and provide definitive, repeatable outcome.

2.7.2 Questions for Qualitative Research

Conversely, qualitative research questions investigate profound insights, subjective experiences, and intricate processes. These inquiries are more adaptable and open-ended, enabling the researcher to modify their approach as the study advances. Qualitative research often engages smaller, targeted audiences and aims to elucidate the "why" and "how" of events, yielding comprehensive, contextual data. Qualitative research questions are often constructed to facilitate the examination of many dimensions of human experiences or social interactions. Creswell (2013) contends that qualitative inquiries are adaptable and non-prescriptive, enabling the researcher to investigate the significance of occurrences or acts. In contrast to quantitative questions, which focus on measurement, qualitative questions strive to uncover, elucidate, or investigate deeper significance levels. Qualitative research questions encompass several subcategories, each designed for a distinct purpose:

- *Contextual Questions* [60]: These inquiries concentrate on comprehending the milieu in which a phenomenon transpires. For example, "In what manner do urban planners incorporate sustainable materials into urban development initiatives?" Contextual inquiries examine the history and circumstances affecting the subject of investigation.
- *Descriptive Questions* [61]: These inquiries comprehensively elucidate specific procedures or occurrences. For instance, "What are the experiential realities of architects who design green roofs in urban environments?" Descriptive questions seek to provide a thorough explanation of the phenomena, typically via narrative or case study methodologies.
- *Exploratory inquiries* [62]: These examine inadequately explored domains. For example, "What is the perception of engineers regarding the role of artificial intelligence in sustainable construction?" Exploratory inquiries seek to furnish initial insights on novel or developing subjects.
- *Evaluative Questions* [63]: These concentrate on appraising the efficacy of a policy or action. For example, "What is the efficacy of green walls in lowering indoor temperatures in office structures?" Evaluative inquiries assess results and effects.

Qualitative research inquiries offer the adaptability to explore intricate social, cultural, or environmental matters. They facilitate a deeper comprehension of the subject, providing elucidations that transcend primary numerical data or statistical correlations [64]. Comprehending the different research topics is crucial for

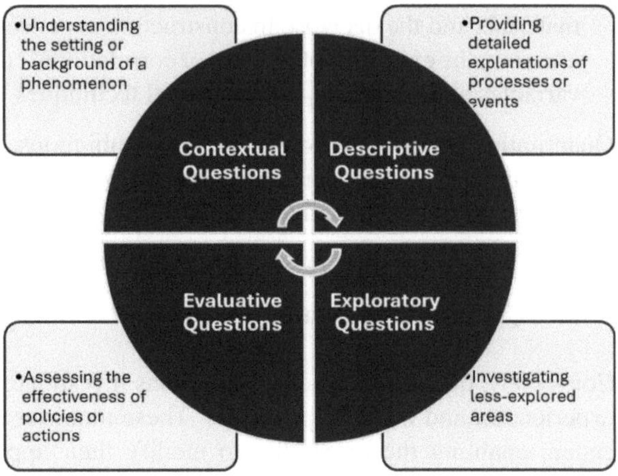

Fig. 3 Taxonomy of research questions: contextual, descriptive, exploratory, and evaluative categories

adequately structuring a study. There are four basic categories of research questions, which are illustrated in Fig. 3. These categories include contextual, descriptive, exploratory, and evaluative. In each quadrant, the focus of these inquiries is highlighted, ranging from the comprehension of the background of a phenomenon (contextual) to the investigation of less-explored regions (exploratory), the provision of thorough explanations (descriptive), and the evaluation of the success of policies or actions (evaluative).

Quantitative research questions provide organized, quantitative insights suitable for hypothesis testing, whereas qualitative research questions provide a more comprehensive examination of intricate occurrences [65]. Both categories of inquiries play essential roles in the research process, contributing to the expansive development of knowledge in distinct yet complementary manners.

2.7.3 Questions for Mixed-Methods Research

Mixed-methods study inquiries integrate quantitative and qualitative methodologies, utilizing the advantages of each to offer a more thorough comprehension of intricate occurrences. Mixed-methods research seeks to collect and evaluate quantitative and qualitative data, enabling researchers to examine research problems from many viewpoints. By amalgamating qualitative and quantitative data, mixed methods can yield a more comprehensive and nuanced response to research questions. Mixed-methods research topics are particularly advantageous in investigations requiring statistical analysis and comprehensive contextual insight. Creswell and Plano Clark (2017) emphasize that mixed-methods research often adheres to one of three primary designs: convergent, explanatory sequential, and exploratory sequential [66]. Each design utilizes distinct ways to amalgamate qualitative and quantitative aspects, although they aim to integrate data from both methodologies to derive more nuanced findings. Mixed-methods research inquiries can be classified as follows:

- *Convergent Mixed-Methods Questions* [67]: These inquiries simultaneously collect qualitative and quantitative data to compare or correlate the results. An example of a convergent inquiry may be: "How do architects' perceptions of sustainability correspond with the actual energy-saving results in green buildings?" Qualitative data (architects' impressions) and quantitative data (energy-saving outcomes) are gathered simultaneously and subsequently compared to evaluate their congruence.
- *Explanatory Sequential Mixed-Methods Inquiries* [68]: This strategy initially collects quantitative and qualitative data to elucidate or further investigate the preliminary findings. For instance, a question may be: "Following the assessment of cost savings from sustainable building materials, what qualitative factors elucidate architects' material selections?" The quantitative data offers a comprehensive assessment of cost reductions, while the ensuing qualitative analysis elucidates the aspects affecting material selection.

The exploratory sequential mixed-methods design starts with the qualitative investigation to guide subsequent quantitative analysis [69]. An example of an exploratory sequential inquiry is: "What themes arise from interviews regarding sustainable architecture, and how can these themes be quantified across a wider demographic?" Initially, qualitative insights are collected to investigate themes or concepts, followed by applying quantitative tools to evaluate or quantify these patterns in a broader sample. Mixed-methods research questions are particularly advantageous when neither a solely qualitative nor solely quantitative approach is enough. They provide the freedom to examine overarching trends (quantitative) and fundamental causes (qualitative), tackling intricate research problems more thoroughly. These questions enable researchers to triangulate data, corroborating results from several perspectives and yielding more robust, dependable conclusions. For example, in research on sustainable construction practices, quantitative data may uncover statistical trends in energy savings associated with various materials, and qualitative data might elucidate the decision-making processes of engineers and architects. Collectively, these methodologies would offer a comprehensive understanding of the results and the motives underlying sustainable design decisions. Formulating research questions is crucial for directing any study since quantitative, qualitative, and mixed methodologies inquiries provide unique yet complementary techniques. Quantitative questions yield organized, quantifiable information suitable for hypothesis evaluation. Qualitative questions provide a comprehensive and adaptable examination of intricate occurrences. Mixed-methods questions amalgamate the advantages of both methodologies, facilitating a more comprehensive and nuanced comprehension of study issues. Scholars can employ diverse research questions to explore the breadth and depth of their themes, thus enhancing the complexity and strength of the knowledge base [70]. A complete overview of the many forms of study and the question categories that correspond to them is shown in Table 1. The study is divided into quantitative, qualitative, and mixed methods. It also provides a detailed explanation of how each type of question handles specific research objectives. Mixed methods incorporate qualitative and quantitative techniques to provide comprehensive knowledge of complex events. Quantitative questions concentrate on comparison and measurement, while qualitative questions investigate context and experiences.

Table 1 Research types and categories

Research type	Category	Description
Quantitative research [71]	Descriptive questions	It focuses on describing the characteristics of a population or phenomenon. For example, "What proportion of architects use sustainable materials in architectural design?"
	Comparative questions	Compares two or more groups to identify differences or similarities. Example: "What are the differences in energy consumption between sustainable and traditional buildings?"
	Relational questions	Examines relationships or correlations between variables. Example: "What is the relationship between using recycled materials and reduced construction costs?"
Qualitative research [72]	Contextual questions	Explores the context or environment surrounding a phenomenon. Example: "How do urban planners integrate sustainable materials into city projects?"
	Descriptive questions	Provides detailed, in-depth descriptions of specific experiences or processes. For example, "What are the experiences of architects designing green roofs in urban areas?"
	Exploratory questions	Investigate new or underexplored topics. Example: "What do engineers think about the role of artificial intelligence in sustainable construction?"
	Evaluative questions	Evaluates the effectiveness of a policy or intervention. Example: "How effective are green walls in reducing indoor temperatures in office buildings?"
Mixed-methods research [73]	Convergent mixed-methods questions	Combines qualitative and quantitative data collected simultaneously to compare results. Example: "How do perceptions of sustainability align with energy-saving outcomes?"
	Explanatory sequential questions	Uses quantitative results to explain or explore qualitative findings. Example: "After measuring the cost savings of green buildings, what factors influence material choices?"
	Exploratory sequential questions	Begins with qualitative data to inform subsequent quantitative research. Example: "What themes emerge from interviews on sustainable architecture that can be measured quantitatively?"

2.8 *Methodological Approaches to Formulating Research Questions*

2.8.1 Formulating Effective "How" Research Questions for In-Depth Analysis

Research questions centered on "How" are crucial for investigating processes, methods, or systems. These sorts of inquiry aim to comprehend the mechanics or operations that underlie certain phenomena or activities by posing "How," the researcher encourages a more thorough examination of the procedures, stages, or techniques that result in specific outcomes [74]. This investigation is precious in disciplines that include intricate interconnections or systems, such as engineering, medicine, social sciences, and management. In sustainable design, a research topic such as "What is the impact of green roofs on energy efficiency in urban buildings?" explores how green roofs serve as a means of conserving energy. This topic necessitates a comprehensive consideration of several elements, including insulation effects, heat absorption, and overall building performance, rather than a simple yes or no response. A well-constructed and efficient "How" inquiry is formulated with clear, pertinent, practical, and precise elements. Clarity guarantees that the question is easily comprehensible and free from any ambiguity. Relevance refers to the extent to which a query tackles a genuine issue or adds value to a substantial pool of information. Practicality guarantees that the question can be resolved using the researcher's existing resources, time, and tools. Ultimately, clarity helps to cut down the scope of the issue, facilitating the collection of precise and significant data [75]. In general, research inquiries about the "How" aspect yield profound discoveries by concentrating on comprehending the complexities of procedures, rendering them exceedingly beneficial for research endeavors focused on innovation, enhancement, or optimization. The relevance of "How" research topics, which concentrate on processes, techniques, and systems, is highlighted in Table 2. It emphasizes their application in intricate disciplines to gain a more profound understanding of processes and events. To conduct an efficient investigation, it is necessary to have critical aspects such as clarity, relevance, and practicality. Because of this, these questions are critical for doing complete research.

2.8.2 Formulating Effective "What" Research Questions for In-Depth Analysis

Research inquiries commencing with "What" are frequently utilized to investigate certain facets of a subject or field of research, to deepen comprehension, and to furnish essential information for subsequent decision-making [76]. These questions generally concentrate on recognizing and delineating technology, materials, processes, or

Table 2 Formulating "How" research questions: investigating processes and systems in complex fields

Aspect	Description	Deep explanation
Focus	Investigating processes, methods, or systems	The focus on processes and methods highlights the necessity of dissecting systems to identify underlying patterns, interconnections, and operational dynamics. This is particularly useful in multidisciplinary studies
Purpose	To understand the mechanics or operations behind phenomena	Understanding the mechanics behind phenomena allows for exploring causal relationships, which can uncover the reasons behind system efficiency or inefficiency, driving improvements and innovation
Application	Useful in disciplines with complex interconnections like engineering, medicine, social sciences, and management	In complex fields, investigating processes helps reveal the intricacies of how interconnected systems function and interact, enabling informed decision-making and practical advancements
Example	Example in sustainable design: "How do green roofs impact energy efficiency in urban buildings?"	This example highlights the importance of asking questions that address real-world challenges, such as urban sustainability and energy efficiency. Green roofs provide insights into integrating ecological designs into city infrastructure
Key elements	Clarity, relevance, practicality	The focus on clarity, relevance, and practicality ensures that any research question remains feasible, addresses real issues, and provides useful insights for application in both theoretical and practical contexts

other essential elements pertinent to the examined topic. "What" queries, by concentrating on specifics, can yield essential data that form the foundation for subsequent investigation. "What" inquiries are predominantly descriptive, focusing on the attributes or traits of a thing that wants elucidation [77]. This renders them particularly advantageous in disciplines such as engineering, architecture, technology, and science, where comprehending the elements or behaviors of systems or processes is essential for informed decision-making. For instance, in eco-friendly technologies or sustainable construction methods, a researcher may ask, "What are the thermal insulation characteristics of materials derived from recycled concrete?" This inquiry seeks to elucidate recycled concrete characteristics that may guide material selection for energy-efficient structures. These findings may enhance industry practices, facilitating sustainability in building. In a comparable situation with green roofs, a researcher may inquire, "What are the principal factors that affect vegetation growth on green roofs in urban environments?" This inquiry aims to ascertain the fundamental elements, including soil composition, plant species, and environmental circumstances, that influence the efficacy of green roofs. By addressing this inquiry, researchers may suggest enhancing green roofs' efficacy and longevity, impacting design and installation methodologies. "What" inquiries serve not just to provide descriptions but also to pinpoint deficiencies or issues in current understanding or practices. Upon the collection and analysis of data, the results can elucidate the topic being examined or provide answers [78]. This methodology is very prevalent in scientific inquiry. For example, in energy-efficient technologies, a researcher would ask, "What are the principal obstacles to implementing solar energy systems in residential structures?" This investigation aims to identify specific barriers, including financial, regulatory, or technological limits, that impede the use of solar energy. The findings may inform solutions to surmount these obstacles, therefore offering a resolution to the issue at hand. By addressing "What" inquiries, researchers can enhance understanding within a particular domain, providing comprehensive insights that facilitate more intricate analyses or practical implementations [79]. In urban planning, a researcher may inquire, "Which green building certifications are most efficacious in fostering environmentally sustainable practices?" The results from this inquiry might assist policymakers and developers in making more informed judgments concerning adherence to environmental regulations. "What" research questions are crucial for obtaining thorough knowledge. They assist in identifying issues, improving comprehension of information or technology, and facilitating informed decision-making. Regardless of whether the objective is to enhance current understanding, resolve a problem, or affect practical results, "What" inquiries are essential to the research process, especially in domains where technical and material innovation propels advancement. Table 3 demonstrates the relevance and significance of "What" research questions in several domains, including engineering and sustainability, highlighting how these inquiries yield practical insights and facilitate informed decision-making.

2.8.3 Formulating Effective "Why" Research Questions for In-Depth Analysis

Research questions commencing with "Why" are often formulated to clucidate the fundamental reasons, motives, or elements influencing an inevitable occurrence. These inquiries transcend superficial observations, enabling researchers to investigate the underlying causes of specific outcomes or behaviors [80]. The inquiry "Why do specific building materials deteriorate more swiftly in coastal regions?" aims to elucidate the environmental conditions, such as salt exposure, humidity changes, or temperature fluctuations, that contribute to accelerated material deterioration. By identifying these aspects, the research can yield insights that increase building methodologies and material choices in coastal areas, resulting in improved durability and performance. In disciplines such as architecture and design, "why" inquiries are essential for examining the reasoning behind decisions and practices. The question "Why do architects favor open floor plans in contemporary office buildings?" explores the rationale behind this prevalent design preference, perhaps uncovering factors such as adaptability, visual attractiveness, promotion of cooperation, or economic efficiency. Comprehending these elements may clarify the values and goals behind current architectural choices, providing profound insights into the design trends that define modern office environments. Moreover, "why" inquiries are crucial for comprehending the determinants affecting structural behavior and design

Table 3 Using "What" questions to clarify systems and drive decision-making

Aspect	Description	Depth explanation
Focus	Grasping particulars within a study area (technologies, materials, procedures)	Breaking down complex systems allows for precisely understanding each component's role and behavior within the broader system
Purpose	Reveal information to improve understanding and aid decision-making	A clearer picture of system operations helps formulate decisions that enhance performance and sustainability
Application Example 1	Investigating thermal insulation qualities of recycled concrete in sustainable buildings	Explores how recycled materials can meet insulation standards, addressing sustainability and energy efficiency in modern construction
Application Example 2	Investigating factors influencing vegetation growth on green roofs in urban areas	Addresses how environmental and material factors can optimize green roofs, enhancing urban sustainability and biodiversity
Using analysis to find solutions	Identify gaps or issues and offer solutions, such as obstacles to solar energy implementation	Analyzing challenges (e.g., cost, technical issues) in solar energy adoption to provide feasible and innovative solutions
Knowledge growth and decision-making	Enhance knowledge and provide insights for informed decisions, e.g., green building certifications	Facilitates better decision-making in green building projects by connecting research insights with certification requirements and sustainability goals

trends [81]. The inquiry "Why are cross-laminated timber structures gaining popularity in sustainable building practices?" aims to elucidate the factors contributing to this trend. The findings may indicate that sustainability, cost-effectiveness, structural integrity, and a diminished carbon footprint are primary factors contributing to the heightened utilization of cross-laminated lumber. Such findings can further the evolution of sustainable building techniques by pinpointing the fundamental elements that facilitate their acceptance. Research questions structured with "Why" facilitate the exploration of profound insights into phenomena, decisions, and behaviors, significantly contributing to theoretical comprehension and practical implementations [25].

- *Tips for Crafting Effective "Why" Research Questions*

Developing impactful "Why" research inquiries is essential for directing and concentrating a study. These inquiries delve into the fundamental reasons, motivations, and influences behind events, promoting a more profound comprehension of intricate matters. Researchers go beyond superficial observations by asking "Why" and delve into the underlying causes of events, trends, or behaviors [82]. Formulating these inquiries necessitates meticulous deliberation about lucidity, feasibility, receptiveness to other viewpoints, and congruity with the empirical and theoretical underpinnings of the study. They thoughtfully formulated "Why" inquiries that offer perceptive elucidations and enhance broader scholarly dialogs and pragmatic implementations. The advanced strategies for designing good "Why" research questions are outlined in Table 4. These techniques emphasize clarity, empirical answerability, and openness to various views. The necessity of contextualizing the subject within socio-economic and technical contexts while simultaneously supporting theoretical advancement and practical applications is brought to light by this. Furthermore, the handbook emphasizes the necessity of critical explanation, which ensures that the study contributes to theory and practice and larger academic discourses within the academic community.

"Why" questions often allow for multiple explanations, encouraging research to remain open to various perspectives rather than focusing on a single hypothesis. This approach ensures a comprehensive and objective analysis, whether investigating causes or motivations and asking "why" is crucial for deepening understanding of the factors shaping a phenomenon, providing a structured framework to address complex issues, and generating data-driven insights that inform decisions and practices.

In conclusion, Fig. 4 depicts the research process following a cyclical model of four stages: describing, explaining, intervening, and predicting. After describing a phenomenon, researchers identify its underlying causes, test interventions, and make predictions, fostering an ongoing cycle of inquiry and learning.

On the other hand, formulating good research questions is essential to direct the topic and methods of a study. A good research question should be well defined, succinct, and particular; it should also reduce the scope of the issue so that it may be investigated in greater depth [8]. It has to be viable for research within the limits that have been established, and it must have practical significance to the area. In addition, it should inspire critical examination, typically phrased with the words "how" or "why," to drive deeper inquiry and provide essential insights to the field. Figure 5 demonstrates the critical components of constructing a solid research question. Practical research questions must be focused on a specific issue or problem, ensuring clarity and precision. They should also be feasible, meaning they can be realistically addressed within the constraints of time and resources. Additionally, such questions must be sufficiently complex, allowing for in-depth analysis and discussion throughout a paper or thesis and ensuring they foster substantive academic inquiry.

3 Articulating the Research Aims and Objectives

One of the most critical aspects of the research process is the establishment of defined research objectives [83]. These objectives serve as a road map that directs the study from formulating research questions to collecting and analyzing that data.

Table 4 Guidelines for formulating effective "Why" research questions: principles, explanations, and example

Guiding principle	In-depth explanation	Illustrative example
Be clear and specific	Formulate the question precisely, ensuring that it focuses on a singular, well-defined issue within the research scope. Clarity helps avoid ambiguity, ensuring that the question effectively directs the investigation toward specific variables	Instead of posing, "Why do construction materials degrade?" the more targeted question could be: "Why do specific types of polymer-based insulation degrade under high UV exposure?"
Ensure it is empirically answerable	The question must be structured so that empirical data can be collected and analyzed to provide concrete answers. Avoid speculative or overly philosophical "Why" questions that real-world data cannot support	A researchable question would be: "Why do certain sustainable building materials demonstrate improved thermal resistance under varying climatic conditions?"
Maintain an open-ended approach	A "Why" question should not imply a singular or predetermined answer. It must allow for exploring multiple perspectives, fostering a comprehensive and multifaceted analysis. Avoid framing the question around one expected outcome	For instance, instead of asking, "Why do architects prioritize energy efficiency?" consider, "Why do architects adopt differing approaches to energy-efficient building design?"
Consider contextual complexity	Ensure that the "Why" question considers the broader context, including socio-economic, technological, and environmental factors, allowing the researcher to account for external influences on the phenomenon	Example: "Why do low-income urban areas in tropical climates adopt fewer energy-efficient building practices despite the availability of sustainable materials?"
Support theoretical and practical inquiry	"Why" questions should support both theoretical development and practical applications. The answers generated can inform practical decision-making and policy recommendations by grounding research in theoretical frameworks	Example: "Why do certain urban planning policies promote the integration of green building technologies while others lag despite economic incentives?"
Guide toward critical explanation	A well-crafted "Why" question should enable the researcher to move beyond mere description, guiding them toward critically explaining the investigated phenomenon's causal relationships or underlying mechanisms	Example: "Why do specific prefabricated materials perform better in modular construction under extreme weather conditions than traditional building materials?"
Inform broader research agendas	The "Why" question should contribute to larger academic or professional discourses. It should align with the current gaps in the literature and potentially inform future research, practice, or policy initiatives	Example: "Why has adopting AI-driven sustainable architecture practices varied significantly between developed and developing countries in recent years?"

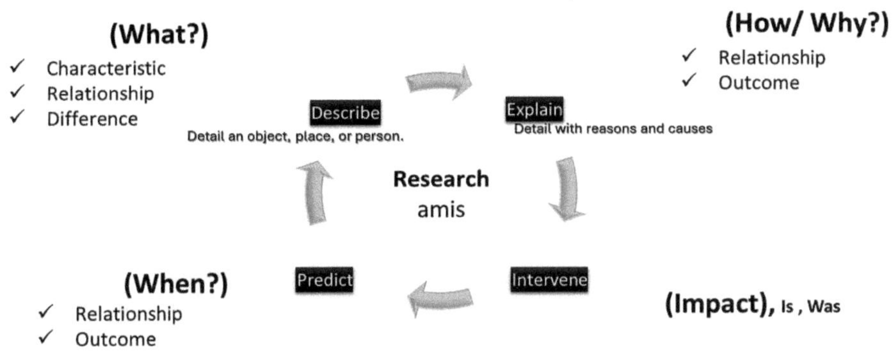

Fig. 4 The research process cycle (*source* authors)

Fig. 5 Constructing effective research question

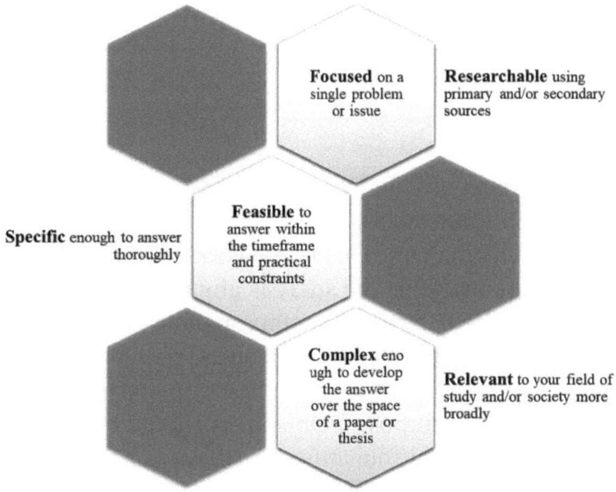

They ensure an ordered and methodical approach by transforming broad research queries into concrete, practical measures. More importantly, they maintain coherence and concentration throughout the study. This section investigates the value of research aims, differentiates between primary and objectives, investigates how well they connect with research questions, and provides an outline of ways for effectively formulating objectives. The research objectives clarify how the study will answer specific research issues. These objectives are typically called "This research aims to examine/analyze/reveal…" The researchers can influence the structure of their dissertation by setting these objectives, with each aim driving the stages of the study and matching chapters with relevant duties. Formulating objectives requires the study to be broken down into crucial stages to maintain coherence and concentration [84]. In most cases, each stage focuses on a different research task, which assists in keeping the study traveling on the right path. The most critical activities should be presented without excessive fragmentation, and the beginning of the description

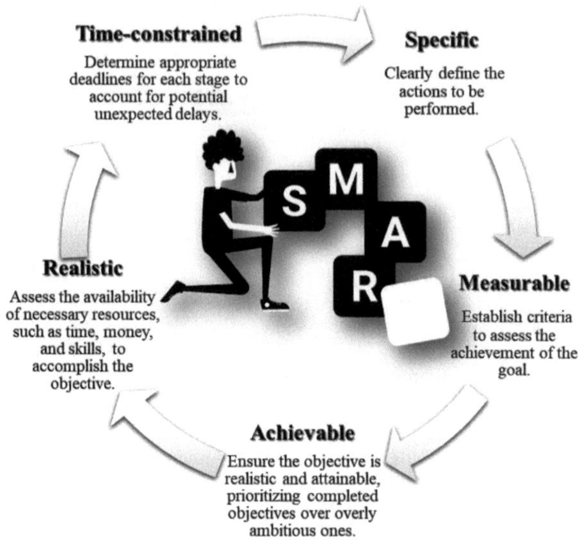

Fig. 6 Cyclical model of iterative processes, highlighting the active and engaging nature of the research process (*source* authors)

should frequently include terms like "investigate," "analyze," "clarify," or "justify." A coherent framework is created by researchers when they structure research objectives in this manner. This framework not only specifies the purpose of the study but also guarantees that the study is aligned with its broader purposes, making it easier for researchers to conduct a thorough and theoretically sound inquiry. The project's objectives are the intended outcomes, stated in general terms, that it wants to achieve upon completion. On the other hand, objectives are the actions that will be performed to achieve the purposes. Objectives should be formulated as precise statements that outline measurable outcomes and the actions necessary to achieve the desired results. When constructing objectives, use forceful positive statements with verbs such as collect, create, classify, develop, devise, measure, produce, revise, select, and synthesize. Avoid using weak verbs such as appreciate, contemplate, inquire, learn, know, comprehend, be aware of, listen, and perceive. In addition, goals should adhere to the **S.M.A.R.T** principle [85], as shown in Fig. 6.

The number of purposes and objectives to be included in the research project can vary. It should be discussed with the project manager, who should provide the audience with support and guidance [86]. Some administrators favor a single well-defined objective, while others anticipate a primary objective with at least two subsidiary objectives. Similarly, the number of objectives is not fixed; however, a substantial number should be developed to measure progress toward achieving the goal(s).

3.1 *Thematic Definition*

- *What exactly is the research goal?*
- *Is it to solve one or several proposed research questions?*
- *Or through analyzing and researching these issues to promote a particular aspect of theoretical understanding?*

Research objectives are formulated for research questions. The primary task of the research objective is to answer the research question; that is, the research question is the "objective" in the research objective, which is its proper meaning [87]. Why should there be both research questions and research objectives? This is because, although we have raised research questions, our answers are not necessarily correct, nor can we obliterate other scholars' existing or possible contributions to this question. When constructing research goals, a safe strategy is to claim that we have given a tentative answer to a specific research question. On the contrary, if we only talk about the research objectives but not the research questions, it will not work! In other words, research questions must be first, followed by research objectives. As far as a specific study is concerned, its objectives have size and degree considerations. Size means that, for a significant goal in a research field, the goal of a topic or an article is small in this big goal. A degree indicates that, as far as a specific study is concerned, it only answers the research question to a certain extent, not 100%; it may be only 70%, 80%, or other degrees [88]. Research objectives should have predictable outcomes that are expected to advance theoretical understanding. In many researchers' opinions, as far as the research objective part is written, both research questions and theoretical understanding must be involved, and both are indispensable. It is necessary to clearly explain the problem to be studied and what theoretical understanding may be obtained [89]. In other words, if the research objective of an application cannot be seen at all, then the research objective can be considered a failure. Similarly, if the research objectives of an application do not see any substantial expected theoretical advances, then the research value will be significantly reduced.

3.1.1 Features

- **Specificity**: Research objectives must clearly define the problem, methods, and expected results. Specific objectives help readers assess the study's rationality, feasibility, and value. The more precise the objectives, the better the overall direction of the research [90].
- **Clarity**: Objectives should clearly outline the research's direction and boundaries [63]. Clarity involves clearly defined actions (research methods) and objects (the research question), a preliminary hypothesis, a direction for theoretical advancement, and clear distinctions between this research and similar studies.
- **Conciseness**: Research objectives should be stated succinctly [92]. Each objective must be expressed in a clear sentence to avoid confusion. Over explanation can indicate uncertainty about the research goals.

3.1.2 Function

The research objectives serve the following three purposes:

- **Specificity**: Research objectives must clearly define the problem, methods, and expected results [90]. Specific objectives help readers assess the studies and avoid confusion; each objective must be expressed clearly with nationality, feasibility, and value. The more precise the objectives, the better the overall direction of the research.
- **Clarity**: Objectives should clearly outline the research's direction and boundaries. Clarity involves clearly defined actions (research methods) and objects (the research question), a preliminary hypothesis, a direction for theoretical advancement, and clear distinctions between this research and similar studies [91].
- **Conciseness**: Research objectives should be stated succinctly [93]. Each objective must be expressed in a clear sentence to avoid confusion. Over explanation can indicate uncertainty about the research goals.

Research objectives are an essential study component, delineating the steps needed to accomplish the overarching goal [94]. These objectives are typically presented as a sequential series of tasks organized within the research work's structure. The objectives provide a systematic approach to the research process, beginning with the theoretical framework and extending to the chapter dedicated to addressing the scientific problem. Here are some examples of research objectives in the fields of architecture and engineering:

Research Goal: To investigate using green building materials to reduce energy consumption in residential construction.

Research Objectives (RO): To analyze the existing literature on green building materials, their properties, and their impact on energy efficiency in residential buildings.

Example: Conduct a comprehensive review of research articles, technical reports, and industry publications to gather information on various green building materials and their energy-saving potential.

RO: To evaluate the thermal performance of selected green building materials through laboratory experiments and numerical simulations.

Example: Conduct controlled experiments to measure different green building materials' thermal conductivity, heat capacity, and insulation properties. Utilize computer simulations to analyze their performance in various climatic conditions.

RO: To compare the energy consumption of residential buildings constructed with traditional materials versus those built with green building materials.

Example: Collect energy consumption data from residential buildings in different locations, compare the data between buildings with traditional materials and those with green building materials, and analyze the energy savings achieved by the latter.

RO: To assess the environmental impact of green building materials throughout their lifecycle, including raw material extraction, production, use, and disposal.

Example: Conduct a lifecycle assessment (LCA) of selected green building materials to evaluate their embodied energy, greenhouse gas emissions, and potential for recycling or disposal at the end of their service life.

RO: To provide recommendations and guidelines for incorporating green building materials in residential construction practices to optimize energy efficiency.

Example: Develop a set of best practices and guidelines for architects, engineers, and construction professionals to effectively integrate green building materials into residential projects, including design considerations, material selection criteria, and construction techniques.

In this example, the research objectives outline the specific tasks that need to be undertaken, such as literature review, laboratory experiments, data collection and analysis, lifecycle assessment, and recommendation development. These objectives guide the researcher in conducting a systematic investigation into using green building materials to reduce energy consumption in residential construction.

3.2 Formulating Goals and Objectives in Research Work

This section aims to clarify the proper approach for formulating goals and objectives in research papers. It is essential to introduce these elements after identifying the study's object and subject and discussing its theoretical and practical significance. Examples of proper goal and objective formulation in architectural engineering to illustrate how to accurately formulate goals and objectives within the fields of architecture and engineering, consider the following examples:

Example I:

Goal: Design an environmentally sustainable and aesthetically pleasing residential building that promotes community interaction.

- **Objective 1**: Research and incorporate sustainable building materials into the design.
- **Objective 2**: Create functional spaces that foster social engagement and communal living.
- **Objective 3**: Optimize energy efficiency and minimize environmental impact through innovative design strategies.

Example II:

Goal: Develop a cost-effective and efficient urban transportation system that reduces congestion and improves commute times.

- **Objective 1**: Analyze traffic patterns and study existing infrastructure to identify bottlenecks and areas for improvement.
- **Objective 2**: Design and implement intelligent transportation systems to enhance traffic flow and reduce congestion.
- **Objective 3**: Assess the environmental impact of the system and integrate sustainable practices, such as promoting public transportation and alternative fuel sources.

These examples demonstrate how to formulate specific, actionable goals and objectives, clearly outlining desired outcomes and the necessary steps to achieve them within the context of architectural engineering research.

3.3 Refining Research Objectives Through a Rigorous Academic Approach

How research objectives are formulated dramatically influences the direction and extent of scholarly investigation [95]. This procedure guarantees that research stays targeted, meaningful, and in line with the overall objectives of the study. By applying rigorous academic methodology to refine these aims, the research becomes more feasible, relevant, and straightforward, ultimately strengthening the study's contribution to the area. This strategy combines analytical reasoning with methodical techniques to guarantee thorough and precise research objectives [96].

3.3.1 Ambition to Expand Theoretical Understanding

The goal of research should be to increase the theoretical knowledge base. One frequent drawback, especially for early-career academics, is limiting research to applications or task-oriented studies with little theoretical value [97]. Such study goals frequently provide little more than replications of previous findings. To contribute significantly to the theoretical discourse, a robust research aim should bring new approaches, question pre-existing frameworks, or suggest unique interpretations.

3.3.2 Practicality and Feasibility

Ambitious goals are necessary but must also be based on practicality. Three main components make up feasibility: (1) well-stated research objectives and hypotheses;

(2) thorough methodological techniques that show how these questions will be handled; and (3) the capacity to break down complex issues into smaller, more manageable ones. Given the limitations of the resources at hand, this guarantees that research aims are not just aspirational but also operational and executable [98].

3.3.3 Combining with a Wider Disciplinary Framework

Research aims must be positioned within the discipline's broader context. Even while a particular research aim may focus on a specialized area or subfield, it must also support the overall objectives of the more extensive academic subject. The subfield must be seen as an essential component of the discipline's "edifice," research goals should align with the main ideas propelling the field. Without this broader connection, objectives risk being seen as disconnected and irrelevant.

3.4 Formulating a Clear and Focused Research Objective

To effectively control the course of a study and determine its scope, it is essential to develop research objectives that are both specific and well defined [99]. The specified objectives serve as a road map for the research process, ensuring that efforts are directed in the right direction and are in line with the general goals of the investigation. Researchers can maintain clarity throughout the study by developing explicit, quantifiable, and attainable goals [100]. This enables them to generate relevant results and contribute to their academic area. When delineating research objectives, three essential components must be contemplated to guarantee clarity and efficacy:

3.4.1 Formulating a Cohesive Framework

A coherent study delineates a distinct cause-and-effect connection, wherein certain acts yield expected effects [101]. The formulation of research objectives generally comprises four elements: (1) the researcher as the agent of action (often implicit); (2) research actions or strategies, employing verbs such as "investigate," "analyze," or "verify"; (3) the subject of these actions, which pertains to the specific research problem or hypothesis; and (4) the anticipated outcomes. Integrating these components, the agent, the action, and the expected outcome, clarifies and renders research objectives actionable throughout the study process.

- Example: Examine the influence of natural illumination on user contentment in workplace settings by assessing the correlation among window orientation, daylight accessibility, and occupant welfare.

- Example: Examine the viability of employing 3D printing technology to produce lightweight, high-strength structural components in aerospace engineering by analyzing material qualities, production techniques, and mechanical performance.

3.4.2 Defining a Distinct Course

Clear direction in research aims entails matching long-term goals with resolving a more significant issue while delineating particular objectives [102]. Articulating these aims often necessitates two components: one phrase to delineate the precise goals and another to underscore their significance to the broader study domain. The specific objectives (or "branches") must correspond with the primary research trajectory, which tackles broadly acknowledged problems. Given that the study's aims may be unfamiliar to readers, it is essential to elucidate their connection to mainstream research, so aiding readers in comprehending their relevance [103].

- Example: Establish explicit research objectives for formulating sustainable urban design methodologies, emphasizing green infrastructure, stormwater management, and biodiversity enhancement. Elucidate how these objectives facilitate the progression of sustainable urban planning methodologies.
- Example: Formulate research objectives focused on improving the efficacy and dependability of renewable energy systems, especially solar power generation, and clarify their role in attaining a sustainable energy future.

3.4.3 Defining Explicit Boundaries

Explicit limits delineate the parameters of the investigation by articulating inclusions and exclusions [104]. These parameters refine the study's focus, guaranteeing that significant findings may be derived. The study objectives must possess quantifiable outcomes, enabling readers to evaluate the extent of their accomplishment. Common hazards include establishing objectives that are too broad or poorly specified, which might compromise the research's credibility. Consequently, it is imperative to delineate explicit boundaries that differentiate the study from other pertinent studies.

- Example: Specify the specifications of a research examining the effects of acoustic design in hospitals by identifying the healthcare environments involved (e.g., patient rooms, waiting areas) and the acoustic metrics to be evaluated (e.g., noise levels, speech intelligibility).
- Example: Define the parameters of a research project aimed at enhancing the structural efficacy of bridge systems by incorporating innovative materials, detailing the types of bridges, characteristics of the materials, and quantifiable metrics for assessing performance and longevity.

In conclusion, when formulating research objectives, one must consider two factors: first, the contribution of the analysis to theoretical progression, and second, the extent of progress attained. The significance of research resides in its contribution to these domains.

4 Aligning Research Questions with Research Objectives

For any research project to succeed, it is essential to achieve alignment between the research questions and the organizational goals [105]. A research inquiry is driven by research questions, which outline the precise examination topics. Research goals, on the other hand, are the actual measures that are necessary to solve these research questions. Establishing a logical link between the two is essential to guarantee that the research will remain focused and coherent, which is essential for obtaining relevant and actionable results. The study can veer off course if the research questions and objectives are not correctly linked. This can result in outcomes that are not clear or conclusions that do not directly answer the query that was first submitted. The researchers must ensure that every study question matches a matching aim that gives a clear route toward resolving the issue [106]. This will help them avoid the situation described above. This link results in the formation of a structured strategy, which guarantees that every aim is consciously planned to investigate the subject at hand and that the study is moving in the direction of clearly stated outcomes. Take, for instance, the following study topic: "How do smart house technologies affect energy consumption patterns in residential areas?" As an illustration, examine this question. "Analyzing the energy savings from integrating smart heating, ventilation, and air conditioning (HVAC) systems in residential buildings" might be the research aim that corresponds to this purpose. In this instance, the target is not only in line with the research topic but also directs the investigation toward a certain analysis: the reduction of energy consumption in smart homes. Because of this obvious connection, the research can remain focused on determining how the technologies above influence energy usage, which guarantees that the study will continue to be pertinent to its core question. Similarly, let us look at a study topic about environmentally responsible building practices: "What are the environmental benefits of using green concrete in large-scale construction projects?" The relevant aim may be "to measure the reduction in carbon footprint achieved by replacing traditional concrete with green concrete in urban infrastructure projects." In this case, the aim directly addresses the research topic, providing a concrete method to evaluate the environmental advantages by concentrating on decreasing the carbon footprint. This direct alignment helps the study retain clarity and precision, ensuring that the techniques and measurements are suited to address the fundamental issue. This alignment also helps the study remain clear. The relationship between research questions and objectives becomes even more important when doing increasingly complicated investigations incorporating several interrelated variables or themes. Consider, for example, a study on material passports, in which the research question may be "How can material passports contribute to the circular economy in construction?" This may be "To evaluate how material passports track and facilitate the reuse of materials across different phases of construction projects." By aligning the aim with the research question, the study continues to concentrate on the function that material passports play in the progression of the circular economy. This guarantees that the study methods, data collecting, and analysis are all aimed at gaining knowledge of how these passports operate within the context of a circular economy. Bringing research questions and objectives into alignment gives several benefits overall. First, it is possible to take an organized approach, in which the methodology naturally emerges from the questions asked. This guarantees that the study is systematic and has a goal. Second, it assists in preserving the coherence of the study by eliminating digressions or diversions that may occur when the study's objectives are not obvious or are not related to the research questions [107]. Lastly, a well-aligned research question-objective structure simplifies conveying the study's goal and findings to stakeholders. This is because the logical path from the question to the outcome is readily apparent. Properly implementing research projects must link the research questions with the objectives. Whether the study is analyzing material passports, studying the impacts of smart home technologies, or exploring green concrete in construction, ensuring that the objectives are related to the research questions helps to keep the study structured, focused, and capable of reaching the intended outcomes. Producing coherent and powerful research results that successfully address the main inquiry requires this alignment as a fundamentally important component [108]. The purpose of Table 7 is to investigate the connections between the many types of research questions, the aims of those questions, and the difficulties encountered in each research technique. When doing quantitative research, the emphasis is placed on quantifiable facts and statistical correlations, whereas qualitative research is conducted to gain a more profound understanding of experiences and processes. A holistic knowledge may be attained through mixed-methods research, which blends both methodologies. There are difficulties in ensuring that the data is representative, managing the variables, and integrating the many forms of data.

Benefits of Aligning Research Questions with Research Objectives

- **Clarity and Focus**: The study maintains its coherence and moves methodically toward answering the research inquiries when it is certain that the objectives are closely related to the research questions [109].
- **Purpose-Driven Research**: Objectives that align with research questions constitute a road map for gathering data. This ensures that each stage of the research process contributes to the overarching objective of answering the questions given [110].
- **Methodological Consistency**: When the questions and objectives are aligned, the research techniques are chosen effectively to address issues [111]. This can be accomplished through qualitative, quantitative, or mixed methodologies approaches.
- **Balanced Exploration**: A balanced approach considering qualitative and quantitative components leads to more comprehensive and trustworthy results, particularly in mixed-methods research [112].

Table 7 Overview of research questions, objectives, and challenges

Research type	Research question	Research objective examples	Purpose	Challenges
Quantitative research	Descriptive: "What percentage of architects use sustainable materials in building design?"	To quantify the use of sustainable materials among architects within a specific population	To provide measurable data on the prevalence of sustainable practices in architecture	It is ensuring that the sample size is representative of generalizing the findings
	Comparative: "What are the differences in energy consumption between green and traditional buildings?"	To compare energy consumption across green and traditional buildings over a specified period	To identify differences in energy efficiency between different building types	It is accurately controlling variables such as building size, climate, and usage patterns
	Relational: "What is the relationship between using recycled materials and reduced construction costs?"	To measure the correlation between recycled materials and overall cost reductions in large-scale construction projects	To assess the financial impact of sustainable material use in construction	Establishing clear causality between the two variables amid other influencing factors
Qualitative research	Contextual: "How do urban planners integrate sustainable materials into their design processes?"	To explore how urban planners incorporate sustainable materials in their workflows and decision-making processes	To understand the practical and theoretical integration of sustainability in urban planning	Capturing the depth and complexity of planners' decision-making processes across varying contexts
	Descriptive: "What are the experiences of architects designing green roofs in urban environments?"	To describe the lived experiences of architects when implementing green roof designs in cities	To provide a narrative or case study account of architects' challenges and successes in green design	Ensuring the subjective experiences collected are relevant and applicable to the broader research context

(continued)

Table 7 (continued)

Research type	Research question	Research objective examples	Purpose	Challenges
	Exploratory: "What are engineers' perceptions of AI's role in sustainable construction?"	To explore engineers' perceptions and insights into how AI is being or can be used in sustainable building practices	To uncover new ideas or insights about the integration of AI in construction	Avoiding bias in responses and ensuring openness to new or unexpected findings
	Evaluative: "How effective are green walls in reducing indoor temperatures in office buildings?"	. To evaluate the effectiveness of green walls in lowering indoor temperatures during summer months	To assess the functional and energy-saving benefits of green wall installations in corporate environments	Measuring effectiveness in varying conditions such as building orientation, wall size, and location
Mixed-methods research	Convergent: "How do perceptions of sustainability align with measured energy-saving outcomes in buildings?"	To compare qualitative perceptions with quantitative energy-saving data in sustainable building projects	To integrate qualitative insights with quantitative data to understand sustainability impacts comprehensively	Balancing the integration of diverse data types and ensuring consistent interpretation across methods
	Explanatory Sequential: "What factors explain the cost savings measured from green building technologies?"	To investigate qualitative factors (e.g., decision-making, material selection) that influence quantitative cost savings in projects	. To identify qualitative reasons that explain the quantitative results of cost savings	Ensuring that the qualitative follow-up adequately explains the quantitative data collected in the first phase
	Exploratory Sequential: "What themes emerge from interviews on sustainable architecture that can be measured?"	To develop themes from qualitative interviews that can later be quantitatively measured in larger populations or surveys	To generate themes and test their prevalence or relevance across broader samples	Identifying themes that are both relevant to the qualitative data and measurable

5 Conclusion

Developing well-organized research questions, objectives, and goals is essential to rigorous scholarly investigation. These elements are the cornerstone of an organized and targeted research process, guaranteeing that academic work significantly contributes to the field's body of knowledge. Research questions are the conceptual framework that organizes an inquiry, directing the study's scope and approach. Well-crafted research questions are essential to concentrate on the study and develop the analytical strategy. Without this kind of focus, research projects risk becoming dispersed and producing results that are unclear or unrelated to the main problems. Furthermore, the integrity of the research process depends on aligning the objectives and research questions. Research questions are operationalized by objectives, reducing general inquiries into focused, doable activities that organize the study. This alignment promotes consistency and coherence across the whole study by guaranteeing a methodical approach to research. When research objectives are written with the S.M.A.R.T criteria (specific, measurable, achievable, relevant, and time-bound) in mind, they give direction and guarantee that the study is feasible and sound regarding methodology given the time, money, and data availability limitations. Furthermore, defining the research objectives is essential to placing the study in a larger theoretical or practical framework. The study aims are concerned with the broader ramifications and contributions of the study, whereas the research objectives concentrate on actions to address the research issues. Well-defined objectives provide the study with a forward-looking perspective by connecting it to more general academic discussions or real-world problems in the subject. The coherence of research questions, goals, and targets promotes a rigorous and meaningful research trajectory, guaranteeing the study's continued relevance and influence. For academic research to succeed, carefully crafting study questions, goals, and purposes is a procedural effort and a strategic requirement. These elements offer a structured framework that promotes methodological precision and conceptual clarity. Without this framework, research risks becoming disjointed or out of alignment, hindering its ability to advance theoretical understanding and real-world applications. Consequently, it is essential to carefully and deliberately align research questions, objectives, and goals to produce practically and academically meaningful research.

References

1. Pratt MG, Sonenshein S, Feldman MS (2022) Moving beyond templates: a bricolage approach to conducting trustworthy qualitative research. Organ Res Methods 25(2):211–238
2. Kazi MR, Chowdhury N, Chowdhury M, Turin T (2022) Knowledge synthesis in health, wellness and social care research: the fundamentals of conducting comprehensive reviews. Borneo J Med Sci (BJMS) 16(1):5–20
3. Schwab S, Janiaud P, Dayan M, Amrhein V, Panczak R, Palagi PM, Hemkens LG, Ramon M, Rothen N, Senn S, Furrer E (2022) Ten simple rules for good research practice. PLoS Comput Biol 18(6):e1010139

4. Ganesh HR, Aithal PS (2022) How do you choose an appropriate research data collection method and method choice among various research data collection methods and method choices during a Ph. Ph.D. program in India? Int J Manage Technol Soc Sci (IJMTS) 7(2):455–489
5. Sundar R, Suguna A, Begum A, Mishra NL (2024) Introductory research methodology. Academic Guru Publishing House
6. Bammer G, O'Rourke M, O'Connell D, Neuhauser L, Midgley G, Klein JT, Grigg NJ, Gadlin H, Elsum IR, Bursztyn M, Fulton EA (2020) Expertise in research integration and implementation for tackling complex problems: when is it needed, where can it be found, and how can it be strengthened? Palgrave Commun 6(1):1–16
7. Glushchenko VV (2023) Ergo design and "specific", measurable," achievable," relevant," and" time-bound"(SMART) technologies as tools for the formation of innovative leadership programs. ASEAN J Econ Econ Educ 2(1):23–34
8. Pandey P, Pandey MM (2021) Research methodology tools and techniques. Bridge Center
9. Levitt HM, Morrill Z, Collins KM, Rizo JL (2021) The methodological integrity of critical qualitative research: principles to support design and research review. J Couns Psychol 68(3):357
10. Hatch JA (2023) Doing qualitative research in education settings. State university of New York press
11. Khoa BT, Hung BP, Hejsalem-Brahmi M (2023) Qualitative research in social sciences: data collection, data analysis and report writing. Int J Public Sector Perform Manage 12(1–2):187–209
12. Hosseini et al (2024) They are formulating research questions for evidence-based studies. J Med Surg Public Health 2:100046
13. Lim WM (2024) What is qualitative research? An overview and guidelines. Austr Market J 144
14. Lynch J, Ramjan LM, Glew PJ, Salamonson Y (2024) How to embed a conceptual or theoretical framework into a dissertation study design. Nurse Res 32(1)
15. Normile IH (2024) Critical being: a philosophical approach to understanding and expanding the scope of critical thinking in UK higher education with attention to cross-cultural diversity
16. Johnson JL, Adkins D, Chauvin S (2020) A review of the quality indicators of rigor in qualitative research. Am J Pharm Educ 84(1):7120
17. Kadasah SF, Abd Al Galil FM, Kolhe B, Shinde SM (2022) Scientific research methodology principles, methods, and techniques. Book Rivers
18. Tracy SJ (2024) Qualitative research methods: collecting evidence, crafting analysis, communicating impact. John Wiley & Sons
19. Rincón-Guio C, Sarache W (2024) Mapping the multifaceted landscape of resilience in manufacturing strategy: a systematic literature review and future research avenues. J Indus Integr Manage 9(2)
20. Almusaed A, Almssad A, Yitmen I (2024) Sustainable built environment and its implications on real estate development: a comprehensive analysis
21. Ruane AC (2024) Synthesis report of the IPCC sixth assessment report (AR6)
22. Shukla BK, Bharti G, Sharma PK, Sharma M, Rawat S, Maurya N, Srivastava R, Srivastav Y (2024) Sustainable construction practices with recycled and waste materials for a circular economy. Asian J Civil Eng 1–22
23. Laudati DA (2024) Tailored implementation of SAFe and scrum in a real-world project: a case study. Doctoral dissertation, Politecnico di Torino
24. Kindermann B, Wentzel D, Antons D, Salge TO (2024) Conceptual contributions in marketing scholarship: patterns, mechanisms, and rebalancing options. J Mark 88(3):29–49
25. Lim WM (2024) What is qualitative research? An overview and guidelines. Aust Market J 14413582241264619
26. Alvesson M, Sandberg J (2024) The art of phenomena construction: a framework for coming up with research phenomena beyond 'the usual suspects.' J Manage Stud 61(5):1737–1765

27. Charef R (2024) A digital framework for the implementation of the circular economy in the construction sector: expert opinions. Sustainability 16(14):5849
28. Bansal et al (2012) Bridging the research–practice gap
29. Barroga E, Matanguihan GJ (2022) A practical guide to writing quantitative and qualitative research questions and hypotheses in scholarly articles. J Korean Med Sc 37(16)
30. Grant C, Osanloo A (2014) Understanding, selecting, and integrating a theoretical framework in dissertation research: creating the blueprint for your "house." Admin Issues J 4(2):4
31. Johnson RB, Christensen LB (2024) Educational research: quantitative, qualitative, and mixed approaches. Sage publications
32. Nguyen HL, Kanbach DK (2024) Toward a view of integrating corporate sustainability into strategy: a systematic literature review. Corp Soc Responsib Environ Manag 31(2):962–976
33. Jin H, Qu X, Liu B, Liu C (2024) Examining the success drivers and barriers of modular building projects using qualitative comparative analysis. Int J Constr Manag 24(6):601–609
34. Sauer PC, Seuring S (2023) How to conduct systematic literature reviews in management research: a guide in 6 steps and 14 decisions. RMS 17(5):1899–1933
35. Clarysse B, Andries P, Boone S, Roelandt J (2023) Institutional logics and founders' identity orientation: why academic entrepreneurs aspire lower venture growth. Res Policy 52(3):104713
36. Wu PY (2024) Data-driven approaches for predicting hazardous substances in the building stock
37. Stiksma F (2023) Lifetime extension assessment of collected washing machines
38. Matt M, Robinson DK, Joly PB, Van Dis R, Colinet L (2023) ASIRPAReal-time in the making or how to empower researchers to steer research towards desired societal goals. Res Eval 32(2):412–425
39. Agarwal SA, Chauhan NKS, Ravikumar S (2024) Basic concepts of research methodology & statistics. Academic Guru Publishing House
40. Beckstead L, McGregor H, Cook IM (2024) Podcast or perish: peer review and knowledge creation for the 21st century
41. Fernando D, Navaratnam S, Rajeev P, Sanjayan J (2023) Study of technological advancement and challenges of façade system for sustainable building: current design practice. Sustainability 15(19):14319
42. Rockmann KW, Vough HC (2023) Using quotes to present claims: practices for the writing stages of qualitative research. Organ Res Methods 10944281231210558
43. Dwivedi YK, Jeyaraj A, Hughes L, Davies GH, Ahuja M, Albashrawi MA, Al-Busaidi AS, Al-Sharhan S, Al-Sulaiti KI, Altinay L, Amalaya S (2024) "Real impact": challenges and opportunities in bridging the gap between research and practice–Making a difference in industry, policy, and society. Int J Inform Manage 102750
44. Almusaed A, Almssad A, Alasadi A, Yitmen I, Al-Samaraee S (2023) Assessing the role and efficiency of thermal insulation by the "Bio-Green Panel" in enhancing sustainability in a built environment. Sustainability 15(13):10418
45. Peel KL (2020) A beginner's guide to applied educational research using thematic analysis. Pract Assess Res Eval 25(1)
46. Giner-Sorolla R, Montoya AK, Reifman A, Carpenter T, Lewis NA Jr, Aberson CL, Bostyn DH, Conrique BG, Ng BW, Schoemann AM, Soderberg C (2024) Power to detect what? Considerations for planning and evaluating sample size. Pers Soc Psychol Rev 28(3):276–301
47. Vollmer HJ (2024) Powerful educational knowledge through subject didactics and general subject didactics. Recent developments in German-speaking countries. Towards Powerful Educ Knowl 87–104
48. Benbrahim FZ, Frichi Y, Benabdelhadi A, Jawab F (2024) The qualitative exploratory study: a necessary prerequisite to the quantitative study. In: Data collection and analysis in scientific qualitative research. IGI Global, pp 57–86
49. Orsini LS, Berger M, Crown W, Daniel G, Eichler HG, Goettsch W, Graff J, Guerino J, Jonsson P, Lederer NM, Monz B (2020) Improving transparency to build trust in real-world secondary data studies for hypothesis testing—why, what, and how: recommendations and a road map from the real-world evidence transparency initiative. Value in Health 23(9):1128–1136

50. Skivington K, Matthews L, Simpson SA, Craig P, Baird J, Blazeby JM, Boyd KA, Craig N, French DP, McIntosh E, Petticrew M (2021) Framework for the development and evaluation of complex interventions: gap analysis, workshop and consultation-informed update. Health Technol Assess (Winchester, England) 25(57):1
51. Flynn SV (ed) (2021) Research design for the behavioral sciences: an applied approach. Springer Publishing Company
52. Tabatabaei F, Tayebi L (2022) Writing a research proposal. Research methods in dentistry. Springer International Publishing, Cham, pp 87–114
53. Dokter G (2023) Circular design through co-creation
54. Evardo Jr OJ, Abina ILS (2023) Research competence of out-of-field teachers in teaching practical research: input to capability building series. Int J Multidiscip Res 5(2)
55. Doody O, Bailey ME (2016) Setting a research question, aim and objective. Nurse Res 23(4):19–23. https://doi.org/10.7748/nr.23.4.19.s5
56. Marshall C, Rossman GB (2011) Designing qualitative research, 5th edn. Sage Publications, Thousand Oaks, CA
57. Mõttus R, Wood D, Condon DM, Back MD, Baumert A, Costantini G, Epskamp S, Greiff S, Johnson W, Lukaszewski A, Murray A (2020) Descriptive, predictive and explanatory personality research: different goals, different approaches, but a shared need to move beyond the Big Few traits. Eur J Pers 34(6):1175–1201
58. Stausberg M (2021) Comparison. The Routledge handbook of research methods in the study of religion, pp 15–33
59. Birhane A (2021) Algorithmic injustice: a relational ethics approach. Patterns 2(2)
60. Levitt HM, McLeod JOHN, Stiles WB (2021) The conceptualization, design, and evaluation OF. Bergin and Garfield's handbook of psychotherapy and behavior change, p 51
61. Van Der Waldt G (2021) Elucidating the application of literature reviews and literature surveys in social science research. Administratio Publica 29(1):1–20
62. Ng DTK, Leung JKL, Chu SKW, Qiao MS (2021) Conceptualizing AI literacy: an exploratory review. Comput Educ: Artif Intell 2:100041
63. Fynn JF, Hardeman W, Milton K, Jones AP (2020) A scoping review of evaluation frameworks and their applicability to real-world physical activity and dietary change programme evaluation. BMC Public Health 20:1–16
64. Neale B (2021) The craft of qualitative longitudinal research: the craft of researching lives through time
65. Gilad S (2021) Mixing qualitative and quantitative methods in pursuit of richer answers to real-world questions. Public Perform Manag Rev 44(5):1075–1099
66. Creswell JW, Clark VLP (2017) Designing and conducting mixed methods research. Sage Publications, Thousand Oaks, CA
67. Busari I (2024) Examining high school mathematics teachers' research-based instructional practices and their relationship to students' learning outcomes: a convergent parallel mixed-method study
68. Birgili B, Demir Ö (2022) An explanatory sequential mixed-method research on the full-scale implementation of flipped learning in the first years of the world's first fully flipped university: departmental differences. Comput Educ 176:104352
69. Dawadi S, Shrestha S, Giri RA (2021) Mixed-methods research: a discussion on its types, challenges, and criticisms. J Pract Stud Educ 2(2):25–36
70. Resch C, Kock A (2021) The influence of information depth and breadth on brokers' idea newness in online maker communities. Res Policy 50(8):104142
71. Hodge SR (2020) Quantitative research. In: Routledge handbook of adapted physical education. Routledge, pp 147–162
72. Mays N, Pope C (2020) Quality in qualitative research. Qualitative research in health care, pp 211–233
73. Lall D (2021) Mixed-methods research: why, when and how to use. Indian J Continuing Nursing Educ 22(2):143–147

74. Sharma A, Kosasih E, Zhang J, Brintrup A, Calinescu A (2022) Digital twins: state of the art theory and practice, challenges, and open research questions. J Ind Inf Integr 30:100383
75. Mezmir EA (2020) Qualitative data analysis: an overview of data reduction, data display, and interpretation. Res Humanit Soc Sci 10(21):15–27
76. Katie B, Mitchell S (2021) Loving what is, revised edition: four questions that can change your life; the revolutionary process called "The Work". Harmony
77. Lee CH, Liu CL, Trappey AJ, Mo JP, Desouza KC (2021) Understanding digital transformation in advanced manufacturing and engineering: a bibliometric analysis, topic modeling and research trend discovery. Adv Eng Inform 50:101428
78. Alam MK (2021) A systematic qualitative case study: questions, data collection, NVivo analysis and saturation. Qual Res Organ Manage Int J 16(1):1–31
79. Ramanadhan S, Revette AC, Lee RM, Aveling EL (2021) Pragmatic approaches to analyzing qualitative data for implementation science: an introduction. Implementation Sci Commun 2:1–10
80. Willis LD (2023) Formulating the research question and framing the hypothesis. Respir Care 68(8):1180–1185
81. Mello PA (2021) Qualitative comparative analysis: an introduction to research design and application. Georgetown University Press
82. Palys T, Atchison C (2021) Research methods in the social and health sciences: making research decisions. Sage Publications
83. Vaughn LM, Jacquez F (2020) Participatory research methods–choice points in the research process. J Participatory Res Methods 1(1)
84. Tomaszewski LE, Zarestky J, Gonzalez E (2020) Planning qualitative research: design and decision making for new researchers. Int J Qual Methods 19:1609406920967174
85. Ogbeiwi O (2021) General concepts of goals and goal-setting in healthcare: a narrative review. J Manag Organ 27(2):324–341
86. Zina O (2021) The essential guide to doing your research project. Sage
87. Zahle J (2021) Objective data sets in qualitative research. Synthese 199:101–117
88. Fink A (2002) How to sample in surveys. Sage Publications
89. Gregor S, Benbasat I (1999) Explanations from intelligent systems: theoretical foundations and implications for practice. MIS Quart 497–530
90. Hennink M, Hutter I, Bailey A (2020) Qualitative research methods. Sage
91. Colquhoun HL, Levac D, O'Brien KK, Straus S, Tricco AC, Perrier L, Kastner M, Moher D (2014) Scoping reviews: time for clarity in definition, methods, and reporting. J Clin Epidemiol 67(12):1291–1294
92. Summers JO (2019) Guidelines for conducting research and publishing in marketing: from conceptualization through the review process. In: How to get published in the best marketing journals. Edward Elgar Publishing, pp 17–27
93. Morkes J, Nielsen J (1997) Concise, scannable, and objective: how to write for the Web. Useit.com 51(1):1–17
94. Hall KL, Vogel AL, Stipelman BA, Stokols D, Morgan G, Gehlert S (2012) A four-phase model of transdisciplinary team-based research: goals, team processes, and strategies. Transl Behav Med 2(4):415–430
95. Leithwood K, Seashore K, Anderson S, Wahlstrom K (2004) Review of research: how leadership influences student learning
96. Dubois A, Gadde LE (2014) "Systematic combining"—a decade later. J Bus Res 67(6):1277–1284
97. Oplatka I (2010) Principals in late career: toward conceptualizing principals' tasks and experiences in the pre-retirement period. Educ Adm Q 46(5):776–815
98. Groat LN, Wang D (2013) Architectural research methods. John Wiley & Sons
99. Wacker JG (1998) A definition of theory: research guidelines for different theory-building research methods in operations management. J Oper Manag 16(4):361–385
100. Yilmaz K (2013) Comparison of quantitative and qualitative research traditions: epistemological, theoretical, and methodological differences. Eur J Educ 48(2):311–325

101. Touborg CT (2018) The dual nature of causation: two necessary and jointly sufficient conditions. Doctoral dissertation, University of St Andrews
102. Grant AM (2020) An integrated model of goal-focused coaching: an evidence-based framework for teaching and practice. Coaching researched: a coaching psychology reader, pp 115–139
103. Locke LF, Silverman SJ, Spirduso WW (2009) Reading and understanding research. Sage
104. Smith HE (2002) Exclusion versus governance: two strategies for delineating property rights. J Leg Stud 31(S2):S453–S487
105. Al-Haddad S (2014) Successful organizational change: aligning change type with methods
106. Hancock DR, Algozzine B, Lim JH (2021) Doing case study research: a practical guide for beginning researchers
107. Taylor ME (2006) A text-linguistic investigation into the discourse structure of James
108. Fullan M, Quinn J (2015) Coherence: the right drivers in action for schools, districts, and systems. Corwin Press
109. Pitney WA, Parker J (2009) Qualitative research in physical activity and the health professions. Human Kinetics, Champaign, IL, pp 63–65
110. Grant G (2024) Purpose-driven innovation leadership for sustainable development: a qualitative case study approach. Taylor & Francis
111. Alavi M, Archibald M, McMaster R, Lopez V, Cleary M (2018) Aligning theory and methodology in mixed methods research: before design theoretical placement. Int J Soc Res Methodol 21(5):527–540
112. Onwuegbuzie AJ, Collins KM (2017) The role of sampling in mixed methods-research. Kölner Zeitschrift für Soziologie und Sozialpsychologie 69(Supplement 2):133–156

Developing a Research Title and Introduction

Abstract Creating a compelling study title and introduction is essential to the process of doing academic research. As the first point of interaction, the title gives readers a brief overview of the study's goals, significance, and scope. A well-written title should not be unduly complicated or ambiguous but precise, pertinent, and descriptive enough to communicate the study topic and approach. The introduction presents the context, importance, and research aims and serves as a solid basis for the study, complementing the title. Conciseness and clarity are essential as they lead readers from a broad summary of the subject to a particular research question. A strong introduction also emphasizes the study's significance in the larger academic conversation by highlighting how it fills knowledge gaps and advances current scholarly discussions. This part ensures the study is well positioned within its academic area and interests readers by giving important contextual information and a clear justification for the research. Ultimately, the research title and introduction play a crucial role in demonstrating the study's applicability, grabbing the reader's attention, and laying the groundwork for an in-depth examination of the research issue.

1 Introduction to the Research Title

The research title, a crucial element in academic inquiry, is the primary engagement point for readers and researchers assessing the research [1]. It serves as a designation and succinct encapsulation of the whole study, providing critical insights into its fundamental topics, breadth, and significance. More than that, an effectively constructed title encapsulates the core of the research and establishes the research's tone, guiding the reader's anticipations concerning the content and methodology. A compelling study title must achieve a balance between conciseness and specificity. It should clearly and simply summarize the principal research topic, technique, and theoretical framework. The title should be straightforward, not excessively unclear or generic [2]. It should furnish sufficient detail to comprehend thc study's emphasis, the environment in which it is embedded, and the wider academic or practical ramifications. Furthermore, the title must precisely represent the research questions and

A. Almusaed et al., *Practice of Research Methodology in Civil Engineering and Architecture*, https://doi.org/10.1007/978-3-031-97393-2_5

objectives, assuring alignment with the study's broad scope. The text should articulate the research issue or topic being addressed and provide insight into the methodological technique employed to address it [3]. This is crucial for assisting potential readers in assessing the relevance of the research to their interests or academic discipline. The title situates the study within more comprehensive educational and professional discussions. By judiciously choosing keywords and phrases that align with contemporary discussions and trends, the title can augment the research's exposure, facilitating its discoverability in academic databases and enriching the continuing intellectual discourse [4]. A well-constructed research title is essential for framing and conveying the significance of the study.

1.1 Reflecting the Content of the Research

The primary function of a research title is to encapsulate the essential subject addressed in the study [5]. A well-constructed title balances conciseness and specificity, succinctly encapsulating the key ideas, research problem, and approach [6]. For example, in studies about energy efficiency in buildings, a title like "Enhancing Energy Efficiency in Commercial Buildings through Passive Design Strategies" succinctly articulates the central theme, the study issue, and the methodological emphasis. This term encapsulates the amalgamation of energy efficiency and design ideas, providing readers with insight into the study's particular focus. For example, in the domain of green buildings, a title such as "Assessing the Influence of Green Roofs on Urban Energy Efficiency: A Comparative Analysis" offers conciseness and precision. The title delineates the factors examined (green roofs and energy performance) and the setting (urban environments) [7]. Potential readers can swiftly comprehend the essence of the research and its broader implications for sustainable construction methods. In other examples, a title may incorporate fundamental elements, such as passive design or green roofs, and allude to theoretical frameworks or approaches [8]. In both instances, the titles function as navigational aids, assisting readers in situating the study within the broader academic discourse on sustainability, architecture, and energy management. A well-crafted title enhances the accessibility and scholarly relevance of research, allowing readers to readily grasp the scope and significance of the study within the broader discourse on energy efficiency and sustainable building practices, while simultaneously establishing a strong initial impression that engages prospective readers.

1.2 Creating a Strong First Impression for Readers

An effectively crafted research title is crucial for conveying an instant and precise comprehension of the research's emphasis [9]. Initially, it must encapsulate the core of the research, ensuring it is readily identifiable and pertinent to the intended audience.

In research domains like energy efficiency in buildings and the recycling of construction materials, a compelling title can promptly convey the significance and breadth of the study, enabling the reader to evaluate its pertinence to their endeavors swiftly. A title such as "Optimizing Energy Efficiency in High-Rise Buildings Through Smart Technology Integration" explicitly conveys the study's emphasis on energy optimization and intelligent technology [10]. It indicates the building type (high-rise) and suggests the methodological approach (technology integration). This title is academically rigorous and fascinating, emphasizing the increasing significance of energy efficiency in modern architectural methods. The title captivates the interest of readers or experts focused on sustainable architecture and intelligent energy solutions through its specificity. In the domain of recycling construction materials, a title like "Assessing the Environmental and Economic Benefits of Recycled Concrete in Urban Construction" clearly conveys the dual emphasis on environmental and economic implications, particularly for recycled concrete [11]. This explicit distinction aids prospective readers or researchers in urban building or sustainability in swiftly determining the relevance of the research to their area of expertise. An effectively constructed title improves the research's visibility in academic databases, augmenting its accessibility and expanding its influence among pertinent research communities. Furthermore, explicitly delineating the study topic and emphasis directs readers through the educational contribution of the research, affirming its scholarly significance and pertinence to current discussions in areas such as sustainable building practices and energy-efficient design [12]. A successful study title should include the following characteristics illustrated in Fig. 1. A well-constructed title must be brief, specific, and pertinent. These three features guarantee that the title appropriately reflects the essential emphasis of the study while preserving clarity and conciseness, which assists readers in rapidly understanding the research goal.

The title of the research project must be carefully prepared to reflect the substance of the research accurately and provide the audience with a profound and instant

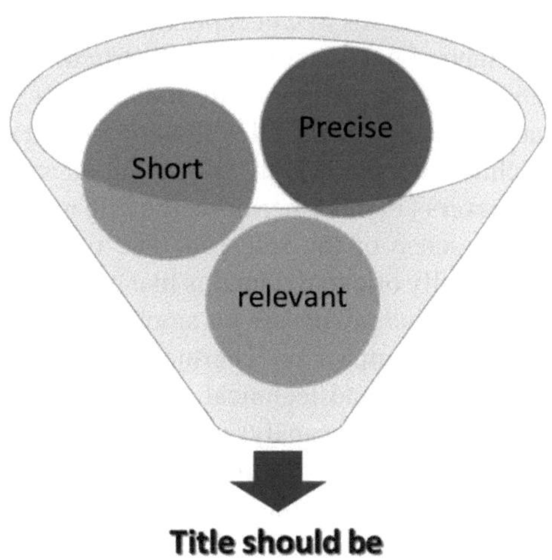

Fig. 1 Key characteristics of an effective research title: precision, relevance, and conciseness (*Source* Authors)

comprehension of the study [13]. A powerful title captures the reader's attention. It serves as a solid academic basis upon which the research is constructed, guaranteeing that it is both relevant and makes a significant scholarly contribution.

1.3 Developing Clear and Engaging Research Titles in Academic Research

The research title is an essential element of academic research, serving as the initial engagement point between the reader and the study. It encapsulates the substance of the study, articulating its scope, emphasis, and importance succinctly and comprehensively [14]. In domains like green buildings and architectural concept design, the title holds heightened significance, as it must encapsulate the transdisciplinary essence of the study while simultaneously signaling rising trends in sustainability and innovation. An effectively crafted title is crucial for determining the research's significance in academic dialog and shaping the reader's anticipations regarding the study's depth and methodology. Clarity and informativeness are essential for a good research title [15]. A title should convey the study emphasis clearly and succinctly, avoiding ambiguity and excessive complexity. For example, in research about green buildings, a title like "Optimizing Energy Efficiency in Urban Green Buildings through Passive Design Strategies" succinctly conveys the central topic, highlighting the emphasis on energy efficiency and the methodological focus on passive design strategies. This clarity emphasizes the research challenge and situates the study within broader conversations on sustainability and energy management, allowing readers to see its significance to their academic or practical pursuits. A research title must be clear, intellectually engaging, and academically rigorous. In fields such as architectural concept design, where originality and innovation are paramount, a title that embodies these aspects will inherently attract the academic community's interest [16]. "Innovative Approaches to Conceptual Architectural Design: Integrating Sustainability and Aesthetic Functionality" clearly articulates the research's dual emphasis on sustainability and aesthetics while indicating a progressive methodology, augmenting its attractiveness. Such titles illustrate the research's relevance to current concerns in architecture and design, enhancing its appeal to academics, practitioners, and policymakers [17]. Furthermore, the title must accurately represent the overarching tone and essence of the research. The title must convey a formal and precise tone in technically oriented domains like green buildings, where empirical investigation and technical judgments are paramount. The title "Quantitative Assessment of Recycled Materials' Impact on Thermal Performance in Sustainable Building Design" has an empirical and technical tone, indicating that the study would encompass thorough quantitative analysis. This strategy guarantees that the title conveys the study subject while establishing expectations for the methods and depth of information to be examined in the research. A robust research title must have vital keywords that accurately represent the primary ideas and concepts of the research, in addition

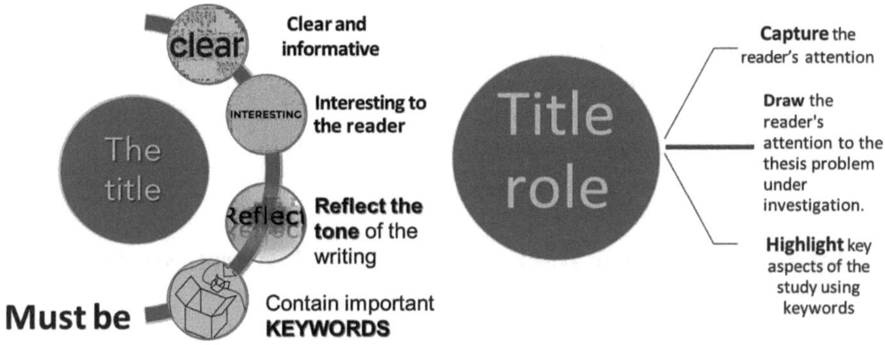

Fig. 2 Core elements and functions of an effective research title in academic research (*Source* Authors)

to clarity, engagement, and tone. This improves the study's visibility in academic databases, guaranteeing accessibility to a broader scholarly audience. For instance, in studies related to sustainable architecture, a title such as "The Role of Recycled Materials in Enhancing Energy Efficiency in Green Building Design" effectively integrates keywords like "recycled materials," "energy efficiency," and "green building [18]." These keywords promptly indicate the research's subject issues and assist in its retrieval by researchers engaged in adjacent sustainability, materials science, and environmental design fields. The title must be both simple and sufficiently specific. A well-constructed research title comprises several fundamental components and functions, illustrated in Fig. 2. In addition to underlining the necessity of capturing the reader's attention, reflecting essential parts of the study, and containing relevant keywords for academic discoverability, it emphasizes the significance of clarity, engagement, and reflection of the research tone.

A title should ideally consist of 5–15 words, enabling the incorporation of essential information without inundating the reader with extra content [19]. A concise and informative title, such as "Sustainable Urban Design through Adaptive Reuse of Materials," effectively conveys the conceptual and methodological focus of the research without superfluous detail. This guarantees that the title stays concentrated while effectively communicating the essential components of the study. The title should exclude superfluous terms that do not contribute meaningful value. Phrases like "An Investigation of" or "A Study on" are sometimes redundant, as they do not enhance the clarity or specificity of the title. For example, instead of "An Examination of Sustainable Practices in Architectural Design," a more concise alternative like "Sustainable Practices in Architectural Design" communicates the same information more efficiently [20]. Eliminating superfluous words facilitates a more direct and concentrated study presentation, thus augmenting its academic and professional significance. A meticulously formulated research title is also essential in educational research, particularly in intricate domains such as green buildings and architectural concept design. It should be clear, engaging, and representative of the research's tone while integrating crucial keywords to enhance discoverability. By ensuring conciseness and clarity, the title functions as a clear and instructive reference for the study, aiding its incorporation into more comprehensive academic and professional dialogs. A compelling title contextualizes and situates the research within current discussions, affirming its academic significance and contribution [21].

2 Introduction to the Background of the Research Study

To give a rational and objective rationale for the issue being investigated, selecting a research topic is an essential fundamental step that must be taken. A well-articulated research subject not only demonstrates the intellectual rigor of the researcher but also paves the way for a systematic approach, which includes an exhaustive assessment of the academic sources pertinent to the investigation [22]. Research must be based on a reputable and dependable framework to guarantee objectivity. To evaluate the credibility and significance of a study, it is necessary to investigate the procedures utilized to validate the study's results within a specific setting. Maintaining adherence to the established academic quality standards is required to ensure that the survey is genuine and relevant [23]. By adhering to these objective stages, researchers considerably increase the possibility that the results will be evaluated positively. A beginning that has been carefully designed can attract reviewers and evaluators, which in turn encourages concentrated engagement with the remainder of the parts. Given that the introduction is the study's entry point, a significant amount of focus must be spent writing an engaging opening. A preliminary justification for the topic that has been selected is provided in this part, which not only serves as an introduction to the subject matter but also as a "prelude" to the more considerable debate that will follow. Starting with a clear and compelling statement is essential when beginning a decent introduction. This can be especially tough for rookie researchers. A well-constructed initial phrase establishes the tone for the whole research project, directing readers to the issue's relevance and defining the aims and objectives of the study outlined in the introduction [24]. There may be unnecessary and critical queries during the research defense, leading to issues with no such opening. The following essential components should be included in an introduction that has been written well:

- Clarification of fundamental ideas and challenges within the current body of research: The introduction must clearly explain the significant difficulty the study is addressing and illuminate the core concepts being discussed. Because of this, the study is built on a solid foundation, and there is no room for confusion about the primary emphasis of the analysis [25].
- The following examples illustrate how the research connects with larger societal or environmental contexts: Demonstrating the study's relevance within a larger framework cannot be overstated [26]. This involves showing how the selected issue corresponds with external influences. This component reflects the

researcher's comprehension level, placing the study within the context of its ramifications for the actual world.

In architectural engineering, for example, research centered on constructing environmentally friendly buildings, the introduction can start by providing an overview of fundamental ideas like sustainability, energy efficiency, and environmental effects [27]. It then investigates how these ideas interact with the surrounding environment, focusing on passive design tactics, incorporating renewable energy, and conserving resources. This establishes a contextual framework for the remainder of the introduction, enabling readers better to grasp the study's broader importance and location.

A practical introduction is accomplished by the utilization of an organized strategy, which generally consists of three essential steps:

- The introduction should clearly describe the links among the study's primary topics, allowing readers to understand how they are related [28]. This is the first step in establishing a clear correlation between the essential aspects of the research.
- This entails quickly detailing the dynamics between the elements and providing a picture of their interrelationships within the research context. The second step is to portray their interaction [29].
- Emphasizing the fundamental variables that differentiate one control domain from another [30]. The introduction should emphasize the various qualities that distinguish one domain of control from another, showing the distinctive characteristics of each sphere within the research.

Adhering to this systematic strategy can guarantee that the introduction is comprehensive and solid, reducing the likelihood of potential problems that may occur later in the research. In addition, the introduction must have a paragraph that traces the historical development of the topic being investigated [31]. This might give the readers a vital background that helps them comprehend the problem's relevance over time. The following is an example of a structure that might be followed by research that investigates the integration of Building Information Modeling (BIM) with sustainable design in architectural engineering:

- Establishing a parallel between the essential components: The introduction would emphasize the significance of sustainable design principles in supporting environmentally friendly construction and highlight the benefits of Building Information Modeling (BIM) in improving design coordination.
- Illustrating their relationship: The introduction may discuss how Building Information Modeling (BIM) technologies enable architects and engineers to analyze environmental implications, improve energy performance, and make educated design decisions. This would illustrate the synergy that exists between BIM and sustainable design.
- Differentiating the control spheres: The introduction would differentiate between Building Information Modeling (BIM), which focuses on digital modeling and information management, and sustainable design, which involves energy efficiency, water conservation, material selection, and occupant comfort.

The introduction could include a section on the historical development of Building Information Modeling (BIM) and sustainable design practices. This section will investigate how incorporating BIM has transformed the design and construction industry, improved efficiency, and enabled a more comprehensive approach to sustainability. This section would further contextualize the research. This background information enriches the introduction, leading to the study's aims and relevance in improving sustainable building practices in architectural engineering. Table 1 summarizes the essential elements in formulating a study introduction, emphasizing obstacles and prospects. Every part underscores the significance of clarity, methodological precision, and contextual pertinence. Researchers must address the intricacies of topic selection, framework formulation, and historical background while capitalizing on possibilities to provide a robust foundation and demonstrate the study's significance within broader academic and practical contexts.

The introduction establishes a solid basis for the research by adhering to these criteria. It piques the attention of supervisors and evaluators, making it easier to conduct a comprehensive review of the study's contribution to the field [39].

3 Crafting an Effective Introduction to Research

The introduction is a crucial component of any research endeavor, laying the foundation for the study and directing readers through the principal themes and objectives [40]. This part is critical for elucidating the research's significance and pertinence within its field. For example, biophilic design and sustainable Building Information Modeling (BIM) have garnered considerable focus in building and architecture. Both domains illustrate the increasing emphasis on sustainability and the amalgamation of nature and technology within the constructed environment. This section explores integrating these topics into an introduction and the necessity of clarity and controlled progression.

3.1 Role and Significance of the Introduction

The introduction establishes the groundwork for any research, especially in fields such as architecture and construction, where practical applications and theoretical advancements frequently converge. An excellent introduction articulates the study's purpose by identifying critical concerns within the research area, emphasizing deficiencies in current knowledge, and delineating the research questions that will direct the investigation. Biophilic architecture, which incorporates natural elements into the built environment, tackles a significant challenge in urban planning: the increasing estrangement between humans and nature in developed regions. The introduction may start with a striking figure: 68% of the global population is anticipated to reside in metropolitan areas by 2050, as reported by the United Nations [41]. This fact

Table 1 Obstacles and prospects in formulating a research introduction

Section	Description	Challenges	Opportunities
Selection of research topic [32]	The initial step in the research process is crucial for ensuring that the chosen topic is articulated with intellectual rigor, laying the foundation for a systematic inquiry and a comprehensive literature review	Identifying an innovative and feasible research topic while balancing academic relevance and practical applicability can be challenging	It presents the possibility of addressing underexplored areas in the field, contributing original research, and establishing the foundation for future investigations
Research framework [33]	A reputable and dependable research framework is essential for ensuring objectivity, rigor, and adherence to academic quality standards, enhancing the credibility and reliability of the study's findings	Establishing a robust theoretical or methodological framework that is sufficiently comprehensive while remaining widely accepted within the discipline may be complex	It enables the development of a solid foundation for the research, ensuring that the results are reliable and valid, thus increasing the study's academic contribution
Introduction importance [34]	A well-constructed introduction provides a rationale for the study, capturing evaluators' interest and contextualizing the research within its broader intellectual and practical implications	Crafting a concise and compelling introduction while clearly outlining the relevance of the research can be particularly difficult for early-career researchers	It offers an opportunity to immediately engage the audience, guiding them through the significance of the research and setting the stage for in-depth exploration
Critical components of an introduction [35]	The introduction must clearly outline the key concepts, research challenges, and the study's relevance within a broader societal, environmental, or academic context, ensuring the research is anchored effectively	The risk of failing to adequately clarify key concepts or misaligning the study's relevance with larger frameworks may diminish the perceived value of the research	It allows the researcher to frame their study within a meaningful context, enhancing its impact and broadening its appeal to a broader academic audience

(continued)

Table 1 (continued)

Section	Description	Challenges	Opportunities
Step-by-step structure of introduction [36]	The introduction should establish logical connections between primary research themes, briefly explain their interrelationships, and highlight the distinctive variables or control domains relevant to the study	Developing a coherent and systematic narrative that effectively links abstract concepts while maintaining a logical flow can pose significant challenges	Ensures a structured, systematic approach to the research, aiding the reader in comprehending the study's focus and methodological coherence
Historical background [37]	Tracing the historical development of the research topic provides essential context, allowing for a deeper understanding of the study's relevance and positioning within the field's evolution	Conducting a thorough historical analysis of the research topic may require extensive time and resources, particularly in under-researched areas	It offers depth and a rich contextual foundation for the study, reinforcing its academic and practical significance
Final relevance [38]	The introduction must underscore the study's significance, clearly articulate its academic and practical contributions, and ensure a comprehensive understanding of its aims and objectives	Communicating the study's relevance without overstating it can be challenging, especially when positioning the research as novel and impactful	It strengthens the groundwork for demonstrating the research's contribution, facilitating a clear justification for its academic and professional importance

presents considerable problems to environmental and human welfare, a disparity that biophilic architecture aims to address. Conversely, sustainable BIM (building information modeling) provides a technological approach to improve sustainability in building endeavors. Integrating sustainability concepts into BIM enables architects and engineers to simulate energy efficiency, conduct lifecycle evaluations, and analyze carbon footprints, establishing a definitive approach to minimizing environmental effects. An introduction on this topic may provide case studies illustrating that BIM decreased material waste by as much as 30%, highlighting the research's practical significance [42].

3.2 *Structural Elements of an Introduction*

The hook, background information, and research statement must be incorporated to construct an engaging opening. These features collaborate to engage readers in the research while providing a concise overview of the study's scope and aims.

3.2.1 The Introductory Hook

The hook constitutes the initial element of the introduction and aims to engage the reader's attention [43]. It must pertain to the research subject and offer a compelling rationale for the significance of the investigation. In studies concerning biophilic design, a captivating focal point may center on the growing acknowledgment of nature's mental health advantages in metropolitan environments. Research indicates that people exposed to natural surroundings have reduced stress levels, excellent cognitive function, and improved well-being. These advantages underscore the necessity of biophilic ideas in modern urban design. In sustainable BIM, the introduction might emphasize the environmental catastrophe confronting the building sector, which accounts for around 39% of worldwide carbon emissions [44].

3.2.2 Contextual Information

After the hook, the introduction must include pertinent background information that situates the study within its broader context. In the domain of biophilic architecture, background material may encompass a discourse on early pioneers such as E.O. Wilson, who promoted the biophilia hypothesis, the notion that people possess an intrinsic affinity to nature [45]. This section will examine the integration of biophilic design into contemporary architectural practices, referencing real-world examples such as Singapore's Gardens by the Bay. This biophilic initiative incorporates abundant greenery within a vibrant urban environment, promoting biodiversity and mitigating the heat island effect. For sustainable BIM, for example, the background material must encompass the development of BIM technology and its growing significance in promoting sustainable practices within the building industry. The capacity of BIM to model energy consumption, material use, and project lifecycle expenses renders it an essential instrument for architects and engineers aiming to achieve environmental sustainability objectives. An exemplary case is the application of BIM in the design and construction of the Bullitt Center in Seattle, recognized as one of the most sustainable commercial buildings worldwide, where BIM technology facilitated optimizing resource efficiency and energy performance.

3.2.3 Research Declaration

The introduction centers on the research statement, delineating the main study's emphasis and objectives [46]. The research statement of a survey on biophilic architecture may suggest examining the impact of incorporating natural features into urban structures on residents' ecological and psychological well-being. This may involve examining case studies to illustrate the enduring advantages of biophilic principles in enhancing air quality, decreasing energy usage, and fostering mental well-being. The study statement for a project on sustainable BIM may concentrate on assessing the role of BIM in promoting the incorporation of sustainability practices over the whole lifetime of a building project, from design to demolition [47]. It might also examine the economic ramifications of implementing BIM for sustainability, evaluating its capacity to decrease building expenses while concurrently mitigating environmental effects.

3.3 Academic Accuracy and Lucidity in the Introduction

The introduction must achieve a balance between conciseness and academic rigor. It must articulate the principal concerns succinctly while eschewing superfluous complications [48]. For example, in biophilic architecture, the introduction should concentrate on the most relevant concepts and examples that align with the study objective rather than providing a comprehensive discussion of all biophilic design theories. Clarity in a sustainable BIM introduction is essential in elucidating how BIM fosters sustainability without inundating the reader with technical terminology. The introduction must furnish sufficient material to illustrate the study's depth while reserving more intricate analysis for later sections. Launching a research study in biophilic architecture and sustainable BIM in construction must be a meticulously organized and lucid roadmap. The text must captivate the reader with an appropriate hook, furnish essential background information, and articulate a well-defined research statement that prepares for future investigation [49]. The introduction effectively contextualizes the study by incorporating examples like biophilic design in Singapore's Gardens by the Bay and sustainable BIM in the Bullitt Center, illustrating its practical implications.

4 Essential Principles for Developing a Compelling Introduction in Academic Writing

To create a clear structure, purpose, and connection to the subject, crafting an excellent beginning when writing for academic purposes is vital. Not only does a well-structured introduction establish the tone of the study, but it also emphasizes the relevance of the investigation, so giving the required context for the research questions and objectives. The purpose of this document is to provide the reader with a road map that outlines the scope and direction of the study. In addition, a well-crafted introduction guarantees that the research topic and the suggested results are consistent, which helps to develop a logical flow that improves the overall argument of the study. By establishing a connection between the introduction and the findings and conclusions of the study, researchers can boost the credibility and effect of their work. Following these criteria will not only enhance the clarity of the introduction but also help establish a strong foundation for academic inquiry, making the study both engaging and accessible.

4.1 Developing the Introduction in a Sequential Manner

A staged structure is a critical tactic in writing a practical introduction. Rather than crafting a comprehensive introduction from the outset, starting with a summary highlighting the study's most essential elements, such as its objectives and purposes, is recommended [50]. For example, when researching the integration of artificial intelligence and the Internet of Things in smart buildings to improve energy efficiency, a concise introduction might focus on the growing importance of digital technologies in achieving sustainability goals. The introduction creates a foundation that may subsequently be developed to represent more particular parts of the study, such as specific technologies, procedures, or case studies of intelligent buildings. This foundation begins with an overview, which is the introduction's basis. It is possible to revise the introduction after the survey has been finished to bring it closer to the primary results it contains. Because it enables the researcher to concentrate on broad elements first, this stage-wise technique benefits subjects using digital technology. It is possible, for example, to provide a quick introduction to the role that automation plays in minimizing construction waste or optimizing resource use in environmentally responsible structures. As the investigation progresses, the introduction may be modified to incorporate particulars such as energy use patterns in smart homes or digital twin models utilized to manage the lifetime of materials.

4.2 Maintaining Proper Length and Proportion

It is of the utmost importance to ensure that the introduction's length does not correspond to the whole length of the study. According to a standard rule of thumb, the introduction should comprise around five to ten percent of the material [51]. This principle applies to a wide range of fields; however, when it comes to digitalization in sustainable construction, an excessively lengthy introduction may overwhelm the reader with unneeded technological details or theoretical discussions, whereas an overly brief introduction may fail to frame the research effectively. Suppose the research is centered on how Building Information Modeling (BIM) makes it easier to design and construct energy-efficient structures, for instance. In that case, it is essential to avoid overloading the introduction with excessive amounts of technical jargon right from the beginning. In its place, the introduction should concisely emphasize the significance of Building Information Modeling (BIM) to sustainability, emphasizing its capacity to forecast energy performance and cut down on construction waste. It is recommended that the methodology and analysis sections be saved for parts that provide more in-depth explorations of its technological applications. A well-balanced introduction will prevent the reader from feeling overwhelmed and will maintain the emphasis on the more general implications of the study without distracting them [52].

4.3 Coherence Between Introduction and Conclusion

Increasing the overall coherence of the study may be accomplished by establishing a distinct connection between the introduction and the conclusion [53]. By doing so, it is ensured that the study objectives that were stated at the beginning are mirrored in the results and outcomes that are offered at the termination. In the context of sustainable building, for instance, if the introduction claims that the research will investigate how digitalization impacts the reduction of carbon emissions in the construction sector, the conclusion should include a summary of the findings that either support or dispute this assertion. The introduction may discuss the utilization of artificial intelligence and drones in site inspections to minimize the amount of material that is wasted. The conclusion should then discuss how the study's findings either demonstrate that these technologies help promote sustainability or highlight limits that were not anticipated. Because it illustrates a logical development from hypothesis to conclusion, this coherence is vital in academic research because it highlights the trend. As an illustration, in the field of environmentally friendly buildings, the introduction can include a hypothesis concerning how digital twins enhance the operational efficiency of green buildings. As a result, the conclusion would be connected to this hypothesis by summarizing the essential facts, such as the reduction in energy consumption accomplished through predictive maintenance made possible by digital twins.

4.4 Field-Specific Variations and Adaptations

Even though these principles provide a broad framework, it is crucial to acknowledge that the form of an introduction may need to be modified depending on the discussed academic topic [54]. Within sustainable building, for instance, the introduction could be required to address multidisciplinary concerns, such as the junction of digital technology, environmental legislation, and architectural design [55]. When writing about the digitalization of sustainable buildings, the introduction might include a discussion of the government's policies, the standards of the industry, or international agreements that impact the use of intelligent technology in the construction sector. Researchers concentrating on digital innovation in the building industry may also need to discuss changing trends. An example that may be addressed in the introduction is the growing dependence on AI-driven decision-making in constructing energy-efficient buildings. As digitalization in environmentally responsible buildings continues to advance, the frameworks researchers employ to present their findings must also continue developing. When the most recent developments in the sector are incorporated into an introduction, it will guarantee that the study will continue to be pertinent and well grounded in the present environment. When designing a successful beginning, it is necessary to take a strategic approach that unfolds in phases, stays proportional to the overall length of the research, maintains consistency with the conclusion, and accommodates changes peculiar to the area. Researchers can communicate their findings in a well-organized, pertinent, and convincing manner if they adhere to these standards, particularly in the dynamic environment of digitalization in sustainable buildings and construction [56]. Table 2 includes essential guidelines for academic introduction writing. These guidelines stress the need for an organized and sequential approach, the use of an acceptable length, the maintenance of coherence between the introduction and conclusion, and the incorporation of field-specific modifications. Focus, detail balance, narrative continuity, and keeping up with developments in ever-changing disciplines are all challenges.

Table 2 Key principles for crafting a scholarly introduction

Essential principles for developing a compelling introduction in academic writing	Description	Challenges
Developing the introduction in a sequential manner [57]	Begin with a concise overview, expanding it as the research evolves	Ensuring adaptability without deviating from the central research focus or prematurely introducing irrelevant elements
Maintaining proper length and proportion [58]	Ensure the introduction comprises 5–10% of the overall text, providing essential context	Striking a balance between sufficient background detail and avoiding overwhelming the reader with excessive technical information
Coherence between introduction and conclusion [59]	Align the research objectives in the introduction with the outcomes in the conclusion	They effectively incorporate unexpected findings while maintaining a consistent narrative and research trajectory
Field-specific variations and adaptations [60]	Tailor the introduction to the specific demands of the field, including current trends and interdisciplinary considerations	The introduction should be aligned with rapidly evolving developments, particularly in digitalization, to ensure ongoing relevance and accuracy

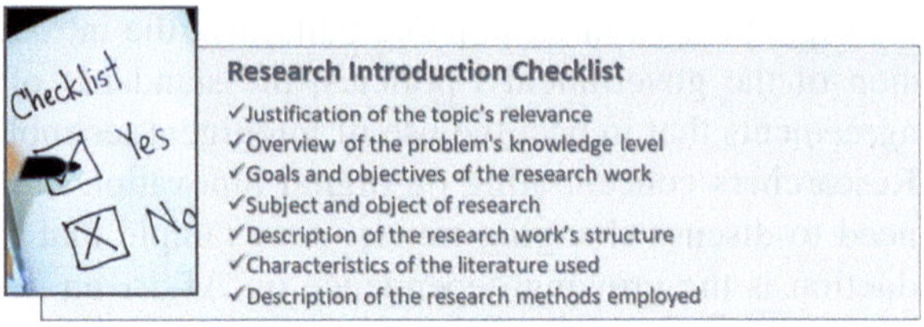

Fig. 3 Research introduction checklist (*Source* Authors)

Figure 3 is an illustration of a "Research Introduction Checklist," which highlights crucial components that are necessary for the creation of a well-structured introduction in academic research. The checklist includes defining the subject and object of the study, providing an overview of the problem's knowledge level, describing the aims and objectives, and providing a justification for the topic's relevance. In addition, it highlights the need to outline the organization of the research activity, detailing the literature utilized and explaining the methodologies used. Following these stages ensures the introduction is comprehensive and cohesive, effectively guiding the reader through the research background.

5 Strategies for Academic Precision and Engagement in Crafting a Research Introduction

When establishing the context for the whole research study, the beginning of a research study is of the utmost importance [61]. It is essential because it can capture the reader's attention, frame the research topic, summarize the subject matter, and identify the study's objectives. To ensure that the rest of the study is clear, coherent, and relevant, it is essential to have a solid beginning that serves as the foundation around which others are constructed. A well-known organizational strategy that is commonly referred to as the "funnel" or "inverted pyramid" technique is frequently utilized by researchers to achieve this level of success [62]. This method enables the researcher to lead the reader from a general viewpoint to the specific subject of the study, assuring smooth and logical development from the beginning to the end of the research. At the start of the funnel approach, a general overview of the subject matter is typically positioned within a more comprehensive sociological, scientific, or global framework. As an illustration, research centered on incorporating renewable energy sources into high-rise buildings can begin by addressing the current worldwide energy problem and the pressing requirement for environmentally responsible design and construction techniques. Having such a comprehensive introduction helps to place the findings into the context of a more general topic that is of widespread relevance. After providing this overarching background, the introduction can gradually reduce its scope, bringing the reader closer to the research issue or problem the study intends to solve [54]. An excellent research introduction follows the steps outlined in Fig. 4. To make significant contributions, it emphasizes crucial features

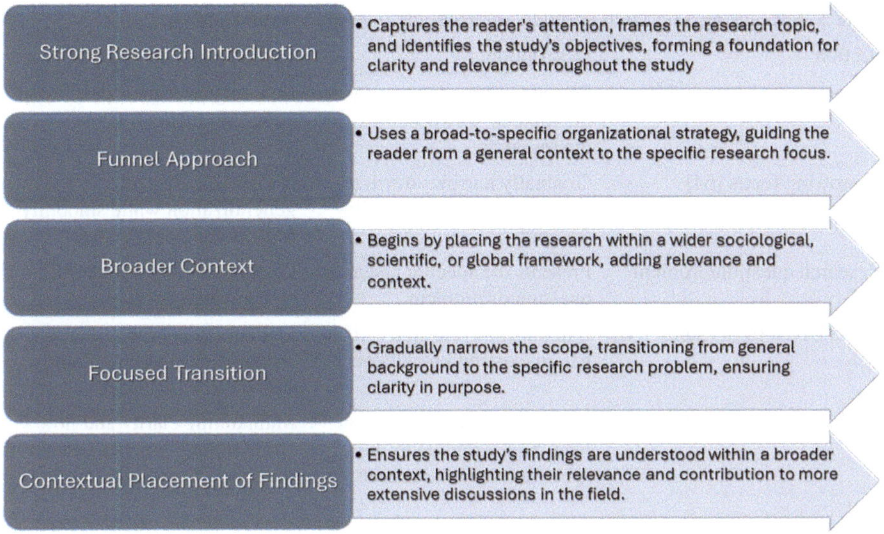

Fig. 4 Strategies for academic precision and engagement in crafting a research introduction (*Source* Authors)

such as a strong introduction, logical flow using a funnel method, a broader context, and a focus, and it places results in a larger academic discourse.

5.1 The Inverted Pyramid Approach: Structuring the Research Introduction

Because it constructs the introduction as a gradual narrowing of focus, the funnel method, also known as the inverted pyramid, is particularly successful [63]. This provides the reader with a logical flow of information presented sequentially. Typically, the first paragraph of a research study will introduce the general topic matter, which may include a global or theoretical issue, before focusing on the specific study challenge. Through this sequence, the reader can comprehend the far-reaching implications of the research while also identifying the void that the study intends to address. Suppose the researcher is investigating using renewable energy systems in high-rise buildings. In that case, they may start by having a conversation about the consumption of energy on a global scale and the effects that traditional energy sources have on the environment. The purpose of this extensive introduction is to emphasize the significance of the subject matter on a more general level. As the introduction becomes more particular, the attention may change to a more specific concern: the technological and architectural obstacles in integrating renewable energy solutions into urban high-rise structures. The research is placed within the more significant academic and practical issues of sustainable urban development due to this shift, which offers context for the

Table 3 Structuring a research introduction using the inverted pyramid approach

Section	Description	Challenges
General introduction [54]	Provides a broad overview of the topic to set the context for the research	Ensuring relevance and capturing reader interest
Narrowing focus [64]	Gradually narrows down to specific concerns related to the study	Avoiding excessive generalization while smoothly transitioning to specifics
Research question/problem [65]	Presents the specific research question or problem	Clearly defining a focused, researchable question
Significance of the study [66]	Explains how the study fills a gap in existing knowledge	Demonstrating the research's unique contribution without overstating its impact
Logical transition [67]	Ensures a smooth transition from broad context to the specific research question	Maintaining clarity and flow while linking broader context to the specific research question

research. The introduction should continue to limit even more by outlining the exact research topic or target as it progresses. An example of a question the researcher may ask is, "Is it possible for high-rise buildings to achieve energy self-sufficiency through the integration of solar photovoltaic systems?" The introduction provides a framework for the study by articulating the research issue clearly and concisely. Still, it also illustrates the space in the existing body of knowledge that research endeavors to fill. To guarantee that the reader can follow the logical transition from a broad context to a specific, targeted research topic, this narrowing process ensures that the introduction is entertaining and instructive. Essential parts of a research introduction are highlighted in Table 3, along with explanations of the sections and the difficulties that go along with them. The text highlights the significance of narrowing down from an overarching overview to a particular research subject while addressing challenges such as maintaining relevance, clarity, and seamless section transitions.

5.2 Background Information and Literature Review in the Introduction

In addition, the introduction must include pertinent background information that assists the reader in comprehending the study subject [68]. This typically consists of a concise assessment of the most critical literature on the topic, a summary of the studies that have already been conducted, and an identification of the gaps the present research intends to fill [69]. For example, a survey incorporating renewable energy sources may refer to earlier research on environmentally responsible construction techniques, highlighting the limits of these studies when applied to high-rise buildings. By doing so, the researcher not only indicates that they have a solid comprehension of the subject matter, but they also justify the necessity of the present study. In addition, providing background information guarantees that the reader is provided with the essential context required to comprehend the relevance of the research. This part doesn't need to be thorough; nonetheless, it should contain sufficient information to indicate the subject matter's significance and the study's originality. As a result of going through this procedure, the researcher can establish their credibility and demonstrate the importance of the work within the context of the more considerable academic debate.

5.3 *The Checklist for the Construction of an All-Encompassing Introduction*

Researchers might use a checklist to verify that they have addressed the essential components when writing the introduction. The following are integral parts:

- Include a clear explanation of why the research issue is significant, not just in the context of the field in which it is being conducted but also in the perspective of a broader social or scientific setting [70]. Consequently, this guarantees the reader knows the study's theoretical or practical consequences.
- Overview of the existing knowledge: Provides a synopsis of the status of research in the field, focusing on the most important studies and pointing out any gaps in the existing body of literature [71]. This helps to contextualize the present study within the academic conversation that is currently taking place.
- The research goals and objectives should include a concise summary of the study's objectives and an explanation of how these objectives relate to the research topic [61]. This part should include an overview of the study's goals and why they are essential.
- Topic and object of study: To begin the research process, define the topic (the primary emphasis) and the object (the more significant phenomena or system within which the subject is positioned) [72]. This will clarify the scope of the investigation.
- Structure of the research work: Summarizes the arrangement of the document to provide the reader with a road map of what to anticipate in the coming sections of the study [73].

5.4 *The Application of Keywords in a Strategic Manner in the Introduction*

Keywords are a significant factor in increasing the exposure and discoverability of a research study [74]. When adequately employed initially, keywords signal the study's emphasis on human readers and digital search algorithms. In a survey of the acoustic performance of sustainable building materials, for instance, it is possible to incorporate keywords such as "acoustic performance," "sustainable materials," and "environmental design" at an early stage in the introduction. This will ensure the research is aligned with relevant research areas and can be easily searched in academic databases. Additionally, the incorporation of keywords into the title of the research article is an additional method that further enhances the possibility that other scientists will discover the publication. As an illustration, a piece of writing titled "Daylighting Strategies for Energy Efficiency in Office Buildings" identifies important topics and contributes to the classification of the study stored in digital repositories. Researchers can guarantee that their work is effectively communicated to the intended audience by including pertinent keywords throughout the introduction [75].

5.5 Choosing the Appropriate Time to Write the Introduction

It is common practice to write the introduction toward the conclusion of the writing process, even though it is often produced at the beginning of a research work. Because of this, the researcher can ensure that the introduction agrees with the results and conclusions of the study, which may change throughout the research process. The introduction should be written last since doing so ensures consistency with the remainder of the work and gives an accurate picture of the researcher's findings [61]. Suppose the researcher waits until the research is finished. In that case, they will have a more excellent knowledge of the study's consequences, allowing them to communicate its importance better. This technique guarantees that the introduction aligns with the research's objectives, scope, and results. The beginning of a research study is an essential piece that must be meticulously [76]. Researchers can design an introduction that is not only interesting but also useful if they make use of the inverted pyramid technique, provide background information, strategically use keywords, and take into consideration the time of the writing process. A well-written introduction offers a framework for the study and increases the article's visibility, influence, and contribution to the academic community [77].

6 Common Mistakes in Formulating an Introduction in Academic Writing

Referring to the "most incorrect and erroneous aspects in the introductory chapter of research" entails an examination of prevalent deficiencies that frequently compromise the quality of research introductions, particularly within the field of architectural engineering. Common pitfalls, often observed at the outset of research, undermine the clarity and coherence necessary for establishing a robust foundation. This analysis will provide specific examples from architectural engineering to illustrate the nature of these errors and their impact on the overall integrity of the research.

- **Lack of Clear Purpose**: The introduction should immediately introduce the research topic and explain why it's essential [78].

  Example

  - *Wrong: "Many buildings are constructed every year."*
  - *Right: "With the increasing demand for sustainable architecture, exploring innovative materials and design strategies has become imperative."*

- **Being Too Vague**: Specifics help set the stage for your research [79].

  Example

  - *Wrong: "Many structures are affected by various factors."*
  - *Right: "Coastal structures in Florida are particularly vulnerable to sea-level rise and saltwater intrusion."*

- **Need to State the Problem**: Could you clearly state the problem you're addressing? [80].

  Example

  - *Wrong: "Energy consumption in buildings is an issue."*
  - *Right: "Current HVAC systems, which account for nearly 40% of a building's energy consumption, often lack efficiency and adaptability."*

- **Not Highlighting the Significance**: Why should anyone care about your study? [81].

  Example

  - *Wrong: "This research looks at green roofs."*
  - *Right: "Green roofs have the potential to mitigate urban heat islands, reduce energy costs, and provide urban habitats, making their optimization critical."*

- **Too Broad or Narrow Scope**: The introduction should balance providing a broader context and narrowing down your research topic [82].

  Example

  - *Wrong: "This study will explore all aspects of urban planning and architectural innovations from the past century."*
  - *Right: "This research will focus on integrating passive solar design strategies in urban residential buildings in the Pacific Northwest."*

- **Failure to State Assumptions or Limitations**: Every research has its constraints. Could you make sure to state them early? [83]

  Example

  - *Wrong: "All buildings benefit from natural ventilation."*
  - *Right: "While many structures can gain advantages from natural ventilation, this research focuses on those in temperate climates, given their unique weather conditions."*

- **Plagiarism**: This is a grave error. Always cite sources when presenting background information [84].

  Example

  - *Wrong: Using a quote or information without citing its source.*
  - *Right: "According to Smith and Jones (2010), urban areas with more green spaces have shown a 15% reduction in ambient temperature."*

- **Poor Structure and Flow**: The introduction should have a logical flow that guides the reader from the broader context to the specifics of your research [85].

  Example

  - *Wrong: Jumping haphazardly between unrelated topics.*
  - *Right: Start with the broader context of sustainable architecture, narrow down to energy efficiency, and specify green roofs' role.*

- **Not Stating the Objectives or Hypothesis**: By the end of the introduction, the reader should know your primary research objectives or hypotheses [86].

  Example

  - *Wrong: "Green roofs are becoming more popular."*
  - *Right: "This study aims to evaluate the thermal performance of green roofs in temperate climates and hypothesizes that their integration can reduce a building's heating and cooling loads by at least 20%."*

- **Ignoring Recent Developments**: Being up to date is essential, especially in fields like architectural engineering [87].

  Example

  - *Wrong: Citing only studies from the 1980s and 1990s.*
  - *Right: Recent literature and technological advancements from the last 5 years.*

The introduction serves as the reader's first impression of the research, setting the stage for everything that follows. Ensuring it is clear, compelling, and error-free is essential for the study's success.

7 Conclusion

Developing a research title and introduction is a crucial aspect of the academic writing process since these elements provide the framework for the whole study. The research title serves as the initial point of interaction between the researcher and prospective readers, functioning as an entryway to the study's content. It offers a concise yet thorough depiction of the research's topic and methods. It aims to serve an essential function in captivating readers and communicating the study's significance to academic and professional audiences. An effectively constructed title encapsulates the primary subject of the research while aligning with current trends and discourse in the pertinent field, augmenting the study's visibility and influence.

Formulating a study title necessitates a balance between brevity and precision. The title should be concise while sufficiently conveying the study's breadth, research inquiries, and theoretical framework. The title can provide a brief overview of the study by emphasizing essential topics and approaches, enabling prospective readers to assess its pertinence to their interests or academic endeavors. Additionally, the title must encapsulate the research's wider academic or practical significance, indicating the study's contribution to the continuing conversation. In disciplines like architecture, energy efficiency, and sustainability, titles that emphasize difficulties and new solutions effectively situate the research within essential dialogs regarding future development. The introduction functions as the official entrance point into the research, offering background and justification for the study while presenting the research topic and objectives. A successful introduction transitions from a broad overview of the study domain to a concentrated emphasis on the tackled issue. This development is crucial for navigating the intricacies of the research issue while clearly articulating the study's objectives and relevance. The introduction serves as a guide, providing readers with an organized overview of the content, methodology, and anticipated research contributions. Essential components of a practical introduction comprise a precise delineation of the research topic, the justification for the investigation, and a brief overview of the study's aims and methods. These elements guide readers and illustrate the researcher's comprehension of the wider academic or practical environment around the research. The introduction contextualizes the work within current debates or new trends, affirming its relevance and prospective contributions. For instance, in studies concerning sustainable construction methodologies, an introduction may commence by highlighting the escalating imperative of tackling climate change and resource exhaustion, subsequently focusing on the contribution of innovative materials or design strategies to enhance energy efficiency. This framework guarantees that readers may comprehend the logical progression of the research and its wider ramifications. Furthermore, the introduction must captivate the reader by providing insight into the study's uniqueness and addition to knowledge. This is frequently accomplished by discussing the research gap the study aims to remedy. Recognizing deficiencies in the current literature enables the researcher to situate their work as a significant contribution to the field, addressing an absence or providing novel insights on known subjects. The introduction must explicitly delineate how the research contributes to knowledge in the selected field, regardless of whether the study seeks to establish a new theoretical framework, implement an innovative technique, or provide practical answers to real-world issues. This is especially crucial in domains like architecture and sustainability, where the fast advancement of technology and design methodologies necessitates continual innovation and reevaluation of conventional paradigms. The clarity and correctness of the research title and introduction are essential for making the study accessible to its target audience. Academic research, particularly in technical disciplines, sometimes encompasses intricate concepts and procedures that may be difficult for readers to comprehend without appropriate context. An effectively crafted title and introduction furnish essential context and clarity, rendering the study accessible while preserving its scholarly integrity. The introduction functions as a screening tool, assisting readers in determining the study's relevance to their research interests or requirements. Moreover, connecting the research title and introduction with the overarching research objectives is paramount. A discrepancy among these factors may result in misunderstanding and reduce the study's efficacy. Both elements must collaborate to provide a unified narrative representing the research's breadth and purpose. This coherence guarantees that the reader's expectations are fulfilled throughout the study, from the initial interaction with the title and introduction to the ultimate presentation of data and conclusions.

Moreover, the introduction is crucial in establishing the tone for the investigation. The tone of the study may fluctuate based on its nature, ranging from technical and formal in quantitative research to more exploratory and discursive in qualitative research. The introduction must balance offering sufficient material to enlighten the reader and avoiding needless complications that may obscure the study's emphasis. A significant factor is the purposeful incorporation of keywords in the title and introduction. Keywords augment the study's exposure in academic databases and guarantee accurate categorization within its discipline. By using pertinent and precise keywords, the researcher enhances the probability that their work will be found by those seeking research on related subjects. This is especially important in multidisciplinary domains such as sustainability and architecture, where research frequently encompasses many areas of expertise. In conclusion, developing a study title and introduction involves meticulous thinking and intentional organization. Both parts convey the research's emphasis, significance, and contributions to the discipline. A well-constructed title and introduction establish the foundation for a successful and meaningful research study by blending conciseness with specificity, clarity with academic rigor, and broad context with aims. These elements captivate the reader and situate the study within more comprehensive educational and practical discussions, guaranteeing a significant contribution to the relevant subject.

References

1. Lee Y, Jang BG, Conradi Smith K (2021) A systematic review of reading engagement research: What do we mean, what do we know, and where do we need to go? Read Psychol 42(5):540–576
2. Purssell E, McCrae N (2020) How to perform a systematic literature review: a guide for healthcare researchers, practitioners, and students. Springer Nature
3. Zou PX, Xu X (2023) Research methodology and strategy: theory and practice. John Wiley & Sons
4. Torres Salinas D, Arroyo Machado W, Robinson García N (2024) Learning materials for the Erasmus+Project RAQMYAT: Parcours SOFTS Skills
5. Cárdenas-Ramírez C, Jaramillo F, Gómez M (2020) Systematic review of encapsulation and shape-stabilization of phase change materials. J Energy Storage 30:101495
6. Wibawa AP, Kurniawan F (2024) A survey of text summarization: techniques, evaluation, and challenges. Nat Lang Process J 7:100070
7. Almusaed A, Yitmen I, Almssad A (2024) Contemporary innovations and sustainable practices in the application of clay materials within architectural design and construction methodologies
8. Perivoliotis D, Arvanitis I, Tzavali A, Papakostas V, Kappou S, Andreakos G, Fotiadi A, Paravantis JA, Souliotis M, Mihalakakou G (2023) Sustainable urban environment through green roofs: a literature review with case studies. Sustainability 15(22):15976
9. Sepasgozar S (2024) Research proposal crucible: crafting with AI awareness: strategies to communicate strong academic arguments. Samad Sepasgozar
10. Hilal NA, Haggag M, Saleh AD (2023) Optimizing energy efficiency in high-rise residential buildings in Abu Dhabi's hot climate: exploring the potential of double skin façades. Buildings 13(9):2148
11. Almusaed A, Yitmen I, Almsaad A, Akiner İ, Akiner ME (2021) Coherent investigation on a smart kinetic wooden façade based on material passport concepts and environmental profile inquiry. Materials 14(14):3771
12. Agbajor FD, Mewomo MC (2024) Green building research in South Africa: a scoping review and future roadmaps. Energy and Built Environment 5(2):316–335
13. Lê JK, Schmid T (2022) The practice of innovating research methods. Organ Res Methods 25(2):308–336
14. Lim WM (2024) What is qualitative research? An overview and guidelines. Australas Mark J 14413582241264619
15. Yadav D (2022) Criteria for good qualitative research: a comprehensive review. Asia Pac Educ Res 31(6):679–689
16. Salama AM (2020) Shaping the future of research in art, architecture, and spatial design fields. J Arts Arch Res Stud 1(2):14–31
17. Dwivedi YK, Jeyaraj A, Hughes L, Davies GH, Ahuja M, Albashrawi MA, Al-Busaidi AS, Al-Sharhan S, Al-Sulaiti KI, Altinay L, Amalaya S (2024) "Real impact": challenges and opportunities in bridging the gap between research and practice–making a difference in industry, policy, and society. Int J Inf Manage 102750
18. Fico D, Rizzo D, Casciaro R, Esposito Corcione C (2022) A review of polymer-based materials for fused filament fabrication (FFF): focus on sustainability and recycled materials. Polymers 14(3):465
19. Weerawardane D, Byrne P (2021) An Introduction to critical thinking, academic writing, and Harvard referencing
20. Almusaed A, Yitmen I (2023) Architectural reply for smart building design concepts based on artificial intelligence simulation models and digital twins. Sustainability 15(6):4955
21. Grove K, Rickards L (2022) Contextualizing Geography's past, present, and future narratives: synthesis, difference, and cybernetic control. Environ Plan F 1(1):26–40
22. Bui HP (ed) (2024) Applied linguistics and language education research methods: fundamentals and innovations: fundamentals and innovations. IGI Global
23. Sotiriadou P, Logan D, Daly A, Guest R (2020) The role of authentic assessment is to preserve academic integrity and promote skill development and employability. Stud High Educ 45(11):2132–2148

24. Bain M, Drinkwater J, Bates J, Bell T, Cullen D, Noone B, Owen D, Redmond P, Samson R, Vázquez L (2021) Interuniversity style guide for writing end-of-degree projects in English
25. Nayak JK, Singh P (2021) Fundamentals of research methodology problems and prospects. SSDN Publishers & Distributors
26. Schoch K (2020) Case study research. Res Des Methods Appl Guid Sch Pract 31(1):245–258
27. Almusaed A, Almssad A, Yitmen I (2024) Sustainable built environment and its implications on real estate development: a comprehensive analysis
28. Vandenbroucke JP, Elm EV, Altman DG, Gøtzsche PC, Mulrow CD, Pocock SJ, Poole C, Schlesselman JJ, Egger M, Strobe Initiative (2007) Strengthening the reporting of observational studies in epidemiology (STROBE): explanation and elaboration. Ann Intern Med 147(8):W-163
29. Tracy SJ (2024) Briefly. Qualitative research methods: Collecting evidence, crafting analysis, communicating impact. John Wiley & Sons
30. Montgomery DC (2019) Introduction to statistical quality control. John Wiley & sons
31. Siddaway AP, Wood AM, Hedges LV (2019) How to do a systematic review: a best practice guide for conducting and reporting narrative reviews, meta-analyses, and meta-syntheses. Annu Rev Psychol 70(1):747–770
32. Kunisch S, Denyer D, Bartunek JM, Menz M, Cardinal LB (2023) Review research as scientific inquiry. Organ Res Methods 26(1):3–45
33. Johnson JL, Adkins D, Chauvin S (2020) A review of the quality indicators of rigor in qualitative research. Am J Pharm Educ 84(1):7120
34. Prochner I, Godin D (2022) Quality in research through design projects: recommendations for evaluation and enhancement. Des Stud 78:101061
35. Van der Waldt G (2020) Constructing conceptual frameworks in social science research. TD J Transdiscipl Res South Afr 16(1):1–9
36. Rashid A, Rasheed R, Amirah NA, Yusof Y, Khan S, Agha AA (2021) A quantitative perspective of systematic research: easy and step-by-step initial guidelines. Turk Online J Qual Inq 12(9)
37. Argyres NS, De Massis A, Foss NJ, Frattini F, Jones G, Silverman BS (2020) History-informed strategy research: The promise of history and historical research methods in advancing strategy scholarship. Strateg Manag J 41(3):343–368
38. Wickert C, Post C, Doh JP, Prescott JE, Prencipe A (2021) Management research that makes a difference: broadening the meaning of impact. J Manage Stud 58(2):297–320
39. Aluko KA (2024) Influence of work ethics and supervisory strategies on effective teaching of business education courses in colleges of education, North-Central, Nigeria (Doctoral dissertation, Kwara State University (Nigeria))
40. Zina O (2021) The essential guide to doing your research project. Sage
41. Hoornweg D, Pope K (2017) Population predictions for the world's largest cities in the 21st century. Environ Urban 29(1):195–216
42. Guerra BC, Leite F, Faust KM (2020) 4D-BIM to enhance construction waste reuse and recycle planning: case studies on concrete and drywall waste streams. Waste Manage 116:79–90
43. Cron L (2012) Wired for story: the writer's guide to using brain science to hook readers from the very first sentence. Ten Speed Press
44. Musolino M, Massimo DE, Marino C, De Paola P, Errigo R, Malerba A, Del Giudice FP (2023) Appraisal of energy saving in green buildings, supported by new BIM platforms. Valori e Valutazioni 34
45. Casmiri IV (2023) Toward a biophilic insurgency through climate fiction and urban planning
46. Benchimol EI, Smeeth L, Guttmann A, Harron K, Moher D, Petersen I, Sørensen HT, von Elm E, Langan SM, Working Committee RECORD (2015) The reporting of studies conducted using observational routinely-collected health data (RECORD) statement. PLoS Med 12(10):e1001885
47. Wong JKW, Zhou J (2015) Enhancing environmental sustainability over building life cycles through green BIM: a review. Autom Constr 57:156–165
48. Turnbull D, Krips H, Dusek V, Fuller S, Sokal A, Bricmont J, Frost A, Chalmers A, Salleh A, Tauber AI, Luxford Y (2000) Clarity, charity and criticism, wit, wisdom and worldliness: avoiding intellectual impositions

49. Krause B (2016) Wild soundscapes: discovering the voice of the natural world. Yale University Press
50. Becker HS (2008) Writing for social scientists: how to start and finish your thesis, book, or article. University of Chicago Press
51. Igwenagu C (2016) Fundamentals of research methodology and data collection. LAP Lambert Academic Publishing
52. Gazzaley A, Rosen LD (2016) The distracted mind: Ancient brains in a high-tech world. MIT Press
53. Abbasian GR, Parvizi Moridani R (2022) Incorporation of discourse coherence strategies into writing skills instruction. Interdiscip Stud Engl Lang Teach 1(2):53–73
54. DePoy E, Gitlin LN (2019) Introduction to research e-book: understanding and applying multiple strategies. Elsevier Health Sciences
55. Almusaed A, Almssad A (2019) City phenomenon between urban structure and composition. Sustain Urban Plan Des 3
56. Jiang Y, Su S, Zhao S, Zhong RY, Qiu W, Skibniewski MJ, Brilakis I, Huang GQ (2024) Digital twin-enabled synchronized construction management: a roadmap from construction 4.0 towards future prospect. Dev Built Environ 100512
57. Barbour RS (2013) Introducing qualitative research: a student's guide
58. Lamon SJ (2020) Teaching fractions and ratios for understanding: essential content knowledge and instructional strategies for teachers. Routledge
59. Madrunio MR (2012) Move sequences in graduate research paper introductions and conclusions. J Engl Lang Teach 2(1):1–14
60. Capron E, Delacour H (2024) Field-configuring places: unpacking the role of place in field evolution. Organ Stud 01708406241280010
61. Hancock DR, Algozzine B, Lim JH (2021) Doing case study research: a practical guide for beginning researchers
62. Helmold M (2022) Performance management in sales. Strategic performance management: achieving long-term competitive advantage through performance excellence. Springer International Publishing, Cham, pp 109–121
63. Hjelm T (2024) Elements of research design. Policy Press.
64. Fischer KW, Farrar MJ (1987) Generalizations about generalization: How a skill development theory explains generality and specificity. Int J Psychol 22(5–6):643–677
65. Quashie S (2022) Formulating Ph. D. construction management research question or problem for mixed method (Qualitative and Quantitative). In: Proceedings of 2021 4th international conference on civil engineering and architecture, pp 537–553. Singapore: Springer Nature Singapore
66. Kieser A, Nicolai A, Seidl D (2015) The practical relevance of management research: turning the debate on relevance into a rigorous scientific research program. Acad Manag Ann 9(1):143–233
67. Hennink M, Hutter I, Bailey A (2020) Qualitative research methods. Sage
68. Grabe W, Stoller FL (2019) Teaching and researching reading. Routledge
69. Fink A (2019) Conducting research literature reviews: from the internet to paper. Sage publications
70. May T, Perry B (2022) Social research: issues, methods and process. McGraw-Hill Education (UK)
71. Liao Y, Deschamps F, Loures EDFR, Ramos LFP (2017) Past, present and future of Industry 4.0-a systematic literature review and research agenda proposal. Int J Prod Res 55(12):3609–3629
72. Mohajan HK (2018) Qualitative research methodology in social sciences and related subjects. J Econ Dev Environ People 7(1):23–48
73. Wilson J (2014) Essentials of business research: a guide to doing your research project
74. Cabonero DA, Jandoc JAP, Cabalo JT, Dominguez WB (2021) Digging the issues and concerns in subject access towards discoverability of information and resources. Libr Philos Pract 1–20

75. National Academies of Sciences, Division of Behavioral, Social Sciences, Committee on the Science of Science Communication and A Research Agenda, 2017. Communicating science effectively: A research agenda
76. Tracy SJ (2024) Qualitative research methods: collecting evidence, crafting analysis, communicating impact. John Wiley & Sons
77. Bajwa NUH, König CJ, Kunze T (2020) Evidence-based understanding of introductions of research articles. Scientometrics 124:195–217
78. Martinich AP (2024) Philosophical writing: an introduction. John Wiley & Sons
79. Wallwork A (2024) English for academic research: vocabulary exercises. Springer Nature
80. Peercy MM, Troyan FJ, Fredricks DE, Hardy-Skeberdis M (2024) Calling for a humanizing turn in language teacher education: problematizing content and language instruction. TESOL Q
81. Giray L (2023) Prompt engineering with ChatGPT: a guide for academic writers. Ann Biomed Eng 51(12):2629–2633
82. Siponen M, Klaavuniemi T, Xiao Q (2024) Splitting versus lumping: Narrowing a theory's scope may increase its value. Eur J Inf Syst 33(4):561–570
83. De Oliveira B (2023) Participatory action research as a research approach: advantages, limitations and criticisms. Qual Res J 23(3):287–297
84. Yadav SK (2023) Referencing and avoiding plagiarism. Research and publication ethics. Springer International Publishing, Cham, pp 205–227
85. Putra JWG, Teufel S, Tokunaga T (2023) Improving logical flow in English-as-a-foreign-language learner essays by reordering sentences. Artif Intell 320:103935
86. Kumari SKV, Lavanya K, Vidhya V, Premila GADJS, Lawrence B (2023) Research methodology, vol. 1. Darshan Publishers
87. Osadcha I, Jurelionis A, Fokaides P (2023) Geometric parameter updating in digital twin of built assets: a systematic literature review. J Build Eng 73:106704

Philosophical Foundations of Research

Abstract This chapter investigates the critical role of philosophical frameworks in developing rigorous, ethical, and socially impactful academic inquiry. Based on ontology and epistemology, these foundations give research into its intellectual structure. They direct the formulation of questions, methods selection, and results interpretation. They bridge the divide between theoretical paradigms and real-world challenges by ensuring coherence and precision. The talk carefully examines how the three main philosophical approaches—positivism, constructivism, and pragmatism—affect how we learn new things. Pragmatism promotes methodological adaptability and practical outcomes. Constructivism investigates the subjective and contextual aspects of human experience. Positivism stresses the importance of being logical and factual in your research. Through case studies in building energy efficiency, urban equity, and climate-responsive architecture, the analysis shows how these paradigms affect different methodological choices and give rich, situation-aware insights. Ethical considerations from deontology, utilitarianism, and virtue ethics seamlessly infuse the discussion. This integration underscores their importance in developing societal trust, equity, and transparency. The work also addresses the changing requirements of modern research, including artificial intelligence and sustainability, and promotes ethically sound and adaptive methodologies. By incorporating philosophical principles, this contribution makes sure that research stays relevant and has the power to change things. It does this by promoting a strong framework that is intellectually rigorous, morally grounded, and responsive to complex societal imperatives.

1 Introduction

Philosophical foundations inform all facets of research, changing the researcher's viewpoint and affecting the development of research topics, hypotheses, procedures, and recommendations [1]. They function as the cognitive structure that delineates the comprehension of knowledge and the interpretation of reality within a particular

A. Almusaed et al., *Practice of Research Methodology in Civil Engineering and Architecture*, https://doi.org/10.1007/978-3-031-97393-2_6

study. Furthermore, governance structures are essential in instilling specific philosophical perspectives, frequently influencing prevailing paradigms and methodologies inside academic and institutional frameworks [2]. Governance-driven ideologies influence researchers' issue conceptualization, inquiry prioritization, and findings interpretation, hence affecting knowledge generation and the application of research outcomes across many domains. Philosophy is the foundation of research, offering the critical conceptual framework for understanding knowledge creation, the essence of reality, and its significant influence on research methodologies. Sun and Zuo [3] assert that research philosophy is a crucial component of the research paradigm, influencing how researchers envision, approach, and conduct their investigations. It is the epistemological framework through which researchers analyze the world, formulate research inquiries, and choose suitable approaches. Also, knowing how research philosophy has changed over time is essential for understanding modern research because it shows the intellectual paths that led to the creation of current methods and paradigms [3]. It becomes more apparent how necessary philosophical foundations are when studying architecture, where creative expression and technical accuracy meet [4]. For example, researchers need solid philosophical knowledge to ground their work, whether examining the social effect of urban settings, the sustainability of building materials, or the cultural symbolism encoded in architectural forms [5]. Researchers can better address complex concerns about knowledge and reality by directing their techniques and ensuring they relate to the study situation. This foundation also increases rigor and clarity. Madden [6] emphasizes that inexperienced researchers frequently need clarification while interacting with technique texts because of the intricacy and inconsistencies of philosophical language [6]. This could also lead to epistemic cliques among new researchers, making it challenging for people from different fields to work together effectively. The study of philosophy provides researchers with essential tools to examine and elucidate the fundamental assumptions of their research, promoting a more unified and integrated approach to knowledge creation. The study of philosophy provides researchers with essential tools to examine and elucidate the fundamental assumptions of their research, promoting a more unified and integrated approach to knowledge creation [7]. For example, an architectural researcher interested in how people interact with public spaces must choose if their study assumes an objective reality or accepts the subjectivity of human experience for their research. For example, a positive approach might concentrate on quantifiable data, such as patterns of foot traffic, swings in temperature, or noise levels, and consider these measurements to be universal indications of space use. On the other hand, an interpretive viewpoint would investigate users' individual experiences by conducting interviews or using ethnographic methodologies to capture the subjective meanings and emotional connections that users have with the spatial environment. These various philosophical viewpoints influence the methodology selection and the interpretation of the findings. The philosophical underpinnings of research are based on two fundamental issues: what is knowledge, and what is the nature of reality? Both concerns are integral to the research process. From the perspective of architecture research, epistemology is concerned with how researchers comprehend and interpret the constructed world. For instance, research on sustainable building methods may explore whether empirical testing, such as evaluating the thermal performance of novel materials or tacit knowledge from experienced practitioners in the field, is the most effective method of acquiring information [8]. An important decision that researchers need to make is whether they believe knowledge to be objective, quantifiable, and generalizable or whether it is subjective and affected by individual or cultural factors. On the other hand, ontology is concerned with the essential characteristics of reality. In architectural studies, this frequently entails discussions regarding whether architectural forms and spaces have intrinsic meanings or whether their importance is produced through contemporary culture and society lenses. An investigator can inquire as to whether a cathedral possesses a holy quality in and of itself or whether the holiness of the cathedral is the result of the beliefs of society and the historical environment in which it was built. The ontological factors researchers look at can change how they structure their questions and interpret the results of their studies [9]. Could you take, for instance, the case of researching green spaces in urban settings? A positive researcher may collect quantitative data on air quality, biodiversity, and temperature regulation. We would assume an objective world where we can quantify and generalize the advantages of green areas. On the other hand, an interpreter would concentrate on the residents' perceptions and emotional ties to these places, applying qualitative approaches to investigating how humans perceive and assign meaning to their environment. Similarly, studying architectural legacy frequently faces challenges in the form of epistemological and ontological concerns. Oral histories and the collective memory of a community are also essential in determining the historical relevance of a structure; does the significance of a building's history depend on material examination and archival records? The legacy value of a structure is either inherent to the structure itself or something developed by society that values it. For example, architectural scholars need a solid understanding of these philosophical dimensions. Due to this, they can express their ideas, connect their methodologies with their research aims, and reply to criticisms with clarity and coherence. For instance, a researcher investigating urban heat islands must decide whether to concentrate on objective temperature data or city people's experiences [10]. Alsaigh and Coyne [11] assert that phenomenology serves as both a philosophical tradition and a methodological framework that includes several techniques. Hermeneutic phenomenology provides a non-prescriptive and adaptive technique suitable for many study situations [11]. Deep engagement with philosophical ideas is necessary to encourage large-scale collaboration across disciplines. This is especially important in architectural research, which often involves interacting with sociology, environmental science, and cultural studies. Researchers with a strong understanding of their philosophical perspective are more adept at interdisciplinary engagement, increasing the likelihood of integrative and creative results. Philosophical underpinnings are essential in architectural study, as the concrete reality of constructed spaces intersects with the abstract realms of meaning, culture, and perception. These foundations offer a vital perspective for researchers to investigate essential concerns of knowledge and reality, facilitating the development of cohesive and thoroughly supported investigations. Investigations based on clear, well-defined philosophical concepts produce significant findings and connect theoretical frameworks with practical applications. This method improves the academic and practical contributions of architectural research by making it easier to understand and combine theory and practice, which is essential for dealing with the complexity of the built environment.

2 Ontology and Epistemology: Fundamental Ideas in Research

Research methods and theoretical frameworks are based on two pillars of philosophy: ontology and epistemology [12]. Ontology study categories of being, relationships between entities, and what exists. It's a way to explore nature's reality [13]. By doing so, an ontology can define its scope and make assumptions about reality. Meanwhile, epistemology delves into how people gain, defend, and confirm knowledge. The main discussion points are knowledge, its origins, and the limits of knowing [14]. Despite their similarity, these concepts center on separate lines of investigation. For instance, a possible ontological question for climate change research is, "is climate change a real phenomenon in and of itself, or is it a product of human meaning and social circumstances?" According to realist ontology, climate change is happening regardless of how humans feel about it and is already affecting ecosystems and weather patterns [15]. Conversely, a constructivist ontology posits that political, cultural, and social frameworks influence our perception and response to climate change. Ontology seeks to define what is considered accurate, whereas epistemology aims to comprehend how we know and understand reality. Taken as a whole, they shape researchers' worldviews, affecting how they approach their work, what methodologies they utilize, and how they interpret their results [16]. Engaging with these philosophical aspects is vital for consistent, thorough, and philosophically grounded academic research. To illustrate, within the same climate change research, one may raise the epistemological question, "how do we gain knowledge about climate change?" An optimistic epistemology would prioritize empirical data such as temperature records and atmospheric carbon dioxide levels as objective facts. In contrast, an interpretivist epistemology may look at how various cultures understand and describe the effects of shifting weather patterns, considering that knowledge is inherently subjective and situational.

2.1 Knowing Epistemology: The Character and Process of Knowledge Acquisition

To answer fundamental problems at the heart of intellectual inquiry, epistemology, which is the philosophical study of nature, origins, and limitations of knowledge, aims to answer the following question: what constitutes knowledge? What kinds of processes led to its acquisition? And how can one differentiate between a justified belief and a subjective opinion? These inquiries are the foundation of philosophical and scholarly investigations and significantly impact the procedures and viewpoints in various academic fields. Several important schools of thought have emerged due to philosophical engagement with epistemology. Each school of thought provides distinctive perspectives on knowledge's origins, validity, and use. Empiricism is a philosophical school of thought that stresses the importance of sensory experience as the primary source of knowledge. It strongly emphasizes observation, measurement, and experimentation as vital means for gaining empirical evidence [17]. This methodology serves as the foundation for the natural sciences, founded on rigorous testing and systematic observation to confirm hypotheses and provide evidence for theoretical claims. Reason and logical inference, on the other hand, are significant sources of information, according to the rationalist movement. René Descartes and other proponents of the idea that certain truths are intrinsic or accessible via logical thinking, irrespective of empirical observation, are among those who argue for this idea [18]. For the most part, rationalist ideals are aligned with disciplines, such as mathematics and theoretical philosophy, which give precedence to abstract thinking as a way of discovering universal truths. By stating that knowledge is not passively acquired but instead actively built via human experiences and interactions with the environment, constructivism presents a dynamic worldview based on the idea that information is not passively received [19]. The fact that this framework emphasizes the contextual and subjective aspects of knowledge makes it particularly relevant in subjects, such as sociology and education, where the emphasis is placed on the interaction between social context and individual comprehension. According to pragmatism, knowledge is evaluated according to its practical utility. Pragmatism contends that the validity of a concept is decided by its capacity to confront issues encountered in the actual world and accomplish the objectives sought. Pragmatism is a philosophical school of thought that emphasizes the relevance of ideas in practical circumstances by shifting the focus from the abstract truth to the functional applicability of ideas. Table 1 delineates essential epistemological approaches—empiricism, rationalism, constructivism, and pragmatism—and their impact on research viewpoints. It emphasizes their definitions, practical uses, and related issues across several fields. Empiricism prioritizes sensory experience and observation, whereas rationalism depends on reason and logic. Constructivism and interpretivism emphasize subjective settings, whereas pragmatism prioritizes usefulness. The difficulties column highlights constraints such as reconciling subjectivity, generalizability, and theoretical depth in the study.

For example

Epistemological viewpoints significantly impact the methodology and results of artificial intelligence [27]. Empiricism emphasizes data-driven methods, such as machine learning, which enables exact predictions and pattern identification. However, empiricism is not infallible due to the need for good-quality datasets and possible bias [28]. Rationalism is the foundation of organized reasoning in

Table 1 Essential epistemological methods: meanings, uses, and difficulties

Aspect	Description	Example/ application	Challenges
Definition [20]	The study of nature, sources, and boundaries of knowledge. It seeks to answer: what is knowledge? How is it acquired? What separates opinion from justified belief?	Foundational for all intellectual inquiry and research design	Balancing philosophical coherence with practical research needs
Empiricism [21]	Knowledge comes from sensory experience. Empirical evidence is gathered through observation, measurement, and experimentation	Natural sciences rely on empirical methods to test hypotheses (e.g., experiments in physics or biology)	Limited in addressing abstract concepts or phenomena not observable through sensory means
Rationalism [22]	Reason and logical deduction are the primary sources of knowledge. Truths can be self-evident or intrinsic, independent of sensory experience	Mathematics and theoretical sciences use rationalist approaches (e.g., proving a theorem in geometry)	Can overlook empirical evidence or real-world applicability in favor of abstract reasoning
Constructivism [23]	Individuals actively construct knowledge through their experiences and interactions with their environment. Emphasizes subjective and contextual information	Sociological studies focusing on cultural narratives or educational frameworks emphasize constructivism	Subjectivity can lead to challenges in establishing generalizable or widely accepted conclusions
Pragmatism [24]	The value of knowledge lies in its practical application. An idea's worth is measured by its usefulness in solving problems or achieving goals	Practical disciplines like engineering or business strategy prioritize pragmatism for real-world solutions	Can prioritize utility over theoretical depth, risking oversimplification of complex issues
Positivist epistemology [25]	It aligns with empiricism, holding that knowledge is objective and can be discovered through systematic observation and experimentation	A scientific experiment measuring the effect of a drug on health outcomes	It may dismiss subjective experiences or social contexts critical for understanding complex phenomena
Interpretivist epistemology [26]	It aligns with constructivism, emphasizing understanding subjective experiences and the meanings people assign to their actions	Ethnographic research examines how different cultures perceive and respond to natural disasters	It can struggle with reproducibility and objectivity, making it harder to validate findings universally

symbolic artificial intelligence and knowledge graphs. While it helps explainability, it has difficulties in dynamic and unstructured contexts. Adaptive learning approaches, such as reinforcement learning, are supported by constructivism [29]. These methods perform very well in changing situations but need many computer resources. The pragmatic approach emphasizes actionable insights, such as those found in decision-support systems, to resolve issues encountered in the real world, often at the expense of theoretical accuracy. The evolution of artificial intelligence is guided by these frameworks, which strike a balance between inherent capabilities and limits [30]. Table 2 highlights how epistemological perspectives shape AI methodologies. Empiricism drives data-centric approaches like machine learning, enabling pattern recognition but requiring extensive datasets. Rationalism emphasizes logical reasoning in systems like knowledge graphs, offering explainability but limited adaptability. Constructivism inspires adaptive learning, excelling in dynamic contexts yet computationally demanding. Pragmatism focuses on real-world utility, exemplified by decision-support systems, delivering practical solutions but sometimes sacrificing theoretical depth. These perspectives guide AI design by balancing strengths and limitations across applications.

2.2 *Analyzing Ontology: The Essence of Reality*

Ontology, the philosophical examination of existence and reality, addresses essential inquiries such as: what entities are present? What types of entities are they? And what is the nature of the interrelationships among these entities? Ontology addresses these inquiries, establishing the conceptual basis for comprehending reality and how it is seen and examined [34]. It influences both theoretical frameworks and the techniques utilized across many fields. Two dominant ontological viewpoints in philosophical and academic debate are realism and constructivism. These viewpoints present divergent interpretations of reality, shaping the pursuit and understanding of knowledge.

Realists contend that reality exists autonomously from human perception or interpretation. From this viewpoint, the world has an intrinsic, objective structure that may be revealed via observation and analysis [35]. Realists contend that the natural sciences are primarily based on this ontological perspective, as they seek to reveal universal rules and principles that govern events. The discipline of physics aims to comprehend unchanging principles such as gravity, regarding them as objective facts that exist independently of human perception or belief [36]. In research, realism frequently drives the application of quantitative approaches aimed at identifying patterns, testing ideas, and generating repeatable outcomes. A realism approach in environmental research may concentrate on quantifying pollution levels throughout ecosystems to ascertain universal criteria for ecological health [37]. This viewpoint supports the idea that methodical experimentation and observation enable researchers to uncover the world's objective reality, devoid of subjective bias. Conversely, constructivism, or relativism, asserts that reality is not static or autonomous but

Table 2 Epistemological perspectives and AI development

Epistemological perspective	Core principles	AI methodologies	Key benefits of AI	Limitations in the AI context	Examples in AI
Empiricism [31]	Knowledge arises from sensory experience, emphasizing observation, measurement, and experimentation	Machine learning (supervised/unsupervised), data mining, neural networks, and statistical models	Enables data-driven insights, accurate predictions, and pattern recognition based on large datasets	It requires vast amounts of high-quality data and is susceptible to biases in data collection	Computer vision for facial recognition and natural language processing (NLP) for sentiment analysis
Rationalism [32]	Knowledge is derived from reason and logical deduction, emphasizing intrinsic truths independent of experience	Symbolic AI, rule-based systems, knowledge graphs, theorem-proving algorithms, and formal logic systems	It facilitates structured reasoning, enables the discovery of abstract patterns, and supports explainable AI models	Limited adaptability to dynamic, unstructured environments; struggles with uncertainty in inputs	AI theorem provers in mathematics, ontology-based knowledge representation for semantic search engines
Constructivism [29]	Knowledge is actively constructed through interaction with the environment, emphasizing context and adaptability	Reinforcement learning, interactive systems, adaptive algorithms, contextual machine learning models	Promotes learning through feedback, enhances adaptability to changing environments, and is context-sensitive	Learning can be slow and computationally intensive, prone to overfitting in complex, dynamic systems	Robotics for adaptive navigation, AI in personalized education platforms that adjust students' progress
Pragmatism [33]	Knowledge is evaluated by its practical utility and ability to solve real-world problems effectively	Decision-support systems, optimization algorithms, and applied machine learning for real-world challenges	Focuses on actionable insights, practical applications, and outcome-driven design	It may sacrifice long-term accuracy or theoretical depth for immediate utility; it can over-prioritize usability	AI in healthcare (e.g., diagnostics), recommendation systems (e.g., Netflix, Amazon recommendations)

is shaped by human experiences, perceptions, and cultural circumstances. Constructivists argue that reality is socially constructed, allowing for the coexistence of numerous realities, each influenced by individual and communal viewpoints [38]. In the social sciences, constructivist techniques examine how people and organizations generate meaning within their cultural and historical settings. A constructivist ontological perspective frequently results in qualitative research methodologies prioritizing subjective experiences and interactions between people and their surroundings [39]. Constructivist research on workplace dynamics may examine employees' perceptions and interpretations of company culture, uncovering varied realities that affect behavior and decision-making. The ontological perspective of a researcher significantly affects the formulation of research topics and the selection of methodologies to investigate them [40]. A realistic viewpoint may prompt investigations aimed at revealing objective facts via controlled experiments and statistical analysis. In contrast, a constructivist view emphasizes the comprehension of subjective experiences and context-specific phenomena, frequently utilizing interviews, ethnographic investigations, or participatory action research. A realist researcher examining climate change may concentrate on discovering measurable patterns in global temperatures and carbon emissions to establish universal rules that govern environmental deterioration [41]. In contrast, a constructivist researcher may investigate how communities interpret and react to climate change, analyzing the subjective reality influenced by cultural, economic, and political issues. Ontology is essential for defining the comprehension and examination of reality. Realism and constructivism provide contrasting yet complimentary viewpoints that address distinct areas of investigation. Realism propels the quest for universal, objective truths, whereas constructivism underscores the significance of context, perception, and human experience. Collectively, these ontological perspectives enhance academic comprehension by providing varied methodologies for examining and comprehending the intricacies of life [42]. Table 3 contrasts realism with constructivism across fundamental dimensions. Realism perceives reality as objective, emphasizing universal truths through quantitative means appropriate for the sciences but frequently neglecting subjective experiences and societal influences. Constructivism posits that reality is socially produced, highlighting context-dependent comprehension using qualitative approaches, which are suitable for social sciences but susceptible to biases and restricted generalizability. Both methodologies need help with study scope and methodological application.

For example

Table 4 examines the impact of various ontological viewpoints on the approaches, advantages, and problems associated with artificial intelligence (AI) applications. The realist school of thought operates on the assumption of an independent reality and employs supervised learning for tasks such as illness diagnosis through medical imaging. However, it needs help dealing with contextual complexities. The constructivist perspective sees reality as something that is socially built. It uses participatory design for systems such as culturally sensitive chatbots, but it also runs the danger of

Table 3 Ontological perspectives in research and applications: realism versus constructivism

Aspect	Description	Example/application	Challenges
Core principles [43]	Realism views reality as existing independently of perception, while constructivism sees reality as socially constructed	Realism: laws of nature (e.g., gravity) Constructivism: cultural interpretations of historical events	Realism: overlooks subjective experiences Constructivism: struggles with generalizability
Nature of knowledge [44]	Realism emphasizes universal truths; constructivism focuses on context-dependent, subjective understanding	Realism: scientific laws in physics Constructivism: community-specific knowledge in anthropology	Realism: assumes universality across contexts Constructivism: susceptible to cultural and personal biases
Research questions [39]	Realism seeks universal principles; constructivism investigates meaning-making in specific contexts	Realism: what are the universal causes of climate change? Constructivism: how do communities interpret climate change impacts?	Realism: ignores localized insights Constructivism: need help to produce predictive models
Methodologies [45]	Realism employs quantitative approaches; constructivism uses qualitative techniques	Realism: experiments and statistical analysis Constructivism: ethnographic studies and interviews	Realism: limited adaptability to subjective nuances Constructivism: resource-intensive data collection
Applications [46]	Realism excels in natural sciences; constructivism thrives in social sciences	Realism: measuring pollution levels in ecosystems Constructivism: studying workplace culture perceptions	Realism: can miss sociocultural factors Constructivism: results may not apply broadly

bias in datasets. Even though it requires many resources, critical realism blends objective and social elements by employing mixed methodologies for social policy research powered by artificial intelligence [47]. While postmodernism challenges solitary realities, it fosters innovation through generative models such as DALL·E. However, it encounters issues in terms of validation. The pragmatic approach emphasizes practical outcomes, such as applying reinforcement learning to real-time systems such as autonomous vehicles; nevertheless, constant updates may result in a loss of theoretical depth. Every point of view forms AI in its unique way.

Table 4 Ontological perspectives in AI development

Ontological perspective	Core principles	AI methodologies	Key benefits of AI	Limitations in the AI context	Examples in AI
Realism [48]	Reality exists independently of human perception, emphasizing universal truths and objective structures	Supervised learning, statistical models, and predictive analytics	High accuracy in well-defined, objective domains Scalable across large datasets with minimal adjustments	Struggles with subjective and contextual nuances It may overlook human factors such as ethics or fairness	Training image recognition models to detect diseases (e.g., AI-powered radiology)
Constructivism [49]	Reality is socially constructed and shaped by human experiences and cultural contexts	Context-aware systems, participatory design, and interpretive AI algorithms (e.g., explainable AI)	Flexible in adapting to diverse cultural and social contexts Provides insights into subjective human experiences	Dataset biases can affect outcomes Higher complexity in training and contextual validation	Chatbots are designed for culturally sensitive communication Personalized AI tutors in education systems
Critical realism [50]	Acknowledges objective reality but emphasizes its socially mediated and layered understanding	Hybrid approaches combine quantitative models with qualitative data (e.g., mixed-methods AI)	Bridge's objective analysis with subjective context Provides nuanced insights into complex problem-solving	It can be resource-intensive due to dual methodological demands Challenges in balancing objectivity and subjectivity	AI systems for public policy analysis, combining statistical predictions with community input

(continued)

Table 4 (continued)

Ontological perspective	Core principles	AI methodologies	Key benefits of AI	Limitations in the AI context	Examples in AI
Postmodernism [51]	Questions the existence of a single, objective reality, emphasizing pluralism and multiple perspectives	Generative models, exploratory AI systems, and adaptive neural networks	Encourages innovation through diverse perspectives Suitable for creative AI applications	It is challenging to validate outcomes Risk of fragmented or inconsistent interpretations	AI-generated art systems (e.g., DALL·E) AI-driven narrative building for personalized entertainment
Pragmatism [52]	Focuses on practical outcomes and actionable reality rather than theoretical absolutes	Reinforcement learning, adaptive AI systems, and real-time decision-making algorithms	Highly effective in dynamic, real-world applications Promotes iterative improvement and user-centered design	May sacrifice theoretical robustness for practical utility Requires continuous fine-tuning and updating	Self-driving cars using real-time sensor data AI for supply chain optimization in logistics

2.3 How Research Views are Shaped by Ontology and Epistemology?

The relationship between epistemology and ontology significantly impacts the theoretical stance of a researcher, the decisions they make about methodology, and the interpretations they get from their findings [53]. When taken as a whole, these concepts provide the foundation of a researcher's perspective, guiding each stage of the study by giving direction. Assembling questions for an investigation: how scientists define the nature of the phenomena that they are examining is influenced by the ontological presumptions that they make. A realist might ask, "what is the effect of X on Y?" assuming an objective link between the two variables. On the other hand, constructivists may study the question, "how do individuals perceive and experience X?" emphasizing the importance of subjective reality. Picking the right methodologies: The tools and methods utilized for data gathering and analysis are determined by the epistemological perspectives held [54]. Optimistic researchers grounded in empirical evidence usually employ quantitative methods, such as conducting experiments, surveys, and statistical analysis. Interpretative researchers may investigate the complexity of human experiences using qualitative methods such as ethnography, discourse analysis, and interviews. Interpreting the results: The researcher's ontological and epistemological viewpoints impact the data's analysis and interpretation. Compared to a realist, a constructivist would emphasize that context, meaning, and human perspectives are more critical than generalizability and absolute facts [25]. By critically scrutinizing epistemological and ontological presuppositions, researchers guarantee that their work is methodologically sound and philosophically consistent without compromising these aspects. They are connected to this, making their findings and the knowledge they contribute more trustworthy. The concepts of epistemology and ontology are fundamental to inquiry's intellectual foundations. The study of ontology is concerned with reality, whereas epistemology is concerned with the nature of knowledge and how it is acquired [55]. Designing investigations, selecting methods, and interpreting data provide the philosophical underpinnings that give the basis. A profound understanding of these concepts contributes to improving academic inquiry. It ensures that research continues to be rigorous, reflective, and significant across all fields of scientific study [56].

3 Philosophical Paradigms in Research

Within the realm of research, philosophical paradigms serve as crucial frameworks that direct researchers' approach, plan, and interpret their investigations [57]. The many existing paradigms indicate unique ways of comprehending reality, knowledge, and methodologies suitable for conducting investigations into the world. Positivism, interpretivism, critical theory, and pragmatism are some of the paradigms that have recently gained the most prominence. Each of these paradigms provides researchers with a distinct perspective through which they might see their work, influencing the selection of study design, technique, and interpretation of findings. A fundamental tenet of positivism is the conviction that reality is objective and exists apart from the subjective viewpoints of individuals [58]. The focal points of this way of thinking are empirical observation and the discovery of universal rules via quantifiable facts and procedures that can be replicated. Employing quantitative approaches like experiments and statistical analysis, researchers working under this paradigm concentrate on causality and generalizability as their primary areas of inquiry [59]. Positivism is the dominant philosophical school, especially in the scientific sciences, where the pursuit of observable and objective facts is the foundation for inquiry. The concepts of determinism, objectivity, and the idea that only one reality can be observed are fundamental to positive philosophy. The interpretivist viewpoint, on the other hand, contends that reality is a social construction shaped by human perception and the surrounding environment. This paradigm's objective is to understand the subjective experiences and interpretations that individuals accord to their activities and the environment in which they find themselves. Interpretivist researchers often use qualitative research techniques to investigate intricate social phenomena, such as interviews, drawing ethnographies, and conducting case studies. In contrast to positivism, which seeks to formulate universal generalizations, interpretivism places a premium on in-depth comprehension and contextual awareness. When it comes to fields like sociology, engineering, and education, where the emphasis is placed on capturing the richness and diversity of human experiences, this paradigm is frequently connected with such fields [60].

By emphasizing the role that power, history, and social institutions play in the formation of knowledge and societal norms, critical theory poses a challenge to both positivism and interpretivism. The paradigm in question is intrinsically activist, with the objective of comprehending the world and bringing about change in it. It is common for researchers who adopt a critical theory viewpoint to investigate issues of oppression, marginalization, and inequality, particularly regarding gender, race, and social class. Critical ethnography, discourse analysis, and feminist research are examples of methodologies within this paradigm [61]. These approaches are inherently reflective and actively engage participants in the research process. The objective is to unearth and question prevalent beliefs to advance social justice and strengthen underrepresented groups. Pragmatism, on the other hand, is a paradigm that is problem-centered and practical, and it places a higher priority on the utility of research outputs than it does on strictly adhering to philosophical theories. Pragmatists contend that the value of research comes in its capacity to solve problems encountered in the real world and offer answers that can be implemented [62]. Methodological diversity is encouraged by this paradigm, which frequently combines qualitative and quantitative techniques to give a thorough knowledge of complicated situations. Pragmatism is essential in applied sectors like healthcare, business, and education, where the primary focus is finding solutions to real issues rather than conforming to rigid theoretical frameworks.

Selecting a philosophical paradigm has significant repercussions on the layout of research projects and the technique used [63]. When doing positivist research, the design is often organized and regulated, guaranteeing that the findings will be objective and reproducible. When testing hypotheses and determining the existence of causal linkages, researchers who adhere to this paradigm frequently depend on experimental designs, surveys, or longitudinal studies. On the other hand, interpretive research uses flexible and exploratory designs, such as ethnography or narrative inquiry, to capture the nuanced and context-dependent character of human experiences [64]. According to the very definition of critical theory research, it frequently incorporates action-oriented or participatory designs to address and transform societal injustices. In contrast, pragmatic research is characterized by its ability to modify its design in response to the situation. It usually employs mixed methods, combining qualitative and quantitative research methodologies.

In addition, the paradigm that is selected forms the procedures that are utilized in the study. Positivism emphasizes quantification to provide findings that can be applied to a broader population [65]. These methodologies employ methods such as statistical analysis and experimental testing. Interpretivist approaches, on the other hand, emphasize qualitative methods such as participant observation and in-depth interviews to understand subjective meanings. Critical theory approaches frequently require introspective and dialogical techniques, such as critical discourse analysis, to bring attention to power dynamics and advocate for change. Because of their intrinsic adaptability, pragmatic methods allow the selection of instruments based on their relevance and utility to the study aims [66].

How researchers interpret their findings is impacted by the philosophical frameworks they use. Research that adheres to positive philosophy emphasizes objectivity, dependability, and generalizability. Interpretivist scholars, on the other hand, prioritize comprehending meanings and experiences that are distinctive to the context. Those who engage in critical theorizing examine their results through the lens of social criticism to bring to light challenging inequalities. On the other hand, pragmatists are concerned with practical applications and interpret discoveries in terms of their ramifications in the actual world and their significance to various stakeholders [62]. Philosophical paradigms serve as a basis for all areas of the research process, from conceptualizing the problem to interpreting the data gained from the investigation. Researchers can ensure coherence and methodological rigor by aligning it with a particular paradigm. This allows them to meet their research objectives while doing their study. Whether the focus is on objective facts, subjective experiences, social critique, or practical solutions, these paradigms provide a variety of valuable and diverse avenues for generating knowledge and driving meaningful development [67]. Figure 1 depicts the link between ontology and epistemology, indicating how both disciplines affect the research process. The questions researchers ask are influenced by ontology, which studies the nature of reality. For example, ontology determines whether the focus should be on objective realities or subjective experiences. Epistemology, concerned with the nature of knowledge, is the guiding principle behind selecting techniques. These methodologies can range from quantitative methods, such as surveys, to qualitative approaches, such as interviews. When taken as a whole, these philosophical foundations govern how findings are understood, whether the emphasis is placed on generalizability or context and meaning. When researchers

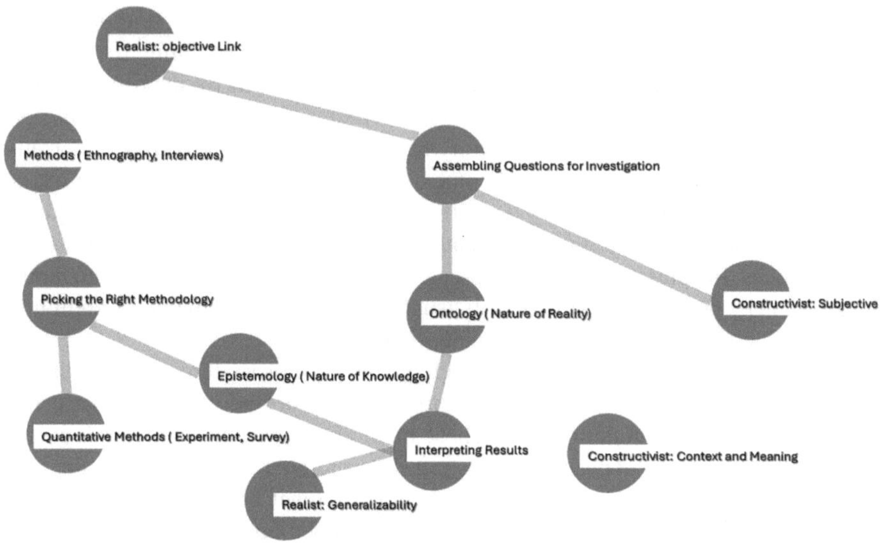

Fig. 1 The interconnection between ontology and epistemology in research (*Source* Authors)

ensure that these notions are consistent, they can construct a methodologically sound framework, increasing the dependability and relevance of their work.

4 Ethics and Philosophical Implications: Integration of Philosophy into Research Practice

Philosophy is the foundation of ethics, offering a comprehensive framework for comprehending and implementing moral concepts in various situations, including research and social settings [68]. Researchers should perform their work responsibly, with transparency and respect for everyone involved. Ethical principles, originating in philosophical traditions, assist researchers in this endeavor [69]. Not only are these values necessary for preserving the honesty of research, but they also play a significant part in establishing trust and responsibility not just inside the academic community but also outside it. By incorporating philosophy into research activity, one can better comprehend the ethical theories that are the foundation for ethical rules. Different points of view can solve possible ethical problems [70]. These include ideas from deontology, which focuses on duty and following the law, and utilitarianism, which focuses on maximizing overall benefits. Virtue ethics, emphasizing moral virtues and individual character, adds complexity to the decision-making process. Researchers can address complex problems with clarity and moral foresight if they match their research procedures with these philosophical traditions [71]. This conversation will examine ethics' philosophical foundations, research application, and broader implications for academia and society. An intellectual approach enables critical thought on the societal effect of research, including concerns of equality, inclusion, and social justice. This is in addition to the fact that it ensures conformity with professional ethical standards. Ethical research is not a stand-alone phenomenon; it influences the public's faith in scientific findings, contributes to public policy formulation, and tackles social issues [71]. Including philosophy in research procedures ensures academic work is rigorous, reputable, and socially responsible. This alignment helps cultivate a culture of empathy, justice, and responsibility, enabling researchers to produce meaningful contributions beyond academia. These contributions benefit society and promote the responsible progress of knowledge.

4.1 Ethical Principles Rooted in Philosophy

In addition to providing formal frameworks to address concerns of morality, justice, and responsibility, philosophical traditions have significantly impacted the formation of ethical standards; this influence has been profound [68]. Using these concepts as essential guidelines is necessary to navigate complicated situations involving right and evil successfully. Important schools of philosophy, such as deontology, virtue ethics, and utilitarianism, have laid the groundwork for modern ethical practices, particularly in research and professional behavior. Deontology emphasizes duty and conformity to laws, virtue ethics emphasizes character and moral qualities, and utilitarianism focuses on seeking the most significant benefit for most people [72]. When taken as a whole, they provide a powerful ethical compass that guides conduct and decision-making.

4.1.1 The Concept of the Greater Good and Utilitarianism

Jeremy Bentham and John Stuart Mill are two philosophers who contributed to developing the philosophical theory known as utilitarianism [73]. This theory argues for acts that maximize overall benefit while minimizing detrimental effects. This notion of the "greater good" is highly pertinent in engineering and architecture, where decisions can significantly influence communities broadly. For instance, the design and building of sustainable infrastructure prioritizes protecting the environment and the long-term benefits to society, even if the initial expenditures are higher. Similarly, safety innovations in building bridges or skyscrapers consider the dangers to workers and possible hazards while also considering the requirement to produce a sturdy and dependable structure for public use [74]. To create solutions that improve the quality of life for many people while also addressing issues regarding safety, cost, and the environment, architects and engineers embody utilitarian concepts. These principles include prioritizing designs that are energy-efficient, accessible spaces, or structures that are resistant to disasters. The examples presented here have insight into utilitarianism's role in influencing ethical practices in various areas.

4.1.2 Ethics That are Based on Duties and Deontology

The philosophical approach known as deontology, based on Immanuel Kant's ideas, emphasizes the observance of moral and ethical principles and obligations [75]. Rather than focusing on the results of activities, this ethical paradigm emphasizes the actions themselves and their inherent significance. In engineering and architecture, deontology emphasizes professional commitment to enforce safety standards, defend public welfare, and guarantee justice. For instance, engineers must build structures emphasizing safety and longevity, even if that means taking shortcuts that might lower costs or boost profits. In the same vein, architects are obligated to take accessibility and inclusivity into consideration, making sure that designs respect the dignity and requirements of all users [76]. In the research context, this approach is expressed in the obligation to protect participants' autonomy, ensure confidentiality, and get informed permission. Deontology emphasizes the significance of preserving the rights and dignity of people over prospective advantages or advances in efficiency. This ensures that ethical concerns remain prominent in professionals' decision-making and practice [77].

4.1.3 Integrity of Character and Ethics of Virtue

Aristotle's philosophy is the foundation of virtue ethics, which emphasizes the development of moral character and qualities such as honesty, integrity, empathy, and responsibility during one's lifetime [78]. In contrast to ethical frameworks founded on rules, virtue ethics emphasizes the individual's character and the motivations behind acts [79]. Taking this point of view empowers experts in engineering and architecture to demonstrate virtues in their work, which in turn helps to cultivate trust and accountability. As an illustration, engineers who exhibit integrity by maintaining openness in structural evaluations can develop confidence with stakeholders. Similarly, architects who demonstrate empathy may build inclusive places that cater to the requirements of a wide range of users. Regarding research, this method goes beyond merely adhering to ethical norms; instead, it promotes a culture in which ethical action results from personal virtue rather than necessity [80]. Researchers who comply with this idea cultivate trust, dependability, and respect in their work, contributing to a more comprehensive culture of integrity that encourages significant breakthroughs while preserving public confidence and ethical responsibility [81].

4.1.4 Fairness and Impartiality

Fairness and impartiality are two values that are firmly ingrained in research ethics. These standards originate in philosophical justice concerns, notably those that John Rawls has investigated [82]. To ensure that no group is subjected to a disproportionate amount of burden or exploitation, these guidelines emphasize the equal treatment of all people who participate in research. When doing research that adheres to ethical standards, recruiting volunteers using fair criteria is necessary, avoiding any biases or methods that might potentially exploit vulnerable or disadvantaged communities [83]. When applied to engineering and architecture, this notion means ensuring everyone has equal access to technological developments or infrastructural initiatives. As an illustration, urban development plans should avoid marginalizing neighborhoods with lower incomes and instead prioritize inclusive designs that benefit all individuals present. Fairness needs increased attention to prevent exploitation among participants in global research settings, which frequently involve power asymmetries and discrepancies in resource availability. By adhering to these values, researchers and professionals can support ethical procedures that protect justice, equity, and respect for all stakeholders participating [84]. For example, in passive building research, equity and inclusion may be manifested by emphasizing designs that are both energy-efficient and accessible, as well as affordable for all economic brackets. When constructing passive house prototypes, researchers should prioritize integrating locally sourced, economical materials and technologies that do not disproportionately advantage wealthier areas to the detriment of underprivileged populations. An illustration would be a research project focused on developing passive cooling systems for residential structures in economically disadvantaged metropolitan regions. Researchers can develop egalitarian and scalable solutions by involving local stakeholders, comprehending regional climate circumstances, and utilizing cost-efficient solutions (e.g., natural ventilation techniques and reflective roofing materials) [85]. These designs guarantee that energy efficiency and comfort are available to a broader demographic, promoting equity in technical progress and alleviating inequalities in living conditions.

4.2 Integrating Ethical Considerations into the Research Process

Ethical norms are not just abstractions but vital and practical elements integrated into every phase of the research process [86]. Their integration guarantees that research methodologies and results conform to ethical standards and societal norms, promoting trust, transparency, and responsibility. At the commencement of every research project, ethical concerns inform the formulation of research questions, ensuring they tackle pertinent and significant topics without damaging participants or communities. In the design phase, approaches are examined to ensure they are inclusive, respectful of multiple viewpoints, and do not exploit or marginalize any group. During the implementation phase, researchers follow guidelines that safeguard privacy, secure informed permission, and ensure equity in participant selection [87]. Data analysis and interpretation are performed with integrity, eschewing manipulation or distortion of results to conform to preconceived narratives. The dissemination of research outputs is conducted ethically, providing accessibility and preventing the establishment of knowledge hierarchies that exclude specific audiences. By integrating ethical considerations consistently, researchers comply with regulatory mandates and honor the broader obligation of fostering equitable and social progress [88]. This demonstrates a commitment to integrity and respect for all stakeholders engaged in the research process. In the realm of building energy efficiency research, ethical integration may entail devising solutions that are both inexpensive and accessible to economically disadvantaged areas. Instead of concentrating exclusively on sophisticated, expensive technology for wealthy markets, researchers should emphasize the creation of affordable energy-efficient retrofits, such as enhanced insulation with locally produced materials. During the implementation phase, these ideas will be evaluated by various community stakeholders to guarantee practical application and fair advantages. The study and distribution of outcomes will emphasize scalable and inclusive methodologies, circumventing knowledge monopolies. By integrating ethical issues consistently, researchers fulfill their obligation to promote equitable and social progress, demonstrating a commitment to honesty and respect for all parties engaged [89].

4.2.1 Ethical Planning and Design of Projects

It is necessary to conduct a thorough analysis of ethical concerns throughout the beginning stages of the research process since ethical planning and design are the core components that ensure the integrity and rigor of the research process. Researchers must conduct thorough risk–benefit assessments; establish protocols to safeguard participants' rights, safety, and confidentiality; and secure official approval from institutional review boards (IRBs) to ensure adherence to ethical and legal standards [90]. The development and execution of informed consent forms are essential to this process. These forms offer participants unequivocal and comprehensive information regarding the aims of the study, the methodology, and any risks or benefits involved with the research. This method upholds fundamental concepts such as autonomy, transparency, and respect, fostering ethical accountability and protecting participants' rights.

4.2.2 Accumulation of Information and Interaction with Participants

Ethical rigor is essential in participant contact and data collection, as both phases significantly impact the integrity and legitimacy of research efforts [91]. Researchers must adhere to fundamental ethical norms, including respect for participants' autonomy, personal information confidentiality, and communication transparency. Data-gathering approaches should be crafted to reduce intrusion and mitigate harm, ensuring the handling of sensitive information aligns with the highest responsibility and security standards. Coercion and undue influence must be meticulously avoided to maintain the voluntariness of participation and ensure fully informed consent. Thorough communication about the research's aims, techniques, and implications is crucial for fostering trust and understanding. Compliance with these ethical standards protects participants' rights and welfare while enhancing the academic integrity of research, ensuring results are both morally valid and scholarly credible [92].

4.2.3 Interpretation and Analysis of the Data

Keeping the legitimacy and integrity of academic research intact requires considering the ethical considerations involved in interpreting and presenting findings [93]. Researchers must portray the results accurately and ensure that interpretations remain impartial and free from personal bias. Because transparency is of the utmost importance, disclosing any potential conflicts of interest that might undermine the idea that the study is unbiased is necessary. In addition to violating professional norms, ethical violations, which include fabricating, falsifying, or selectively representing results, also undermine the credibility of the scientific endeavor. When these ethical values are adhered to, the dependability of conclusions is strengthened, the rigor of scholarly contributions is enhanced, and the integrity of the academic community is preserved [94]. Adhering to such norms is necessary to generate methodologically strong and morally sound information.

4.2.4 Spreading and Influence on Others

Underscoring the necessity of fair and responsible communication is the fact that the ethical duties of researchers extend to the dissemination and application of their results [95]. To democratize information and increase its influence on society, it is vital to provide accessibility through open-access efforts and to develop inclusive public involvement. Researchers must thoroughly evaluate the more significant implications of their work, especially in morally complicated fields like genetics, public health, and artificial intelligence, where discoveries may bear substantial and far-reaching effects. An assessment of this nature is critical. Regarding dissemination strategies, intellectual integrity, methodological transparency, and accessibility to a wide range of audiences should be prioritized. This will ensure that alignment with ethical norms is not compromised. Making this commitment improves research's practical value and societal relevance and reduces the dangers of misinterpretation, abuse, or unintentional harm in crucial application areas [96].

4.3 *Implications for Philosophy and Their Wider Scope of Influence*

Not only can the incorporation of ethical and philosophical ideas into research processes have a significant impact, but it can also shape individual studies and the larger academic and societal frameworks that research functions within. To guarantee that research procedures and results are up to the highest intellectual and ethical integrity standards, these foundations offer a framework for scientific rigor and moral accountability [97]. This alignment enhances the legitimacy of academic activities, fostering confidence among participants, stakeholders, and the public. Philosophical concepts urge researchers to analyze the influence that their work has on society critically. This helps researchers match their goals with broader objectives such as social equality, environmental sustainability, and public health. This alignment encourages the development and implementation of inclusive, egalitarian, and transparent approaches, thereby fostering the production of knowledge that is both responsible and influential. Philosophical frameworks provide vital tools for resolving new ethical problems in rapid evolution domains, such as artificial intelligence, genomics, and data analytics [98]. These guiding principles allow the responsible growth of innovation while simultaneously retaining ethical monitoring. They achieve this by providing a systematic approach to evaluate and address innovative challenges. Ultimately, incorporating philosophy into research benefits society by opening new intellectual conversations and helping make essential discoveries [99]. All these things contribute to the continued growth of knowledge and the progress of society. For example, philosophical issues in building research affect the balance between quantitative assessments, such as how much less energy is used after retrofitting, and qualitative insights into how people perceive and use energy-saving technologies. Integrating these dimensions ensures that technical solutions are effective, socially acceptable, and widely adoptable, advancing equity in sustainable practices.

4.3.1 Increasing One's Credibility and Conviction

Establishing a foundation of trust among those who participate in research, those who collaborate with researchers, and the public is essential to accomplishing academic activities [100]. The strict commitment to ethical norms in research activity is a crucial factor contributing to the considerable enhancement of this confidence. When philosophical ideas are included in the planning and execution of research, ethical considerations are incorporated as core aspects rather than being considered as concerns that are extraneous or superficial. Such an approach increases scientific inquiry's epistemic and moral validity to ensure scientific investigation is based on larger community ideals. In addition, the researcher's reputation and professional integrity are improved due to this profound dedication to ethical rigor, which fosters confidence among academic colleagues and stakeholders. Ultimately, this alignment of ethical and methodological rigor contributes to advancing the reputation and effect of science [101].

4.3.2 Fostering a Sense of Social Responsibility

Ethically grounded philosophical frameworks provide researchers with a basis to critically assess the societal implications of their work critically, ensuring that research goals are consistent with overarching societal needs [102]. This connection underscores accountability and prioritizes resolving urgent global issues like social justice, environmental sustainability, and public health fairness. In ecological research, advancing energy-efficient building solutions must encompass technological innovation and equal access for marginalized people. This strategy mitigates inequities while promoting ecological objectives. By integrating social responsibility into research frameworks, researchers facilitate generating information that supports inclusive growth and communal welfare [103]. This integration guarantees academic progress is academically demanding, socially transformational, and ethically robust.

4.3.3 Serving as a Guide for New Fields

The fast progression of technology and scientific developments has generated intricate ethical challenges, requiring robust philosophical frameworks for successful resolution [104]. These frameworks facilitate a systematic and consistent methodology for addressing ethical problems in developing fields such as artificial intelligence, gene editing, and big data. In these domains, algorithmic bias, genetic equality, and data privacy require careful consideration. In artificial intelligence, the ethical ramifications of decision-making algorithms necessitate examination to guarantee fairness, transparency, and accountability. By anchoring these investigations in established philosophical traditions, scholars may traverse the complex ethical landscape with intellectual rigor and critical depth. Integrating philosophical inquiry with technical innovation promotes responsible research methods and guarantees that breakthroughs accord with social ideals and ethical principles [105].

4.3.4 Promoting Reflective Practice Encouragement

Researchers are encouraged to engage in reflective practice, which involves questioning their motives, assumptions, and obligations when ethics and philosophy are integrated into their working practices. Both the academic community and society stand to profit from this reflection, which ultimately results in more careful research [106].

To do responsible research, it is essential to understand the link between ethics and philosophy. Researchers are provided with the ethical framework that guides them in creating, conducting, and disseminating their study. Philosophical concepts such as utilitarianism, deontology, and virtue ethics are examples of such principles. By incorporating these principles into research practice, one may ensure that studies are methodologically sound and responsible from a moral and social standpoint. Researchers contribute to the progress of knowledge in a manner that respects the rights of individuals, promotes the well-being of society, and confronts the pressing concerns of our day by cultivating a culture of ethical inquiry and action [107]. Table 5 comprehensively summarizes philosophy's multidimensional role in determining ethical research techniques. For example, it shows how fundamental ethical theories like utilitarianism, deontology, virtue ethics, and justice frameworks can solve moral problems in various research settings. Specific practical applications, like assuring inclusion, fostering sustainability, and protecting participant rights in research and engineering, align with the many philosophical foundations. The table also emphasizes how these frameworks affect upcoming issues, notably in fields such as genetic research, artificial intelligence, and global sustainability. It stresses that ethical issues should be considered at all stages of research, from planning to sharing the results, to ensure that it is responsible, fair, and in line with society. Broader implications underscore the transformative potential of ethics-motivated research. These implications include the promotion of reflective practice and the facilitation of global collaboration. The table generally indicates that incorporating philosophical rigor into ethical research processes helps cultivate trust, inclusion, and social advancement. This helps to ensure that research continues to be relevant and responsible in the face of growing global concerns.

5 Applying Philosophical Foundations to Research

The conceptualization of research issues and selection of acceptable procedures are significantly influenced by philosophical foundations, which play a crucial role in both processes [118]. From the perspective of epistemology, which studies the nature and extent of knowledge, and ontology, which investigates the nature of reality, these foundations offer a logical and systematic framework for academic investigation. Academics can guarantee their work's methodological rigor, consistency, and ethical integrity by aligning their study with these philosophical ideals. The alignment in question makes it easier to comprehend the research setting more deeply. It encourages critical reflection on the assumptions that form the basis of the investigation [119]. Furthermore, this method assures that the study is relevant to society by addressing practical requirements while conforming to theoretical and ethical criteria for research. As a result, incorporating philosophical underpinnings contributes to increased legitimacy, validity, and overall influence of academic research.

Table 5 Foundations, uses, and social impact of ethical aspects

Dimension	Description	Philosophical foundation	Practical applications	Broader implications
Philosophical basis of ethics [108]	Philosophy offers the intellectual foundation for comprehending and implementing ethical values in study and practice	It is grounded in deontological ethics, utilitarianism, virtue ethics, and theories of justice, such as Rawls' theory of fairness	Frameworks direct ethical rules for research, public policy, engineering standards, and professional conduct	Establishes a framework for resolving moral challenges, guaranteeing consistency in ethical thinking across several fields
Utilitarian ethics [109]	Stresses the optimization of society's advantages while mitigating damage; choices are assessed according to their consequences	Utilitarianism of Jeremy Bentham and John Stuart Mill	Engineering designs centered on sustainability (e.g., green infrastructure) and public health initiatives that equilibrate resources with social demands	Emphasizes the importance of long-term societal and environmental well-being, tackling issues related to resource distribution and global equality
Deontological ethics [110]	Emphasizes compliance with regulations, responsibilities, and ethical principles irrespective of outcomes, highlighting inherent moral duties	Immanuel Kant's deontology	Professional obligations in engineering (e.g., structural integrity), inclusivity-oriented architectural methodologies, and informed consent in research endeavors	Strengthens procedural integrity and upholds individual rights, frequently acting as the ethical foundation for regulatory norms and legal systems
Virtue ethics [111]	Emphasizes moral character, underscoring honesty, empathy, and integrity as the cornerstone of ethical behavior	Aristotle's virtue ethics	Enhancing trust by honest assessment in scientific inquiry, creating inclusive environments through empathic interaction, and cultivating responsibility in academic partnerships	Fosters a culture of ethical excellence by associating personal virtue with professional honesty, so establishing enduring trust in academic and societal environments

(continued)

Table 5 (continued)

Dimension	Description	Philosophical foundation	Practical applications	Broader implications
Justice and fairness [112]	Advocates for treating people and communities fairly and dividing up the advantages and disadvantages	John Rawls' theory of justice	Equitable recruitment in research, inclusive urban planning to prevent socio-economic marginalization and accessibility in infrastructural initiatives	Alleviates social imbalances and promotes inclusiveness, ensuring progress does not intensify existing disadvantages
Integration in research processes [113]	Incorporating ethical issues across all phases of research to guarantee integrity, equality, and societal unity	Interdisciplinary ethical theories (e.g., bioethics, research ethics)	Processes for informed consent, safeguards for data privacy, fair involvement of participants, and distribution via open-access platforms	Enhances research credibility, conforms to regulatory criteria, and guarantees relevance to various stakeholders
Emerging ethical challenges [114]	Emerging technical and scientific fields require novel, ethical frameworks to tackle unexpected challenges	Pragmatic and adaptive ethical theories (e.g., applied ethics)	Ethics of artificial intelligence (e.g., algorithmic bias mitigation), ethical parameters in genomic research (e.g., CRISPR), and privacy concerns about big data	Fosters innovation in ethical governance and policymaking, facilitating proactive answers to swiftly changing issues in science and technology
Reflective practice in research [102]	Researchers must rigorously evaluate their ethical assumptions, obligations, and social implications	Reflexive methodologies in social and natural sciences	Continuously assess research objectives against ethical standards, promote method accountability, and guarantee inclusiveness in research results	Promotes adaptable and responsible research methodologies, establishing a basis for ethical knowledge creation amid social and technological change

(continued)

Table 5 (continued)

Dimension	Description	Philosophical foundation	Practical applications	Broader implications
Ethics-driven dissemination [115]	It guarantees that discoveries are disseminated judiciously, considering accessibility, societal implications, and the risk of abuse	Ethics of knowledge dissemination (e.g., open science principles)	Publications with open access, transparent techniques, and involvement with impacted communities to contextualize findings	Publications with open access, transparent techniques, and participation with impacted communities to contextualize findings
Global implications [116]	Harmonizing ethical frameworks across cultures to tackle global concerns such as climate change, pandemics, and technology upheaval	Global ethics and cross-cultural philosophical discourse	Multilateral frameworks for equitable climate solutions, inclusive global health policies, and ethical artificial intelligence for different user demographics	Bridges cultural and disciplinary gaps, promoting a cohesive strategy for global concerns while honoring local settings and values
Societal transformation [117]	Ethical research promotes societal transformation by tackling equality, sustainability, and inclusion via multidisciplinary cooperation and philosophical precision	Critical theory and transformative ethics	Employing research to confront structural injustices, such as gender and racial biases in design, while promoting participatory approaches to elevate disadvantaged perspectives	Facilitates enduring social advancement by incorporating ethical foresight into implementable policies and technology, emphasizing inclusion and fairness

5.1 Influence on the Development of Research Questions and Methods' Analysis

The formulation of research questions and the selection of methodologies to answer those questions are profoundly impacted by philosophical perspectives during the research process [1]. For example, researchers operating under a positive paradigm will likely structure their queries around discovering causal linkages, forecasting outcomes, or measuring phenomena. This is because the positive paradigm argues that there is an objective reality that can be measured. These questions are specific and are intended to produce insights that may be applied to a broader context. The production of empirical evidence is the goal of most of the procedures utilized in positivist research [120]. These approaches include experimental designs, surveys, and statistical analyses. On the other hand, researchers are anchored in a constructivist paradigm, which holds that reality is socially created and subjective, and they concentrate on investigating lived experiences, social processes, or the interpretations individuals assign to events. They ask intentionally open-ended and interpretive questions to capture complexity rather than reduction. Interviews, ethnography, and case studies are examples of qualitative research methods frequently used in constructivist methodology. These methods enable researchers to gain comprehensive knowledge of circumstances. A pragmatic paradigm that bridges the gap between positivism and constructivism emphasizes practical results and research issues addressed through mixed techniques [121]. Researchers who adhere to the sensible approach construct their studies to balance quantitative rigor and qualitative depth, ensuring a general comprehension of complicated phenomena. This paradigm highlights the flexibility and adaptability of research, which allows for methodologies to be tailored to the requirements of the investigation.

5.2 *Studies of Real-World Applications Illustrated Through Case Studies*

5.2.1 Research on the Energy Efficiency of Buildings: Case Study No. 1

The use of philosophical principles is readily apparent within the realm of study about the energy efficiency of buildings. A positive researcher working in this field might concentrate on measuring the amount of energy savings that can be realized through retrofitting techniques such as enhanced insulation or installing solar panels. The question "what is the percentage reduction in energy consumption achieved through retrofitting?" is an example of a research question that might be asked. Methodologies such as controlled experiments, computer modeling, and statistical analysis will be considered to produce reproducible and generalizable findings [122].

On the other hand, a constructivist approach to the same subject matter can investigate how individuals adapt to energy-saving technology in their houses and how they perceive these technologies. The question "how do residents perceive the usability and effectiveness of new energy technologies?" might be investigated by investigating research questions. Qualitative approaches, such as focus groups or ethnographic observations, would highlight societal or cultural hurdles to adoption and capture everyone's subjective experiences [123]. In addition to the technical focus of positive research, this method offers insights into the human aspects of energy efficiency to enhance energy efficiency.

5.2.2 Urban Development and Social Equity: A Case Study of the Second Case

When researching urban development, philosophical concerns impact the formulation of equity-related issues and the methodologies utilized to study such questions. A researcher who adheres to the pragmatic approach could use a combination of qualitative and quantitative research approaches to investigate the effects of urban redevelopment programs on underserved neighborhoods [124]. Some questions might include "what are the changes in housing affordability following urban renewal?" and "how do residents experience the social and cultural impacts of these projects?" Qualitative interviews or focus groups would provide a comprehensive viewpoint that considers measured consequences and lived experiences with residents impacted by the housing market trends. These interviews or groups would supplement quantitative assessments of housing market trends.

5.2.3 The Third Case Study: Adaptation in Architecture to Climate Change Analysis

Philosophical concepts also guide the study of climate-responsive architecture [125]. When it comes to lowering the temperatures within a building during periods of extreme weather, a positive approach can concentrate on determining whether design interventions, such as green roofs or passive cooling systems, are effective. The following research questions might be asked: "what percentage reduction in indoor temperature achieved by green roofs under heatwave conditions?" Empirical evidence might be obtained for design optimization by conducting performance simulations and experimental setups. On the other hand, a constructivist researcher can investigate how conventional architectural methods are modified to meet climate hazards in economically fragile areas. Ethnographic approaches would be guided by questions such as "how do cultural beliefs shape adaptive architectural strategies?" This would allow for a more nuanced understanding of the practices that are currently in place. This type of study brings the relevance of incorporating indigenous knowledge into sustainable design, which also underlines the interaction between cultural and environmental elements.

The questions shape the research process researchers ask and the techniques they employ to answer them. Philosophical underpinnings are vital for directing the research process [126]. The positivist, constructivist, and pragmatist perspectives provide complementary theoretical approaches when comprehending complicated occurrences. Case examples in building energy efficiency, urban development, and climate-responsive architecture highlight how philosophical issues impact the research design and the study results [127]. Scholars can generate scientifically robust, morally sound, and socially beneficial work by connecting their research with these core principles. This allows them to contribute to essential breakthroughs in both knowledge and practice.

6 Conclusion

The chapter analyzes the philosophical foundations that influence research, illustrating their essential role in molding the research process, from question formulation to interpretation of findings. These foundations, grounded in epistemology and ontology, offer an intellectual framework that guides methodological precision, ethical integrity, and societal significance in academic research. Integrating philosophical concerns allows researchers to synchronize methodology and outcomes with overarching goals, such as enhancing knowledge and tackling intricate social issues. Philosophical frameworks like positivism, constructivism, and pragmatism provide unique methods for comprehending reality and obtaining knowledge. Positivism prioritizes objectivity and generalizability via empirical observation and quantitative methods, rendering it especially useful in engineering and the natural sciences. Constructivism, conversely, emphasizes subjective experiences and contextual interpretations, utilizing qualitative methodologies to investigate human connections and cultural events. Pragmatism connects various paradigms, promoting methodological adaptability and practical relevance, offering a balanced framework for multidisciplinary and applied research.

Case studies demonstrated the practical applicability of these paradigms. In building energy efficiency research, positivist methodologies accurately measure energy savings, whereas constructivist techniques investigate perceptions and experiences related to energy-saving devices. In urban planning, pragmatic methodologies integrate quantitative research with qualitative insights to tackle social fairness concerns, uncovering measurable advantages and intricate lived experiences. In climate-responsive architecture, philosophical frameworks inform adaptive design inquiries, reconciling empirical thermal performance evaluations with research of anthropological, cultural, and environmental dynamics. These examples illustrate how philosophical underpinnings shape research strategies and foster thorough and contextually aware understandings. The integration of philosophical principles also guides ethical research methodologies. Ethical theories based on deontology, utilitarianism, and virtue ethics establish a basis for defining researchers' obligations to participants, communities, and society. These principles guarantee that research is performed with respect, openness, and equity, cultivating trust and credibility in academic and wider societal contexts. Incorporating ethical issues into study design and implementation allows researchers to maintain the integrity of their work while responding to social requirements. Philosophical investigation is essential in tackling modern difficulties. Global challenges like urbanization, climate change, and technological progress require multidisciplinary collaboration, while philosophical frameworks offer coherence and flexibility. Critically engaging with epistemological and ontological inquiries enables researchers to formulate methodologically rigorous studies that consider stakeholders' varied realities and requirements.

The philosophical foundations are dynamic, adapting to new ethical concerns and methodological obstacles. These frameworks provide crucial instruments for traversing novel ethical and conceptual landscapes in artificial intelligence, genomics, and environmental sustainability. Anchoring research in established philosophical tenets enables scholars to confront difficulties with intellectual rigor and ethical foresight, assuring their work's relevance and effect. Philosophical underpinnings are essential to the research process, providing clarity, coherence, and ethical direction across several fields. These paradigms direct the formulation and implementation of methodologically rigorous and socially significant research through the positivist quest for objective truths, the constructivist analysis of subjective experiences, or the pragmatic amalgamation of diverse methodologies. Aligning research with these ideals allows researchers to improve knowledge, tackle critical societal issues, and maintain the ethical standards essential for transformational inquiry. Integrating these principles into research methodology cultivates a strong, egalitarian, and progressive corpus of knowledge adept at tackling contemporary difficulties.

References

1. Dehalwar K, Sharma SN (2023) Fundamentals of research writing and uses of research methodologies. Edupedia Publications Pvt Ltd
2. Wiranto EB, Sule MM, Gafoordeen N (2023) The baseline of multicultural education: an examination from Islamic and Buddhist standpoints. Multicult Islam Educ Rev 1(2):96–108
3. Sun P, Zuo X (2024) Evolution and history of research philosophy. J Manag Res 24(1):09725814
4. Nia HA, Rahbarianyazd R (2020) Aesthetics of modern architecture: a semiological survey on the aesthetic contribution of contemporary architecture. Civ Eng Arch 8(2):66–76
5. Hazbei M, Cucuzzella C (2023) Revealing a gap in parametric architecture's address of "Context." Buildings 13(12):3136
6. Madden AD (2022) A review of essential research tools without the confusing philosophy. High Educ Res Dev 41(5):1633–1647
7. Moshood TD, Rotimi FE, Rotimi JO (2022) An integrated paradigm for managing efficient knowledge transfer: towards a more comprehensive philosophy of transferring knowledge in the construction industry. Constr Econ Build 22(3):65–98
8. Sohail M, O'Donovan A, O'Sullivan PD (2024) Design frameworks for ventilative cooling: perceptions from building design practitioners in Ireland and the United Kingdom. Build Environ 263:111860
9. McCrudden MT, Marchand G (2020) Multilevel mixed methods research and educational psychology. Educ Psychol 55(4):197–207
10. Bhadsavle MM (2023) Climate change communication for enhanced urban resilience: communicating urban heat stress through visual narratives
11. Alsaigh R, Coyne I (2021) Doing a hermeneutic phenomenology research underpinned by Gadamer's philosophy: a framework to facilitate data analysis. Int J Qual Methods 20:16094069211047820
12. Manik KN, Nababan IA, Sai YH, Ginting SA, Sari W (2024) Study of ontological, epistemological and axiological principles in final academic assignments for dissertation scientific papers: philosophical values in scientific writing. Int J Multidiscip Appl Bus Educ Res 5(1):330–338
13. Van Inwagen P (2022) Being: a study in ontology. Oxford University Press
14. Bennett C (2021) Evolving epistemology: a reconsideration of evolutionary debunking arguments against religion. Doctoral dissertation, University of Oxford
15. Ayer J (2022) Ontological security in the new climate regime. Doctoral dissertation, University of British Columbia

16. Wilson D, Mikahere-Hall A, Sherwood J (2022) Using indigenous kaupapa Māori research methodology with constructivist grounded theory: generating a theoretical explanation of indigenous womens realities. Int J Soc Res Methodol 25(3):375–390
17. Pekala RJ (2013) Quantifying consciousness: an empirical approach. Springer Science & Business Media
18. Atwebembeire S (2023) Empiricism and rationalism: kant's reconciliatory project and its heritage
19. MacLeod A, Burm S, Mann K (2022) Constructivism: learning theories and approaches to research. Res Med Educ 25–40
20. Grant R, Phene A (2022) The knowledge based view and global strategy: past impact and future potential. Glob Strateg J 12(1):3–30
21. Cobb D (2022) Empiricism in the Philosophy of Science
22. Ezebuilo HC (2020) The rationalist and empiricist epistemological strategies and their implications in ethics. IGWEBUIKE Afr J Arts Hum 6(4)
23. Larison KD (2022) On beyond constructivism: Using intersubjective approaches to promote learning in the science classroom. Sci Educ 31(1):213–239
24. Wills J, Lake RW (2020) Introduction: the power of pragmatism. ATISM, p 1
25. Shrestha BK, Sharma LR (2024) Discerning the distinctive characteristics of key research paradigms and their constituents. Nepal J Multidiscip Res 7(2):30–44
26. Burns M, Bally J, Burles M, Holtslander L, Peacock S (2022) Constructivist grounded theory or interpretive phenomenology? Methodological choices within specific study contexts. Int J Qual Methods 21:16094069221077758
27. Finn VK (2020) Exact epistemology and artificial intelligence. Autom Doc Math Linguist 54:140–173
28. Akinosho TD (2024) Investigating a deep learning approach to real-time air quality prediction and visualisation on UK highways. Doctoral dissertation, University of the West of England, Bristol
29. Kim M, Adlof L (2024) Adapting to the future: ChatGPT as a means for supporting constructivist learning environments. TechTrends 68(1):37–46
30. Rayhan A, Rayhan R, Rayhan S (2023) Artificial general intelligence: roadmap to achieving human-level capabilities
31. Maass W, Storey VC (2021) Pairing conceptual modeling with machine learning. Data Knowl Eng 134:101909
32. Wang W, Yang Y, Wu F (2022) Towards data-and knowledge-driven artificial intelligence: a survey on neuro-symbolic computing. arXiv:2210.15889
33. Jebreili S, Goli A (2024) Optimization and computing using intelligent data driven. In: Optimization and computing using intelligent data-driven approaches for decision-making: optimization applications, p. 90
34. De Bianchi S, Giovannetti L (2023) Introduction: exploring the dynamics between ontology and epistemology. THAUMÀZEIN 11(1):1–27
35. Lawani A (2021) Critical realism: what you should know and how to apply it. Qual Res J 21(3):320–333
36. Stezhko Z, Rymar S (2023) Philosophical problems of scientific cognition
37. Li Y, Huang Y, Zhang X, Jiang Y, Jing X, Deng Y, Li R, Jiang D, Mu J (2024) Ecosystem health assessment: a PSR analysis combining AHP and EW methods for Sansha Bay, China. Front Mar Sci 11:1396026
38. Kaczmarczyk MR (2024) The myth of agreement. J Theory Soc Behav
39. Pilarska J (2021) The constructivist paradigm and phenomenological qualitative research design. In: Research paradigm considerations for emerging scholars, pp 64–83
40. Toyon MAS (2021) Explanatory sequential design of mixed methods research: phases and challenges. Int J Res Bus Soc Sci (2147–4478), 10(5):253–260
41. Maclear J (2021) The international norm dynamics of climate change in global health governance: a constructivist analysis. Doctoral dissertation, Stellenbosch: Stellenbosch University

42. Srikanth M (2024) Exploring the perceptions and experiences of e-learning in consultancy organizations: a qualitative study with a focus on the Indian context. Doctoral dissertation, University of Essex
43. Bøe TD (2021) Ethical realism before social constructionism. Theory Psychol 31(2):220–236
44. Machado LMO, Martínez-Ávila D, Almeida MB, Borges MM (2023) Towards a moderate realist foundation for ontological knowledge organization systems: the question of the naturalness of classifications. J Inf Sci 01655515231160031
45. Bogna F, Raineri A, Dell G (2020) Critical realism and constructivism: merging research paradigms for a deeper qualitative study. Qual Res Organ Manag Int J 15(4):461–484
46. Samoraj JL (2022) Does realism still dominate the theory and practice of international relations? Lessons on the 21st century grand chessboard
47. Sonoda T (2020) The politicisation of evidence: a critical realist approach to data-driven education planning, practice and exclusion in Syria. University of London, University College London (United Kingdom)
48. Aberšek B, Flogie A, Pesek I (2023) Philosophical and social realm. AI and Cognitive Modelling for Education. Springer International Publishing, Cham, pp 7–117
49. Hodorog A (2023) The role of semantics in enhancing user experience in building and city web applications. Doctoral dissertation, Cardiff University
50. Ryu S (2020) The role of mixed methods in conducting design-based research. Educ Psycholo 55(4):232–243
51. Rijos A (2024) A posthumanist approach to artificial intelligence research—theory and Praxis
52. Rahmani, S., Aghalar, H., Jebreili, S. and Goli, A., Optimization and computing using intelligent data-driven approaches for decision-making. In Optimization and Computing using Intelligent Data-Driven Approaches for Decision-Making (pp. 90–176). CRC Press
53. Ugwu CI, Ekere JN, Onoh C (2021) Research paradigms and methodological choices in the research process. J Appl Inf Sci Technol 14(2):116–124
54. Godwin A, Benedict B, Rohde J, Thielmeyer A, Perkins H, Major J, Clements H, Chen Z (2021) New epistemological perspectives on quantitative methods: an example using topological data analysis. Stud Eng Educ 2(1):16–34
55. Otoo BK (2020) Declaring my ontological and epistemological stance. J Educ Thought (JET)/ Revue de la Pensée Éducative 53(1):67–88
56. Kunisch S, Denyer D, Bartunek JM, Menz M, Cardinal LB (2023) Review research as scientific inquiry. Organ Res Methods 26(1):3–45
57. Kumatongo B, Muzata KK (2021) Research paradigms and designs with their application in education. J Lexicogr Termin 5(1):16–32. Online ISSN 2664-0899. Print ISSN 2517-9306
58. Comte A (2022) A general view of positivism: or, summary exposition of the system of thought and life. DigiCat
59. Mohajan HK (2020) Quantitative research: a successful investigation in natural and social sciences. J Econ Dev Environ People 9(4):50–79
60. Bentalha B, Alla L (2024) Revealing the subtleties: the art of qualitative studies in science and management. Applying qualitative research methods to management science, pp 1–21. IGI Global
61. Lowe RJ (2021) Frame analysis in critical ethnography: applications for ELT research. Lang Cult Curric 34(3):307–320
62. Allemang B, Sitter K, Dimitropoulos G (2022) Pragmatism as a paradigm for patient-oriented research. Health Expect 25(1):38–47
63. Khatri KK (2020) Research paradigm: a philosophy of educational research. Int J Engl Lit Soc Sci 5(5):1435–1440
64. Omodan BI (2024) Research paradigms and their methodological alignment in social sciences: a practical guide for researchers. Taylor & Francis
65. Park YS, Konge L, Artino AR Jr (2020) The positivism paradigm of research. Acad Med 95(5):690–694
66. Racine E, Kusch S, Cascio MA, Bogossian A (2021) Making autonomy an instrument: a pragmatist account of contextualized autonomy. Hum Soc Sci Commun 8(1):1–15

67. Phillips MJ (2023) Towards a social constructionist, criticalist, Foucauldian-informed qualitative research approach: opportunities and challenges. SN Soc Sci 3(10):175
68. Bhat RM, Silllalee A, Kandasamy LS (2023) Concepts and contexts: the interplay of philosophy and history in understanding human society. East Asian J Multidiscip Res 2(6):2581–2590
69. Brown C, Spiro J, Quinton S (2020) The role of research ethics committees: friend or foe in educational research? An exploratory study. Br Edu Res J 46(4):747–769
70. Arnold DG, Beauchamp TL, Bowie NE (2020) Ethical theory and business. Cambridge University Press
71. Floridi L, Strait A (2021) Ethical foresight analysis: what it is and why it is needed? The 2020 yearbook of the digital ethics lab, pp 173–194
72. Morrell K, Dahlmann F (2023) Aristotle in the Anthropocene: the comparative benefits of Aristotelian virtue ethics over Utilitarianism and deontology. Anthropocene Rev 10(3):615–635
73. Ghanbarian A (2023) Exposition and analysis of the formation and shaping of the foundations of Jeremy Bentham's Utilitarianism (1748–1832). Int J Multicult Multireligious Underst 10(10):202–215
74. Bader SM (2023) VR-based safety training program for high-rise building construction. Master's thesis, The American University in Cairo (Egypt)
75. Kwemarira G, Ntamu DN, Nsereko I, Balunywa W (2023) Deontological ethical orientations and public interest in government schools. Public Organ Rev 23(4):1455–1476
76. Kapsalis ET (2024) Contemporary use of co-design and computational design methods for inclusive built environments: a case study of platform-train interfaces. Doctoral dissertation, University of Nottingham
77. Doherty RF (2020) Ethical dimensions in the health professions-e-book: ethical dimensions in the health professions-e-book. Elsevier Health Sciences
78. Wang G, Hackett R (2023) Moral virtues: philosophical origins and contemporary understandings. Leadership and Virtues, pp 7–24. Routledge
79. Taggart G, Zenor J (2022) Evaluation as a moral practice: the case of virtue ethics. Eval Program Plann 94:102140
80. Vandenberghe F (2021) What's good about the good life? Action theory, virtue ethics and modern morality. Philos Soc CritIsm 47(7):814–830
81. Konadu K, Opoku Mensah A, Koomson S, Abraham EM, Amuzu J, Agyapong JAM (2023) A model for improving the relationship between integrity and work performance. Int J Ethics Syst
82. Nnodim P (2020) Beyond justice as fairness: rethinking rawls from a cross-cultural perspective. Rowman & Littlefield
83. Ferreira RJ, Buttell F, Ferreira S (2015) Ethical considerations for conducting disaster research with vulnerable populations. J Soc Work Values Ethics 12(1):29–40
84. Bussu S, Lalani M, Pattison S, Marshall M (2021) Engaging with care: ethical issues in participatory research. Qual Res 21(5):667–685
85. Topham S, Boscolo P, Mulquin M (2023) Personal data-smart cities: how cities can utilise their citizen's personal data to help them become climate neutral, p 356. Taylor & Francis
86. Burr C, Leslie D (2023) Ethical assurance: a practical approach to the responsible design, development, and deployment of data-driven technologies. AI Ethics 3(1):73–98
87. Newman PA, Guta A, Black T (2021) Ethical considerations for qualitative research methods during the COVID-19 pandemic and other emergency situations: navigating the virtual field. Int J Qual Methods 20:16094069211047823
88. Mobarak MH, Hasan MJ, Rahman M, Rana M, Saikat AS, Tasnim R (2024) Ethical and social consideration in mechatronics. Mechatronics: fundamentals and applications, pp 161–196. Singapore: Springer Nature Singapore
89. Al-kfairy M, Mustafa D, Kshetri N, Insiew M, Alfandi O (2024) Ethical challenges and solutions of generative AI: an interdisciplinary perspective. Informatics 11(3):58. MDPI

90. Lapid MI, Ouellette Y, Drake MT, Clarke BL (2023) Institutional review board (IRB): US perspectives. Handbook of bioethical decisions. Volume II: scientific integrity and institutional ethics, pp 219–240. Cham: Springer International Publishing
91. Levitt HM, Morrill Z, Collins KM, Rizo JL (2021) The methodological integrity of critical qualitative research: principles to support design and research review. J Couns Psychol 68(3):357
92. Ten Cate O (2022) The ethics of health professions education research: protecting the integrity of science, research subjects, and authorship. Acad Med 97(1):13–17
93. Konidena BK, Malaiyappan JNA, Tadimarri A (2024) Ethical considerations in the development and deployment of AI systems. Eur J Technol 8(2):41–53
94. Giray L (2024) Academic citation partnership: a predatory practice that undermines scholarly integrity. Ann Biomed Eng 1–5
95. Field SM, Thompson J, De Rijcke S, Penders B, Munafò MR (2024) Exploring the dimensions of responsible research systems and cultures: a scoping review. R Soc Open Sci 11(1):230624
96. Garcini LM, Barrita A, Cadenas GA, Domenech Rodríguez MM, Galvan T, Mercado A, Moreno O, Paris M, Rojas Perez OF, Silva M, Venta A (2023) A decolonial and liberation lens to social justice research: upholding promises for diverse, inclusive, and equitable psychological science. Am Psychol
97. Kolstoe SE, Pugh J (2024) The trinity of good research: distinguishing between research integrity, ethics, and governance. Account Res 31(8):1222–1241
98. Lescrauwaet L, Wagner H, Yoon C, Shukla S (2022) Adaptive legal frameworks and economic dynamics in emerging tech-nologies: navigating the intersection for responsible innovation. Law Econ 16(3):202–220
99. Lin Z (2023) Why and how to embrace AI such as ChatGPT in your academic life. R Soc Open Sci 10(8):230658
100. Christopher S, Watts V, McCormick AKHG, Young S (2008) Building and maintaining trust in a community-based participatory research partnership. Am J Public Health 98(8):1398–1406
101. National Academies of Sciences, Policy, Global Affairs and Committee on Responsible Science (2017) Fostering integrity in research. National Academies Press
102. May T, Perry B (2022) Social research: issues, methods and process. McGraw-Hill Education (UK)
103. Aziza OR, Uzougbo NS, Ugwu MC (2023) Legal frameworks and the development of host communities in oil and gas regions: balancing economic benefits and social equity. World J Adv Res Rev 19(3):1582–1594
104. Vallor S (2016) Technology and the virtues: a philosophical guide to a future worth wanting. Oxford University Press
105. Medel-Ramírez C, Medel-Lopez H (2024) Exploring the multidimensional aspects of open knowledge: philosophical foundations, educational innovations, and scientific research. Educ Innov Sci Res
106. Senge PM, Scharmer CO (2008) Community action research: learning as a community of practitioners, consultants and researchers. Handbook of action research: the concise paperback edition, pp 195–206
107. Abma T, Banks S, Cook T, Dias S, Madsen W, Springett J, Wright MT (2019) Participatory research for health and social well-being. Springer International Publishing, Cham
108. Bleazby J (2020) Fostering moral understanding, moral inquiry and moral habits through philosophy in schools: a Deweyian analysis of Australia's ethical understanding curriculum. J Curric Stud 52(1):84–100
109. Jedličková A (2024) Ethical approaches in designing autonomous and intelligent systems: a comprehensive survey towards responsible development. AI & SOCIETY, pp 1–14
110. Akinyele D (2024) Ethical dilemmas in leadership: navigating between utilitarian and deontological perspectives
111. Hossain KA (2023) Evaluation of ethical values to develop global human resource. J Lib Arts Humlties 4(4):1–29

112. Fanton M, Valdevino Oliveira Silva W (2023) Rawls' theory of justice. Encyclopedia of business and professional ethics, pp 1539-1544. Cham: Springer International Publishing
113. Gusliana HB, Ardiyanto SY, Widiyanto WE, Setyawan H, Putro BN, Darmawan A, Radulovic N (2024) Ensuring fairness and transparency in sports through compliance to legal principles and ethical standards. Retos: nuevas tendencias en educaci6n física, deporte y recreaci6n (60):204–216
114. Vidal C, Heylighen F (2022) Ethics and complexity: why standard ethical frameworks cannot cope with socio-technological change. Humanism and its discontents: the rise of transhumanism and posthumanism. Springer International Publishing, Cham, pp 197–216
115. Halim Z, Durya NPMA, Kraugusteeliana K, Suherlan S, Alfisyahrin AL (2023) Ethics-based leadership in managing information security and data privacy. Jurnal Minfo Polgan 12(2):1819–1828
116. Rošker JS (2023) Confucian relationism and global ethics: alternative models of ethics and axiology in times of global crises, vol 13. Brill
117. O'Gara K (2022) Equitable transformation, transforming equality: a grounded theory study on the role of equality and equity in research and policies for transformative change (Master's thesis)
118. Naeem M, Ozuem W, Howell K, Ranfagni S (2023) A step-by-step process of thematic analysis to develop a conceptual model in qualitative research. Int J Qual Methods 22:16094069231205788
119. Ho YR, Chen BY, Li CM (2023) Thinking more wisely: using the socratic method to develop critical thinking skills amongst healthcare students. BMC Med Educ 23(1):173¨
120. Asad MM, Hussain N, Wadho M, Khand ZH, Churi PP (2021) Integration of e-learning technologies for interactive teaching and learning process: an empirical study on higher education institutes of Pakistan. J Appl Res High Educ 13(3):649–663
121. Sim CL, Wong TA, Sin KY, Sivakumaran VM (2024) Pragmatism as a paradigm for quality management research in bridging academic-practitioner gaps. Int J Qual Serv Sci 16(2):330–342
122. Usui T, Macleod MR, McCann SK, Senior AM, Nakagawa S (2021) Meta-analysis of variation suggests that embracing variability improves both replicability and generalizability in preclinical research. PLoS Biol 19(5):e3001009
123. Mrabti L, Alaoui ZB (2024) Balancing qualitative and quantitative research methods: insights and applications. Data collection and analysis in scientific qualitative research, pp 87–118. IGI Global
124. Mulisa F (2022) When does a researcher choose a quantitative, qualitative, or mixed research approach? Interchange 53(1):113–131
125. Lin H, Yin S, Xie C, Lin Y (2022) Integrated pedagogy with climate-responsive strategies: vernacular building renovation design. Buildings 12(9):1294
126. Cuthbertson LM, Robb YA, Blair S (2020) Theory and application of research principles and philosophical underpinning for a study utilizing interpretative phenomenological analysis. Radiography 26(2):e94–e102
127. Mahar WA (2021) Methodology for the design of climate-responsive houses for improved thermal comfort in cold semi-arid climates

Research Design and Methodology

Research, in its most rigorous form, represents a methodical quest for knowledge supported by a defined framework rather than merely an act of questioning. Central to this framework are design and technique, two interrelated components that shape the investigative process's epistemological basis, consistency, and dependability. This book emphasizes the importance of careful preparation, conceptual clarity, methodological accuracy, and ethical integrity in academic research while exploring these fundamental components. The chapters in this part sequentially and integratively address the evolution of research frameworks and the formulation of formal research proposals. They provide a coherent discussion on transforming abstract research issues into organized studies that yield legitimate, generalizable, and significant results. These contributions reflect a broader commitment to academic rigor and practical relevance, crucial for advancing knowledge and guiding practice. The architectural blueprint of any empirical or theoretical study is its research design, which outlines the logical chain linking the development of research questions to data collection, analysis, interpretation, and conclusion drawing. A well-conceived design ensures that every aspect of the research process is methodically aligned with the overall goals and objectives of the study. It serves as a roadmap for operationalizing abstract ideas, selecting appropriate scientific tools, and ensuring consistency throughout the empirical process. The type of research challenge, the researcher's epistemological perspective, and the desired contribution to the academic field will determine the choice of a suitable research design. Some studies may require a positivist, quantitatively oriented framework emphasizing measurement and causation. In contrast, others may need a constructivist or interpretivist approach better suited for exploring meanings, experiences, and contextual events. Increasingly, research is being conducted under a mixed-methods paradigm that intentionally integrates the advantages of both qualitative and quantitative traditions. This inclusive strategy captures statistical generalizability while considering the richness and depth of human and social phenomena. Mixed-methods research is particularly advantageous for addressing multi-dimensional research questions that resist reductionist approaches. Such mixed-approach designs enhance result robustness and offer a more comprehensive understanding of the topic by triangulating data sources, analytical methods, and theoretical perspectives. Consequently, integrating approaches must be guided by a unifying framework that maintains internal consistency and methodological congruence throughout the research. The absence of such congruence significantly raises the risk of analytical fragmentation or epistemological conflict. Establishing validity and reliability stands as the primary concern of research design. These standards ensure that results can be replicated or generalized across contexts and accurately represent the phenomena under investigation. Among the various methods employed to enhance these aspects are instrument pilot testing, adopting consistent procedures, cross validation, and ensuring inter-rater reliability in qualitative coding. Triangulation, the process of converging information from multiple sources to support conclusions, strengthens methodological rigor. Furthermore, adherence to ethical criteria is equally fundamental to the credibility of research. From question development to result dissemination, ethical considerations permeate every stage of the research process. Issues such as informed consent, voluntary participation, confidentiality, and the ethical use of data require special attention. Researchers must adhere to institutional and disciplinary ethical guidelines while honoring each participant's dignity, autonomy, and well-being. Conducting ethical research is both a legal requirement and a critical aspect of academic legitimacy and social responsibility. Beyond guiding the conduct of studies, this part of the book also examines the formulation of the research proposal, an essential initial step in academic research. A research proposal is a written document outlining the rationale, objectives, methodological framework, and expected contributions of the proposed study. It fulfills various purposes: securing institutional approval, attracting research funding, and articulating an implementation strategy. Therefore, the proposal must demonstrate conceptual clarity, methodological accuracy, and feasibility regarding both time and budget. The key components of a well-researched proposal include a clearly defined research question, a specific set of objectives or hypotheses, a thorough review of relevant literature, and a robust methodological approach. The literature review is particularly crucial for situating the study within current academic debates, identifying theoretical and empirical gaps, and showcasing the originality and relevance of the proposed research. A detailed methodology outlines the research design, sampling strategy, data collection methods, analytical processes, and ethical safeguards. Including a timetable and budget further enhances the credibility and viability of the study project. While the structure of research proposals may differ across disciplines and institutions, the fundamental principles remain constant: clarity of purpose, alignment of design with objectives, methodological integrity, and ethical accountability. A well-conceived proposal not only charts the course of research but also anticipates potential challenges, acknowledges limitations, and articulates the academic and practical significance of the study. This part emphasizes the dual imperative of academic research: to generate methodologically sound and socially relevant information. It illustrates that a scholarship's quality, legitimacy, and impact are primarily determined by research design and methods rather than peripheral factors. Research can illuminate theory, inform practice, and contribute to societal development by adhering to ethical, transparent, and systematic processes. Rather than offering a prescriptive template, this part presents conceptual tools and methodological strategies that can be adapted for various disciplinary contexts and research inquiries. The ideas articulated here foster the qualities of academic reflexivity, methodological creativity, and epistemological awareness essential for conducting exemplary research in dynamic and complex environments. By applying these principles, research achieves academic rigor and lasting relevance.

Research Design and Strategy

Abstract This chapter shows a research design and strategy meticulously constructed and organized to investigate the primary research topics in depth and precisely. The chapter comprehensively analyzes complex, multi-dimensional phenomena using a mixed-methods approach, which blends quantitative rigor with qualitative richness. By ensuring that discoveries are empirically based and contextually nuanced, this methodological approach contributes to the advancement of theoretical insights while also supporting practical implementations. Each methodological option is painstakingly linked with the study's objectives, enhancing the research process's credibility, trustworthiness, and transparency. Implementing strategic procedures such as triangulation, pilot testing, and inter-rater evaluations strengthens the validity and reliability of the results. These techniques all contribute to the consistency and correctness of the results presented. Ethical standards are strictly adhered to ensure that participant's rights are protected. This includes obtaining informed permission and maintaining data confidentiality. This helps ensure the study maintains its moral integrity and academic rigor. In addition to being flexible and all-encompassing, this research methodology can accommodate ever-evolving issues within dynamic domains and generate significant findings with a wide range of potential applications. This chapter offers substantial results in theoretical breadth and may be applied to various practical circumstances. It does so by adhering to a methodologically sound framework, which provides a firm platform for future study.

1 Introduction

According to Javali [1], research, derived from the term "recerchier" (meaning "to seek again"), is the systematic and empirical pursuit of knowledge. It involves a structured methodology for investigating topics, enhancing understanding, and expanding a field's knowledge base. According to the methodological framework, the Research Design and Strategy chapter forms the foundation for the entire investigation. This chapter systematically outlines the chosen approach, techniques, and procedures used for data collection, analysis, and interpretation, aligned with the study's objectives.

A. Almusaed et al., *Practice of Research Methodology in Civil Engineering and Architecture*, https://doi.org/10.1007/978-3-031-97393-2_7

Presenting the rationale and justification for these methods establishes a coherent framework to guide the research process and ensure methodological rigor [1]. This chapter is crucial for establishing the rigor, transparency, and validity of the study, meticulously addressing these aspects to ensure the credibility and trustworthiness of the findings. It is a comprehensive blueprint detailing the methods and techniques employed to achieve the study's objectives and answer its research questions. By providing a structured framework, this chapter enables consistency and methodological rigor throughout the research process. Its straightforward presentation of data collection and evaluation procedures minimizes bias, thereby maximizing the reliability and validity of the results. The chapter underscores the importance of aligning the research design and approach with the study's objectives. All methodological choices, whether related to research methodology, data collection methods, or data analysis techniques, should directly support achieving the study's goals and answering its research questions [2]. For example, if the study examines the effectiveness of environmentally responsible building techniques in improving energy efficiency, a mixed-methods approach might be most suitable. The research can provide statistical evidence alongside detailed contextual insights by combining quantitative and qualitative methods. Therefore, methodological selections are deliberate and purpose-driven, ensuring that the data collected is highly relevant to addressing the research questions and fulfilling the study's overarching goals.

- *What is the connection between research design, questions, and objectives?*

The meticulous alignment of the study design with the research questions and objectives established at the outset is fundamental to maintaining research rigor. According to Abusaada and Elshater [3], an essential aspect of engaging with the research literature is systematically including pertinent sources while rigorously excluding irrelevant ones. This process requires a careful, methodologically sound approach to selecting appropriate references, ensuring the literature reviewed is relevant and contributes meaningfully to the study's foundational objectives [3]. These research questions and objectives directly inform the selection of suitable methodologies, actively shaping methodological decisions. Qualitative approaches, such as semi-structured interviews, provide specific insights and a comprehensive understanding of the subject. According to Lim [4], qualitative research must clearly articulate its focus, rigor, and relevance, underscoring its role in analyzing complex social phenomena and generating rich, human-centered insights. The rationale for employing qualitative methods should be carefully examined, emphasizing their critical importance in revealing nuanced, contextually grounded perspectives. While this approach offers strengths, such as contextual depth and holistic viewpoints, it also presents limitations, including challenges in generalizability and potential researcher bias. Practical strategies to mitigate these limitations, such as reflexivity and methodological transparency, further enhance the robustness of qualitative research [4]. For example, suppose the research aims to explore viewpoints on integrating artificial intelligence in urban planning. In that case, qualitative methods are employed if the research aims to explore viewpoints on integrating artificial intelligence in urban planning methods. In contrast, if the purpose is to test a hypothesis regarding the energy efficiency of new building materials under climatic circumstances, then a quantitative experimental design would be appropriate [5]. This approach would allow for statistical analysis as well as the testing of hypotheses in an objective manner. By connecting each methodological option back to the research issues, this careful selection of techniques helps to develop coherence and consistency in the study [6]. Using this methodology, the data are guaranteed to be trustworthy and genuine and directly contribute to addressing the research questions. Therefore, a well-structured design supports the study's objective, enhances the methodology's transparency, and demonstrates that the research is reliable and organized. Because of this deliberate alignment, the findings are solid from an academic standpoint and meaningful from a practical standpoint, contributing to advancing knowledge and application within the sector [7].

2 Research Paradigm and Approach

2.1 The Optimistic Viewpoint

The optimistic worldview is built on the idea that reality is objective and exists apart from human intellect [8]. The positive theory, founded on empiricism, asserts that knowledge is obtained by observing and measuring phenomena that can be quantified and submitted to statistical analysis [9]. This paradigm aims to uncover generalizable rules or principles by analyzing the correlations between variables and relating them to quantitative approaches. For example, in passive building research, a positive approach might involve measuring the energy efficiency of buildings with specific insulation types, window placements, and building orientations [10]. Researchers can statistically analyze which design factors lead to optimal thermal performance by collecting empirical data on energy usage under controlled conditions. This data-driven approach enables researchers to establish generalizable principles on energy efficiency, aiding architects and engineers in creating more sustainable buildings based on objective, measurable results. Assuming that knowledge is best acquired from empirical data that can be duplicated and validated independently of subjective influence, positivist research is conducted, with the researcher maintaining a detached and objective posture throughout the investigation [11].

2.2 The Interpretive Perspective

Abductive methodology is a hybrid approach that combines elements of deductive and inductive reasoning, enabling researchers to move fluidly between theory and empirical evidence. This approach provides enhanced flexibility within the research process, accommodating evolving insights. Proudfoot [12] suggests that hybrid thematic analysis, which integrates inductive and deductive reasoning, presents substantial opportunities for researchers; however, its potential within integrative mixed-methods research still needs to be explored, indicating a promising area for further methodological development [12]. In abduction, the researcher aims to identify the most plausible explanations for observed phenomena, continuously refining and reassessing theoretical constructs as new data emerges. Havinga and Rae [13] argue that ethnographic fieldwork addresses many critical needs in safety research, yet it introduces distinct challenges. Specifically, ethnography in safety studies often leans toward either a sympathetic portrayal of safety practices from the practitioners' perspectives or provides a sociological critique of existing practices. This dual focus highlights ethnography's potential to enrich understanding while also necessitating careful consideration of its interpretative limitations in applied safety contexts [13]. This strategy is beneficial for examining complex or ambiguous phenomena because it enables a thorough understanding of the subject under study.

For example, an abductive approach would allow researchers to iteratively refine energy efficiency or indoor environmental quality theories in sustainable building design as new case studies or building performance data become available [14]. Rather than adhering strictly to pre-existing theories on sustainable construction, such as those focused solely on minimizing energy use, an abductive approach might explore how occupant behavior influences energy consumption patterns or how green certifications impact long-term building performance [15]. By dynamically updating hypotheses with new findings, researchers can more effectively construct theories that account for the multifaceted challenges inherent in sustainable building practices. This flexibility fosters continuous theory construction and enhancement, making abduction a valuable exploratory and explanatory research methodology. In sustainable architecture, for example, abductive reasoning can assist in developing adaptive building designs that are energy-efficient and responsive to varying environmental and social contexts. Through the iterative, reflective nature of abduction, researchers can thus create more robust and context-sensitive frameworks that contribute to a deeper understanding of sustainability in the built environment [16].

2.3 Chosen Paradigm and Justification

An inductive methodology has been selected for this study due to its unity with the interpretivist paradigm and the exploratory aims of the research. The inductive approach facilitates a flexible and adaptive investigative process, crucial for understanding participants' subjective experiences and the meanings they attribute within a specific context [17]. Unlike deductive methodologies that confine the analysis to established theoretical frameworks, the inductive approach allows themes and insights to emerge organically from the data. This methodological openness is particularly valuable for capturing the complexity and depth of participant experiences, enabling a nuanced, grounded comprehension of the phenomenon in question [18]. For example, in research focused on daylighting within research buildings, an inductive approach permits collecting and interpreting participant insights without imposing predetermined frameworks related to energy efficiency or optimal lighting configurations. Participants such as architects, facility managers, and building occupants may provide unique perspectives on their experiences with natural lighting, uncovering unanticipated benefits like enhanced cognitive performance, well-being, or satisfaction, which extend beyond conventional lighting metrics. These insights contribute to a more user-centered understanding of daylighting strategies that might not surface through deductive, hypothesis-driven inquiry alone [19]. The inductive methodology suits this study because it emphasizes exploring under-researched or nuanced aspects of daylighting design and occupant experience. By commencing with detailed observations and allowing patterns and meanings to develop organically, this approach fosters the generation of insights that are contextually resonant and deeply relevant to the specific research environment. Additionally, the inductive approach aligns with the interpretive commitment to viewing knowledge as socially and culturally constructed, underscoring that participants' experiences with daylighting are influenced by their unique interactions with the built environment [20]. Ultimately, by embracing the evolving interaction between data and interpretation, this inductive approach enables the research to yield theoretically significant and empirically grounded insights, thus contributing to an enriched understanding of sustainable daylighting practices in architectural contexts. This approach supports the development of contextually valid findings that hold both academic and practical significance, advancing knowledge within sustainable building research. An overview of the interpretative paradigm is presented in Table 1. This paradigm emphasizes comprehending participants' subjective meanings and experiences within the context of their own experiences. As a result of the fact that researchers might get different insights from the same data, this method acknowledges the difficulty of reaching agreement on interpretations. Even though the inductive technique offers flexibility and makes it possible to undertake an adaptable approach to exploring experiences, it may run into problems with rigor because systematic analysis without a formal framework can be challenging to carry out. Exploratory research is aligned with the inductive method since it allows for the organic development of ideas. However, the open-ended nature of the inductive approach may result in excessive data.

2.4 The Anticipated Results

Using an interpretative perspective and an inductive technique, this research aims to provide extensive insights into the phenomena being studied, taking context into account. The anticipated results of this methodological decision are as follows:

- The study's inductive analysis focuses on patterns and themes that emerge from the qualitative data and represent the participants' varied viewpoints and experiences [24]. The results will align with the participant's experiences, as themes

Table 1 Analytical dimensions and challenges in interpretive paradigm and inductive methodology

Aspect	Explanation	Challenges
Chosen paradigm [21]	The interpretive paradigm emphasizes understanding participants' subjective meanings and experiences within their context	This may lead to difficulties in achieving consensus on interpretations, as different researchers may derive varied meanings from the same data
Methodology [22]	The inductive approach facilitates a flexible, adaptive investigative process crucial for exploring participant experiences	Flexibility can lead to issues of rigor; ensuring systematic analysis without a structured framework can be challenging
Justification for methodology [23]	The inductive approach aligns with the exploratory nature of the research, allowing insights to develop organically	The open-ended nature may result in data overload, where the researcher struggles to identify relevant themes amid a wealth of qualitative information
Comparison with deductive approach [24]	Unlike deductive methods, the inductive approach allows for emergent themes, promoting a comprehensive understanding	The absence of a guiding hypothesis can lead to the risk of findings being perceived as anecdotal rather than grounded on a clear theoretical basis
Value of inductive methodology [18]	Captures the complexity and depth of experiences, enabling a nuanced understanding of the studied phenomenon	The subjective nature of the data may introduce bias, as the researcher's interpretations could influence the representation of participant experiences
Application example [25]	In research on daylighting, the inductive approach allows for collecting participant insights without imposing predetermined frameworks, fostering authentic responses	Participants may have difficulty articulating their experiences, potentially leading to incomplete or unclear data that complicates analysis
Participant insights [26]	It enables diverse perspectives from architects, facility managers, and occupants, uncovering unexpected benefits such as improved well-being and cognitive performance	Variability in participant understanding and articulation of their experiences can lead to inconsistent data collection
Research suitability [27]	It is ideal for exploring under-researched aspects of lighting design and occupant experience, allowing themes to develop organically	Exploring nuanced topics may require considerable time and effort to gather and analyze data, potentially delaying research outcomes

(continued)

Table 1 (continued)

Aspect	Explanation	Challenges
Alignment with interpretivism [28]	It recognizes knowledge as socially constructed, highlighting the influence of participants' interactions	The socially constructed nature of knowledge can make generalizing findings beyond the study's specific context difficult
Outcome of inductive approach [29]	Facilitates the generation of theoretically significant and empirically grounded insights	There may be challenges in translating qualitative findings into actionable recommendations, particularly for practitioners who prefer quantitative data
Academic and practical significance [30]	Produces contextually valid findings that advance sustainable building research, offering academic and practical implications	Balancing academic rigor with practical applicability can be challenging, as insights may only sometimes align with established practices or industry standards

will emerge organically from the data rather than be tested against preconceived notions. The emergence of these themes is crucial because it will allow us to investigate intricacies and subtleties that might otherwise go unnoticed in a more rational, organized method.

- Insights tailored to the study's setting: The interpretative paradigm allows researchers to zero in on the exact social, cultural, and environmental factors that shape how experiences play out [31]. The findings are particularly relevant within that context since this method recognizes that participants' experiences are impacted by their surroundings, social structures, and cultural dynamics. The findings will enhance our comprehension of the phenomena in a meaningful way that considers the specifics of the research context; nonetheless, they should not be applied universally [32].
- Although the results might not apply to everyone, they should provide a thorough and empirically rich description of the phenomena, leading to valuable theoretical insights. This research will give theoretically sound findings that enhance our understanding of the phenomena under study by meticulously analyzing participants' experiences. These findings have the potential to pave the way for more studies by identifying recurring themes and patterns and providing theoretical and empirical inspiration for additional research.
- The study is expected to provide theoretically significant and practically relevant insights, offering a paradigm to guide future practice or policy in comparable settings [30]. Researchers in adjacent areas can use context-specific findings to inform their practice by providing them with empirically grounded understandings that can be applied to similar situations. Adaptation in comparable situations can be guided by the contextual application of the insights, even if they may

not be immediately transferable. For example, in the field of energy efficiency, the study might examine the adoption of smart meters in residential buildings, revealing both theoretical benefits, such as reduced energy use and cost savings, and practical challenges, like user acceptance and setup costs. While this model may not be directly replicable in all regions due to regulatory or infrastructure differences, other areas could adapt core principles—like community engagement or focusing on cost–benefit analyses—to fit their specific contexts. Additionally, insights gained could help researchers and policymakers in related fields, such as water conservation or waste reduction, by providing a framework to guide similar resource-focused initiatives [33].

- Choosing an interpretative paradigm and an inductive approach lays a solid groundwork for investigating complicated, subjective experiences. By making sure the research aims are compatible with the selected strategy, this methodological alignment improves the study's capacity to produce insights that are solid conceptually and have real-world applications. Because the research technique is tailored to the question, enabling a level of understanding that could not be attained using more organized, hypothesis-driven methods, this congruence also increases the trustworthiness of the results. The research is anticipated to offer highly descriptive, theoretically significant, and practically applicable insights, thereby significantly adding to the current corpus of knowledge.

The study's methodologically sound design will do more than add to current knowledge; it will provide the groundwork for further theoretical and empirical investigations into the topic, shedding light on its unique setting and ensuring its steady advancement. Based on an interpretative perspective and inductive approach, Table 2 presents the expected outcomes of a research project. Essential parts include finding new subjects, developing insights particular to the situation, and adding to our theoretical knowledge. Methodological congruence enhances credibility, while practical relevance is demonstrated via application in related domains. Striking a balance between insights unique to context and those with broader relevance are two of the many challenges. Another is dealing with data overload. The study's overarching goal is to produce theoretically and pragmatically essential results.

3 Research Strategy and Methodology

3.1 Research Methodology

The foundational step in accomplishing the goals and assuring the validity of the findings is selecting a suitable research approach [39]. According to the research objectives, this study uses a case study methodology, which closely aligns with the investigation. The research project's goals drive the selection of a case study because it can offer comprehensive insights, contextual comprehension, or empirical proof. For example, consider a study examining energy efficiency improvements

Table 2 Interpretive inductive research methodology expected outcomes and obstacles

Aspect	Description	Significance	Challenges
Emergent themes and patterns [34]	The inductive analysis will focus on discovering organic themes and patterns that reflect participants' varied experiences and viewpoints. These themes will emerge naturally from the data, providing a nuanced exploration of complex insights that might remain hidden in structured methodologies	Allows for a deeper understanding of intricacies in participant experiences, enhancing the richness and relevance of findings	It requires careful management to avoid data overload and focus on relevant themes. Without a structured framework, ensuring rigor in identifying significant patterns can be challenging
Context-specific insights [35]	The interpretative paradigm will highlight the specific social, cultural, and environmental factors that shape participants' experiences. Findings will emphasize the influence of contextual elements, ensuring that the insights are highly relevant within the study's unique setting	This contributes to a meaningful understanding of the phenomenon in its specific context, acknowledging that results may not be universally applicable but are highly valuable within the research environment	Insights may lack broader generalizability, making applying findings to other contexts challenging. Customizing analysis to each unique setting requires extensive contextual understanding
Theoretical contributions [36]	By examining participants' experiences in detail, the study aims to offer a rich, empirically grounded description of the phenomenon. This approach reveals theoretical insights that deepen understanding and inspire future research, identifying recurring themes contributing to existing theories	It provides a foundation for theoretical development and paves the way for additional studies inspired by the study's grounded, contextually relevant insights	Balancing depth and scope of theoretical insights can be challenging, as extensive qualitative data may lead to theoretical ambiguity. Ensuring insights remain relevant to broader theory requires meticulous analysis
Practical relevance [37]	The study will offer context-specific insights that practitioners in related fields can use to inform practice. While not directly transferable, the insights provide a framework that can guide adaptation in similar contexts, offering practical relevance and the potential to inform future policies	It equips practitioners and policymakers with adaptable knowledge, offering empirical insights that inform practice and policymaking in similar environments	Translating qualitative insights into actionable recommendations may prove difficult, as practitioners may require more quantifiable data for direct application

(continued)

Table 2 (continued)

Aspect	Description	Significance	Challenges
Methodological alignment [38]	Selecting an interpretive paradigm and inductive approach builds a strong foundation for exploring complex, subjective experiences. This alignment between research goals and methodology ensures conceptually sound and practically applicable insights, highly trustworthy and relevant to real-world applications	Enhances the study's credibility, offering theoretically sound and applicable findings well suited to both the research goals and the broader implications of the insights	Maintaining methodological alignment demands consistency, which can be difficult when adapting to emergent findings. Balancing flexibility with rigor in the analysis process is essential yet challenging
Contribution to knowledge [30]	The study is expected to deliver descriptive, theoretically significant, and practically relevant findings that extend the current knowledge body. The research's robust methodological design ensures it adds to existing literature and forms a basis for ongoing theoretical and empirical investigations into the phenomenon	Advances the field by providing foundational insights that support further research and understanding, emphasizing the study's role in enriching the knowledge base and contributing to the steady progress of the discipline	Ensuring contributions are academically rigorous and practically practical requires significant refinement of insights. Adapting findings to broader research contexts without diluting their depth poses a challenge

in office buildings. By selecting a case study approach, researchers could focus on a specific building or set of buildings to observe energy-saving technologies, occupant behavior, and building performance over time. One notable example is the research conducted on the U.S. Department of Energy's "Energy Smart" buildings, where energy-saving measures were implemented, monitored, and analyzed over several years. This approach enabled the researchers to identify the direct effects of energy-saving technologies (like LED lighting and smart thermostats) and how occupants interacted with these technologies, which proved crucial in understanding the full scope of energy savings. Such a case study could reveal behavior patterns, uncover challenges, and show specific savings outcomes, thus providing a rich, context-sensitive understanding of energy efficiency initiatives. The logic for this decision resides in its capacity to investigate phenomena in real-world situations. This is particularly helpful in research projects that need an understanding of complex and dynamic interactions within environmental settings. Case studies are beneficial when examining a restricted number of units (such as organizations, communities, or groups) [40]. They also make it possible to investigate the underlying mechanisms, linkages, or patterns they uncover. Because it offers a comprehensive investigation of phenomena that occur within natural ecosystems, the case study methodology is particularly well suited for research questions that address "how" and "why" inquiries. For this investigation, the case study approach makes it possible to conduct a detailed examination of [certain situations or organizations], capturing the numerous specifics and one-of-a-kind dynamics pertinent to the research aims. On the other hand, the justification for selecting a survey approach is that it is highly effective at collecting standardized information from a large sample, making it easier to generalize the results [41]. In research that seeks to evaluate attitudes, behaviors, or traits across a population, surveys are an excellent data collection method. The data acquired through surveys is often gathered through questionnaires or structured interviews. Surveys are an efficient method to obtain quantitative data that may be used to discover patterns or test ideas. Taking this strategy makes it possible to gather a wide range of insights, making it possible to do statistical analysis and perhaps extrapolate to a larger population. This approach aligns with the research's objective to evaluate attitudes, behaviors, or trends. In the case of an experimental approach, the reasoning is based on the strategy's effectiveness in establishing causality through manipulating variables and controlling circumstances. Researchers can witness the direct influence of independent variables on dependent variables through experiments, which are suitable for investigations that need exact measurements and control over extraneous elements. The testing of hypotheses is supported by experimental research, which, when constructed with great care, provides a high level of internal validity. The experimental design is handy for answering "what works" questions by illustrating the causal links between the variables [42]. An experimental method for this study would provide a controlled environment to test specific hypotheses regarding [a particular intervention or variable], which would align with knowledge of the causal linkages between the variables. Starting with the study objectives, which influence the technique chosen, Fig. 1 shows the research methodology selection process. Case studies, surveys, and experiments are some methods that could emerge from the technique. The study environment determines the specific benefits of each strategy. The three main research methods are case studies, surveys, and experiments. The former two allow researchers to delve deeply into complicated real-world situations, while the latter collects more broad data from large samples.

3.2 The Methodology of Research

This study employs a mixed-method approach, which combines qualitative and quantitative research methods. This strategy was chosen as the methodology because it is consistent with the data and analysis required to meet the study objectives [43]. Examples of complex dynamics that may be explored using qualitative methods include the attitudes of building occupants about energy-efficient designs in residential or commercial spaces. For instance, if a qualitative methodology is used, it is beneficial for examining these dynamics. Through in-depth interviews or focus groups with residents, facility managers, and architects, researchers can obtain nuanced insights into how various stakeholders perceive and interact with sustainable building aspects.

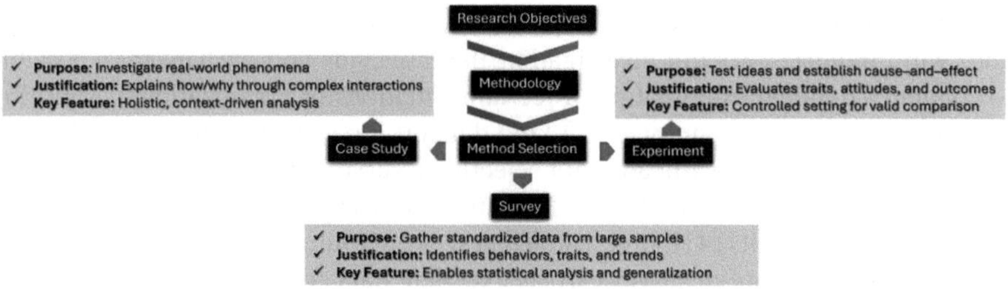

Fig. 1 Research methodology selection process (*Source* Authors)

Qualitative research would be helpful in this situation because it enables a more in-depth knowledge of elements such as comfort, perceived worth of energy savings, or obstacles in everyday operations. This is because quantitative data alone would not be able to reveal all of these factors. Taking such an approach enables researchers to not only determine "what" is working or not but also to dive into the "how" and "why" behind these encounters. This ultimately results in a deeper understanding of user behavior and the efficacy of sustainable construction techniques in real-world contexts. On the other hand, if the objective is to assess specific results, such as the reduction in energy consumption because of various sustainable building technologies (for example, solar panels, energy-efficient HVAC systems, or enhanced insulation), then a quantitative approach could be the most appropriate method to use. Using this strategy, the researchers could create patterns, uncover correlations, or determine causal linkages between architectural elements and energy efficiency levels. This would require gathering numerical data on energy consumption across several buildings. An investigation, for instance, might investigate how particular insulating materials contribute to a reduction in heating and cooling expenses, therefore delivering objective conclusions supported by statistical evidence that can be extrapolated to buildings of a similar nature. Alternatively, a mixed-methods approach is optimal when qualitative and quantitative data are essential to address the research questions comprehensively. For instance, initial quantitative surveys could assess energy savings across diverse buildings in research on sustainable buildings, providing measurable data on the effectiveness of energy-efficient measures. Following this, qualitative interviews with building occupants could explore their experiences and satisfaction with the implemented energy-saving features. This approach enables researchers to contextualize quantitative findings on energy efficiency improvements with qualitative insights into occupant satisfaction and perceptions of sustainability features. As Almusaed [44] noted, combining statistical evidence of energy savings with an in-depth understanding of user perspectives enriches the analysis, offering a more holistic evaluation of the effectiveness of sustainable building practices [44]. Figure 2 illustrates the options of research methodologies: qualitative, quantitative, and mixed methods. Each methodology fulfills distinct research requirements: qualitative research elucidates insights in natural contexts, and quantitative research methodically examines hypotheses and quantifies variables, while mixed-methods

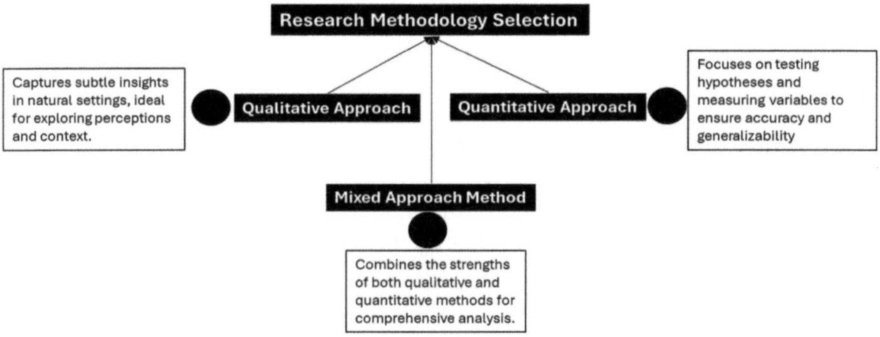

Fig. 2 Selection framework for qualitative, quantitative, and mixed-methods approaches (*Source* Authors)

amalgamate both for an exhaustive examination. Arrows denote relationships and appropriateness according to research aims.

4 Data Collection and Analysis Techniques

Investigating recycling techniques within building materials requires a systematic approach to gathering and analyzing data [45]. This methodology guarantees the production of solid and dependable findings that significantly contribute to the construction area. It is vital for a research study that investigates this subject to use a mixture of data-gathering instruments and analytical approaches that are specialized in qualitative, quantitative, or mixed-methods research. This part aims to provide an overview of the data-collecting tools, sampling tactics, and analytical frameworks utilized to collect and evaluate stakeholder insights and monitor trends within the construction sector.

4.1 Data Collection

Efficient data collection is essential for research, as it gathers material that directly influences the cornerstone of research. Collecting data efficiently synthesizes evidence directly informing the study results [46]. The selection of data-collecting techniques for a survey of recycling building materials is based on the tools' ability to capture quantitative measurements of material reuse and qualitative insights into industry practices, obstacles, and stakeholder views [47]. This ensures the study is thorough, with both the selected methodologies and the overall objective supporting a complete understanding of the recycling landscape in the building industry.

4.1.1 Data Collection Tools and Methods

To get both quantitative and qualitative insights, the study makes use of a variety of data-gathering methods, each of which is by the overall research design:

- **The Interviews**

Interviews are a qualitative research method used to collect primary data by questioning one or more individuals about their thoughts [48], experiences, or viewpoints on a specific topic or subject matter from the interviewee's perspective. The three primary forms of interviews are structured, unstructured, and semi-structured interview formats. For example, there are issues linked to using recycled materials and considerations of cost and practicality. Semi-structured interviews are very beneficial since they enable participants to discuss their experiences, such as integration. When researching recycling building materials, conducting in-depth interviews with construction workers, architects, and environmental consultants may give significant insights into recycled steel or concrete. These interviews can also indicate restrictions such as challenges with sourcing or compliance with regulatory requirements. This qualitative data provides insights into the reasons and obstacles associated with using recycled materials, which give nuanced insights [49].

- **The Use of Questionnaires and Surveys**

A questionnaire is any written collection of questions, but a survey is both the set of questions and the process of collecting, aggregating, and evaluating the replies from those questions. A questionnaire can look like any written set of questions. One example is that quantitative surveys can collect information from many participants, providing organized statistics on recycling adoption rates and industry trends [50]. For example, a study directed toward construction companies may determine the percentage of projects that use recycled concrete or lumber, or it could evaluate the perceptions of the cost-effectiveness and performance of these materials. While close-ended survey questions simplify the process of data processing and enhance the comparability of results, they also make it possible to quantify patterns that may be used to support conclusions that can be generalized.

- **Group Discussions**

Focus groups are a research approach that involves gathering a small group of individuals to answer questions in a monitored environment [51]. The group is selected based on the demographic characteristics that have been predetermined, and the questions are intended to provide light on a subject that is of interest. One example is that focus groups elicit collective viewpoints on recycling methods and sustainability regulations. These groups include stakeholders such as construction managers, suppliers, and legislators. These conversations have the potential to provide more in-depth insights into aspects such as subsidies offered by the government for recycled materials and perceived quality problems. Focus groups provide a dynamic environment for exploring complex themes, particularly identifying common perspectives on the challenges and advantages of recycling methods in the construction sector [52].

- **Methods Based on Observations**

As a method of qualitative research, observation research is a methodology in which researchers watch the continuing activity of participants in a natural setting. The degree of engagement that the market researcher will have in the study will vary depending on the characteristics of the observational research being conducted and the objectives of the investigation [53]. One example is that to collect data through observation, it is necessary to observe recycling operations on building sites directly. Researchers can watch the processes of managing, storing, and reusing demolition trash to obtain insight into the absolute recycling methods and protocols that are in place. Inconsistencies or innovations in recycling methods can be brought to light by observational data, which provides a crucial framework for understanding how sustainable policies are applied or altered [54].

- **Analyzing Documents and Data from the Archives**

This study is more comprehensive by incorporating pre-existing data sources, including municipal reports, regulatory papers, and project specifications [54]. These sources offer historical context or benchmarks in recycling procedures. An example of this would be examining government legislation concerning the recycling of construction materials over time, which may assist in mapping changes in industry standards. Furthermore, case studies on previous building projects that used recycled materials can offer insight into long-term results, such as durability and environmental effects, adding to a more comprehensive understanding of the topic.

4.1.2 Sampling Approach

The sampling approach, for example, needs to be meticulously crafted to guarantee a representative and pertinent dataset, covering a wide range of opinions from employees working in the construction sector about using recycled materials.

- **A Sampling at Random**

Using random sampling in quantitative surveys helps improve generalizability by ensuring the sample represents various construction entities during the survey [55]. Participants were randomly chosen to create a diverse dataset that accurately reflects recycling practices across the sector. These participants included both small contractors and significant enterprises.

- **Surveying with a Purpose**

Using purposive sampling in qualitative data collection allows for the targeted selection of participants with specialized knowledge [56]. For example, architects who specialize in sustainable design or managers who are active in green construction projects are examples of participants who can be chosen. This methodology enhances the relevance and depth of qualitative data, ensuring that the insights gained are directly related to the concentration of the study on recycled materials.

- **Stratified Samples**

Using stratified sampling, it is possible to guarantee that participants with prior experience working with various recycled materials, such as concrete, metals, or lumber, are represented proportionally [57]. This technique makes it easier to do comparative research across different types of materials, improving the study's ability to generalize its findings across various recycling applications.

- **A Sample Taken at Convenience**

Convenience sampling can be used in exploratory phases, even though its generalizability is limited [58]. This is because access to a larger sample is restricted during these phases. For example, preliminary data from local building sites may be collected to understand recycling procedures. This understanding can then be used to influence further systematic sampling.

4.1.3 Sources of Information

This research incorporates both primary and secondary sources of information, so offering a holistic perspective on recycling practices:

- **Data Collected Through Primary Sources**

Primary data, which includes interviews, surveys, focus groups, and observations, provides firsthand insights into building professionals' recycling techniques.

- **Data from Secondary Sources**

Secondary sources, which include government publications and industry reports, offer context, support comparative research, and allow the findings to be placed within a more comprehensive industrial or regulatory framework.

4.2 Data Analysis

The data analysis is organized to address the study's objectives successfully. It does this by employing procedures adapted to the features of the data and the overall research strategy [59]. To arrive at findings that align with the research objectives, several different analytical methods are utilized, and these methods vary according to whether the data are qualitative or quantitative.

4.2.1 Qualitative Data Analysis

The interpretation of qualitative data, which frequently originates from interviews, focus groups, and observations, calls for specific analytical approaches [60]. This is because qualitative data is comprised of information that is not numerical.

- **Thematic Analysis**

Thematic analysis is a method that helps identify recurrent themes within qualitative data. This method is beneficial in recycling materials [61]. When conducting interviews with participants, for instance, it is possible that themes such as "cost limitations," "quality perceptions," and "regulatory hurdles" may emerge. These themes will assist researchers in comprehending the prevalent challenges associated with using recycled materials.

- **Content Analysis**

Content analysis is a method that allows systematic investigation of text data, which in turn permits the quantification of specific phrases or topics [62]. An example of this would be the analysis of interview transcripts with architects, which may demonstrate that "sustainability" and "client demand" are often discussed variables. This would provide a structured method for evaluating the elements driving or impeding recycled material usage [63].

- **Narrative Analysis**

The narrative analysis process, frequently utilized in the context of stakeholder accounts, assists in interpreting how participants communicate their experiences [64]. When it comes to procuring recycled steel, for instance, construction managers could discuss their experiences, emphasizing the difficulties associated with dependability and compliance. Regarding recycling procedures in the construction industry, this investigation gives insight into how personal experiences and professional ideals impact recycling behaviors. Software applications like NVivo and Atlas.ti are frequently utilized to aid with qualitative analysis [65]. These applications make it easier to code and visualize data, ensuring the discovery of themes in an organized and systematic manner.

5 Validity, Reliability, and Ethical Considerations in Research

When doing research in the academic world, achieving high standards of validity, reliability, and ethical integrity is of the utmost importance. While validity assures that the research effectively represents what it seeks to assess, reliability focuses on the consistency of findings across various trials or circumstances [66]. Validity and reliability are two related concepts. These components, when used together, increase the dependability of the outcomes of a study, making them reliable for both practical applications and future research due to their reliability. Ethical considerations, which are equally important, guide researchers when they conduct studies responsibly [67]. These factors ensure that participants and their data are protected and respected. In the following discussion, we will cover the procedures followed to maintain the validity and reliability of research and the ethical norms used to inform and control these behaviors.

5.1 Validity in Research

Validity, at its core, is a term that relates to the honesty and correctness of the data and conclusions in connection to the research questions that were presented [68]. When research properly measures what it promises to measure, it is said to have a high degree of measuring accuracy. It is usual practice for academics to focus on several types of validity, including content, construct, and external.

- **Content Validity**

Content validity pertains to the extent to which a measure adequately represents the concept it aims to assess. This type of validity is essential in both qualitative and quantitative researches. In survey-based research, ensuring content validity requires designing questions that comprehensively address the topic's scope, avoiding omissions and irrelevant items [69]. Researchers can accomplish this by engaging with field experts or employing established frameworks for item selection, thereby minimizing the potential for bias or misrepresentation in data collection.

- **Construct Validity**

Construct validity assesses the extent to which an instrument accurately measures the theoretical construction it aims to evaluate [70]. For example, when a study seeks to determine "social anxiety," it is crucial to ensure that the survey items or interview questions precisely reflect this particular construct rather than a related but distinct concept such as "general anxiety." This is typically accomplished through using established scales validated by previous studies or by performing exploratory factor analysis to confirm that the items cluster as anticipated.

- **External Validity**

External validity denotes the degree to which research findings can be generalized beyond the conditions of the study [71]. This concern frequently arises in studies involving highly specific or narrow participant samples, as it may restrict the broader applicability of the findings. Methods to improve external validity involve employing random sampling, conducting studies in various contexts, and executing meta-analyses to evaluate overarching trends [72].

- **Internal Validity**

Internal validity evaluates the extent to which observed effects can be ascribed to the variables under investigation rather than external factors [73]. Experimental studies typically require the control of confounding variables to ensure that observed changes result from manipulating the independent variable rather than extraneous factors. Random assignment and control groups are established techniques for improving internal validity [74].

- **Triangulation**

Triangulation determines a point's location by forming triangles from known points [75]. This technique is widely used in various fields, including surveying, navigation, and geolocation, to enhance measurement accuracy and reliability. Triangulation is a method employed to confirm findings by cross verifying data across various methods or sources. This method is particularly beneficial in qualitative research, as subjectivity may lead to biases. A study may utilize interviews, surveys, and document analysis to collect varied perspectives and validate findings. Integrating multiple viewpoints minimizes bias, enhancing the conclusions' validity [76].

- **Pilot Testing**

Pilot testing is a preliminary study aimed at refining data collection instruments, identifying potential issues, and confirming the feasibility of the methodology [77]. Testing research tools allows researchers to identify and rectify ambiguous or problematic items, ensuring that the main study captures accurate and relevant data. This step is especially advantageous in survey research, as ambiguous questions may result in inconsistent responses.

5.2 *Reliability in Research*

Reliability pertains to the consistency and repeatability of results across various study instances [78]. A reliable study yields consistent results when replicated under identical conditions, indicating robustness and dependability.

- **Test–Retest Reliability**

Test–retest reliability is evaluated by administering the identical measurement instrument to the same participants at two distinct time points and analyzing the results for comparison [79]. Consistency in responses reflects high test–retest reliability, indicating that the instrument produces stable measurements over time. This method is commonly employed in longitudinal research or to validate new assessment instruments.

- **Inter-Rater Reliability**

Inter-rater reliability refers to the degree of agreement among independent observers or raters assessing the same phenomenon [80]. It is a crucial metric in research to ensure consistency and objectivity in measurements. Inter-rater reliability evaluates the extent of concordance among researchers or raters examining the same phenomenon. This is essential in qualitative research, as subjective interpretations of data may differ. To ensure inter-rater reliability, researchers train raters in standardized criteria and calculate agreement scores, such as Cohen's kappa, to quantify the consistency level among observers [81].

- **Internal Consistency**

Internal consistency assesses the degree to which items in a test measure a unified construct. Cronbach's alpha is frequently utilized to evaluate the extent of correlation among items within a scale [82]. High internal consistency signifies that all items assess the same underlying construction, thereby improving the reliability of the results.

- **Replicability**

Replicability refers to the ability to reproduce the results of a study or experiment under the same conditions [83]. It is a fundamental aspect of scientific research, ensuring that findings are reliable and can be verified by others. Replicability is a fundamental principle in scientific research, denoting the ability to reproduce a study's conclusions under identical conditions. Researchers who publish detailed methodology and data collection instruments enable other scholars to replicate the study, thereby verifying the reliability of the findings [84]. Reproducible results strengthen confidence in research conclusions, thereby contributing to the accumulation of reliable knowledge.

5.3 Ethical Considerations

Researchers are responsible for conducting studies responsibly, assuring participants' protection and the integrity of the research process [85]. Research ethics comprises a set of principles that guide researchers in this endeavor. Many other aspects fall under the category of ethical issues, such as informed consent, confidentiality, and data protection.

- **Consent Given After Being Informed**

Informed consent is a critical ethical concept that requires researchers to offer participants clear and thorough information about the study, including its aim, methods, risks, and benefits. This information must be provided to participants before they grant their permission. The participants must consent to participate in the activity willingly and be allowed to raise questions [86]. Through this procedure, participants are guaranteed to make an educated decision, respecting their liberty and well-being. It is possible to take extra precautions in situations involving vulnerable groups to ensure that the consent given is appropriately informed.

- **Privacy and Protection of Personal Information**

Confidentiality refers to the commitment to maintaining the privacy of the information provided by participants and safeguarding their identities and other personal details [87]. To protect the confidentiality of sensitive information, researchers anonymize the data, which is often accomplished by providing participants with pseudonyms or codes. Compliance with regulatory requirements, such as the General Data Protection Regulation (GDPR) in Europe, can be ensured using data protection measures such as safe storage and encryption [88]. These procedures prohibit unauthorized access to protected data. As a result, the participants' anonymity is safeguarded, which helps to cultivate trust and encourages truthful replies.

- **An Approval of Ethics**

Research institutes that include human participants must undergo ethics assessment before any study involving human subjects [89]. Ethics committees evaluate ideas to identify whether they comply with ethical norms, protecting participants from potential damage and maintaining the credibility of the research. This permission procedure is especially critical in sensitive research fields, such as healthcare or social sciences, where participants may be exposed to psychological hazards or emotional suffering throughout the research [90].

- **Possession of the Right to Withdraw**

Participants should be able to withdraw from the research at any moment without incurring any penalties or being discriminated against [91]. It is essential to recognize that participants may alter their minds or experience discomfort during the study process. This concept acknowledges and respects the participants' autonomy. Researchers must convey this right clearly and concisely, ensuring participants can withdraw from the study. Researchers guarantee that their studies are trustworthy, relevant, and respectful of the participants by adhering to ethical norms and meticulously addressing validity and reliability in their studies. When these principles are included in the research process, it is possible to confidently use research findings in academic and practical contexts. This contributes to the credibility and ethical soundness of the study undertaken. By taking these precautions, researchers ensure that their work is honest and trustworthy, contributing to the progress of knowledge responsibly [92]. Figure 3 visually represents the fundamental aspects that fall under validity, dependability, and ethical issues in research. The core aspects essential for preserving academic investigations' integrity, consistency, and ethical soundness are

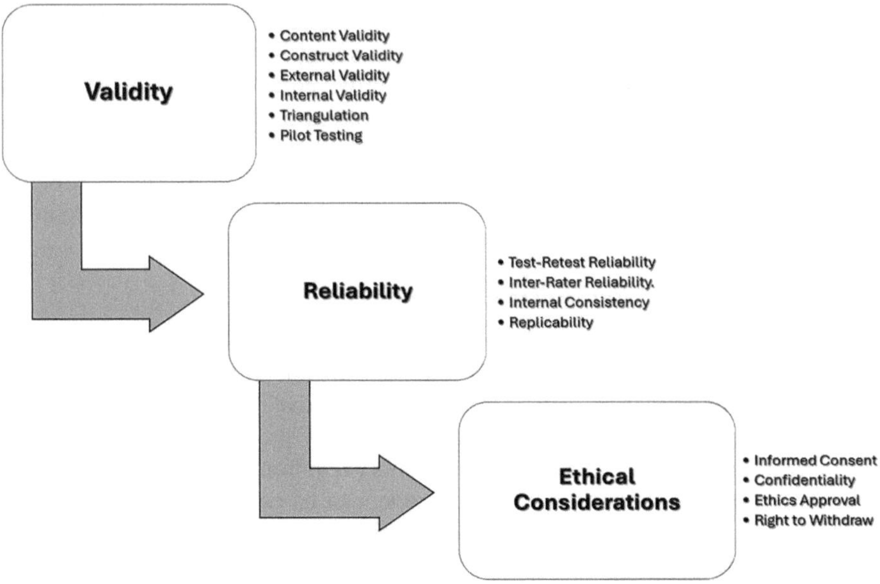

Fig. 3 Fundamental components of validity, reliability, and ethical considerations in research (*Source* Authors)

outlined in each area. Particular attention is paid to the stringency of the methodology and the adherence to the requirements for participant protection.

6 Conclusion

A research design constitutes a structured framework for systematically addressing a research problem through collecting and analyzing empirical data. Constructing a research design involves articulating the study's core objectives and selecting a methodological approach that aligns with these aims. Central to this process is the decision between employing primary research—where original data is gathered—or secondary research, which involves the analysis of pre-existing data. Also, a good research design spells out the sampling strategy, including the ways and standards for choosing participants. This ensures that the sample population accurately reflects the characteristics necessary to the study's goals, improving the validity and reliability of the results. This study's research design and strategy conclude with reaffirming the methodology's significance as a fundamental component. The methodology is essential in accomplishing the research objectives and answering the questions it poses. The research offers a balanced framework for rigorous empirical analysis while capturing the phenomenon's subtle and complex characteristics. Coordinating quantitative and qualitative methodologies precisely provides for this.

This all-encompassing method not only guarantees the dependability and trustworthiness of the study but also optimizes its value by delivering theoretical and practical insights, which are vital in increasing knowledge within the field of sustainable building research. In this research design, one of the most important aspects is the seamless alignment between the study questions, objectives, and methodological choices chosen during the process. To encourage openness and clarity, every methodological decision, whether in sampling, data collection, or analysis, was selected to connect with the study's objectives closely. Not only does this meticulous alignment assist in constructing a logical path for the research, but it also guarantees that every step makes a significant contribution to the overall objective of the scientific investigation. In this way, the alignment between research questions and techniques helps to create both rigor and transparency, ensuring that the study remains firmly rooted in its aims.

The study's mixed-methods methodology, which blends positive and interpretative perspectives, highlights its comprehensive character and makes it possible to conduct an all-encompassing investigation of sustainable construction techniques. This method strengthens the robustness of the findings by including both data-driven analysis and contextual insights. It makes it easier to validate the conclusions statistically and provides more nuanced knowledge that contributes to the enrichment of the area. When evaluating complicated subjects, such as sustainable practices that entail human interactions, such as the function of occupant behavior in enhancing energy efficiency, this dual perspective is particularly essential since it allows two perspectives to be considered simultaneously. Using this mix of quantitative and qualitative insights, the study generates relevant results from an academic standpoint. It makes discoveries applicable in the real world and connects them with their implications. Ultimately, the methodological design guarantees that the research findings are relevant and accessible to academic and industrial audiences, eventually contributing to better educated and efficient sustainable construction practices. Another thing that makes the study flexible and open to participant-centered insights is the inductive methodology, which lets themes come up naturally from the data without predefined frameworks. When researching areas with minimal past research, such as daylighting methods in sustainable buildings, where current theories may be insufficient or partial, this is especially beneficial since it allows for a comprehensive exploration of the subject matter. By fostering a process in which patterns, themes, and linkages are found via extensive observations, the inductive technique creates a process that encourages in-depth knowledge that is anchored in the experiences of the participants. In addition to the focus that the interpretative paradigm places on collecting subjective meanings and comprehending the socially constructed knowledge that participants bring to the research, this technique complements that paradigm. The inductive method exposes hidden aspects of sustainable practices, such as how different architectural characteristics affect occupant well-being and pleasure, which could otherwise go unnoticed in stricter, hypothesis-driven research methods. This is because the inductive method allows for emergent insights to be uncovered.

The study's design is anticipated to advance theoretical frameworks and practical applications within sustainable architecture significantly. By synthesizing participants' diverse experiences and perspectives through emergent themes, this research provides an in-depth examination of sustainable building practices. This approach is intended to yield context-specific insights that address the intricate social, cultural, and environmental factors influencing sustainable behaviors. Such findings are pivotal for academic discourse and practical implementation, offering nuanced perspectives on complex topics like the interplay between occupant behavior and energy efficiency. The results underscore the critical role of sustainable design in enhancing occupant well-being and performance, thereby contributing substantively to both the theoretical foundations and practical strategies of sustainable architecture. Table 3 presents a comprehensive review of essential elements within a study design framework, elucidating the purpose of each component, its contribution to research rigor and relevance, and the problems encountered. It emphasizes the strategic equilibrium required to connect methodological selections with research goals while confronting potential challenges in obtaining reliable, significant results.

The study methodology rigorously adheres to ethical standards and incorporates measures to uphold validity and reliability, ensuring credible and significant findings. Ethical protocols, such as informed consent, confidentiality, and respect for participant autonomy, are embedded within the research process, safeguarding participant rights while strengthening the integrity and reliability of the findings. The study enhances its validity through triangulation and pilot testing, ensuring the results accurately capture the intended assessment. To further ensure data reliability, measures such as test–retest reliability and inter-rater reliability assessments are employed, promoting consistency throughout data collection and analysis. These efforts collectively provide a robust foundation for reliable and relevant findings applicable across academic and professional domains. The study's design also advances the understanding of evaluating sustainable practices and their application in architecture. By integrating quantitative findings and contextual insights, the methodology aligns with the unique demands of sustainable building research. Consequently, the investigation produces empirically grounded, practically oriented insights adaptable to future research needs. With an inductive strategy and a mixed-methods approach, you can get comprehensive and adaptable results that help you develop the best sustainable design methods. This adaptability is especially valuable in the dynamic and rapidly evolving field of sustainable design, where new technologies and practices continuously emerge. Thus, this study ensures that its contributions are relevant to contemporary academic discussions and practical applications supporting sustainable construction methods' progression. Through a methodologically rigorous yet flexible approach, this research establishes a valuable framework for future advancements in sustainable architecture. This study's research design and strategy create a coherent and thorough platform for investigating sustainable construction techniques throughout the study. By aligning each methodological decision with the study's aims, the research can acquire a degree of coherence that enhances the dependability and transparency of its findings. A thorough study that balances quantitative and qualitative insights may be achieved through the mixed-methods approach,

Table 3 Key aspects, significance, and challenges in research design and methodology

Aspect	Description	Significance	Challenges
Research design framework	A structured framework addresses research problems via empirical data collection and analysis, integrating primary (original data) and secondary (pre-existing data) research	Ensures study objectives align with a chosen methodology for reliable, valid outcomes	Balancing depth and breadth of data without compromising either can be difficult
Sampling strategy	Defines participant selection methods and criteria, ensuring the sample represents essential characteristics aligned with study goals	Enhances validity and reliability by aligning the sample with study objectives, improving result relevance	Maintaining sample representativeness while managing participant accessibility
Mixed-methods approach	Combines quantitative and qualitative methodologies to capture both empirical data and contextual insights	Strengthens study robustness; balances quantitative rigor with qualitative depth, allowing exploration of complex, multifaceted phenomena	Integrating diverse data sources and aligning findings can be complex and time-intensive
Methodological alignment	Ensures seamless connection between research questions, objectives, and methodological choices, promoting rigor and transparency	Maintains coherence in the study, creating a logical research progression and enhancing credibility	Requires continuous refinement to keep alignment as the study progresses and findings evolve
Ethical standards	Adheres to informed consent, confidentiality, and participant autonomy	Protects participants' rights, enhancing ethical integrity and credibility	Balancing thorough ethical protocols with practical limitations, such as time constraints
Validity and reliability measures	Implements triangulation, pilot testing, test–retest, and inter-rater assessments to ensure consistent and accurate results	Reinforces the study's consistency and credibility, contributing to reliable findings	Testing measures without overextending resources or adding complexity to the study
Inductive approach	Allows themes and insights to emerge naturally, fostering participant-centered insights without predefined frameworks	Provides depth in under-researched areas, facilitating the discovery of novel insights	Risk of excessive data or analysis overload due to the open-ended nature of the approach

(continued)

Table 3 (continued)

Aspect	Description	Significance	Challenges
Comprehensive perspective	It blends positive and interpretive paradigms, creating an inclusive approach to analyzing data-driven and contextual insights	Broadens understanding by exploring topics from multiple viewpoints, enhancing theoretical and practical implications	Managing the integration of opposing paradigms without bias or compromising consistency
Adaptability for dynamic fields	Methodological flexibility allows for responsiveness to new insights and evolving research needs	Supports research relevance in rapidly changing areas, such as fields where new techniques and technologies continuously emerge	Risk of methodological drift, potentially reducing focus or coherence of research objectives
Academic and practical contributions	Generates context-specific insights beneficial for academic inquiry and practical applications, contributing to sustainable architecture discourse and decision-making frameworks in various fields	Advances knowledge base and informs best practices, promoting practical application in academic and professional contexts	Ensuring findings are both rigorously academic and accessible for practical application

which offers a contextually rich understanding in addition to giving empirical data. Adding ethical concerns, validity and reliability measurements, and other metrics to the inductive method gives the study more depth and credibility. This ensures that the results are academically sound and helpful in real life. These choices in methods contribute to advancing research on sustainable building practices and provide essential contributions that promote informed decision-making in architecture and urban planning.

References

1. Javali S, Javali MSB (2024) A text book of research methodology and applied statistics. Academic Guru Publishing House
2. Tracy SJ (2024) Qualitative research methods: collecting evidence. John Wiley & Sons, Crafting Analysis, Communicating Impact
3. Abusaada H, Elshater A (2024) Beyond keywords: Effective strategies for building consistent reference lists in scientific research. Publications 12(3):25
4. Lim WM (2024) What is qualitative research? An overview and guidelines. Australas Mark J 14413582241264619
5. Almusaed A, Yitmen I, Almssad A, Myhren JA (2024) Construction 5.0 and sustainable neuro-responsive habitats: integrating the brain–computer interface and building information modeling in smart residential spaces. Sustainability 16(21):9393
6. Fellows RF, Liu AM (2021) Research methods for construction. John Wiley & Sons

7. Dwivedi YK, Jeyaraj A, Hughes L, Davies GH, Ahuja M, Albashrawi MA, Al-Busaidi AS, Al-Sharhan S, Al-Sulaiti KI, Altinay L, Amalaya S (2024) "Real impact": challenges and opportunities in bridging the gap between research and practice–making a difference in industry, policy, and society. Int J Inf Manag 102750
8. Segev M (2022) The value of the world and of oneself: philosophical optimism and pessimism from Aristotle to modernity. Oxford University Press
9. Kedron P (2024) Logical positivism. The encyclopedia of human geography. Springer International Publishing, Cham, pp 1–6
10. Almusaed A, Yitmen I, Almsaad A, Akiner İ, Akiner ME (2021) Coherent investigation on a smart kinetic wooden façade based on material passport concepts and environmental profile inquiry. Materials 14(14):3771
11. Dehalwar K, Sharma SN (2023) Fundamentals of research writing and uses of research methodologies. Edupedia Publications Pvt Ltd
12. Proudfoot K (2023) Inductive/deductive hybrid thematic analysis in mixed methods research. J Mixed Methods Res 17(3):308–326
13. Havinga J, Rae A (2024) Pluralist function interrogation: a methodology for analysing workplace practices. Saf Sci 173:106440
14. Weerasinghe A (2023) The impacts of occupant behaviours on energy consumption in New Zealand office buildings: a thesis submitted in partial fulfilment of the requirements for the degree of Doctor of Philosophy in Construction at Massey University, Albany, New Zealand (Doctoral dissertation, Massey University)
15. Christensen PH, Robinson SJ, Simons RA (2018) The influence of energy considerations on decision-making by institutional real estate owners in the US. Renew Sustain Energy Rev 94:275–284
16. Andersson E, Borgström S, Haase D, Lanemeyer J, Mascarenhas A, McPhearson T, Wolff M, Laszkiewicz E, Kronenberg J, Barton D, Herreros-Cantis P (2021) A context-sensitive systems approach for understanding and enabling ecosystem service realization in cities. Ecol Soc 26(2)
17. Susanto PC, Yuntina L, Saribanon E, Soehaditama JP, Liana E (2024) Qualitative method concepts: literature review, focus group discussion, ethnography and grounded theory. Siber J Adv Multidiscip 2(2):262–275
18. Bentalha B, Alla L (2024) Revealing the subtleties: the art of qualitative studies in science and management. Applying qualitative research methods to management science, pp 1–21. IGI Global
19. Dissaux T (2024) The impact of information retrieval in architectural design: cognitive strategies for navigating parametric design environments. Doctoral dissertation, Universite de Liege (Belgium)
20. Ngeno FC (2020) Situating the concept of sustainable design in Nairobi, Kenya. Doctoral dissertation, University of Sheffield
21. Léveillé DJ (2019) Knowledge creation within the boundaries of physical spaces: a case study of individual knowledge creators within an innovative firm. Northcentral University
22. Sibeoni J, Verneuil L, Manolios E, Révah-Levy A (2020) A specific method for qualitative medical research: the IPSE (inductive process to analyze the structure of lived experience) approach. BMC Med Res Methodol 20:1–21
23. Jebb AT, Parrigon S, Woo SE (2017) Exploratory data analysis as a foundation of inductive research. Hum Resour Manag Rev 27(2):265–276
24. Azungah T (2018) Qualitative research: deductive and inductive approaches to data analysis. Qual Res J 18(4):383–400
25. Animashaun BR (2023) Building lighting and heating systems and effect on building performance and occupants. Doctoral dissertation, The University of Huddersfield
26. Knight C, Haslam SA (2010) The relative merits of lean, enriched, and empowered offices: an experimental examination of the impact of workspace management strategies on well-being and productivity. J Exp Psychol Appl 16(2):158
27. Weise A, Büchter R, Pieper D, Mathes T (2020) Assessing context suitability (generalizability, external validity, applicability or transferability) of findings in evidence syntheses in

healthcare—an integrative review of methodological guidance. Research Synthesis Methods 11(6):760–779
28. Packard MD (2017) Where did interpretivism go in the theory of entrepreneurship? J Bus Ventur 32(5):536–549
29. Sarker S, Lau F, Sahay S (2000) Using an adapted grounded theory approach for inductive theory building about virtual team development. ACM SIGMIS database: the DATABASE for advances in information systems 32(1):38–56
30. Sovacool BK, Axsen J, Sorrell S (2018) Promoting novelty, rigor, and style in energy social science: towards codes of practice for appropriate methods and research design. Energy Res Soc Sci 45:12–42
31. Yanow D, Schwartz-Shea P (2015) Interpretation and method: empirical research methods and the interpretive turn. Routledge
32. Kivunja C, Kuyini AB (2017) Understanding and applying research paradigms in educational contexts. Int J High Educ 6(5):26–41
33. Ejigu AK, Yeshitela K (2023) Integrating resource oriented sanitation technologies with urban agriculture in developing countries: measuring the governance capacity of Arba Minch city, Ethiopia. Front Sustain Cities 5:1153502
34. Qaissi A (2024) Exploring thematic analysis in qualitative research. Data collection and analysis in scientific qualitative research, pp 253–294. IGI Global
35. Burns M, Bally J, Burles M, Holtslander L, Peacock S (2022) Constructivist grounded theory or interpretive phenomenology? Methodological choices within specific study contexts. Int J Qual Methods 21:16094069221077758
36. Tracy SJ (2024) Qualitative research methods: collecting evidence, crafting analysis, communicating impact. John Wiley & Sons
37. Reed JE, Howe C, Doyle C, Bell D (2018) Simple rules for evidence translation in complex systems: a qualitative study. BMC Med 16:1–20
38. Lal JP, Bedekar JS, Athawale BK (2023) Research methodology: theory. Academic Guru Publishing House, Practices and tools
39. Morse JM, Barrett M, Mayan M, Olson K, Spiers J (2002) Verification strategies for establishing reliability and validity in qualitative research. Int J Qual Methods 1(2):13–22t
40. Elsawah S, Pierce SA, Hamilton SH, Van Delden H, Haase D, Elmahdi A, Jakeman AJ (2017) An overview of the system dynamics process for integrated modelling of socio-ecological systems: lessons on good modelling practice from five case studies. Environ Model Softw 93:127–145
41. Denscombe M (2017) EBOOK: the good research guide: for small-scale social research projects. McGraw-Hill Education (UK)
42. Hurry J (2016) Experimental design. Designing research in education: concepts and methodologies, p 99
43. Mulisa F (2022) When does a researcher choose a quantitative, qualitative, or mixed research approach? Interchange 53(1):113–131
44. Almusaed A (2010) Biophilic and bioclimatic architecture: analytical therapy for the next generation of passive sustainable architecture. Springer Science & Business Media
45. Munaro MR, Tavares SF, Bragança L (2020) Towards circular and more sustainable buildings: a systematic literature review on the circular economy in the built environment. J Clean Prod 260:121134
46. Saharan VA, Kulhari H, Jadhav H, Pooja D, Banerjee S, Singh A (2020) Introduction to research methodology. Principles of research methodology and ethics in pharmaceutical sciences, pp 1–46. CRC Press
47. Abdullah Osman Mohammed L (2024) Improving the supply chain of building materials. Doctoral dissertation, Politecnico di Torino
48. Austin Z, Sutton J (2014) Qualitative research: getting started. Can J Hosp Pharm 67(6):436
49. Schultz FC, Reinhardt RJ (2022) Facilitating systemic eco-innovation to pave the way for a circular economy: a qualitative-empirical study on barriers and drivers in the European polyurethane industry. J Ind Ecol 26(5):1646–1675

50. Hou D, Guthrie P, Rigby M (2016) Assessing the trend in sustainable remediation: a questionnaire survey of remediation professionals in various countries. J Environ Manage 184:18–26
51. Krueger RA (2014) Focus groups: a practical guide for applied research. Sage publications
52. Giorgi S, Lavagna M, Wang K, Osmani M, Liu G, Campioli A (2022) Drivers and barriers towards circular economy in the building sector: Stakeholder interviews and analysis of five European countries policies and practices. J Clean Prod 336:130395
53. Morgan SJ, Pullon SR, Macdonald LM, McKinlay EM, Gray BV (2017) Case study observational research: a framework for conducting case study research where observation data are the focus. Qual Health Res 27(7):1060–1068
54. Varadarajan R (2017) Innovating for sustainability: a framework for sustainable innovations and a model of sustainable innovations orientation. J Acad Mark Sci 45:14–36
55. Abowitz DA, Toole TM (2010) Mixed method research: Fundamental issues of design, validity, and reliability in construction research. J Constr Eng Manag 136(1):108–116
56. Campbell S, Greenwood M, Prior S, Shearer T, Walkem K, Young S, Bywaters D, Walker K (2020) Purposive sampling: complex or simple? Research case examples. J Res Nurs 25(8):652–661
57. Balador Z (2020) Increasing the use of reclaimed and recycled building materials in New Zealand: Stakeholder perceptions. Doctoral dissertation, Open Access Te Herenga Waka-Victoria University of Wellington
58. Schreier M (2018) Sampling and generalization. The SAGE handbook of qualitative data collection, pp 84–97
59. Benbasat I, Goldstein DK, Mead M (1987) The case research strategy in studies of information systems. MIS Q 369–386
60. Onwuegbuzie AJ, Dickinson WB, Leech NL, Zoran AG (2009) A qualitative framework for collecting and analyzing data in focus group research. Int J Qual Methods 8(3):1–21
61. Agyekum K, Adinyira E, Baiden B, Ampratwum G, Duah D (2019) Barriers to the adoption of green certification of buildings: a thematic analysis of verbatim comments from built environment professionals. J Eng, Des Technol 17(5):1035–1055
62. Krippendorff K (2018) Content analysis: an introduction to its methodology. Sage publications
63. Govindan K, Hasanagic M (2018) A systematic review on drivers, barriers, and practices towards circular economy: a supply chain perspective. Int J Prod Res 56(1–2):278–311
64. Beattie V (2014) Accounting narratives and the narrative turn in accounting research: issues, theory, methodology, methods and a research framework. Br Account Rev 46(2):111–134
65. Woods M, Paulus T, Atkins DP, Macklin R (2016) Advancing qualitative research using qualitative data analysis software (QDAS)? Reviewing potential versus practice in published studies using ATLAS. ti and NVivo, 1994–2013. Soc Sci Comput Rev 34(5):597–617
66. Mohajan HK (2017) Two criteria for good measurements in research: validity and reliability. Annals of Spiru Haret University. Eco Series 17(4):59–82
67. Sieber JE, Tolich MB (2013) Planning ethically responsible research, vol 31. Sage
68. Anderson ST, Lin KK (2024) Scientific method. Translational orthopedics, pp 13–15. Academic Press
69. Green JP, Dalal RS, Fyffe S, Zaccaro SJ, Putka DJ, Wallace DM (2023) An empirical taxonomy of leadership situations: development, validation, and implications for the science and practice of leadership. J Appl Psychol 108(9):1515
70. Ahmed I, Ishtiaq S (2021) Reliability and validity: importance in medical research. Methods 12(1):2401–2406
71. Findley MG, Kikuta K, Denly M (2021) External validity. Annu Rev Polit Sci 24(1):365–393
72. Rauch A, Saeed S, Frese M (2024) Decoding evidence-based entrepreneurship: a systematic review of meta-analytic choices and reporting. J Small Bus Manag pp 1–47
73. Slocum TA, Pinkelman SE, Joslyn PR, Nichols B (2022) Threats to internal validity in multiple-baseline design variations. Perspect Behav Sci 45(3):619–638
74. Karunarathna I, De Alvis K, Gunasena P, Jayawardana A (2024) Designing and conducting clinical research: methodological approaches

75. Henry S, Christian JA (2023) Absolute triangulation algorithms for space exploration. J Guid Control Dyn 46(1):21–46
76. Khoa BT, Hung BP, Hejsalem-Brahmi M (2023) Qualitative research in social sciences: data collection, data analysis and report writing. Int J Public Sect Perform Manag 12(1–2):187–209
77. Aschbrenner KA, Kruse G, Gallo JJ, Plano Clark VL (2022) Applying mixed methods to pilot feasibility studies to inform intervention trials. Pilot Feasibility Stud 8(1):217
78. Aithal A, Aithal PS (2020) Development and validation of survey questionnaire and experimental data–a systematical review-based statistical approach. Int J Manag Technol Soc Sci (IJMTS) 5(2):233–251
79. Clark K, Birch-Hurst K, Pennington CR, Petrie AC, Lee JT, Hedge C (2022) Test-retest reliability for common tasks in vision science. J Vis 22(8):18–18
80. Cole R (2023) Inter-rater reliability methods in qualitative case study research. Sociol Methods Res 00491241231156971
81. Jacobson E, Conboy L, Tsering D, Shields M, McKnight P, Wayne PM, Schnyer R (2019) Experimental studies of inter-rater agreement in traditional chinese medicine: a systematic review. J Altern Complement Med 25(11):1085–1096
82. Goodboy AK, Martin MM (2020) Omega over alpha for reliability estimation of unidimensional communication measures. Ann Int Commun Assoc 44(4):422–439
83. Gundersen OE (2021) The fundamental principles of reproducibility. Phil Trans R Soc A 379(2197):20200210
84. Carcary M (2020) The research audit trail: methodological guidance for application in practice. Electron J Bus Res Methods 18(2):166–177
85. Kang E, Hwang HJ (2021) Ethical conducts in qualitative research methodology: participant observation and interview process. J Res Publ Ethics 2(2):5–10
86. Morris C, Detrick JJ, Peterson SM (2021) Participant assent in behavior analytic research: Considerations for participants with autism and developmental disabilities. J Appl Behav Anal 54(4):1300–1316
87. Hulkower R, Penn M, Schmit C (2020) Privacy and confidentiality of public health information. Public Health Inform Inf Syst 147–166
88. Kolasa K, Ken Redekop W, Berler A, Zah V, Asche CV (2020) Future of data analytics in the era of the general data protection regulation in Europe. Pharmaco Eco 38(10):1021–1029
89. White MG (2020) Why human subjects research protection is important. Ochsner J 20(1):16–33
90. Pearlman W (2023) Emotional sensibility: exploring the methodological and ethical implications of research participants' emotions. Am Polit Sci Rev 117(4):1241–1254
91. Fernandez Lynch H (2020) The right to withdraw from controlled human infection studies: justifications and avoidance. Bioethics 34(8):833–848
92. Kjeldaas S, Dassler T, Antonsen T, Wikmark OG, Myhr AI (2023) With great power comes great responsibility: why 'safe enough' is not good enough in debates on new gene technologies. Agric Hum Values 40(2):533–545

Developing a Research Proposal

Abstract Formulating a research proposal is crucial in promoting scholarly investigation as the basis for obtaining financing and academic endorsement. A well-organized proposal converts a novel concept into a meticulously designed research project, offering a definitive framework that delineates the study question, objectives, technique, and expected contributions. The core of this method is formulating a specific research question and conducting a comprehensive literature assessment, positioning the study within the current knowledge framework, and highlighting the gaps the proposed research intends to address. A consistent methodological framework is crucial, outlining the data collection procedures, analysis, and interpretation to ensure the research's viability and academic rigor. The proposal must demonstrate methodological rigor and address ethical issues, especially with human subjects, assuring adherence to norms of informed consent, confidentiality, and data protection. A detailed budget and timeframe augment the proposal's credibility, indicating that the project is financially and chronologically feasible. Furthermore, the proposal must highlight the research's importance, clarifying its possible contributions to scholarly dialog and practical implementations. A comprehensive, rational, and compelling argument in a research proposal successfully connects intellectual innovation with practical application, enhancing the study's potential for successful execution and significant impact.

1 Introduction to the Thematic Area

The most cutting-edge research concept, backed by a reliable methodology and a persuasive explanation, has the potential to significantly contribute to the development of innovative solutions and real-world applications [1]. Thousands of research concepts have the potential to transform industries, introduce new philosophies, and solve current difficulties that are occurring on a worldwide scale. Nevertheless, the capacity to acquire support and money is a significant factor in determining whether a research concept will be successful, regardless of how ground breaking it may be [2]. The process of transforming a visionary idea into a fully established research project

A. Almusaed et al., *Practice of Research Methodology in Civil Engineering and Architecture*, https://doi.org/10.1007/978-3-031-97393-2_8

is fraught with obstacles, one of the most important of which is the acquisition of financial backing. Obtaining financial support is a crucial stage in the process of doing research. When no financial resources are available, even the most well-formulated ideas are impossible to test, verify, or implement. This is because they stay theoretical. The provision of funding enables the acquisition of vital resources, like cutting-edge machinery and technology, data-gathering tools, and the workforce required to carry out a project [3]. In addition, it helps to disseminate the research findings, making it possible for the results to be communicated to a broader audience through channels such as journals, conferences, and collaborative efforts. This financial support is essential to the vitality of the research ecosystem, which permits discoveries that can potentially change both businesses and society. Acquiring finance is frequently tricky and competitive, even though it is critical. An overwhelming number of financing applications are submitted to research institutions, private investors, and government agencies each year; nevertheless, only a small percentage of these projects are successful. Not only does the level of novelty or relevance of the study matter in this highly competitive climate, but the degree to which the concept is effectively communicated is also the most critical factor [4]. Regardless of the promise that lies under the surface, it is unlikely that a proposal will be given serious consideration if it does not possess clarity, coherence, or a clear description of the possible impact that it may have. As a result, the presentation of a study proposal becomes equally important as the idea itself. A well-prepared research proposal must bridge the gap between a creative idea and the financial assistance necessary for its actualization [5]. This proposal aims to act as a formal document that explains the research strategy in an organized and compelling manner. The document provides an overview of the project's goals and relevance and exhibits the project's wider worth and capacity to be applied. The proposal serves as a road map for potential funders, demonstrating how the study will be carried out, the techniques utilized, and how the findings will address specific issues or expand knowledge in each subject. The capacity to emphasize the practical ramifications of the study is one of the most essential features of a proposal that is more likely to be successful. Funders often are interested in understanding how their monetary contributions will result in quantifiable and significant outcomes. A solid proposal presents a compelling case for funding by explicitly expressing how the study adds to advancing technology, science, or social well-being. It also tackles real-world concerns by defining how the research solves such issues [6]. Having the capacity to transform a theoretical concept into a strategy that can be implemented and is motivated by outcomes is frequently what differentiates successful research proposals from those that are not. Additionally, a study proposal establishes whether the project can be carried out successfully. What is more critical to funders than the originality of the idea itself is whether the project can be carried out in a manner that is feasible within the constraints of the planned timetable and budget. The trust of the funders in the effective implementation of the project is increased when they are provided with a comprehensive explanation of the methodology, resources, and persons engaged in the project. It assures them that the research team is adequately equipped to meet the project's challenges and generate significant results. Overall, the quality of the study proposal is the most critical factor in determining whether financing can be obtained. A proposal that has been carefully constructed not only communicates the relevance of the research concept but also instills confidence in the practicality of the idea and the potential effect it may have [7]. The research proposal is a document that takes a good concept and develops it into a concrete plan that has the potential to attract the required financial backing to bring it to reality. This is accomplished through a precise, clear, and compelling presentation. It is possible that even the most forward-thinking research ideas will not be realized if they do not receive such funding, which would result in their potential benefits to society being unmet.

2 Essential Components of a Coherent Research Proposal

A well-organized research proposal is essential for the success of any academic undertaking, be it a dissertation or a thesis [8]. The proposal functions as a comprehensive guide for the research process, delineating the study's scope and objectives while showcasing the researcher's capacity for logical organization of ideas. While the precise structure of research proposals may differ among institutions, the majority conform to a series of fundamental components that guarantee clarity, logical coherence, and scholarly rigor [5]. These fundamental elements provide a definitive framework, directing the researcher and the review committee through the proposed study's aims, methods, and significance within the academic discipline. Incorporating these essential parts guarantees that the proposal is organized thoroughly and cohesively. Using a consistent structure, the researcher may convey the importance of their study while demonstrating its viability. This document examines the essential elements of a well-structured research proposal and analyzes the tactics required to enhance each component for a successful study submission.

2.1 The Role and Importance of the Title or Title Page in a Research Proposal

A research proposal's title or title page is the initial formal impression, articulating the central subject of the study. It comprises critical details such as the research title, author's name, institutional affiliation, and submission date [9]. The title page for a research proposal on Smart Buildings and Energy Efficiency may state: "Enhancing Energy Efficiency in Smart Buildings through Advanced AI Systems." The researcher's name, institution, and submission date are explicitly indicated beneath the title. This page establishes the proposal's tone, providing a brief summary of the study's emphasis. An effectively crafted title should concisely convey the core of the research, offering explicit guidance on the study's topic. Research on Materials Passports in Building Construction may be titled "Enhancing Sustainability through

Materials Passports: A Lifecycle Approach to Building Materials Management." This term conveys the research's focus on sustainability and implementing materials passports. This system records the material characteristics and possibilities for reuse or recycling in construction elements. The title offers a concise and helpful summary and directs the reader's expectations concerning the study's scope and aims. The headline regarding materials passports would emphasize sustainable construction techniques and lifecycle management, underscoring the research's contribution to the increasing need for environmental accountability in the building industry [10]. The title page often contains the researcher's name, institutional affiliation, and submission date, formalizing and situating the proposal within an academic context. This information is crucial, especially in research that connects with policy and industry norms, such as initiatives concerning integrating materials passports. The title and title page define the study's identity, establishing a framework for a coherent and academically pertinent proposal.

2.2 Explanation in Extensive Detail of the Proposed Research and Historical Context

A detailed explanation of the proposed research and its historical context is essential to every research proposal [11]. This part necessitates a comprehensive examination of the research issue, highlighting its importance in academic and societal contexts. In a study on biophilic design, the researcher may emphasize how biophilic principles incorporating nature into architectural spaces address the increasing need for well-being in urban settings. A historical analysis may delineate the evolution of biophilic design from the first ecological design theories to its contemporary implementation in sustainable construction. The proposal situates the study within the larger context of environmental design, illustrating its significance to current challenges like mental health, environmental sustainability, and urban resilience [12]. In a study on ecological design, the investigator may examine the evolution of sustainable construction practices in response to significant environmental issues. This may involve analyzing the redefinition of early green construction methods in response to global climate change and resource depletion issues. A thorough literature assessment would reveal deficiencies in the existing discourse, particularly about the inadequate emphasis on incorporating adaptable design solutions in low-income housing, which may serve as the focal point of the proposed study [13]. The researcher differentiates their approach from other studies by pinpointing deficiencies in the current literature, notably the absence of empirical investigations about the long-term advantages of biophilic and environmental design in varied urban contexts. By citing essential literature, the researcher displays a profound comprehension of the domain, affirming their proposed contribution's intellectual and practical importance. This foundation establishes the research as a significant contribution to the current academic discourse and a prospective remedy for tangible environmental and societal issues [14].

2.3 *Evaluation of the Research Project*

The research project assessment critically synthesizes essential elements to provide a concise yet thorough summary [15]. This part summarizes the research question or hypothesis, a brief methodology, and anticipated outcomes, offering an overview of the project's primary emphasis. A study investigating building energy efficiency may present a research topic such as, "In what ways can smart insulation technologies diminish energy consumption in urban residential structures?" The technique may encompass a comparative comparison of conventional insulating materials with novel, energy-efficient alternatives. A brief overview would delineate the comparison methodology and emphasize expected results, including a substantial decrease in energy usage and operational expenses. In a project centered on recycled building materials, the hypothesis may examine the environmental advantages of using recycled materials in contemporary construction methodologies. The summary may delineate a process in which varying percentages of recycled materials are integrated into architectural structures and evaluated for performance, durability, and sustainability [16]. Expected outcomes may encompass increased sustainability, less carbon footprint, and optimized material lifecycle management. The summary must succinctly convey the project's importance, illustrating its approach to tackling essential sustainability concerns. The research enhances both academic debate and the actual progress of sustainable construction by prioritizing energy efficiency and the utilization of recyclable resources [17]. Precise and targeted examples make the summary fascinating and appealing, prompting further investigation.

2.4 *Key Scientific Contributions and Emerging Gaps in the Field*

The literature review, an overview of significant scientific research within a discipline, is essential to a research proposal [18]. In the Sustainability in Building Information Modeling (BIM), this level entails a comprehensive assessment and critical evaluation of significant contributions to incorporating sustainable practices within BIM frameworks. The researcher must initially delineate pivotal investigations, including those examining BIM's potential to enhance energy efficiency in architectural designs or its application in minimizing waste during construction [19]. These studies establish a basis for comprehending how BIM has furthered sustainability objectives in the building sector. A researcher may examine seminal studies that investigate the application of BIM in energy modeling, illustrating how BIM tools may replicate building performance and energy usage during the lifespan [20]. Through the analysis of these contributions, the researcher can elucidate the effectiveness of BIM in advancing sustainability while also pinpointing areas requiring further investigation, such as the scarcity of studies on integrating recycled materials within BIM models or utilizing BIM in circular economy frameworks. This overview enables the researcher to situate their work as essential to the discipline. The researcher illustrates their work's significance and imperative nature by pinpointing deficiencies such as inadequate emphasis on how BIM might improve sustainability in smaller scale projects or disadvantaged areas. Critically interacting with existing research enables the researcher to propose new avenues, providing insights that may advance the frontiers of sustainable practices in BIM [21].

2.5 Detailed Explanation of the Proposed Methodology for the Research Study

The way the research will be carried out is described in the section under "study design" or "methodology." It should offer a clear and complete strategy of the methodology that will be employed, including methods for data collection, strategies for analysis, and any equipment or instruments that will be necessary [22]. In addition, this part should discuss the practicability of the planned study, considering aspects such as the already available resources, the limitations of time, and the potential difficulties. A robust methodology section assures reviewers that the research has been carefully considered and that the researcher is outfitted with the necessary resources to conduct the investigation properly. Incorporating these fundamental elements into a study proposal enables researchers to develop a robust basis for their inquiry. Every component of the planned research project is essential for maintaining clarity, coherence, and viability [23]. A meticulously organized proposal substantiates the study topic's legitimacy and facilitates a successful and influential academic pursuit. Table 1 delineates the fundamental elements of a coherent study proposal, emphasizing critical components such as the title, study context, project appraisal, literature review, and methodology. Every element enhances the coherence and clarity of the idea. The chart delineates prevalent issues at each step, providing insights into possible hurdles and ideas for improving the proposal's overall quality.

Table 1 Essential components and challenges in a coherent research proposal

Component	Description	Challenges
Title or title page [24]	It clearly articulates the research subject, aims, and parameters. It contains the researcher's name, affiliation, and submission date for official context	Formulating a brief title that accurately encapsulates the study's substance
Research explanation and context [25]	This section comprehensively examines the research topic, its historical context, and its importance in academic or societal frameworks	It guarantees comprehensive background research while eschewing superfluous information that obscures understanding
Research project evaluation [26]	The summary encapsulates the research topic, methods, and anticipated outcomes, highlighting the study's significance, viability, and prospective contributions	The goal is to maintain brevity while providing enough detail to demonstrate the study's significance and viability
Key scientific contributions and gaps [18]	Review critical studies in the field, identify gaps in existing research, and position the proposed study to address these gaps or advance current knowledge	Find out where there are gaps in the literature and fill them with your suggested study
Methodology and research design [27]	It describes the techniques for gathering data, the approaches for analyzing it, and the resources available to handle issues with practicality, obstacles, and morality	It is foreseeing pragmatic obstacles, like resource restrictions or methodological constraints

3 Efficient Introduction to a Research Proposal

The introduction of a research proposal is essential for contextualizing the study and providing preliminary insight into the research topic, aims, and importance [28]. This part often encompasses 1–2 pages and aims to captivate the reader by elucidating the study's significance, timeliness, and merit. An effectively organized introduction meets the standards of clarity, relevance, and academic rigor, ensuring it engages curiosity while situating the topic within the broader academic and societal framework. For example, the proposal may focus on using artificial intelligence (AI) in smart buildings to enhance energy efficiency, a vital issue in the architecture, engineering, and construction (AEC) industries [29]. The core research topic of this proposal is: how might AI-driven systems be employed to improve energy efficiency in smart buildings? This inquiry directly pertains to a fundamental difficulty in contemporary urban development—the necessity to diminish energy usage while preserving operational efficiency. The introduction must effectively delineate the study challenge and its significance to contemporary technology breakthroughs and sustainability initiatives, ensuring engagement and academic rigor. AI in smart buildings provides possible solutions for real-time monitoring and optimization of energy consumption, markedly decreasing energy waste by autonomously regulating systems like heating, ventilation, and lighting according to occupancy and ambient factors [30]. The introduction highlights the importance of the study by stressing new solutions, rendering it appealing and influential. A good introduction must delineate the methods to examine the study challenge. This study will employ a mixed-methods approach to collect quantitative and qualitative data. Quantitative data will be obtained from case studies of intelligent buildings utilizing AI-driven energy management systems, assessing key performance factors like energy savings, cost reduction, and environmental impact [31]. This will be enhanced by qualitative data from expert interviews, providing profound insights into the difficulties and possibilities of integrating AI technology in building systems. The introduction establishes the viability and academic rigor of the proposed study through a rigorous methodological framework, augmenting its credibility. A fundamental condition for a robust academic introduction is to articulate the importance of the study, both inside the scholarly field and in its broader social context. This research possesses significant promise for enhancing smart infrastructure and artificial intelligence applications in sustainable construction methods. It enhances the expanding corpus of knowledge about the transformative potential of AI in energy management within the built environment. It tackles pressing global issues, like energy conservation and reducing carbon footprints. Predictive AI algorithms can effectively facilitate energy use in smart buildings, leading to substantial cost reductions and less environmental impact [32]. The introduction aligns the research with urgent global challenges, highlighting the study's significant significance and attracting the attention of funding organizations and academic committees. Ultimately, an introduction must not only articulate the research issue but also substantiate the importance of the study, demonstrate its relevance to current academic and social demands, and elucidate its contributions to theoretical progress and practical implementation [33]. By fulfilling these criteria, the introduction transforms into a crucial component of a successful research proposal, offering a clear, engaging, and academically rigorous entry into the project. The essential elements for writing a practical introduction to a research proposal are outlined in Table 2. A straightforward research question, an outline of the study's relevance and scope, a description of the research aims, and an explanation of the methodology are some of the main components highlighted in this section. In addition, the table emphasizes highlighting the repercussions for society and ensuring the introduction is engaging. One challenge is explaining the study's realistic impact while balancing information and brevity.

Table 2 Criteria for a good introduction in a research proposal

Criterion	What it needs to include?	Role	Challenges
Clear research question [34]	A brief, clearly articulated research topic that delineates the primary issue or problem the study investigates	Defines the objective and scope of the investigation, directing the research framework and technique	Crafting a specific inquiry appropriately scoped and aligned with the research objectives
Relevance and scope [35]	Contextual information on the significance of the topic and its alignment with the field of study	Demonstrates the importance of the research within the academic field and its relevance to current societal or technological issues	Ensuring the introduction sufficiently explains the scope without overwhelming the reader with too much background detail
Research objectives [36]	The study's precise objectives are often congruent with the research issue	Clarifies the researcher's objectives, establishing explicit goals for the investigation	Formulate attainable, quantifiable objectives corresponding to the study issue and the broader discipline
Methodology overview [37]	Formulate attainable, quantifiable objectives corresponding to the study issue and the broader discipline	Provides an overview of the study methodology, ensuring the reader comprehends the strategy and its practicality	Attaining equilibrium between brevity and detail—providing adequate information to demonstrate the comprehensiveness of the procedure without overwhelming the introduction
Significance of research [38]	Elucidation of how the research will augment scholarly comprehension and tackle pragmatic challenges	The study's scholarly and practical value is demonstrated, increasing the likelihood of receiving approval or finance	Without making assumptions or broad assertions about the study's significance, it lays out the possible effects
Societal implications [39]	Analysis of the study's potential societal, industrial, or policy impacts	It highlights the broader implications, showing that the study addresses real-world problems and has the potential to make a difference	Forecasting the effects reasonably without exaggerating the breadth of the research's political significance

(continued)

Table 2 (continued)

Criterion	What it needs to include?	Role	Challenges
Engagement and attraction [40]	Expressions that convey the author's excitement for the subject and captivate the reader about it	Hooks readers and shows how seriously the researcher takes the subject	Maintaining the reader's interest by combining an academically rigorous tone with an appealing narrative approach
Scope and relevance to field [41]	Direct correlation to contemporary research trends and the broader academic dialog within the discipline	Positions the topic within current academic discourse and addresses contemporary challenges	Exhibiting originality while acknowledging previous studies and articulating contributions to the current knowledge

4 Background and Significance of the Study

A study proposal's "Background and Importance" section is often succinct yet intensely concentrated. Depending on the research's extent and intricacy, this part may vary from one to two pages. The objective is to convey the fundamental reasoning and importance without inundating the reader with extra data. An organized and coherent methodology is crucial to sustain the reader's engagement and successfully emphasize the critical elements of the study [42]. To enhance the proposal's appeal, many essential methods must be implemented:

- **Clarity and Precision**: The terminology employed in this section must be unequivocal and devoid of ambiguity [43]. Each line must contribute meaningfully to the argument, and jargon should be employed sparingly, only when essential. This enables individuals from other academic disciplines to interact with the material and comprehend its significance.
- **Engagement via Relevance**: Highlighting the study's practical ramifications and significance guarantees that the proposal connects with the reader [44]. Connecting the study to contemporary global issues—such as sustainability, climate change, or technological innovation—establishes a direct link to broader academic and social discussions.
- **Logical Progression**: This part must advance coherently from issue identification to delineating the suggested solution, the research inquiries, and the prospective contributions to the existing body of knowledge [45]. A structured framework facilitates the reader's comprehension of the story and highlights the consistency of the study strategy.

- **Emphasizing Innovation**: To attract attention, it is essential to accentuate the research's distinctive or original aspects [46]. Highlighting an innovative approach, a fresh theoretical perspective, or an underexplored field of study would distinguish the proposal.
- **Equilibrium of Detail**: The section must be concise yet sufficiently detailed to demonstrate the profundity of the investigation. Essential concepts must be succinctly elucidated, and significant terminology must be specified [47]. Refrain from inundating the section with excessive details while ensuring enough substance to pique attention.
- **Visual Appeal**: While not necessarily essential in text-dense proposals, using headers, subheadings, or illustrations, where suitable, may enhance readability and increase engagement [48]. A visually organized layout improves readability, particularly when conveying intricate concepts or inquiries.

While the "Background and Importance" section should be concise, it should also be compelling and designed to pique interest in the proposal while delineating the research's objective, importance, and contribution to the field. Concentrating on clarity, involvement, and unique input will make the proposal appealing and academically persuasive [49]. A visually organized layout, with headers, subheadings, or illustrations where suitable, is not just a bonus but a necessity that improves readability, particularly when conveying intricate concepts or inquiries. Table 3 is a checklist for enhancing the clarity, engagement, and organization of research proposals' "Background and Importance" part. It delineates issues like clarity, logical coherence, and relevance, offering pragmatic answers and illustrations. Adhering to these guidelines—such as eschewing jargon, emphasizing originality, and sustaining an appropriate degree of detail—enhances the section's impact and accessibility for readers.

5 Review of Prior Literature and Research

Brevity and concentration should be maintained while presenting the background and importance of the suggested study in a clear and organized literature review [56]. In a typical three- to five-page review, it is crucial to emphasize critical studies, approaches, and literature gaps supporting the new research. The review needs to be concise, lucid, and persuasive. This format strikes a balance between being transparent and detailed:

- **Citing Previous study**: Closely to the five to eight most pertinent papers addressing the study topic. Reviewing every survey on the subject is unnecessary; instead, focus on the studies that indicate significant advances or gaps in the field [57]. For instance, cite studies that look at energy-efficient building practices and emphasize the gaps your study fills in sustainable building materials. Explain everything succinctly and ensure it relates directly to the study's goals.

Table 3 Checklist for enhancing the "Background and Importance" section

Challenges	Solutions	Examples
Clarity and precision [50]	Ensure vocabulary is exact, eliminate excessive jargon, and render each statement significant to the argument	Clarify essential words such as "sustainability" and refrain from employing needlessly convoluted academic terminology
Engagement via relevance [51]	Connect the study to pertinent global challenges such as sustainability, climate change, or technology to enhance its relevance in broader discourse	For example, if the research examines sustainable construction materials, it can be connected to addressing climate change
Logical progression [52]	Systematize the segment from issue identification to delineating solutions and contributions, ensuring clarity and coherence	Ensure the section transitions seamlessly, starting with the issue (e.g., deficiency of energy-efficient materials) and progressing to the suggested resolution (e.g., developing innovative sustainable materials)
Emphasizing innovation [53]	To attract attention, emphasize distinctive or new elements of the study, such as an original approach or viewpoint	An illustration may involve introducing an innovative method for utilizing recycled resources in building, which remains mostly unexamined
Equilibrium of detail [54]	Offer sufficient information to demonstrate depth while preventing reader overload. Articulate fundamental terminology and elucidate critical topics concisely	For example, elucidate phrases such as "carbon footprint" without inundating the section with extraneous information or superfluous technicalities
Visual appeal [55]	Utilize headers, subheadings, or diagrams to segment text and enhance the comprehensibility of intricate subjects	Utilize titles like "Problem Statement" and "Proposed Solution" to direct the reader or incorporate a basic flowchart to elucidate the study methodology

- **Analyzing Approaches and Results**: Choose three to four essential papers illustrating various approaches [58]. Examine how they tackled the subject, mainly how they investigated related problems. For example, in studies on biophilic design, qualitative interviews may be used in one study, while energy performance models are used in another. Briefly describe these approaches, highlighting how they apply to your research and why some are more suited.
- **Comparing and Contrasting**: List the main points where earlier research has agreed and disagreed [59]. Spend two to three points pertinent to the study instead of discussing every dispute. For instance, although some research emphasizes the energy efficiency of green roofs, others contend that they improve biodiversity. Clearly state these differences and demonstrate how they support or contradict the present study's findings. For clarity, use bullet points or brief paragraphs.

- **Reviewing Literature**: Summarize two to three studies that have significantly impacted or are exceptionally convincing in the subject [60]. Emphasize the advantages of their approaches or the significance of their conclusions. Mention, for example, whether the research combined biophilic concepts uniquely or employed sophisticated thermal models. This section should highlight why these studies are particularly pertinent to the investigation.
- **Concerning the Proposed Research**: Describe in this last section how the research contributes to, synthesizes, or departs from the body of current literature [61]. For instance, highlight how this research fills a need if no studies have paired biophilic design with energy-efficient materials. This part should be brief and clarify, without going into much detail, the unique contribution the new research provides to the discipline.

Creating a compact and attractive review:

- Use distinct headings: Headings like "Methodological Approaches" or "Key Studies" help maintain the review's structure and assist the reader in navigating it [62].
- Make choices: Steer clear of going into too much information. Discard any information unrelated to the study topic and concentrate on what does [63].
- Brief paragraphs or bullet points: Use them to emphasize important concepts or simplify comparisons when appropriate [64].
- Visual components: Include tables or charts if possible since they might help summarize methods or findings. This keeps the review visually structured and lessens the need for lengthy explanations [65].

By adhering to this format, the literature review may be condensed into three to five pages while offering a precise and concentrated explanation of the significant studies, approaches, and gaps that bolster the suggested investigation [66]. A systematic strategy for structuring a literature review is presented in Table 4. This approach focuses on crucial aspects such as citing previous research, methodology, comparisons, evaluations, and linking to the planned study. The document draws attention to possibilities for clarity and creativity, problems that should be avoided, and includes examples that apply to each area.

6 Research Design and Methods

The primary purpose of the research design and methodology section is to clarify how the research will be done and how it intends to address the research questions, which act as an essential foundation for any study [37]. This part needs to show confidence in the rigor and relevance of the planned study, even though the research might not have been started yet. The objective is to persuade the reader that the study

Table 4 Structure of literature review with opportunities, challenges, and examples

Section	Content focus	Opportunities	Challenges	Examples
Citing previous research [67]	Summarize five to eight pivotal studies pertinent to the study topic	Establishes context and creates a robust framework for the study	The potential exists for overwhelming the reader with many citations, so hindering concentration	Reference research on energy-efficient structures to underscore deficiencies in material innovation
Examining methodologies and outcomes [68]	Examine three–four principal approaches employed in prior studies	Facilitates a critical evaluation of various research methodologies, enhancing the selection process	A brief presentation is necessary to prevent inundating the reader with technical specifics	Contrast qualitative interviews in biophilic design with quantitative energy performance models
Comparing and contrasting [69]	Identify two or three notable areas of consensus or contention within the literature	It offers explicit insights into the scholarly discourse and aids in pinpointing research deficiencies	It must eschew superficial comparisons and need meticulous investigation to be effective	Comparative analysis of the ecological advantages of green roofs versus their thermal insulation characteristics
Evaluating the literature [70]	Concise evaluation of two—three significant studies about their techniques or results	Provides a concentrated analysis of the most credible studies, underscoring the significance of the new findings	Critically evaluating research impartially may be complicated, particularly in contentious fields	Assess research employing modern energy modeling software beside another utilizing conventional methodologies
Relating to the proposed research [71]	Elucidate how the planned research extends, deviates from, or integrates prior studies	It elucidates the distinctive contribution of the research, establishing it as innovative	It must distinguish the novel study from previously established concepts	Elucidate how the amalgamation of energy-efficient materials with biophilic design constitutes an innovative contribution

design and techniques that have been selected are not only suitable for solving the research topic but that they will also successfully interpret and evaluate the outcomes that are anticipated. To guarantee that the research objectives and methodology are consistent, it is necessary to develop a logical alignment between the research objectives and the respective techniques of data collecting and analysis [72]. The research proposal should use the most effective procedures currently in use in the industry, including specific examples from an exhaustive literature analysis. This strategy not only helps to establish the study's foundation in the existing body of scholarly work, but it also contributes to the justification of the selected methods by demonstrating how they have been effectively implemented in earlier research. Specification of the methods that will be used to gather data, the types of analyses that will be carried out, and the methods that will be used to evaluate the validity of the findings are all essential components of a robust study design. Specifically, qualitative techniques like interviews or case studies may be used to investigate in-depth viewpoints. At the same time, quantitative means like surveys or experiments may be utilized to quantify factors and test hypotheses [73]. For example, qualitative methods may be used to explore in-depth perspectives. Each technique needs to be meticulously explained, with explicit explanations of why it is the most appropriate strategy for addressing the research question about the research question. When detailing the techniques that were selected, it is essential to include comprehensive plans for the collection of data, analysis, and interpretation of the data. Outlining the timeframe for conducting the study, the methods or instruments used to gather data, and the statistical or thematic analysis applied to the results are all components included in this step. The possible difficulties that may appear while putting the study plan into action are another critical factor that must be considered [74]. To demonstrate adequate planning and to boost the credibility of the study, it is important to anticipate any difficulties, whether they be logistical, ethical, or methodological. The research design section guarantees that the study can give important insights into the field by matching research aims with acceptable methodologies and relying upon established approaches from relevant literature. This ensures that the study can contribute to the field. Not only does a well-crafted methodology enhance the basis of the research, but it also instills confidence that the study will be able to traverse complexity and provide legitimate and dependable results. Table 5 illustrates the situation. A practical study design synchronizes objectives with suitable data gathering and analysis methodologies. By adhering to established protocols, anticipating obstacles, and presenting transparent methods, researchers may guarantee the validity and dependability of their results while successfully and confidently answering the research issues.

7 Contribution and Relevance of Knowledge

This section explains how the proposed study would contribute to the advancement, expansion, or critical transformation of the current body of knowledge on the relevant topic [81]. The contribution can be conceptual, analytical, or methodological, and it should reflect the broader academic and practical consequences. For instance, research on sustainable building materials might present innovative materials and show how their application decreases energy consumption in buildings. This would give substantial gains to both theoretical knowledge and market practices in the construction sector. In addition, the contribution of the proposed study is based on the possibility that it may either question or refine already established hypotheses. For example, suppose the research studies the structural performance of recycled

Table 5 Research design and methods overview

Key component	Description	Purpose	Challenges
Research clarity [75]	The research design must provide clarity on the methodology of the study	To establish trust in the research's rigor and significance	Ensuring the study design is sufficiently explicit to eliminate uncertainty
Coherence of research goals and methods [76]	Ensuring that research aims are logically congruent with methodologies	To substantiate the application of methodologies that successfully tackle the research issue	Adjustments to research objectives and methods may be necessary as the study advances
Data collection methods [77]	Detailing the methodologies for data acquisition (e.g., interviews, surveys)	To gather data that effectively responds to the research inquiries	Obtaining dependable data sources or volunteers may prove challenging
Data analysis techniques [78]	Detailing the methodologies for data analysis and evaluating validity	To understand and analyze results with accuracy and dependability	Selecting appropriate analytical methods for intricate data or outcomes
Detailed research plan [79]	Formulating a comprehensive strategy for executing and analyzing research	To guarantee that the study is methodologically rigorous and consistent with its objectives	Unexpected alterations to the research strategy, including delays or resource deficiencies
Anticipation of challenges [80]	Foresee any impediments in the research and strategize accordingly	To augment the study's credibility and readiness for difficulties	Addressing unexpected logistical or ethical obstacles during data collecting and processing

building materials. In that case, it may call into question the traditional assumptions about the durability and efficacy of these materials, adding to discussions within the area. Identifying holes in existing approaches or discovering aspects that had not been addressed before might serve as a source of inspiration for future study, which could lead to more empirical studies and theoretical advancements. Additionally, it is essential to emphasize the practical ramifications of the research. For instance, developing innovative thermal insulation materials might directly impact residential and commercial construction, improving energy efficiency and reducing operational costs. These findings would not only be of academic interest, but they also have the potential to affect governmental decisions on building codes, sustainability rules, and urban growth plans. It is also essential to consider the influence of planned research on society. Suppose the findings of the study show both ecological and economic benefits, such as materials that are both cost-effective and ecologically sustainable. In that case, the research can improve quality of life, particularly in inadequately resourced areas. It is also possible that this will affect political discourse, leading to the implementation of stricter environmental legislation and supporting innovation in both the public and commercial sectors about sustainable practices. This research has the potential to significantly contribute to the understanding of complicated issues in the area by addressing topics that have not yet been answered and by filling crucial gaps in the current literature. This would pave the way for future academic investigation and practical application. Table 6 delineates the fundamental components of research contributions, encompassing knowledge enhancement and societal influence. It underscores the impact of research on further studies, actual implementations, policy formulation, and prospective economic advantages. Furthermore, it tackles issues that may emerge, like opposition to innovation, difficulties in applying theory to reality, and uncertainty regarding policy or market acceptance, all essential factors in academic research.

Table 6 Key elements of research contribution and potential challenges

Key element	Description	Potential challenges
Advancement of knowledge [82]	In what ways does the research augment, elucidate, or contest established theories or concepts?	Opposition to novel concepts and challenges in achieving acceptance within the scholarly community
Future research implications [27]	How the findings may inspire or guide future studies or inquiries?	Ambiguity in the repeatability of findings or absence of prompt further investigations
Practical applications [83]	The applicability of the results in practical contexts to enhance procedures or solutions	Challenges in converting theoretical findings into practical, scalable implementations
Impact on society [84]	Possible societal advantages encompass enhancing quality of life or tackling social issues	Assessing social effects may be intricate, and advantages may need years to materialize
Influence on policies and practices [85]	The capacity to affect policy choices, industry standards, or intervention strategies	Opposition from policymakers or stakeholders, particularly if the results contest prevailing standards
Commercial and public benefits [86]	The study results may have economic, public, or commercial advantages	Ambiguity in market acceptance, possible financial liabilities, or insufficient industry engagement
Unresolved questions and research gaps [87]	Identify deficiencies in the existing literature and elucidate how the research solves these deficiencies	It must enhance efficiency in acquiring complete data or ensure the study adequately addresses the highlighted shortcomings
Potential for breakthroughs [88]	The potential for the study to provide discoveries or innovations upon implementation	The success of breakthroughs may hinge on external circumstances, like technological preparedness and wider commercial acceptability

8 Formulating a Research Budget

A well-structured budget is essential for research projects, especially when pursuing external financing. While not all academic institutions need meticulous budget preparation for research incorporating historical resources or scientific publications, several funding agencies demand a thorough financial strategy. The budget delineates the anticipated expenses related to the study and guarantees compliance with the funding organization's stipulations [89]. When formulating a research budget, it is crucial to account for specific fundamental components to furnish a comprehensive rationale for each expenditure. The subsequent elements must be explicitly delineated:

- Overall estimated cost: The financing necessary for the effective execution of the research must be explicitly delineated [90]. This encompasses a detailed enumeration of all main elements, guaranteeing that no substantial costs are excluded.
- Rationale required for each budget item: Every budget item must be justified [91]. This section must elucidate how the allocated money would facilitate the proper execution of the study, ensuring that the expenses are both essential and justifiable.
- Origin of financial projections: It is essential to provide specifics on the methodology used to ascertain each cost [92]. This section must explicitly outline the process employed to estimate each expenditure, whether sourced from supplier quotations, historical data, or expert consultations.

Accurately assigning prices to specific aspects in historical or scientific literature research can be difficult, especially when traditional pricing methods are inapplicable. Researchers must examine the following factors:

- Access to specialized tools and programs: Is access to specific software, databases, or analytical instruments necessary for research? Allocate funds for requisite acquisitions, subscriptions, installation expenses, or training fees related to these instruments [93].
- Assessing the time commitment and opportunity costs: Accurately estimating the time necessary for research is essential [94]. This may involve considering time off work, which might have financial repercussions, or the necessity to employ replacements to uphold other professional responsibilities throughout the study phase.
- Travel and data collection expenses: Should the research require fieldwork or travel to specific sites for data gathering or meetings with participants, the related travel, lodging, and logistical expenses must be included in the budget [95].
- Personnel and outsourcing: The budget must delineate the roles, duties, and remuneration of research assistants or expert consultants when necessary [96]. Furthermore, all jobs eligible for outsourcing must be distinctly specified, along with their corresponding expenses.

Establishing a contingency fund to manage unexpected bills or crises is advisable. A precisely crafted budget exemplifies fiscal discipline and enhances the entire research proposal by ensuring donors of the project's thorough planning and financial feasibility. Table 7 presents a detailed analysis of the essential required financial components for research, encompassing expenses for software, time allocation, travel, staff, and outsourcing. The chart also underscores possible difficulties in cost estimation, controlling travel variability, and securing the prompt engagement of research assistants or consultants. This systematic method aids in forecasting financial risks and guarantees that the research budget is feasible and well substantiated.

Table 7 Detailed research budget and associated challenges

Budget item	Description	Estimated cost (currency)	Challenges
Software/Tools [97]	Expenditures associated with acquiring software or accessing databases and tools required for research	Depending on the specifications of the software or tool	Challenges in accurately assessing software or tool expenses without preliminary study or estimates
Time investment (research leave) [98]	Projected duration of absence from work for research purposes, encompassing costs for substitute personnel or foregone earnings	Determined by time investment and compensation framework	Ambiguity over the duration necessary for research and possible interruptions to routine tasks
Travel and accommodation [99]	Expenditures for travel related to site visits, interviews, and data collecting, encompassing transportation and accommodation	Contingent upon distance, duration, and trip frequency	Variations in travel and lodging expenses resulting from unexpected events
Research assistants/ consultants [100]	Remuneration for research assistants or consultants, specifying their responsibilities and compensation framework	Determined by hourly or project-based remuneration for assistants/consultants	Identifying competent assistants or consultants at affordable prices while adhering to budgetary constraints
Outsourcing (if applicable) [101]	It involves delegating specialized tasks or services that need external expertise	Contingent upon the extent of outsourced services	Overseeing outsourced tasks while adhering to budget constraints and avoiding delays
Contingency fund [102]	Resources are allocated for unexpected costs or crises during the research process	A typical allocation is a percentage (5–10%) of the whole budget	Precisely forecasting unforeseen expenses or establishing a suitable contingency sum

9 Developing a Schedule (Timeline)

Developing a realistic and concise research timetable is essential for ensuring the assignment is finished within the designated time range, as dictated by your academic program or calendar [103]. For example, if you are enrolled in a 2-year master's degree, your research must be completed within that timeframe, encompassing periods for writing, editing, and any necessary presentations. The timetable must have a comprehensive calendar of actions to be executed, commencing with fundamental research, succeeded by data collecting, analysis, composition, and final editing. The initial milestone involves finalizing your literature research within the first 3 months, which is succeeded by data gathering commencing in the fourth month and extending for 2 months. Each phase must have designated completion dates to sustain development and guarantee the study continues schedule. Monitoring advancement is of similar significance. Key events in the timetable include submitting a draft paper to your supervisor by the sixth month. The timeframe must be sufficiently adaptable to permit necessary modifications. If data gathering exceeds the anticipated duration, the timetable may require revision to accommodate this delay. Moreover, pertinent events associated with the study, such as conference presentations or progress updates, must be incorporated into the chronology. If you must present a poster at a conference in the ninth month, this will act as a provisional deadline to ensure the project remains on track. Establishing a dynamic, frequently updated timetable allows researchers to monitor their work efficiently, adjust to problems, and assure timely completion of all milestones. A structured research timetable is presented in Table 8. This timeline illustrates significant steps of the research process, including preliminary research, literature evaluation, data collecting, and analysis, ending in the submission of the proposal. Examples of milestones are included with each phase so that progress can be tracked and potential obstacles may come up. The purpose of this timetable is to act as a thorough guide for properly managing time and ensuring that all critical tasks are accomplished within the time range that has been allowed.

10 Ethical Considerations in Research

Ethical issues are essential to the integrity and rigor of any research effort, especially when involving human subjects [111]. Incorporating ethical criteria in a study proposal guarantees respect for participants' rights, dignity, and well-being while ensuring compliance with recognized academic and professional norms. Ethical precautions are essential for preserving the integrity of the research process and safeguarding persons from possible damage [112].

Table 8 Research timeline with challenges

Phase	Description	Estimated time frame	Milestone example	Challenges
Preliminary research [104]	Preliminary examination of the study subject and identification of essential theoretical frameworks and sources	Month x	Finalization of subject investigation and submission of a written research proposal	Possible difficulties in refining the study scope and selecting the most pertinent academic sources
Literature review [105]	Comprehensive analysis of current scholarly literature to contextualize the study and pinpoint deficiencies	Months x-x	Submission of the literature review to the supervisor for evaluative input	Challenges in obtaining extensive and pertinent literature or reconciling contradictory study results
Data collection [106]	Whether primary or secondary, data is gathered by surveys, experiments, or archive studies	Months x-x	Finalization of data acquisition and preliminary arrangement of results	Postponements resulting from participant accessibility, logistical complications, or unexpected technical difficulties
Data analysis [107]	Methodical examination of aggregated data to identify significant patterns and relationships	Months x-x	Finalization of the data analysis and formulation of the findings section	Obstacles in analyzing intricate datasets or deriving precise conclusions from confusing information
Draft writing [108]	Prepare the preliminary research manuscript, integrating all gathered data, analyses, and conclusions	Months x-x	Submission of the comprehensive research document for supervisor evaluation	Achieving a balance between analytical depth and clarity while ensuring compatibility with research objectives
Editing and revisions [109]	Amendment of the research document by input, guaranteeing cohesion, clarity, and scholarly rigor	Months x-x	Finalization of all amendments and meticulous refinement of the research material	Thoroughly addressing specific criticism under time limitations while maintaining accuracy in updates
Final submission [110]	Completing the research document, encompassing formatting and compliance with submission criteria	Month x	Submitting the completed research work for evaluation or grading	Possible last-minute format problems, submission requirements, or unanticipated administrative hold-ups

10.1 Informed Consent

Informed consent is a fundamental principle of ethical research involving human participants. It guarantees that participants are well apprised of the research's nature, purpose, methods, and potential dangers before participating [113]. Furthermore, informed consent must delineate participants' rights, including their entitlement to withdraw from the research at any moment without repercussions. The study proposal must provide a comprehensive account of the consent procedure, indicating whether it will be secured in written or spoken format. The wording included in permission forms must be clear and understandable to all parties [114]. Specific ethical concerns must be implemented when engaging with vulnerable populations, like minors, the elderly, or persons with cognitive disabilities. In such instances, further precautions must be instituted, encompassing the acquisition of agreement from legal guardians or caretakers, guaranteeing voluntary involvement, and safeguarding against coercion. The study proposal must explicitly outline these procedures to ensure adherence to ethical norms, especially regarding protecting vulnerable populations [112].

10.2 Confidentiality and Anonymity

Safeguarding participants' confidentiality and anonymity constitutes a significant ethical problem [115]. The plan must delineate the methods for managing, storing, and protecting personal and sensitive data. This may entail using encrypted files, password protection, and various safe data management methodologies. The plan must delineate the individuals permitted access to the data, guaranteeing that only authorized staff are included. To maintain anonymity, researchers should use pseudonyms or employ coded identities instead of participants' names in study data, reports, and publications [116]. When handling sensitive information—such as health, financial, or personal data—the proposal must provide supplementary precautions to guarantee data confidentiality, including restricted access and improved security processes. Furthermore, the proposal must outline the protocols for data disposal upon the conclusion of the research. A definitive plan for the timing and method of data destruction must be incorporated to guarantee that participant information is not maintained needlessly, bolstering the research's ethical standards. It is important to understand that ethical concerns are essential for the competent execution of research. Addressing informed permission, confidentiality, and anonymity safeguards participants and bolsters the study's legitimacy and academic rigor [117]. Informed permission, protecting vulnerable groups, confidentiality, anonymity, data protection, and data disposal are some critical ethical issues outlined in Table 9, which provides an overview of the study methodology. The importance of protecting the rights of participants and maintaining compliance with ethical and regulatory standards is emphasized by the fact that each ethical issue is accompanied by a description and the obstacles connected with it.

Table 9 Ethical considerations in research

Ethical aspect	Description	Key challenges	Example
Informed consent [118]	Ensuring that participants are adequately informed about the study's methods, dangers, and withdrawal rights	Ensuring that intricate study facts are communicated to all participants simply and understandably	Giving participants thorough consent documents that outline the purpose, methods, and ability to withdraw from the research
Vulnerable populations [119]	Put in place extra protections for children, the elderly, and others who are cognitively impaired	Preserving voluntary involvement while giving vulnerable populations sufficient legal protections	Gaining guardian approval before participating in an educational performance survey if a minor
Confidentiality [120]	Restricting exposure to authorized parties and safeguarding data access to protect personal information	Responsible for handling sensitive data to stop illegal access and data breaches	Only the study team can access encrypted files containing health survey data
Anonymity [121]	Using codes or pseudonyms to protect participant anonymity in study papers and data	Weighing the necessity of participant anonymity against the possibility of participant re-identification for additional study	Protecting the identity of interviewees in a social stigma study by using pseudonyms
Data protection [122]	Using password security, encryption, and restricted access procedures to store data securely	Establishing and upholding safe procedures for data preservation during the study	Encrypting participant financial data to guard against unwanted access in economic behavior research
Data disposal [123]	Predetermined procedures and deadlines destroy data once the study is completed to preserve participant privacy	Creating a precise data disposal strategy that complies with legal and ethical constraints	After a project, digital data is securely destroyed, and paper records are shredded to protect privacy

11 Concluding Research Statement

The final paragraph of a research proposal is essential in which the researcher discusses anticipated results and justifies the conclusions the investigation intends to accomplish. Because it frequently includes practical applications with substantial social, economic, or environmental ramifications, this might be of utmost significance for projects that fall within the engineering discipline of civil engineering [124]. To clarify the scope and dependability of the study for decision-makers, funders, or academic assessors, it is essential to outline the desired results and potential repercussions of the research. While the findings presented at this stage are based on preliminary assumptions and currently available data, they may be adjusted following the completion of the study [125]. On the other hand, by predicting the possible outcomes, you provide a framework for evaluating the significance and dependability of the study. For example, in the context of a civil engineering project to enhance the structural integrity of bridges in seismic zones, you can anticipate that your study will develop new materials or design processes capable of withstanding even greater earthquake pressures. The assessors are provided with an understanding of the study's practical significance; nevertheless, this preliminary finding may be susceptible to alteration after testing has been completed [126]. An essential component of the concluding statement is the discussion of the potential repercussions or wider ramifications that the study may have. For instance, if research in civil engineering investigates the utilization of intelligent materials in building construction, the outcomes may include increased sustainability, longer building lifespans, or decreased maintenance costs. These predicted results reflect a comprehensive understanding of the subject matter and demonstrate that the study will contribute to the existing body of academic knowledge and practical applications in the real world. Additionally, the concluding statement should include a section for applications if necessary. Civil engineering research may require supplemental resources such as technical data, design blueprints, or compliance forms according to various safety standards and construction regulations. For example, suppose the research involves testing new road materials for improved durability. In that case, the application section might include data on material performance tests, environmental impact assessments, or instructions on road safety compliance [127]. These documents strengthen the proposal by showing thorough preparation and attention to detail regarding the technical and regulatory aspects of the study. The concluding statement combines the most critical components of the proposal by predicting anticipated results, acknowledging possible adjustments, and illustrating the study's broader implications. The real-world significance and impact of the work are highlighted through examples from civil engineering, such as seismic-resistant structures or intelligent materials. Including supplementary documents in the designated application section further enhances the credibility of the research and strengthens the case for approval or funding. The research closing statement should encompass the key components outlined in Table 9. These components include the forecasted outcomes, anticipated conclusions, broader implications, and the application section. As detailed in Table 10, significant challenges involve the unpredictability of research results, the complexities of evaluating broader societal impacts, and the critical need to ensure compliance with relevant regulatory and ethical standards, particularly within the application section.

Table 10 Key components and challenges of the research closing statement

Research aspect	Description	Example (civil engineering)	Challenges
Forecasted outcomes [128]	Presenting the probable outcomes of the investigation based on existing facts and assumptions	Creating innovative materials for earthquake-resistant bridges	Challenges in precisely forecasting results owing to fluctuating factors in real-world circumstances
Anticipated conclusions [129]	Revealing the prospective results derived from the research, which are subject to modification	In conclusion, intelligent materials may extend the longevity of structures	Conclusions may change due to the variability of data in subsequent study phases
Broader implications [14]	Emphasizing the social, environmental, or practical implications of the study findings	Highlighting effects such as enhanced sustainability or decreased maintenance expenses	The broader consequences may be challenging to assess once the study is finalized and implemented
Application section [130]	Supplementary resources must be supplied, including data tables, consent forms, and technical recommendations	Incorporating material performance test outcomes or road safety compliance standards	Ensuring all supplemental documents are comprehensive and adhere to compliance and regulatory standards

12 Typical Errors to Steer Clear of in Research Proposals

Developing a persuasive research proposal is essential for securing approval or financing for academic study [131]. Due to the very competitive nature of research submission procedures, especially in esteemed journals with rejection rates over 90%, even little mistakes can significantly reduce a proposal's likelihood of success [132]. Frequent errors, including ambiguous scope, overly lengthy sentences, and insufficient relevant literature, frequently result in rejection. Furthermore, excessive attention to little details, the presentation of feeble arguments, or the oversight of typographical errors can significantly diminish the proposal's credibility. To enhance the likelihood of approval, researchers must guarantee clarity, conciseness, and rigor in their writing [133]. This introduction delineates common hazards in research proposals. It offers strategies to circumvent them, empowering researchers to write focused, logical, and persuasive proposals, enhancing their chances of acceptance.

12.1 Overly Prolonged Sentences

Proposals must be brief and direct. Long sentences make reading the text more complex and obfuscate meaning [134]. The argument is more straightforward to comprehend when sentences are brief and concise.

- For example, "this study applies quantitative methods in exploring various environmental factors influencing urban sustainability across major cities, particularly renewable energy, green building, and climate change."
- Enhanced: "This study examines environmental factors affecting urban sustainability, focusing on renewable energy, green building, and climate change, using quantitative methods."

12.2 Imprecise Duration and Range

Many proposals regarding the technique and research topic are vague [135]. To effectively communicate the direction and viability of the study, a clear and concise statement of aim is essential. For instance:

- Initial title: "This study explores cultural, social, political, and economic factors influencing urban development over time."
- Enhanced: "This study examines economic and social factors affecting urban development in post-industrial cities over the next 12 months."

12.3 Insufficient Review of Literature

While conciseness is valued, leaving out crucial works tarnishes the argument. A solid foundation requires citations for critical research. For example:

- Original statement, "Urban development literature offers various perspectives."
- Rewritten: "Brown (2018) and Smith (2015) investigate the role of economic policies on urban development. This study evaluates the contribution of local government to sustainable growth, building on earlier results."

12.4 Excessive Focus on Small Details

Frequently, proposals overemphasize unimportant details, casting shadows over the main research topics [136]. Less attention must be paid to details, and more attention must be paid to essential concerns.

- For example: "Attention will be given to survey formatting and question font size, as they may influence responses."

- Enhanced: "The study emphasizes survey question clarity and relevance to ensure accurate data collection."

12.5 *Feeble or Non-persuasive Arguments*

A proposal must persuasively argue the significance of the research and connect it to the study's more enormous scholarly or social implications [137]. For example:

- Original statement, "This research is important for urban planning."
- Enhanced: "This research addresses sustainability challenges in urban planning, providing data that can inform policy decisions."

12.6 *Errors and Bad Style of Writing*

Grammar and style mistakes detract from a proposal's professionalism. Thorough proofreading is necessary to preserve credibility and academic standards. For example:

- Originally stated: "The research will answer key questions in urban sustainability."
- Enhanced: "The research will answer key questions in urban sustainability."

Research proposals that address these typical mistakes—unambiguous sentences, a clearly defined scope, the inclusion of relevant literature, a concentration on essential themes, compelling arguments, and error-free writing—are more likely to be successful in a highly competitive academic setting.

13 Conclusion

Creating a research proposal is a procedural obligation and a fundamental component of the academic research process that influences a scholarly investigation's feasibility, coherence, and ultimate success. Chapter 8 has examined the essential elements required to develop a coherent research proposal, emphasizing the pivotal function of this document in guiding the research process, obtaining financing, and enhancing the broader academic dialog. The success of every research proposal hinges on the formulation of a precise and focused research topic. The research question functions as the conceptual foundation for the study, directing the formulation of objectives, the choice of methodology, and the proposal's organization. The research topic identifies gaps in the current knowledge, defining the inquiry's scope and significance to ongoing academic discussions and social issues. The capacity to formulate and enhance a research topic that is both meaningful and investigable is an essential talent, emphasized in this chapter as a crucial factor in assessing a proposal's academic value. Literature evaluation is equally important, as it establishes the educational framework for the research. A comprehensive and analytical literature review transcends mere summarization of previous studies; it interacts with the contemporary knowledge landscape, delineates dominant trends, discussions, and deficiencies, and contextualizes the proposed research within this framework. This synthesis of current material not only substantiates the necessity for the research but also illustrates the researcher's proficiency in the topic and capacity to make a significant contribution. This Chapter emphasizes that a well-conducted literature review establishes the groundwork for the study, connecting previous knowledge with forthcoming investigations. Methodological rigor constitutes a critical element of a comprehensive research proposal. The methodology section delineates the research plan, encompassing the precise techniques for data collecting, analysis, and interpretation. A methodologically rigorous plan guarantees the research's viability and bolsters its credibility by illustrating that it will be executed methodically and by accepted academic norms. The chapter underscores that the selected methodologies must correspond with the research question and objectives, guaranteeing that the study can produce valid and trustworthy outcomes. The technique must foresee probable obstacles and constraints, including contingency strategies to address these risks. This foresight is essential in convincing reviewers of the research's viability and the researcher's readiness.

Ethical issues, especially with human participants, are emphasized as essential to the integrity of the research process. A research proposal must include measures to protect participants' rights, dignity, and welfare, assuring adherence to ethical norms, including informed consent, confidentiality, and data protection. Compliance with these ethical norms is not only a procedural obligation but is essential for preserving the credibility and legitimacy of the study. Consequently, ethical rigor is critical to a proposal's integrity and must be included in the methodological framework. Incorporating a comprehensive budget and timeframe enhances the research proposal by demonstrating its practical viability. A well-formulated budget offers a detailed analysis of the expected expenditures related to the study, encompassing equipment, supplies, labor, and travel costs. The proposal provides a transparent financial plan, assuring potential investors of the project's fiscal viability and the optimal allocation of resources. A delineated timetable, incorporating milestones and progress-monitoring systems, guarantees the project's chronological viability and compliance with requisite deadlines.

As indicated in the chapter, the budget and timeframe are essential in demonstrating to funders and reviewers that the study is theoretically valid and practically feasible. Additionally, an effective research proposal must explicitly delineate the overarching importance of the topic, encompassing both scholarly contributions and prospective practical applications. The proposal must illustrate how the study will enhance knowledge in its domain by providing novel insights, ideas, or methodologies that rectify current deficiencies. Furthermore, it should emphasize the study's possible social, economic, or technical ramifications, demonstrating how the results may guide policy, stimulate innovation, or aid in addressing urgent global issues. This chapter emphasizes that the capacity to convert academic research into tangible results is frequently a crucial determinant in obtaining financing and institutional backing. Ultimately, clarity and consistency in writing are essential for the success of a study proposal. A meticulously constructed, logically coherent, and error-free proposal improves its readability and persuasiveness, guaranteeing that its points are effectively conveyed to reviewers. The chapter underscores the need for accuracy and conciseness, indicating that convoluted or confusing wording may diminish the proposal's effectiveness. The capacity to articulate intricate concepts clearly and accessibly signifies intellectual rigor and is crucial for obtaining endorsement from academic committees or funding organizations. Formulating a research proposal is a complex process necessitating meticulous preparation, critical analysis of existing literature, methodological precision, and pragmatic foresight. Every element of the proposal, from developing the research topic to the technique specification, ethical issues, funding, and timeline, enhances its robustness and feasibility. This chapter illustrates that a meticulously constructed research proposal establishes the foundation for a successful study and is essential for obtaining the requisite support and resources to realize the research objectives. By following the concepts delineated in this chapter, researchers can improve the quality of their ideas and augment their likelihood of making a significant contribution to their area. Beginning with the title and abstract, the important components of a research proposal are depicted in Fig. 1. These sections range from contributions to knowledge to the title and abstract. This section provides a comprehensive overview of the content flow, highlighting how each area, beginning with the introduction and continuing through the methodology, literature review, and objectives, contributes to a unified research plan that addresses ethical concerns, budget, timetable, and anticipated outcomes.

Fig. 1 Key components of a research proposal

References

1. Hanington B, Martin B (2017) The pocket universal methods of design: 100 ways to research complex problems, develop innovative ideas and design effective solutions. Rockport
2. Aagaard K, Schneider JW (2016) Research funding and national academic performance: examination of a Danish success story. Sci Public Policy 43(4):518–531

3. Moloi GT (2024) Funding challenges encountered by technology startups in South Africa (Doctoral dissertation, Regenesys Business School).
4. Darmawan D (2024) Distribution of six major factors enhancing organizational effectiveness. J Distrib Sci 22(4):47–58
5. Sepasgozar S (2024) Research proposal crucible: crafting with AI awareness: strategies to communicate strong academic arguments. Samad Sepasgozar
6. Hazari A (2024) Developing research proposal. Research methodology for allied health professionals: a comprehensive guide to thesis & dissertation, pp 59–71. Singapore: Springer Nature Singapore
7. Alawag AM, Alaloul WS, Liew MS, Musarat MA, Baarimah AO, Saad S, Ammad S (2023) Critical success factors influencing total quality management in industrialised building system: a case of Malaysian construction industry. Ain Shams Eng J 14(2):101877
8. Muasya JN, Mulwa PM (2023) Pilot study, a neglected part of qualitative and quantitative research process: evidence from selected PhD thesis and dissertations. High Educ Res 8(4):115–123
9. Bhaskar R, Ola M (2024) Chapter objectives. Principles of research methodology and ethics in pharmaceutical sciences: an application guide for students and researchers, p 271
10. Almusaed A, Almssad A, Yitmen I, Wallhagen M, Yang YF, Iyer-Raniga U (2024) Integrative approaches in urban sustainability: architectural design, technological innovations and social dynamics in global contexts. BoD–Books on Demand
11. Pandey P, Pandey MM (2021) Research methodology tools and techniques. Bridge Center
12. Sartorio FS, Aelbrecht P, Kamalipour H, Frank A (2021) Towards an antifragile urban form: a research agenda for advancing resilience in the built environment. Urban Des Int 26(2):135
13. Sharmin T, Khalid R (2022) Post occupancy and participatory design evaluation of a marginalized low-income settlement in Ahmedabad, India. Build Res Inf 50(5):574–594
14. Dwivedi YK, Jeyaraj A, Hughes L, Davies GH, Ahuja M, Albashrawi MA, Al-Busaidi AS, Al-Sharhan S, Al-Sulaiti KI, Altinay L, Amalaya S (2024) "Real impact": challenges and opportunities in bridging the gap between research and practice–making a difference in industry, policy, and society. Int J Inf Manage 102750
15. Ketchen DJ Jr, Craighead CW (2023) What constitutes an excellent literature review? Summarize, synthesize, conceptualize, and energize. J Bus Logist 44(2):164–169
16. Almusaed A, Almssad A, Alasadi A, Yitmen I, Al-Samaraee S (2023) Assessing the role and efficiency of thermal insulation by the "Bio-Green Panel" in enhancing sustainability in a built environment. Sustainability 15(13):10418
17. Zhang K, Qing Y, Umer Q, Asmi F (2023) How construction and demolition waste management has addressed sustainable development goals: Exploring academic and industrial trends. J Environ Manage 345:118823
18. Chigbu UE, Atiku SO, Du Plessis CC (2023) The science of literature reviews: searching, identifying, selecting, and synthesising. Publications 11(1):2
19. AlJaber A, Alasmari E, Martinez-Vazquez P, Baniotopoulos C (2023) Life cycle cost in circular economy of buildings by applying building information modeling (BIM): a state of the art. Buildings 13(7):1858
20. Ciccozzi A, de Rubeis T, Paoletti D, Ambrosini D (2023) BIM to BEM for building energy analysis: a review of interoperability strategies. Energies 16(23):7845
21. Abanda FH (2023) Informed teaching for assessing BIM courses during COVID-19 and beyond. J Appl Res High Educ 15(2):568–598
22. Schlussel MM, Sharp MK, de Beyer JA, Kirtley S, Logullo P, Dhiman P, MacCarthy A, Koroleva A, Speich B, Bullock GS, Moher D (2023) Reporting guidelines used varying methodology to develop recommendations. J Clin Epidemiol 159:246–256
23. Enworo OC (2023) Application of Guba and Lincoln's parallel criteria to assess trustworthiness of qualitative research on indigenous social protection systems. Qual Res J 23(4):372–384
24. Haghani M (2023) What makes an informative and publication-worthy scientometric analysis of literature: a guide for authors, reviewers and editors. Transp Res Interdiscip Perspect 22:100956

25. Kristóf T, Nováky E (2023) The story of futures studies: An interdisciplinary field rooted in social sciences. Soc Sci 12(3):192
26. Matt M, Robinson DK, Joly PB, Van Dis R, Colinet L (2023) ASIRPAReal-time in the making or how to empower researchers to steer research towards desired societal goals. Res Eval 32(2):412–425
27. Tracy SJ (2024) Qualitative research methods: collecting evidence, crafting analysis, communicating impact. John Wiley & Sons
28. Halme M, Piekkari R, Matos S, Wierenga M, Hall J (2024) Rigour vs. reality: contextualizing qualitative research in the low-income settings in emerging markets. Br J Manag 35(1):36–51
29. Yitmen I, Almusaed A, Alizadehsalehi S (2023) Investigating the causal relationships among enablers of the construction 5.0 paradigm: integration of operator 5.0 and society 5.0 with human-centricity, sustainability, and resilience. Sustainability 15(11):9105
30. Hassan Q, Sarhan N, Awwad EM, Al-Musawi TJ, Ghazaly NM, Viktor P, Fodor M, Iqbal A, Zhiltsov S, Makhmudov A, Al-Jiboory AK (2024) Optimizing smart building energy systems for sustainable living: a realistic approach to enhance renewable energy consumfaption and reduce emissions in residential buildings. Energy Build 114354
31. Ali DMTE, Motuzienė V, Džiugaitė-Tumėnienė R (2024) AI-driven innovations in building energy management systems: a review of potential applications and energy savings. Energies 17(17):4277
32. Farzaneh H, Malehmirchegini L, Bejan A, Afolabi T, Mulumba A, Daka PP (2021) Artificial intelligence evolution in smart buildings for energy efficiency. Appl Sci 11(2):763
33. Luft JA, Jeong S, Idsardi R, Gardner G (2022) Literature reviews, theoretical and conceptual frameworks: An introduction for new biology education researchers. CBE—Life Sci Educ 21(3):rm33
34. Hancock DR, Algozzine B, Lim JH (2021) Doing case study research: a practical guide for beginning researchers
35. Walkington C, Bernacki ML (2020) Appraising research on personalized learning: definitions, theoretical alignment, advancements, and future directions. J Res Technol Educ 52(3):235–252
36. Hohenschurz-Schmidt DJ, Cherkin D, Rice AS, Dworkin RH, Turk DC, McDermott MP, Bair MJ, DeBar LL, Edwards RR, Farrar JT, Kerns RD (2023) Research objectives and general considerations for pragmatic clinical trials of pain treatments: IMMPACT statement. Pain 164(7):1457–1472
37. Jacobsen KH (2020) Introduction to health research methods: a practical guide. Jones & Bartlett Publishers
38. Tai HY, Chen YS (2021) The effects of L2 proficiency on pragmatic comprehension and learner strategies. Edu Sci 11(4):174
39. Muhonen R, Benneworth P, Olmos-Peñuela J (2020) From productive interactions to impact pathways: understanding the key dimensions in developing SSH research societal impact. Res Eval 29(1):34–47
40. Edwards AL (2022) The paratexts of audience engagement: cover matter that draws in and keeps readers
41. Lawrence MG, Williams S, Nanz P, Renn O (2022) Characteristics, potentials, and challenges of transdisciplinary research. One Earth 5(1):44–61
42. Rahimi M, Zhang LJ (2022) Effects of an engaging process-genre approach on student engagement and writing achievements. Read Writ Q 38(5):487–503
43. Hills RM Jr (2023) Strategic ambiguity and article VII: why the framers decided not to decide. J Am Const Hist 1:379
44. Lee Y, Jang BG, Conradi Smith K (2021) A systematic review of reading engagement research: what do we mean, what do we know, and where do we need to go? Read Psychol 42(5):540–576
45. Chen S, Sharma G, Muñoz P (2023) In pursuit of impact: from research questions to problem formulation in entrepreneurship research. Entrep Theory Pract 47(2):232–264
46. Xiao T, Makhija M, Karim S (2022) A knowledge recombination perspective of innovation: review and new research directions. J Manag 48(6):1724–1777

47. Sundar R, Suguna A, Begum A, Mishra NL (2024) Introductory research methodology. academic Guru Publishing House
48. Kardgar A (2024) Effectiveness of automatically advanced vs. manually advanced infographics in health awareness (Doctoral dissertation, Purdue University Graduate School)
49. Hyland K, Zou H (2022) Pithy persuasion: engagement in 3 minute thesis presentations. Appl Linguis 43(1):21–44
50. Willoughby SD, Johnson K, Sterman L (2020) Quantifying scientific jargon. Public Underst Sci 29(6):634–643
51. Hügel S, Davies AR (2020) Public participation, engagement, and climate change adaptation: a research literature review. Wiley Interdiscip Rev Clim Chang 11(4):e645
52. Petrevska Nechkoska R, Petrevska Nechkoska R (2020) Problem investigation and solution requirements for tactical management. Tactical management in complexity: managerial and informational aspects, pp 27–63
53. Lê JK, Schmid T (2022) The practice of innovating research methods. Organ Res Methods 25(2):308–336
54. Saunders L, Wong MA (2020) Instruction in libraries and information centers: an introduction
55. Elahi E (2023) Role of images on World Wide Web Readability, Doctoral dissertation, Universidad Carlos III de Madrid
56. Mpofu FY (2021) Review Articles: A Critical Review Of The Pitfalls And Guidelines To Effectively Conducting And Reporting Reviews. Technium Soc Sci J 18:550
57. Zhu M, Sari AR, Lee MM (2020) A comprehensive systematic review of MOOC research: research techniques, topics, and trends from 2009 to 2019. Educ Tech Res Dev 68:1685–1710
58. Jaakkola E (2020) Designing conceptual articles: four approaches. AMS Rev 10(1):18–26
59. Uma AN, Fornaciari T, Hovy D, Paun S, Plank B, Poesio M (2021) Learning from disagreement: a survey. J Artif Intell Res 72:1385–1470
60. Anderson MH, Lemken RK (2023) Citation context analysis as a method for conducting rigorous and impactful literature reviews. Organ Res Methods 26(1):77–106
61. Lim WM, Kumar S, Ali F (2022) Advancing knowledge through literature reviews:'what', 'why', and 'how to contribute.' Serv Ind J 42(7–8):481–513
62. Karunarathna I, Bandara S, Jayawardana A, De Alvis K, Gunasena P, Hapuarachchi T, Ekanayake U, Rajapaksha S, Gunawardana K, Aluthge P, Gunathilake S Navigating the literature: rules for systematic review writing
63. Sunstein CR (2020) Too much information: understanding what you don't want to know. MIT Press
64. Devaraj A, Wallace BC, Marshall IJ, Li JJ (2021) Paragraph-level simplification of medical texts. In Proceedings of the conference. Association for computational linguistics. North American Chapter. Meeting, vol 2021, p 4972. NIH Public Access
65. Mishra P, Kumar S, Chaube MK (2023) Graph interpretation, summarization, and visualization techniques: a review and open research issues. Multimed Tools Appl 82(6):8729–8771
66. Wan Y, Liu Y, Chen Z, Chen C, Li X, Hu F, Packianather M (2024) Making knowledge graphs work for smart manufacturing: research topics, applications and prospects. J Manuf Syst 76:103–132
67. Szucs D, Ioannidis JP (2020) Sample size evolution in neuroimaging research: an evaluation of highly-cited studies (1990–2012) and of latest practices (2017–2018) in high-impact journals. Neuroimage 221:117164
68. Heck R, Hallinger P (1999) Conceptual models, methodology, and methods for studying school leadership. The 2nd handbook of research in educational administration, pp 141–162
69. Logullo P, van Zuuren EJ, Winchester CC, Tovey D, Gattrell WT, Price A, Harrison N, Goldman K, Chisholm A, Walters K, Blazey P (2024) Accurate consensus reporting document (ACCORD) explanation and elaboration: guidance and examples to support reporting consensus methods. PLoS Med 21(5):e1004390
70. Turnbull D, Chugh R, Luck J (2023) Systematic-narrative hybrid literature review: a strategy for integrating a concise methodology into a manuscript. Soc Sci Hum Open 7(1):100381

71. Welch C, Paavilainen-Mäntymäki E, Piekkari R, Plakoyiannaki E (2022) Reconciling theory and context: how the case study can set a new agenda for international business research. J Int Bus Stud 53(1):4–26
72. Jaskó S, Skrop A, Holczinger T, Chován T, Abonyi J (2020) Development of manufacturing execution systems by industry 4.0 requirements: a review of standard-and ontology-based methodologies and tools. Comput Ind 123:103300
73. Mohajan HK (2020) Quantitative research: A successful investigation in natural and social sciences. J Econ Dev Environ People 9(4):50–79
74. Clark JS, Porath S, Thiele J, Jobe M (2020) Action research. New Prairie Press
75. Imran A, Ghufran F, Shah T, Ahmad B, Anwar A (2024) Optimizing research article structure and style: best practices for ensuring clarity and impact in scholarly publications. J Asian Dev Stud 13(2):1432–1444
76. Pratt MG, Sonenshein S, Feldman MS (2022) Moving beyond templates: a bricolage approach to conducting trustworthy qualitative research. Organ Res Methods 25(2):211–238
77. Jain N (2021) Survey versus interviews: comparing data collection tools for exploratory research. Qual Rep 26(2):541–554
78. Popenoe R, Langius-Eklöf A, Stenwall E, Jervaeus A (2021) A practical guide to data analysis in general literature reviews. Nord J Nurs Res 41(4):175–186
79. Morciano C, Errico MC, Faralli C, Minghetti L (2020) An analysis of the strategic plan development processes of major public organisations funding health research in nine high-income countries worldwide. Health Res Policy Syst 18:1–16
80. Urueña S (2021) Responsibility through anticipation? The 'Future Talk'and the quest for plausibility in the governance of emerging technologies. NanoEthics 15(3):271–302
81. Pereira V, Bamel U (2021) Extending the resource and knowledge-based view: a critical analysis into its theoretical evolution and future research directions. J Bus Res 132:557–570
82. Seidman S (2016) Contested knowledge: social theory today. John Wiley & Sons
83. Schmidt NA, Brown JM (2024) Evidence-based practice for nurses: appraisal and application of research. Jones & Bartlett Learning
84. Rony MKK, Parvin MR, Wahiduzzaman M, Akter K, Ullah M (2024) Challenges and advancements in the health-related quality of life of older people. Adv Public Health 2024(1):8839631
85. Nguyen J (2024) Beyond policy impacts: internal strategic capabilities as determinants of industrial energy efficiency implementation. Energy Policy 184:113898
86. Xu A, Song M, Xu S, Wang W (2024) Accelerated green patent examination and innovation benefits: an analysis of private economic value and public environmental benefits. Technol Forecast Soc Chang 200:123105
87. Law R, Lei SSI, Zhang K, Lau A (2024) Bridging the theory-practice gap: a critical reflection on information and communication technology research. Int J Contemp Hosp Manag 36(6):1980–1990
88. Héraud JA, Popiolek N (2024) A model of breakthrough innovation: simultaneity of discovery and invention. J Innov Econ Manag 43(1):159–186
89. Fantu M (2024) Assessment of budget practice in NGO's, in the case of Grarbet Tehadiso Mahber. Doctoral dissertation, St. Mary's University
90. Rodrigues SL, da Silva EA (2024) Implementation of environmental life cycle costing: procedures, challenges, and opportunities. Int J Life Cycle Assess 29(5):803–837
91. Jayanti P, Jayanti S, Yavuz U, Hernandez L (2024) A universal, sound, and complete forward reasoning technique for machine-verified proofs of linearizability. Proc ACM Program Lang 8(POPL):2456–2484
92. Al Wahid SA, Mohammad N, Islam R, Faisal MH, Rana MS (2024) Evaluation of information technology implementation for business goal improvement under process functionality in economic development. J Data Anal Inf Process 12(2):304–317
93. Ateş NT, Tekin N (2024) Digital and online tools employed by prospective teachers to access information. Int J Assess Tools Educ 11(1):171–194

94. Woods B, Lomas J, Sculpher M, Weatherly H, Claxton K (2024) Achieving dynamic efficiency in pharmaceutical innovation: identifying the optimal share of value, the payments required and evaluating pricing policies
95. Kapiszewski D, MacLean L, Smith L (2024) Digital fieldwork. Doing good qualitative research, p 323
96. Quinones A (2024) Outsourcing auxiliary services in higher education: an integrated risk management model
97. Michael EI, Etukudo OB (2024) Operating system software and Cd-Rom databases as tools for information delivery in university e-libraries in south-south zone, Nigeria. Glob J Educ Res 23(2):231–240
98. Elliott R (2024) The economics of the UK health and social care labour market. Oxford University Press
99. Shafiqah A (2024) Travel smart, spend smart: the intersection of tourism and financial planning. Spend smart: the intersection of tourism and financial planning
100. Vetoshkin, E. and Rozhnov, I., 2024. Impact of AI on Game Development: opium des Volkes or Panacea
101. Mutongerwa NF, Juqu ZF, Matebese H (2024) A paradigm shift in outsourcing functions to external service providers in the public sector: a liability for the state. Am J Supply Chain Manag 9(1):13–32
102. Mangarin RA, O'Loughlin TW (2024) Covid-19 policy responses and implementation for a contingency-based financial planning for school heads. TWIST 19(2):9–16
103. Felder RM, Brent R (2024) Teaching and learning STEM: a practical guide. John Wiley & Sons
104. Lynch J, Ramjan LM, Glew PJ, Salamonson Y (2024) How to embed a conceptual or theoretical framework into a dissertation study design. Nurse Res 32(1)
105. Zhao P, Zhang X, Cheng MM, Yang J, Li X (2024) A literature review of literature reviews in pattern analysis and machine intelligence. arXiv:2402.12928
106. Ohme J, Araujo T, Boeschoten L, Freelon D, Ram N, Reeves BB, Robinson TN (2024) Digital trace data collection for social media effects research: APIs, data donation, and (screen) tracking. Commun Methods Meas 18(2):124–141
107. Riwayat AAP, Susilawati AD, Naqiah Z (2024) Purchasing patterns analysis in e-commerce: a big data-driven approach and methodological. Int J Softw Eng Comput Sci (IJSECS) 4(1):148–164
108. Moschis GP (2024) Writing manuscripts for academic journals. Academic research in business and the social sciences: a guidebook for early career researchers, pp 209–232. Cham: Springer Nature Switzerland
109. Heinemann F, Asatryan Z, Bachtrögler-Unger J, Birkholz C, Corti F, Ehrlich MV, Fratesi U, Fuest CLEMENS, Lang V, Weber M (2024) Enhancing objectivity and decision relevance: a better framework for evaluating cohesion policies. ZEW-centre for european economic research discussion paper, pp 24–034
110. Enciso RE Implementing a robust monitoring system for research submissions: a committee-based approach
111. Guelmami N, Ben Ezzeddine L, Hatem G, Trabelsi O, Ben Saad H, Glenn JM, El Omri A, Chalghaf N, Taheri M, Bouassida A, Ben Aissa M (2024) The ethical compass: establishing ethical guidelines for research practices in sports medicine and exercise science. Int J Sport Stud Health 7(2):31–46
112. Brittain S, Ibbett H, de Lange E, Dorward L, Hoyte S, Marino A, Milner-Gulland EJ, Newth J, Rakotonarivo S, Veríssimo D, Lewis J (2020) Ethical considerations when conservation research involves people. Conserv Biol 34(4):925–933
113. Punnaram A, Arya PB (2021) Ethics in biomedical research: with special reference to psychiatric research
114. Millum J, Bromwich D (2021) Informed consent: what must be disclosed and what must be understood? Am J Bioeth 21(5):46–58

115. Kang E, Hwang HJ (2021) Ethical conducts in qualitative research methodology: participant observation and interview process. J Res Publ Ethics 2(2):5–10
116. Pascale J, Lineback JF, Bates N, Beatty P (2022) Protecting the identity of participants in qualitative research. J Surv Stat Methodol 10(3):549–567
117. Slattengren EH (2020) Institutional integrity: perceptions of organizational legitimacy and organizational virtuousness in a research university setting. Doctoral dissertation, University of Minnesota
118. Xu A, Baysari MT, Stocker SL, Leow LJ, Day RO, Carland JE (2020) Researchers' views on, and experiences with, the requirement to obtain informed consent in research involving human participants: a qualitative study. BMC Med Ethics 21:1–11
119. Gordon BG (2020) Vulnerability in research: basic ethical concepts and general approach to review. Ochsner Journal 20(1):34–38
120. Keshta I, Odeh A (2021) Security and privacy of electronic health records: concerns and challenges. Egypt Inform J 22(2):177–183
121. Vorhölter J (2021) Anthropology anonymous? Pseudonyms and confidentiality as challenges for ethnography in the twenty-first century. Ethnoscripts 23(1):15–33
122. Sauber AM, El-Kafrawy PM, Shawish AF, Amin MA, Hagag IM (2021) A new secure model for data protection over cloud computing. Comput Intell Neurosci 2021(1):8113253
123. Maritsch F, Cil I, McKinnon C, Potash J, Baumgartner N, Philippon V, Pavlova BG (2022) Data privacy protection in scientific publications: process implementation at a pharmaceutical company. BMC Med Ethics 23(1):65
124. Josa I, Aguado A (2021) Social sciences and humanities in the education of civil engineers: current status and guidelines proposal. J Clean Prod 311:127489
125. Zhang MF, Dawson JF, Kline RB (2021) Evaluating the use of covariance-based structural equation modelling with reflective measurement in organizational and management research: a review and recommendations for best practice. Br J Manag 32(2):257–272
126. Scientific Committee EFSA, More S, Bambidis V, Benford D, Bragard C, Hernandez-Jerez A, Bennekou SH, Koutsoumanis K, Machera K, Naegeli H, Nielsen SS (2020) Draft for internal testing scientific committee guidance on appraising and integrating evidence from epidemiological studies for use in EFSA's scientific assessments. EFSA J 18(8):e06221
127. Guerrieri M, Casto BML, Peri G, Rizzo G (2020) Smart vs conventional motorways: environmental impact assessment under realistic traffic conditions. Sci Total Environ 727:138521
128. Dessaint O, Foucault T, Frésard L (2024) Does alternative data improve financial forecasting? The horizon effect. J Financ 79(3):2237–2287
129. Ikram MA, Kieboom BC, Brouwer WP, Brusselle G, Chaker L, Ghanbari M, Goedegebure A, Ikram MK, Kavousi M, de Knegt RJ, Luik AI (2024) The rotterdam study. Design update and major findings between 2020 and 2024. Eur J Epidemiol 39(2):183–206
130. Manyara AM, Davies P, Stewart D, Weir CJ, Young AE, Blazeby J, Butcher NJ, Bujkiewicz S, Chan AW, Dawoud D, Offringa M (2024) Reporting of surrogate endpoints in randomised controlled trial reports (CONSORT-Surrogate): extension checklist with explanation and elaboration. BMJ 386
131. Hyland K, Jiang F (2024) Hyping the REF: promotional elements in impact submissions. High Educ 87(3):685–702
132. Moschis GP (2024) Evaluating, accepting, and rejecting manuscripts. Academic research in business and the social sciences: a guidebook for early career researchers, pp 233–263. Cham: Springer Nature Switzerland
133. Kornieva Z, Lukianenko V, Baklazhenko Y (2024) Enhancing manuscript quality: common mistakes and best practices for educational research papers. Adv Educ 4–11
134. Fisher J, Hallinan S, Lu X, Gordon M, Harchaoui Z, Choi Y (2024) StyleRemix: interpretable authorship obfuscation via distillation and perturbation of style elements. arXiv:2408.15666
135. Koed Madsen A (2024) Digital methods as 'experimental a priori'–how to navigate vague empirical situations as an operationalist pragmatist. Convergence 30(1):94–115

136. Hadzi A (2024) Extended reality in european academic institutions: interdisciplinary research on open justice. Int J Technol Knowl Soc 20(2):67
137. Carnegie GD, Gomes D, Parker LD, McBride K, Tsahuridu E (2024) How accounting can shape a better world: framework, analysis, and research agenda. Meditari Account Res

Qualitative Research Methods

Understanding human experience requires depth, context, and sensitivity to the meanings people and communities assign to their actions and interactions; it transcends numerical patterns, statistical correlations, and generalizable laws. The strength and potential of qualitative research lie in this. While quantitative approaches aim to verify hypotheses through controlled measurement and analysis, qualitative research is driven by the desire to understand events as they naturally unfold, in all their complexity and depth. From this perspective, qualitative research represents both a philosophy and a methodology; it is a way of engaging with the world that values meaning-making, subjective interpretation, and contextual knowledge. This part combines theoretical frameworks with practical tools, providing a comprehensive foundation in qualitative research techniques to help scholars navigate the dynamic field of human inquiry. Throughout these chapters, readers explore various interconnected approaches: ethnography, case studies, grounded theory, phenomenology, narrative and discourse analysis, and advanced data collection and management techniques grounded in fundamental philosophical beliefs about reality, knowledge, and values. Each chapter offers depth and breadth suitable for beginners and those seeking to refine and expand their methodological toolkit by building on prior knowledge. At its core, qualitative research is grounded in an interpretative and constructivist viewpoint. This perspective posits that reality is perceived as subjective and socially constructed, influenced by the viewpoints, experiences, and backgrounds of the individuals involved. Rather than pursuing a singular "truth," qualitative researchers seek to explore how people understand and experience the world. This interpretative approach allows for examining intricate social processes, cultural patterns, and personal experiences that would be difficult, if not impossible, to access through quantitative methods alone. Therefore, qualitative approaches are well suited for addressing questions about how and why events occur, rather than merely what happens. Emphasizing thick description, a term popularized by anthropologist Clifford Geertz, serves as one of the defining characteristics of qualitative research. It refers to the rich, detailed narratives that unveil behaviors and their significance within specific cultural contexts. Whether studying communities, organizations, families, or individuals, qualitative methods strive to capture the nuances of everyday life, including emotions, intentions, rituals, power dynamics, and social norms. This is especially important in sociology, anthropology, education, health, architecture, and urban planning, where understanding lived experiences can inform theory and practice. Qualitative research employs various data sources and collection methods, each chosen for its potential to yield rich, insightful ideas. These include audiovisual material, documents, diaries, field notes, life stories, participant observation, focus groups, and in-depth interviews. Often large, complex, and typically non-standardized, the resulting data necessitates significant interpretative skill and methodological rigor in both collection and analysis. The researcher plays a critical role throughout these practices, not merely as a detached observer but as an active participant in the meaning-making process. Reflexivity, the practice of critically scrutinizing one's positionality, biases, and impact on the research process, is fundamental to maintaining qualitative integrity. This part emphasizes the diverse methodological techniques available within the qualitative tradition. For example, ethnography allows researchers to observe behaviors, relationships, and cultural practices in their natural setting through deep, long-term engagement with a particular community or environment. In contrast, case study research focuses on specific systems such as organizations, events, or institutions, enabling in-depth, context-specific analysis utilizing multiple data sources. While phenomenology seeks to understand and articulate lived experiences from the perspective of those who have them, grounded theory provides an inductive framework for developing theory directly from data. Focusing on language, meaning, and identity, narrative and discourse analyses examine how stories are articulated and how language shapes reality and power relationships. These methodologies are flexible frameworks, not rigid templates, that can be adapted to address specific research questions and contexts' unique needs. Driven by a commitment to rigor, transparency, and ethical integrity, qualitative researchers often combine methods or traverse various traditions. A shared acknowledgment of the complexity of human experience and the necessity of understanding individuals on their terms unites them. Data management and analysis represent other critical dimensions of qualitative research explored in this part. Unlike quantitative datasets that can be organized into spreadsheets and analyzed using statistical tools, qualitative data tend to be messy, subjective, and context-dependent. Analyzing such data involves identifying themes, patterns, categories, and relationships through coding, memoing, and constant comparison. Recently, there has been a rising dependency on digital tools such as ATLAS.ti, MAXQDA, and NVivo to support these analytical processes. Nevertheless, the human element remains central. While software can facilitate analysis, the essence of meaning-making ultimately relies on the researcher's insight, sensitivity, and interpretative expertise. This part weaves ethical considerations, underscoring the researchers' obligation to treat subjects with dignity, respect, and care. This includes obtaining informed consent, safeguarding confidentiality, and being transparent about the study's goals and potential implications. Ethical practice also entails an awareness of power dynamics, cultural sensitivity, and potential harm. The ethical dimensions of qualitative research become increasingly intricate and significant when studies focus on vulnerable populations or sensitive topics. Ultimately, this part situates qualitative research within broader academic and professional contexts. Far from being a marginal or "soft" approach, qualitative research is widely recognized as essential for addressing pressing real-world issues, from healthcare delivery and policy development to social justice, educational reform, and sustainable urban planning. Its capacity to yield context-rich, actionable insights renders it invaluable in transdisciplinary endeavors where complexity, ambiguity, and human factors are pivotal. Qualitative approaches enhance academic knowledge by encompassing diverse stakeholder perspectives, values, and lived experiences and foster more inclusive, responsive, and humane forms of innovation. This part invites you to actively engage in qualitative research as both a method and a perspective, a way of understanding and relating to the world. Its ethical foundation, practical guidance, and conceptual tools empower researchers to conduct meaningful, high-quality qualitative inquiries that expand knowledge, challenge assumptions, and deepen human understanding in all its richness and diversity.

Introduction to Qualitative Research

Abstract This chapter introduces qualitative research and emphasizes its importance in understanding the depth, complexity, and context of human and societal experiences. The main goal is to give scholars a solid theoretical and methodological framework for rigorous qualitative research, addressing its limitations and potential. Interpretive and constructivist qualitative research emphasizes reality's subjective and social construction. Instead of numerical precision, it seeks rich, contextual insights to address "why" and "how" issues. The chapter discusses ethnography, grounded theory, phenomenology, and narrative analysis, each with distinct techniques for studying human behavior, social processes, and cultural dynamics. This chapter examines ontology, epistemology, and axiology, which influence qualitative research's framework and goals. Data-gathering methods, including interviews, focus groups, and observations, are explained and used to get honest participant opinions. Methodological integrity involves reducing researcher bias, ensuring rigor, and improving dependability and validity. The chapter shows how qualitative research may finesse complicated events and improve academic and professional knowledge via theoretical and applied examples. Qualitative research is essential for solving transdisciplinary problems, encouraging creative inquiry, and expanding knowledge across domains.

1 Introduction

To investigate and comprehend the nuanced breadth and depth of human experiences, qualitative research is a methodological technique that focuses on examining and understanding these aspects. Specifically, Lim (2024) stresses its merits, which include its capacity to deliver rich contextual insights and complete and holistic viewpoints. Despite these qualities, there are significant obstacles that offset them. These issues include low generalizability and vulnerability to researcher bias. Lim emphasizes the significance of strong mitigation techniques, guaranteeing qualitative research's rigor, dependability, and validity [1]. These tactics are essential for addressing the constraints that arise from qualitative research. Qualitative research

A. Almusaed et al., *Practice of Research Methodology in Civil Engineering and Architecture*, https://doi.org/10.1007/978-3-031-97393-2_9

emphasizes investigating subjective reality using textual or visual evidence, setting itself apart from quantitative methods concentrating on numerical data and statistical analysis. Ahmad et al. (2019) characterize qualitative research as a naturalistic inquiry focused on comprehending social phenomena in their authentic circumstances. This paradigm prioritizes exploring the "why" above the "what" of social events, depending on the interpretative meaning-making processes intrinsic to human experience. Qualitative researchers utilize many approaches, such as biography, case studies, historical analysis, discourse analysis, ethnography, grounded theory, and phenomenology, to investigate human phenomena instead of depending on logical or statistical frameworks [2]. This technique seeks to discover the meanings that people or groups attach to events to provide significant insights into the complexities of human behavior and social phenomena. Qualitative research's interpretive and constructivist aspects are at the heart of the field [3]. Through qualitative research, the objective is to understand events within the specific social, cultural, and historical contexts in which they occur. The positive notion that the cosmos is objective and quantifiable is challenged by this alternative theory, which is founded on the theoretical premise that reality is socially produced and intrinsically subjective. Within the realm of research technique, Susanto et al. (2024) contend that qualitative research procedures, which include literature reviews, focus group discussions, ethnography, and grounded theory, constitute an essential component of scientific literature. This literature evaluation takes a descriptive qualitative approach, employing data-gathering procedures, including exhaustive literature research and reviewing previous studies pertinent to the topic [4]. The purpose of this chapter is to present a comprehensive analysis of qualitative research, including a description of its methodology, philosophical basis, and the practical issues that drive its implementation. This chapter emphasizes the distinctive contributions that qualitative research makes to the understanding of complex events. This highlights the vital role that qualitative research plays in both academic and professional investigations [5]. Both systematic and interpretive approaches are utilized in qualitative research. In doing so, it emphasizes the significance of context and prioritizes understanding the meanings that individuals attach to their experiences and behaviors. The philosophical underpinnings of qualitative research are predominantly interpretive and constructivist, emphasizing the subjective and socially constructed interpretations of reality. According to Paudel (2024), research is a systematic process involving collecting, analyzing, interpreting, and presenting data through appropriate methodologies and tools to investigate a subject or question and enhance understanding. Philosophical perspectives inherently influence the selection of research frameworks and methods. Specifically, the intellectual paradigm—ontology, epistemology, axiology, and methodology—shapes the research process's foundational assumptions, principles, beliefs, and values [6]. Historical contributions from sociology, anthropology, and psychology have significantly shaped its evolution, shifting from positivist traditions to more interpretive and critical viewpoints [7]. This evolution highlights that qualitative research can address the intricacies of human experiences. Interpretivism, for example, emphasizes understanding behavior from the viewpoints of people experiencing it, whereas phenomenology is principally concerned with the experiences that individuals have in their lives [8]. On the other hand, critical theory criticizes power systems and amplifies excluded voices. This is in line with the emancipatory potential expressed by qualitative research. There are a few fundamental qualities that separate qualitative research from quantitative methodologies. These characteristics characterize qualitative research. It is more likely that results will correctly represent situations in the actual world when data is collected in natural settings rather than controlled ones. According to Karunarathna et al. (2024), choosing the proper data-gathering technique is essential for solid research, but it is insufficient to ensure a project's success. To reduce mistakes and improve the validity and reliability of research findings, it is crucial to guarantee the quality and correctness of the gathered data [9]. The participants' opinions and voices are prioritized, frequently requiring more extended involvement and conversation with the participants. Research designs are adaptable and can accommodate discoveries as the investigation progresses [10]. The data often consists of items that are either unstructured or semi-structured, such as interviews, observations, and visual artifacts, which enables thorough comprehension.

Researchers must actively integrate reflexivity into their methodologies to critically assess their biases during data analysis, enabling them to discern patterns and generate meaningful insights. As Jamieson (2023) argues, reflexivity is deeply tied to an academic's identity and influence on their work. It is pivotal in debates surrounding objectivity, subjectivity, and knowledge production within social scientific research. A fundamental distinction between qualitative and quantitative research is the emphasis on reflexivity. Qualitative research centers on participants' lived experiences and necessitates researchers to critically examine how their positionalities shape the research process [11]. Consequently, reflexivity is a cornerstone of qualitative research training, fostering a deeper awareness of the interplay between the researcher and the research context. Because these characteristics underline qualitative research's inductive and exploratory nature, it is particularly well suited for investigating complex social dynamics, cultural phenomena, and individual or collective actions [12].

The ontological and epistemological foundations of qualitative and quantitative research methodologies serve as the primary basis for distinguishing between these two approaches. Dieronitou (2014) posits that Content Analysis (CA) and Critical Discourse Analysis (CDA) share similar ontological underpinnings across both quantitative and qualitative paradigms. This observation challenges the rigid methodological distinctions paradigmatic purists advocate and supports adopting a more integrative, hybrid approach to research methodologies. The study concludes that qualitative dimensions inherently inform quantitative methods and, conversely, that quantitative elements underpin qualitative approaches, highlighting their interdependence [13]. The constructivist paradigm, which views reality as socially produced and subjective, is the framework for qualitative research [14]. On the other hand, quantitative inquiry is consistent with positive traditions since it presupposes the existence of an objective world that can be investigated and quantified. The nature of the data is one of the most significant contrasts between qualitative and quantitative researches [15]. Qualitative research analyzes non-numerical data, such as words and narratives, while quantitative research primarily employs numerical data and statistical models. According to Odiadi (2024), qualitative research emphasizes exploring subjective experiences and contextual complexity, offering valuable insights into intricate social processes and organizational behavior. In contrast, quantitative research prioritizes objectivity and statistical precision to test hypotheses and identify patterns within extensive datasets. Integrating these methodologies provides a more comprehensive perspective, combining the depth and nuance of qualitative insights with quantitative approaches' generalizability and predictive power. This synthesis enriches academic inquiry and practical application [16]. Qualitative research methods aim to investigate meanings, experiences, and processes. In contrast, the primary goal of quantitative research methods is to test hypotheses and determine the links between causes and effects [17]. The qualitative research method emphasizes depth and is characterized by techniques such as interviews, ethnography, and case studies. Quantitative research utilizes standardized methodologies, including questionnaires and experiments, to emphasize breadth, impartiality, and generalizability. In contrast, Dehalwar and Sharma (2024) emphasize the necessity of comprehensively grasping the differences between quantitative and qualitative research approaches. This information strengthens researchers' capacity to make informed methodological choices and significantly adds to the broader academic conversation on research paradigms and practices [18]. When conducting qualitative research, the researcher is actively immersed and thoughtful, using their point of view to interpret the findings. Quantitative research aims to reduce the amount of influence that researchers must preserve impartiality. Despite these distinctions, qualitative and quantitative approaches are not in competition with one another but rather complement one another. The qualities of both methodologies are used in mixed-methods research, which results in a more thorough grasp of the research problems being investigated [19].

When doing qualitative research, context and subjective interpretation are paramount. In contrast to quantitative approaches, which frequently strive to generalize their findings, qualitative research endeavors to fully comprehend events within the unique circumstances in which they occur [20]. This contextual emphasis guarantees that the findings are in depth and pertinent. To ensure that findings are culturally grounded, it is necessary to acknowledge that actions and experiences are inextricably linked to their respective settings [21]. When qualitative research is conducted, it avoids the imposition of external biases by locating the findings within the cultural frameworks of the participants [22]. The knowledge of social processes and the development of those processes requires a thorough comprehension of contextual elements such as time and place. When doing qualitative research, interpretation is necessarily subjective since it involves the researcher's interaction with the data it uses. This subjectivity is not a constraint but rather a strength since it enables a more profound connection with the nuances of human life. The discipline of reflexivity, which involves critically assessing one's biases, is one of the most critical factors in ensuring that interpretations remain trustworthy and authentic [23]. An example would be qualitative research investigating energy efficiency in building design. This study might examine how architects and engineers perceive and prioritize sustainability within certain socio-economic circumstances. The research may highlight, through interviews and case studies, the difficulties that professionals experience when attempting to strike a balance between creative design and regulatory compliance, as well as the cultural and organizational elements that influence the decision-making processes of these experts [24]. This perspective, which is contextualized, provides a greater understanding of the subjective and practical dynamics that shape energy-efficient construction techniques. Qualitative research offers its issues, including retaining rigor, managing vast amounts of data, and balancing depth and focus [25]. Researchers must overcome these obstacles while ensuring reliable and significant findings [26]. To have a proper appreciation for the contributions that qualitative research makes to the study of human behavior and social phenomena, it is necessary to have a solid understanding of the philosophical underpinnings, methodological characteristics, and contextual importance of qualitative research [27]. Through its emphasis on depth, subjectivity, and context, qualitative research provides us with vital insights into the intricacies of human experiences, thus improving our knowledge of the world [28]. Figure 1 visually represents the fundamental components and interactions involved in qualitative research. Qualitative research explores human experiences, focusing on the subject matter's depth, complexity, and subjective reality. The technique emphasizes literary and visual evidence to investigate cultural contexts, time, and location. Among the most essential characteristics are data collection in natural situations, the elevation of participant voices, and the utilization of adaptable designs. Interviews, observations, and visual materials are a few varied data sources used in qualitative research to provide a thorough knowledge of intricate social and cultural phenomena. Because of this abundance of data, researchers can capture the complexities of human experiences and the interaction of social interactions in their natural settings. Qualitative research is methodologically versatile and can adjust to complex and dynamic phenomena because it prioritizes context, interpretation, and the subjective aspects of reality. Because of these characteristics, qualitative research may provide depth and understanding of links and processes that quantitative studies could miss. By dividing these complex procedures into interconnected parts, the figure successfully streamlines them and offers a fundamental and understandable depiction of qualitative research techniques and their uses. It draws attention to how interpretative and context-sensitive qualitative research is by highlighting the connections between context, interpretation, and data sources. This representation highlights the methodological adaptability of qualitative research, demonstrating its capacity to tackle complex issues and unearth subtle insights across disciplines. In addition to making qualitative concepts more approachable, this kind of depiction highlights how important they are to improving our comprehension of intricate social and cultural phenomena, especially those that require in-depth and contextual examination.

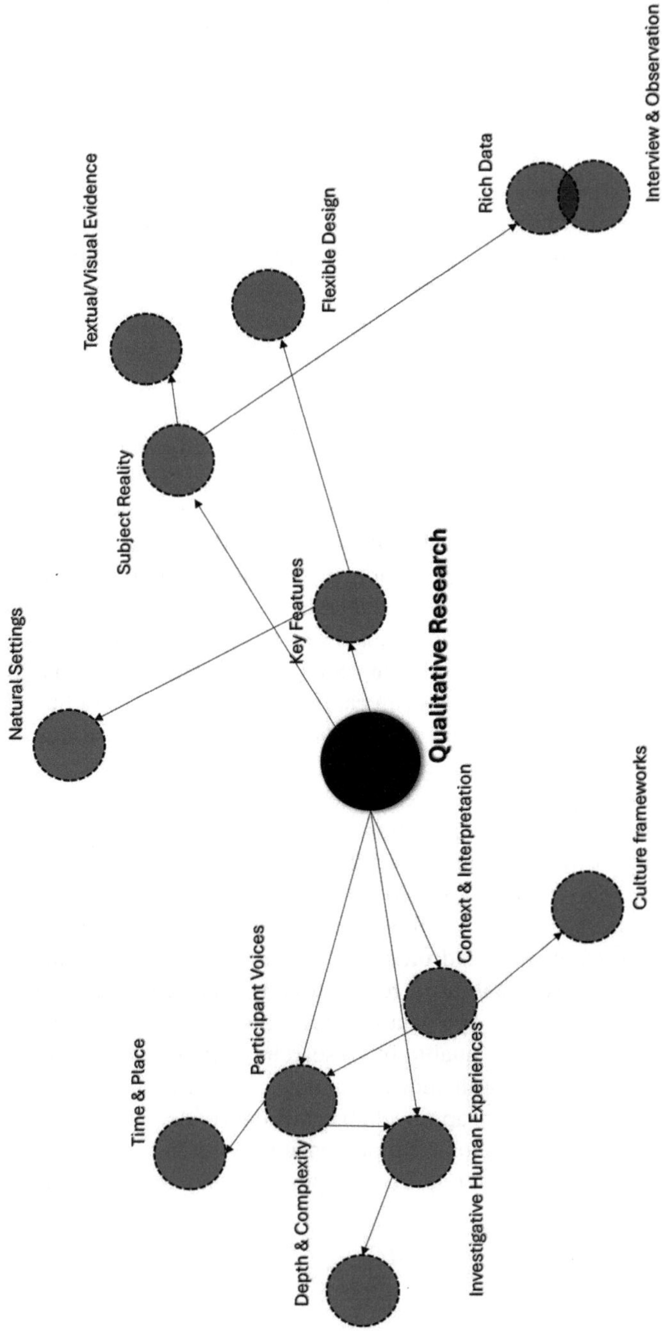

Fig. 1 Simplified conceptual framework of qualitative research

2 The Foundations and Methodological Nuances of Qualitative Research

2.1 *A Synopsis of Qualitative Research Goals and Purposes*

Investigating the breadth, depth, and complexity of human experiences is the objective of qualitative research, which aims to achieve a more profound comprehension of events through the study process [29]. This method is both methodical and interpretative, with the primary objective being to discover the meanings people or groups attach to the events they have participated in. In contrast to quantitative research, which is based on numerical data and statistical analysis, qualitative research emphasizes textual or visual evidence, which provides a wealth of insights into subjective reality. This chapter delves into the concept of qualitative research, examining its defining characteristics and distinctions. It provides a comprehensive analysis of the unique methodological framework that underpins qualitative research. Additionally, it explores the historical evolution of qualitative research and investigates the philosophical foundations that shape its approach, emphasizing its interpretative and constructivist orientation [30]. Furthermore, the chapter emphasizes the significance of context and subjective interpretation within qualitative research approaches. These components are necessary for comprehending the various layers of social and behavioral phenomena, enabling researchers to place their results within their cultural, social, and historical contexts. This section highlights the importance of qualitative research in addressing the intricacies of human experiences by emphasizing the factors above. A systematic review of the fundamental features of qualitative research and the obstacles connected with them is shown in Table 1. It sheds insight into the basic aspects constituting the qualitative research paradigm, such as the aims, technique, data focus, philosophical foundations, and contextual importance. The table also discusses issues associated with each element, such as preserving rigor, handling enormous amounts of data, and merging discoveries particular to the context with findings that apply to a broader range of situations. The table highlights the depth and complexity of qualitative research by showing this dual focus. At the same time, it acknowledges the methodological considerations required for the efficient execution and interpretation of qualitative research.

2.2 *Key Characteristics and Methodological Frameworks of Qualitative Research*

According to Khoa et al., 2023 [35], qualitative research may be characterized as an inquiry method that aims to comprehend social or human problems by constructing complex and comprehensive pictures using words, reporting the specific perspectives of informants, and carrying out the study in a natural context. An interpretative and

Table 1 Overview of qualitative research aspects and challenge

Aspect	Description	Challenges
Objective [31]	Investigates the breadth, depth, and complexity of human experiences to understand the meanings people or groups attach to events	Ensuring depth without losing focus on key objectives
Methodology [17]	Both methodical and interpretative, emphasizing the exploration of subjective realities through non-numerical methods	Maintaining rigor and reducing researcher bias
Data emphasis [32]	Focuses on textual or visual evidence, contrasting with quantitative methods based on numerical data and statistical analysis	Balancing the richness of textual data with the difficulty of analyzing large volumes
Philosophical underpinnings [33]	Grounded in interpretative and constructivist paradigms, exploring reality's subjective and social construction	Addressing subjective interpretation without compromising reliability
Contextual significance [34]	It highlights the importance of cultural, social, and historical contexts in interpreting findings, enabling a deeper understanding of social and behavioral phenomena	Integrating context-specific insights with broader applicability

constructivist technique prioritizes participants' subjective and lived experiences [36]. As a result, it is beneficial for investigating topics that are difficult to quantify.

Among the most important aspects of qualitative research are the following:

- Naturalistic inquiry gathers data in real-world settings rather than controlled environments, ensuring findings are grounded in actual conditions [37]. For example, in architectural engineering research, this approach can be used to study how occupants interact with green building features, such as natural ventilation, to evaluate their impact on comfort and energy efficiency.
- The participant-centric approach prioritizes participants' viewpoints and voices, often requiring prolonged engagement and interaction [38]. For example, in architectural engineering research, this method can be applied by involving building occupants to gather their feedback on daylighting design, ensuring spaces are both functional and user friendly based on their experiences.
- Emergent study designs are flexible, allowing adjustments as new insights are gained during the research process [39]. For example, in architectural engineering, a study on energy-efficient building materials might evolve to include occupant feedback on thermal comfort, ensuring that the research addresses performance metrics and user satisfaction.

- Rich and holistic data often consist of unstructured or semi-structured formats, such as interviews, observations, and visual materials, enabling a comprehensive understanding of the studied phenomena [40]. For example, this approach might be used in civil engineering research to analyze construction site workflows by combining worker interviews, site observations, and time-lapse photographs to identify efficiency bottlenecks and improve project management.

Researchers perform interpretive analysis, which involves analyzing data to generate interpretations and patterns. They frequently incorporate their reflexivity into the process. These features highlight qualitative research's inductive and exploratory nature, making it particularly well suited for studying complex social dynamics, cultural phenomena, and individual or collective actions.

2.3 A Paradigmatic View Comparing Qualitative and Quantitative Methods

The ontological and epistemological assumptions that go into qualitative and quantitative research are a significant point of differentiation between the two types of study [41]. Comparatively, quantitative research is found on a positive framework that presupposes an objective world that can be measured and quantified. In contrast, qualitative research frequently functions under a constructivist paradigm, which postulates that reality is socially produced and subjective. One of the most important distinctions that can be made between qualitative and quantitative research paradigms depends on the characteristics of the data. What distinguishes qualitative research from quantitative research is its emphasis on non-numerical data, including narratives, photographs, textual descriptions, and other information rich in context. A profound and nuanced comprehension of phenomena may be attained via this methodology, which is intended to dive into the complexities of human experiences, social processes, and meanings. On the other hand, quantitative research is based on numerical data and uses statistical models and variables that can be measured to assess patterns, test hypotheses, and identify causal correlations. This methodology emphasizes impartiality, accuracy, and generalizability, which enables researchers to recognize linkages and patterns that are more general. Incorporating these methodologies into a framework that utilizes mixed methods improves the research process by combining qualitative depth with quantitative rigor to provide comprehensive and robust findings across various fields of study [42]. Qualitative research aims to investigate and comprehend meanings, experiences, and social processes, emphasizing the complexity of human behavior and the subtleties influenced by context. Insights into the subjective and complicated features of social phenomena that cannot be measured are provided by qualitative research, which is founded on approaches such as interviews, ethnography, and case studies. The positivist paradigm is utilized in quantitative research, which emphasizes the testing of hypotheses, the identification of causal linkages, and the generalizability of findings. Researchers can recognize trends and develop statistical connections through this methodology, which involves the utilization of standardized instruments such as surveys and experiments to collect objective numerical data [43]. There are substantial differences between these paradigms regarding the researcher's function. The researcher is fully involved and thoughtful when conducting qualitative research. They interpret the data from their positionality while preserving reflexivity to reduce the likelihood of any adverse biases occurring. Because of this immersive engagement, it is possible to get a deep knowledge of events within the contexts in which they appear in nature. Quantitative research, on the other hand, emphasizes the researcher's neutrality, which helps ensure impartiality and minimizes the researcher's effect on the data, improving replicability and dependability. Even though these methods are founded on different ontological and epistemological principles, they are not fundamentally incompatible with one another but rather complimentary. By synthesizing the depth and contextual richness of qualitative research with the statistical rigor, generalizability, and precision of quantitative analysis, mixed-methods research offers a robust framework for addressing complex and multifaceted research problems. This integrative approach leverages the unique strengths of each methodology, providing a more comprehensive and nuanced understanding than either method could achieve independently. For example, in intelligent building research, qualitative methods can investigate the subtle experiences of occupants interacting with advanced technologies such as adaptive lighting or thermal control systems. These methods uncover critical insights into usability, satisfaction, and behavioral responses, revealing the subjective dimensions of human–technology interaction. Simultaneously, quantitative techniques deploy sensors and advanced monitoring systems to collect empirical data on performance metrics such as energy efficiency, indoor air quality, and operational reliability. The convergence of these approaches enables researchers to align user-centered design considerations with technological optimization, fostering functional and contextually relevant solutions. This synergy is particularly vital in addressing the challenges posed by contemporary built environments, where technological advancements must integrate seamlessly with human needs. Mixed-methods research thus serves as a robust paradigm for generating actionable insights that inform policy, enhance design practices, and promote sustainable innovation. By bridging the gap between subjective experiences and measurable outcomes, this approach ensures a holistic perspective, advancing theoretical knowledge and practical applications in intelligent building systems and beyond [44]. A visual representation of the complementing strengths of mixed-methods research is presented in Fig. 2. While quantitative techniques emphasize generalizability and observable facts, qualitative approaches give depth by concentrating on subjective experiences and context. On the other hand, qualitative approaches focus on context and provide breadth. Collectively, they improve research outcomes, particularly in complicated disciplines such as studies of smart buildings, by combining specific findings with broader applications.

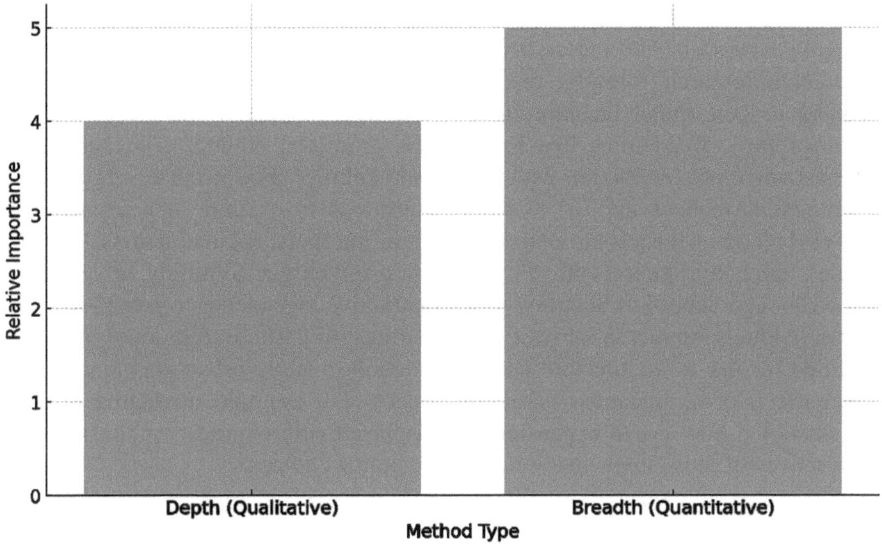

Fig. 2 Strengths of mixed-methods integration. *Source* authors

Qualitative approaches like interviews or focus groups might examine how residents use smart devices like adaptive lighting and temperature controls. Qualitative insights may show user irritation with complicated interfaces, while quantitative data shows energy savings. This comprehensive approach combines technical performance measurements with user experience to improve smart building design.

2.4 *Comparing Qualitative and Quantitative Approaches: A Paradigmatic Perspective*

2.4.1 The Development of Qualitative Research and Its Philosophical Underpinnings

Over time, qualitative research has changed dramatically and become essential for comprehending intricate social and behavioral phenomena. Sociological, anthropological, and psychological interdisciplinary contributions have influenced their evolution by moving the emphasis from positivist traditions emphasizing objectivity and quantification to interpretative and critical paradigms placing a higher value on subjectivity and contextual depth [45]. This development is especially seen in industries like building and construction, where a growing number of subjects, including stakeholder engagement, construction site dynamics, and sustainability practices, are being explored via qualitative approaches.

2.4.2 Early Development and Ethnography's Function

Qualitative research formally emerged when anthropology and sociology saw tremendous conceptual breakthroughs in the late nineteenth and early twentieth centuries [46]. Academics like Franz Boas promoted ethnographic approaches emphasizing close interaction with people and cultures. For instance, ethnographic techniques have been applied in building research to examine how construction workers behave on the job site, emphasizing how interpersonal interactions, language hurdles, and cultural norms affect collaboration and output. Similarly, urban studies in the Chicago School of Sociology used qualitative approaches to provide light on how individuals interact in intricate urban settings [47]. These techniques have been modified for use in architecture and construction to study urban planning and the social effects of significant construction projects. For example, qualitative research has examined how urban expansion has displaced communities, emphasizing the perspectives of those impacted to help guide policy choices.

2.4.3 Philosophical Foundations of Qualitative Research

Interpretivism, phenomenology, and critical theory are philosophical traditions that have strong roots in qualitative research [48]. These paradigms have added something special to qualitative research's theoretical and methodological foundations. Interpretivism, greatly influenced by Wilhelm Dilthey and Max Weber, strongly emphasizes interpreting human behavior via everyone's subjective perspectives [49]. This method is especially pertinent in building research, where stakeholder views are crucial. Interpretivist techniques, for instance, have been used to comprehend how contractors and architects view project goals differently, frequently resulting in misunderstandings and delays. Researchers can suggest methods to align objectives and enhance cooperation by investigating these individualized viewpoints. By exploring the core of human experiences, phenomenology—which Edmund Husserl developed—complements interpretivism [50]. Phenomenological studies in building study frequently concentrate on how people perceive constructed surroundings. To further understand how design features like lighting and ventilation affect comfort and productivity, researchers may, for example, look at how residents of green buildings see their interior spaces. The Frankfurt School is the source of critical theory, which gives qualitative research a transformational aspect [51]. It challenges structural injustices and power dynamics while promoting the participation of underrepresented voices. Critical theory has been applied in construction research to examine labor conditions on construction sites, particularly concerning migrant worker exploitation, dangerous working conditions, and wage inequities. Critical theory-driven research can help shape legislation that improves safety standards and labor rights by elevating the voices of these workers.

2.4.4 Broadening Perspectives: Postmodern and Poststructuralist Impacts

Qualitative research expanded in the second half of the twentieth century because of postmodern and poststructuralist ideas [52]. These schools of thought accepted reality's fractured, fluid aspect and questioned conventional grand narratives. Postmodern viewpoints have been used in the building industry to examine the changing connection between cultural identity and architectural design. Qualitative research has examined, for example, how postmodern architecture embraces modern aesthetics while incorporating a variety of cultural influences to produce places that are in tune with local traditions [53]. Like this, poststructuralist methods have been used to dismantle prevailing sustainability narratives, highlighting the necessity of context-specific and locally relevant solutions in green construction techniques. Focusing on context and subjective interpretation is one characteristic that distinguishes qualitative research. Qualitative research aims to comprehend phenomena within cultural, social, and historical contexts, unlike quantitative techniques, which frequently seek to generalize findings across populations. Context is essential in building research to comprehend project dynamics [54]. Research on building delays, for instance, may look at how stakeholder relationships, cultural attitudes toward deadlines, and local regulatory contexts affect project timeframes. The idea of embeddedness emphasizes how crucial it is to place these results in their more extensive settings to make sure they are both pertinent and nuanced. Cultural sensitivity is another vital qualitative research component, especially in global building projects. For example, cross-cultural collaboration on international construction teams has been studied using qualitative approaches, demonstrating how variations in communication styles and decision-making processes affect project outcomes.

2.4.5 The Role of Subjectivity and Reflexivity

Subjectivity, which positivist paradigms frequently view as a shortcoming, is acknowledged as a methodological strength in qualitative research [55]. A more thorough investigation of the nuances and complexity of human experiences is made possible by the interpretative character of qualitative research. Researchers can use their insights to unearth deep, context-specific understandings of social phenomena by focusing on meanings and viewpoints that numerical techniques might miss [56]. However, including subjectivity necessitates a strict adherence to reflexivity, especially when discussing delicate or divisive subjects. In qualitative research, reflexivity is essential because it forces researchers to critically examine their positionality, underlying presumptions, and potential biases. This is particularly important in domains such as construction research, where scholars frequently traverse the complex dynamics of community opposition to infrastructure initiatives. Such opposition usually results from various social, cultural, and economic issues that call for careful discussion with those who may be impacted. For example, examining resistance to extensive urban expansion entails comprehending how these initiatives affect regional customs, community structures, and environmental stability [57]. Reflexivity guarantees that researchers stay sensitive to these disadvantaged people's voices and lived experiences by avoiding the imposition of outside interpretations or prejudices. Qualitative techniques' rigor, legitimacy, and ethical accountability are improved by including reflective activities within the research process [58]. The viewpoints of underrepresented or marginalized groups may be genuinely reflected in the findings thanks to this reflective approach, which promotes more inclusive, contextually aware, and workable solutions to the intricate problems that arise in building and infrastructure projects. For example, a reflexive approach may entail working with community members to confirm results and co-interpret data.

2.4.6 Qualitative Research's Implications and Contributions

Exploring complex social and behavioral processes in construction and beyond requires qualitative research because of its emphasis on depth, context, and interpretive insights. Qualitative research encourages inclusion and respect for diversity by prioritizing participants' views and perspectives [59]. Because of its natural flexibility, qualitative research is ideally suited to dealing with construction projects' complex and dynamic character, allowing researchers to negotiate challenges in real-world settings successfully. Because of their adaptability, qualitative approaches may change as the study environment does, considering unanticipated factors and new information. Qualitative research bridges the gap between technical and human-centered concerns by emphasizing the investigation of social reality's complex and contextualized features. This leads to a more thorough knowledge of study topics. For example, qualitative studies frequently incorporate perspectives from various stakeholders, such as architects, engineers, and tenants, in sustainable building techniques. These viewpoints show how design features like natural ventilation, energy-efficient systems, and material choices affect usability and user happiness. Focus groups, in-depth interviews, and case studies provide researchers with insightful input on the practicality and efficacy of sustainable solutions. The integration of sustainability objectives with practical, occupant-friendly applications is advanced by these insights, which guide the creation of more effective, context-sensitive, and user-centered solutions [60]. The development of qualitative research reflects a rising knowledge of the significance of subjectivity, context, and interpretive depth in comprehending human experiences. Qualitative research, which has its roots in transdisciplinary contributions and is enhanced by many philosophical frameworks, is still essential to expanding our understanding of social reality [61]. Its significance in tackling intricate, practical issues and providing insights that guide policy, advance practices, and improve the built environment is demonstrated by its use in building and construction research. Figure 3 shows qualitative research's key features, highlighting its subjective and contextual nature. Reflectivity in addressing biases and their consequences in offering depth, inclusiveness, and personalized solutions for

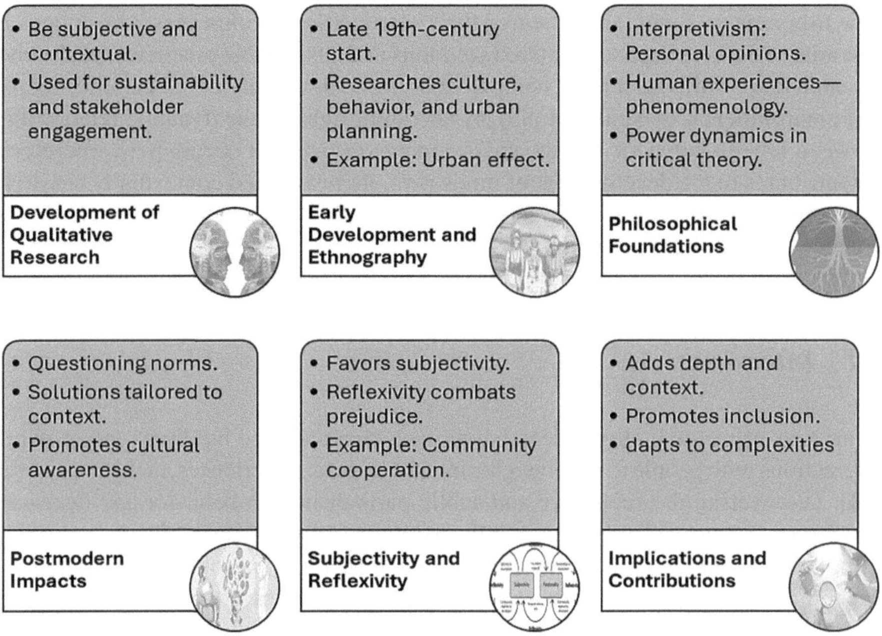

Fig. 3 Key aspects of qualitative research. *Source* Authors

complicated scenarios are highlighted, as are early ethnographic foundations, philosophical traditions like interpretivism, the influence of postmodern philosophy, and the importance of early ethnography.

3 A Critical Review of Qualitative Research and Its Importance

The qualitative research technique is a rigorous methodological approach aimed at investigating and understanding human experiences, social activities, and contextual phenomena in more depth. Qualitative research seeks to provide nuanced knowledge of complicated reality [17]. In contrast to quantitative research, which emphasizes numerical data and statistical methodologies, qualitative research emphasizes narrative data and interpretative analysis. Several types of data collecting, including in-depth interviews, participant observations, case studies, focus groups, historical analysis, and visual data interpretation, are the foundation of this methodology [62]. Thanks to these methodologies, researchers can capture the subjective sensations and contextual complexities sometimes overlooked by quantitative techniques. When it comes to fields like architecture, design, and urban planning, qualitative research is fundamental because of the profound impact it has on these fields. Knowing how users interact with and perceive their surroundings within these disciplines is essential to developing user-centered solutions and sustainable practices. Qualitative research helps cultivate a more comprehensive understanding of the cultural, social, and environmental dynamics at play by shedding light on the dynamic relationship between human behavior and the environments in which it occurs [63]. Therefore, it contributes to the development of innovative, inclusive, and contextually sensitive frameworks, ultimately leading to the advancement of academic and professional practices within these categories of study.

3.1 Interviews

Interviews are a qualitative research approach characterized by direct and in-depth interactions with people to gain insights into their ideas, experiences, and perspectives [64]. Discovering the subjective and subtle parts of human behavior and decision-making processes is challenging, but this strategy is highly successful. It is possible to conduct interviews in an organized, semi-structured, or unstructured manner, depending on the research purpose and the flexibility necessary to investigate emergent topics effectively. Regarding architecture and engineering, interviews are essential for grasping intellectual and practical aspects of design elements. For example, engaging with architects about their design ideas may shed light on the cognitive and creative processes behind their work, the obstacles they experience during the design development process, and the unique solutions they utilize. These encounters give information rich in context and include specifics beyond surface-level observations. They also provide vital insights into the dynamic relationship between theory and practice.

How Can Research Interviews Be Conducted Effectively?

Researchers can collect rich, in-depth insights from various participants through research interviews, a fundamental component of qualitative research. They are handy for investigating complicated subjects, capturing one-of-a-kind points of view, and revealing delicate nuances that may otherwise be unreachable through other means. It is necessary to possess rigorous preparation, clear communication, and an approach that strikes a balance between structure and flexibility to guarantee the effectiveness of interviews. A well-defined research topic is essential to properly connecting interviews with the study's objectives. It is crucial to develop interview questions carefully to ensure they are relevant and provide sufficient depth in addressing the topic being investigated. Questions such as "What factors influence your decision-making process in a project?" serve as a clear entrance point and an invitation for participants to provide relevant insights. For example, this question may be used to describe the decision-making process. A flexible checklist may guarantee that essential themes are addressed while allowing follow-up questions to investigate unexpected replies or developing themes. Questions with open-ended responses are beneficial for generating responses that are both thorough and complete. The question "Can you describe your approach to problem-solving in challenging situations?" is an example of a question that invites participants to communicate their experiences and perspectives more in depth. Furthermore, having a neutral tone during the whole interview is equally important. Using neutral language, such as "What are your thoughts on this method?" rather than "Don't you think this method is the best option?" is an effective way to guarantee that the comments provided by participants are genuine and objective. Additionally, at the end of the interview, clarifying any unclear responses and summarizing the most essential elements increases the quality and dependability of the data obtained. One example of a researcher's ability to summarize would be to state, "To confirm, you mentioned that collaboration plays a central role in achieving your goals." Could you explain how this is put into practice? Not only does this technique validate the information, but it also allows for more elaboration if required. The research interviews provide rich, contextually relevant material, contributing to a more profound knowledge of the topic investigated. Researchers can increase the quality and depth of the insights obtained through this method by utilizing a methodical approach, enabling open discourse, and maintaining neutrality. For example, developing frameworks and rules that improve architectural practices may also be informed by interviews, which can be used to gather information [65]. This technique contributes to the advancement of knowledge, the promotion of critical thought, and the promotion of integrating theoretical concepts with practical applications in design and construction. It does this by capturing a variety of views. An illustration of how interviews can be used as a qualitative research method to investigate ideas, experiences, and decision-making is provided in Fig. 4. Their ability to provide in-depth insights, improve comprehension of design processes, and bridge the gap between theoretical knowledge and actual implementations is extremely valuable.

The three essential characteristics that establish research interviews as an effective qualitative method are the ability to adapt questions to participants' experiences and perspectives, the elaborative potential to clarify responses, and the ability to capture contextual factors influencing responses.

- Researchers can change and adjust their questions based on the replies they receive from participants, allowing for examining emergent themes or unexpected subjects throughout the interview process [66].
- Elaborative potential: Participants can expound on complicated concepts or express their thoughts in greater detail. This helps limit misunderstandings, allows researchers to investigate subtleties, and effectively resolves gaps in participants' viewpoints [67].
- Contextual understanding: Interviews offer insights into the social, cultural, and environmental elements that shape participants' experiences. This enables a better grasp of the participants' perspectives and the larger context influencing their replies [68].

In qualitative research, interviews are an essential tool for collecting data that is rich in detail, relevant to the situation, and specific to the individual. Because of their adaptability, capacity for elaboration, and capacity to capture contextual

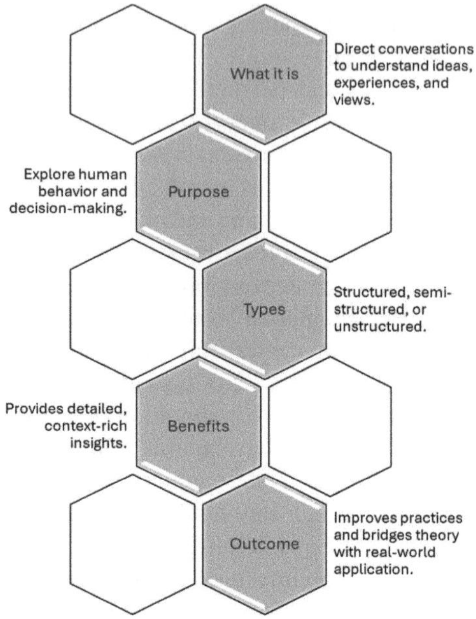

Fig. 4 The role of interviews in investigating design processes and decision-making. *Source* authors

awareness, they are handy for investigating complicated phenomena and gaining nuanced insights.

3.2 Feedback

Feedback is a crucial element of iterative enhancement in design and idea development. It involves gathering and examining individuals' perspectives, assessments, and responses to specific concepts or results [69]. In architecture and engineering, for example, collecting feedback on a building's design provides essential insights into public perceptions, user happiness, and operational deficiencies, eventually enabling improvements that meet the functional and experiential requirements of end-users. The feedback acquisition procedure must be methodologically rigorous and strategically organized to guarantee thorough and impartial data collection [70]. This entails utilizing several methodologies, such as structured surveys, comprehensive interviews, focus group discussions, and observational studies. Each method may be customized to target facets of the project, encompassing aesthetic preferences and usability issues. User satisfaction surveys can evaluate aesthetic appeal, ergonomic functionality, and accessibility, whereas stakeholder interviews offer qualitative insights into intricate experiences and expectations. Feedback is essential for aligning design results with user requirements and social standards. In architectural contexts, this alignment guarantees that structures surpass their functional purpose by promoting a harmonic relationship between inhabitants and the constructed environment. Feedback systems reveal preferences for essential design components, including natural lighting, spatial fluidity, material selection, and energy efficiency [71]. Integrating these preferences into the design process improves the project's relevance, usefulness, and sustainability. Moreover, feedback functions as a diagnostic instrument for detecting hidden faults and opportunities for enhancement inside a project. A systematic investigation of user reactions might uncover significant undetected deficiencies, like suboptimal spatial arrangement, insufficient signage, or environmental inefficiencies [72]. Proactively addressing these concerns reduces possible unhappiness and enhances the project's overall quality and functionality. In addition to enhancing projects, feedback enriches the overall corpus of architectural knowledge and professional practice. Interacting with varied stakeholders throughout the feedback process equips architects with improved insights into social trends, changing user behaviors, and novel design methodologies. These insights promote a cyclical learning process that guides future initiatives and supports adaptive, evidence-based practices within the profession. Feedback enhances cooperation and stakeholder involvement by highlighting participatory design concepts. By engaging users and communities in the assessment and development process, architects exhibit attentiveness to their concerns, cultivating trust and a feeling of collective ownership [73]. This inclusion improves the product's relevance and creates a foundation for ongoing communication between professionals and the communities they serve. In conclusion, feedback is a fundamental component in promoting architectural quality. Engaging systematically with stakeholders enables the development of responsive, user-centered designs that meet functional needs and social ambitions. Thus, feedback serves as a mechanism for enhancement and a fundamental element of creativity and accountability in architectural engineering practice [74]. Important feedback features, such as its goals, techniques, uses, and difficulties, are summarized in Table 2. It emphasizes the role of feedback in bringing designs closer to user demands, finding areas of improvement, expanding expertise, and encouraging teamwork to achieve better architectural results.

3.3 Case Studies

Case studies are an essential research tool, particularly in architecture and engineering. In these fields, analyzing specific phenomena, such as creating iconic structures or applying passive building principles, requires a methodology that is nuanced and sensitive to the environment in which it is being conducted. Researchers and practitioners can uncover effective strategies, critically analyze design techniques, and extract insights that might guide and inspire future architectural undertakings when they conduct systematic analyses of instances [82]. This approach offers a solid foundation for comprehending the dynamic relationship between historical, cultural, and contextual elements that play a role in forming architectural practices. By way of

Table 2 Crucial elements of input in design and assessment

Aspect	Description	Applications	Challenges
Definition of feedback [75]	Process of collecting and analyzing evaluations to improve ideas or designs	Feedback on building design for user satisfaction and issues	Ensuring unbiased and informed responses
Methods [76]	Use of structured approaches for reliable data collection	Surveys, interviews, focus groups, and observations	Choosing suitable methods for diverse users
Alignment with needs [77]	Ensures projects meet user preferences and functional demands	Incorporating preferences for lighting, materials, and space	Balancing user needs with technical limits
Diagnostic tool [78]	Identify deficiencies for targeted improvements	Addressing poor layouts, signage, or inefficiencies	Detecting early issues with limited data
Knowledge contribution [79]	Enhance practice by integrating user insights and trends into future designs	Using trends for adaptive and innovative practices	Applying abstract trends effectively
Collaboration [80]	Build trust and ownership through inclusive stakeholder engagement	Engaging communities from diverse perspectives	Managing conflicting priorities
Purpose and value [81]	Improves quality by aligning designs with user and societal goals	Creating functional, sustainable, and user-centered projects	Balancing short-term and long-term needs

illustration, case studies make it possible to conduct an in-depth analysis of architectural approaches and the consequences of these methodologies in terms of aesthetics, sociocultural relevance, and functional performance. This is accomplished by carefully examining design decisions, material applications, and structural innovations. This method bridges the gap between theoretical paradigms and their practical implementations across various historical and geographical settings. It sheds light on the adaptation techniques responsible for the life and significance of architectural works. Case studies are defined by several fundamental qualities, including their holistic approach, reliance on numerous sources of evidence such as interviews, papers, observations, and artifacts, and flexibility for various research purposes [32]. Individually, they may be classified into three distinct types: exploratory, descriptive, and explanatory, each fulfilling a distinct research aim. Explanatory investigations identify causal mechanisms and correlations, descriptive studies give extensive contextual assessments, and exploratory case studies construct theoretical frameworks for understudied issues [83]. Exploratory case studies are a type of theoretical framework development. Case studies, for example, play an essential role in passive building research because they make it possible to conduct in-depth examinations of energy-efficient design methods and the performance of these strategies in the actual world. Analyzing thermal insulation systems, ventilation strategies, and energy use patterns may be performed to solve obstacles in obtaining net-zero energy objectives. In addition, case studies shed light on the interplay between building materials, climate, and the behavior of occupants, which further helps implement sustainability principles. Beyond academic inquiry, case studies bridge the gap between theory and practice, providing insights that may be put into reality for innovation, policy formulation, and decision-making [84]. Not only do their results appeal to academic audiences, but they also resonate with professional audiences, which helps to stimulate critical engagement, evidence-based decision-making, and the growth of architectural practices that are culturally sensitive and sustainable. Due to their versatility and depth, case studies are vital for tackling the intricacies of current architectural and engineering difficulties. Table 3 illustrates the sequential stages involved in conducting case study research, beginning with defining the phenomena and continuing through the analysis of data and drafting findings. Each phase, which includes the aim, the collecting of data, and the applications, blends theory and practice, providing an organized strategy for examining complicated phenomena within the settings of their real-world applications.

3.4 Observations

Observational research is a method that focuses on analyzing how different stakeholders interact with one another and work together during different processes [94]. By observing team dynamics, decision-making tactics, and problem-solving procedures in real-world contexts, researchers can better comprehend the mechanisms that govern the development of practices. It is possible to identify significant characteristics that influence cooperation via this technique. These factors include communication patterns, role distribution, and adaptive tactics. This approach offers useful insights into the intricacies of teamwork and creativity across various organizational contexts and sectors [95]. An excellent illustration of this would be the utilization of observational research in architecture to investigate how architects, engineers, and contractors collaborate on environmentally friendly construction projects. Researchers can determine the process by which decisions are made regarding materials and energy-efficient systems by observing meetings and conversations that take place on-site [96]. They can also discover communication difficulties and identify potential solutions to improve cooperation and the overall effectiveness of the project. A key approach in qualitative research is observation, which enables the systematic study of behaviors, interactions, and processes within their natural settings. Consequently, observation is an essential component of qualitative research. Researchers can get direct and genuine insights into real-world dynamics' intricacies by observing things as they occur [97]. As a result of its ability to bridge the gap between theoretical frameworks and actual realities, this approach is a crucial instrument in qualitative research. It is not enough to view something passively; observation is a process that is both purposeful and methodical. A definition of the phenomenon that will be investigated, selecting suitable instruments for data collection, and recording observations

Table 3 Design process in case study research

Aspect	Explanation
Definition [85]	Detailed investigation of a phenomenon within its real-life context using multiple data sources
Purpose [86]	To understand complex phenomena and refine theoretical frameworks
Key characteristics [32]	– Holistic description and explanation – Flexible design and data collection – Use of multiple evidence sources – Emphasis on context
Types of case studies [87]	– Exploratory: Identify trends or develop frameworks – Descriptive: Provide detailed analysis of phenomena – Explanatory: Uncover causal relationships
Applications [88]	– Investigating challenges in sustainable architecture and passive design – Analyzing complex systems and their contexts
Benefits [89]	– Detailed insights – Methodological flexibility – Real-world applicability – Robust findings from multiple evidence sources
Limitations [90]	– Limited generalizability – Resource-intensive – Complex data analysis – Potential for subjective interpretation
Data collection [91]	Includes interviews, observations, documents, artifacts, and records to provide comprehensive insights
Analysis [92]	– Organize and code data – Identify patterns/themes – Interpret findings in context – Ensure accuracy and consistency
Significance [93]	Connects theory with practice, offering actionable insights and advancing knowledge in various fields

in an organized or unstructured way are all required steps. It is necessary to have proper preparation and execution to guarantee the dependability and authenticity of the collected data. When doing research, researchers must identify significant variables, provide circumstances for observation, and educate observers to guarantee consistency in their methodology [98]. It is possible to classify observation techniques according to their structure, the method of observation, and the function that the researcher plays in the process. Structured observation is a method that adheres to predetermined categories and applies to hypothesis testing and systematic description. The opposite of structured observation is unstructured observation, which is more exploration and flexible, enabling the discovery of patterns or behaviors that were not anticipated. The researcher may take on various roles, including those of a participant, in which they fully immerse themselves in the group being researched, and those of a non-participant, in which they watch from a distance to limit their effect

[99]. There are two types of observation: direct observation, in which researchers are present at the events being observed, and indirect observation, in which instruments such as cameras or records are used to collect data. Furthermore, it can be overt, in which the participants are aware that they are being watched, or covert, in which the observation is carried out without the participants' awareness to reduce the likelihood of their behavior being altered. Within the realm of qualitative research, using observation has several benefits. In addition to providing rich data in context and typical real environments, it also captures behaviors and interactions as they occur. This is especially helpful in studying non-verbal communication, spontaneous interactions, and processes that participants might struggle with articulating. As an additional benefit, observation enables a full knowledge of complex systems by illuminating the interaction of several components in real time. Nevertheless, there are difficulties associated with observation. It may be time-consuming and resource-intensive, and it requires careful preparation and execution to be successful. One further concern is the possibility of observer bias, which occurs when the researcher's point of view tends to impact how observed actions are interpreted [100]. Especially when it comes to covert observations, it is necessary to consider ethical factors such as protecting individuals' privacy and obtaining informed consent. In addition, the information that is gathered through observation is frequently dependent on the immediate environment and may not be easily generalized. Although there are many obstacles to overcome, observation continues to be an essential approach in qualitative research. When carried out methodically and ethically, it offers important insights into human behavior and social relationships, enhancing our comprehension of complicated phenomena and advancing theoretical and practical knowledge [101]. Table 4 presents a succinct overview of observation methods, emphasizing their principal characteristics. Observation is a methodical method for examining behaviors and processes in real-world and controlled environments. The table classifies observation kinds, roles, and procedures, including their benefits, drawbacks, and ethical implications. It underscores the importance of observation in qualitative research for acquiring actionable insights and connecting theory with practical applications.

3.5 Focus Group

Focus groups are a well-known qualitative research tool that has gained widespread recognition for their ability to offer a structured platform to obtain various opinions via organized conversations [110]. This method is beneficial if you are interested in researching multidimensional themes, such as sustainable architectural engineering and urban planning, where it is essential to integrate technical competence, community views, and environmental considerations. Focus groups are great for revealing nuanced ideas, encouraging collaborative problem-solving, and generating insights that are frequently difficult to obtain through surveys or individual interviews [111]. Focus groups do these things by stimulating interaction among members. As an

Table 4 Important features of methods for observation

Aspect	Details
Definition [9]	A method to systematically observe and record behaviors or events in natural or controlled settings
Purpose [102]	To understand real-world behaviors, interactions, and processes as they naturally occur
Types [103]	– Structured: Uses predefined categories – Unstructured: Flexible and exploratory
Roles [99]	– **Participant:** Observer is part of the group – **Non-participant:** Observer watches from outside
Techniques [104]	– **Direct:** Observer watches in real time – **Indirect**: Uses tools like cameras or recordings
Visibility [105]	– **Overt**: Subjects know they are being observed – **Covert**: Subjects are unaware of observation
Advantages [106]	– Real-world insights – Captures non-verbal cues – Data in natural settings
Limitations [107]	– It takes a lot of time – Inherent prejudice in the observer – Moral considerations
Ethics [108]	Ensure privacy and informed consent and minimize the impact on participants
Significance [109]	Provides actionable insights and connects theory to practice

illustration, in sustainable architectural engineering, focus groups function as a platform that brings together professionals from various fields to discuss the difficulties and possibilities associated with implementing environmentally aware methods. In the case of research concerning the incorporation of solar energy in residential buildings, participants can include homeowners, architects, civil and electrical engineers, and urban planners. A full investigation of the issue is ensured by this variety, which combines technical understanding with experiences gained from real-world situations [112]. It is possible that architects would emphasize the aesthetic and structural consequences of solar panels, while engineers may stress the installation's complexity or the maintenance requirements. At the same time, homeowners could simultaneously express concerns over prices, perceived advantages, or compatibility with the infrastructure that is already in place. Through these kinds of conversations, technical obstacles are identified, and social and economic aspects that influence adoption are uncovered. The insights acquired from these focus groups provide politicians and designers with information that can be used to improve the adoption of solar energy and the public's acceptance of it. Similarly, focus groups play an essential role in evaluating sustainable development projects and the consequences those initiatives have for communities in various cases that pertain to urban planning [113]. Multiple stakeholders, such as environmental experts, urban planners, construction professionals, and citizens, are frequently involved in these conversations. As an illustration, a discussion group focusing on green construction methods may investigate energy efficiency, environmental effects, and community well-being. In addition to discussing the advantages of green infrastructure, such as decreased carbon footprints and improved living conditions in metropolitan areas, participants might also discuss the problems associated with green infrastructure, such as locating sustainable materials, achieving a cost–benefit balance, and overcoming regulatory obstacles. Urban planners might use these insights to establish policies encouraging environmentally responsible growth while addressing economic restrictions [114]. The emphasis groups place on engagement and collaborative discourse, which frequently results in the emergence of new ideas, is one of the distinguishing characteristics of these groups. During a discussion on modular building, for instance, an engineer would emphasize the environmental efficiency of modular construction, inspiring an architect to comment on how modular systems are in harmony with design goals. This lively interchange contributes to the enrichment of the conversation. It encourages the development of creative ideas that approach sustainability from various perspectives [115].

Nevertheless, there are several difficulties associated with focus groups. Dominant individuals can obscure the voices of more reserved others, and groupthink may restrict the expression of various ideas. A well-executed moderation strategy is necessary to preserve equilibrium, guarantee inclusiveness, and direct conversations while allowing for the natural exploration of different subjects [116]. Focus groups continue to be a practical approach for gathering rich, qualitative data that may be used to influence sustainable architectural and urban planning practices despite the limitations that have been mentioned. Focus groups provide an excellent platform for tackling sustainability challenges and creating actionable insights that drive significant change. This is accomplished via careful planning, expert facilitation, and the engagement of various stakeholders. Table 5 lists the benefits and drawbacks of using focus group interviews. It highlights their capacity to investigate group dynamics, find new ideas, and provide varied and rich data through engagement. On the other hand, there are obstacles such as participant domination, groupthink, limited generalizability, and practical problems. These qualities are demonstrated in real-world examples.

Focus group interviews are powerful qualitative research tools that allow for exploring collective opinions and experiences [125]. By assembling diverse participants and fostering interactive discussions, researchers can uncover nuanced insights contributing to more informed decision-making. Although focus groups have limitations, their ability to generate rich, qualitative data makes them invaluable in urban planning and product development. When carefully planned and skillfully moderated, focus group interviews offer a unique and dynamic approach to understanding complex topics [126].

Table 5 Advantages and challenges of focus group interviews with practical examples

Category	Subcategory	Description	Example
Advantages	Rich data collection [117]	The interactive nature generates detailed and diverse qualitative data, sparking new ideas and uncovering deeper insights	Participants discuss user experience with a sustainable housing project, generating diverse qualitative data
	Efficient feedback [118]	Multiple participants provide various perspectives in a single session, saving time and resources	A single session gathers feedback from architects, planners, and homeowners, reducing the time spent on individual interviews
	Emergent insights [119]	Group discussions lead to new perspectives and ideas that may not surface in individual interviews	An architect introduces modular design, inspiring engineers and planners to explore practical applications
	Exploration of group dynamics [120]	Observing participant interactions reveals attitudes, opinions, and decision-making processes influenced by group dynamics	Homeowners debating solar energy adoption provide insights into cost vs. benefit considerations
Challenges	Dominance of certain participants [121]	Strong personalities may dominate discussions, potentially overshadowing quieter participants. Moderators must ensure balanced participation	A dominant homeowner monopolizes the discussion on green building practices, requiring moderator intervention
	Groupthink [122]	Participants may conform to majority opinions, suppressing their true thoughts. Encouraging diverse viewpoints is essential	Participants hesitate to criticize a proposed policy for fear of opposing the group consensus
	Limited generalizability [123]	Small, non-random samples may not represent the broader population. Combining focus groups with other methods strengthens conclusions	Findings from focus groups discussing solar adoption might not apply to all regions
	Logistical challenges [124]	Coordinating schedules, finding suitable locations, and ensuring technical reliability (for virtual groups) can be time-intensive	The virtual focus group encounters technical issues, which delay data collection and increase the administrative workload

4 Comprehensive Framework for Qualitative Research: A Systematic Approach

Even though the qualitative research process is intrinsically adaptable to suit a wide range of techniques and project contexts, it must adhere to fundamental principles to guarantee consistency and effectiveness. This guide provides a comprehensive and organized explanation of the basic steps of qualitative research, focusing on its use in environmentally responsible construction methods.

4.1 Determining the Subjects of the Research

When doing qualitative research, the first step is clearly articulating the study's goals. Identifying knowledge gaps and opportunities requires analyzing relevant inputs, including strategic papers, stakeholder comments, and policies already in place. Considering the community's concerns and the stakeholders' objectives, for example, researchers could evaluate the environmental effect of green roofing systems in a sustainable construction project. The potential impact of the project, the availability of resources, and the timeline limits are all considered while developing a prioritized research plan [127]. For example, if a green roofing program decreases the number of urban heat islands, it will be prioritized over less critical renovations. Researchers pose essential questions, such as "How do residents perceive the trade-offs between the costs and benefits of green roofs?" Researchers formulate these questions. In selecting participants, attention is placed on stakeholders such as urban planners, architects, and citizens who have direct control over the project or are directly affected by it.

4.2 The Process of Selecting an Appropriate Methodology

The research focus is taken into consideration when selecting the method. In sustainable design, secondary analysis may entail examining previous research on the effectiveness of green buildings in terms of energy performance or materials management [44]. Methods might be used to investigate attitudes (for example, the acceptance of sustainable materials by the community) or actions (for example, the adoption of environmentally friendly practices by contractors) depending on the purpose of the research. In the context of exploratory research, semi-structured interviews can potentially disclose attitudes on integrating solar panels. In the context of research evaluation, usability testing might be used to evaluate how users interact with environmentally friendly building systems. Distinguished sources, such as the recommendations provided by the United States Green Building Council, assist in refining methodological choices [128].

4.3 The Realignment of Stakeholders

The research objectives are guaranteed to fulfill practical demands through alignment with stakeholders while maintaining relevance to the strategic context [129]. Detailed plans that outline the scope and timetable are offered to solicit stakeholder feedback. In a sustainable building program, for instance, construction managers could propose removing regulatory impediments or delays in the supply chain. The researchers consider this feedback and modify the designs of the studies so that they integrate practical factors such as timetables for the acquisition of materials. Research frameworks are further improved by peer criticisms, which also contribute to the development of consensus and methodological rigor [130].

4.4 Preparation and Carrying Out the Plan

As part of the preparation process, things like arranging logistics, such as test sites or laboratories for environmentally friendly buildings, and organizing interviews with participants, such as environmental engineers and architects, are essential [131]. The conformity with ethical standards is ensured by legal documents, which may include confidentiality agreements. As an illustration, researchers may create prototypes for hands-on participant interaction during a focus group investigating energy-efficient building materials. The process of execution entails the facilitation of meetings in which stakeholders offer input on innovations, such as solar-integrated facades. At the same time, researchers document observations and record sessions for analysis.

4.5 Insights Analysis and Reporting as the Fifth Step

Data analysis reveals patterns and insights connected to the research aims [130]. For example, the outcomes of a study that investigates the practicability of passive cooling systems can emphasize that inhabitants place a higher value on energy savings than they do on the installation price. In addition to assisting in synthesizing data, collaborative approaches such as affinity mapping also include stakeholders in the process of thematic analysis. To successfully engage policymakers, architects, and community groups, deliverables such as sustainability impact evaluations, lifetime cost studies, and green building personas are adapted to meet specific needs.

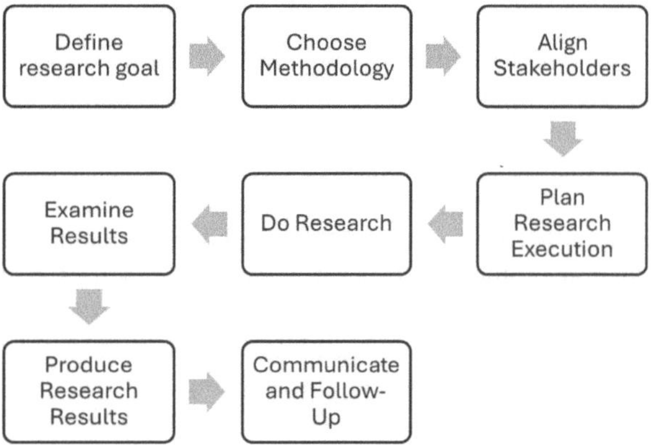

Fig. 5 The qualitative research process. *Source*: authors

4.6 *Implementation and the Spreading of Information*

When disseminating findings, it is necessary to offer outputs adapted to different audiences' needs. Detailed studies for urban planners may address zoning concerns, while executive summaries may concentrate on the return on investment (ROI) from investments in environmentally responsible construction practices [131]. When it comes to urban contexts, for instance, the findings of a study on green wall systems might be used to drive legislative changes that encourage the cultivation of vertical gardens. During the follow-up process, it is ensured that the findings are included in strategic frameworks, such as municipal sustainability plans or architectural design guidelines, which ultimately lead to achieving practical results.

The qualitative research method is iterative and collaborative, with careful preparation, implementation, and communication. By following this systematic method, researchers may provide actionable insights that successfully guide strategic decisions and promote organizational success. As shown in Fig. 5, qualitative research is broken down into eight distinct processes. These steps include setting research objectives, selecting methodology, aligning stakeholders, preparing for execution, conducting research, assessing findings, generating outputs, and distributing results with follow-up. Each step emphasizes methodical preparation and execution to ensure efficient qualitative inquiry and the distribution of findings that have a significant impact.

5 Conclusion

This chapter emphasizes the vital role of qualitative research in enhancing our comprehension of the intricacies associated with human and societal processes. Qualitative research emphasizes depth, context, and interpretative insight rather than numerical accuracy, facilitating the exploration of subtle inquiries about "why" and "how" events transpire. Its dependence on interpretative and constructivist paradigms underscores reality's socially created and subjective essence, providing a methodological framework adept at examining cultural, social, and historical components sometimes neglected by quantitative methods. This methodological focus guarantees that qualitative research encompasses the interaction between personal experiences and broader societal circumstances, enhancing the investigation depth. The chapter focuses on qualitative procedures, including ethnography, grounded theory, phenomenology, and narrative analysis, each offering unique yet complementary methods for examining human behavior and social dynamics. These approaches are supported by a strong philosophical framework encompassing ontology, epistemology, and axiology, guaranteeing that the research process fits its aims and maintains methodological rigor. Qualitative research prioritizes context and subjective interpretation, yielding significant and practical insights, especially in disciplines where standardization and objectivity may mask essential elements of lived experiences. Furthermore, the chapter delineates essential data-gathering methodologies—namely, interviews, focus groups, and observations—that facilitate researchers acquiring comprehensive, nuanced insights from participants. These methodologies have problems, such as generalizability concerns, researcher bias, and the labor-intensive characteristics of qualitative analysis. These constraints are effectively addressed by reflexivity, methodological triangulation, and adherence to ethical norms, which collectively improve qualitative results' rigor, dependability, and validity. Although qualitative research is grounded in subjectivity, this does not undermine its scholarly significance. Instead, it enhances its significance by enabling the investigation of emerging patterns, revealing the fundamental meanings behind observable occurrences, and offering theoretical insights that further knowledge. Its inductive and exploratory characteristics render it an exceptionally efficient instrument for examining intricate social and cultural processes, guaranteeing that results are profoundly rooted in their surrounding reality. The chapter emphasizes the synergy between qualitative and quantitative researches, promoting an integrated approach using mixed techniques. Mixed-methods research integrates the contextual depth of qualitative insights with the generalizability and accuracy of quantitative data, providing a holistic viewpoint that can tackle complex research inquiries. This synthesis enhances theoretical exploration and practical implementation, broadening the applicability of research outcomes in academic and professional fields. Qualitative research is a crucial and necessary element of scholarly investigation, especially in fields that need a profound comprehension of human experiences and social processes. Its methodological rigor, versatility, and dedication to contextual sensitivity offer exceptional chances to investigate intricate topics that defy reductionist methods. Qualitative research enhances our understanding of lived experiences and significantly aids in tackling real-world issues. Its function in cultivating transdisciplinary comprehension and guiding new methodologies highlights its lasting significance as a fundamental element of scholarly and professional inquiry.

References

1. Lim WM (2024) What is qualitative research? An overview and guidelines. Australasian Market J 14413582241264619
2. Ahmad S, Wasim S, Irfan S, Gogoi S, Srivastava A, Farheen Z (2019) Qualitative v/s. quantitative research summarized review. Population 1(2):2828–2832
3. Denicolo P (2016) Constructivist approaches and research methods: A practical guide to exploring personal meanings
4. Susanto PC, Yuntina L, Saribanon E, Soehaditama JP, Liana E (2024) Qualitative method concepts: literature review, focus group discussion, ethnography, and grounded theory. Siber J Adv Multidis 2(2):262–275
5. Hennink M, Hutter I, Bailey A (2020) Qualitative research methods. Sage
6. Paudel P (2024) Examining paradigmatic shifts: unveiling the philosophical foundations shaping social research methodologies. J Univ Ruhuna 12(1)
7. Sun P, Zuo X (2024) Evolution and history of research philosophy. J Manag Res 24(1):28–61
8. Acharya R (2024) Examining interpretivism in social science research. Education
9. Karunarathna I, Gunasena P, Hapuarachchi T, Gunathilake S (2024) Comprehensive data collection: methods, challenges, and the importance of accuracy
10. Martin F, Chen Y, Moore RL, Westine CD (2020) A systematic review of adaptive learning research designs, context, strategies, and technologies from 2009 to 2018. Education Tech Research Dev 68(4):1903–1929
11. Jamieson MK, Govaart GH, Pownall M (2023) Reflexivity in quantitative research: a rationale and beginner's guide. Soc Pers Psychol Compass 17(4):e12735
12. Merriam SB, Grenier RS (eds) (2019) Qualitative research in practice: examples for discussion and analysis. Wiley
13. Dieronitou I (2014) The ontological and epistemological foundations of qualitative and quantitative methods to research. Int J Econ 2(10):1–17
14. Pilarska J (2021) The constructivist paradigm and phenomenological qualitative research design. Research paradigm considerations for emerging scholars, pp 64–83
15. Emon MMH (2024) Research approach: a comparative analysis of quantitative and qualitative methodologies in social science research
16. Odiadi J (2024) Qualitative and quantitative research in business studies: managing. Available at SSRN 4876332
17. Pregoner JD (2024) Research approaches in education: a quantitative, qualitative and mixed methods comparison. IMCC J Sci 4(2):31–36
18. Dehalwar K, Sharma SN (2024) Exploring the distinctions between quantitative and qualitative research methods. Think India Journal 27(1):7–15
19. Guetterman TC, Manojlovich M (2024) Grand rounds in methodology: designing for integration in mixed methods research. BMJ Qual Saf 33(7):470–478
20. Omodan BI (2024) Research paradigms and their methodological alignment in social sciences: a practical guide for researchers. Taylor & Francis
21. Campos F, Nguyen H, Ahn J, Jackson K (2024) Leveraging cultural forms in human-centered learning analytics design. Br J Edu Technol 55(3):769–784
22. Ademolu E (2024) Birds of a feather (don't always) flock together: Critical reflexivity of 'Outsiderness' as an 'Insider ' doing qualitative research with one's 'Own People.' Qual Res 24(2):344–366
23. Lim WM (2024) What is qualitative research? An overview and guidelines. Australasian Market J, 14413582241264619
24. Kulkov I, Kulkova J, Rohrbeck R, Menvielle L, Kaartemo V, Makkonen H (2024) Artificial intelligence-driven sustainable development: examining organizational, technical, and processing approaches to achieving global goals. Sustain Dev 32(3):2253–2267
25. Burns M, Bally J, Burles M, Holtslander L, Peacock S (2022) Constructivist grounded theory or interpretive phenomenology? Methodological choices within specific study contexts. Int J Qual Methods 21:16094069221077758

26. Ankeny RA, Leonelli S (2024) Investigating research practices: how qualitative methods enhance philosophical understandings of science—qualitative Psychology 11(2):247
27. Limna P (2023) The impact of NVivo in qualitative research: perspectives from graduate students. J Appl Learn Teach 6(2)
28. Fossey E, Harvey C, McDermott F, Davidson L (2002) Understanding and evaluating qualitative research. Aust N Z J Psychiatry 36(6):717–732
29. Hlady-Rispal M, Jouison-Laffitte E (2014) Qualitative research methods and epistemological frameworks: a review of publication trends in entrepreneurship. J Small Bus Manage 52(4):594–614
30. Prosek EA, Gibson DM (2021) Promoting rigorous research by examining lived experiences: a review of four qualitative traditions. J Couns Dev 99(2):167–177
31. Tracy SJ (2024) Qualitative research methods: collecting evidence, crafting analysis, communicating impact. Wiley
32. William FKA (2024) Interpretivism or constructivism: navigating research paradigms in social science research. Interpretivism Constructivism: Navigating Res Paradigms Soc Sci Res 143(1):5–5
33. Meydan CH, Akkaş H (2024) The role of triangulation in qualitative research: converging perspectives. In Principles of conducting qualitative research in multicultural settings, pp 98–129. IGI Global
34. Khoa BT, Hung BP, Hejsalem-Brahmi M (2023) Qualitative research in social sciences: data collection, data analysis and report writing. Int J Public Sector Perform Manag 12(1–2):187–209
35. Gillani D (2021) Can and" should" qualitative research be value-free? understanding the epistemological tussle between positivists and interpretivists. J Pol Stud 28:181
36. Mandia D (2023) A naturalistic inquiry of the experiences of women diagnosed with gestational diabetes (Doctoral dissertation)
37. Oyinloye B (2024) Reimagining the researcher-participant ethics relationship: a participant-centred, values-based approach in comparative and international education. Compare J Comparat Int Educ 54(4):646–660
38. Thompson Burdine J, Thorne S, Sandhu G (2021) Interpretive description: a flexible qualitative methodology for medical education research. Med Educ 55(3):336–343
39. Shrivastava A, Singh BK, Gurjar ES, Kushawaha PP (2024) Research methodology: concepts, techniques, and applications. Academic Guru Publishing House
40. Mulisa F (2022) When does a researcher choose a quantitative, qualitative, or mixed research approach? Interchange 53(1):113–131
41. Murto J (2021) Non-numerical investor communication–Does the depth of non-numerical communication in the financial outlook of an equity story impact investment decision?
42. Pandey P, Pandey MM (2021) Research methodology tools and techniques. Bridge Center
43. Almusaed A, Yitmen I, Myhren JA, Almssad A (2024) Assessing the impact of recycled building materials on environmental sustainability and energy efficiency: a comprehensive framework for reducing greenhouse gas emissions. Buildings 14(6):1566
44. Chirkov V (2024) Statistical positivism, psychoanalysis, and the theory of sociocultural models: a review of three approaches to acculturation. Int J Intercult Relat 101:102009
45. Masullo G (2024) The phases of qualitative and quantitative methods in Italian sociology: institutionalisation, social engagement, and emerging problems. The American Sociologist, pp 1–26
46. Zhou J, Wang P (2024) Unveiling urban interactions: Incorporating the elements of knowledge, community, and innovation in the built environment's choreography. J Knowl Econ 15(3):12121–12151
47. Mathotaarachchi KP, Thilakarathna KAAN (2021) Philosophy of approaches in social sciences: a review of positivism, phenomenology and critical social sciences in qualitative research. Technium Soc Sci J 20:944
48. Osbaldiston N (2023) Understanding social research methods and perspectives in sociology. Exploring Sociology in the Antipodes: Introduction to Sociology–1st Australian Edition

49. Kuchinke KP (2023) Phenomenology and human resource development: philosophical foundations and implication for research. Hum Resour Dev Rev 22(1):36–58
50. Giroux HA (2023) Critical theory and educational practice. In: The critical pedagogy reader, pp 50–74. Routledge
51. Stewart GT, St. Pierre E, Devine N, Kirloskar-Steinbach M (2021) The end of the dream: Postmodernism and qualitative research. Qual Inquiry 27(8–9):1051–1058
52. Rashdan W, Ashour AF (2024) Heritage-inspired strategies in interior design: balancing critical regionalism and reflexive modernism for identity preservation. Heritage 7(12):6825
53. Awan U, Sroufe R (2022) Sustainability in the circular economy: insights and dynamics of designing circular business models. Appl Sci 12(3):1521
54. Ali IM (2024) A guide for positivist research paradigm: from philosophy to methodology. Idealogy J 9(2)
55. Mirgissa K (2024) Qualitative research methodology and methods: researcher's guide. ZAIRAICHI, pp 7–220
56. Almusaed A, Yitmen I, Almssad A (2024) Contemporary innovations and sustainable practices in the application of clay materials within architectural design and construction methodologies
57. Berkovich I, Grinshtain Y (2023) A review of rigor and ethics in qualitative educational administration, management, and leadership research articles published in 1999–2018. Leadersh Policy Sch 22(3):549–564
58. Levitt HM, Morrill Z, Collins KM, Rizo JL (2021) The methodological integrity of critical qualitative research: principles to support design and research review. J Couns Psychol 68(3):357
59. Almusaed A, Almssad A (eds) (2023) Sustainable smart cities: a vision for tomorrow. BoD–Books on Demand
60. Pipere A, Lorenzi F (2021) The dialogical potential of transdisciplinary research: challenges and benefits. World Futures 77(8):559–590
61. Shackleton S, Bezerra JC, Cockburn J, Reed MG, Abu R (2021) Interviews and surveys. The Routledge handbook of research methods for social-ecological systems, pp 107–118. Routledge
62. Peng L, Jia R (2023) Exploring the interplay of the physical environment and organizational climate in innovation. Sustainability 15(20):15013
63. Osborne N, Grant-Smith D (2021) In-depth interviewing. In: Methods in urban analysis, pp 105–125. Singapore: Springer Singapore
64. Ullah F (2021) A beginner's guide to developing review-based conceptual frameworks in the built environment. Architecture 1(1):5–24
65. DeJonckheere M, Vaughn LM (2019) Semistructured interviewing in primary care research: a balance of relationship and rigour. Family Med Commun Health 7(2)
66. Garcia-Varela F, Bekerman Z, Nussbaum M, Mendoza M, Montero J (2025) Reducing interpretative ambiguity in an educational environment with ChatGPT. Comput Educ 225:105182
67. Clegg S, Stevenson J (2013) The interview reconsidered: Context, genre, reflexivity and interpretation in sociological approaches to interviews in higher education research. High Educ Res Dev 32(1):5–16
68. Cennamo K, Kalk D (2019) Real world instructional design: an iterative approach to designing learning experiences. Routledge
69. Aguinis H, Ramani RS, Alabduljader N (2018) What you see is what you get? Enhancing methodological transparency in management research. Acad Manag Ann 12(1):83–110
70. Hammes S, Geisler-Moroder D, Hauer M, Weninger J, Obleitner M, Miller J, Plörer D, Van Karsbergen V, Weitlaner R, Pfluger R (2024) Concepts of user-centred lighting controls for office applications: a systematic literature review. Build Environ, 111321
71. Kiss B, Sekulova F, Hörschelmann K, Salk CF, Takahashi W, Wamsler C (2022) Citizen participation in the governance of nature-based solutions. Environ Policy Gov 32(3):247–272

72. Song K, Chen Y, Duan Y, Zheng Y (2023) Urban governance: a review of intellectual structure and topic evolution. Urban Governance
73. Banihashem SK, Noroozi O, Van Ginkel S, Macfadyen LP, Biemans HJ (2022) A systematic review of the role of learning analytics in enhancing feedback practices in higher education. Educ Res Rev 37:100489
74. Taherdoost H (2021) Data collection methods and tools for research; a step-by-step guide to choose data collection technique for academic and business research projects. Int J Acad Res Manag (IJARM) 10(1):10–38
75. Popoola OA, Adama HE, Okeke CD, Akinoso AE (2024) Conceptualizing agile development in digital transformations: Theoretical foundations and practical applications. Eng Sci Technol J 5(4):1524–1541
76. Venkatasubramanian V, Rengaswamy R, Kavuri SN (2003) A review of process fault detection and diagnosis: Part II: Qualitative models and search strategies. Comput Chem Eng 27(3):313–326
77. Peine A, Herrmann AM (2012) The sources of use knowledge: Towards integrating the dynamics of technology use and design in the articulation of societal challenges. Technol Forecast Soc Chang 79(8):1495–1512
78. Favretto N, Shackleton S, Sallu SM, Hoffman T (2021) Editorial for special issue:"collaboration and multi-stakeholder engagement in landscape governance and management in Africa: lessons from practice." Land 10(3):285
79. Reich BH, Benbasat I (2000) Factors that influence the social dimension of alignment between business and information technology objectives. MIS Quart, 81–113
80. Razavian M, Paech B, Tang A (2019) Empirical research for software architecture decision making: an analysis. J Syst Softw 149:360–381
81. Welch C, Paavilainen-Mäntymäki E, Piekkari R, Plakoyiannaki E (2022) Reconciling theory and context: how the case study can set a new agenda for international business research. J Int Bus Stud 53(1):4–26
82. Kuran O (2024) Closing the theory-practice gap: socio-economic approach and action research in management. Syst Pract Action Res, 1–28
83. Reis HT, Gable SL, Maniaci MR (2014) Methods for studying everyday experience in its natural context. Handbook of research methods in social and personality psychology 2, pp 373–403
84. Ravitch SM, Riggan M (2016) Reason & rigor: how conceptual frameworks guide research. Sage Publications
85. Casula M, Rangarajan N, Shields P (2021) The potential of working hypotheses for deductive exploratory research. Qual Quant 55(5):1703–1725
86. Mba EJ, Okeke FO, Igwe AE, Ozigbo CA, Oforji PI, Ozigbo IW (2024) Evolving trends and challenges in sustainable architectural design; a practice perspective. Heliyon
87. Oberprieler NG, Pladevall-Vila M, Johannes C, Layton JB, Golozar A, Lavallee M, Liu F, Kubin M, Vizcaya D (2024) FOUNTAIN: a modular research platform for integrated real-world evidence generation. BMC Med Res Methodol 24(1):224
88. Karunarathna I, Gunasena P, Hapuarachchi T, Ekanayake U, Rajapaksha S, Gunawardana K, Aluthge P, Bandara S, Jayawardana A, Kapila De Alvis S, Navigating the complexities of data collection: a critical analysis of methods and challenges
89. Onwuegbuzie AJ, Leech NL, Collins KM (2010) Innovative data collection strategies in qualitative research. Qualitative report 15(3):696–726
90. Michelen M, Phan M, Zimmer A, Coury N, Morey B, Montiel Hernandez G, Cantero P, Zarate S, Foo MA, Tanjasiri S, Billimek J (2024) Practical qualitative data analysis for public health research: a guide to a team-based approach with flexible coding. Int J Qual Methods 23:16094069241289280
91. Mohrman SA, Lawler E (2011) Useful research: advancing theory and practice. Berrett-Koehler Publishers
92. Watson C, Husband G, Ireland A (2021) Opening the 'black box': what does observational research reveal about processes and practices of governing? J Manage Governance 25:189–221

93. Shaibu OG, Mbaegbu R, Echetama FC. Zion-ngoka CK (2024) Effect of team creativity on the sustainability of oil lubricant companies in imo state. Soc Sci 1(2):1–15
94. Hasselsteen L, Lindhard SM, Kanafani K (2024) Resource management at modern construction sites: bridging the gap between scientific knowledge and industry practice and needs. J Environ Manage 366:121835
95. Bhaskar P, Gupta PKK (2024) Delving into educators' perspectives on ChatGPT in management education: a qualitative exploration. Interactive technology and smart education
96. Barries HEM (2024) Observation as a research technique.(reflections, types, recommendations and examples). Russian Law J 10(2):792y
97. Cheek C, Austin E, Richardson L, Testa L, Ransolin N, Francis-Auton E, Safi M, Murphy M, De Los Santos A, Vukasovic M, Clay-Williams R (2024) Non-participant observations in experience-based codesign: an example using a case study research approach to explore emergency department care. Int J Qual Methods 23:16094069241289278
98. Donapati RR, Shukla A, Bapi RS (2024) Action-outcome delays modulate the temporal expansion of intended outcomes. Sci Rep 14(1):2379
99. DePoy E (2024) Introduction to research-e-book: understanding and applying multiple strategies. Elsevier Health Sciences
100. Vigliocco G, Convertino L, De Felice S, Gregorians L, Kewenig V, Mueller MA, Veselic S, Musolesi M, Hudson-Smith A, Tyler N, Flouri E (2024) Ecological brain: reframing the study of human behaviour and cognition. Royal Soc Open Sci 11(11):240762
101. de Haan E, Padigar M, El Kihal S, Kübler R, Wieringa JE (2024) Unstructured data research in business: toward a structured approach. J Bus Res 177:114655
102. Solano MT, Cerecedo-Nuñez HH, Ovando MAP, Faubert J, Misaghian K, Lugo JE (2024) A method for directly observing mechanical oscillations in photonic structures based on porous silicon nanostructures. Micro 4(1):80–96. MDPI
103. Luoma T, Kosunen S (2024) Participation and educational conservatism among lower secondary school students in a disadvantaged neighborhood in Finland: a request for visible pedagogies. Peabody J Educ 99(1):23–41
104. Zhou Q, Suraworachet W, Cukurova M (2024) Detecting non-verbal speech and gaze behaviours with multimodal data and computer vision to interpret effective collaborative learning interactions. Educ Inf Technol 29(1):1071–1098
105. Hua S, Jin S, Jiang S (2024) The limitations and ethical considerations of chatgpt. Data intelligence 6(1):201–239
106. World Medical Association (2024) World medical association declaration of Helsinki: ethical principles for medical research involving human participants. JAMA
107. Dwivedi YK, Jeyaraj A, Hughes L, Davies GH, Ahuja M, Albashrawi MA, Al-Busaidi AS, Al-Sharhan S, Al-Sulaiti KI, Altinay L, Amalaya S (2024) "Real impact": challenges and opportunities in bridging the gap between research and practice–Making a difference in industry, policy, and society. Int J Inf Manag, 102750
108. Morgan DL (2018) Basic and advanced focus groups. Sage Publications
109. Rinkus MA, Donovan SM, Hall TE, O'Rourke M (2021) Using a survey to initiate and sustain productive group dialogue in focus groups. Int J Soc Res Methodol 24(3):327–340
110. Liu F, Panagiotakos D (2022) Real-world data: a brief review of the methods, applications, challenges and opportunities. BMC Med Res Methodol 22(1):287
111. Zhang Z, Yu J, Tian J (2024) Community participation, social capital cultivation and sustainable community renewal: a case study from Xi'an's southern suburbs. China. J Knowl Econ 15(3):11007–11040
112. Al-Asadi A, Almusaed A, Al-Asadi F, Almssad A (2024) Enhancing urban sustainability through industrial synergy: a multidisciplinary framework for integrating sustainable industrial practices within urban settings–The case of Hamadan industrial city. Open Eng 14(1):20240033
113. Bondar I, Malaniuk V, Dubovyi O, Birillo I, Skovronskyi B (2023) Environment design and cultural innovations: analysis of creative approaches

114. Nederhand J, Edelenbos J (2023) Legitimate public participation: AQ methodology on the views of politicians. Public Adm Rev 83(3):522–536
115. Engström A, Johansson A, Edh Mirzaei N, Sollander K, Barry D (2023) Knowledge creation in projects: an interactive research approach for deeper business insight. Int J Manag Proj Bus 16(1):22–44
116. Curelaru M, Curelaru V, Cristea M (2022) Students' perceptions of online learning during COVID-19 pandemic: a qualitative approach. Sustainability 14(13):8138
117. Rapp A (2021) In search for design elements: a new perspective for employing ethnography in human-computer interaction design research. Int J Human-Comput Interact 37(8):783–802
118. Jones C, Volet S, Pasternak DP, Heinimäki OP (2022) Interpersonal affect in groupwork: a comparative case study of two small groups with contrasting group dynamics outcomes. Frontline Learn Res 10(1):46–75
119. Leese B, Rosen AM (2024) Increasing inclusion in classroom discussion: the raised block as a classroom response system in international studies. Int Studies Perspect, ekae004
120. Güner Gültekin D (2024) Groupthink in contemporary decision making: a failure to dissent
121. Raifman S, DeVost MA, Digitale JC, Chen YH, Morris MD (2022) Respondent-driven sampling: a sampling method for hard-to-reach populations and beyond. Current Epidemiology Reports 9(1):38–47
122. Suchman L, Gitome S, Nyando M, Kwena ZA, Wekesa P, Okumu S, Ndunyu L, Okoli C, Tijani A, Jegede A, Idiodi I (2023) Many cooks in the Kitchen: iterating a qualitative analysis process across multiple countries, sites, and teams. Global Health Sci Pract 11(6)
123. Chai HH, Gao SS, Chen KJ, Duangthip D, Lo ECM, Chu CH (2021) A concise review on qualitative research in dentistry. Int J Environ Res Public Health 18(3):942
124. Prosser AM, Heung LN, Blackwood L, O'Neill S, Bolderdijk JW, Kurz T (2024) 'Talk amongst yourselves': designing and evaluating a novel remotely-moderated focus group methodology for exploring group talk. Qual Res Psychol 21(1):1–43
125. Sahoo SK, Goswami SS (2023) A comprehensive review of multiple criteria decision-making (MCDM) Methods: advancements, applications, and future directions. Decision Making Adv 1(1):25–48
126. Li X, Feng W, Liu X, Yang Y (2023) A comparative analysis of green building rating systems in China and the United States. Sustain Cities Soc 93:104520
127. Hristov I, Appolloni A (2022) Stakeholders' engagement in the business strategy as a key driver to increase companies' performance: Evidence from managerial and stakeholders' practices. Bus Strateg Environ 31(4):1488–1503
128. Kowalski CP, Nevedal AL, Finley EP, Young JP, Lewinski AA, Midboe AM, Hamilton AB (2024) Planning for and assessing rigor in rapid qualitative analysis (PARRQA): a consensus-based framework for designing, conducting, and reporting. Implement Sci 19(1):71
129. Save P, Terim Cavka B, Froese T (2021) Evaluation and lessons learned from a campus as a living lab program to promote sustainable practices. Sustainability 13(4):1739
130. Saura JR (2021) Using data sciences in digital marketing: Framework, methods, and performance metrics. J Innov Knowl 6(2):92–102
131. Mendes JFC (2023) Is there a real benefit from investing in Smart Cities? Measuring return on investment: a conceptual framework

Ethnography and Case Study Methods

Abstract This chapter examines ethnography and case study methodologies as foundational approaches in qualitative research, highlighting their distinct yet complementary roles in exploring complex social, cultural, and organizational phenomena. Ethnography, rooted in anthropology, emphasizes immersive fieldwork and participant observation to capture the lived experiences, beliefs, and practices within their natural contexts. On the other hand, case studies focus on detailed examinations of specific phenomena, integrating diverse data sources to provide context-specific and actionable insights. The chapter explores the theoretical underpinnings of both methodologies, grounded in interpretivist paradigms that prioritize contextual depth and subjective understanding over positivist generalizations. The integration of these approaches is emphasized to address gaps in traditional research designs, particularly in capturing cultural and systemic dynamics. Ethical considerations, such as informed consent, confidentiality, and reflexivity, are discussed as critical for maintaining rigor and participant dignity throughout the research process. While ethnography provides richly detailed narratives illuminating cultural practices, case studies deliver structured analyses conducive to theory development and practical application. Despite challenges such as the time-intensive nature of ethnography and the limited generalizability of case studies, both methods are essential for advancing multidisciplinary research and fostering nuanced understandings of human experiences and organizational systems.

1 Introduction

Ethnography and case study methodologies are essential qualitative techniques for investigating intricate social, cultural, and organizational phenomena. Jones-Hooker and Tyndall (2023) contend that although case studies offer comprehensive insights, they may overlook cultural aspects. Integrating targeted ethnography as an adjunct element helps bridge this gap by revealing the values, beliefs, and traditions that shape the case [1]. This integration improves the study design, providing a more comprehensive grasp of the subject. Qualitative research methodologies such as ethnography and

A. Almusaed et al., *Practice of Research Methodology in Civil Engineering and Architecture*, https://doi.org/10.1007/978-3-031-97393-2_10

case study techniques are necessary to thoroughly investigate and comprehend intricate human experiences in their authentic settings [2]. These methodologies provide a solid basis for studying complex interactions among people, groups, and more prominent systemic frameworks by analyzing their theoretical underpinnings, methodological frameworks, and practical applications. To maintain the study's honesty and the participants' dignity, these methods also stress the significance of attending to ethical concerns [3]. Ethnography and case study methods, key approaches in qualitative research, share a focus on contextual depth and interpretive analysis. Zarestky (2023) highlights that selecting an approach, designing accordingly, and detailing methods remain ongoing challenges for qualitative researchers in HRD and beyond [4]. Ethnography, rooted in anthropology, prioritizes intensive involvement in cultural and social processes. For example, in construction 6.0 research, ethnographic methodologies may elucidate the interactions between workers and sophisticated automation technology on construction sites, offering insights into cultural adaptation to digital tools and procedures. Through participant observation and extended fieldwork, ethnography uncovers core societal values, customs, and rituals. Construction 6.0 research explores worker interactions with advanced automation, providing insights into cultural adaptation to digital tools and processes [5]. By adopting an emic perspective capturing the insider's view, ethnographers generate comprehensive descriptions of the lived experiences of people and communities, significantly enhancing the knowledge of many socio-economic situations [6]. The case study technique employs a more defined and systematic approach, thoroughly investigating phenomena, including persons, organizations, events, or processes [7]. This intrinsically adaptable methodology allows researchers to amalgamate many data sources and utilize various analytical techniques. For example, a case study may investigate the use of digital twins in large-scale infrastructure projects to analyze the influence of these digital twins on the project's efficiency, the collaboration among the team members, and the decision-making procedures. In situations like this, case studies are beneficial for answering research questions that need an in-depth investigation of the intricacies of the actual world. They provide insights that are both context-specific and readily transferable to different situations. Comparative investigation of many situations enhances their applicability, offering a foundation for theory development and improvement across diverse fields [8]. The theoretical foundations of these approaches are based on the interpretivist paradigm, which contests positivist concepts of objectivity and generalizability [9]. According to interpretivism theory, reality is socially created and intrinsically subjective. As a result, methodological techniques must give priority to participants' viewpoints and experiences. An example would be an ethnographic study on construction 5.0 advances, investigating how project teams perceive and assimilate innovative technology into their processes. On the other hand, a case study might be used to track the progress of a specific digital twin project during its lifespan to identify successful strategies and challenges. Methodologies such as ethnography and case studies match this epistemological approach because they emphasize qualitative data like narratives, observations, and artifacts that represent the complex reality of human and organizational behavior. An investigation of how project teams perceive and use novel technologies in their operations might be the subject of an ethnographic study of construction 5.0 innovations [10]. As an alternative, a case study might be used to trace the lifetime of a particular project to discover both successful practices and challenging obstacles. This approach brings to light the relevance of these approaches in producing contextually relevant and insightful findings. Across a wide range of academic disciplines, the usefulness of ethnography and case study methodologies is immediately apparent. Ethnography has significantly contributed to advancing research on underrepresented people in sociology and anthropology. It has helped shed insight into the social and cultural forces driving inequality and resistance. In business and organizational studies, case studies have been shown to be invaluable for dissecting the complexities of decision-making, leadership techniques, and innovative routes. These methodologies are indispensable in construction research, as they have been important in studying project delivery processes and identifying critical elements that influence stakeholder engagement and site management. As a result, they have made it possible to make targeted adjustments and structural developments. By analyzing the impact of these technologies on stakeholder engagement and site management, researchers may determine methods to optimize their integration. Although ethnography and case study methodologies possess revolutionary potential, they are fraught with methodological and ethical issues. Ethnography necessitates substantial temporal and emotional commitment, frequently compelling researchers to traverse intricate social contexts and power relations [11]. Moral difficulties, including the assurance of informed permission, the preservation of secrecy, and the management of the dual roles of observer and participant, are inherent to the ethnographic process. Although providing a more organized framework, case studies encounter criticisms over their restricted generalizability and the possibility of researcher bias. A case study of a successful digital twin implementation in a particular construction project may not be immediately relevant to projects in disparate regulatory or cultural environments. A case study of a successful digital twin implementation in a particular building project may not be immediately relevant to other projects in differing regulatory or cultural environments [12]. Depending on comprehensive case analysis requires methodological precision to guarantee the reliability and applicability of results. Ethical issues play a pivotal part in the implementation of both techniques. Researchers must employ a reflexive methodology, critically examining their positionality, assumptions, and possible impact on the research process. Maintaining openness in participant communication, securing informed permission, and protecting anonymity and confidentiality are crucial for preserving ethical norms. Cultural sensitivity is crucial, particularly in cross-cultural and international research contexts, where power imbalances and differing normative frameworks can affect interactions and interpretations [13]. By mastering these methodologies, scholars can advance knowledge across disciplines, fostering a deeper understanding of the social, cultural, and organizational dimensions. Table 1 summarizes the most critical components of ethnography and case study approaches, including their definitions, advantages, disadvantages, examples, and ethical implications. Specifically, it illustrates how ethnography provides insights into culture and society, while case studies

Table 1 Comparative overview of ethnography and case study methodologies

Aspect	Ethnography	Case Study
Definition [14]	A qualitative research method focusing on cultural and social processes through in-depth fieldwork	A qualitative method involves thoroughly examining specific instances or processes to understand phenomena
Strengths [15]	Provides deep insights into societal values, beliefs, and cultural adaptations	Offers comprehensive, context-specific insights by combining multiple data sources
Challenges [16]	Requires significant time, emotional investment, and ethical considerations like consent and dual roles	Limited generalizability and potential researcher bias require precision for reliable findings
Examples of use [17]	Investigating worker interactions with automation or studying marginalized communities	Analyzing project lifecycles or organizational decision-making processes
Ethical considerations [18]	Emphasizes cultural sensitivity, reflexivity, and protection of anonymity in cross-cultural research	Ensures participant communication, informed consent, and transparent methodologies
Relevance across disciplines [19]	Highly relevant in sociology, anthropology, and studies requiring cultural insights	Widely applied in business, organizational studies, and applied research

provide thorough analyses relevant to the setting. As a result, both methods are vital tools across a wide range of academic and applied research fields.

2 Ethnography: A Comprehensive Exploration of Methodology, Applications, and Challenges

Ethnography, an empirical research methodology rooted in anthropology, is primarily concerned with the systematic investigation and in-depth understanding of individuals and the social groups to which they belong [20]. This approach emphasizes the importance of contextualizing human behavior and social practices within their natural environments, allowing for a nuanced exploration of cultural and social dynamics. Mahadevan (202) underscores the relevance of a multi-sited ethnographic approach, mainly when it incorporates a sensitivity to power dynamics and a historical perspective. Such an approach is deemed highly suitable for examining the multifaceted phenomenon of global migration [21]. This methodology enables scholars to uncover the interconnected and overlapping processes that shape migration patterns by engaging with multiple research sites. In qualitative research, ethnography systematically investigates people's behaviors, relationships, and cultural traditions within their local contexts [22]. This approach necessitates the researcher to become actively involved in the participants' day-to-day activities, which helps to cultivate a first-hand view of the events they have lived through. In engineering, researchers aim to uncover patterns and subtleties that characterize a system's operational dynamics and design practices through comprehensive analysis, simulations, and hands-on experimentation. The immersive aspect of ethnography makes it possible to gain a thorough grasp of the intricacies of human experiences [23]. This helps bridge the gap between the views of outsiders and the realities of those familiar with the subject matter. Researchers can further their understanding of the unwritten norms, shared meanings, and collective values that contribute to forming group identity when they live and participate in the community. This technique gives a voice to individuals frequently excluded from standard research paradigms, which is particularly useful in comprehending marginalized communities with very little documentation [24]. In addition, ethnography presents researchers with the task of being reflective and understanding their impact on the study process and the findings. With its ability to cultivate empathy and analysis rich in context, ethnography significantly contributes to a more profound knowledge of cultural variety. Figure 1 depicts ethnography as a research paradigm grounded in anthropology, highlighting methodical inquiry and contextual comprehension. Principal applications encompass multi-sited investigations concerning global migration and underprivileged populations. Ethnography incorporates active engagement, facilitating understanding of cultural processes. Its significance encompasses sociology, education, healthcare, and engineering, enabling an in-depth examination of individual and systems behaviors.

Fig. 1 Key concepts, applications, and interdisciplinary relevance of ethnography. *Source* Authors

2.1 *Fundamentals of Ethnography: Investigating Culture and Society*

Ethnography, a fundamental aspect of anthropological study, developed as a systematic and rigorous approach for investigating and recording cultures' cultural practices and social dynamics [25]. This methodology aims to elucidate human behavior via examination in its natural setting. In contrast to other techniques that frequently isolate factors in controlled settings or depend predominantly on quantitative data, ethnography underscores the need to engage deeply with the everyday lives of communities. This immersion enables ethnographers to watch and document the nuances of human interaction and cultural expression, elucidating how people and communities construct, maintain, and negotiate their social realities. Ethnographers are motivated by revealing the norms, beliefs, and behaviors that constitute the core of a community's cultural framework [26]. These factors, frequently overlooked by community members, provide significant insight into how individuals comprehend their environment and engage with each other. Through meticulous observation of rituals, customs, and daily interactions, ethnographers endeavor to encapsulate the fundamental nature of a community's lifestyle. They aim to record visible actions and analyze and comprehend their underlying meanings and motives. This process necessitates sensitivity, patience, and a readiness to confront complexity and paradoxes intrinsic to human society. The roots of ethnography are intimately linked to the evolution of anthropology as a field of study. Initially, anthropology focused on examining "exotic" or "primitive" communities, frequently situated in remote regions. Early ethnographers, like Bronisław Malinowski, significantly advanced the discipline by formulating methodologies for extended fieldwork and participant observation [27]. Malinowski notably resided among the Trobriand Islanders in the early twentieth century, chronicling their social structures, rituals, and economic systems. His methodology, which underscored the significance of community engagement and involvement in daily activities, established a benchmark for ethnographic research and continues to have influence today [28].

Fundamentally, ethnography pertains to narrative construction. Ethnographers endeavor to narrate the experiences of the populations they examine, frequently elevating voices that may remain unacknowledged. In doing so, they enhance a more profound and intricate comprehension of human variation. Ethnographers may investigate sustainable building practices in urban settings, analyzing how local populations modify traditional construction methods to conform to contemporary sustainability objectives. The ethnographer may elucidate the interaction between global sustainability initiatives and local customs by chronicling how these communities reconcile cultural values with environmental issues [29]. This research underscores the inventive strategies utilized by these communities and advises policymakers and architects on culturally sensitive methods to advance sustainability development. The ethnographic approach is very participatory, depending on creating trust and rapport between the researcher and the community. Ethnographers must manage the intricate balance of being both an outsider and a participant, watching critically yet connecting honestly with their subjects. This dual function has distinct issues, since ethnographers must be aware of their prejudices and the possible influence of their presence on the community. Reflexivity, the critical examination of one's position and impact as a researcher, is an essential component of ethnographic research [6]. Ethnography necessitates a dedication to prolonged involvement. In contrast to studies that depend on short surveys or restricted contacts, ethnographic research sometimes extends across months or even years. This prolonged duration enables ethnographers to better comprehend the rhythms and patterns of communal existence, along with the fundamental beliefs and assumptions that influence conduct. Prolonged interaction may yield insights that are not evident in brief investigations. For example, ethnographers examining rural communities may reveal nuanced power dynamics that affect decision-making or the manifestation of social hierarchies in everyday encounters. The insights derived from ethnography extend well beyond the academic sphere. Ethnographic research enhances our theoretical comprehension of culture and society, informs policymaking, enhances education, and facilitates the implementation of more prosperous social initiatives [30]. Ethnographic studies of healthcare practices have shown the obstacles marginalized populations encounter in obtaining medical treatments, resulting in the formulation of solutions more aligned with the needs of these communities. Likewise, ethnographic research inside organizational contexts has offered significant insights into workplace dynamics, assisting organizations in cultivating more inclusive and productive cultures. Ethnography is an effective instrument for examining the intricacies of human existence. Through immersion in the social environments of their subjects, ethnographers reveal the complex interconnections among culture, behavior, and meaning [31]. Their research deepens our comprehension of human experience and offers pragmatic insights applicable to real-world situations. By prioritizing meticulous observation, profound participation, and insightful interpretation, ethnography remains essential in exploring humanity.

2.2 Ethnography: Features, Approaches, and Uses

Ethnography is a qualitative research method distinguished from other methods by unique characteristics that set it apart specifically [32]. Participant observation, in which the researcher actively engages with the investigated community, is essential to its technique. By adopting the twin roles of participant and observer, ethnographers can collect genuine material rich in context while preserving the objectivity required for critical analysis [33]. The blending of etic and emic viewpoints is a significant characteristic of ethnography. The emic viewpoint represents the insider's perspective of the culture, capturing the understanding and experiences of the community. In contrast, the etic viewpoint provides the researcher with an analytical interpretation. Considering these viewpoints together results in a more well-rounded and all-encompassing comprehension of cultural dynamics. The construction of detailed descriptions, which are thorough and nuanced narratives that represent the depth and complexity of social phenomena within their respective settings, is another aspect emphasized in ethnography. In addition to this, another distinguishing feature of ethnography is its adaptability. Adapting to new circumstances and insights is an essential component of the anthropological process because fieldwork's unpredictability frequently requires researchers to modify their approaches in response to new obstacles and discoveries [34]. A methodical approach is taken in the ethnographic research process, which consists of numerous essential components. The process starts with defining the research topic, followed by developing research questions that are specific and targeted to direct the investigation. These inquiries must be tightly associated with the cultural or social phenomenon being investigated. As the next step in preparing for fieldwork, it is necessary to identify an appropriate location, acquire access to the community, and address issues regarding administration and ethics. One of the most critical aspects of this phase is attempting to build trust and rapport with the participants. When collecting data, ethnography uses a wide range of methods, such as participant observation over an extended period, in-depth interviews, focus groups, and examining artifacts and documents [35]. Many methods guarantee the creation of a complete and extensive dataset. Researchers must consistently reflect on their role, biases, and effect on the study to maintain reflexivity, an essential process component. By maintaining this level of self-awareness, the research can keep its credibility and integrity intact. The analysis step entails synthesizing data to recognize patterns, themes, and cultural structures, illuminating the concealed meanings of the material gathered [36]. Representation in engineering research involves presenting findings accurately reflecting participants' perspectives while situating them within relevant theoretical frameworks. For example, civil engineering may include understanding workers' experiences with safety protocols on construction sites and aligning these insights with safety engineering models. Similarly, in mechanical engineering, ethnographic methods can explore operators' interactions with machinery to inform ergonomic design improvements. However, ethnography in engineering research faces challenges, such as the extensive time and resources required for field studies. Additionally, mitigating biases from close researcher-participant interactions necessitates continuous self-reflection to ensure objective and valid interpretations [37]. For ethical problems, such as striking a balance between protecting participants' privacy and conducting in-depth research, adhering to standards such as confidentiality and informed permission is necessary. In addition, the fact that the ethnographic research findings are context-specific frequently restricts their capacity to generalize to other groups or locations. The method of study known as ethnography is a rigorous and flexible methodology that offers profound insights into the cultures and behaviors of human beings [38]. The researchers can unearth the complexity of social life due to its emphasis on immersion, participant observation, and thorough narratives. Ethnography continues to be a beneficial method to investigate and analyze cultural and social phenomena, even though it is fraught with difficulties such as time constraints and ethical problems. In low-energy building research, ethnography investigates how occupants adjust their behaviors in response to energy-efficient equipment. For example, it investigates how cultural norms impact heating preferences or appliance usage, showing the difficulties of developing novel building designs that align with user behaviors. Architects and policymakers may use these findings to help them develop housing solutions that are culturally sensitive and sustainable [39]. Table 2 provides an academic overview of ethnography's function in investigating cultural and social phenomena. It summarizes ethnography's most important characteristics, research methodologies, applications, strengths, and limits.

Table 2 A Synopsis of ethnographic features, methods, and their use

Aspect	Description	Examples/Applications
Core characteristics [40]	– Active engagement through participant observation – Combines emic (insider) and etic (outsider) perspectives – Produces thick descriptions	Documenting cultural rituals and their social significance within a community
Flexibility [41]	Ethnography is adaptable, evolving with the dynamics of fieldwork to address unexpected developments	Adjusting research focuses on unforeseen cultural practices that are discovered during fieldwork
Research process [42]	– Defining the Research Problem: Formulating specific and relevant questions – Preparation for Fieldwork: Gaining access and establishing trust – Data Collection: Using participant observation, interviews, and artifact analysis – Reflexivity: Continuously reflecting on biases and influence – Data Analysis: Synthesizing patterns and cultural constructs – Representation: Presenting findings within broader theoretical contexts	Investigating social hierarchies, cultural practices, or healthcare dynamics
Strengths [16]	– Provides deep, contextually grounded insights – Captures the complexity of human interactions – Blends qualitative and interpretative approaches	Identifying how local traditions interact with global sustainability initiatives
Limitations [43]	– Time-intensive fieldwork – Potential for researcher bias – Ethical challenges (privacy, informed consent) – Limited generalizability due to context-specific findings	Difficulty generalizing findings from one community to a broader population

2.3 *A Step-by-Step Guide to Ethnographic Research for In-Depth Cultural Understanding Research*

2.3.1 Defining the Research Problem

Identifying a specific and well-described research issue is the cornerstone of each ethnographic research project [44]. This phase necessitates a comprehensive background awareness and a thorough assessment of the relevant literature. A particular community should be the focus of the study questions to elucidate the intricacies of cultural phenomena, social behaviors, or relationships within that group [45]. For example, a researcher may ask, "How do residents of passive houses perceive the impact of energy-efficient designs on their daily lives?" alternatively, "What cultural adaptations do communities make when implementing passive house standards in varying climatic conditions?" Using these specific questions, the framework for a significant and practicable investigation within the available resources' constraints and the time limit is established [46].

2.3.2 Fieldwork Preparation (Detailed Version with Examples from Building Energy Research)

Thorough planning is essential for successful ethnographic fieldwork, guaranteeing the achievement of study objectives while upholding ethical and practical viability [3]. This phase consists of many important elements:

- Choosing the Research Location

Selecting an appropriate study site is crucial for addressing the unique research inquiries. The site must conform to the objectives and parameters of the study [47]. For example, in building energy research, if the aim is to investigate energy efficiency in urban residential structures, the researcher may choose a highly populated metropolitan region that combines high-rise and low-rise edifices. Research on off-grid renewable energy systems may need engagement with isolated rural populations that employ solar microgrids.

- Obtaining Access

Access to the field site necessitates establishing rapport with essential stakeholders, like gatekeepers, building managers, or community leaders, who can enable admission and participation [48]. Fostering trust and mutual respect is vital to acquiring significant insights. For example, Researchers studying energy consumption behavior in commercial office environments may partner with facility managers or sustainability officers. Preliminary discussions may encompass elucidating the research aims, addressing any apprehensions, and assuring participants of the study's non-intrusive character.

- Meeting Ethical Obligations

Ethics are fundamental to ethnographic research, safeguarding the rights and welfare of participants. Researchers must comply with ethical guidelines and protocols [49]. For example, stakeholders like homeowners or tenants may disclose sensitive information about their energy use patterns or budgetary limitations in building energy research. Researchers are required to:

- Obtain Informed Consent: Clearly and comprehensively communicate the study's objectives, procedures, and the intended use of the collected data to participants [50]. Ensure that consent is voluntary and documented, allowing participants to make informed decisions about their involvement.
- Ensure Confidentiality: Protect participant privacy by anonymizing data, removing identifying information, and securely storing sensitive details [51]. Take proactive measures to prevent the unintended disclosure of information during data collection, analysis, and publication.
- Comply with Ethical Standards: Follow the guidelines and protocols established by ethical review boards. For studies involving human participants, secure approval from an Institutional Review Board (IRB) or equivalent authority, ensuring the research adheres to local, institutional, and international ethical standards [52].

- Logistical Coordination

Comprehensive logistical preparation guarantees the seamless implementation of fieldwork and reduces interruptions [53]. This entails coordinating lodging, transportation, and necessary equipment while preserving the adaptability to respond to unexpected obstacles. For example, in a study examining thermal comfort in residential structures, researchers may need to Organize extended accommodations near the designated structures to enable consistent monitoring and data acquisition. Acquire apparatus, including thermal imaging cameras, temperature sensors, and data recorders for surveilling interior environments. Prepare for contingencies, including delays in data collection caused by participant unavailability or technical malfunctions with equipment [54].

- Formulating Fieldwork Instruments and Protocols

Preparing suitable tools and guidelines improves the efficiency and consistency of data collection. This may encompass organized or semi-structured interview protocols, observation checklists, and digital data management systems [55]. For example, fieldwork instruments may encompass surveys to assess user happiness and usability

Fig. 2 Essential ethical obligations for researchers. *Source* Authors

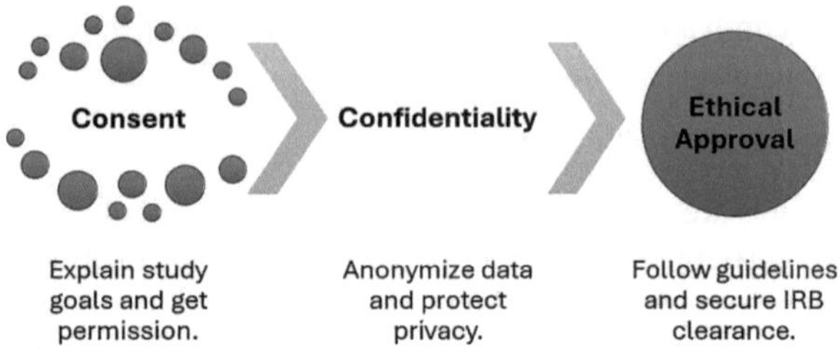

in studies investigating user engagement with smart home energy systems. Observational procedures to record resident interactions with thermostats or energy management applications. A secure database for organizing survey findings, observational notes, and gadget usage data.

- Cultural Awareness and Instruction

Comprehending the field location's cultural, social, and environmental background facilitates respectful participant interaction and improves data quality [55]. For example, researchers examining energy conservation methods in a multicultural metropolitan community may find homes with varying cultural perspectives on energy consumption. Training in cultural sensitivity mitigates the risk of misinterpreting actions or encountering communication obstacles. Thorough fieldwork preparation connects theoretical research goals with practical execution. This entails a sophisticated methodology for site selection, access acquisition, ethical consideration, and logistical preparation in building energy research. Addressing hurdles in this field—such as technological intricacies or varied participant demographics—guarantees the results' validity and comprehensiveness [56]. Receiving informed permission, protecting the confidentiality of participants through the anonymization of data, and complying with ethical norms by obtaining the requisite clearance from the Institutional Review Board (IRB) are some of the critical ethical criteria in Fig. 2.

2.3.3 Collection of Data

Ethnographic research is the meticulous, methodical, and comprehensive data gathering via various complementary methodologies [57]. These methodologies reveal social and cultural behaviors' complexities, interrelations, and subtle dynamics. Researchers utilize several methods to grasp the group or phenomena under investigation comprehensively. The subsequent procedures are frequently employed to guarantee the richness, depth, and dependability of the collected insights.

- *Participant Observation*

The researcher engages in the community's everyday activities and interactions, balancing active engagement and objective observation [58]. This method reveals nuanced behaviors and interactions that may not be evident using alternative techniques, and for example, observing people in a net-zero energy housing complex yields insights into their interactions with energy-efficient devices, including smart thermostats, solar panels, and energy storage systems. Engagement in community gatherings on energy use or recycling activities exposes shared perspectives and obstacles [59].

- *Comprehensive Interviews*

Interviews, whether organized, semi-structured, or unstructured, provide a medium for participants to convey their experiences, opinions, and viewpoints [60]. These interviews are beneficial for examining personal motives and obstacles to adopting new practices. For example, interviews with homes that have adopted renewable energy systems might disclose personal motives, such as financial savings or ecological considerations, as well as problems, such as initial installation expenses or technical obstacles related to CO_2 emission reduction [61].

- *Focus Groups*

Group talks facilitate the exchange of ideas among members, allowing researchers to gather collective perspectives and analyze group dynamics [62]. This approach is essential for comprehending community-level choices and collective activities. For example, conducting focus groups in a community transitioning to a zero-carbon program can elucidate prevalent problems, such as reluctance to change or the perceived intricacy of new technology, while simultaneously uncovering collective objectives for minimizing CO_2 emissions.

- *Analysis of Artifacts and Documents*

Examining objects and papers enhances the research by offering historical and contextual profundity [63]. This approach involves analyzing physical artifacts, policies, or media that impact or represent the community's practices. For example, examining municipal energy policies, environmental initiatives, or advertising content for zero-carbon technology can elucidate the influence of policies on community habits. Analyzing smart home usage records or thermal images of buildings may provide insights into energy efficiency and heat loss patterns, aiding in achieving CO_2 reduction objectives [64].

- *Repetitive Characteristics of Data Acquisition*

Ethnographic data collecting is intrinsically iterative. Researchers consistently enhance their methodologies and modify their emphasis as they discover new insights, ensuring the study remains attuned to the community's dynamics [65]. For example, when examining a net-zero housing initiative, preliminary observations may reveal insufficient interaction with energy conservation systems. This discovery may encourage researchers to perform subsequent interviews or focus groups to investigate obstacles to utilization, such as user interface design or insufficient training in CO_2-conscious technology.

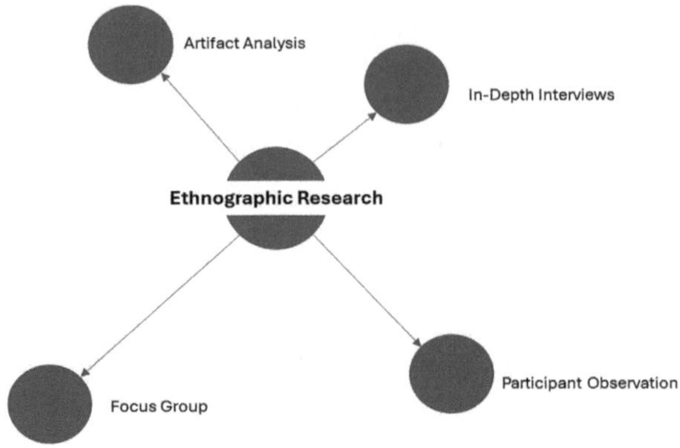

Fig. 3 Principal methods in Ethnographic research. *Source* Authors

Using these complementary methodologies, ethnographic research provides a nuanced understanding of the interactions between social dynamics, technology, and community behaviors [66]. For example, observing how individuals utilize smart home devices can reveal patterns in energy management, while conducting interviews can uncover motivations and challenges associated with energy conservation. Analyzing local policies and regulations offers additional insights into external influences on these practices. Together, these approaches create a detailed picture of the factors driving or hindering measures to mitigate CO_2 emissions [67]. This comprehensive methodology ensures that findings are grounded in the lived experiences of the community, making them highly relevant and actionable. Figure 3 illustrates the primary techniques employed in ethnographic research, such as artifact analysis, focus groups, participant observation, and in-depth interviews, all essential for exploring complex social and cultural practices.

2.3.4 Reflexivity in Ethnographic Research

Throughout the study process, ethnographers reflexively examine their roles, biases, and impacts [68]. Researchers can gain genuine insights reflecting the participants' reality when they acknowledge personal influence, preserve ethical integrity, and ensure transparency. This allows for a more profound understanding of intricate social dynamics and reduces the likelihood of potential interpretation errors. Essential elements comprise:

- *Recognizing Positionality*

Researchers must acknowledge the impact of their own identity, background, and biases on their interactions and interpretations [69]. This insight is essential for comprehending how their role within the study framework influences data collection and processing. A researcher examining material passports in sustainable building may possess a past advocate for circular economy principles. This viewpoint may inadvertently affect their data interpretation or the prioritization of specific participant perspectives, such as prioritizing architects who advocate for reuse above contractors who emphasize pragmatism.

- *Sustaining a Reflexive Journal*

Maintaining a comprehensive journal enables researchers to record observations, choices, and emotional reactions during study [70]. For example, this record ensures openness and aids in recognizing instances when biases or preconceptions may have impacted the study process. During site visits to evaluate the implementation of building material passports, a researcher may see sentiments of irritation when stakeholders oppose new documentation methods. Contemplating this answer in a journal may uncover inherent prejudices about the reception of innovation, necessitating modifications in strategy.

- *Participating in Analytical Discourse*

Frequent dialogs with colleagues, mentors, or peers might assist researchers in recognizing and mitigating any biases or blind spots [71]. These discussions offer novel viewpoints and interrogate assumptions that would otherwise remain unexplored. A researcher presenting preliminary data on material passport adoption may get comments from peers indicating an excessive emphasis on the perspectives of significant construction businesses [72]. This discovery may necessitate a reassessment to guarantee that the perspectives of small contractors are adequately reflected.

- *Reflexivity in Practice*

In ethnographic research, reflexivity may encompass:

- Acknowledge that stakeholders possess diverse goals influenced by their jobs, including the equilibrium of sustainability, cost, and practical issues.
- Considering how the researcher's educational background may result in prioritizing specific measurements or viewpoints over others.
- Incorporating feedback from many stakeholders to enhance research questions and methodologies, guaranteeing thorough and inclusive outcomes.

By employing reflexivity, researchers guarantee that their findings accurately reflect the perspectives and reality of all stakeholders, mitigating the impact of their biases [73]. This method is especially vital when examining intricate and interdisciplinary subjects such as the implementation of building material passports, fostering a comprehensive awareness of both difficulties and potential. Ethnographic research relies heavily on reflexivity, as seen in Fig. 4. Recognizing positionality, keeping a reflective diary, engaging in analytical discourse, and putting reflexivity into practice are the four main points it emphasizes. These interdependent components promote ethical integrity, reduction of prejudice, and genuine portrayal of participants' reality.

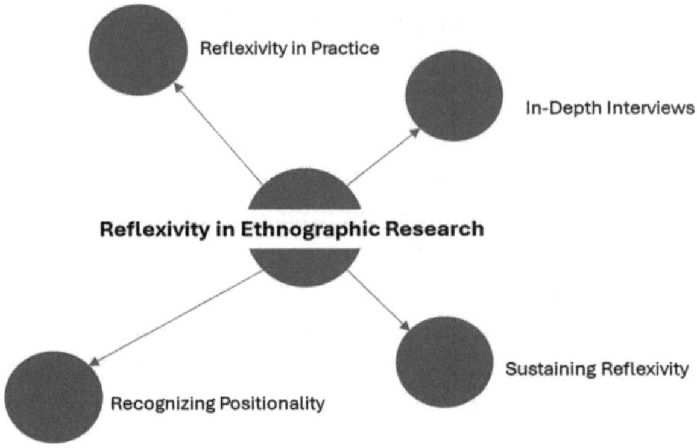

Fig. 4 Essential elements of reflexivity in ethnographic research. *Source* Authors

2.3.5 Data Analysis in Ethnographic Inquiry

Analyzing ethnographic data is a thorough and iterative method to profoundly comprehend the cultural and social processes at play [74]. This entails using various data sources, including interviews, observations, field notes, and other qualitative materials, to recognize patterns, themes, and insightful realizations. Because of the iterative nature of the procedure, researchers can adjust their analysis as new information becomes available, which guarantees a wholly comprehensive and nuanced interpretation. Data classification, identifying repeated themes, and contextualizing results within the larger cultural framework are all critical aspects of the process. The objective is to develop an all-encompassing story that represents the intricacies of the community being researched. This narrative will provide significant insights into the group's beliefs, behaviors, and interactions. Collaboration and reflexivity are crucial throughout this analytical process [75].

- *The Organization of Data*

The first step in the analysis is to organize the data systematically. To assemble artifacts and documents logically, researchers must transcribe interviews, organize field notes, and categorize them [2]. An example would be the organization of interview transcripts from architects, contractors, and policymakers researching recycled materials in sustainable buildings [76]. These transcripts are structured with field notes detailing site observations of material reuse techniques and images of construction processes.

- *Inductive Analysis, Number Two*

Inductive analysis is a method for directly recognizing patterns and themes from the data without relying on preexisting ideas or frameworks [77]. Findings can arise naturally through this method. For example, while analyzing the data collected from a recycling program inside a building project, themes such as "challenges in material sorting" and "stakeholder collaboration" may naturally emerge. These themes reflect the participants' actual experiences.

- *Third, coding*

The process of methodically applying labels or categories to data segments to categorize similar ideas and recognize underlying trends is known as coding [78]. In the context of a sustainable building project, for example, codes such as "cost barriers," "policy incentives," and "design innovations" may be applied to interview data. This enables researchers to monitor how these factors impact the use of recycled materials.

- *An examination of themes*

When doing a thematic analysis, it is necessary to synthesize recurrent themes encompassing the research setting's cultural, social, and practical dynamics [79]. For example, ethnographic research on recycling materials may disclose topics such as "perceptions of material durability," "aesthetic concerns with reused materials," and "impact of regulatory frameworks on adoption rates," which would provide insights into the attitudes of the community as well as the impacts of the system.

- *Triangulation*

Triangulation is a method that involves cross-validating the findings by comparing data from numerous sources to improve their reliability and validity significantly [80]. During an investigation into environmentally friendly construction materials, for instance, the results obtained from interviews with contractors can be checked with site observations and policy papers to guarantee that the conclusions drawn regarding the obstacles to recycling adoption are consistent.

- *Iterative Re-Analysis*

When conducting ethnographic data analysis, researchers often revisit the data when new insights arise, expanding their understanding of the cultural and social context for the data [81]. A second assessment, for instance, would emphasize the importance of community education and awareness initiatives, motivating further investigation into these aspects. This contrasts with the first research, which focuses on the financial difficulties associated with recycling. Analyzing ethnographic data requires organization, categorization, topic synthesis, and triangulation to unearth complex insights [82]. This technique enables researchers to uncover practical obstacles, stakeholder dynamics, and cultural attitudes about sustainable construction and recycling materials. This helps ensure that the findings are credible and based on the realities of the community. Figure 5 depicts an optimized framework for the analysis of ethnographic data. Commencing with data organization, it advances through pattern recognition, systematic coding, and thematic synthesis. Validation by triangulation and repeated re-analysis guarantees thorough and dependable results. The technique incorporates cultural and practical insights, highlighting a comprehensive understanding of social dynamics.

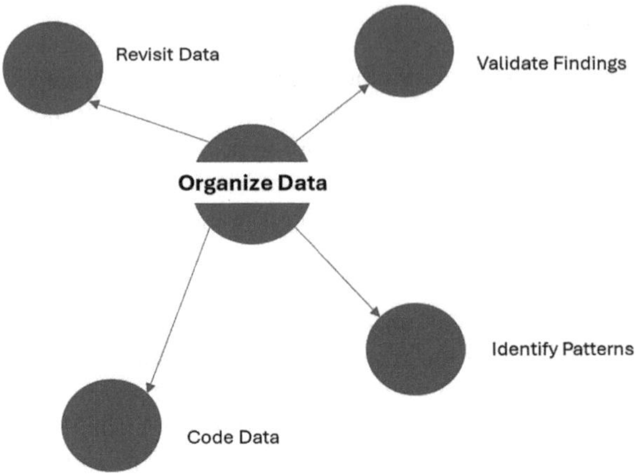

Fig. 5 Streamlined methodological framework for ethnographic data analysis. *Source* Authors

2.3.6 Representation of Ethnographic Findings

The portrayal of ethnographic discoveries necessitates authenticity, scholarly precision, and good communication to guarantee significant and believable insights [83]. Researchers must prioritize precisely representing participants' lived experiences through comprehensive articulation while contextualizing findings within broader theoretical frameworks and current literature to underscore their relevance and significance. Utilizing captivating tales renders the findings accessible while maintaining scientific rigor, enhancing comprehension of intricate social processes. Moreover, ethical issues are crucial, encompassing the preservation of participants' anonymity, the avoidance of misrepresentation, and the acquisition of feedback to preserve integrity [84]. Using examples for these primary ways enhances the comprehension and implementation of these tactics.

- *Verify Authenticity*

Findings must accurately represent participants' actual experiences. Comprehensive descriptions, narratives, and direct quotations are vital [85]. For example, in the examination of rural education, citing a teacher who states, "We frequently lack resources; thus, we devise our teaching materials from commonplace items" may underscore the community's resilience and ingenuity while maintaining its voice.

- *Situate Discoveries*

To emphasize their importance, insights must be connected to overarching theoretical frameworks and established literature [86]. For example, research on urban commuting may link participants' dissatisfaction with unreliable public transportation to conceptions of mobility justice, contextualizing their experiences within a worldwide discourse on fair urban infrastructure.

- *Embrace a Captivating Writing Style*

Integrating storytelling with analytical depth renders results approachable while preserving scholarly rigor [87]. For example, narrating the daily routine of an architectural engineer working on a high-rise project, which includes everything from early morning meetings to on-site problem-solving and challenges in implementing sustainable materials, provides a realistic and approachable perspective on the complexities of design, construction, and environmental considerations, while also reinforcing the more general principles of architectural innovation and sustainability.

- *Tackle Ethical Issues*

Ethical representation guarantees credibility and respect for participants. This entails preserving privacy, preventing misrepresentation, and soliciting comments on the results [88]. For example, in building research, designating a participant as "Engineer A" and quoting their statement, "Balancing cost constraints with the implementation of energy-efficient designs in new construction projects presents significant challenges, ensures anonymity while emphasizing systemic issues related to sustainability and budgetary limitations in architectural practices.

- *Methods of Dissemination*

Ethnographic discoveries may be disseminated via several channels, such as academic journals, books, conference presentations, and multimedia formats [89]. For example, a researcher examining traditional crafts may publish their results through videos showcasing artists in action, supplemented by interviews and theoretical analysis. The essential elements outlined in Fig. 6 for documenting ethnographic discoveries begin with accurate representation, relating insights to theoretical frameworks, using compelling narratives, addressing ethical issues, and communicating results. Rigid and robust research communication is emphasized systematically.

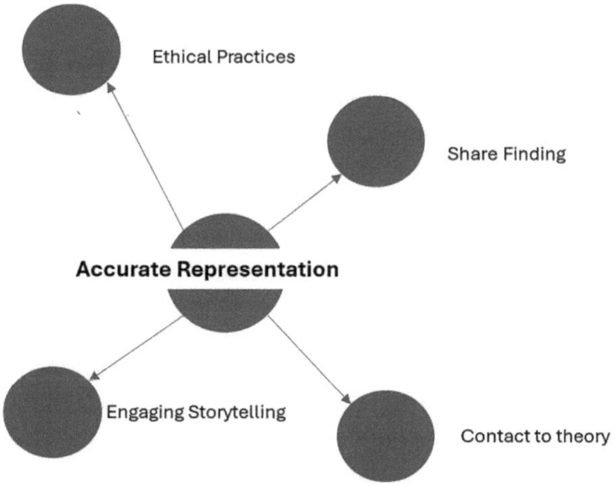

Fig. 6 Essential steps in depicting ethnographic findings

3 Case Study Method

The case study method is an example of a rigorous qualitative research strategy that enables an in-depth evaluation of real-world occurrences within their natural and complex environments [90]. It provides a rich and complete knowledge of dynamic interactions and contributing elements. It is particularly well-suited for investigating complex circumstances where the boundaries between the phenomena and their surroundings are hazy. This methodological technique is widely used in various fields, including the social sciences, education, healthcare, and management, all of which are environments in which contextual variables play an essential role in interpreting results. Case studies can promote the detection of subtle patterns and unearth nuanced insights that are frequently unreachable via more general research approaches [91]. This is because case studies prioritize holistic perspectives. Although qualitative, the case study approach can combine quantitative data to give a far more comprehensive analysis. It is a flexible instrument that may describe, compare, evaluate, and analyze many elements of complicated research topics. For example, a case study may investigate the financial and environmental consequences of utilizing recycled building materials in sustainable construction projects. Alternatively, it could explore the operational performance of zero-energy buildings across various climatic situations. The ability of the case study technique to capture the richness and complexity of real-world events is one of its strengths [92]. This method offers researchers the chance to connect theory and practice. Thanks to this methodological framework, researchers can address complex research issues and provide relevant, context-specific findings with academic and real-world value, substantially contributing to theoretical breakthroughs and practical implementations.

3.1 Types of Case Studies

There are many case studies, each intended to address a different set of research objectives and contribute to the larger body of knowledge [7]. There are several types of case studies, including exploratory, descriptive, explanatory, and multiple case studies. At the beginning of the research process, exploratory case studies are typically utilized to investigate unique occurrences and develop ideas for later study. They intend to recognize developing patterns and offer early insights that they have uncovered [93]. In urban planning, for instance, an exploratory case study may investigate how inhabitants engage with newly built public spaces to establish hypotheses regarding the influence these places have on social cohesion and well-being. Detailed and contextually rich reports of specific cases are the primary focus of descriptive case studies, emphasizing the unique characteristics of the reviewed cases. This research aims to create a comprehensive narrative encapsulating the core of the investigated phenomena. For instance, a descriptive case study may contain documentation of implementing an innovative educational method in a specific classroom, stressing the distinctive characteristics of the approach and the contextual forces that influenced it. Case studies that explain the situation examine causal mechanisms and linkages within settings [94]. To advance our understanding of the elements that lead to specific outcomes, this research addresses the "how" and "why" questions that have been generated. The successful adoption of an organizational change effort is one example of what may be analyzed in an explanatory case study. The research would concentrate on the interaction between leadership, employee engagement, and structural factors. In contrast, multiple case studies entail the investigation of several examples to detect cross-case trends, enhancing the robustness and generalizability of the researchers' results. Numerous case studies in healthcare research might, for instance, evaluate the adoption of telemedicine across many institutions to find common obstacles and best practices to provide more comprehensive policy recommendations. The case study technique uses these many categories to create a flexible framework that may be utilized to answer various research issues [95]. As a rigorous and influential methodological instrument, its relevance is highlighted by its ability to bridge the gap between theoretical study and practical implementation. Table 3 delineates the principal categories of case studies—exploratory, descriptive, explanatory, and multiple, alongside their research objectives, intrinsic obstacles, and related opportunities. It emphasizes the adaptability of case study methodologies in fulfilling various research aims while acknowledging its constraints and capacity for enhancing understanding through context-specific insights.

Table 3 Overview of case study types, difficulties, and possibilities

Type of case study	Purpose	Challenges	Opportunities
Exploratory [96]	Investigate unique occurrences and develop ideas for further research	Difficulty in defining the initial scope and formulating hypotheses without prior data	Provides early insights and hypotheses that guide future research directions
Descriptive [97]	Provide detailed and contextually rich reports of specific cases, emphasizing uniqueness	Potential for bias in emphasizing unique aspects over generalizability	Enables detailed understanding of unique cases and contextual nuances
Explanatory [98]	Examine causal mechanisms and relationships, addressing '"how" and '"why" questions	Complexity in identifying causal relationships due to confounding variables	Facilitates deeper comprehension of causal mechanisms and their applications
Multiple [99]	Investigate several examples to detect cross-case trends and enhance generalizability	High resource requirements and difficulty in managing multiple cases simultaneously	Enhances generalizability and robustness by identifying commonalities across cases

3.2 Evaluating the Appropriateness of the Case Study Method

When the goal is to obtain a comprehensive and context-specific understanding of phenomena within their real-world environment, the case study technique is beneficial since it allows for achieving this goal [100]. The one described below best investigates new or complicated problems that surface-level observations cannot fully address. Researchers can explore the dynamic interplay of numerous elements that influence phenomena using case studies because they offer a perspective that is both comprehensive and nuanced. An example would be using a case study technique to investigate the use of recycled concrete in urban building projects [101]. This would provide significant insights into sustainability practices, regulatory constraints, and cost–benefit dynamics specific to urban areas. This kind of work not only contributes to the academic discourse on environmentally responsible building techniques but also provides practical knowledge that can be used to direct both industry practices and government policies. Case studies are commonly used in academic research, notably in theses and dissertations, because of their capacity to keep the scope of the investigation reasonable while giving significant and thorough conclusions [102]. There are two basic methods in which a case study technique may be employed, and the research objectives and the essence of the study determine these approaches. To grasp the phenomena deeply, a single case study is conducted, which entails in-depth research of a single, complicated instance. Exploring one-of-a-kind or uncommon incidents that call for in-depth investigation is a particularly fruitful use of this method [103]. The process of doing several case studies entails looking at several different examples to permit comparative analysis. The capacity of researchers increases the generalizability and robustness of their findings while simultaneously exposing deeper insights about the phenomena that can be achieved by identifying patterns, variations, or inconsistencies. Researchers can handle complex research topics using a case study technique, which helps them bridge the gap between theoretical understanding and practical application. Table 4 illustrates the practical uses of case studies by combining research topics with specific instances. The presentation of a wide range of themes, such as environmentally responsible building practices, marketing techniques, and sustainable construction, demonstrates the adaptability of the case study methodology.

3.3 Steps in Conducting a Case Study

A case study is a methodical procedure aimed at gaining an all-encompassing comprehension of a particular event within the context of its actual occurrence in the real world [109]. Because of its remarkable adaptability, this technique provides significant insights across various industries. Researchers may guarantee that their conclusions are rigorous and relevant by following a series of systematic stages, which include selecting cases, constructing theoretical frameworks, collecting data, and

Table 4 Research applications for case studies

Research question	Case study example
How can recycled materials enhance sustainable construction practices?	Analysis of recycled concrete usage in eco-friendly residential developments [104]
What are effective design strategies for zero-energy buildings in cold climates?	Evaluation of a zero-energy building prototype in a Scandinavian region [105]
How do regulations shape the implementation of energy-efficient building projects?	Comparison of energy efficiency standards and their outcomes in EU member states [106]
What are the socio-economic benefits of integrating renewable energy systems into residential buildings?	Impact assessment of solar panel installations on household energy savings in rural communities [107]
How do passive design strategies impact energy consumption in urban housing developments?	Study of passive cooling techniques in high-density urban apartments [108]

analyzing the data. Each stage creates a comprehensive narrative, connecting practical observations and general theoretical and intellectual debates [78]. An organized strategy like this unearths new information. It tackles real-world problems, generating ideas that can be put into practice and contribute to advancing both theory and practice.

3.3.1 Step 1: Select a Case

When doing research using a case study, one of the most essential and basic steps is to choose a suitable example. It is necessary for the selected case to be well aligned with the study's aims and to provide considerable potential for generating valuable insights.:

- **Generate novel insights** into the subject matter.
- **Challenge existing theories** or assumptions.
- **Identify practical solutions** to specific problems.
- **Offer directions for future research.**

For example, investigating a zero-energy residential building in Sweden could illuminate practical strategies for achieving energy efficiency in cold climates. Alternatively, examining a project involving recycled construction materials might reveal challenges in material sourcing, quality assurance, and public acceptance.

Selecting Outlier vs. Representative Cases:

Outlier Case: The researcher may select a creative initiative, such as a net-positive energy residential complex in Norway [110]. This complex fulfills its energy requirements while generating excess energy. By examining this outlier, the researcher might reveal atypical architectural elements, sophisticated technology, or community-oriented strategies that enhance the building's remarkable energy efficiency. Integrating advanced solar panels with energy-efficient insulating methods may provide innovative solutions to energy issues in extreme regions.

Representative Case: Alternatively, the researcher may examine a conventional zero-energy residential structure in Germany, exemplifying a typical instance of energy-efficient construction in temperate climes [111]. This case study enables the researcher to analyze overarching patterns and discern universally applicable energy conservation techniques, including incorporating passive heating systems and intelligent energy management technology. Insights acquired here may drive legislative recommendations or industry standards relevant to broader initiatives. By selecting an outlier or a representative instance, the researcher can yield significant insights by contesting established assumptions through atypical examples or by corroborating and enhancing generalizable tactics through typical situations [112]. This fundamental stage guarantees that the case study will enhance scholarly understanding and practical progress in energy-efficient construction methods.

3.3.2 Step 2: Build a Theoretical Framework

One of the most critical steps is constructing a theoretical framework to ensure that a case study is academically rigorous [113]. Case studies emphasize real world, context-specific features, yet anchoring them in applicable theories helps to relate the study to previously established knowledge and strengthens its contribution to larger academic discourse. A correctly designed framework provides clarity and guides the investigation and interpretation of findings. In case study research, academic rigor is ensured by establishing a strong theoretical foundation [114]. By connecting the case study to current understanding, the theoretical framework allows the results to be included in more extensive scholarly debates. It shows how the study relates to accepted ideas and provides a basis for data analysis and interpretation. A strong theoretical framework can accomplish several goals:

- *Exemplify of a Theory*

Case studies can show how current ideas function in practical situations. For instance, a project utilizing recycled bricks may be the subject of a case study looking at the circular economy framework in urban development [115]. The research examines the project's material flows, waste reduction tactics, and economic effects to show how circular economy concepts may be used for sustainable urban development.

- *Extending Theory*

Case studies can enhance preexisting theoretical frameworks by adding additional elements or aspects [116]. For example, researching zero-energy buildings in tropical regions may uncover special design modifications, such as improved shading or creative ventilation techniques, that deepen our knowledge of energy-efficient architecture in harsh weather [117]. Considering various environmental conditions, these findings might improve current theories on sustainable building design.

- *Challenge a Theory*

Case studies can also challenge and disprove preconceived notions, advancing theory [118]. For example, a case study might demonstrate that, in some circumstances, recycled concrete outperforms conventional materials in terms of durability, calling into question accepted notions about material performance and leading to a reassessment of industry norms.

- *Literature Review*

A comprehensive literature study is necessary to find essential concepts, earlier studies, and pertinent hypotheses while developing a theoretical framework [119]. This might involve examining research on circular economy models, adaptive design for climate resilience, or the relative performance of conventional and recycled materials in the context of sustainable urban and architectural studies. The literature review guides the framework, filling in any gaps or restrictions in the body of current research while guaranteeing that the analysis is based on academic knowledge [120]. Researchers can increase their results' educational and practical value by providing examples, extending or questioning ideas, and ensuring that the study is firmly based on a sound theoretical framework.

3.3.3 Step 3: Collect and Analyze Data

The data collection phase is among the most essential steps in a case study since it lays the groundwork for meaningful analysis and interpretation [121]. Researchers often apply a combination of qualitative and quantitative methodologies to capture the intricacies of real-world occurrences. Integrating both views provides a more thorough picture of the topic investigated.

Qualitative Methods

Qualitative data collection may gain a wealth of contextual and deep insights into individuals' and groups' lived experiences, attitudes, and actions [122]. Typical methods include conducting interviews, participating in focus groups, making observations, and analyzing documents. One example of qualitative research methods would be observing construction sites that use recycled building materials and conducting interviews with project managers and contractors. This would be done in the context of a case study that investigates recycled building materials. These interviews may investigate difficulties such as managing supply chains, guaranteeing material quality, and the acceptability of stakeholders.

Quantitative Methods

Adding quantitative data gives the research further depth and accuracy, providing insights that can be measured and statistically analyzed [123]. Statistics on the durability of materials, energy consumption metrics, cost savings, and waste reduction numbers are examples of the kind of numerical data researchers could collect. When it comes to recycling building materials, quantitative analysis may concentrate on contrasting the lifespan costs of recycled materials with those of traditional materials and assessing the amount of trash reduced throughout the construction process.

Mixed-Methods Approach

Combining qualitative and quantitative research methodologies may provide a robust and multifaceted view [124]. For example, a case study of a wind farm development project, for instance, might involve the following activities by researchers:

- Quantitative Data: Energy output, employment data, and revenue numbers must be examined to evaluate economic and environmental effects.
- For qualitative data, it is recommended that interviews be conducted with local people and stakeholders to investigate their perspectives on the wind farm's influence on the neighborhood.

The following is an example of a comparable strategy researchers may use while studying zero-energy buildings: Tracking energy consumption patterns, operational efficiency, and financial savings over time is essential to collecting quantitative data. Documenting resident experiences with energy-efficient devices and observing architectural adjustments to local climatic circumstances are parts of the qualitative data collection process. Researchers can develop a comprehensive and well-balanced narrative that satisfies the study objectives while maintaining credibility and rigor by carefully gathering and integrating data using a variety of methodologies. New insights are uncovered during this phase, and the study's contributions to academic and practical knowledge are simultaneously strengthened.

3.3.4 Step 4: Describe and Analyze the Case

The final stage of a case study is to produce a complete, cohesive, and analytically rich narrative [125]. This is accomplished by integrating the collected data. This step is crucial because it places the findings within a larger academic and practical framework and transforms raw data into relevant insights. The description and the analysis must adhere to a systematic methodology to guarantee clarity, relevance, and substantialness.

Key Components of Description and Analysis

- Provide Contextual Details

To begin, provide a comprehensive explanation of the situation, which should include the geographical location of the case, the extent and scope of the project, the major actors engaged, and the precise objectives being addressed [126]. For example, a case study on recycling building materials can contain information about the construction location, the sort of recycled materials utilized, for example, recycled steel, and the stakeholders, including architects, contractors, and legislators.

– *Incorporate both theoretical and literary elements.*

It is essential to place the findings within the context of previously conducted research and current theoretical frameworks to emphasize their contribution to academic knowledge [127]. For example, examining a high-rise building project that uses recycled steel may be connected to ideas on the circular economy or sustainable design. This would demonstrate how the findings agree with, expand upon, or contradict established conceptions.

• *Discuss Broader Implications*

Consider how the findings contribute to a better understanding of the phenomena and the significance of these discoveries to broader academic and social issues. For instance, research on recycled materials may give insights into lowering carbon footprints in the building sector or overcoming regulatory hurdles, impacting business practices and policy development.

• *Reporting Formats*

Based on the study aims and the target audience, case studies can take on a variety of formats, including the following:

– Parts structured for techniques, findings, and discussion are called scientific structures. These parts are typically utilized in subjects such as engineering and environmental science.
– Narrative style is characterized by a thematic investigation of the relevance of the case, which is often chosen in the social sciences and the humanities.

• *A Real-World Illustration*

For example, Recycling Materials Used in Construction. A case study that focuses on the use of recycled steel in the construction of high-rise buildings might:

– Include a detailed context description specifying the project's location, its scale [128], e.g., a twenty-story office building, and the stakeholders involved, such as environmental authorities, construction companies, and other relevant entities.
– Conduct a rigorous evaluation of the results by presenting a quantitative analysis of the environmental benefits, such as reductions in embodied carbon, alongside the economic advantages, including cost savings achieved through material.
– A Discussion on Obstacles: It is essential to investigate obstacles such as satisfying regulatory criteria, ensuring that the quality of the material remains constant, and overcoming public cynicism regarding recycled materials.

3.3.5 Considerations on an Ethical Level

Research methods such as ethnography and case studies must adhere to stringent ethical standards to protect the participants' rights, dignity, and well-being [129]. These criteria serve as the fundamental building blocks for ensuring the legitimacy and integrity of the research process. This context is guided by several important factors that drive ethical activities. Participants must be provided with a comprehensive explanation of the study's objectives, procedures, and potential outcomes so that researchers can get informed consent, which is of the utmost importance [130]. Through this procedure, participation is ensured to be entirely voluntary and founded on a comprehensive comprehension of the scope of the research. In a case study on sustainable urban development, for example, participants, such as architects or community members, must be informed of how their contributions will be used and the possible ramifications of their participation [131].

When it comes to protecting sensitive information and maintaining the identity of participants, confidentiality is essential. To prevent unwanted access, researchers must adopt steps such as anonymizing data and keeping the information in a safe location [132]. In research projects, including urban planners or construction businesses, maintaining secrecy protects important business information while encouraging employees to reply honestly. Researchers are required to respect the cultural norms, values, and power dynamics that exist within the context of the study to demonstrate cultural sensitivity. For example, while conducting ethnographic research on Indigenous housing patterns, researchers must modify their approaches to respect the local community's customs and avoid imposing their prejudices. To minimize harm, it is necessary to refrain from taking any acts that could result in participants experiencing emotional, social, or professional harm [133]. Researchers are responsible for using extreme caution to guarantee that the data collection and interpretation techniques do not jeopardize the participants' health or reputations. The final point is that Ethical Reflexivity necessitates continual thought-on and reaction to ethical problems while researching [134]. With this continued commitment, we guarantee that ethical concerns will play a significant role in each study phase. Researchers not only safeguard the participants in their studies by following these principles but also improve the dependability and significance of their results during the research process. Ethical rigor is, therefore, essential for conducting both practical and responsible research.

4 Comparing Ethnography and Case Study Methods

In qualitative research, two unique methodologies offer diverse frameworks for investigating social phenomena. These are the ethnography and case study methods [90]. There are some parallels between these methods in terms of methodology; nevertheless, there are considerable differences in their emphasis, scope, data-gathering procedures, outputs, and inherent constraints. Through an immersive and comprehensive perspective, the primary objective of ethnography is to discover and analyze patterns of cultural and social activities. Participant observation, in-depth interviews, and other qualitative methods are frequently utilized in this approach, which is founded on the sustained engagement of the researcher within the area [122]. Through this immersive technique, the researcher can develop rich, contextually grounded narratives and provide detailed knowledge of the beliefs, behaviors, and relationships of the community or group being researched. On the other hand, empirical research requires significant time because it depends on extensive fieldwork. Furthermore, the intimately involved role of the researcher may allow for the introduction of bias since the researcher's interpretations and positionality will invariably impact the findings. On the other hand, the case study technique is used to analyze examples or phenomena within the confines of their contexts. Compared to the broad framework of ethnography, the scope of a case study is more limited and clearer [114]. Case studies use a wide range of data sources, such as interviews, documents, and observations, often gathered within a minimal time. The methodology aims to develop analytical frameworks and conclusions supported by evidence that will give insights into the particularities and consequences of the case being investigated. Although case studies are beneficial for investigating complicated phenomena in more depth, they have drawbacks. These drawbacks include difficulties in generalizing findings beyond the unique environment in which they were conducted and the possibility of subjective interpretations of data [135]. Further evidence of the differences between the two techniques may be seen in each strategy's results. There is a tendency for ethnography to provide stories that are narrative-driven and intensely detailed, with an emphasis on depth and cultural understanding. On the other hand, case studies are concerned with the generation of actionable insights via the use of systematic analysis and interpretation. They frequently contribute to developing theoretical models or suggestions that can be implemented in the real world. In qualitative research, ethnography and case studies are complementary approaches but separate from one another [2]. Even though it demands a significant amount of time and may be subject to prejudice, ethnography is an excellent method for gaining a thorough and culturally ingrained understanding of social processes. On the other hand, case studies provide concentrated analyses relevant to the setting in which they are conducted. These analyses produce practical results, but their generalizability is restricted. The selection of one of these approaches needs to be based on the study's objectives, the characteristics of the phenomena being investigated, and the required depth and breadth of inquiry [16].

5 Conclusion

Ethnography and case studies are fundamental to qualitative research because they provide unique yet complementary methods to investigate complex social phenomena. Ethnography strongly emphasizes immersion in cultural contexts, enabling researchers to record complex understandings of behaviors, relationships, and the meanings individuals attach to their surroundings or surroundings. Ethnography uses prolonged engagement, participant observation, and qualitative research methodological approaches to provide in-depth and contextually grounded insights into cultural and social dynamics. On the other hand, it depends on extensive fieldwork and the subjective engagement of the researcher, which involves several obstacles, such as the time commitment, the possibility of bias, and the ethical complications involved. On the other hand, the case study technique is a method that methodically investigates phenomena within the settings of their real-world occurrences. Researchers can investigate trends, formulate hypotheses, and investigate the linkages between causes and effects via case studies using various data sources such as interviews, documents, and observations. This strategy successfully tackles multidimensional problems where the borders between the phenomenon and its surroundings are poorly defined. Even though they provide practical insights and organized analytical frameworks, case studies have limits in terms of their generalizability and the risk of subjective interpretation. Therefore, to guarantee dependability, a rigorous methodological design is required. Both techniques require academic rigor, reflexivity, and ethical awareness. When doing research, researchers must handle ethical factors such as informed permission, confidentiality, and cultural sensitivity while preserving reflexivity to reduce the likelihood of bias. When choosing among these methodologies, the decision should be based on the study's objectives, the complexity of the phenomenon being investigated, and the desired depth and breadth of the investigation. On the other hand, case studies are more suited for providing practical insights and evidence-based conclusions within limited contexts. In contrast, ethnography excels in discovering profound cultural knowledge and social dynamics. The outcomes of these approaches reflect the different focuses that they utilize. In contrast to case studies, which provide systematic analyses that add to theoretical models or practical suggestions, ethnography frequently results in the production of tales that are rich in detail and shed light on cultural practices and social and organizational systems. Despite their differences, both approaches significantly advance knowledge across various fields, including education, healthcare, business, and urban planning. They achieve this by addressing complex issues and giving findings relevant to the context in which they are being used. Both the ethnographic and case study methodologies are crucial instruments for qualitative research. Each of these methodologies possesses its advantages and caters to specific research requirements. Case studies offer more focused and practical insights, whereas ethnography provides a more in-depth and culturally relevant perspective. Understanding and addressing the intricacies of human behavior and social systems provides a solid foundation that bridges the gap between theory and practice. To guarantee that their contributions continue to be academically rigorous and practically significant, their practical application depends on methodological precision, ethical integrity, and a precise alignment with research aims.

References

1. Hooker C, Tyndall DE (2023) Application of case study research and ethnography methods: lessons learned. Appl Nurs Res 73:151713
2. Tracy SJ (2024) Qualitative research methods: collecting evidence, crafting analysis, communicating impact. Wiley
3. Taquette SR, da Matta B, Souza LM (2022) Ethical dilemmas in qualitative research: a critical literature review. Int J Qual Methods 21:16094069221078732
4. Zarestky J (2023) Navigating multiple approaches to qualitative research in HRD. Hum Resour Dev Rev 22(1):126–138
5. Almusaed A, Yitmen I, Almssad A (2023) Reviewing and integrating aec practices into industry 6.0: strategies for smart and sustainable future-built environments. Sustainability 15(18):13464
6. Roshan A, Elhami A (2024) Ethnographic research methods and the role of reflexivity in qualitative research. In: Principles of conducting qualitative research in multicultural settings, pp 1–14. IGI Global
7. Hancock DR, Algozzine B, Lim JH (2021) Doing case study research: a practical guide for beginning researchers
8. Kheradmandi N, Mehranfar V (2022) A critical review and comparative study on image segmentation-based techniques for pavement crack detection. Constr Build Mater 321:126162
9. Pervin N, Mokhtar M (2022) The interpretivist research paradigm: a subjective notion of a social context. Int J Acad Res Progress Educ Dev 11(2):419–428
10. Almusaed A, Yitmen I, Almssad A, Myhren JA (2024) Construction 5.0 and sustainable neuro-responsive habitats: integrating the brain-computer interface and building information modeling in bright residential spaces. Sustainability 16(21):9393
11. Deng C (2023) When ethnography meets scientific aspiration: a comparative exploration of ethnography in anthropology and accounting. Qual Res Account Manag
12. Ammar A, Nassereddine H, AbdulBaky N, AbouKansour A, Tannoury J, Urban H, Schranz C (2022) Digital twins in the construction industry: a perspective of practitioners and building authority. Frontiers Built Environ 8:834671
13. Anjum G, Aziz M (2024) Advancing equity in cross-cultural psychology: embracing diverse epistemologies and fostering collaborative practices. Front Psychol 15:1368663
14. Mohajan HK (2018) Qualitative research methodology in social sciences and related subjects. J Econ Dev Environ People 7(1):23–48
15. Khalil S (2024) Understanding society: an introduction to social science. Pakistan J Soc Sci Rev 2(01):57–68
16. Mrabti L, Alaoui ZB (2024) Balancing qualitative and quantitative research methods: insights and applications. In: Data collection and analysis in scientific qualitative research, pp 87–118. IGI Global
17. Marabelli M, Newell S, Handunge V (2021) The lifecycle of algorithmic decision-making systems: organizational choices and ethical challenges. J Strateg Inf Syst 30(3):101683
18. Pelzang R, Hutchinson AM (2022) Cultural sensibility in accessing participants in cross-cultural research. In: Handbook of qualitative cross-cultural research methods, pp 100–120. Edward Elgar Publishing
19. Ringel L (2024) Organizational sociology and organization studies: Past, present, and future. In: Sociological thinking in contemporary organizational scholarship, pp 55–83. Emerald Publishing Limited
20. Verd JM, Quaranta G (2024) Case study and ethnography. In: Handbook of case study research in the social sciences, pp 172–192. Edward Elgar Publishing
21. Mahadevan J (2024) How language power, white subalternity, and compressed modernity frame highly-skilled non-Western migrants in an East-German company: insights from multi-sited ethnography. J Global Mobility: Home Expatriate Manag Res 12(3):394–416

22. Kelly GJ (2023) Qualitative research as culture and practice. Handbook of research on science education, pp 60–86
23. Ceuterick M, Ingraham C (2021) Immersive storytelling and affective ethnography in virtual reality. Rev Commun 21(1):9–22
24. Bharadwaj S, Howard J, Narayanan P (2021) Using participatory action research methodologies for engaging and researching with religious minorities in contexts of intersecting inequalities
25. Madden R (2022) Being ethnographic: a guide to the theory and practice of ethnography
26. Hilser HHB (2021) Collective stewardship and pathways to change: Understanding pro-social values, connectedness to nature and empathic capacity to cultivate ecocentrism in rural communities of North Sulawesi. University of Exeter (United Kingdom), Indonesia
27. Kubica G, Brzeziński D (2024) Revisiting the life and work of Bronisław Malinowski. Bronisław Malinowski and his legacy in contemporary social sciences and humanities: on the centenary of Argonauts of the western pacific
28. Smith E (2023) Queer desire and the anthropological imagination: Randolph stow and Bronislaw Malinowski in the Trobriand Islands. Modernism/Modernity 30(1):21–36
29. Yu H (2024) Sense of place and sustainable development: the case of a Tibetan luxury enterprise. Organ Environ, 10860266231226143
30. Jovanovic J (2024) Exploring educational equity: an ethnographic case study of non-profit initiatives in early childhood education (Master's thesis, University of South Florida).]
31. Dogan Akkaya F (2023) The power of ethnographic toolkit in understanding transnational culture. Ethnography, 14661381231220212
32. Sari EN, Wandanaya AB, Dewa DMRT, Hadayanti D, Kusnadi IH (2024) Ethnographic analysis to identify customer experience and customer behavior in the food and beverage industry. Jurnal Sistim Informasi dan Teknologi, 54–59
33. Shin SYSSY, Miller SMS (2022) A review of the participant observation method in journalism: designing and reporting. Rev Commun Res, 10
34. Schoon EW (2023) Fieldwork disrupted: how researchers adapt to losing access to field sites. Sociol Methods Res, 00491241231156961
35. Hammouchi MS, El Idrissi AB (2024) Unusual qualitative research practices in management sciences: the case of ethnography and dyadic approach. In: Data collection and analysis in scientific qualitative research, pp 163–192. IGI Global
36. Gunio MJ (2021) Determining the influences of a hidden curriculum on students' character development using the Illuminative evaluation model. J Curriculum Studies Res 3(2):194–206
37. Nardon L, Hari A, Aarma K (2021) Reflective interviewing—Increasing social impact through research. Int J Qual Methods 20:16094069211065232
38. Ameli K (2022) Multispecies ethnography: methodology of a holistic research approach of humans, animals, nature, and culture. Rowman & Littlefield
39. Iwuanyanwu O, Gil-Ozoudeh I, Okwandu AC, Ike CS (2024) Cultural and social dimensions of green architecture: designing for sustainability and community well-being. Int J Appl Res Soc Sci 6(8):1951–1968
40. Bausch M, Barmeyer C, Guttormsen DS (2024) An outsider's insights from the inside: implications of emic concepts on qualitative international management research. Qual Res Organ Manag Int J 19(4):256–281
41. McArdle R (2022) Flexible methodologies: a case for approaching research with fluidity. Prof Geogr 74(4):620–627
42. Cebada-Relea AJ, López M, Claus R, Aenlle M (2023) Short-term analysis of extreme wave-induced forces on the connections of a floating breakwater. Ocean Eng 280:114579
43. Douglas M (2021) Systems of support for researchers experiencing stress in difficult contexts: challenges and visions for improvement

44. Wanat M, Weller S, Borek AJ, Newhouse N, McNiven A (2024) Crafting tempo and timeframes in qualitative longitudinal research: case studies from health research. Int J Qual Methods 23:16094069241270400
45. Verma S (2023) Exploring the intersections of community and cross-cultural psychology: enhancing well-being and understanding diversity. Shodh Sari-An Int Multidis J 2(4):419–424
46. Almusaed A, Almssad A (2022) Blending human ware with software and hardware in the design of smart cities. In: Sustainable smart cities-a vision for tomorrow. IntechOpen
47. Karunarathna I, De Alvis K, Gunasena P, Jayawardana A (2024) Designing and conducting clinical research: methodological approaches
48. Oladayo AM, Odukoya O, Sule V, Molobe I, Busch T, Akodu B, Adeyemo WL, Gowans LJ, Eshete M, Alade A, Awotoye W (2024) Perceptions and beliefs of community gatekeepers about genomic risk information in African cleft research. BMC Public Health 24(1):507
49. Hair N, Akdevelioglu D, Clark M (2023) The philosophical and methodological guidelines for ethical online ethnography. Int J Mark Res 65(1):12–28
50. Gisladottir U, Nakikj D, Jhunjhunwala R, Panton J, Brat G, Gehlenborg N (2022) Effective communication of personalized risks and patient preferences during surgical informed consent using data visualization: qualitative semistructured interview study with patients after surgery. JMIR Hum Factors 9(2):e29118
51. Crossfield SS, Zucker K, Baxter P, Wright P, Fistein J, Markham AF, Birkin M, Glaser AW, Hall G (2022) A data flow process for confidential data and its application in a health research project. PLoS ONE 17(1):e0262609
52. Resnik DB (2021) Standards of evidence for institutional review board decision-making. Account Res 28(7):428–455
53. Bekishev Y, Pisarenko Z, Arkadiev V (2023) FMEA model in risk analysis for the implementation of AGV/AMR robotic technologies into the internal supply system of enterprises. Risks 11(10):172
54. Thörnblom A, Svensson Cabral S (2024) Navigating the unexpected: causes of deviations in turnaround maintenance: a qualitative case study of a Swedish pulp and paper mill
55. Karunarathna I, Gunasena P, Hapuarachchi T, Gunathilake S (2024) Data collection fundamentals: a guide to effective research methodologies and ethical practices
56. Khan MM (2024) Optimizing web surveys in research: methodological considerations and validity aspects. Int J Res Sci Innov 11(4):75–105
57. Pretnar Žagar A, Podjed D (2024) Ethnography beyond thick data. Ann Anthropological Practice 48(2):272–288
58. Javaid ZK, Ijaz S, Latif E, Azam S (2024) A qualitative study on student-teacher interaction: student engagement and teacher feedback. Int J Contemporary Issues Soc Sci 3(2):1670–1683
59. Helmefalk M, Palmquist A, Rosenlund J (2023) Understanding the mechanisms of household and stakeholder engagement in a recycling ecosystem: the SDL perspective. Waste Manage 160:1–11
60. Croucher SM, Cronn-Mills D (2021) Interviewing. In: Understanding communication research methods, pp 113–128. Routledge
61. Almusaed A, Yitmen I, Myhren JA, Almssad A (2024) Assessing the impact of recycled building materials on environmental sustainability and energy efficiency: a comprehensive framework for reducing greenhouse gas emissions. Buildings 14(6):1566
62. Cole MB (2024) Group dynamics in occupational therapy: the theoretical basis and practice application of group intervention. Taylor & Francis
63. Bhat RM, Rajan P, Gamage L (2023) Redressing historical bias: exploring the path to an accurate representation of the past. J Soc Sci 4(3):698–705
64. Merabet GH, Essaaidi M, Haddou MB, Qolomany B, Qadir J, Anan M, Al-Fuqaha A, Abid MR, Benhaddou D (2021) Intelligent building control systems for thermal comfort and energy-efficiency: a systematic review of artificial intelligence-assisted techniques. Renew Sustain Energy Rev 144:110969
65. Sanchez HS, Eski M, Batlle IC (2024) Bricolage for innovative qualitative social science research: a perspective on its conceptual hallmarks. Qual Inq 1:15

66. Siru C (2022) Societal shifts: exploring dynamics in sociology and anthropology. Int J Res Rev Techn 1(1):8–14
67. Hoang AT, Foley AM, Nižetić S, Huang Z, Ong HC, Ölçer AI, Nguyen XP (2022) Energy-related approach for reduction of CO_2 emissions: a critical strategy on the port-to-ship pathway. J Clean Prod 355:131772
68. White E, Davis M (2023) Exploring subjectivity: a critique of reflexivity in qualitative research. Qual Res Rev Lett 1(02):96–103
69. Secules S, McCall C, Mejia JA, Beebe C, Masters AS, L Sánchez-Peña M, Svyantek M (2021) Positionality practices and dimensions of impact on equity research: a collaborative inquiry and call to the community. J Eng Educ 110(1):19-43
70. Khoa BT, Hung BP, Hejsalem-Brahmi M (2023) Qualitative research in social sciences: data collection, data analysis and report writing. Int J Public Sector Perform Manag 12(1–2):187–209
71. Cheng SM, McKinney CC, Hurtado-de-Mendoza A, Chan S, Graves KD (2024) Confidence, connection & collaboration: Creating a scalable bias reduction improvement coaching train-the-trainer program to mitigate implicit bias across a medical center. Teach Learn Med 36(3):381–398
72. Widyasningrum A (2021) Blockchain for a circular economy an explorative research towards the feasibility of blockchain in enhancing the implemented material passport and life cycle assessment methodology within construction sector in The Netherlands
73. Guttormsen DS, Moore F (2023) 'Thinking about how we think': using Bourdieu's epistemic reflexivity to reduce Bias in international business research. Manag Int Rev 63(4):531–559
74. Markham A, Pronzato R (2024) A critical (theory) data literacy: tales from the field. Inf Learn Sci 125(5/6):293–320
75. Cayir E, Felder TM, Nkwonta CA, Jackson JR, Dawson R (2022) Discovering new connections: Insights from individual and collective reflexivity in a mixed methods study. Int J Qual Methods 21:16094069221105708
76. Engelbrecht A, Yitmen I, Almusaed A (2024) Industry 4.0 transformation towards industry 5.0 paradigm-challenges, opportunities and practices: challenges, opportunities and practices
77. Christou PA (2022) How to use thematic analysis in qualitative research. J Qual Res Tourism 3(2):79–95
78. Naeem M, Ozuem W, Howell K, Ranfagni S (2023) A step-by-step process of thematic analysis to develop a conceptual model in qualitative research. Int J Qual Methods 22:16094069231205788
79. Ozuem W, Willis M, Howell K (2022) Thematic analysis without paradox: sensemaking and context. J Cetacean Res Manag 25(1):143–157
80. Meydan CH, Akkaş H (2024) The role of triangulation in qualitative research: converging perspectives. In: Principles of conducting qualitative research in multicultural settings, pp 98–129. IGI Global
81. Rapp A (2021) In search for design elements: a new perspective for employing ethnography in human-computer interaction design research. Int J Human-Comput Interaction 37(8):783–802
82. Costa K, Mfolo L (2024) Qualitative evidence synthesis using the COSTAQDA online software
83. Bang TC (2024) Ensuring credibility and trustworthiness in qualitative inquiries. In: Applied linguistics and language education research methods: fundamentals and innovations, pp 70–85. IGI Global
84. Gashu KD, Guadie HA (2024) Ethics in public health informatics. In: Public health informatics: implementation and governance in resource-limited settings, pp 225–262. Cham: Springer Nature Switzerland
85. Scheffelaar A, Janssen M, Luijkx K (2021) The story as a quality instrument: developing an instrument for quality improvement based on narratives of older adults receiving long-term care. Int J Environ Res Public Health 18(5):2773
86. Luft JA, Jeong S, Idsardi R, Gardner G (2022) Literature reviews, theoretical frameworks, and conceptual frameworks: An introduction for new biology education researchers. CBE—Life Sci Educ 21(3):rm33

87. Ryan L (2023) Visual analytics fundamentals: creating compelling data narratives with Tableau. Addison-Wesley Professional
88. Markov K, Ivanova K (2023) Sample code of ethics for an international scientific society in informatics. Int J Inf Theories Appl 3(3):266–299
89. Luzón MJ, Pérez-Llantada C (2022) Digital genres in academic knowledge production and communication: perspectives and practices
90. Priya A (2021) Case study methodology of qualitative research: Key attributes and navigating the conundrums in its application. Sociol Bull 70(1):94–110
91. Omodan BI (2024) Research paradigms and their methodological alignment in social sciences: a practical guide for researchers. Taylor & Francis
92. Fitton R (2021) Building energy performance assessment based on in-situ measurements: challenges and general framework
93. Baron RA (2006) Opportunity recognition as pattern recognition: How entrepreneurs "connect the dots" to identify new business opportunities. Acad Manag Perspect 20(1):104–119
94. Beach D, Pedersen RB (2016) Causal case study methods: Foundations and guidelines for comparing, matching, and tracing. University of Michigan Press
95. Harrison H, Birks M, Franklin R, Mills J (2017) Case study research: foundations and methodological orientations. In: Forum qualitative Sozialforschung/Forum: qualitative social research, vol 18, No 1
96. Swedberg R (2020) Exploratory research. Prod Knowl Enhancing Progress Soc Sci 2(1):17–41
97. Simons H (2015) Interpret in context: generalizing from the single case in evaluation. Evaluation 21(2):173–188
98. Bentouhami H, Casas L, Weyler J (2021) Reporting of "theoretical design" in explanatory research: a critical appraisal of research on early life exposure to antibiotics and the occurrence of asthma. Clin Epidemiol, pp.755–767
99. Rokade V (2024) Innovative strategies and applications in case study methodology-a review. AG Volumes, pp 124–136
100. Warren V, Bell R (2022) The role of context in qualitative case study research: understanding service innovation. SAGE Research Methods
101. Aguacil S, Duque S, Lufkin S, Rey E (2024) Designing with building-integrated photovoltaics (BIPV): a pathway to decarbonize residential buildings. J Build Eng 96:110486
102. Zhou H, Jiang FK (2023) 'The study has clear limitations': presentation of limitations in conclusion sections of PhD dissertations and research articles in applied linguistics. Engl Specif Purp 71:34–47
103. Hagawe HM, Mobarek A, Hanuk A, Jamal A (2023) A unique business model for micro-finance institution: the case of assadaqaat community finance (ACF). Cogent Bus Manag 10(1):2135202
104. Zeyad AM (2023) Sustainable concrete production: incorporating recycled wastewater as a green building material. Constr Build Mater 407:133522
105. Gibbons L, Javed S (2022) A review of HVAC solution-sets and energy performace of nearly zero-energy multi-story apartment buildings in Nordic climates by statistical analysis of environmental performance certificates and literature review. Energy 238:121709
106. Björklund M, von Malmborg F, Nordensvärd J (2023) Lessons learnt from 20+ years of research on multilevel governance of energy-efficient and zero-carbon buildings in the European Union. Energ Effi 16(8):98
107. Khan SA, Al-Ghamdi SG (2021) Renewable and integrated renewable energy systems for buildings and their environmental and socio-economic sustainability assessment. Energy Systems Evaluation (Volume 1) Sustainability Assessment. Springer International Publishing, Cham, pp 127–144
108. Lopez-Ordoñez C, Garcia-Nevado E, Crespo-Cabillo I, Calzada JR, Coch H (2024) Reducing residential cooling demand in a sprawling desert city through vertical urban densification. J Build Eng 95:110089
109. Agarwal SA, Chauhan NKS, Ravikumar S (2024) Basic concepts of research methodology & statistics. Academic Guru Publishing House

110. Vavouris A, Guasselli F, Stankovic L, Stankovic V, Gram-Hanssen K, Didierjean S (2024) A complex mixed-methods data-driven energy-centric evaluation of net-positive households. Appl Energy 367:123404
111. Jelves DB (2023) Towards nearly zero-energy buildings
112. Cutolo D, Ferriani S (2024) Now it makes more sense: How narratives can help atypical actors increase market appeal. J Manag 50(5):1599–1642
113. Halkias D, Neubert M, Thurman PW, Harkiolakis N (2022) The multiple case study design: methodology and application for management education. Routledge
114. Cleland J, MacLeod A, Ellaway RH (2021) The curious case of case study research. Med Educ 55(10):1131–1141
115. Appendino F, Roux C, Saadé M, Peuportier B (2021) The circular economy in urban projects: a case study analysis of current practices and tools. Transactions of AESOP
116. Verbrugghe N, Rubinacci E, Khan AZ (2023) Biomimicry in architecture: a review of definitions, case studies, and design methods. Biomimetics 8(1):107
117. Almusaed A, Almssad A, Yitmen I, Wallhagen M, Yang YF, Iyer-Raniga U (2024) Integrative approaches in urban sustainability: architectural design, technological innovations and social dynamics in global contexts. BoD–Books on Demand
118. Hanisch M (2024) Prescriptive theorizing in management research: a new impetus for addressing grand challenges. J Manage Stud 61(4):1692–1716
119. Karunarathna I, De Alvis K, Gunasena P, Jayawardana A (2024) Creating value through literature reviews: techniques for identifying research gaps
120. Mancin S, Sguanci M, Andreoli D, Soekeland F, Anastasi G, Piredda M, De Marinis MG (2024) Systematic review of clinical practice guidelines and systematic reviews: a method for conducting comprehensive analysis. MethodsX 12:102532
121. Alam MK (2021) A systematic qualitative case study: questions, data collection, NVivo analysis and saturation. Quali Res Organ Manag Int J 16(1):1–31
122. Rana K, Poudel P, Chimoriya R (2023) Qualitative methodology in translational health research: current practices and future directions. In: Healthcare Vol 11, No 19, p 2665. MDPI
123. Karunarathna I, Gunasena P, Hapuarachchi T, Gunathilake S (2024) Comprehensive data collection: methods, challenges, and the importance of accuracy
124. Pregoner JD (2024) Research approaches in education: a comparison of quantitative, qualitative and mixed methods. IMCC J Sci 4(2):31–36
125. Lundin NB, Cowan HR, Singh DK, Moe AM (2023) Lower cohesion and altered first-person pronoun usage in the spoken life narratives of individuals with schizophrenia. Schizophr Res 259:140–149
126. Urbinati A, Landoni P, Cococcioni F, De Giudici L (2021) Stakeholder management in open innovation projects: a multiple case study analysis. Eur J Innov Manag 24(5):1595–1624
127. Thomas A, Gupta V (2022) Tacit knowledge in organizations: bibliometrics and a framework-based systematic review of antecedents, outcomes, theories, methods and future directions. J Knowl Manag 26(4):1014–1041
128. Taherdoost H (2021) Data collection methods and tools for research; a step-by-step guide to choose data collection technique for academic and business research projects. Int J Acad Res Manag (IJARM) 10(1):10–38
129. Khan N (2024) Research ethics in education and training. Ethical frameworks in special education: a guide for researchers (Vol 2nd), p 79
130. Millum J, Bromwich D (2021) Informed consent: What must be disclosed and what must be understood? Am J Bioeth 21(5):46–58
131. White M, Langenheim N (2021) A ladder-truss of citizen participation: re-imagining Arnstein's ladder to bridge between the community and sustainable urban design outcomes. J Design Res 19(1–3):155–183
132. Zuo Z, Watson M, Budgen D, Hall R, Kennelly C, Al Moubayed N (2021) Data anonymization for pervasive health care: systematic literature mapping study. JMIR Med Inform 9(10):e29871t

133. Scheuerman MK, Jiang JA, Fiesler C, Brubaker JR (2021) A framework of severity for harmful content online. Proc ACM Human-Comput Interact 5(CSCW2):1–33
134. Meskens G (2024) The ethical motivation for post-normal science. Futures 163:103457
135. Maxwell JA (2021) Why are qualitative methods necessary for generalization. Qual Psychol 8(1):111

Grounded Theory and Phenomenology

Abstract This chapter delivers a powerful, in-depth exploration of grounded theory and phenomenology, two cornerstone qualitative research methodologies. Grounded theory offers a systematic, data-driven approach to theory development, while phenomenology delves into the richness of lived experiences. Whether uncovering emerging patterns or capturing human perception, these methods provide invaluable insights across social sciences, engineering, healthcare, and more. Designed for researchers, scholars, and students, this chapter cuts through complexity, contrasting epistemological foundations, methodological frameworks, and analytical techniques with clarity. From coding processes and theoretical sampling to bracketing and thematic analysis, it equips readers with practical strategies to enhance rigor and impact. By bridging philosophy and research application, this chapter empowers scholars to make informed methodological choices, ensuring innovation, depth, and credibility in their studies. Whether developing ground-breaking theories or uncovering profound human insights, this is your essential roadmap to qualitative research excellence.

1 Introduction

Grounded theory and phenomenology are essential qualitative research methods shaping academic studies. Urcia (2021) states they are among the most popular qualitative inquiry methods. Over time, both theory and technique have been refined, leading to diverse adaptations for various epistemological perspectives, philosophical foundations, and research methodologies. Urcia (2021) further notes that novice researchers often confuse the methodological principles of these techniques or inadvertently incorporate elements from their adaptations, leading to conceptual and procedural inconsistencies [1]. Qualitative research demands an appropriate technique and adaptation that aligns with the research question and the researcher's worldview to ensure methodological rigor and coherence. Although both methodologies are rooted in the qualitative research tradition, their epistemological and ontological orientations differ significantly [2]. This distinction greatly impacts

A. Almusaed et al., *Practice of Research Methodology in Civil Engineering and Architecture*, https://doi.org/10.1007/978-3-031-97393-2_11

how researchers comprehend and interpret social phenomena. Inductive grounded theory develops theories through repeated data collection and analysis. According to McCall and Edwards (2021), grounded theory is one of social science research's most significant qualitative methods. However, its application in engineering education has been inconsistent, exacerbating debates about methodological rigor and grounded theory's epistemology. This heterogeneity raises the risk of methodological inconsistencies and limits its theoretical contributions and practical application, particularly for novice researchers unfamiliar with its principles [3]. In contrast, phenomenology prioritizes subjectivity and the significance that individuals assign to their interactions with the environment. It aims to explore and elucidate the essence of lived experiences [4]. The broader philosophical debates regarding the nature of reality, the significance of interpretation in research, and how knowledge is generated become clearer through these methodological differences. To better understand the positioning of grounded theory and phenomenology within broader methodological discourses, it is crucial to recognize qualitative research paradigms' role in knowledge production. Qualitative research challenges the traditional positivist paradigm based on the notion that an objective, observable reality exists independently of human experience. Instead, qualitative methods, such as grounded theory and phenomenology, align more closely with interpretivist and constructivist epistemologies [4]. These epistemologies acknowledge that knowledge is co-constructed and highlight the influence of context, perspective, and subjectivity. Grounded theory is a method that facilitates a systematic approach to developing conceptual frameworks that emerge organically from data. This approach allows for the identification of patterns and social processes. Conversely, phenomenology emphasizes depth over breadth. It was developed to capture the underlying essence of experiences by examining how individuals perceive, understand, and interpret the realities they have lived through [5]. These research approaches have conceptual foundations deeply rooted in intellectual traditions that have significantly influenced the evolution of social scientific studies. Auguste Comte's work spurred positivism's rise as a dominant paradigm in the nineteenth century [6]. Positivism maintains that knowledge is acquired through actual observation and observable facts. This paradigm was essential in establishing a framework for broadly applying quantitative research methods. Conversely, as scholars began to recognize the limitations of purely empirical techniques in capturing complex human behaviors and social systems, alternative paradigms, such as constructivism and interpretivism, started to gain traction [7]. Max Weber and Wilhelm Dilthey were key thinkers who contributed to the development of interpretivism [8]. Interpretivism highlights the significance of understanding social behavior within its subjective and cultural context. Constructivism further developed this idea, which argues that reality is not simply found but actively constructed through human interaction. This notion is particularly relevant in grounded theory research [9]. The study of human experience is approached differently from phenomenology, which is rooted in the philosophical inquiries of Edmund Husserl and later expanded by Martin Heidegger [10]. Husserl's transcendental phenomenology aimed to uncover the fundamental structures of consciousness through a method known as epoché or bracketing [11]. This technique suspends preconceived beliefs and biases to describe experiences accurately. In contrast, Heidegger introduced hermeneutic phenomenology, which proposed that understanding is inherently linked to interpretation and that human existence (Dasein) is intrinsically connected to historical and cultural contexts [12]. These philosophical perspectives have significantly influenced the methodological applications of phenomenology in contemporary research. The Discovery of Grounded Theory (1967) is the source for the formalization and methodological evolution of Grounded Theory [13]. This can be traced back to the theory's initial publication. Their method was ground breaking because it challenged the focus on deductive reasoning and hypothesis-driven research. Instead, they promoted an inductive, data-driven process where theoretical discoveries arise directly from actual observations. This represented a significant departure from prior practices. Over time, various versions of Grounded Theory have emerged. These include the Glaserian method, which emphasizes emergent theory through an unstructured coding process, and the Straussian approach, which incorporates more defined analytical techniques. Reflexivity and researcher positionality have been integrated into the analytical process through Kathy Charmaz's constructivist Grounded Theory, which has been introduced more recently [14]. This further highlights the interpretative dimensions of qualitative investigation. Similarly, phenomenology has evolved, adapting to various philosophical perspectives and academic needs. Husserlian phenomenology remains influential because of its focus on descriptive analysis and the intentionality of consciousness. However, Heideggerian and later existential phenomenologies have expanded the scope of phenomenology by incorporating elements of interpretation, embodiment, and temporality [15]. For example, the contributions of Maurice Merleau-Ponty underscore the importance of perception and bodily experience in the formation of consciousness [16]. This illustrates the intricate connection between human comprehension and existence's sensory and physical aspects. Scholars can investigate more complex human experiences, like trauma, identity, and emotional involvement, with greater depth and richness thanks to these theoretical breakthroughs, which have enhanced the relevance of phenomenology across various research topics. In contemporary academic research, the case for adopting these approaches hinges on their potential to provide rich, context-sensitive insights that are often unattainable through quantitative methods [17]. In exploratory research contexts, grounded theory is essential when existing theoretical frameworks fall short or novel social patterns and behaviors need detailed characterization. These situations require systematic conceptualization. The application of this theory can be observed across diverse sectors such as sociology, business, and healthcare, which all recognize the significance of understanding social processes, decision-making, and behavioral dynamics. Conversely, phenomenology is an excellent choice for research to illuminate the range of human experiences [18]. It offers a methodological framework that can be used to explore consciousness, perception, and meaning-making. This approach is especially prevalent and important in fields like psychology, nursing, and anthropology, where documenting individuals' lived realities contributes to a more comprehensive understanding of social and psychological processes. Grounded Theory and Phenomenology, in addition to their contributions to methodology, engage with broader epistemological questions regarding the nature of knowledge, objectivity, and the role of the researcher in the research process [19]. The iterative and comparative method employed in grounded theory challenges the notion that theories must emerge from predetermined assumptions. Instead, it promotes a dynamic and evidence-driven model for generating knowledge. In contrast, phenomenology reaffirms that understanding is fundamentally subjective and situated within personal and cultural contexts. This raises significant questions about how meaning is constructed and understood. Emphasizing the importance of carefully selecting research methodologies that align with the research questions' nature and the study's epistemic assumptions, these methodological and philosophical issues play a crucial role. The conceptual foundations and philosophical bases of Grounded Theory and Phenomenology highlight the significant contributions of each approach to qualitative research. For example, grounded theory facilitates identifying trends in domestic energy usage by analyzing user behavior and decision-making in energy conservation research, leading to more effective conservation techniques. Conversely, phenomenology in urban planning examines how individuals perceive and engage with public spaces, offering insights into walkability, green areas, and overall well-being [20]. These qualitative methodologies provide valuable insights for formulating sustainable energy solutions and human-centric urban settings. Phenomenology offers a means to explore the depth and complexity of lived experiences, while grounded theory aims to produce theoretical constructs through systematic analysis. Although distinct, both methods serve as valuable tools for expanding knowledge in the social sciences and humanities [21]. Researchers can ensure their investigations remain rigorous, contextually relevant, and theoretically insightful by critically engaging with their epistemological assumptions, methodological approaches, and practical applications. Table 1 systematically compares grounded theory and phenomenology, two prominent qualitative research approaches. It highlights their key differences in epistemological perspective, methodological framework, data collection, and analytical methods. Grounded theory focuses on developing theories through systematic coding, while phenomenology explores lived experiences through theme analysis. Understanding these characteristics is essential for researchers in selecting the most suitable approach for their study.

Table 1 Comparison of grounded theory and phenomenology

Aspect	Grounded theory	Phenomenology
Definition [22]	A qualitative research method that develops theories through iterative data collection and analysis	A qualitative research method that explores the essence of lived experiences
Epistemological orientation [23]	Constructivist and interpretivist epistemology emphasizes theory development from empirical data	Interpretivist epistemology focuses on individual perceptions and meanings
Ontological assumption [24]	Assuming social reality is constructed through interactions and processes	Assuming reality is subjective and shaped by personal experiences
Methodological approach [25]	Systematic and inductive; relies on iterative coding and constant comparison	Descriptive and interpretative; emphasizes rich, in-depth exploration
Data collection [26]	Interviews, observations, documents, and open-ended surveys	In-depth interviews, personal narratives, and participant observations
Data analysis [27]	Open, axial, and selective coding leads to theory generation	Thematic analysis through epoché, bracketing, and reduction
Outcome [28]	Develops conceptual frameworks and theories grounded in data	Reveals deep insights into human experience and perception
Challenges [29]	Methodological inconsistencies, particularly among novice researchers	Subjectivity and researcher bias in interpretation
Key theorists [30]	Barney Glaser, Anselm Strauss, Kathy Charmaz	Edmund Husserl, Martin Heidegger, Maurice Merleau-Ponty

2 Conceptual Foundations and Philosophical Underpinnings

2.1 Methodological Distinctions and Applications

Both ontology and epistemology are essential in shaping research techniques and influencing how scholars construct and interpret knowledge. Bracken (2010) emphasized the importance of aligning ontological perspectives, epistemological attitudes, and data collection methods to ensure coherence in research. This alignment is vital in studies investigating identity formation, where various methodological approaches can yield nuanced insights [31]. Grounded Theory (GT) and phenomenology are recognized as qualitative methods with distinct epistemological and ontological foundations. Glaser and Strauss (1967) introduced GT based on constructivist and post-positivist epistemologies. Lee (2024) outlines key characteristics of GT, including contemporaneous data collection, continual comparison, theoretical sampling, phased coding, mamboing, and theoretical saturation, concluding with suggestions for its broader application [32]. Rather than relying on pre-formed hypotheses, it promotes an emergent, inductive approach to developing ideas grounded in rigorously acquired and analyzed data. Pragmatism and relativism align with GT, which acknowledges that human interactions form numerous realities. Honda (2024) draws attention to the ontological extension of STGT and the diversity of epistemic options it offers [33]. Phenomenology focuses on how humans construct reality through their subjective experiences, rooted in interpretivist and idealist epistemology. Husserl developed an organized and analytical method using mathematics and logic, prioritizing intentionality and the removal of assumptions about the external world. Heidegger expanded phenomenology by incorporating existential and hermeneutic elements, drawing inspiration from religion and poetry. In contrast to Heidegger's esprit de finesse, which adopts a more interpretative and existential approach, Husserl's esprit de géométrie exemplifies his methodical rigor, as noted by Sass (2021). This distinction is clear because their approaches and writing styles differ [34]. Conversely, Heidegger's hermeneutic phenomenology offers an existentialist viewpoint, highlighting the relationship between experience and context. Husserl's transcendental phenomenology aligns with idealism, suggesting that human consciousness shapes reality [35]. The choices made by researchers across various fields, such as sustainable building studies, are influenced by these fundamental characteristics [36]. For instance, a Grounded Theory approach may be employed in sustainable building research to develop concepts regarding how architects and engineers implement sustainability principles in building design. By examining qualitative data collected from experts in the field, researchers can gain insights into the decision-making processes and iterative changes that occur during a project's design and construction phases. Conversely, phenomenology can be applied to investigate how inhabitants experience and interact with sustainable environments. They can express their thoughts on indoor air quality, energy efficiency, and available material options. This comparison underscores the methodological flexibility of both approaches in addressing sustainability-related issues.

2.2 Philosophical Foundations and Evolution of Research Approaches

Different philosophical traditions are the source of Grounded Theory and Phenomenology, which in turn define the methodological frameworks of these two approaches. Among these customs are including

- Positivism: The traditional techniques of positivism emphasize objectivity, measurement, and empirical generalization. Constructivist orientations have emerged in GT, notably in Charmaz's Constructivist Grounded Theory, where meaning arises via interactions between researchers and participants [37]. Although GT initially had post-positivist influences, particularly in its organized coding procedures, it has progressed toward constructivist orientations.
- Interpretivism, a philosophical school of thought, is aligned with phenomenology, emphasizing subjective meaning and lived experiences [23]. Since reality is a social construction, researchers must fully immerse themselves in participants' views to acquire genuine awareness.
- Constructivism: Later advances in GT, especially Charmaz's Constructivist Grounded Theory, emphasize that ideas are co-constructed rather than found [38]. This paradigm is aligned with contemporary phenomenology, which likewise emphasizes interpreting meaning in context and the depth of understanding.

These philosophical ideas influence how research on sustainable construction is carried out. For instance, positivist traditions may emphasize measuring energy efficiency indicators in green buildings. In contrast, interpretivist and constructivist paradigms emphasize how individuals and communities perceive and interact with sustainability elements.

- An examination of the development of phenomenology and grounded theory:
  - The Development of Grounded Theory according to Glaser and Strauss (1967) [39].
  - The discovery of grounded theory occurred in 1967 when Glaser and Strauss established grounded theory as a methodical approach to producing data-driven theories [40].
  - Glaser developed the GT approach in 1978, reinforcing a rigorous yet flexible analytical procedure. This was the year when theoretical sensitivity was introduced [41].
  - In the 1990s, Strauss and Corbin offered a more structured version of the process known as axial coding, whereas Glaser maintained a purist and emergent approach. This was the moment when divergence occurred [42].
  - Charmaz's Constructivist GT, completed in 2006—Charmaz stressed the researcher's involvement in co-constructing meaning with participants, which marked a paradigm change in the research process [43].
- The history of phenomenology, beginning with Husserl and current applications:
  - Husserl's transcendental phenomenology centered on bracketing prejudices to gain access to pure consciousness [44].
  - Husserl's detached method was rejected by Heidegger's Hermeneutic Phenomenology, which emphasized interpretation and existential context [45].
  - Embodied phenomenology, developed by Merleau-Ponty, emphasizes the body's role in perception and experience, which is significant to how people engage with sustainable environments [46].

- In sustainable building research, phenomenology may investigate how inhabitants experience well-being in green buildings, while the field can generate ideas on how sustainability principles grow in construction [47].

2.3 Contemporary Relevance and Justifications for Employing These Methodologies

The versatility of both techniques will ensure that they continue to be relevant:

- Addressing complexity is an appropriate method for researching transdisciplinary issues related to sustainability [48].
- In contrast to phenomenology, which focuses on recording human-centered experiences, GT is flexible enough to construct theories [49].
- Improved human-centered research is essential for determining the usability of green buildings and the level of satisfaction experienced by their occupants [50].
- Theory and practice are brought together through phenomenology, which gives contextual knowledge, and GT, which provides practical examples [23].
- Using phenomenology and GT in mixed methods, research frequently improves quantitative research in evaluating sustainable building practices [51].

Regarding research focused on sustainability, Grounded Theory and Phenomenology are essential as they provide unique perspectives on the built environment [23]. Phenomenology offers profound insights into human experiences, such as user interactions with green buildings, while Grounded Theory drives the development of sustainability models. Together, these methodologies bridge the gap between theory and practice, fostering a more holistic approach to sustainable architecture. By integrating experiential understanding with data-driven theory generation, they contribute to the evolution of ecologically responsible architectural practices, ensuring human-centric design and long-term environmental sustainability. Figure 1 visually represents the connection between Grounded Theory (GT) and Phenomenology, illustrating both approaches' philosophical roots, historical development, and contemporary significance. The circular organization of the graphics creates the impression that various techniques engage in a dynamic and evolving conversation. Key thematic areas include the development of phenomenology and general theory, their philosophical foundations, and their applications in today's world. This design emphasizes the interrelated nature of different qualitative research methodologies by showcasing their connectedness.

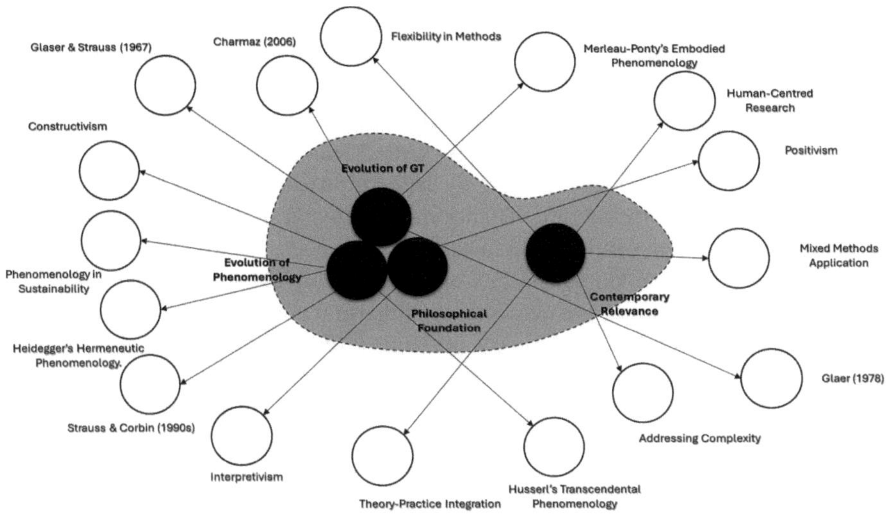

Fig. 1 Conceptual framework of grounded theory and phenomenology. *Source* Authors

3 Conceptual Foundations and Philosophical Underpinnings

3.1 Understanding Phenomenology in Engineering

Phenomenology and Grounded Theory are two significant research approaches that give contrasting views on engineering [52]. They bridge the gap between theoretical inquiry and practical application by providing a bridge between the two. Phenomenology, grounded in Edmund Husserl's philosophical tradition and further developed by thinkers such as Martin Heidegger and Maurice Merleau-Ponty, explores human experience within its contextual framework. Cheeli (2024) asserted that research methodology is an independent scientific discipline that fosters systematic inquiry across various domains. The literature highlights its multifaceted and theoretical nature, integrating experimental elements and philosophical frameworks. It encompasses qualitative and quantitative methodologies, spanning a range of research strategies from basic to highly complex [53]. This methodological approach is critical in engineering research, where human interactions with the built environment are central to evaluating design effectiveness. By emphasizing individuals' subjective experiences, the research method known as phenomenology enables a thorough investigation of how people perceive green buildings, interact with them, and are affected by them. Mathews (2021), however, identifies a theoretical gap in understanding these technologies' impact on professionals in the field. This underscores the need for further academic inquiry [54]. The significance of phenomenology in engineering research is further highlighted by its ability to provide valuable insights into user experiences. Green buildings often feature natural lighting, ventilation, energy-efficient technology, and biophilic elements, all designed with ecological principles. According to Aj-Bolouri (2023), immersive virtual reality (IVR) is becoming an increasingly important component of information systems (IS) research [55]. Current research has not sufficiently clarified how virtual reality experiences gain meaning for end-users, as their effectiveness largely depends on user engagement and the interpretation of design elements [56]. Traditional architectural assessment frameworks, such as energy efficiency ratings and sustainability indices, often overlook occupants' lived experiences in residential spaces. A phenomenological approach addresses this gap by investigating subjective perceptions of thermal comfort, spatial organization, air quality, and natural light [57]. By utilizing in-depth interviews, observational studies, and immersive methods, researchers can better understand the emotional, cognitive, and sensory responses provoked by the built environment.

3.2 Models that Are Dependent on Data and Grounded Theory

On the other hand, grounded theory offers a systematic method for building theories based on empirical data, contrasting with phenomenology, which emphasizes lived experiences [58]. This method is crucial for research subjects characterized by dynamic environmental, social, and technical factors that necessitate evidence-based and adaptable solutions. Through grounded theory, researchers can create conceptual models directly from observable patterns [59]. By employing this methodology, researchers can draw theoretical conclusions from empirical observations. This approach follows a systematic strategy that includes data collection, coding, classification, and theory formulation. Research in ecological building design heavily relies on grounded theory to establish frameworks for integrating ecological design principles with human behavioral dynamics in sustainability initiatives, ensuring these strategies are grounded in real-world interactions. For example, one study used grounded theory to explore feedback mechanisms, energy usage patterns, and tenant behaviors in residential green buildings. Thematic analysis and iterative coding facilitated the identification of key themes, such as occupant adaptation techniques, barriers to sustainable behavior, and the effect of technological advances on user participation. These findings informed strategies that enhance energy efficiency and encourage environmentally conscious actions, deepening our understanding of how users interact with sustainable design elements. This approach improves the practicality of sustainable architectural solutions by aligning technical performance with user-centered concerns and establishing theoretical models based on empirical evidence [60]. These discoveries formed a foundation for models integrating technical performance and human-centered considerations, contributing to a more holistic view of sustainable design. This research method guarantees that theoretical frameworks are developed not in isolation but are firmly anchored in observable, real-world events by emphasizing empirical evidence. Grounding research conclusions in empirical data boosts their significance, making them more pertinent to the challenges faced in environmentally responsible architecture and design. Furthermore, merging theoretical concepts with empirical data fosters innovation by allowing researchers and practitioners to detect emerging trends, adapt methods based on user behavior, and refine sustainable design processes. This iterative approach ensures that architectural solutions remain dynamic and responsive, enabling effective responses to shifts in social conditions, environmental factors, and technological advancements. For example, this approach bolsters the integration of ecological principles with user experience in ecological buildings, leading to designing models that are more responsive and resilient. By bridging the gap between theory and practice, grounded theory-driven research supports the development of frameworks that address both performance indicators and user engagement, ultimately improving the functionality and sustainability of built environments. The ongoing advancement of environmental and architectural research is driven by the synergy between empirical data and theoretical innovation, ensuring that design solutions are increasingly practical and contextually relevant. Consequently, grounded theory is vital for enhancing sustainable design practices, advancing architectural research, and connecting theoretical understanding with practical application. In architectural engineering research, the core components of grounded theory are illustrated in Fig. 2, showcasing the systematic approach employed by this methodology. The figure illustrates how theory originates from facts through systematic methods, linking theoretical research to real-world applications. It underscores the effectiveness of grounded theory in sustainable design, integrating ecological principles with insights into human behavior, ultimately bridging the gap between theory and practice.

3.3 A Holistic Approach to Design Through the Integration of Phenomenology and Grounded Theory

Incorporating phenomenology and grounded theory into engineering research results in establishing a synergistic framework that improves both the theoretical rigor and the practical application of the engineering study [61]. In capturing the breadth and depth of human experiences, phenomenology significantly contributes to research development. On the other hand, grounded theory integrates these insights into consistent and generalizable models. A thorough understanding of design that considers both qualitative and quantitative aspects may be achieved by utilizing this mix of instructional methods. This methodological integration is critical in modern architectural discourse, where there is a growing emphasis on sustainability, which needs research tactics that cross disciplinary boundaries and utilize several methodologies [62]. Researchers can investigate how humans feel and interact with sustainable architectural aspects, such as energy-efficient technology, spatial configurations,

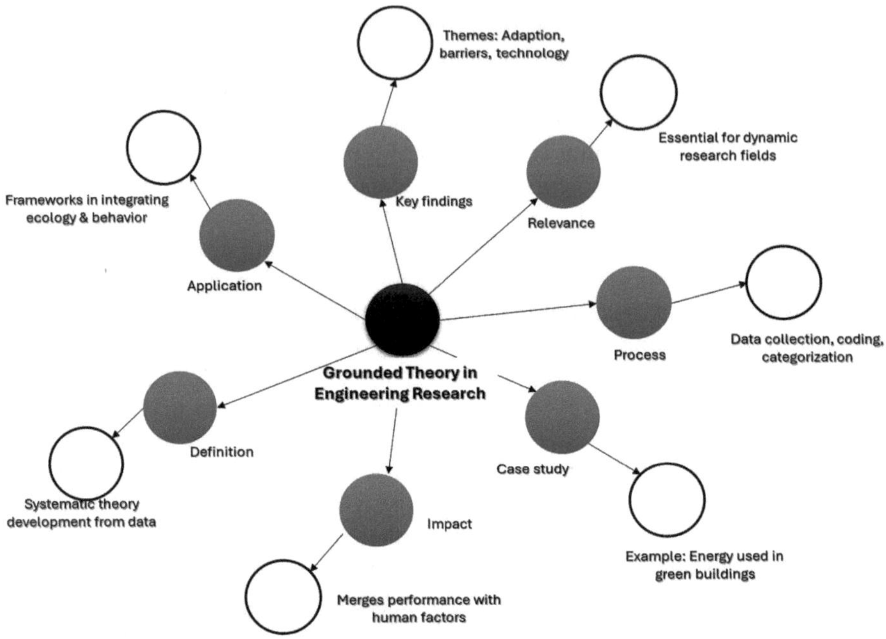

Fig. 2 Conceptual framework of grounded theory application in architectural research. *Source* Authors

and biophilic design components, by employing the methodology of phenomenology while conducting their investigations. These ideas are then refined by grounded theory, which methodically finds emerging themes and arranges them into conceptual models that support evidence-based design methods. When it comes to green buildings, for example, this dual approach is beneficial because it allows for the collection of subjective evaluations of comfort and well-being while simultaneously allowing for the development of data-driven frameworks that maximize energy performance and occupant satisfaction [63]. Through this methodological synthesis, it is ensured that sustainability initiatives go beyond solely technical concerns and incorporate user-centered viewpoints, hence generating design solutions that are more adaptable, robust, and human centered. To offer a firm foundation for advancing engineering and engineering design research, this framework integrates phenomenological investigation with the systematic approach grounded theory takes [64]. This combination method is beneficial in the post-occupancy evaluation (POE) of green buildings, one of the most critical areas. Quantitative performance measurements, such as energy savings, indoor air quality, and water efficiency, are frequently the primary emphasis of traditional performance evaluation techniques (POE) [65]. The human element of sustainability is not entirely captured by these measures, even though they are essential. Researchers can explore how occupants feel and adapt to architectural features by incorporating a phenomenological inquiry into POE [66]. This approach allows them to uncover insights that might otherwise be overlooked. Subsequently, grounded theory enables the integration of these insights into more comprehensive conceptual frameworks, which can influence future design practices and policy proposals [67]. Furthermore, combining these approaches aligns with the broader paradigm shift in engineering research toward human-centered sustainability. Unlike past technocratic methods prioritizing environmental performance over users' well-being, modern engineering recognizes that true sustainability requires balancing ecological imperatives with human needs. This shift is facilitated by merging phenomenology and grounded theory, ensuring that models are technically robust and socially relevant [63]. Integrating phenomenology and grounded theory establishes a solid foundation for advancing architectural practices by connecting theoretical insights with empirical evidence [68]. Phenomenology fosters a deeper understanding of user experiences, while grounded theory provides a systematic framework for developing models based on real-world data. Their combined use bridges the gap between theoretical constructions and practical application, promoting a comprehensive approach to ecological design [69]. This methodological synergy ensures that engineering is technically sensible and experientially enriching by emphasizing human-centered and environmental considerations. By concurrently incorporating experiential and evidence-based perspectives, this integration supports the creation of innovative, adaptable, and contextually relevant design solutions in response to evolving global challenges [70]. As sustainability, climate adaptation, and user well-being increasingly become central to architectural discourse, this dual approach enhances the capacity to create built environments that are both resilient and responsive. Beyond its contributions to architectural research, this methodological integration informs broader discussions on sustainable development, ensuring that design strategies align with ecological imperatives and human needs [71]. By refining theory and practice, this approach plays a crucial role in shaping future engineering methodologies and strengthening the relationship between the built environment and its social, environmental, and psychological impacts. Here is a systematic summary of how phenomenology and grounded theory are integrated in practice, as shown in Table 2. This summary highlights how different approaches complement each other and enhance design in theory and practice. The table emphasizes their impact on sustainability, user-centered methods, and multidisciplinary research, ensuring a strong foundation for engineering studies.

Table 2 Integration of phenomenology and grounded theory in research

Key concept	Description	Impact
Integration of phenomenology and grounded theory [72]	Merging phenomenology with grounded theory boosts theoretical rigor and practical application in engineering research	Enhances engineering and architectural research frameworks
Role of phenomenology [73]	Captures the richness of human experiences, providing qualitative insights into research and design	Enhances the understanding of user behavior in sustainability environments
Role of grounded theory [73]	Refines phenomenological insights into structured, generalizable models that support evidence-based design	Develop organized models that enhance architectural design practices
Methodological importance [74]	Encourages interdisciplinary research, which is crucial for tackling sustainability and complex design challenges	Encourages innovative, adaptable, and contextually relevant approaches and solutions
Post-occupancy evaluation (POE) [75]	Improves traditional POE by adding subjective user experiences along with quantitative performance metrics	Enhances building performance evaluations with a user-centered approach focus
Human-centered sustainability [76]	Ensures that sustainability models balance ecological imperatives with user well-being and adaptability	Aligns sustainability with human experience, promoting resilience design
Theoretical and practical integration [77]	Promotes an integrated approach that aligns engineering methodologies with environmental and social considerations	Shapes future engineering and design methodologies with empirical and theoretical grounding

4 Grounded Theory: Systematic Methodology and Analytical Rigor

4.1 Reasoning Based on Inductive Reasoning and the Development of Theory

Grounded Theory (GT) is a qualitative research technique that develops theory from systematically gathered data, emphasizing interaction, action, and processes (Bischof & Freyb, 2022). Discursive psychology, also rooted in symbolic interactionism, highlights the role of language in meaning-making. Its incorporation into GT can enhance analytical depth by introducing discourse-oriented approaches, thereby expanding GT's applicability in language-focused research [78]. GT employs an inductive reasoning process, which allows theoretical ideas to emerge naturally from facts. This contrasts with deductive methods, which are based on pre-existing assumptions. This technique effectively captures complex social and behavioral phenomena by enabling researchers to formulate theories grounded in real-world evidence [79]. It facilitates the development of such theories. Researchers can refine new concepts through iterative data collection and analysis, ensuring that theories remain anchored in empirical evidence rather than speculative assumptions. Grounded Theory (GT) emphasizes developing theory through methodical engagement with facts to ensure that theoretical categories and connections are derived rather than predetermined. GT is often described as a purely inductive, qualitative approach that involves collecting and interpreting data without any theoretical biases, according to Themelis et al. (2023), who highlights a common misconception in the research methodology literature. Due to this misunderstanding, researchers postpone conducting literature reviews until the end of the study to prevent prior knowledge from influencing the analysis [80]. Consequently, they overlook the complex interaction between data and existing theoretical frameworks. For example, GT can be applied in architectural engineering research to explore sustainable design methods. This is achieved by analyzing qualitative data gathered from architects and engineers. Through iterative data analysis, researchers uncover emerging themes, such as adaptive reuse strategies or passive cooling techniques, rather than enforcing preconceived sustainability frameworks. This enables them to refine theoretical concepts that are rooted in practice. This aspect of GT is critically essential. Researchers begin with broad questions and allow themes and patterns to emerge naturally while analyzing the data. This methodology ensures that the ideas generated are contextually relevant and can be adapted to various research situations [81]. GT is exceptionally valuable when addressing topics requiring thorough research into dynamic social, organizational, and behavioral processes. This is because GT allows researchers to delve into uncharted territory, unlike methods that enforce strict theoretical frameworks or guidelines.

4.2 *Analytical Procedures and Procedures for the Methodological Framework*

A structured framework ensures the methodological rigor of Grounded Theory (GT). This framework includes essential components such as theoretical sampling, ongoing comparative analysis, and systematic coding techniques. Al-Eisawi (2022) presents a design framework aimed at novice qualitative researchers. This framework visually represents GT as a methodological toolbox rather than just a procedure for producing theories. It facilitates the extraction of conclusions supported by empirical evidence and allows for thorough data classification, ensuring analytical depth and coherence in qualitative research [82]. By emphasizing methodological structure, the framework enhances the reliability and validity of GT, thereby assisting researchers in systematically gaining insights from data. Theoretical sampling is a method that directs data collection flexibly and iteratively. This approach allows researchers to select participants or data sources based on evolving conceptual ideas. Consequently, this ensures that each new data point significantly contributes to the development of the emerging theory. Unlike conventional sampling techniques, which aim for representativeness, theoretical sampling focuses on identifying variance and expanding the conceptual framework. This is achieved by continually reworking the study topic in response to new conceptual developments. Constant comparative analysis is a crucial component of the analytical process [83]. A continual comparison of freshly obtained data with data collected in the past is required for this analysis. This comparison aims to uncover similarities, differences, and developing trends. As a result of this ongoing process, researchers can make conceptual links between various data points. This procedure guarantees the developing theory will remain internally consistent and firmly rooted in empirical facts [84]. The refinement of theoretical categories, the clarification of links between ideas, and the strengthening of the general analytical framework are all achieved through the methodical comparison of facts. This iterative interaction with data allows ideas to grow spontaneously, boosting their validity and usefulness. Rather than depending on established assumptions, this approach allows for the evolution of hypotheses [85]. At the heart of grounded theory, the continual comparison approach guarantees that theoretical ideas grow inductively rather than being imposed from the outside. While simultaneously collecting and analyzing data, researchers refine categories and ideas based on observations made in the actual world [86]. This recurrent strategy helps to encourage theoretical saturation, which is the process by which categories become well defined and sufficiently developed via frequent engagement with data. In addition, the rigorous approach ensures that ideas develop directly from empirical evidence, which contributes to an increase in the dependability and credibility of the findings of the research projects. The dynamic interaction between the acquisition of data and its analysis makes it possible to arrive at nuanced interpretations that reflect the intricacies of the phenomena being examined [87]. Researchers can design robust, adaptable, and contextually relevant frameworks when they ground theoretical development in facts [88]. These frameworks are responsible for driving significant discoveries and actual improvements in practice. Open coding, axial coding, and selective coding are the three basic steps that make up the coding process in GT. Additionally, the coding process is essential to the data interpretation process. Open coding is a method that includes breaking down qualitative data into discrete pieces and assigning labels to emergent concepts. This method enables researchers to find various themes without imposing limiting categorizations on the data. The next step in the process is called axial coding, which involves building linkages between multiple topics and organizing them into cohesive theoretical constructs. During this stage, the primary focus is determining the causal links, contextual impacts, and interactions within the data collection [89]. The Grounded Theory (GT) phase, known as selective coding, is important since it synthesizes the most important theoretical categories into a cohesive framework. This phase ensures that the framework is both comprehensive and consistent within itself. In this step, essential concepts are methodically integrated, and theoretical constructs are refined. This process builds upon the foundations formed during open and axial coding. The process of selective coding helps to construct a structured, empirically based theoretical model that strengthens the explanatory power of the emerging theory. This is accomplished by building links between basic categories. This organized coding procedure considerably strengthens the methodological rigor of GT, which in turn makes it possible to analyze complicated qualitative data in a way that is both systematic and transparent. In this method, theoretical conceptions are not formed by predetermined assumptions but develop inductively from empirical data. This technique assures that theoretical constructions evolve inductively. The iterative structure of GT makes it possible for theories to develop dynamically through constant engagement with data, ultimately resulting in resilient, adaptive, and contextually relevant perspectives. Figure 3 illustrates that the interrelated processes within GT, including theoretical sampling, comparative analysis, and coding approaches, reveal that data refining is an iterative rather than a linear process. The basis is open coding, which divides raw data into distinct ideas. Axial coding, on the other hand, is responsible for establishing links between these concepts to create organized categories. These categories are subsequently incorporated into a full theoretical framework through selective coding, guaranteeing theoretical saturation and internal coherence through its implementation. As a result of this systematic methodology, the created theory is strengthened in terms of its validity, reliability, and application. This further solidifies GT as a rigorous and successful strategy for constructing empirically driven theoretical models in qualitative research.

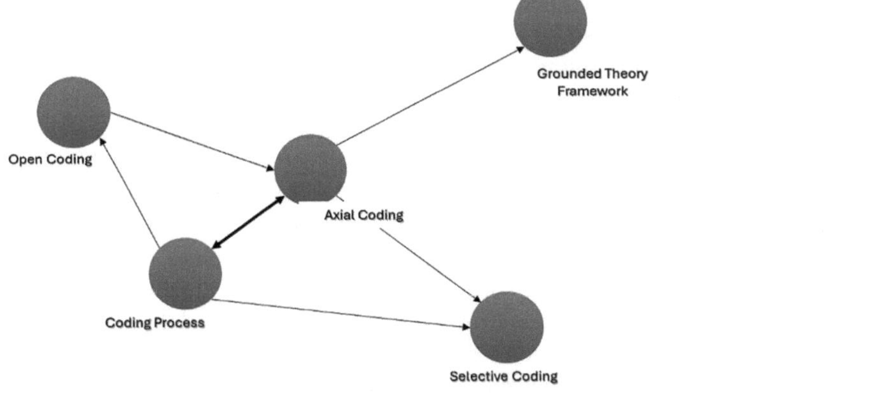

Fig. 3 Complex diagram of grounded theory analytical procedures. *Source* Authors

4.3 Various Applications, Obstacles, and Methodological Improvements

Grounded theory (GT) is standard across various fields of study, including sociology, organizational studies, and behavioral research [90]. Collecting lived experiences and examining the underlying social structures have proven invaluable tools in sociological research, allowing for exploring social interactions, identity construction, and cultural phenomena. GT has been employed in research organizations to analyze workplace behaviors, leadership styles, and institutional transformation [91]. This has provided researchers with significant insights into how individuals navigate organizational contexts. The approach has also been applied in behavioral research to understand decision-making processes, psychological reactions, and the impact of social circumstances on the behavior of individuals and groups [92]. Researchers across diverse fields can construct theoretical models that accurately reflect the complexities of the real world due to the versatility of GT. However, despite its many benefits, GT has its drawbacks, especially regarding the preservation of impartiality and the reduction of researcher bias. Since the approach relies on the researcher's interpretive role, subjective viewpoints may influence data coding and theory development. Thus, reflexivity is crucial, and researchers must critically evaluate their assumptions and transparently document their analytical judgments. Additionally, reaching theoretical saturation, the point at which no new conceptual insights emerge requires thorough evaluation to ensure that findings are comprehensive and to avoid premature conclusions [93]. Managing large volumes of qualitative data also presents challenges, necessitating rigorous organization and systematic documentation of analytical techniques. Implementing GT is further complicated by balancing inductive emergence with theoretical sensitivity. Although the method encourages the development of theories grounded in evidence, researchers must remain aware of existing theoretical frameworks and avoid allowing them to shape emerging discoveries prematurely. To achieve this balance, it is essential to maintain ongoing engagement with both empirical data and relevant literature [94]. This ensures that the theories developed remain original and well informed. Researchers can harness the full potential of GT to create theories that are both empirically grounded and theoretically rigorous by refining methodological strategies, incorporating reflexivity, and upholding analytical transparency [95]. For example, grounded theory (GT) can be employed in construction research to explore the acceptance and implementation of digital twins and digital replicas of actual buildings used for real-time monitoring, simulation, and performance enhancement. Researchers utilize GT techniques to collect qualitative data from industry experts, including engineers, project managers, and site operators, to identify emerging themes in digital twin integration. Significant challenges, such as data interoperability, real-time accuracy, and interactions with legacy systems, are systematically addressed through theoretical sampling. These findings are refined via comparative analysis, establishing a link between technological constraints and broader organizational and workflow inefficiencies. In contrast to axial coding, which develops connections between factors such as operational efficiency, predictive analytics, and technical limitations, open coding helps identify the most critical user experiences. Ultimately, selective coding provides a conceptual framework that synthesizes these findings to enhance digital twin applications in the construction sector. This methodological approach ensures that theoretical advancements remain empirically grounded, fostering innovation, improving project efficiency, and advancing methodological rigor in digital twin research.

5 Phenomenology: Exploring Subjectivity and Lived Experience

5.1 The Foundations of Phenomenology, Both Ontological and Epistemological

The rigorous qualitative research method known as phenomenology aims to understand reality as individuals experience it. Giorgi (1994) states that the phenomenological approach to qualitative research is compared to various other qualitative methodologies based on distinct paradigms. The thesis argues that a deeper understanding of phenomenological philosophy can provide the alternative framework many authors have sought [96]. Phenomenology is a phenomenological approach. Using a constructivist ontological framework as its foundation, it proposes that reality is not an absolute objective but a dynamic construct molded by human consciousness and perception. The epistemological foundation of phenomenology emphasizes the importance of first-person experiences, acknowledging that these accounts are necessary for expressing the essence of human knowledge [97]. Husserlian phenomenology, which is descriptive, and Heideggerian phenomenology, which is interpretive, are fundamentally different. In his descriptive phenomenology, Edmund Husserl strongly emphasizes the importance of bracketing, also known as epoché, to suspend prejudices and make it possible to examine actual events without intermediaries [98]. Through this rigorous methodology, researchers can uncover the fundamental components of experience, referred to as "essences." On the other hand, Martin Heidegger's interpretive (hermeneutic) phenomenology opposes the concept of bracketing by stating that understanding is inherently entwined with interpretation. Heideggerian phenomenology recognizes the researcher's participation in the process of co-constructing meaning through analysis, and as a result, it embraces subjectivity as an essential component of inquiry [99]. These contrasting viewpoints contribute to the advancement of phenomenological research by providing techniques that are complementary to one another while still being unique from one another.

5.2 *Methodological Approaches in Phenomenological Research*

To ensure depth, coherence, and analytical precision, phenomenological research often adheres to an organized and variable methodological trajectory [100]. The first step in the process is called epoché, and it involves making a conscious effort to put aside any preconceived notions and approach the participants' experiences with empathy. Following this comes the reduction process, which involves reducing experiences to their underlying structures [101]. This process culminates in essence identification, which involves uncovering fundamental theme aspects. Data gathering in phenomenological inquiry is often carried out through in-depth interviews in a semi-structured format. These interviews provide participants with the opportunity to explain their experiences in narratives that are rich and descriptive. Complementary methods include documenting experiential narratives and thematic analysis, making recognizing repeating themes in several reports easier. This research approach is not designed to provide conclusions that can be generalized; rather, it is intended to shed light on the profound and subjective aspects of human experience. One of the most critical analytical approaches, phenomenological hermeneutics, emphasizes the interpretive element of meaning-making [102]. Understanding is fundamentally influenced by context, language, and cultural influences, which this method understands and accepts. Researchers use a technique that is known as the hermeneutic circle, which is an iterative procedure. This method involves dynamic interaction between the individual components of the data (participant narratives) and the larger conceptual whole (emerging themes). Through this iterative process, one can arrive at a consistent and contextually pertinent interpretation of lived experience [103].

5.3 *Ensuring the Strictness of Phenomenological Research Technique*

A methodological commitment to trustworthiness, integrity, and authenticity is essential for establishing rigor in phenomenological research before it can be deemed rigorous [104]. Unlike quantitative research, which relies on statistical validation, phenomenological inquiry employs qualitative criteria to enhance validity and epistemological strength [105]. Thorough documentation of methodological procedures enables the establishment of trustworthiness, ensuring transparency from data collection to thematic analysis. Researchers must critically evaluate and acknowledge their influence on the interpretive process to fulfill reflexivity requirements, which are crucial for maintaining methodological integrity. Sustained engagement with participants bolsters credibility, deepening the understanding of their lived experiences [106]. Specific methods, such as member checking, where participants review and validate interpretations—further enhance trustworthiness by aligning the researcher's insights with the participants' realities. Triangulation, involving multiple data sources and analytical perspectives, also improves the reliability of conclusions [107]. In research, authenticity refers to the extent to which it accurately and completely represents the participants' experiences without distortion. Researchers must resist imposing extraneous theoretical frameworks to preserve the authenticity of their findings, grounding their interpretations in the words and perspectives of the participants themselves. Employing detailed descriptions, extensive data, and direct quotations from participants maintains the richness and depth of subjective experiences, contributing to increased authenticity [100]. By meticulously addressing these methodological issues, phenomenological research sustains its academic legitimacy while providing valuable insights into the nature of experience. This methodological approach improves qualitative research and enhances phenomenology's capacity to illuminate human consciousness, perception, and the process of meaning-making in a profoundly nuanced way. For example, a phenomenological study on energy efficiency in buildings examined how architects and engineers perceive the adoption of passive design strategies. Through in-depth interviews, researchers identified key experiential themes, such as challenges in regulatory compliance, cost concerns, and the integration of renewable energy systems, revealing professionals' lived experiences and decision-making processes in sustainable construction. By addressing these methodological issues, phenomenological research maintains its academic legitimacy while offering essential insights into the nature of lived experience. This meticulous approach enhances qualitative research and strengthens phenomenology's ability to illuminate human consciousness, perception, and the intricacies of meaning-making.

6 Comparative Methodological Analysis, Theoretical Implications, and Research Applications

6.1 Distinctions Between Epistemological and Methodological Approaches

There are two distinct epistemological orientations: theory generation and phenomenological insight. Phenomenology and Grounded Theory (GT) are separate qualitative research approaches based on differing epistemological foundations. Bakker (2019) asserts that these procedures are prevalent in qualitative research, each evolving into many adaptations with distinct philosophical underpinnings and methodological approaches. Nevertheless, inexperienced researchers frequently combine or misapply various methodologies, resulting in methodological discrepancies. The choice of a suitable technique and its tailored modification should be directed by the research question and the researcher's epistemological position to guarantee methodological rigor. Notwithstanding their prevalent application, a paucity of literature methodically examines and differentiates the philosophical foundations of their many adaptations [108]. The primary focus of GT is the creation of theories that originate in pragmatic and symbolic interactionist traditions. In conducting systematic analysis, researchers utilizing GT aim to develop middle-range hypotheses based on empirical facts [109]. Through this iterative process, hypotheses emerge from the facts rather than being imposed from the outset. Conversely, phenomenology, heavily influenced by Husserlian and Heideggerian philosophy, emphasizes understanding lived experiences from a first-person perspective [110]. Rather than concentrating on developing theories, phenomenology seeks to comprehend the fundamental nature of human experiences and how individuals make sense of the world around them. Instead of creating abstract theoretical frameworks, the emphasis is on exploring subjectivity, perception, and meaning [111]. This underlying epistemological difference leads to the development of diverse techniques, guiding researchers in various directions regarding knowledge production.

- An Examination of the Differences Between Coding Schema and Thematic Interpretation in Concerns to Data Analysis Techniques

Concerning data analysis and epistemological foundations, phenomenology and grounded theory (GT) couldn't be more different [112]. Their distinct methods of data analysis are a key point of differentiation. By employing a descriptive or interpretative approach, phenomenology seeks to understand lived experiences by identifying key themes and meanings. In contrast, GT develops a theory based on data through a systematic coding process. This framework typically includes open, axial, and selective coding to identify patterns and connections systematically. Both approaches offer unique perspectives on qualitative research and complement each other well. A process known as open coding involves the fragmentation of data into discrete components and the assignment of codes to newly emerging concepts [113]. Following this, axial coding creates links between categories, and selective coding is used to synthesize these findings into a cohesive theoretical framework. The procedure is iterative and reflexive, and it frequently involves the utilization of continuous comparison approaches to enhance recently developed ideas. On the other hand, phenomenology takes a more flexible and interpretive approach, emphasizing theme analysis. Researchers immerse themselves in participants' narratives, during which they look for repeating themes that indicate the substance of the events that people have experienced [114]. In descriptive phenomenology, bracketing, or epoché, is employed to exclude the researcher's preconceived notions, guaranteeing an objective portrayal of the participants' viewpoints. Within the framework of interpretive (hermeneutic) phenomenology, which was developed by Heidegger, the researcher's participation in the process of meaning-making is acknowledged, which helps to cultivate a co-constructed understanding. In contrast to GT, which seeks to isolate generalizable concepts through coding, phenomenological analysis preserves the complexity and variety of tales [115].

6.2 *Methodological Considerations and Contributions to the Research Discussion*

Comparing the expansion of theoretical frameworks to deepening phenomenological understanding might contribute to academic discourse [116]. The contributions that GT and phenomenology provide to academic discourse complement one another. Through the development of empirically based ideas that are relevant across a variety of settings, GT broadens the scope of theoretical frameworks. This methodological approach is beneficial in fields like sociology, psychology, and organizational studies, which are all areas in which explanatory theories have the potential to influence both policy and practice. The findings of GT studies frequently impact later research because they give rise to flexible frameworks that direct further empirical investigations [117]. The study of phenomenology, on the other hand, contributes to a more profound understanding by providing a wealth of context-specific insights into human experiences. Its contributions to academic discourse include illuminating complex viewpoints that may be neglected in research and are motivated by various theories. Because they lighten individual experiences with sickness, learning processes, and therapeutic interactions, phenomenological investigations benefit multiple fields, including healthcare, education, and counseling [118]. Phenomenology enriches areas that emphasize human-centered inquiry because it captures these subjective qualities. The decision between phenomenology and GT is not random; instead, it is determined by the aims of the research, the contextual aspects, and the alignments between the disciplines. The study that seeks to construct explanatory frameworks and theoretical generalizations is the type of research that gravitates most toward GT. In situations where emergency theories are used to guide practical solutions, such as when researching social processes, organizational behaviors, or policy-related concerns, it is especially beneficial from this perspective [119]. It is common for researchers working in fields that emphasize empirical validation and theoretical contributions to gravitate toward GT. In contrast, phenomenology is ideally suited for research that aims to comprehend the subjective experiences that people have in their lives. Phenomenology is a powerful tool for researchers interested in studying phenomena such as sorrow, trauma, identity, or consciousness because it captures how individuals establish meaning. The alignment of disciplines also plays a significant part. However, GT is commonly used in sociology and management studies, and a substantial amount of research in psychology, philosophy, and healthcare supports phenomenology [120]. Researchers must conduct an in-depth analysis of their research questions and objectives to ascertain which methodological approach is most suitable for achieving their academic aims.

6.3 Considerations Regarding Ethical Issues and Future Research

Regarding subjectivity, researcher reflexivity, and participant representation, GT and phenomenology need to consider ethical issues [121]. Academics working in GT must stay attentive to the practice of pushing data into pre-existing theoretical frameworks. This will ensure that ideas arise from empirical evidence in an accurate manner. Reflexivity is critical in minimizing biases that may impact coding and interpretation. Because of the interpretive character of phenomenological research, it demands an even higher sensitivity to subjectivity than other types of study. Researchers are responsible for realizing their positionality and recognizing how their various points of view impact the analysis [122]. The extent of ethical duty extends to the portrayal of participants to ensure that lived experiences are presented accurately and without distortion. Including direct quotations and detailed descriptions is recommended to enhance credibility and ensure that the participants' voices are accurately portrayed. The acquisition of informed permission, the maintenance of confidentiality, and the adherence to ethical research standards are of the utmost importance in guaranteeing the integrity of both approaches [123]. In interdisciplinary research, the intersection of grounded theory and phenomenology is presented as a potential future direction. GT and phenomenology have long been considered separate approaches; nevertheless, recent research drawing from a variety of disciplines reveals that there may be potential connections between the two. Researchers have investigated hybrid techniques that combine GT's methodical coding with phenomenology's rich experiential focus. This confluence holds excellent promise in professions like education and healthcare, where theoretical growth and profound subjective discoveries are essential [124]. For example, in healthcare research, the combination of GT and phenomenology enables the development of patient-centered care models that are theoretically sound and based on an individual's experiences. Researchers can provide holistic insights into patient care by applying GT to recognize patterns in patient narratives and concurrently using phenomenology to investigate the extent to which those experiences are experienced. In a similar vein, this integration may be helpful in educational research by helping to bridge the gap between pedagogical theories and student experiences. This approach ensures that empirical trends and personal learning journeys guide curriculum creation. Looking toward the future, various methodological improvements hold the potential to enhance the interplay between phenomenology and GT methods. Exploring mixed-methods designs that incorporate qualitative and quantitative methodologies can help balance theory development and the depth of lived experiences. Additionally, advancements in digital ethnography and AI-driven qualitative analysis have created new opportunities to combine the analytical rigor of GT with the interpretive richness of phenomenology. Phenomenology and grounded theory represent two distinct yet complementary qualitative research approaches. While GT excels in theory formation through structured data analysis, it is a valuable tool for constructing explanatory frameworks [125]. In contrast, phenomenology provides profound insights into human experiences,

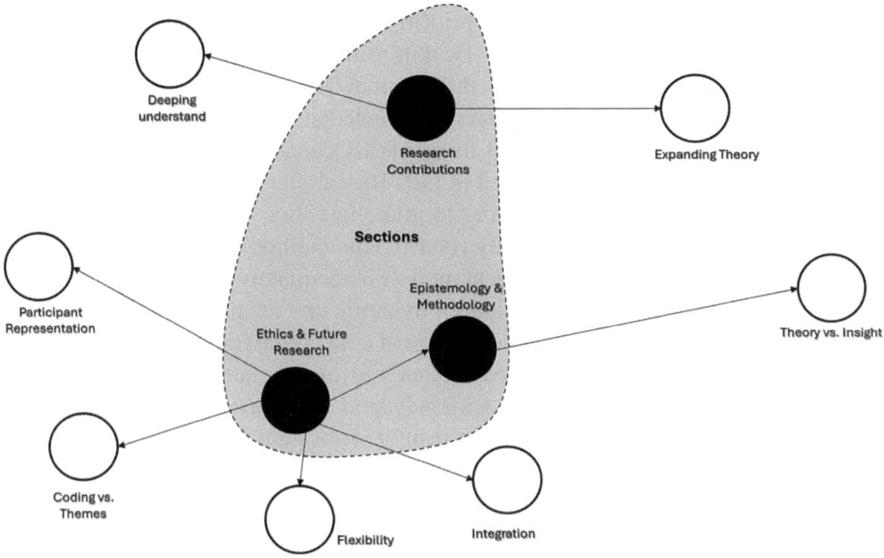

Fig. 4 Comparative methodological framework and research implications. *Source* Authors

enriching fields of study that prioritize subjective understanding. These methodologies have traditionally been considered separate, yet interdisciplinary research offers a pathway to integrating their strengths. Acknowledging both methods' epistemological, analytical, and ethical dimensions enables researchers to make informed methodological choices that align with their academic objectives. Future research should continue exploring intersections between GT and phenomenology to foster methodological innovations that advance qualitative inquiry across multiple disciplines [126]. Figure 4 presents a hierarchical summary of key methodological features and their applications in scholarly research. The article is structured into three main sections: Research Contributions, Ethics and Future Research, and Epistemology and Methodology. These components offer valuable insights for social sciences, psychology, healthcare, education, and interdisciplinary research by emphasizing theoretical perspectives, data analysis techniques, and ethical considerations.

7 Conclusion

The synthesis of important topics and ideas from the conversation reveals that grounded theory and phenomenology are separate but compatible techniques in qualitative research. Both methodologies offer significant frameworks for examining intricate social processes, yielding profound, subtle insights beyond superficial interpretations. Grounded theory, by its repeated data-gathering and analysis process, enables researchers to build theories that arise organically from empirical evidence. Conversely, phenomenology prioritizes lived experiences to reveal the core of human perception and awareness. Despite differing epistemological underpinnings and methodological approaches, their mutual dedication to qualitative research underscores their importance in enhancing knowledge across several fields. One of the primary advantages of grounded theory is its capacity to provide substantive and formal theories securely anchored in fact. Researchers utilize ongoing comparative analysis and theoretical sampling to guarantee their conclusions are contextually grounded and methodologically robust. This method is particularly advantageous in areas where current theoretical models inadequately account for new social phenomena. Grounded theory facilitates a dynamic and adaptable research process by permitting evidence to guide the formation of conceptual categories. Nonetheless, issues such as theoretical saturation, researcher bias, and the labor-intensive processes of coding and memo writing must be acknowledged as intrinsic complications of this technique. Phenomenology offers deep insights into human experiences by highlighting intentionality and subjectivity. By employing bracketing, researchers seek to suspend preconceived assumptions, allowing for exploring participants' actual experiences in their most authentic form. This methodological approach is essential in psychology, nursing, education, and philosophy, where a profound comprehension of subjective experiences propels theoretical progress and practical implementations. Nonetheless, phenomenology possesses inherent limits. The necessity for stringent reflexivity, possible researcher bias, and difficulties in accurately recording individual views provide methodological and interpretive issues. Notwithstanding these obstacles, its capacity to elucidate the intricacies of consciousness and meaning-making renders phenomenology an essential instrument in qualitative research. An essential factor in qualitative research is the epistemological stance of the researcher. Grounded theory and phenomenology, although both qualitative methodologies, fundamentally diverge in their approaches to data engagement and meaning construction. Grounded theory functions from a pragmatic and constructivist perspective, highlighting the interaction and collaborative creation of meaning between the researcher and the participant. Phenomenology, by contrast, employs an interpretative and descriptive methodology aimed at uncovering the essential structures of experience. These epistemological discrepancies influence data-gathering techniques, analytical approaches, and the final presentation of results. Identifying these distinctions enables researchers to synchronize their methodological selections with their study aims and philosophical perspectives. The relevance of grounded theory and phenomenology extends across several fields, improving scholarly and practical practices. In health sciences, grounded theory has been important in formulating patient-centered care models by elucidating emergent patterns in healthcare delivery and patient experiences. Conversely, phenomenology has facilitated the comprehension of chronic disease, mental health issues, and the lived realities of oppressed groups. Both techniques in education provide significant insights into student experiences, instructor pedagogies, and institutional dynamics. The flexibility of these methods underscores their enduring significance in modern research. Another significant aspect of discourse is the methodological rigor necessary for both methods. In grounded theory, achieving theoretical saturation demands ongoing data refining, necessitating a balance between depth and breadth in sampling. Researchers must be vigilant about coding procedures, ensuring that personal biases or theoretical inclinations do not unduly impact emergent categories. Phenomenological researchers must conduct precise data interpretation, skillfully balancing descriptive integrity with interpretative depth. This stringent methodology safeguards the integrity and reliability of qualitative research outcomes. Although grounded theory and phenomenology differ methodologically, both are dedicated to enhancing the qualitative research domain. Their contributions surpass academic investigation, influencing policy choices, clinical interventions, and organizational tactics. These techniques enhance the knowledge of social realities by encapsulating the depth and complexity of human experiences. Future research may benefit from methodological triangulation, integrating aspects of both methodologies to improve analytical rigor and interpretative depth. Integrative techniques can mitigate separate methodological constraints while capitalizing on their distinct strengths. An additional examination of these approaches indicates their adaptive characteristics in addressing modern research issues. The rising digitalization of data, the heightened focus on participatory research, and the intersectionality of social identities require methodological adjustments. Grounded theory has progressed with integrating digital coding and data visualization technologies, improving its analytical methodologies' efficiency. Phenomenology has dealt with multidisciplinary viewpoints by integrating ideas from cognitive science, ethnography, and emotion theory. These methodological advancements underscore the flexibility and robustness of qualitative research traditions. The ethical implications of employing grounded theory and phenomenology warrant consideration. In all techniques, the relationship between the researcher and participant is crucial in influencing the data-gathering process. Ethical issues about informed consent, participant confidentiality, and data representation necessitate meticulous attention. Researchers must diligently uphold ethical norms, particularly when dealing with sensitive subjects or vulnerable groups. Transparency in methodological decisions, reflexivity in interpretation, and knowledge of power dynamics augment the ethical integrity of qualitative research. The evolution of qualitative techniques suggests a promising future for grounded theory and phenomenology. Continued theoretical developments, technological advances, and multidisciplinary partnerships will influence scholarly discourse and professional practice. Its ability to encapsulate human intricacy, produce significant discoveries, and aid in knowledge creation guarantees its enduring significance. Researchers must remain involved in essential methodological discussions, developing, and modifying these strategies to address new research inquiries and social issues. Ultimately, grounded theory and phenomenology are essential approaches that enrich the qualitative research paradigm. Their methodological precision, theoretical advancements, and practical implementations underscore their significance in scholarly and professional domains. By recognizing their distinct strengths and limits, researchers may approach the intricacies of qualitative inquiry with clarity and honesty. Doing so enhances a more nuanced, complete, and human-centered comprehension of social processes, affirming that qualitative research is fundamental to scholarly inquiry and practical application.

References

1. Urcia IA (2021) Comparisons of adaptations in grounded theory and phenomenology: selecting the specific qualitative research methodology. Int J Qual Methods 20:16094069211045474
2. Grass K (2024) The three logics of qualitative research: epistemology, ontology, and methodology in political science. Amer J Qual Res 8(1):42–56
3. McCall C, Edwards C (2021) New perspectives for implementing grounded theory. Studies Eng Educ 1(2):93–107
4. Prosek EA, Gibson DM (2021) Promoting rigorous research by examining lived experiences: a review of four qualitative traditions. J Couns Dev 99(2):167–177
5. Gagura G (2023) Brief review of phenomenological research design in education. Ethiopian J Educ Studies 3(1)
6. Hossain MS, Alam MK, Ali MS (2024) Phenomenological approach in the qualitative study: data collection and saturation. ICRRD Qual Index Res J 5(2):148–172
7. Baidi R, Ahmad A, Munib A (2023) Testing the truth of Comte's positivism based on Habermas critical theory. Trunojoyo Law Rev 5(1):39–53
8. Damböck C (2021) Dilthey, historicism, and logical empiricism. In: The Routledge handbook of logical empiricism, pp 34–42. Routledge
9. Stough LM, Lee S (2021) Grounded theory approaches used in educational research journals. Int J Qual Methods 20:16094069211052204
10. Dahlstrom DO (2023) Heidegger, Martin (1889–1976). Encyclopedia of Phenomenology. Springer International Publishing, Cham, pp 1–15
11. Raphael CE (2023) Husserlian phenomenology as foundationalism: critical review of a-world-in-brackets. AMAMIHE J Appl Philos 21(2)
12. Heidegger M (2024) Phenomenology and theology (Vol 27). Minerva Heritage Press
13. Forde AI (2023) The American founding documents and democratic social change: a constructivist grounded theory (Doctoral dissertation, Walden University)
14. Low J, Hyslop-Margison C (2021) Kathy Charmaz: theory, method, and scholarly identity. Symb Interact 44(3):632–664
15. De Santis D, Hopkins BC, Majolino C (eds) (2021) The Routledge handbook of phenomenology and phenomenological philosophy. Routledge, Taylor & Francis Group
16. Godwins J (2024) Philosophy of body: Merleau-Ponty and the philosophy of mind. TWIST 19(1):209–227
17. Homer ST, Lim WM (2024) Theory development in a globalized world: bridging "doing as the Romans do" with "understanding why the Romans do it." Glob Bus Organ Excell 43(3):127–138
18. Kelly NJ, Hallam J, Bignell S (2023) Using interpretative phenomenological analysis to gain a qualitative understanding of presence in virtual reality. Virtual Reality 27(2):1173–1185
19. Folgueiras Bertomeu P, Sandín Esteban MP (2023) The research question in hermeneutic phenomenology and grounded theory research. Qual Report 28(5):1452–1472
20. Lomadze E (2024) Pedestrian perception of walkability in public spaces: example of a historical Neighbourhood In Batumi
21. Meyer ET, Schroeder R (2023) Knowledge machines: digital transformations of the sciences and humanities. MIT Press
22. Locke K, Feldman M, Golden-Biddle K (2022) Coding practices and interactivity: beyond templates for analyzing qualitative data. Organ Res Methods 25(2):262–284
23. Burns M, Bally J, Burles M, Holtslander L, Peacock S (2022) Constructivist grounded theory or interpretive phenomenology? Methodological choices within specific study contexts. Int J Qual Methods 21:16094069221077758
24. Berger P, Luckmann T (2023) The social construction of reality. In: Social theory re-wired, pp 92–101. Routledge

25. Ramanadhan S, Revette AC, Lee RM, Aveling EL (2021) Pragmatic approaches to analyzing qualitative data for implementation science: an introduction. Implementation Sci Commun 2:1–10
26. Osborne N, Grant-Smith D (2021) In-depth interviewing. In: Methods in urban analysis, pp 105–125. Singapore: Springer Singapore
27. Amalanathan S, Reddy-Best KL (2024) Modesty in business, bold in fashion: entrepreneurial experiences of US Muslim women in niche fashion markets. J Innov Entrepreneurship 13(1):57
28. Larsen HG, Adu P (2021) The theoretical framework in phenomenological research: development and application. Routledge
29. Levitt HM, Surace FI, Wu MB, Chapin B, Hargrove JG, Herbitter C, ... Hochman AL (2022) The meaning of scientific objectivity and subjectivity: from the perspective of methodologists. Psychol Methods 27(4):589
30. Butt PM (2024) Feminism, theology and everyday domestic skill within a phenomenological framework: an autoethnographic reflection (Doctoral dissertation, University of Birmingham)
31. Bracken S (2010) Discussing the importance of ontology and epistemology awareness in practitioner research. Worcester J Learn Teach (4)
32. Lee J (2024) Understanding the epistemic diversity and essential components of the grounded theory method: empowering human resource development researchers. Hum Resour Dev Rev 23(2):276–293
33. Hoda R (2024) Research philosophy. Qualitative research with socio-technical grounded theory: a practical guide to qualitative data analysis and theory development in the digital world. Springer International Publishing, Cham, pp 93–112
34. Sass L (2021) Husserl, Heidegger, and the paradox of subjectivity. Cont Philos Rev 54(3):295–317
35. Džanić D (2023) Transcendental Phenomenology as human possibility: Husserl and fink on the Phenomenologizing Subject (Vol. 237). Springer Nature
36. Hauashdh A, Jailani J, Rahman IA (2022) Strategic approaches towards achieving sustainable and effective building maintenance practices in maintenance-managed buildings: a combination of expert interviews and a literature review. J Build Eng 45:103490
37. Chakabwata W (2025) Grounded theory for a doctoral thesis: retrospective and prospective insights. In: Qualitative research methods for dissertation research, pp 251–278. IGI Global Scientific Publishing
38. Rodríguez-Labajos L, Thomson CS, O'Brien G (2021) Applying constructivist grounded theory in co-production research: a case study exploring its potential and lessons for construction management research. Constr Manag Econ 39(5):369–382
39. Heath H, Cowley S (2004) Developing a grounded theory approach: a comparison of Glaser and Strauss. Int J Nurs Stud 41(2):141–150
40. Reichertz J (2019) Abduction: The logic of discovery of grounded theory–an updated review. The Sage handbook of current developments in grounded theory, 259–281
41. Martin VB, Gynnild A (Eds) (2011) Grounded theory: the philosophy, method, and work of Barney Glaser. Universal-Publishers
42. Qureshi HA, Ünlü Z (2020) Beyond the paradigm conflicts: a four-step coding instrument for grounded theory. Int J Qual Methods 19:1609406920928188
43. Priya KR (2019) Using constructivist grounded theory methodology: Studying suffering and healing as a case example. The SAGE handbook of current developments in grounded theory, 392–412
44. Dotts JM (2023) Go On… Shatter It! a qualitative transcendental phenomenological study of gender-based Prejudices of senior-level women administrators at community colleges across Texas
45. Edwards C (2024) Denaturalising phenomenology: husserl and Heidegger's transcendental account of the knowledge problem (Doctoral dissertation, Murdoch University)
46. Du Toit J (2024) A Merleau-Pontian phenomenology of the virtual: disembodied challenges and embodied prospects. South African J Philos 43(4):307–322

47. Almusaed A, Almssad A, Alasadi A, Al-Asadi F (2025) Sustainable reimagining of urban habitats: navigating Iraq's housing crisis through socio-technological and environmental approaches. Ain Shams Eng J 16(2):103280
48. Horn A, Scheffelaar A, Urias E, Zweekhorst MB (2023) Training students for complex sustainability issues: a literature review on the design of inter-and transdisciplinary higher education. Int J Sustain High Educ 24(1):1–27
49. James ML (2023) Lived Experiences of K-12 traditional teachers' preparedness to instruct during emergency remote learning: a phenomenological study (Doctoral dissertation, Northcentral University)
50. Zhang Y, Chen J, Liu H, Chen Y, Xiao B, Li H (2024) Recent advancements of human-centered design in building engineering: a comprehensive review. J Build Eng, 108529
51. Unuigbe M (2021) Influencing factors to sustainability transitioning in commercial buildings in the Nigerian built environment: a grounded theory study [Online]. Leeds Beckett University (Doctoral dissertation, Leeds Beckett University)
52. Hoda R (2024) Qualitative research with socio-technical grounded theory
53. Cheeli BP (2024) Process of phenomenology as qualitative research method in English. Int J Methodol 3(1):1–20
54. Mathews M (2021) A hermeneutic phenomenological study into the impact of BIM on the social dynamics of the AEC professional in the workplace.. A Hermeneutic Phenomenological Study into the impact of BIM on the Social Dynamics of the AEC professional in the workplace
55. Haj-Bolouri A (2023) The experience of immersive virtual reality: a phenomenology-inspired inquiry. Commun Assoc Inf Syst 52(1):782–814
56. Lee JG, Seo J, Abbas A, Choi M (2020) End-Users' augmented reality utilization for architectural design review. Appl Sci 10(15):5363
57. Almusaed A (2010) Biophilic and bioclimatic architecture: analytical therapy for the next generation of passive sustainable architecture. Springer Science & Business Media
58. Masoodi M (2017) A comparative analysis of two qualitative methods: deciding between grounded theory and phenomenology for your research. Profesinis rengimas: tyrimai ir realijos 28:23–40
59. Engward H, Davis G (2015) Being reflexive in qualitative grounded theory: discussion and application of a model of reflexivity. J Adv Nurs 71(7):1530–1538
60. Irizar-Arrieta A, Casado-Mansilla D, Garaizar P, López-de-Ipiña D, Retegi A (2020) User perspectives in the design of interactive everyday objects for sustainable behaviour. Int J Hum Comput Stud 137:102393
61. Johnson RD (2016) Integrated project delivery in architecture, engineering, and construction: an interpretative phenomenological analysis of practice. Colorado Technical University
62. Stock P, Burton RJ (2011) Defining terms for integrated (multi-inter-trans-disciplinary) sustainability research. Sustainability 3(8):1090–1113
63. Chance S, Duffy G, Bowe B (2020) Comparing grounded theory and phenomenology as methods to understand lived experience of engineering educators implementing problem-based learning. Eur J Eng Educ 45(3):405–442
64. Chun Tie Y, Birks M, Francis K (2019) Grounded theory research: a design framework for novice researchers. SAGE Open Med 7:2050312118822927
65. Altomonte S, Kaçel S, Martinez PW, Licina D (2024) What is NExT? A new conceptual model for comfort, satisfaction, health, and well-being in buildings. Build Environ, 111234
66. Xin Y (2024) Phenomenological narratives: exploring space and identity in fiction
67. Guetterman TC, Babchuk WA, Howell Smith MC, Stevens J (2019) Contemporary approaches to mixed methods–grounded theory research: a field-based analysis. J Mixed Methods Res 13(2):179–195
68. Allen N, Davey M (2018) The value of constructivism grounded theory for built environment researchers. J Plan Educ Res 38(2):222–232
69. Baldassarre B, Keskin D, Diehl JC, Bocken N, Calabretta G (2020) Implementing sustainable design theory in business practice: a call to action. J Clean Prod 273:123113

70. Behague D, Tawiah C, Rosato M, Some T, Morrison J (2009) Evidence-based policy-making: the implications of globally-applicable research for context-specific problem-solving in developing countries. Soc Sci Med 69(10):1539–1546
71. Keeler M, Vaidya P (2016) Fundamentals of integrated design for sustainable building. Wiley
72. Luthfiandana R, Santioso LL, Febrian WD, Soehaditama JP, Sani I (2024) Qualitative research concepts: phenomenology, grounded theory, ethnography, case study. Narrative. Siber J Adv Multidisciplinary 2(1):26–36
73. Bornhoff A, Davis EB, Yousey J, Kimball CN, Stier E, Wang E (2024) Patient and provider perspectives on the phenomenon and effective treatment of treatment-resistant depression: a grounded theory. J Affect Disorders Reports 16:100779
74. Christensen J, Ekelund N, Melin M, Widén P (2021) The beautiful risk of collaborative and interdisciplinary research. A challenging collaborative and critical approach toward sustainable learning processes in academic profession. Sustainability 13(9):4723
75. Elsayed M, Pelsmakers S, Pistore L, Castaño-Rosa R, Romagnoni P (2023) Post-occupancy evaluation in residential buildings: a systematic literature review of current practices in the EU. Build Environ 236:110307
76. Almusaed A, Almssad A, Yitmen I, Wallhagen M, Yang YF, Iyer-Raniga U (2024) Integrative approaches in urban sustainability: architectural design, technological innovations and social dynamics in global contexts. BoD–Books on Demand
77. Ibeh CV, Awonuga KF, Okoli UI, Ike CU, Ndubuisi NL, Obaigbena A (2024) A review of agile methodologies in product lifecycle management: bridging theory and practice for enhanced digital technology integration. Eng Sci Technol J 5(2):448–459
78. Bischof A, Freybe K (2022) Grounding theory in digital data: a methodological approach for a reflective procedural framework. J Cultural Analyt 7(4)
79. Cooksey RW (2024) Systemic Contextualisation: guiding assumptions and framing. In: Unity from diversity: pluralist systemic thinking for social and behavioural research, pp 223–324. Singapore: Springer Nature Singapore
80. Themelis C, Sime JA, Thornberg R (2023) Informed grounded theory: a symbiosis of philosophy, methodology, and art. Scand J Educ Res 67(7):1086–1099
81. Sefeni D, Johnson M, Lee J (2024) Game-theoretic approaches for stepwise controllable text generation in large language models. Authorea Preprints
82. Al-Eisawi D (2022) A design framework for novices using grounded theory methodology and coding in qualitative research: organizational absorptive capacity and Knowledge Management. Int J Qual Methods 21:16094069221113552
83. Braun V, Clarke V (2021) Can I use TA? Should I use TA? Should I not use TA? Comparing reflexive thematic analysis and other pattern-based qualitative analytic approaches. Couns Psychother Res 21(1):37–47
84. Li Y, Shang H (2023) How does e-government use affect citizens' trust in government? Empirical evidence from China. Inf Manag 60(7):103844
85. Pérez Bentancur V, Tiscornia L (2024) Iteration in mixed-methods research designs combining experiments and fieldwork. Sociol Methods Res 53(2):729–759
86. Khoa BT, Hung BP, Hejsalem-Brahmi M (2023) Qualitative research in social sciences: data collection, data analysis and report writing. Int J Public Sector Perform Manag 12(1–2):187–209
87. Phillips MJ (2023) Towards a social constructionist, criticalist, Foucauldian-informed qualitative research approach: opportunities and challenges. SN Soc Sci 3(10):175
88. Van Assche K, Beunen R, Duineveld M, Gruezmacher M (2023) Adaptive methodology. Topic, theory, method and data in ongoing conversation. Int J Soc Res Methodol 26(1):35–49
89. Karunarathna I, Gunasena P, Hapuarachchi T, Gunathilake S (2024) Data collection fundamentals: a guide to effective research methodologies and ethical practices
90. Linden KLV, Palmieri PA (2021) Criteria for assessing a classic grounded theory study: a brief methodological review with minimum reporting recommendations. Grounded Theory Rev 20(2)

91. Weru J (2024) A grounded theory study exploring the process of professional identity construction among black corporate professionals in predominately white organizational spaces (Doctoral dissertation, The Chicago School of Professional Psychology)
92. Pandita S, Mishra HG, Chib S (2021) Psychological impact of covid-19 crises on students through the lens of Stimulus-Organism-Response (SOR) model. Child Youth Serv Rev 120:105783
93. Rahimi S (2024) Saturation in qualitative research: an evolutionary concept analysis. Int J Nursing Studies Adv 6:100174
94. Boccoli G, Gastaldi L, Corso M (2023) The evolution of employee engagement: towards a social and contextual construct for balancing individual performance and wellbeing dynamically. Int J Manag Rev 25(1):75–98
95. Peters HC (2023) A-methodological saturation: a grounded theory analysis. Couns Psychol 51(7):933–969
96. Giorgi A (1994) A phenomenological perspective on specific qualitative research methods. J Phenomenol Psychol 25(2):190–220
97. Klinke ME, Fernandez AV (2023) Taking phenomenology beyond the first-person perspective: conceptual grounding in collecting and analyzing observational evidence. Phenomenol Cogn Sci 22(1):171–191
98. Cypress BS (2021) Fundamentals of qualitative phenomenological nursing research. Wiley
99. Emon MMH (2024) Research design in social sciences research: exploring methodological evolution and comprehensive insights
100. Lim WM (2024) What is qualitative research? An overview and guidelines. Australasian Market J, 14413582241264619
101. Chae M, Kim B (2024) Reducing cultural barriers: a grounded theory approach to nursing student attitudes after multicultural education. Risk management and healthcare policy, pp 2241–2253
102. Agarwal A, Sandiford PJ (2021) Fictionalizing dialogue: interpretative phenomenological analysis in organizational research. Qual Res Organ Manag Int J 16(1):218–236
103. Braun V, Clarke V (2022) Conceptual and design thinking for thematic analysis. Qual Psychol 9(1):3
104. Vella J (2024) In pursuit of credibility: evaluating the divergence between member-checking and hermeneutic phenomenology. Research in social and administrative pharmacy
105. Janis I (2022) Strategies for establishing dependability between two qualitative intrinsic case studies: a reflexive thematic analysis. Field Methods 34(3):240–255
106. Hilliard F, Goldstein E, Nervik K, Croes K, Ossorio PN, Zgierska AE (2023) Voices of women with lived experience of substance use during pregnancy: a qualitative study of motivators and barriers to recruitment and retention in research. Fam Community Health 46(1):1–12
107. Rockmann KW, Vough HC (2024) Using quotes to present claims: practices for the writing stages of qualitative research. Organ Res Methods 27(4):621–649
108. Bakker JI (2019) Grounded theory methodology and grounded theory method: Introduction to the special issue. Sociol Focus 52(2):91–106
109. Hadley G, Hadley H (2024). Grounded theory. Less frequently used research methodologies in applied linguistics 6:127
110. Lumma AL, Weger U (2023) Looking from within: Comparing first-person approaches to studying experience. Curr Psychol 42(12):10437–10453
111. Halliday MAK, Matthiessen C (2006) Construing experience through meaning: a language-based approach to cognition. A&C Black
112. Anastasi NA, ProfDoc A (2021) Exploring counselling psychologists' experience of dual practice: a grounded theory (Doctoral dissertation, University of Roehampton)
113. Levine TH, Mitoma GT, Anagnostopoulos DM, Roselle R (2023) Exploring the nature, facilitators, and challenges of program coherence in a case of teacher education program redesign using core practices. J Teach Educ 74(1):69–84
114. Rau A, Coetzee JK (2022) Designing for narratives and stories. Quali Res Design, 700

115. Wiltshire G, Ronkainen N (2021) A realist approach to thematic analysis: making sense of qualitative data through experiential, inferential, and dispositional themes. J Critical Realism 20(2):159–180
116. Liu C, Tseng MY (2021) Paradigmatic variation in hedging and boosting: a comparative study of discussions in narrative inquiry and grounded theory research. Engl Specif Purp 61:1–16
117. Kafetzopoulos D (2023) Talent development: a driver for strategic flexibility, innovativeness and financial performance. EuroMed J Bus 18(2):296–312
118. Liao KC, Peng CH, Snell L, Wang X, Huang CD, Saroyan A (2021) Understanding the lived experiences of medical learners in a narrative medicine course: a phenomenological study. BMC Med Educ 21(1):321
119. Lindquist EA, Wellstead A (2021) The policy cycle: From heuristic to a theory-informed research and advice. In: Handbook of public administration, pp 303–322. Routledge
120. Timonen V, Foley G, Conlon C (2018) Challenges when using grounded theory: a pragmatic introduction to doing GT research. Int J Qual Methods 17(1):1609406918758086
121. Mruck K, Mey G (2007) Grounded theory and reflexivity. The Sage handbook of grounded theory, pp 515–538
122. Massoud MF (2022) The price of positionality: assessing the benefits and burdens of self-identification in research methods. J Law Soc 49:S64–S86
123. Pina E, Ramos J, Jorge H, Váz P, Silva J, Wanzeller C, Abbasi M, Martins P (2024) Data privacy and ethical considerations in database management. J Cybersecurity Privacy 4(3):494–517
124. Ghimire A, Yunjing QIU, Thapa B, Ranabhat M (2024) Exploring the factors affecting undergraduate nursing students' migration intention: a qualitative study. Nurse Educ Today 143:106414
125. van Veggel N, Engward H, Birks M, Mills J, Using AI in Grounded Theory research–a proposed framework for a ChatGPT-based research assistant
126. Ripani G, Zdzinski SF (2024) Beyond the divide between qualitative and quantitative research: a comparison of grounded theory and structural equation modeling. Bull Counc Res Music Educ 241:49–74

Narrative and Discourse Analysis

Abstract Narrative and discourse analysis are essential approaches in modern research, connecting individual experiences with broader cultural norms and power relations. This chapter offers an in-depth examination of these methodologies, illustrating their ability to analyze the formation, negotiation, and distribution of meaning in human communication. Narrative analysis reveals how personal and communal tales impact identity, memory, and cultural values. In contrast, discourse analysis examines the overarching language and ideological structures that uphold power dynamics and form societal perspectives. This approach, rooted in substantial theoretical traditions from Propp, Foucault, and Bruner, clarifies these techniques' historical development and transdisciplinary significance. Narrative analysis elucidates the lived experiences of individuals and societies, uncovering the complex mechanisms of identity development and meaning construction. Simultaneously, discourse analysis dissects language utilization within institutional and cultural frameworks, revealing the foundational beliefs and power dynamics that influence social discourses. This chapter also explores innovative breakthroughs, like artificial intelligence and multimodal tools, which enhance the analytical capacities of these approaches. Including ethical considerations and methodological precision underscores the transformational capacity of story and discourse analysis in tackling intricate global issues. These approaches are essential for enhancing academic research, promoting critical insights, and guiding multidisciplinary studies in a progressively linked world.

1 Introduction

Narrative analysis and discourse analysis are key methodologies across various academic disciplines, serving as essential tools to explore how meaning is constructed, shared, and understood [1]. Originating from the social sciences and humanities, these approaches have become indispensable for examining how individuals and communities articulate, interpret, and represent their experiences [2]. Narrative analysis focuses on the structured stories people use to make sense of

A. Almusaed et al., *Practice of Research Methodology in Civil Engineering and Architecture*, https://doi.org/10.1007/978-3-031-97393-2_12

their lives, emphasizing how personal and collective experiences are organized and communicated [3]. In contrast, discourse analysis examines the broader linguistic, cultural, and social frameworks that shape and influence communication practices, revealing underlying power dynamics, ideologies, and societal norms. This section thoroughly explores both methodologies, outlining their definitions, historical development, and pivotal roles in contemporary research. Narrative analysis traces how individuals construct coherent accounts of their experiences, offering insights into identity formation, memory, and meaning-making processes. Meanwhile, discourse analysis explores patterns of language use, uncovering how texts and conversations are embedded within larger social and cultural contexts [4]. The analysis highlights how these methodologies, while distinct, often intersect in their focus on the interplay between individual agency and structural influences in communication. By examining their distinctions and overlaps, this discussion emphasizes the complementary strengths of narrative and discourse analysis in addressing complex questions about human communication and meaning. These methodologies provide a robust framework for analyzing how meaning is negotiated and shared, contributing to a deeper understanding of contemporary societal and scholarly challenges [5]. Their enduring relevance underscores their utility in bridging theoretical exploration and practical application, offering critical insights into how people and communities navigate an increasingly interconnected and dynamic world. The qualitative research approach, known as narrative analysis, investigates the tales that individuals can generate to comprehend and make sense of the experiences they have lived through [6].

- *What are Narrative and Discourse Analysis, and How Are They Applied?*

Narratives—ordered recounts of events grouped chronologically and thematically—offer a structured lens through which human acts, identities, and relationships may be comprehended. It is essential to note that narratives reflect and actively influence reality, illuminating the process by which identity, agency, and cultural norms are produced. Through narrative analysis, researchers can unearth more profound meanings, values, and social frameworks concealed inside stories. This methodology shows how knowledge and experience are negotiated via social interaction and storytelling [7]. On the other hand, discourse analysis focuses on systematically studying language usage within the context of its more extensive social, cultural, and institutional settings [8]. To investigate spoken, written, and multimodal texts, this technique goes beyond the study of sentences to explore how language forms and reflects power relations, ideologies, and social practices. At the same time as it examines the structural components of language, such as grammar and lexicon, discourse analysis also investigates more major societal issues, such as power dynamics and institutional effects—this type of analysis functions on both the micro and the macro levels [9]. The scope of its application encompasses a wide range of resources, including ordinary conversations and policy texts, as well as discourse in the media and communication inside organizations. Research on building design can benefit from the practical application of narrative and discourse analysis. For example, narrative analysis may be used to investigate interviews with architects, engineers, or tenants. This can show how individuals interpret their experiences, values, and problems about environmentally responsible building practices and energy efficiency. Researchers can learn about cost, comfort, and environmental responsibility by examining personal accounts [10]. On the other hand, discourse analysis may be used to investigate policy papers, literature from the media, or architectural reports to discover prevailing ideologies, power relations, and institutional discourses influencing the adoption of passive building practices. For example, researchers may investigate how governmental laws or industry publications explain sustainability goals, affecting views and driving trends in the construction sector.

- *What is the Development and Importance of Narrative and Discourse Analysis in Modern Research?*

The historical evolution of story and discourse analysis highlights their emergence as independent but interconnected tools for investigating communication, meaning construction, and social processes [11]. These frameworks have developed across several fields, becoming vital instruments for tackling intricate modern challenges. Narrative analysis originated from early twentieth-century anthropology, psychology, and literary research. Innovative scholars like Vladimir Propp and Claude Lévi-Strauss established the groundwork by recognizing repetitive patterns and structural roles in myths, folktales, and cultural narratives [12]. Propp's formalist theory of narrative structure and Lévi-Strauss's structuralist analysis of myths provide essential insights into how tales influence human comprehension and societal unity. In the 1970s and 1980s, narrative analysis saw a resurgence due to the contributions of researchers like William Labov and Jerome Bruner, who highlighted the significance of tales in daily life, identity development, and cognitive functions [13]. Bruner illustrated that narratives are fundamental to forming personal and societal importance, connecting individual experiences with broader social settings. Narrative analysis in the current study has expanded beyond its literary and psychological origins to include sociology, education, healthcare, and anthropology [14]. This has emerged as an essential approach for examining identity development, trauma rehabilitation, and societal transformation. Individual and communal narratives provide crucial insights into how people and groups comprehend daily events, navigate cultural conventions, and address socio-political issues. Narrative analysis has been vital in examining disadvantaged groups, illuminating their challenges and resilience in contexts characterized by systematic injustices. Concurrently, discourse analysis arose in the mid-twentieth century, significantly shaped by the linguistic ideas of Ferdinand de Saussure and Noam Chomsky [15]. Early structuralist linguistics concentrated on language as a system of signs, but discourse analysis marked a substantial transition by integrating interdisciplinary insights from sociology, philosophy, and cultural studies. Academics like Michel Foucault and Norman Fairclough contributed to developing discourse analysis to investigate the complex interconnections among language, power, and ideology [16]. Foucault specifically examined how discourses shape knowledge, govern behavior, and perpetuate power relations, uncovering the inherent ideological roles of language. Fairclough's critical discourse analysis similarly emphasizes how language sustains social injustices while facilitating avenues for resistance and reform. In the twenty-first century, discourse analysis has become an essential instrument for comprehending the function of language in media, politics, and organizational communication. It allows academics to examine how prevailing narratives either uphold or contest societal norms, revealing the processes by which language affects identity, power dynamics, and cultural hegemony [17]. This analytical framework is beneficial for scrutinizing media representation, political discourse, and institutional communication, offering refined perspectives on how language influences public perception and collective behavior. The importance of these approaches in modern research is their capacity to tackle complex societal problems. Narrative analysis emphasizes lived experiences, highlighting the perspectives of people and communities confronting structural obstacles and cultural changes [18]. Discourse analysis concurrently scrutinizes the overarching socio-cultural narratives and power dynamics that influence these encounters. Collectively, these methodologies provide synergistic insights that enhance our comprehension of communication, identity, and meaning construction in a more linked world. Researchers can examine the relationship between personal narratives and dominant discourses by integrating story and discourse analysis [19]. This comprehensive framework facilitates the analysis of the interplay between individual narratives and societal structures, yielding an in-depth comprehension of human communication and its influence on knowledge, behavior, and social transformation. Global issues necessitate sophisticated, multidisciplinary strategies, highlighting the ongoing significance of these tools in promoting essential and transformational research.

- *How Can Narrative and Discourse Analysis Enhance Engineering Research?*

Despite contrasting methodologies and areas of emphasis, narrative, and discourse analysis have considerable similarities that augment their significance in modern studies, particularly engineering studies [20]. Both techniques provide unique yet complimentary viewpoints on the creation and conveyance of meaning, rendering them advantageous across many disciplines. Narrative analysis investigates human experiences via tales or accounts, emphasizing the structure, substance, and themes included in narratives [21]. Rooted in narrative theory and psychology, it emphasizes the significance of storytelling in forming identity and comprehending human experience. Narrative analysis in engineering research can examine how practitioners articulate problem-solving processes or significant events that impact their design decisions, uncovering the human and experiential aspects of engineering practice. Conversely, discourse analysis examines language utilization across contexts and levels, from individual statements to institutional documents [22]. Grounded in post-structuralism and critical theory, it highlights the interaction of language, power, and ideology as essential components for comprehending engineering communication. Discourse analysis frequently examines technical papers, standards, or organizational communications, investigating how language transmits and shapes knowledge, especially within institutional and professional contexts [23]. Notwithstanding their methodological disparities, both theories acknowledge the significance of context in influencing interpretation. Analyzing an engineer's account of a failed design iteration and examining discussions on safety rules recognizes that language is intricately woven within cultural, historical, and professional contexts. A common theoretical foundation is their comprehension of meaning as generated rather than inherent. Both emphasize how language concurrently mirrors and influences human understanding, actively participating in the negotiation of meaning [24]. For example, discourse analysis may elucidate the framing of sustainability within engineering design regulations, whereas narrative analysis can disclose how engineers express their contributions to sustainability objectives. When integrated, these methodologies offer a more thorough understanding of engineering communication and meaning-making processes. Research on extensive infrastructure projects may utilize narrative analysis to understand engineers' viewpoints on cooperation difficulties [25]. In contrast, discourse analysis might investigate how project documentation articulates matters like risk management and stakeholder involvement. Through the integration of story and discourse analysis, researchers may understand how language, context, and experience collaboratively shape communication practices in engineering disciplines [26]. Table 1 discusses the similarities and differences between narrative Analysis and Discourse Analysis, focusing on their definitions, theoretical underpinnings, and applications, notably in architectural study. Personal narratives are the primary subject of narrative analysis, which investigates the construction of identities and the experiences of individuals. Analysis of themes, structure, and the flow of narrative are some of the tools that it employs. For example, it may investigate the inspirations of architects or their views on design issues. On the other hand, discourse analysis examines the context in which language is used within larger systems, therefore revealing ideologies and power relations. Critical theory is the foundation for this approach, which analyzes systemic communication by analyzing professional writings such as sustainability reports or urban planning plans. Both methods provide complementary insights: personal narratives emphasize creative qualities, whereas discourse discloses the effect of institutional factors.

Narrative and discourse analysis effectively examine the intricacies of language and meaning [34]. Their unique yet intersecting techniques provide complementary insights essential to engineering research. By analyzing the personal narratives created by engineers and the overarching discourses influencing technical and organizational contexts, academics can attain a more profound comprehension of professional practices, communication dynamics, and the significance of language in fostering creativity [35]. This dual emphasis guarantees the sustained relevance of both methodologies in the ever-changing realm of engineering research. Figure 1 delineates the principal distinctions and intersections between story analysis and discourse analysis. Narrative analysis examines personal accounts, scrutinizing their structure and content to comprehend experiences and identity development. Conversely, discourse analysis prioritizes examining language within social, cultural, and institutional frameworks, revealing underlying ideologies and power structures. The overlap underscores their mutual emphasis on context and meaning-making, crucial for comprehending communication. Both techniques are utilized in research to investigate how individuals and groups construct, negotiate, and disseminate meaning, providing alternative perspectives on human experiences and societal frameworks.

Table 1 Narrative analysis versus discourse analysis in architectural research

Aspect	Narrative analysis	Discourse analysis	Example in architectural research (Narrative Analysis)	Example in architectural research (Discourse Analysis)
Definition [3]	It focuses on personal stories and experiences, analyzing their structure and content	Examines language use in broader contexts, focusing on power, ideology, and systems	Analyzing architects' personal stories about design inspirations	Studying language in architectural codes or sustainability reports
Theoretical foundations [27]	Based on narrative theory and psychology	Rooted in poststructuralism and critical theory	Exploring how architects' experiences shape their creative process	Investigating language in urban planning policies to uncover hidden ideologies
Primary focus [28]	Individual experiences and identity formation	Patterns of language use across professional and institutional settings	Interviews with architects reflecting on challenges in specific projects	Analyzing architectural critiques in professional journals
Scope [29]	Individual and event-specific; focuses on unique interpretations of experiences	Broad and systemic; examines communication in cultural and institutional contexts	Studying how architects narrate cultural influences on their designs	Analyzing discourse in housing policies or zoning regulations
Key techniques [30]	Analyzes story structure, themes, and narrative flow	Examines language patterns, syntax, and vocabulary	Exploring case studies where architects narrate their storytelling through space	Studying rhetoric in reports on sustainable urban development
Context of application [31]	Focuses on personal, historical, and cultural contexts that influence individuals	Emphasizes how language reflects and shapes broader systems and ideologies	Analyzing architects' reflections on designing post-war rebuilding projects	Examining government discourse promoting architectural mega-projects

(continued)

Table 1 (continued)

Aspect	Narrative analysis	Discourse analysis	Example in architectural research (Narrative Analysis)	Example in architectural research (Discourse Analysis)
Relevance [32]	Provides insight into personal and creative dimensions of architectural practice	Highlights power dynamics and ideologies within professional communication	Studying how architects' experiences with clients influence final designs	Analyzing competition jury reports to identify biases toward design trends
Complementary Use [33]	Captures human perspectives and reflective storytelling	Analyzes broader systemic language and institutional frameworks	Understanding the personal inspirations behind iconic cultural spaces	Investigating discourse in heritage preservation policies to assess values

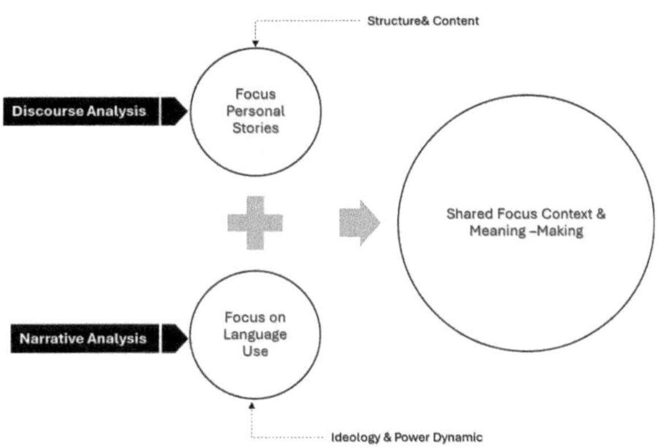

Fig. 1 Narrative analysis versus discourse analysis: emphasis, intersections, and research applications. *Source* Authors

2 Theoretical Foundations

Regarding the social sciences and the humanities, it is necessary to comprehensively investigate the theoretical frameworks responsible for shaping modern understandings of society, identity, and power systems [36]. Human behavior, cultural systems, and power dynamics may be investigated using these frameworks, which offer essential analytical tools [37]. These perspectives are shaped by foundational fields of study such as sociology, philosophy, and critical theory. Several prominent theoretical paradigms, including as structuralism, post-structuralism, and social constructionism, have developed as significant approaches. These paradigms give diverse perspectives on the construction of meaning, the development of identity, and the establishment of societal norms [38]. Several important academics, notably

Michel Foucault and Paul Ricoeur, have contributed substantially to advancing these conversations [39]. The discourse and power relations studies that Foucault provides shed light on the processes responsible for enforcing and resisting societal norms. At the same time, Ricoeur's research on narrative identity highlights the significance of language and storytelling in forming individual and social identities. These theoretical underpinnings offer a powerful lens to investigate the complex interactions between discourse, culture, and power [40]. Therefore, they make it possible to comprehend better the processes by which identities and social structures are generated, sustained, and modified.

2.1 Systems, Structures, and Meaning in the Context of Structuralism

The linguistic ideas of Ferdinand de Saussure served as the foundation for the development of the key philosophical paradigm known as structuralism, which arose in the middle of the twentieth century [41]. Structuralism is based on the notion that meaning is not intrinsic to individual components but rather something that emerges from the interactions of those components within a broader system [42]. The distinction Saussure makes between the signifier, the material form, such as a sound or picture, and the signified, the conceptual notion it represents, illustrates how meaning is formed relationally. The meaning of a sign is not determined by its inherent characteristics; instead, it is determined by its location within the linguistic structure and how it contrasts with other signals [43]. This relational framework emphasizes language as a system of differences in which meaning is derived via opposition rather than isolated elements. By studying myths, family systems, and cultural practices, Claude Lévi-Strauss expanded the scope of structuralist theories beyond the realm of language and into anthropology. He hypothesized that human societies are guided by profound, unconscious structures that can mold ideas and actions. Lévi-Strauss revealed that shared, underlying logic lurks behind the surface of cultural variety by revealing universal patterns in myths and rituals that people from different community's experience [44]. This suggests that human societies work according to systematic principles of acting instinctively, structuring meaning, and social interaction. These structures reflect universal cognitive processes, further implying human society's function. The fact that structuralism places such a strong emphasis on determinism and universalism, however, quickly prompted negative replies. Critics suggested that structuralism neglects the fluidity and contextuality of meaning, reducing human agency and historical uniqueness to inflexible structures [45]. Furthermore, the structuralist concept of permanent systems was not well-equipped to account for change, variety, and the instability of meaning. In response to these constraints, post-structuralism emerged as a theoretical framework to question and dismantle the assumptions underpinning structuralism. Philosophers such as Jacques Derrida and Michel Foucault questioned the stability of structures, emphasizing the contingent, dynamic, and discursive character of meaning [46]. This move signified a transition from structuralist universalism to a view of meaning as continually negotiated, unstable, and influenced by circumstances loaded with power.

2.2 *Deconstructing Stability and Authority Through the Lens of Post-Structuralism*

It was a critical response to the apparent rigidity of structuralism that gave rise to post-structuralism, which challenged the assumptions of fixed meanings and universal structures made by structuralism [47]. Some of the most influential intellectuals in history, like Jacques Derrida, Michel Foucault, and Julia Kristeva, have underlined that meaning is neither fixed nor objective but somewhat fluid, reliant on context, and constantly changing [48]. Post-structuralism emphasizes the situational and created character of language, power, and identity by arguing against the determinism associated with structuralism. Jacques Derrida developed the notion of deconstruction, which lies at the heart of post-structuralism. This concept reveals the inconsistency of meaning that exists inside language. In his work, Derrida presents the idea of difference, which describes how an infinite interaction of differences continually delays meaning. According to his argument, language does not have an intrinsic meaning; instead, it acts as a relational system in which meanings are contingent and unstable. Contradictions, hierarchies, and exclusions inherent in texts are brought to light through deconstruction, undermining claims of a fixed or universal reality. Through rethinking power and discourse, Michel Foucault contributed to expanding post-structuralist philosophy. Foucault criticizes the conventional conceptions of power as hierarchical or centralized in his essential writings, such as Discipline and Punishment and The History of Sexuality [49]. He contends that power is not centralized but distributed through networks ingrained in institutions, behaviors, and knowledge systems. According to Foucault, power is exercised through discourse, consisting of language and knowledge systems that generate "truths," legitimate acts, and establish identities to construct identities. Through examining how discourses influence societal norms and power relations, Foucault shows the processes responsible for maintaining and normalizing authority. Julia Kristeva contributes to further developing post-structuralist concepts by investigating the function of language and subjectivity [50]. Her work emphasizes the dynamic interaction between the semiotic (pre-linguistic impulses) and the symbolic (organized language systems), destabilizing fixed subject positions and portraying identity as fluid and fractured [51]. As a result, post-structuralism challenges the concepts of stability, authority, and objectivity that are the foundation of dominant narratives and thinking systems. By illuminating the constructed and contingent nature of meaning, power, and identity, it provides essential tools for challenging oppressive systems, challenging societal norms, and encouraging a more excellent knowledge of the cultural and social processes at play. For example, Sustainable building design reflects post-structuralist thought by questioning conventional, rigid definitions of architecture and stability. Instead of fixed meanings, sustainability emphasizes adaptability, context-specific solutions, and decentralized power dynamics in material choices and energy systems. Buildings are reimagined as dynamic, responsive entities shaped by cultural, environmental, and social discourses.

2.3 Narrative Identity Refers to Storytelling as a Method of Individual Creation

The notion of narrative identity, defined by philosopher Paul Ricoeur, highlights the relevance of storytelling in building personal and social identities [52]. According to Ricoeur, humans generate meaning and continuity in their lives by arranging events into coherent narratives, which he calls employment [53]. This narrative framework reconciles fragmented experiences, providing a fuller knowledge of oneself and one's role in the world. Ricoeur's perspective extends beyond individual identities, stressing communal narratives created by cultural, historical, and institutional causes.

For example, in the context of sustainable design, a study on the community of Vauban, an eco-district in Freiburg, Germany, reveals how architectural spaces contribute to narrative identity. Residents typically explain their decision to settle in Vauban as part of a more significant tale of environmental action and devotion to ecological principles [54]. These narratives are not only reflections of the physical architecture of the district—which stresses energy efficiency, renewable resources, and walkability but active constructions of identity that link personal beliefs with social aims. By integrating sustainability into their daily experiences, people establish and reaffirm their identities as environmental stewards, ultimately contributing to a collective narrative that presents the district as a global model for sustainable living.

Further, Ricoeur's theory provides a lens for understanding the function of storytelling in reconciling historical and current narratives within architectural design [55]. For instance, in reconstructing Copenhagen's Nordhavn area, narratives of industrial legacy were carefully conserved and incorporated into a vision of sustainable urban development [56]. The design integrates reused materials and industrial elements while infusing green technology and public areas that enable community engagement. This mix of historical continuity and forward-looking sustainability allows communities and stakeholders to develop a standard narrative that honors the past while embracing creative, eco-conscious possibilities [57]. Moreover, narrative identity offers a vital framework for understanding how excluded cultures employ storytelling to exert agency within dominant architectural paradigms. For example, research on indigenous-led sustainable architecture efforts, such as the award-winning Cree Cultural Institute in Quebec, reveals how storytelling may challenge traditional design conventions [58]. By incorporating indigenous cultural tales into the architecture through symbolic shapes, material choices, and spatial arrangements, the design expresses a communal identity that resists assimilation and upholds cultural sovereignty. Through these places, communities can develop powerful narratives and honor heritage while addressing modern environmental issues. Using Ricoeur's idea of narrative identity, sustainable architecture emerges as a functional discipline and a medium for building tales that reflect and shape individual and collective goals [59]. This is because the concept of narrative identity is applied to sustainable architecture. This approach aligns design methods with ethical imperatives, promoting inclusion, sustainability, and cultural resilience.

3 Methodological Approaches

Methodological narrative and discourse analysis approaches demand a rigorous and systematic process that integrates effective data collection strategies, comprehensive analytical frameworks, advanced techniques, and an acute awareness of ethical complexities [60]. The success of such research begins with the strategic collection of data tailored to the aims and context of the study. This involves employing diverse methods, such as in-depth interviews, participant observation, and the gathering of textual or multimodal artifacts, to elicit detailed and authentic narratives [61]. Open-ended, probing questions are critical in fostering an environment where participants can articulate their experiences and perspectives. Simultaneously, field observations and the analysis of naturally occurring discourse provide a robust foundation for interpreting how language operates within specific social and cultural contexts. The analytical phase of narrative and discourse research relies on applying comprehensive frameworks designed to systematize the interpretation of data [62]. Thematic analysis, widely used in qualitative research, identifies patterns and themes within narratives, enabling researchers to distill the complexity of individual stories into coherent, overarching insights. On the other hand, critical discourse analysis (CDA) offers a powerful tool for interrogating the underlying power dynamics, ideologies, and social structures embedded within language use [63]. Both approaches necessitate an iterative engagement process with the data, where interpretations are continuously refined and validated to balance depth and reliability. Such frameworks demand sensitivity to the interplay between micro-level linguistic features and the broader macro-level sociocultural dynamics that shape discourse. Researchers increasingly rely on advanced tools and techniques to enhance the precision and depth of analysis. Coding, whether conducted manually or with software like NVivo or Atlas's, provides an essential foundation for organizing data into meaningful categories [64]. Through iterative coding cycles, researchers can identify patterns and relationships that might remain obscured. Narrative mapping serves as another critical technique, offering a visual representation of temporal, spatial, and relational dimensions within narratives, which helps to uncover underlying trajectories and thematic interconnections. Intertextual analysis further augments the analytical process by examining how texts interact and influence one another, revealing how narratives and discourses are constructed, contested, and transformed across different contexts [65].

For example, in research on smart houses, interviews with residents can yield narratives about their experiences of living with automated systems, emphasizing themes such as convenience, control, or privacy concerns. Narrative mapping might visually chart the interactions between household members and smart devices over time, revealing patterns of adaptation or resistance to technology. Ethical considerations are integral to narrative and discourse research, as the interpretive nature of this work often involves sensitive topics and vulnerable populations. Researchers must ensure that participants' consent is informed and voluntary, and confidentiality must be upheld to protect individuals from harm or exposure. The subjective and interpretive dimensions of narrative and discourse analysis also require researchers to adopt a reflexive stance, continually examining their own biases and positionality to safeguard the authenticity and integrity of the research [66]. Furthermore, the representation of participants' voices in scholarly work raises important questions about power dynamics, necessitating a conscientious effort to amplify, rather than distort or diminish, the narratives of those studied [67]. The methodological rigor required for narrative and discourse analysis encompasses strategic data collection, robust analytical frameworks, the integration of sophisticated techniques, and a vigilant adherence to ethical principles. By employing these interconnected strategies, researchers can produce nuanced, meaningful, and ethically sound analyses that contribute significantly to understanding language, identity, and social structures within their fields [68]. The essential components of qualitative research approaches are outlined in Table 2, emphasizing the importance of methodological rigor and ethical integrity. In-depth interviews and participant observations are two examples of the various methods used in data gathering. These methods are adapted to the study's aims, with informed consent and confidentiality paramount. Techniques for conducting interviews include asking questions that are both open-ended and in-depth to elicit complex responses in a safe setting. Field observations record naturally occurring social interactions to ensure that non-intrusive techniques are followed. Data interpretation that is both systematic and clear may be made easier using analytical frameworks such as theme analysis and Critical Discourse Analysis (CDA). Using software such as NVivo, organized data organization, and visualization may be accomplished through coding and narrative mapping tools. Reflexivity highlights the significance of conducting an in-depth analysis of researchers' biases. Throughout the research process, specific ethical guidelines should be followed to guarantee that the participants' voices are accurately reflected.

Table 2 Key research methodologies, analytical techniques, and ethical considerations

Aspect	Description	Examples/Techniques	Ethical considerations
Data collection [69]	Employ diverse methods tailored to research aims and context, fostering authentic narratives	In-depth interviews, participant observation, textual/multimodal artifact collection	Ensure informed consent, voluntary participation, and confidentiality to protect individuals from harm or exposure
Interview strategies [70]	Uses open-ended, probing questions to elicit detailed experiences and perspectives	Example: Residents discussing smart house experiences emphasize convenience, control, and privacy concerns	Create a safe environment encouraging honest communication
Field observations [71]	Observes naturally occurring discourse within specific social and cultural contexts	Analyzing interactions between household members and automated systems	Respect privacy and avoid intrusive observation methods
Analytical frameworks [72]	Systematic tools to interpret data with precision and depth	Thematic analysis, Critical Discourse Analysis (CDA)	Maintain transparency in analytical procedures to avoid misrepresentation
Thematic analysis [73]	Identifies patterns and overarching themes within narratives	Coding cycles (manual or software-aided), e.g., NVivo or Atlas.ti	Be reflexive about researcher biases influencing theme identification
Critical discourse analysis [74]	Explores power dynamics, ideologies, and social structures in language use	Examine macro-level sociocultural dynamics alongside micro-level linguistic features	Ensure critical perspectives do not overshadow participants' authentic narratives
Coding [75]	Organizes data into meaningful categories for analysis	Iterative coding (manual/software-aided)	Ensure anonymity in coded data to protect participant identity
Narrative mapping [76]	Visually represents temporal, spatial, and relational dimensions within narratives	Example: Charting adaptation/resistance patterns to smart technology among households	Avoid misrepresentation by cross-validating mapped narratives with participant feedback
Intertextual analysis [77]	Examines interactions and influences between texts to reveal discourse construction	Identifies contested or transformative narrative elements across contexts	Acknowledge the origins and context of intertextual references to respect intellectual contributions
Advanced techniques [78]	Enhances precision and depth in analysis	Use software tools (NVivo, Atlas.ti) for narrative mapping	Ensure tools are ethically deployed, preserving participant integrity

(continued)

Table 2 (continued)

Aspect	Description	Examples/Techniques	Ethical considerations
Reflexivity [79]	Researchers continually examine their biases and positionality	Critical reflection journals, team discussions on positionality	Addresses how researcher identity and context influence data interpretation
Participant voice [80]	Amplifies authentic narratives, avoiding distortion or diminishment	Direct quotations and co-constructed narratives	Strive for accurate representation, allowing participants to review portrayals of their input
Ethical adherence [81]	Ensures research integrity through strategic, sensitive approaches	Voluntary consent processes, safeguarding vulnerable populations, robust confidentiality protocols	Continuously revisit ethical principles throughout all stages of research to adapt to emergent challenges

4 Reflections and Future Directions

4.1 Emerging Trends and Innovations in Narrative and Discourse Analysis

In narrative and discourse analysis, emerging trends and cutting-edge innovations are discussed [82]. The academic field of story and discourse analysis is undergoing a continual revolution, pushed by developments in theoretical paradigms and technical improvements. The rising use of multimodal narratives, which go beyond the conventional text-based formats and incorporate visual, audio, and digital media, is a noteworthy development in recent years [83]. Researchers can now record more complex and richer tales reflecting modern communication practices, thanks to the trend toward multimodality. In addition, techniques that involve student participation and co-construction are becoming increasingly popular. These methods emphasize cooperation between researchers and participants in interpreting narratives [84]. This helps to ensure that various voices and experiences are accurately represented. Alterations in narrative analysis are also being brought about by technological advancements, notably in artificial intelligence (AI) and machine learning (ML) [85]. Utilizing these technologies makes it possible to handle enormous amounts of information, recognize patterns, and conduct complex language analysis. For example, natural language processing (NLP) techniques can identify subtle linguistic characteristics and discourse patterns across large corpora [86]. This, in turn, makes it possible to conduct both scalable and complex studies. In addition, interactive platforms and digital storytelling tools are altering the process of collecting, analyzing, and presenting narratives, increasing accessibility and engagement [87]. Within the context of contemporary narrative analysis, Table 3 emphasizes significant characteristics. Theoretical paradigms investigate multimodal storytelling using artificial intelligence and machine learning to evaluate layered tales. However, dangers such as oversimplification need the use of culturally adaptable tools. Participant co-construction emphasizes collaborative storytelling through interactive platforms to achieve balanced computational support without losing authenticity. Natural language processing technologies make real-time theme analysis possible, which can recognize complicated discourse patterns in computational pattern recognition. Nevertheless, ethical considerations are raised when algorithms have prejudices. To assure ethical, egalitarian, and context-aware narrative insights, future directions will emphasize the establishment of inclusive platforms for co-creation and the enhancement of the interpretability of artificial intelligence. Technology, ethics, and narrative innovation all work together to create a synergy reflected in these developments.

Table 3 Key aspects and innovations in narrative analysis

Aspect	Key concepts	Notable innovations	Challenges and ethical considerations	Future directions
Advancements in theoretical paradigms [83]	Exploration of multimodal narratives integrating text, visuals, audio, and digital platforms	AI and ML applications capable of analyzing complex, multilayered narratives at scale	Risks of over-simplification or inadequate contextual representation	Development of adaptive AI/ML frameworks to ensure culturally sensitive and context-aware analysis
Participant Co-construction [88]	Emphasis on the role of participants in shaping narratives collaboratively	Enhanced interactive platforms enabling co-creation of stories and participatory digital tools	Balancing computational capabilities with participant autonomy and authenticity	Designing systems that ensure equitable representation while promoting inclusivity in narrative construction
Computational pattern recognition [89]	Utilization of NLP techniques to detect nuanced discourse patterns across vast textual corpora	Sophisticated tools for real-time discourse mapping and thematic analysis	Ethical concerns about biases inherent in computational tools impacting narrative insights	Increasing the interpretability and transparency of AI-driven tools to enhance their reliability and ethical use

4.2 Challenges and Ethical Dimensions in Computational Narrative Analysis

The area of story and discourse analysis continues to have considerable limitations despite general developments in the subject [1]. One of the most urgent difficulties is the limits of computational tools, namely the tendency of artificial intelligence (AI) and machine learning (ML) algorithms to oversimplify or misunderstand complicated narratives. This is one of the most critical challenges [90]. Even though these technologies improve efficiency and scalability, they frequently need more contextual awareness for nuanced analysis. For example, algorithms trained on enormous datasets may concentrate on surface-level patterns. Still, they may need help to grasp subtleties such as cultural nuances, subtext, or non-linear storytelling structures. Because of this constraint, it is necessary to find a way to strike a balance between the capabilities of computers and the interpretive skills of human analysts [91]. A more holistic view of narratives may be fostered by successfully integrating different techniques, which can reduce the possibility of reductionist interpretations. The ethical considerations in narrative analysis present an additional obstacle, particularly considering the growing dependence on digital technology for data collection. Regarding privacy, permission, and data ownership, online platforms and social media narratives present substantial ethical difficulties [92]. For example, evaluating personal tales published on social media without the proper controls may unwittingly breach users' privacy or misappropriate the material they have provided. Considering this, researchers must implement stringent ethical frameworks to address the issues present in electronic contexts [93]. The collection of informed permission, anonymizing sensitive data, and guaranteeing transparency throughout the study process should all be contained within such frameworks. A strategy that considers several dimensions and combines technical innovation with ethical and methodological rigor is necessary to navigate these challenges successfully [94]. Researchers can more effectively harness computational technologies by addressing these limits, which allows them to maintain the integrity and depth of narrative inquiry [95]. Because of this, the interaction between the effectiveness of machines and the insights of humans, supported by ethical concerns, is essential for advancing the area in a responsible and potentially significant manner. The application of multidisciplinary approaches has substantial prospects for methodological progress in narrative and discourse analysis. Utilizing ideas from sociology, anthropology, linguistics, and computer science facilitates the creation of complete frameworks for narrative investigation [96]. Combining qualitative and quantitative paradigms can be accomplished by merging computational methods, such as natural language processing (NLP), with ethnographic techniques [97]. This combination improves analytical scalability while maintaining the necessary contextual complexity for nuanced interpretation. In civil engineering research, this multidisciplinary integration has demonstrated significant advantages. Studies on project management narratives frequently utilize ethnographic approaches to document team dynamics and decision-making processes while employing computational tools to evaluate communication patterns within extensive datasets, including email exchanges and meeting transcripts. Researchers can discern underlying project success or failure tendencies through sentiment analysis and keyword extraction. Discourse analysis in public consultation for infrastructure projects integrates language analysis with computer methods to evaluate community sentiment and involvement, providing essential insights for stakeholder management [98]. This multidisciplinary method connects human interpretive abilities with computational efficiency, guaranteeing the capture of intricate details and overarching patterns in tales. Thus, these techniques promote a deeper and more comprehensive knowledge of speech in civil engineering situations. An Analysis of the Revolutionary Effects Technology and Digital Tools Have Had on the Field [99]. Narrative and discourse analysis have been dramatically reimagined due to the digital revolution, introducing contemporary tools and approaches that challenge conventional practices. Software such as NVivo, Atlas.ti, and MAXQDA supports systematic arrangement and analysis of qualitative data [100]. This software enables researchers to manage vast and complicated datasets with precision. Additionally, these technologies make it possible to do complex coding, theme analysis, and visualization approaches, which enables the discovery of detailed patterns within tales [101]. Table 4 examines some of the most important advances and problems in story analysis. Computational limits emphasize the difficulty of integrating artificial intelligence and machine learning with human-centric thinking. This challenge can be overcome by utilizing hybrid systems that combine machine learning with qualitative tools. Misinterpretation of cultural subtleties continues to be a significant obstacle, which calls for more nuanced forms of communication. Privacy and anonymization are concerned with ethical issues associated with sensitive narrative data, and recent developments in anonymization techniques aim to prevent data misuse. Contextual depth highlights the necessity of preserving narrative richness through creative strategies that balance scalability and complexity. Future directions advocate for solutions involving multiple disciplines and robust frameworks to ensure ethical, thorough, and context-aware narrative analysis.

4.3 *Digital Platforms and Immersive Technologies in Narrative Analysis*

The study of modern cultural, political, and social discourses has been revolutionized due to the growth of social media and internet platforms as dynamic stores of narrative material [105]. These platforms give researchers unrivaled access to constantly shifting narratives in real-time. This allows researchers to investigate the dynamic relationship between individual expressions and collective voices. Social media platforms enable rapidly disseminating and amplifying personal narratives, shaping collective perspectives, and encouraging dynamic conversations. On the other hand, studying online narratives calls for a critical and reflective methodology

Table 4 Navigating challenges in narrative analysis

Aspect	Key concepts	Notable innovations	Challenges and ethical considerations	Future directions
Computational limitations [102]	Difficulty in integrating AI/ML with human-centric interpretations for a holistic understanding of narratives	Innovative hybrid systems combining machine learning with qualitative analysis tools	Misinterpretation of cultural nuances and subtexts by algorithmic models	Refining hybrid approaches that bridge computational and qualitative methods for comprehensive narrative analysis
Privacy and anonymization [103]	Concerns over narratives derived from sensitive data or social media interactions	Advanced anonymization techniques and ethical data governance protocols	Potential misuse or unauthorized dissemination of personal data	Formulating robust legal and ethical frameworks for the responsible use of narrative data
Contextual depth [104]	Ensuring that computational tools retain the rich, contextual depth of narratives	Techniques for balancing efficiency with the retention of narrative intricacies	Challenges in scaling nuanced analysis to larger datasets without loss of depth	Promoting multidisciplinary approaches to balance scalability and depth in narrative analysis

because of these tales' fluid and context-dependent character and the ethical difficulties they provide. Rigorous ethical frameworks must ensure that research techniques are accountable and transparent [106]. Key considerations include privacy, permission, and ownership of data. The possibilities for narrative exploration are further expanded by the emergence of immersive technologies such as augmented reality (AR) and virtual reality (VR). These technologies enable lifelike and interactive settings, facilitating a more profound interaction with narrative data [107]. For instance, virtual reality (VR) can rebuild historical or cultural narratives, providing users with an immersive platform on which they may relive and reflect on experiences from the past. The preservation of cultural heritage, cultivating empathy, and improving educational endeavors are all areas in which such applications are particularly worthwhile [108]. By recreating events or rebuilding environments that would otherwise be unavailable, augmented reality and virtual reality also make it possible for academics to investigate intricate social dynamics. When these technical breakthroughs are paired with more conventional methods, they make it possible for researchers to break through the limitations traditionally established in narrative analysis. Scholars can obtain a more nuanced and comprehensive comprehension of narratives by integrating immersive technology with existing qualitative and quantitative research approaches [109]. Through the utilization of this multidisciplinary strategy, the gap between technological advancement and cultural sensitivity is bridged. This technique guarantees that narrative inquiry will continue to be ethically grounded while also using the full potential of current technologies. In the end, this synthesis contributes to the advancement of the study of narratives by providing novel avenues for comprehending human experiences in the era of digital technology [110]. Table 5 examines recent developments in storytelling, with particular emphasis on real-time data, immersive methodologies, and cultural legacy preservation. The real-time narrative data is magnified by dynamic platforms such as social media, which enables the transmission of information on a broad scale but raises ethical issues regarding the possibility of wrong depiction. Future technologies will validate and contextualize these tales, which is the goal. To rebuild historical and cultural tales, immersive storytelling includes augmented and virtual reality (AR/VR), even though accessibility and resource demands present problems. Efforts should be made to integrate these technologies with more conventional approaches while democratizing them. Cultural preservation makes use of immersive tools to protect history. However, there is a possibility that these tools might be appropriate. Approaches involving collaboration with local communities guarantee the sharing of cultural tales genuinely and ethically.

Table 5 Innovations and challenges in storytelling technologies

Aspect	Key concepts	Notable innovations	Challenges and ethical considerations	Future directions
Real-time narrative Data [111]	Dynamic online platforms amplify individual stories into collective narratives	Social media as a catalyst for large-scale narrative dissemination and interaction	Ethical implications of context-dependent narratives and misrepresentation	Creation of tools to validate and contextualize narratives shared on dynamic platforms
Immersive storytelling [112]	Augmenting traditional methods with AR/VR for lifelike interaction and cultural storytelling	Use of AR/VR for reconstructing historical and cultural narratives	High resource requirements and accessibility issues for immersive technologies	Democratizing access to immersive tools and combining them with traditional storytelling methods for broader applications
Cultural preservation [113]	Immersive technologies as tools for safeguarding and disseminating cultural heritage	Applications in education and historical reconstructions	Risks of cultural appropriation or misrepresentation in narrative recreation	Collaborative efforts with local communities to ensure authenticity and respect in cultural narrative preservation

4.4 Advancing Narrative Analysis Through Multidisciplinary Collaboration

It is necessary to have robust collaboration across disciplines to advance narrative and discourse analysis [1]. This kind of collaboration enables academics to tackle complex topics using novel methodologies. Scholars may develop cutting-edge strategies and procedures by drawing on insights from various areas. Artificial intelligence (AI) and machine learning (ML) algorithms to analyze large-scale narrative datasets might be improved, for instance, through partnerships with computer scientists [114]. While this happens, working with psychologists can enhance our knowledge of human storytelling by allowing us to investigate cognitive and emotional elements that are included inside narratives. When story analysis is combined with other fields of study, such as public health, education, and policy studies, it opens fascinating possibilities for research that will have a significant impact [115]. Trans narratives play a vital role in creating public views and influencing decision-making. When embedded narratives are analyzed, key themes may be uncovered, pressing societal concerns can be addressed, and findings can guide evidence-based policy-making. Increasing the scope of study to encompass a broader range of cultural and linguistic situations also provides a one-of-a-kind chance for comparative analysis [116]. These kinds of research have the potential to reveal universal narrative patterns in addition to culturally particular variances, promoting global intercultural understanding and aiding the creation of relevant theoretical frameworks across cultures. The development of story analysis will be influenced by such factors as the incorporation of developing technology, viewpoints from several disciplines, and ethical issues. Even though technological constraints and ethical conundrums continue to exist, they present opportunities for invention and refinement [117]. To expand the frontiers of story inquiry and provide significant insights into the complicated links between language, culture, and society, researchers who are willing to face these difficulties have the potential to push the bounds of narrative inquiry. The field's capacity to remain methodologically rigorous and ethically upright while adapting to ever-changing communicative settings is essential to the field's future. Ultimately, the continuous development of story and discourse analysis will be contingent on its ability to balance technical innovation and a dedication to solving social concerns [118]. This will ensure that tale and discourse analysis remain relevant in an always-evolving world. The ever-evolving approaches being utilized in story studies are highlighted in Table 6. Even though ethical issues around data management continue to exist, interdisciplinary integration encourages cooperation across sectors, with collaborations in artificial intelligence and machine learning pushing scalable solutions. An important goal for the future is to increase the amount of cultural and linguistic variety in research. Even though technological obstacles are involved in combining these paradigms, ethnographic methodologies combine natural language processing (NLP) with ethnography to provide frameworks that merge qualitative and quantitative methods. Continued investigation of mixed-method techniques is necessary to develop more comprehensive applications. The practical applications

Table 6 Innovative approaches to narrative research

Aspect	Key concepts	Notable innovations	Challenges and ethical considerations	Future directions
Interdisciplinary integration [119]	Collaboration across disciplines enhances methodological rigor in narrative studies	Partnerships with computer scientists for scalable AI/ML solutions and innovations	Addressing ethical dilemmas arising from cross-disciplinary data handling and interpretations	Expanding research to encompass broader cultural and linguistic diversity for more inclusive insights
Ethnographic techniques [97]	Integration of NLP methods with ethnography for in-depth narrative analysis	New frameworks blend qualitative and quantitative approaches for holistic insights	Technical limitations in combining distinct paradigms into cohesive methodologies	Continued exploration of mixed method approaches for comprehensive application in diverse fields, such as policymaking
Practical applications [120]	Insights are applied in public health, education, and policymaking to drive societal benefits	Development of user-friendly digital tools for direct application of narrative frameworks in various domains	Ensuring that methodologies remain adaptable and relevant across different sectors and audiences	Building translational frameworks to make narrative research accessible and actionable for real-world impact

illustrate the social influence of narrative research in fields such as public health and education. These applications highlight the necessity of adaptive methodologies and user-friendly technologies to improve accessibility and relevance in the real world.

5 Conclusion

Narrative and discourse analysis have become fundamental approaches in modern research, providing advanced frameworks for investigating meaning-making, communication, and social structuration processes. These techniques facilitate a thorough comprehension of the processes influencing human behavior, identity, and societal norms by connecting individual experiences with broader cultural, institutional, and systemic frameworks. This chapter has illustrated the significant capacity of these techniques to examine intricate inquiries about language, power, and agency across many academic contexts. Narrative analysis offers essential insights into how people and groups formulate and convey coherent narratives to comprehend their lived experiences. This method elucidates the mechanics of identity creation, memory production, and cultural expression by concentrating on the narrative's structural, thematic, and chronological components. It emphasizes the importance of storytelling in molding human comprehension, demonstrating how tales reflect and actively affect individual and societal reality. Narrative analysis is applicable across several sectors, including sociology, healthcare, engineering, and education. It highlights its versatility and effectiveness in examining identity, resilience, and the human aspects of professional and social activities. Conversely, discourse analysis broadens the analytical framework by contextualizing language within its extensive sociocultural and institutional environments. This technique rigorously analyzes how language embodies and sustains power dynamics, ideologies, and institutional conventions. Discourse analysis, by examining underlying communication processes, offers vital tools for deconstructing prevailing narratives, contesting institutional frameworks, and revealing mechanisms of control and resistance. Its contributions to media studies, political discourse, and organizational communication illustrate its effectiveness in examining and transforming public views and policy frameworks. Narrative and discourse analysis form a complementary and dynamic paradigm that unites human communication's micro and macro aspects. The interaction of these techniques enables a detailed examination of the connections between individual agency and structural limitations. By integrating insights from both methodologies, researchers can elucidate the reciprocal construction and influence of personal narratives and societal discourses, thereby thoroughly comprehending the processes of meaning negotiation, identity formation, and social change. The use of new technology has greatly enhanced the functionalities of these techniques. Advancements in artificial intelligence, natural language processing, and multimodal analysis have improved narrative and discourse research's scalability, accuracy, and profundity. These instruments enable the methodical study of large datasets, promote the detection of hidden patterns, and permit the evaluation of intricate multimodal texts. Simultaneously, innovative platforms like digital storytelling, augmented reality, and virtual reality have transformed narrative collection, analysis, and presentation methods, enhancing interaction with research subjects and viewers. Nonetheless, these technological improvements pose obstacles, especially in preserving contextual richness, the risk of algorithmic simplicity, and the ethical issues related to data privacy, consent, and intellectual property rights. This chapter emphasizes the importance of ethical rigor and methodological reflexivity in story and discourse analysis. Due to their interpretative and socially contextualized characteristics, these approaches require researchers to manage the intricacies of enhancing participant voices while preserving analytical impartiality. Reflexive techniques, transparency in analytical procedures, and iterative validation with participants are crucial for guaranteeing the credibility, integrity, and ethical soundness of study findings. The chapter emphasizes the importance of multidisciplinary techniques to augment these strategies, including ideas from sociology, linguistics, anthropology, and computational sciences. Narrative and discourse analysis has considerable potential for tackling global and multidisciplinary issues. Their focus on sustainability, public health, education, and social justice illustrates their potential to support evidence-based policymaking, promote intercultural understanding, and enhance equity-oriented research. Comparative research across cultural and linguistic settings can reveal universal patterns and culturally distinct storytelling practices, aiding in creating globally applicable theoretical frameworks and approaches. Narrative and discourse analysis are essential for progressing research in human communication, meaning construction, and social interaction. Their lasting significance stems from their capacity to connect human and structural aspects, amalgamate theoretical and practical knowledge, and adjust to changing technological and multidisciplinary environments. By persistently enhancing these techniques, researchers may intensify their involvement with critical social challenges, contribute to transformational scholarships, and provide practical insights that enlighten and influence the future of human knowledge and collaboration.

References

1. Johnstone B, Andrus J (2024) Discourse analysis. Wiley
2. Eiranen R, Hatavara M, Kivimäki V, Mäkelä M, Toivo RM (2022) Narrative and experience: interdisciplinary methodologies between history and narratology. Scand J Hist 47(1):1–15
3. De Fina A (2021) Doing narrative analysis from a narratives-as-practices perspective. Narrat Inq 31(1):49–71
4. Ariawan DA, Nurkamto J, Sumardi S (2022) Critical discourse analysis of cultural representations in EFL national commercial textbooks. Studies English Lang Educ 9(3):1112–1127
5. Jones SR, Torres V, Arminio J (2021) Negotiating the complexities of qualitative research in higher education: Essential elements and issues. Routledge
6. Nigar N (2020) Hermeneutic phenomenological narrative enquiry: a qualitative study design. Theory Pract Lang Studies 10(1):10–18
7. Jørgensen KM (2022) Storytelling, space and power: an Arendtian account of subjectivity in organizations. Organization 29(1):51–66
8. Uzokova S (2020) Principles of discourse analysis in linguistics. Журнал иностранных языков и лингвистики 1(1):7–12
9. Yahya NQ (2024) A critical discourse analysis of Iraqi EFL university students' performance at micro level. Midad Al-Adab Refereed J 1(36)
10. De Villiers C, Farooq MB, Molinari M (2022) Qualitative research interviews using online video technology–challenges and opportunities. Meditari Account Res 30(6):1764–1782
11. Catalano T, Waugh LR (2020) Critical discourse analysis, critical discourse studies and beyond, vol 11. Springer, Cham
12. Maiolani M (2023) Anthropology and modern Italian Literature: Italo Calvino, Primo Levi, Gianni Celati (Doctoral dissertation)
13. Swargiary K (2024) Language and learning: the crucial role of language in the teaching-learning process. Google
14. Blustein DL, Grzanka PR, Gordon M, Smith CM, Allan BA (2024) The psychology of precarity: a critical framework. American Psychologist
15. Yemelianova OV (2024) Theory of language discourse with the content module: composition of texts of different genres
16. Khan A (2024) The language of power: discourse analysis and the politics of representation. Liberal J Lang Literat Rev 2(01):22–31

17. Alvero AJ, Lee J, Regla-Vargas A, Kizilcec RF, Joachims T, Antonio AL (2024) Large language models, social demography, and hegemony: comparing authorship in human and synthetic text. J Big Data 11(1):138
18. Brotman S, Ferrer I, Koehn S (2020) Situating the life story narratives of aging immigrants within a structural context: The intersectional life course perspective as research praxis. Qual Res 20(4):465–484
19. McMahon R, Kaiser W (2022) Narrative Ju-jitsu: counter-narratives to European union. J Contemp Eur Stud 30(1):1–9
20. Sochacka NW, Walther J, Rich JR, Brewer MA (2021) A narrative analysis of stories told about engineering in the public discourse: Implications for equity and inclusion in engineering. Studies Eng Educ 2(2):54–77
21. Kiss D, Horváth Z, Kassai S, Gyarmathy AV, Rácz J (2022) Folktales of recovery–from addiction to becoming a helper: deep structures of life stories applying Propp's theory: a narrative analysis. J Psychoactive Drugs 54(4):328–339
22. Matthews A, Kotzee B (2022) Bundled or unbundled? A multi-text corpus-assisted discourse analysis of the relationship between teaching and research in UK universities. Br Edu Res J 48(3):578–597
23. Irwin LN, Posselt JR (2022) A critical discourse analysis of mainstream college student leadership development models. J Leadership Educ 21(4):1–22
24. Strobel M (2023) Collaborative dialogue and deliberative communication: reading circles with Young Adult novels and adolescent learners of English as a Second Language (Doctoral dissertation, University of Glasgow)
25. Sergeeva N, WInch GM (2020) Narrative interactions: how project-based firms respond to Government narratives of innovation. Int J Project Manage 38(6):379–387
26. Hennessy S (2024) Analysing educational dialogue around shared artefacts in technology-mediated contexts: a new coding framework. Classroom Discourse, pp 1–35
27. Piper A, So RJ, Bamman D (2021) Narrative theory for computational narrative understanding. In: Proceedings of the 2021 conference on empirical methods in natural language processing, pp 298–311
28. Tomlinson M, Jackson D (2021) Professional identity formation in contemporary higher education students. Stud High Educ 46(4):885–900
29. Su W, Junge S (2023) Unlocking the recipe for organizational resilience: a review and future research directions. Eur Manag J 41(6):1086–1105
30. Berhe A, Guinaudeau C, Barras C (2023) Survey on narrative structure: from linguistic theories to automatic extraction approaches. Revue TAL: traitement automatique des langues 63(1)
31. Causadias JM (2020) What is culture? Systems of people, places, and practices. Appl Dev Sci 24(4):310–322
32. Bhandari MP (2023) The fundamental principles of social sciences. Bus Ethics Leadership 7(2):73–86
33. Bentalha B, Alla L (2024) Revealing the subtleties: the art of qualitative studies in science and management. Applying qualitative research methods to management science, pp 1–21. IGI Global
34. Chavez JV, Del Prado RT (2023) Discourse analysis on online gender-based humor: markers of normalization, tolerance, and lens of inequality. In: Forum for linguistic studies (Transferred) (Vol 5, No 1, pp 55–71
35. Moezabadi DM (2024) Enhancing competence in civil engineering: assessment of international research experience (Doctoral dissertation, Arizona State University)
36. Sovacool BK, Hess DJ, Cantoni R (2021) Energy transitions from the cradle to the grave: a meta-theoretical framework integrating responsible innovation, social practices, and energy justice. Energy Res Soc Sci 75:102027
37. Siru C (2022) Societal shifts: exploring dynamics in sociology and anthropology. Int J Res Rev Techn 1(1):8–14
38. Hughes P (2020) Paradigms, methods and knowledge. In Doing early childhood research, pp 31–55. Routledge

39. Bowden B (2020) The intellectual origins of postmodernism. The Palgrave handbook of management history, pp 645–669
40. Ramasubramanian S, Banjo OO (2020) Critical media effects framework: bridging critical cultural communication and media effects through power, intersectionality, context, and agency. J Commun 70(3):379–400
41. Stawarska B (2020) Saussure's linguistics, structuralism, and phenomenology. Springer International Publishing
42. Tronca L, Forsé M (2022) Towards a sociology of reasonableness: structure and action in the structural interactionist approach. Italian Sociol Rev 12(3)
43. Barham L, Everett D (2021) Semiotics and the origin of language in the Lower Palaeolithic. J Archaeol Method Theory 28(2):535–579
44. Zerubavel E (2020) Generally speaking: an invitation to concept-driven sociology. Oxford University Press
45. Bouchard J, Bouchard J (2021) The structure–agency debate and its relevance to AL. Complexity, emergence, and causality in applied linguistics, pp 95–175
46. Eskandarnezhad O (2024) Deconstructivism and deconstruction: a critical re-evaluation of derrida through heidegger with regard to architecture
47. Woermann M (2016) Bridging complexity and post-structuralism. Switzerland: Springer International
48. Debnath R (2019) Revisiting the tale of Jacques Derridas deconstruction: critical study (Doctoral dissertation, University of North Bengal)
49. Morris A (2018) Watching our weight: an exploration into the politics of the body using intersections between post-structuralist feminism and Foucault's discipline and punish (Doctoral dissertation, University of Birmingham
50. Ghosh A (2021) Disciplinary power and practices of body politics: an evaluation of Dalit women in Bama's Sangati and P. Sivakami's The Grip of Change through Foucauldian discourse analysis. Huma Soc Sci Commun 8(1):1–9
51. Warner C (2024) Multiliteracy play: designs and desires in the second language classroom
52. Dohnalová M (2021) Ricoeur's concept of narrative identity (Doctoral dissertation, Masarykova univerzita, Filozofická fakulta)
53. Ricoeur P (1995) Figuring the sacred: religion, narrative, and imagination. Fortress Press
54. East M (2018) Rising cities: continuity, innovation and deliberation of Vauban District. Ecocycles 4(2):58–71
55. Faulder SJ (2024) Philosophical hermeneutics and the architecture of Álvaro Siza: meaning. Taylor & Francis, Action and Place
56. Journey RK (2021) Future perfect: experimental prototype city of tomorrow (Doctoral dissertation, The University of Chicago)
57. Alzaydi A (2024) Balancing creativity and longevity: the ambiguous role of obsolescence in product design. J Clean Prod 445:141239
58. Maguire M (1994) Cultural stances informing storytelling among bilingual children in Quebec. Comp Educ Rev 38(1):115–143
59. Pérez-Gómez A (2020) Narrative language, architecture and the city. IN_BO. Ricerche e progetti per il territorio, la città e l'architettura 11(15):8–15
60. Alejandro A, Zhao L (2024) Multi-method qualitative text and discourse analysis: a methodological framework. Qual Inq 30(6):461–473
61. Reavey P, Johnson K (2017) Visual approaches: using and interpreting images. The SAGE handbook of qualitative research in psychology, 354–373
62. Herman L, Vervaeck B (2019) Handbook of narrative analysis. U of Nebraska Press
63. Caballero-Mengíbar A (2015) Critical discourse analysis in the study of representation, identity politics and power relations: a multi-method approach. Communi Soc 28(2):39–54
64. Kalpokas N, Radivojevic I (2022) Bridging the gap between methodology and qualitative data analysis software: a practical guide for educators and qualitative researchers. Sociol Res Online 27(2):313–341

65. Souto-Manning M (2014) Critical narrative analysis: the essential interplay of discourse and narrative analyses. Int J Qual Stud Educ 27(2):159–180
66. Consoli S, Ganassin S (2023) Reflexivity in applied linguistics. Routledge
67. Owens J (2007) Liberating voices through narrative methods: the case for an interpretive research approach. Disability Soc 22(3):299–313
68. Cameron D, Frazer E, Harvey P, Rampton MBH, Richardson K (2018) Researching language: issues of power and method. Routledge
69. Onwuegbuzie AJ, Leech NL, Collins KM (2010) Innovative data collection strategies in qualitative research. Qual Report 15(3):696–726
70. Brédart A, Marrel A, Abetz-Webb L, Lasch K, Acquadro C (2014) Interviewing to develop Patient-Reported Outcome (PRO) measures for clinical research: eliciting patients' experience. Health Qual Life Outcomes 12:1–10
71. Denzin N, Lincoln Y (2025) Ethnography and participant observation. Qual Res Methods Media Studies, 181
72. Ncube MM, Ngulube P (2024) Enhancing environmental decision-making: a systematic review of data analytics applications in monitoring and management. Discover Sustain 5(1):290
73. Grubert E (2020) Socialized misinformation: questionable Information and the Influence on entities a qualitative exploratory case study (Doctoral dissertation, University of Phoenix).
74. Butt SK, Ahmad S, Jahan M (2024) Language power, and gender politics: a critical discourse analysis of woman empowerment slogans as tools for social change. Liberal J Lang Literat Rev 2(4):157–173
75. Auerbach C, Silverstein LB (2003) Qualitative data: An introduction to coding and analysis (Vol 21). NYU Press
76. Wainer LS (2022) The informalization of formal housing projects in the global south: policy failure or counterhegemonic city-making? (Doctoral dissertation, Massachusetts Institute of Technology
77. Lupu I, Sandu R (2017) Intertextuality in corporate narratives: a discursive analysis of a contested privatization. Account Auditing Account J 30(3):534–564
78. Ronzani CM, da Costa PR, da Silva LF, Pigola A, de Paiva EM (2020) Qualitative methods of analysis: an example of Atlas. TI™ Software usage. Revista Gestão & Tecnologia (4):284–311
79. Corlett S, Mavin S (2018). eflexivity and researcher positionality. The SAGE handbook of qualitative business and management research methods, 377–399
80. Kennelly IB (2017) Voice Matters: Narratives and perspectives on voice in academic writing (Doctoral dissertation, University of Sheffield)
81. Findley MG, Ghosn F, Lowe SJ (2024) Vulnerability in research ethics: a call for assessing vulnerability and implementing protections. Proc Natl Acad Sci 121(34):e2322821121
82. Dolgopolovas V, Dagiene V (2021) On the future of computational thinking education: moving beyond the digital agenda, a discourse analysis perspective. Sustainability 13(24):13848
83. Al-Saadawi HFT, Das B, Das R (2024) A systematic review of trimodal affective computing approaches: text, audio, and visual integration in emotion recognition and sentiment analysis. Expert Syst Appl, 124852
84. Spector-Mersel G (2010) Narrative research: time for a paradigm. Narrat Inq 20(1):204–224
85. Rane NL, Paramesha M, Choudhary SP, Rane J (2024) Artificial intelligence, machine learning, and deep learning for advanced business strategies: a review. Partners Univ Int Innov J 2(3):147–171
86. Hirschberg J, Manning CD (2015) Advances in natural language processing. Science 349(6245):261–266
87. Alexander B (2017) The new digital storytelling: creating narratives with new media--revised and updated edition. Bloomsbury Publishing USA
88. Hill A (2019) Power to the people: responsible facilitation in co-creative story-making
89. O'Neill L, Anantharama N, Buntine W, Angus SD (2021) Quantitative discourse analysis at Scale—AI, NLP and the transformer revolution (No. 2021–12). Monash University, SoDa Laboratories

90. Shandilya SK, Datta A, Kartik Y, Nagar A (2024) Role of artificial intelligence and machine learning. In: Digital resilience: navigating disruption and safeguarding data privacy, pp 313–399. Cham: Springer Nature Switzerland
91. Rockwell G, Sinclair S (2022) Hermeneutica: computer-assisted interpretation in the humanities. mit Press
92. Taylor J, Pagliari C (2018) Mining social media data: how are research sponsors and researchers addressing the ethical challenges? Research Ethics 14(2):1–39
93. Bente BE, Van Dongen A, Verdaasdonk R, van Gemert-Pijnen L (2024) EHealth implementation in Europe: a scoping review on legal, ethical, financial, and technological aspects. Frontiers in Digital Health 6:1332707
94. Sovacool BK, Axsen J, Sorrell S (2018) Promoting novelty, rigor, and style in energy social science: towards codes of practice for appropriate methods and research design. Energy Res Soc Scie 45:12–42
95. Bryda G, Costa AP (2023) Qualitative research in digital era: innovations, methodologies and collaborations. Social Sciences 12(10):570
96. Kim JH (2015) Understanding narrative inquiry: the crafting and analysis of stories as research. Sage Publications
97. Li Z, Abramson CM (2024) Ethnography and machine learning: synergies and new directions. arXiv preprint arXiv:2412.06087
98. Verma S (2022) Sentiment analysis of public services for smart society: literature review and future research directions. Gov Inf Q 39(3):101708
99. Yu Y, Jiang Y (2024) Understanding the digital revolution and its influences. Springer Books
100. Sampson C, Wong LS (2023) Using qualitative data analysis software to help explore critical research questions: a tool, not a replacement. In: Handbook of critical education research, pp 637–654. Routledge
101. Staley DJ (2015) Computers, visualization, and history: how new technology will transform our understanding of the past. Routledge
102. Kore A (2022) Designing human-centric AI experiences. Apress, Berkeley
103. Saxena AK (2022) Enhancing data anonymization: A semantic K-anonymity framework with ML and NLP integration. Sage Sci Rev Appl Mach Learn 5(2):81–92
104. He Y, Cao S, Shi Y, Chen Q, Xu K, Cao N (2024) Leveraging large models for crafting narrative visualization: a survey. arXiv preprint arXiv:2401.14010
105. KhosraviNik M, Unger JW (2016) Critical discourse studies and social media: power, resistance and critique in changing media ecologies. Methods Critical Discourse Studies 3:205–233
106. Slade S, Prinsloo P (2013) Learning analytics: ethical issues and dilemmas. Am Behav Sci 57(10):1510–1529
107. Asif M, Mondal A, Soumil S, Das A, Sahoo P (2024) Augmented reality and virtual reality in education: a transformative journey into immersive learning environments. Adv Comput Solutions, 185
108. Cooper B (2011) Empathy in education: engagement, values and achievement. Bloomsbury Publishing
109. Lazou C, Tsinakos A (2023) Critical immersive-triggered literacy as a key component for inclusive digital education. Educ Sci 13(7):696
110. Dincelli E, Yayla A (2022) Immersive virtual reality in the age of the Metaverse: a hybrid-narrative review based on the technology affordance perspective. J Strateg Inf Syst 31(2):101717
111. Maghrabi RO (2017) Online social systems, social actions, and politics: a narrative analysis of the role of social media in revolutionary political change. The University of North Carolina at Greensboro
112. Boboc RG, Băutu E, Gîrbacia F, Popovici N, Popovici DM (2022) Augmented reality in cultural heritage: an overview of the last decade of applications. Appl Sci 12(19):9859
113. Oladokun BD, Ajani YA, Ukaegbu BC, Oloniruha EA (2024) Cultural preservation through immersive technology: the metaverse as a pathway to the past. Preservation, Digital Technol Culture 53(3):157–164

114. Nguyen G, Dlugolinsky S, Bobák M, Tran V, López García Á, Heredia I, Malík P, Hluchý L (2019) Machine learning and deep learning frameworks and libraries for large-scale data mining: a survey. Artif Intell Rev 52:77–124
115. Tracy SJ (2024) Qualitative research methods: Collecting evidence, crafting analysis, communicating impact. Wiley
116. Cooper H (2015) Research synthesis and meta-analysis: A step-by-step approach (Vol 2). Sage Publications
117. Van de Poel I, Royakkers L (2023) Ethics, technology, and engineering: an introduction. Wiley
118. De Fina A, Georgakopoulou A (2011) Analyzing narrative: discourse and sociolinguistic perspectives. Cambridge University Press
119. Ejjami R, Integrative literature review 5.0: leveraging AI and emerging technologies to redefine Academic Research
120. Misuraca G, Broster D, Centeno C (2012) Digital Europe 2030: designing scenarios for ICT in future governance and policy making. Gov Inf Q 29:S121–S131

Qualitative Data Collection and Management

Abstract Deep insight into human experiences, social relationships, and institutional systems, qualitative research is a fundamental methodological framework spanning many academic fields. Emphasizing their theoretical roots, methodological rigor, and ethical implications, this chapter comprehensively examines qualitative approaches. Emphasizing structured, semi-structured, unstructured interviews; focus groups; participant and non-participant observations; and document and archive analysis methodically addresses fundamental qualitative data collecting techniques. Regarding their epistemological coherence, pragmatic relevance, and impact on research validity, every method is evaluated severely. Methodological alignment—ensuring data collecting techniques complement the general research design is under great focus. The conversation covers digital technologies, including NVivo, ATLAS.ti, and MAXQDA, for qualitative data analysis, transcribing techniques, and theme coding schemes (open, axial, and selective). It also covers the challenges of data long-term preservation and metadata management. The chapter emphasizes ethical responsibilities of qualitative research, including participant confidentiality, informed permission, adherence to Institutional Review Boards (IRB), ethical data-sharing methods, and methodological problems. Using triangulation, member checking, audit trails, and reflexivity, strategies for creating trustworthiness—credibility, transferability, dependability, and confirmability are exhaustively examined. Scholars trying to improve their qualitative studies' methodological rigor, openness, and ethical integrity will find great value in this chapter. It offers theoretical depth and pragmatic direction for doing intense, morally conscious, epistemologically sound qualitative research.

1 Introduction

Academic research in many disciplines, including engineering, education, healthcare, and transdisciplinary studies, depends critically on qualitative research. Originally in the social sciences and, more recently, engineering, Santiago-Delefosse et al. emphasize the significant expansion of qualitative research approaches over the past thirty

A. Almusaed et al., *Practice of Research Methodology in Civil Engineering and Architecture*, https://doi.org/10.1007/978-3-031-97393-2_13

years [1]. This rise has made thorough review of the quality standards applied to evaluate qualitative research necessary, leading to different assessment policies. These suggestions, however, expose notable differences in structure and vocabulary, which complicates the standardization of qualitative research evaluation [2]. Academic study in many different disciplines depends critically on qualitative research. Originally in the social sciences and more lately in the health sciences, Santiago-Delefosse et al. underline the notable expansion of qualitative research approaches over the past thirty years. This rise has made a thorough review of the quality standards applied to assess qualitative research necessary, leading to the formulation of many evaluation policies. However, these suggestions show quite different structural structures and vocabulary, making it difficult to standardize qualitative research evaluation [3]. Emphasizing meaning-making, subjectivity, and the subtleties of lived events, qualitative research is essentially interpretive. This compares the statistical inference and the numerical data emphasis of quantitative methods. In qualitative research, methodical data collecting, organization, and data management define the integrity and rigor of the study [4]. This guarantees the results are epistemologically coherent, ethically reasonable, and methodologically strong. Therefore, qualitative research's legitimacy and academic worth depend on operationally necessary approaches for gathering and evaluating qualitative data. Esteve and Lee [5] claim that just a tiny fraction of scholarly journal papers in this subject center on qualitative research. Among several strategies, case studies are critical. Moreover, the way methodological decisions are reported varies greatly [5]. The research design's ontological and epistemological paradigms should coincide with the choice of suitable qualitative data collecting techniques, guaranteeing theoretical coherence and methodological rigor. Asrifan et al. [6] underline the need to thoroughly examine qualitative methods like participant observation, focus groups, and interviews. These techniques record rich, analytically complicated, contextually anchored data [6]. The multiple qualitative research approaches—grounded theory, phenomenology, ethnography, and case study research—demand different data collecting techniques. Grounded theory derives theoretical formulations from empirical facts by means of repeated data collecting and continuous comparison. According to Hallberg [7], Glaser and Strauss's "The Discovery of Grounded Theory" (1967) changed qualitative research by bringing methodical methodologies, questioning quantitative methods, and so validating qualitative inquiry as a scientific tool. Previously, qualitative techniques were mostly taught orally and seen as a forerunner of statistical research [7]. Conversely, phenomenology research stresses looking at lived experiences using in-depth interviews and thorough descriptive narratives. Extended field observations are quite important for ethnographic study. On the other hand, case study research builds a whole understanding of a certain event using a range of data sources [8]. Methodological coherence is therefore crucial to ensure that the data collecting processes follow the general study paradigm. One of the main ways to get qualitative data, generally seen as one of the most effective techniques for getting rich and complex insights, is interviews. Three forms may be used to classify them: organized, semi-organized, and unstructured. Every one of these approaches aims at a particular kind of study. Using a pre-made set of questions, structured interviews assure consistency in the data-gathering procedures and simplify comparative analysis [9]. Semi-structured interviews allow one to keep a guiding framework while enabling more freedom to research and analyze new subjects [10]. Conversely, unstructured interviews use a more exploratory and open-ended methodology. Based on their personal experiences and points of view, this will let participants help to guide the direction of the discourse [11]. The effectiveness of interviews in qualitative research depends on the researcher's ability to build rapport with the respondents, lower the possibility of answer biases, and regularly engage in reflexivity all through the data-collecting process [12]. Focus groups are a useful instrument in addition to interviews for compiling data on the group opinions and social interactions. Focus group participants are urged to connect, allowing researchers to probe subjects including consensus, divergence, and how people create meaning via social interaction [13]. Nonetheless, the validity of the results from the focus group mostly relies on the moderator's capacity to control the group dynamics, avoid prominent voices from overshadowing others, and provide an environment suitable for honest and open communication. Since it is not always feasible to ensure participants' confidentiality in a public debating environment, ethical problems are highly relevant in research employing focus groups [14].

Observational methods are another important element of qualitative data collection, especially in ethnographic and case study research [15]. Two kinds of observations are participant and non-participant. Whereas the latter consists of maintaining an external, detached perspective [16], the former category entails the researcher actively engaging in the area. Direct experience of social events made possible by participant observation helps the researcher to better understand the surroundings of such occurrences. Still, this approach raises moral questions concerning the researcher's influence and the likelihood of conflicts of interest. Although non-participative observation lowers the possibility of researcher tampering, this approach might limit access to insider knowledge. To guarantee the trustworthiness of data acquired via surveillance, meticulous documentation—including field notes, reflective notebooks, and, if allowed, audio-visual recordings—is very vital [17]. Policy documents, institutional reports, historical records, and current texts may all be studied very well using documentation and archive analysis and main data collecting techniques. Archival and document analysis are two further approaches of gathering data. Using textual artifacts as main data sources, this approach is useful in policy evaluation, historical research, and discussion analysis. Unlike observations and interviews, document analysis does not call for direct human interaction [17]. Some ethical questions, including informed permission, become easier. Simultaneously, one should consider the validity and reliability of the results by considering the sources' legitimacy, the context's interpretation, and the data triangulation. Managing qualitative data helps to guarantee the integrity, accessibility, and safety of research conclusions [18]. Though gathering qualitative data is a necessary part of research, properly handling such material is also important. Transcription comes first in starting the data organizing process. This process turns observations and recorded interviews into written form for study. Researchers have to decide between automated transcription methods, which provide efficiency but usually need post-processing techniques for error correction, and manual transcribing, which assures a better degree of correctness but takes a lot of time [19]. After transcription, qualitative data is coded a rigorous process wherein textual information is organized into topics of relevance. Basic concepts are identified using open coding; axial coding is used to create theme links; and selective coding is utilized to improve fundamental theoretical structures. The analytical framework applied determines the variation in coding practices. Recent developments in digital technology have fundamentally changed qualitative data management by enabling researchers to effectively store, code, and retrieve [20] specialized software like NVivo, ATLAS.ti, and MAXQDA. These instruments improve research accuracy by helping to organize representation, integrate multimedia elements, and execute sophisticated analytical searches. However, software learning curves, data security issues, and over-reliance on automation require careful thought. Ethical issues remain a priority in qualitative data management, particularly about anonymity, safe storage, and adherence to Institutional Review Board (IRB) policies [21]. Researchers must use anonymization, encrypted storage, and limited access among other protections. Policies for long-term preservation and ethical data-sharing also must balance participant privacy and openness. In qualitative research, establishing credibility calls for following exacting methodological guidelines including credibility, transferability, dependability, and confirmability [22]. Triangulation, member screening, and extended participation with study subjects all help to increase credibility. While dependability is guaranteed by extensive documentation and audit trails [23], detailed contextual descriptions improve transferability. Reflexivity and open presentation of researcher positionality help to establish confirmability. The process of qualitative data collecting and management calls for both ethical integrity and strategic decision-making. Collection methods have to fit the theoretical and epistemological framework of the study if empirical data fairly depicts human experiences [24]. Methodological rigor, ethical data processing, and safe storage help to maintain study credibility and participant anonymity. Researchers must stay flexible while upholding academic rigidity as qualitative approaches change with technological developments and multidisciplinary integration [25]. Figure 1 illustrates a structured approach to qualitative data collection and management, emphasizing key methodological aspects such as interviews, focus groups, and observational techniques, along with essential data management considerations, including technological innovations, ethical safeguards, and research integrity.

2 Key Qualitative Data Collection Methods

Instead of quantitative measurement, qualitative research uses comprehensive, contextualized descriptions to explain occurrences. Despite its quantitative roots, Gelo et al. [26] suggest that engineering has increasingly used qualitative methods in recent decades. Qualitative research is increasingly valued for capturing the intricacies of human, social, and contextual elements in engineering investigations. The developing epistemological landscape of engineering research has led to a dispute

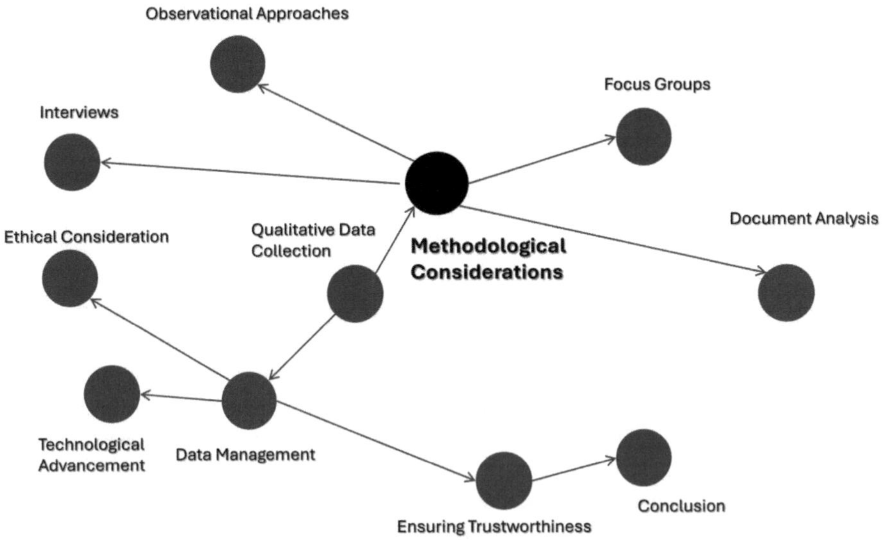

Fig. 1 Streamlined framework for qualitative data acquisition and administration. *Source* Authors

about the relative benefits and application of qualitative and quantitative methods [26]. An essential qualitative research component is the data-gathering process, which utilizes several approaches to provide comprehensive insights [27]. These methodologies enable researchers to investigate human experiences, behaviors, and interactions across varied contexts, offering significant contextual and interpretive richness to study outcomes [28]. Qualitative data-gathering methods are extensively employed in social sciences, education, health studies, and business research to investigate intricate situations when quantitative methods may prove inadequate [29]. These methodologies allow researchers to capture subjective viewpoints, social interactions, and cultural factors influencing individuals' experiences. Furthermore, qualitative research facilitates the exploration of new themes and unforeseen patterns that may not be evident in organized numerical data [30]. The principal qualitative data-gathering methods encompass interviews, focus groups, observations, document analysis, and comprehensive discussions of their methodologies, benefits, and obstacles. Every approach possesses distinct advantages and drawbacks, and the choice of the most suitable methods is contingent upon the research objectives, ethical issues, and the characteristics of the study population [31]. Utilizing a mixture of these methodologies, called triangulation, can augment the credibility and depth of qualitative research outcomes. For example, many qualitative research methodologies are employed in passive building research to explore human interactions with energy-efficient structures [32]. Research studies utilize interviews and observational methods to gather data on occupant experiences, adaptation behaviors, and perceptions of comfort in passive constructions. Ethnographic research findings indicate that occupants adjust ventilation, shading, and insulation to achieve the ideal interior environment. These qualitative insights enhance the quantitative energy performance data, improving our understanding of user behavior, design effectiveness, and long-term sustainability in passive building innovations.

2.1 Interviews

Interviews are a typical qualitative data collection approach. Researchers can learn more about individuals' ideas, feelings, and viewpoints through these interviews. Campion et al. [33] advocate a rigorous six-step employment interview method for reliability and fairness. Interview questions are designed after a thorough investigation to guarantee relevancy. Second, all applicants get standardized questions for homogeneity. Third, rating scales specify objective scoring. Fourth, interview panels record and analyze replies. Fifth, all candidates proceed similarly. Finally, job relevance, fairness, and testing criteria are underlined for optimal hiring and evaluation [34]. They offer insightful perspectives into the subjective perceptions, decision-making processes, and experiences that people have lived through. Interviews can be conducted in various formats, including structured, semi-structured, and unstructured, depending on the degree of structure and flexibility involved [35]. Structured interviews use a predetermined list of questions, ensuring all participants are treated similarly. It is possible to be flexible while conducting semi-structured interviews, which enables a more in-depth investigation of developing topics [36]. Unlike organized interviews, unstructured interviews are open-ended, allowing free-flowing conversations that reveal complex opinions. To guarantee that the data you collect is reliable, an effective interview needs attentive listening, reflexivity, and ethical concerns [37].

2.1.1 Structured Interviews

Through a prepared set of standardized questions, structured interviews ensure that all interviews are conducted consistently and uniformly [33]. When a systematic comparison of responses is necessary, this methodological approach is especially valuable in research. Structured interviews enhance similar data collection and increase reliability by reducing questions' variability. Campion et al. [38] concluded that interviews could be refined by integrating various structural components. They further suggested that strengthening this widely used selection technique should be one of the most significant goals for future research and practice [38]. Consequently, structured interviews are well-suited for large-scale research, surveys, and evaluations. However, one significant drawback of structured interviews is their inherent limitations, which can prevent respondents from providing more detail in their answers [34]. As a result, this method may not capture subtle insights, complex views, or emerging themes that less organized formats might reveal. Unlike semi-structured or unstructured interviews, structured interviews do not allow for spontaneous follow-up questions, which may limit the qualitative data gathered [39]. Despite these limitations, structured interviews remain a popular research method, particularly in fields where collecting objective and repeatable data is crucial for ensuring the validity and reliability of empirical studies [31].

2.1.2 Semi-structured Interviews

When it comes to qualitative research, semi-structured interviews are beneficial because they balance consistency and flexibility in methodology [36]. Interviewers can explore emerging topics based on participant responses while adhering to a predetermined set of key questions prepared in advance [40]. This flexibility allows for a more in-depth exploration of complex social phenomena compared to structured interviews, leading to more prosperous and comprehensive data collection. Additionally, researchers can clarify ambiguous responses and delve deeper into insights through semi-structured interviews, which helps enhance the validity of the findings [41]. Due to their ability to accommodate diverse perspectives while maintaining a structured framework, they are widely used in social sciences, psychology, and organizational research. For example, unstructured interviews have been utilized in passive building research to explore architects' and engineers' perspectives on sustainable design [32]. Researchers investigating passive heating and cooling systems encountered unexpected challenges, including regulatory constraints, client hesitance, and material limitations, which impede energy-efficient building solutions during open-ended discussions.

2.1.3 Unstructured Interviews

Because they are open-ended, unstructured interviews allow for a lively and impromptu conversation between the interviewer and the interviewee. Although much of the literature on interviews and selection procedures sees unstructured interviews as significantly less accurate and reliable than structured interviews, Chauhan [42] notes that a close examination demonstrates their apparent benefits. Unstructured interviews help capture subtle insights, enabling researchers to examine complicated viewpoints and experiences despite worries about impartiality and consistency. Unstructured interviews are a potent technique where rich, exploratory data is crucial because of their flexibility, deepening qualitative research [42]. This method is effective for gaining deep insights into personal experiences and perspectives. Unlike structured ones, the flexibility offered by unstructured interviews allows interviewers to explore unexpected topics [43]. However, the diversity of responses can make analysis challenging. Identifying similar themes demands careful interpretation since everyone may provide different answers [34]. In qualitative research, this approach is especially valuable when it's essential to grasp feelings, motivations, and opinions unique to each person. Despite their challenges, unstructured interviews remain a powerful method for gathering rich and detailed data across various research subjects [42].

2.1.4 Designing Effective Interview Guides

Developing a well-structured guide for conducting interviews is essential to ensure the success of an interview-based study [44]. Maintaining coherence, facilitating the gathering of valuable data, and increasing the depth of replies are all benefits of a guide that has been thoughtfully constructed. It is vital to establish specific objectives to align questions with the study's objectives. The use of open-ended questions enables participants to provide detailed comments while retaining neutrality and helps reduce the amount of prejudice the interviewer may have [45]. A logical succession of questions is essential to increase participant involvement and ensure a natural flow. Using follow-up questions also makes conducting a more in-depth investigation of the primary topics possible. Researchers can systematically gather high-quality qualitative data by following these criteria, which enhances their study's reliability and validity [46].

Table 1 presents a comparative overview of structured, semi-structured, and unstructured interviews. This overview highlights each interview type's advantages, disadvantages, and possibilities. Structured interviews ensure consistency but lack sufficient depth, while semi-structured interviews offer flexibility but demand skilled execution. Although time-consuming, unstructured interviews provide a wealth of qualitative data. Each approach offers unique research opportunities.

Table 1 Interview types—benefits, challenges, and opportunities

Interview type	Benefits	Challenges	Opportunities
Structured [47]	Ensures consistency, facilitates comparison and reduces interviewer bias	Limited depth; it may not capture nuanced insights	Valuable for extensive studies that require standardized responses needed
Semi-structured [36]	It provides flexibility and encourages a more profound exploration of emerging topics	It requires skilled interviewers, and balancing structure with flexibility can be difficult	Can adjust to new information and refine research direction in real-time
Unstructured [47]	Encourages open conversation and captures complex opinions and experiences	It is time-consuming and more challenging to analyze due to unstructured responses	Offers valuable qualitative insights and a deeper understanding of participants' perspectives

2.2 Focus Groups

A qualitative research approach known as focus groups is a method in which a moderate conversation about a particular subject is held with a few people carefully selected for participation [48]. Researchers can investigate collective insights, social interactions, and the formation of shared meanings through this technique, which emphasizes group dynamics to elicit varied views [49]. The interactive nature of focus groups encourages a more profound level of participation by allowing participants to improve and build upon each other's points of view [50]. This results in generating rich qualitative data that is contextually nuanced and crucial for thorough analysis.

2.2.1 Group Dynamics

When leading successful focus groups, a solid understanding of group dynamics is essential [51]. The interactions among participants can enhance conversations, leading to the generation of diverse ideas; however, they can also contribute to conformity bias. While the moderator ensures that more dominant voices do not overshadow others and actively encourages quieter individuals to engage, they play a vital role in maintaining a balanced contribution [52]. Effectively managing these dynamics fosters richer discourse, yielding more comprehensive and credible qualitative data. The fundamental elements of focus group dynamics are depicted in Fig. 2, emphasizing essential aspects such as promoting a diversity of viewpoints, avoiding conformity bias, controlling dominating voices, guaranteeing balanced participation, and encouraging involvement from more reserved people. The quality of discussions is improved, and the validity of qualitative data gathering is reinforced when these dynamics are moderated effectively.

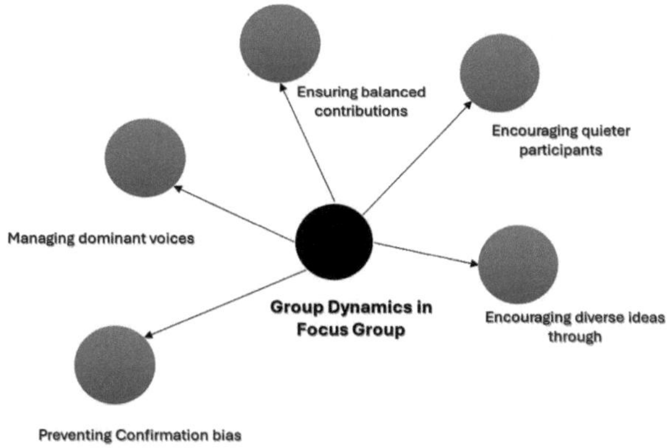

Fig. 2 Group dynamics in focus groups. *Source* Authors

2.2.2 Moderator Roles

The moderator is crucial in guiding discussions within the focus group to ensure systematic and relevant data gathering [48]. One of the moderator's essential duties is ensuring that conversations align with the study's goals while fostering a polite, welcoming atmosphere that encourages candid and meaningful participation. By actively managing group dynamics, making sure no one voice dominates the conversation, and allowing everyone the opportunity to engage, moderators play a crucial role in maintaining a balance in participation. This requires effective facilitation strategies, such as refocusing discussions, posing follow-up questions, and encouraging more reserved participants to share their thoughts. A good moderator must also be skilled in resolving conflicts positively. When handled appropriately, disagreements and differing viewpoints can enrich the conversation by introducing new perspectives and promoting critical thinking skills. Instead of stifling disagreements, the moderator should promote civil discourse, guiding participants toward shared interests or, if necessary, acknowledging opposing views in a manner that fosters critical thought. In addition to resolving conflicts, moderators also need to incorporate new ideas to ensure cohesion and thematic clarity in the discussion's conclusions. They enhance the analytical rigor of the discussion by emphasizing key concepts, identifying trends, and linking participants' contributions to broader study issues. Ultimately, an effective moderator cultivates an environment where conversations are both lively and methodologically sound, ensuring that qualitative data collection remains robust, organized, and ethically conducted [53]. The primary duties of a moderator in promoting productive conversations are graphically depicted in Fig. 3. The central node, "Moderator Roles," is divided into five crucial responsibilities: promoting civil discourse, staying focused, enticing involvement, handling disagreements, and summarizing essential ideas. Qualitative data collection is organized and fruitful thanks to these interrelated functions.

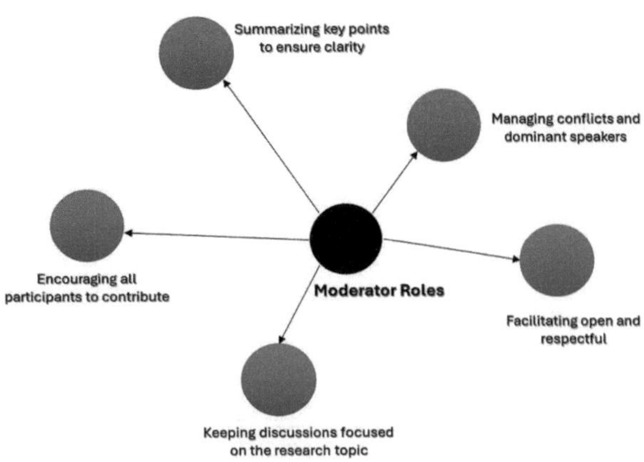

Fig. 3 Moderator roles in focus groups. *Source* Authors

2.2.3 Ethical Considerations in Collective Discussions

Focus groups must be conducted with ethical concerns to ensure participants' safety and welfare [54]. Obtaining informed consent is crucial as it provides participants with an understanding of the study's purpose and that their participation is voluntary. Despite the group setting, confidentiality must be maintained, requiring measures to prevent the disclosure of sensitive information. Moderators should also exercise caution when discussing sensitive topics to minimize potential anxiety and create a safe environment [55]. Addressing power imbalances ensures everyone can express their opinions without fearing retaliation. Adhering to ethical guidelines enhances qualitative research processes' legitimacy, integrity, and moral rigor [18]. Figure 4 visually represents the fundamental ethical concepts that direct talks inside focus groups. To protect participants and preserve ethical integrity, it is vital to ensure that informed permission is obtained, that confidentiality is maintained, that sensitive themes are managed and that power imbalances are avoided. These considerations boost the dependability of qualitative data, enhancing the legitimacy of the research and promoting trust among participants.

2.3 Observations

Observation is a fundamental qualitative research method to analyze behaviors, relationships, and the effects of context in natural settings [56]. This technique enables researchers to collect real-time data without solely relying on self-reported information, providing direct insights into social dynamics. There are two distinct types of observation: participant observation, where the researcher engages in the environment, and non-participant observation, where the researcher observes without direct involvement [57]. Non-participant observation ensures minimal disturbance, while participant observation allows for deeper engagement. Each approach offers its advantages. Researchers can enhance the validity and reliability of qualitative data

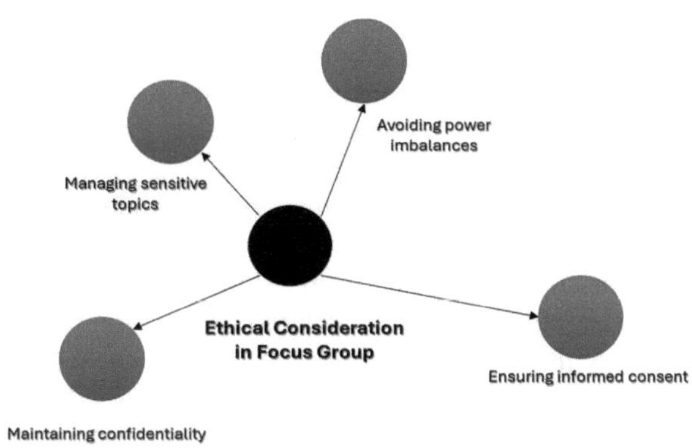

Fig. 4 Ethical considerations in focus groups. *Source* Authors

collection and analysis by systematically documenting the actions and contextual factors they observe.

2.3.1 Participant Observation

Participant observation is a qualitative research methodology in which the researcher immerses themselves in the study setting [58]. This immersive approach provides an insider perspective, offering deep insights into social dynamics, behaviors, and cultural contexts. Researchers can uncover complex nuances that alternative methods might overlook by directly engaging individuals [15]. However, ethical challenges arise, including the researcher's potential impact on the environment and the difficulty of maintaining impartiality. To mitigate bias, reflexivity, and methodological rigor are essential. Despite these challenges, participant observation is crucial for collecting rich, qualitative social and behavioral research data [59].

2.3.2 Non-participant Observation

Non-participant observation is a qualitative research methodology where the researcher observes a group or environment without direct involvement [57]. This detached approach reduces researcher bias, enabling a more impartial examination of behaviors and interactions. By maintaining distance, the researcher can systematically record patterns without disrupting the natural dynamics of the environment [60]. However, a significant limitation is the potential lack of contextual richness, as the observer does not directly engage with the setting. Nevertheless, non-participant observation is essential for studying objective behavioral patterns, significantly when investigator involvement could influence participant reactions or disrupt the normal flow of events [61].

2.3.3 Strategies for Field Notes and Reflexivity

Observations must be thoroughly documented for qualitative research to be reliable and rigorous. Structured field notes are the primary tool for systematically documenting environments, interactions, and behaviors [62]. Researchers must provide comprehensive explanations that include relevant details such as participant dynamics, the surrounding context, and significant trends. Maintaining a clear distinction between objective observations and interpretive comments is essential to prevent factual accounts from being conflated with subjective interpretations [63]. In observational research, reflexivity is crucial, requiring researchers to critically evaluate their preconceptions and biases at each stage of the data-gathering process. This self-awareness enhances the trustworthiness of the findings and minimizes the potential for interpretational biases. In addition to field notes, researchers can reflect on their influence on the observed environment by maintaining a reflective notebook

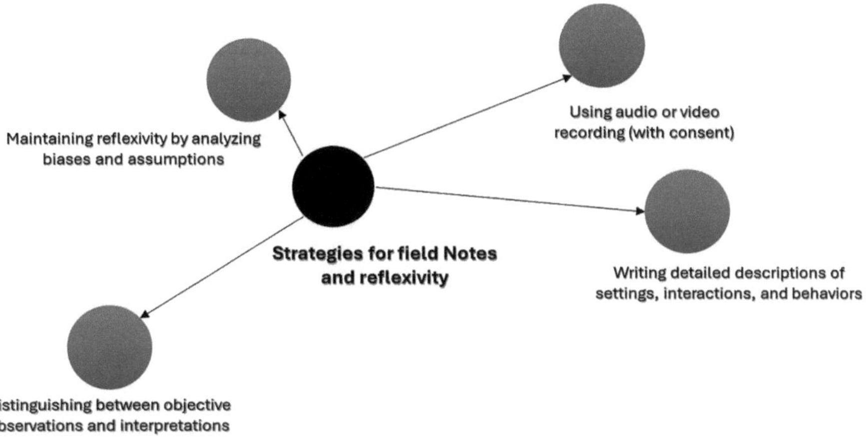

Fig. 5 Strategies for field notes and reflexivity

[64]. Audio or video recordings can also support field notes to offer more prosperous, comprehensive data. Ethical considerations must be prioritized to ensure recordings are made with participants' informed consent [54]. These practices enhance the transparency and reliability of observational research, enabling a more thorough and dependable examination of social interactions and behaviors in their natural settings. Essential techniques for preserving rigor in observational research are shown in Fig. 5. Detailed descriptions, a distinction between observations and interpretations, and reflexivity to reduce bias are all necessary for producing practical field notes. Furthermore, with permission, audio or video recordings can improve data accuracy. These techniques enhance qualitative research's depth and dependability.

2.4 *Document and Archival Analysis*

In qualitative research, document and archive analysis is a methodical study of current texts, including policy documents, reports, newspapers, diaries, and other written materials. According to Kayesa and Shung-King [65], document analysis is widely used in policy analysis research, although its use is sometimes lacking in both goal and methodological rigor. This study searched nine electronic databases using a qualitative systematic review method to find papers from low—and middle-income countries [65]. This approach benefits historical and policy research by enabling scholars to methodically examine historical events, institutional records, and policy changes. Reviewing these papers, scholars can find patterns, trends, and discourses that influence institutional and societal systems [66]. One of its primary benefits is the ability to document and archive research to offer deep contextual insights without necessitating direct participant interaction. Scholars must thoroughly assess

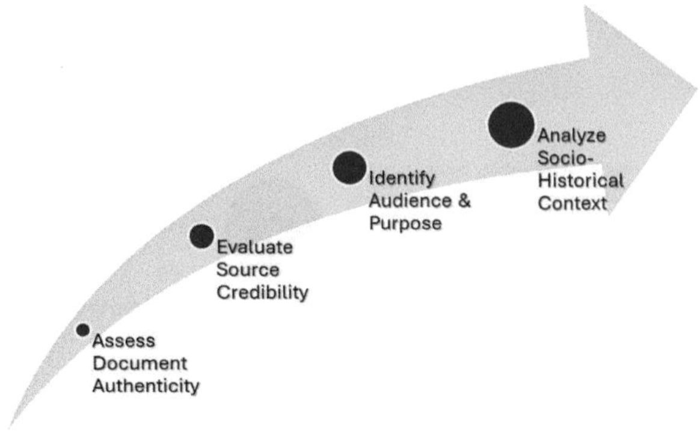

Fig. 6 Systematic approach to text analysis

documents' veracity, legitimacy, and potential biases to guarantee accurate interpretations. Despite its difficulties, this approach is still crucial for methodically examining historical and modern research problems [67].

2.4.1 Examining Existing Texts

To ensure the validity of the analysis, a methodical and sound approach must be employed while conducting a thorough investigation of the available texts. Researchers must first evaluate the document's authenticity, differentiating between original and replicated copies to ascertain its integrity. Ensuring credibility and reducing bias rely on assessing the validity of sources. Albahri [68] emphasizes how artificial intelligence applications are transforming the laws and regulations that govern professionalism, efficiency, and work ethics, in addition to the existing policies and procedures. Trustworthy practices in this context are still nascent [68]. Also, understanding the document's intended audience and purpose is essential for a comprehensive grasp of its communicative goal [69]. Enhancing analytical depth involves carefully considering the socio-historical context in which the work was produced. By incorporating these criteria into their work, researchers can maintain scholarly rigor and provide a complete, critical interpretation of textual sources. An organized framework for text analysis is shown in Fig. 6. Assessing document validity comes first in the process, followed by source trustworthiness. After determining the target audience and goal, researchers examine the socio-historical background. These methodical procedures guarantee a thorough and exact interpretation of the textual elements.

2.4.2 Policy Document Analysis

Ensuring credibility and reducing bias depend on evaluating the validity of sources. Albahri [68] highlights how artificial intelligence applications have transformed the

Table 2 Key aspects of policy analysis

Aspect	Description	Methodological approach	Key challenges
Source credibility and bias in architectural policies [71]	Assessing the reliability, objectivity, and potential biases of sources in policy analysis	Triangulating sources, conducting validity checks, and analyzing policy documents through critical discourse analysis	Subjectivity in selecting sources, risks of misinformation, and underlying biases that influence policy interpretations
Stakeholder influence and policy development [72]	Identifying key stakeholders and their roles in shaping policy decisions	Stakeholder mapping, network analysis, and influence assessment in the construction industry	Power imbalances between policymakers, industry stakeholders, and regulatory bodies
Critical analysis and comparative policy evaluation [73]	Examining language, rhetoric, and ideological influences on policies in various regions	Text mining, discourse analysis, and thematic coding alongside comparative and longitudinal studies	Bias in policy writing, misinterpretation risks, and architectural policy objectives variability

laws and regulations governing professionalism, efficiency, work ethics, and policies and procedures. Although still in their early stages, trustworthy and explainable artificial intelligence (XAI) in healthcare continues to face ongoing challenges related to reliability and transparency [25]. Identifying key stakeholders is the first step for researchers to comprehend the power dynamics and interests influencing policy choices. Critical evaluation of the language and framing used within these texts is essential to uncover policy writings' underlying ideologies, rhetorical techniques, and unconscious biases [70]. Researchers can identify structural and discursive policy variations by performing a comparative study across different periods and geographical locations. Additionally, a thorough understanding of policy effectiveness and its consequences on society can be achieved by conducting a comprehensive evaluation of policy implementation and outcomes based on documented evidence. Table 2 presents an organized and methodical review of essential factors in engineering policy analysis. These elements include the credibility of the source, the effect assessment, the influence of stakeholders, comparative assessments, and the problems that are involved in the execution of the policy.

3 Organizing, Storing, and Securing Qualitative Data: Tools and Techniques

Efficient data management is essential in qualitative research, as it ensures data legitimacy, improves accessibility, and protects long-term security throughout the study. In their research, Karunarathna et al. [74] highlight the importance of the data collection phase. This phase is crucial because it provides the necessary information to address research questions, test hypotheses, and meet study objectives [74]. Due to the inherently unstructured, context-dependent, and complex nature of qualitative data, researchers must develop organization, storage, and security techniques that are methodologically rigorous, transparent, and systematic [75]. Practical data management advantages are improving the credibility of research, encouraging collaboration, and ensuring compliance with ethical and legal standards. This section explores the moral imperatives of data security, confidentiality, and adherence to institutional, national, and international regulations regarding research data protection and responsible data sharing. It highlights the best practices and technological advancements across three key areas: structured data organization and thorough documentation, digital storage solutions and analytical tools tailored for qualitative research, and the ethical principles of data security and confidentiality.

3.1 Data Organization and Documentation

It is necessary to have a technique of managing qualitative data that is both well-documented and organized to retain analytical rigor, guarantee methodological transparency, and improve the reproducibility of research findings [76]. An effective organization systematically allows for data retrieval, makes conducting in-depth analysis more effortless, and maintains ethical research standards. To preserve coherence among datasets, researchers ought to use consistent coding methods, extensive metadata annotations, and established file-naming protocols. The interpretability of findings and the credibility of qualitative research are strengthened by comprehensive documentation, including data collection methods, notes taken by researchers, and information on the context in which the findings were obtained [77]. By implementing systematic data management procedures, researchers can maximize the efficiency of their workflow and guarantee the long-term utility of qualitative observations.

3.1.1 Transcription Techniques

Transcription is a fundamental process in qualitative research, converting spoken data from audio or video recordings into textual format to facilitate systematic analysis [77]. The selection of transcription techniques should be guided by accuracy, efficiency, and the ability to preserve contextual and discursive nuances critical for qualitative inquiry.

- Manual Transcription is widely regarded as the most precise method, allowing for meticulously preserving linguistic subtleties, speaker inflections, and nonverbal cues. This approach ensures a high level of contextual fidelity, which is particularly crucial for discourse and conversation analysis [78]. However, it is resource-intensive, requiring substantial time and researcher involvement.

- Automated Transcription, powered by artificial intelligence (AI)-based tools such as Otter.ai and Rev.com, enhances efficiency by rapidly converting speech into text [79]. Despite its advantages, automated transcription often lacks precision, mainly when dealing with specialized terminology, overlapping dialogue, or diverse dialectal variations. Consequently, post-transcription verification and editing are necessary to ensure reliability.

A hybrid approach integrating automated transcription with manual refinement is often recommended to optimize accuracy and efficiency [78]. This method leverages the speed of AI-driven tools while preserving the contextual and linguistic integrity of the data through human intervention. The selection of an appropriate transcription method should align with the research objectives, the complexity of the dataset, and the ethical considerations surrounding data accuracy and representation.

3.1.2 Coding Systems for Thematic Analysis

Coding is a fundamental component of qualitative data analysis, facilitating systematic categorization, organization, and interpretation of textual data [80]. Thematic coding involves a structured approach to identifying patterns and meaning within qualitative datasets. The process typically begins with open coding, where researchers inductively identify emerging themes and concepts from raw data without predefined categories. This exploratory stage ensures that analysis remains grounded in the data rather than being restricted by prior assumptions. Following this, axial coding establishes relationships between individual codes, grouping them into structured categories and identifying connections that contribute to a more cohesive analytical framework. Lastly, selective coding focuses on identifying overarching themes that integrate various data elements, leading to developing a coherent theoretical model. To enhance accuracy, consistency, and efficiency in coding, researchers often utilize qualitative data analysis software such as NVivo, ATLAS.ti, or MAXQDA, which offer advanced tools for theme identification, data visualization, and structured analysis [81].

3.1.3 Reflexive Documentation: Research Diaries and Memos

Reflexive documentation is paramount in ensuring that qualitative research is conducted transparently and boosting its credibility [23]. This includes research diaries and analytical memoranda. The systematic tracking of analytical judgments is made easier by these records, allowing researchers to critically reflect on their positionality and assist in mitigating the effects of any semantic biases. Researchers can increase the dependability of their results by providing a clear audit trail of their analytical procedures, which may be accomplished by maintaining extensive documentation. Regarding longitudinal research, where emerging insights require a systematic record of analytical trajectories, reflexive documentation is handy. This type of documentation allows for consistency in interpretation and reinforces the depth and rigor of qualitative investigation.

3.2 Digital Tools for Qualitative Data Storage and Analysis

Integrating digital tools into qualitative research workflows significantly enhances data management efficiency, supports systematic analysis, and ensures adherence to institutional and ethical data governance frameworks. Given the complexity and volume of qualitative data, selecting appropriate digital solutions is critical for maintaining data integrity, facilitating secure storage, and enabling sophisticated analytical processes [82]. Researchers must assess available tools based on several key factors, including data security protocols, accessibility for collaborative research, and compliance with regulatory standards such as GDPR and institutional ethical guidelines. Advanced qualitative data analysis software, such as NVivo, ATLAS.ti, and MAXQDA, offers comprehensive coding, theme identification, and visualization features, streamlining the analytical process. Additionally, secure cloud-based storage solutions with encryption and controlled access enhance data protection while ensuring researchers can efficiently retrieve and manage data. A well-structured approach to digital tool selection optimizes workflow efficiency and safeguards the ethical handling of qualitative research data [83].

3.2.1 Qualitative Data Analysis Software

Enhanced analytical rigor and methodological transparency may be achieved by utilizing qualitative data analysis software (QDAS), which offers sophisticated functions for organizing, coding, and analyzing intricate qualitative information [84]. Pattern identification, data visualization, and integration with mixed-method research methodologies are all made easier with the help of these innovative technologies. QDAS platforms that are among the most extensively utilized include:

- NVivo provides a substantial range of tools for organizing massive datasets, undertaking thematic analysis, and combining qualitative and quantitative data [85].
- Since ATLAS.ti emphasizes qualitative data visualization and supports collaborative research workflows, it is well-suited for projects involving teams of people.
- MAXQDA is characterized by its user-friendly code structures and broad interoperability with mixed-method research approaches.

The selection of software needs to be guided by the project's special analytical requirements, budget restrictions, institutional resources, and user competency. The efficient application of QDAS improves the effectiveness of qualitative research by assuring the systematic management of data while preserving the study's methodological rigor and repeatability [86].

3.2.2 Cloud-Based Versus Local Data Storage

Choosing a suitable data storage infrastructure significantly influences data security, accessibility, and compliance with ethical and legal standards, making it an essential decision for researchers [87]. Institutions and individual researchers must carefully evaluate the trade-offs associated with different storage systems to ensure data integrity and confidentiality.

- Cloud-based storage: Cloud systems such as Google Drive, Dropbox, and OneDrive offer significant benefits, including remote data access, automatic backups, and seamless collaboration with other research teams [87]. These features facilitate effective data management, especially in projects involving multiple institutions or disciplines. However, relying on third-party providers raises concerns about data privacy, security vulnerabilities, and compliance with data protection regulations specific to a particular country (for example, the General Data Protection Regulation and the Health Insurance Portability and Accountability Act). As a result, institutions must evaluate the security measures employed by the provider. These protocols should include encryption standards, data residency regulations, and access control measures [88].
- Local Storage: Locally controlled storage options, such as institutional servers, encrypted external hard drives, and secure network-attached storage (NAS), offer a greater degree of control over data security and compliance with institutional regulations. Implementing these solutions decreases reliance on external service providers and diminishes the risks associated with unauthorized third-party access [89]. However, it is essential to consider challenges such as limited remote access, the potential for hardware failures, and increased maintenance needs.

Hybrid storage formats are gaining popularity to enhance both usability and security. These models safeguard sensitive research data while ensuring flexibility [90]. They achieve this by combining cloud-based and local storage systems, utilizing encryption, multi-factor authentication, and role-based access controls. Adopting hybrid methods allows researchers to leverage the advantages of cloud storage for collaboration while keeping crucial data assets in local environments they can manage. This approach helps researchers mitigate the risks of data breaches and regulatory non-compliance.

3.2.3 Metadata Management for Data Reproducibility

Robust metadata management is essential to research data governance to assure qualitative datasets' systematic organization, traceability, and reusability [83]. By establishing stringent metadata rules, researchers can improve the integrity of empirical findings and maximize their repeatability. In the realm of metadata management, one of the most essential aspects is the assignment of unique and permanent identifiers to datasets. This not only helps to maintain referential integrity but also makes it easier to retrieve the data over an extended period. Furthermore, to ensure openness in research procedures, it is vital to have complete documentation of the methodology used to collect data, timestamps, and contextual factors [91]. Standardizing file name standards and coding structures is another factor that helps organize and retrieve datasets effectively. This, in turn, reduces the risk of discrepancies and enhances the interoperability of research endeavors. Adhering to data stewardship principles by implementing organized metadata frameworks makes it possible to do secondary analysis and facilitate comparisons across different studies. For this reason, it is necessary to have a well-defined metadata management strategy to guarantee compliance with open scientific laws and to encourage methodological rigor in qualitative research [83].

3.3 Data Security and Ethical Considerations

When handling sensitive materials related to participants, it is essential to maintain confidentiality, ensure data security, and uphold ethical integrity. This is a crucial requirement for managing qualitative research data. Researchers must adhere to strict ethical frameworks and institutional regulations that protect data throughout their lifespan [92]. Implementing robust data storage solutions, enforcing stringent access controls, and applying secure data transfer protocols are vital. These measures aim to prevent unauthorized access and breaches. Additionally, ethical responsibilities include obtaining informed consent from participants, ensuring transparency about data usage, and protecting participant anonymity when necessary. By following these principles, the study is more likely to be viewed as credible, ethically, thereby ensuring compliance with established academic and legal standards.

3.3.1 Confidentiality and Anonymization Strategies

One of the most fundamental ethical obligations in qualitative research is to maintain the anonymity of participants [56]. This requirement protects individual privacy and ensures adherence to institutional and legal standards. Researchers use various advanced strategies to reduce the risk of disclosing confidential information, including the following:

- Anonymization is a method that involves the irreversible removal or alteration of personally identifiable information (PII) from research datasets [93]. This technique effectively eliminates any potential re-identification of participants.
- Pseudonymization is a technique in which participants are assigned coded IDs that replace their data [94]. An encrypted and secure key file maintains the

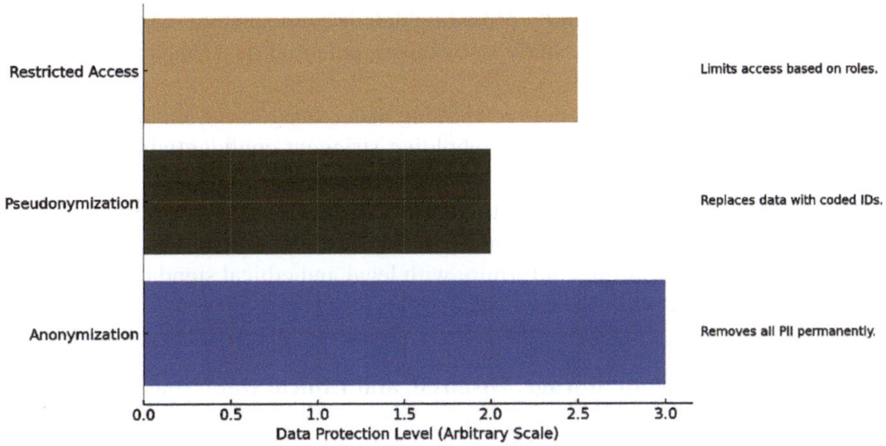

Fig. 7 Comparison of data protection methods based on protection level

connection between the two, accessible only to authorized personnel under certain circumstances.

- Restricted Access Controls: Implementing hierarchical, role-based access systems ensures that only individuals with clearly defined permission levels can access sensitive data. This significantly reduces the risk of unauthorized disclosure or breaches [95].

These procedures align with the requirements of the institutional review board (IRB) and overarching data protection legislation, such as the General Data Protection Regulation (GDPR). They are essential for maintaining the ethical integrity of research. Figure 7 shows how three data security methods Restricted Access, Pseudonymization, and Anonymization stack up against each other on a random scale of data protection. Restricted Access only offers a little bit of protection by limiting user roles. Pseudonymization replaces information that can be linked to a person with coded identifiers, which gives moderate protection. Anonymisation permanently removes all personally identifiable information (PII), which is the best way to protect your privacy. The graphic shows that each method is becoming better at keeping data private.

3.3.2 Institutional Review Board (IRB) Compliance and Informed Consent

Institutional Review Boards (IRBs) serve as essential regulatory bodies responsible for ensuring the ethical integrity of research involving human participants [96]. Researchers are obligated to obtain formal IRB approval before the commencement of any study and must rigorously adhere to established ethical guidelines, which include.

- Obtaining informed consent by providing detailed information, ensuring participants fully comprehend the study's objectives, potential risks, and data usage and withdrawal rights [97].
- Explicitly delineating protocols concerning data management, including storage, access, and dissemination, while upholding stringent confidentiality safeguards to protect participant anonymity.
- Demonstrating full compliance with relevant regulatory frameworks, such as the General Data Protection Regulation (GDPR) and institutional data governance policies, thereby ensuring conformity with legal and ethical standards governing research practices [98].

3.3.3 Long-Term Data Storage, Disposal, and Ethical Data Sharing

Beyond the immediate lifespan of data collecting and analysis, ethical requirements in research cover the long-term management, secure disposal, and appropriate sharing of research data. These obligations extend beyond the typical lifecycle of data gathering and processing. While upholding these obligations, it is necessary to adhere to stringent protocols to preserve participants' privacy and ensure compliance with the criteria set out by the institution, the law, and ethical principles. Some critical factors to consider are.

- Secure Archiving: Data should be placed systematically in institutional repositories that have been certified and that implement restricted access methods [90]. This will ensure that the data is preserved for an extended period and protect sensitive information from unauthorized access. Ethical Disposal: Outdated, redundant, or sensitive datasets must be deleted irreversibly using safe and verifiable processes. This will ensure that the anon
- The amity of the participants is maintained, and the legislation governing data protection is adhered to.
- Data distribution must be carried out precisely with ethical norms and participant permission agreements to ensure full compliance with institutional rules and international data protection frameworks [99]. This is referred to as "responsible data sharing."

These policies protect the rights of participants throughout the data lifespan and ensure research integrity.

4 Ensuring Trustworthiness, Rigor, and Ethical Integrity in Qualitative Data Management

It is essential to ensure the credibility, rigor, and ethical integrity of qualitative research to guarantee the reliability and validity of the findings from this type of study. To establish trustworthiness, qualitative research requires a methodologically sound and transparent strategy. This contrasts with quantitative research, which relies on statistical metrics to assess dependability and validity [100]. To achieve this goal, researchers must employ strategies that enhance rigor, minimize bias, and address the ethical concerns inherent in qualitative research. Key techniques include maintaining reflexivity, ensuring longer participant engagement, and utilizing triangulation to confirm results through diverse data sources or perspectives. Additionally, member checking, where participants review and validate the accuracy of findings or interpretations, can further enhance credibility. Throughout the study process, maintaining ethical integrity involves ensuring confidentiality, obtaining informed consent from participants, and adhering to the moral standards set by the institution. Researchers must provide a thorough audit trail of methodological decisions and analytical techniques to meet the transparency requirement in data reporting and academic writing, which is also fundamentally important. By adhering to these principles, qualitative researchers can ensure their work reflects methodological rigor, ethical responsibility, and academic integrity [101]. This will ultimately enhance the trustworthiness of their research findings and contribute meaningfully to the larger body of scholarly knowledge. In engineering research, guaranteeing trustworthiness, methodological rigor, and ethical integrity calls for applying methodical validation procedures, strong experimental verification, and ethical guideline adherence to improve the accuracy, dependability, and repeatability of computational modeling, finite element analysis, and data-driven engineering simulations.

4.1 Establishing Trustworthiness in Qualitative Research

When doing qualitative research, credibility, transferability, dependability, and confirmability of the findings are the four key criteria to evaluate trustworthiness. These criteria act as qualitative counterparts to the notions of validity and reliability typically utilized in quantitative research [102]. This ensures that the study findings are reliable and consistent with the methodology.

- It is essential to ensure the study's findings are accurate and genuine to establish credibility. This will ensure that the findings accurately represent the participants' experiences and viewpoints.
- Transferability is evaluated to determine how much a research study's findings may apply to different contexts or settings [103]. Thoroughly describing the research environment and the participants makes this evaluation possible.
- The concept of dependability emphasizes the consistency and stability of the research process over time. This necessitates the explicit documentation of the decisions made about the methodology [104].
- An audit record of the research process supports the concept of confirmability, which assures that the conclusions are formed by the participants' experiences rather than by the researcher's bias.

- When taken as a whole, these requirements ensure that qualitative research is conducted rigorously and dependably.

4.1.1 Credibility: Triangulation, Member Checking, and Prolonged Engagement

Credibility refers to qualitative research findings' integrity, accuracy, and authenticity, ensuring they represent participants' perspectives and lived experiences [105]. It reflects the degree to which the results are trustworthy and free from researcher bias or misinterpretation. To enhance credibility, researchers employ several rigorous strategies:

- Methodological Triangulation involves validating results through multiple data sources, research methods, or investigators [106]. By gathering data from diverse techniques such as interviews, observations, and document analysis, this approach ensures that findings are substantiated by various strands of evidence, thereby minimizing reliance on a single source and strengthening internal validity.
- Member Checking enhances the validity of interpretations by inviting participants to review and reflect on preliminary results. This process allows participants to confirm the accuracy of the findings, reducing the potential for researcher-induced bias or misrepresentation and ensuring a more authentic portrayal of participants' experiences.
- Prolonged Engagement refers to sustained immersion in the research setting, which fosters trust and rapport with participants [107]. This extended involvement allows researchers to gain deeper insights into the phenomenon under investigation, mitigating superficial observations or distortions arising from brief interactions or limited exposure. Through these strategies, researchers can ensure that their findings remain methodologically rigorous, comprehensive, and reflective of participants' experiences.

4.1.2 Transferability: Thick Description and Contextualized Findings

Transferability is a term that describes the extent to which the findings of qualitative research may be used or expanded to different contexts, situations, or people [108]. The idea of transferability in qualitative research is established by dense description, in contrast to the concept of generalizability in quantitative research, which is dependent on statistical representation. To do this, you must provide a comprehensive and abundantly contextualized account of the study environment. This will include descriptions of the participants, settings, and pertinent situational elements.

By providing detailed accounts of the social, cultural, and environmental conditions in which the study is conducted, researchers make it possible for others to evaluate the findings regarding their relevance and capacity to be applied to various scenarios [109]. Having access to such a comprehensive level of information gives future researchers, practitioners, or policymakers the ability to ascertain whether the findings obtained from the study may be usefully translated to other comparable situations. As an illustration, a study that investigates the level of satisfaction experienced by occupants in energy-efficient buildings might offer specific descriptions of the weather conditions, the materials used in construction, and the users' behaviors. This would enable researchers from other regions to assess how much the findings apply to their sustainable building projects.

4.1.3 Dependability: Audit Trails and Methodological Consistency

The stability, reliability, and consistency of qualitative research findings over time is what is meant by the term "dependability." [104] Furthermore, it guarantees that the research process is logical, traceable, and well-documented, which enables other individuals to evaluate the study's methodological soundness. Researchers can implement the following measures to improve dependability: Audit trails entail methodical documentation of each research stage. This documentation includes data collection techniques, coding decisions, and analytical methodologies. Thanks to this public record, external reviewers can evaluate the methodological rigor, check the research stages, and ensure that the conclusions are based on a consistent and transparent approach. Utilizing organized and established protocols for conducting interviews, observations, and coding processes is necessary to achieve consistency in the data collection process [110]. This method reduces the likelihood of variances due to the researcher's subjectivity or procedure discrepancies. Researchers can limit the possibility of bias by using consistent techniques across all data collection phases. This helps ensure that the results remain reliable and repeatable in comparable situations. For example, when researching the performance of passive houses, it is essential to maintain consistency in the methods used to gather data. Utilizing similar temperature sensors and standardizing occupant questionnaires are two procedures that assure the reliability of energy efficiency measures across various housing models.

4.1.4 Confirmability: Reflexivity and Transparency in Researcher Positioning

The term "confirmability" relates to the degree to which qualitative research conclusions are influenced by the facts obtained, as opposed to the researcher's personal biases, preconceptions, or subjective interpretations [23]. This guarantees that the results are founded on evidence, enhancing the research's credibility and contributing to its objectivity. Researchers can utilize the following measures to increase confirmability.

Throughout the research process, reflexivity entails continuous and critical self-examination of the researcher's role, potential biases, and underlying epistemological assumptions [111]. A reflective diary allows researchers to methodically document their ideas, shifting viewpoints, and decision-making processes to further their research. This method encourages openness and helps reduce the impact of personal subjectivity on the interpretation of facts from a quantitative perspective.

The positionality statement that the researcher must create must clearly define the researcher's background, motives, and possible power dynamics with participants. Openly disclosing personal experiences, professional positions, and relationships with the study setting is one way researchers may improve the transparency of their experimental methodology. This acknowledgment assists readers in determining the degree to which the researcher's stance impacts on the findings, which ultimately lends credence to the legitimacy and impartiality of the findings [112]. For example, in a study on ecological building design, a researcher with prior experience in sustainable architecture ought to contemplate how their knowledge can impact the interpretation of data about energy-efficient materials and design choices. Outlining four main criteria, credibility, transferability, dependability, and confirmability, Table 3 offers a disciplined framework for guaranteeing trustworthiness in qualitative research. It addresses definitions, primary methodologies, and engineering research applications by emphasizing methodological rigor, openness, and dependability in qualitative studies, especially in sustainable architecture and energy efficiency research.

Table 3 Trustworthiness in qualitative research

Criteria for trustworthiness	Definition	Key Strategies	Examples in engineering research
Credibility	Ensure the findings' accuracy, integrity, and authenticity reflect participants' experiences and minimize bias	Triangulation, Member Checking, and Prolonged Engagement	Validating energy efficiency data using interviews, simulations, and field observations to ensure its reliability
Transferability	Evaluating the relevance of research findings across various contexts with thorough descriptions of settings and participants	Thick Description and Contextualized Findings	Offering thorough insights into sustainable building studies covering materials, climate, and occupant behaviors
Dependability	Ensuring methodological consistency and reliability through thorough documentation of research procedures and maintenance of transparency	Audit Trails and Methodological Consistency	Employing standardized sensors and occupant questionnaires in passive house studies to guarantee reproducibility
Confirmability	Ensuring that data influences findings instead of researcher bias through reflexivity and transparency in research positioning	Reflexivity and Transparency in Researcher Positioning	Keeping reflexive diaries when examining the impact of sustainable architecture principles on energy performance

4.2 Addressing Researcher Bias and Ethical Dilemmas

When doing qualitative research, ethical challenges extend beyond protecting participant anonymity and encompass addressing issues such as the researcher's prejudice, power dynamics, and cultural sensitivity. The researcher's bias can affect the data collection, interpretation, and analysis processes, leading to skewed conclusions depending on the researcher's ideas or preconceived notions [112]. The ability to maintain a reflective mindset and critically examine one's effect on the research process is a vital skill for researchers to possess. In addition, it is necessary to recognize and address the power imbalances between researchers and participants to guarantee genuine and respectful interactions. It is also essential for researchers to demonstrate cultural sensitivity since they must consider and respect the various social circumstances, values, and histories of the participants. The implementation of measures such as reflexivity, informed consent, and open communication assists in minimizing ethical hazards, therefore contributing to the development of a research environment that is more egalitarian and trustworthy [113].

4.2.1 Mitigating Researcher Bias and Ensuring Objectivity

Although qualitative research inherently acknowledges the researcher's subjective role, it is essential to implement strategies that minimize bias and promote objectivity. Several methods help ensure the integrity of the research process: Bracketing is particularly vital in phenomenological research, requiring researchers to set aside personal biases, assumptions, and preconceptions consciously [114]. This practice allows the data to emerge organically, ensuring that participants' experiences are accurately represented without undue influence from the researcher's perspective.

Peer Debriefing involves engaging colleagues or fellow researchers in critical discussions about the research process and findings. These interactions promote reflective thinking, challenge assumptions, and help identify potential biases that could shape data interpretation. Negative Case Analysis requires researchers to seek out and consider data that contradicts emerging patterns or interpretations. Including these discrepancies enhances the validity of the findings by ensuring that conclusions are not selectively biased toward confirmatory evidence, promoting a more balanced and comprehensive understanding of analysis [115]. For example, in a study on sustainable building practices, negative case analysis might involve examining projects where energy-saving measures failed to meet expected outcomes, providing a more nuanced understanding of challenges in implementing green design principles.

4.2.2 Addressing Power Dynamics in Qualitative Research

Power inequalities between researchers and participants may impact the data-collecting process, potentially resulting in skewed or biased findings [116]. Researchers should take the following measures to reduce the impact of these imbalances: Establishing a connection with participants that is one of collaboration and respect is an essential aspect of non-hierarchical interaction practice. Researchers can encourage open, honest, and frank replies from participants if they minimize the use of authoritative positions and establish a sense of mutual respect for the participants. Throughout the study process, participants are given an agency through participant empowerment. Ethical responsibility may be ensured by allowing participants to review interview transcripts, offer interpretation input, and verify the veracity of their remarks. This strategy reduces the possibility of distortions caused by the researcher, and the legitimacy of the findings is substantially improved. The researcher must refrain from asking leading questions while using neutral interview techniques. Instead, they should allow participants to express their opinions and construct their narratives freely [117]. This method reduces the likelihood of interviewer bias and guarantees that the participants' viewpoints remain at the forefront of the study, fostering more reliable and accurate conclusions. For example, when researching digital twins for construction, building managers check the simulation results to help guarantee that the data accurately represents actual operational issues. This, in turn, reduces the amount of bias that researchers impose when evaluating energy efficiency or predictive maintenance solutions.

4.2.3 Ethical Considerations in Cross-Cultural Research

Qualitative research in cross-cultural situations provides a unique set of ethical problems, notably representation, language obstacles, and power relations [118]. The following are some of the best strategies researchers should implement to handle these complexities properly: A profound understanding of the sociocultural settings in which the study is conducted is essential to creating cultural competence. Through this knowledge, researchers can avoid ethnocentric interpretations, which helps guarantee that the findings are respectful and appropriately reflect participants' lived experiences within their cultural frameworks. Translation and interpretation that adhere to ethical standards emphasize preserving language precision throughout the transcription, translation, and analysis. Careful attention is imperative to guarantee that cultural subtleties and conceptual meanings are kept, preventing misinterpretations that might skew the research outcomes [119]. Local stakeholders actively participate in the research process through community-based participatory approaches. The participation of participants in decision-making processes encourages ethical collaboration, honors local knowledge systems, and increases the legitimacy of cultural practices. The study results are strengthened in relevance and credibility because of this inclusive strategy, which also helps eliminate power structures. When researching passive houses, for example, working with local builders and inhabitants ensures that cultural preferences, such as traditional ventilation systems, are considered [120, 121]. This improves the ethical validity of the study as well as the conclusions about energy efficiency.

4.3 Transparency in Data Reporting and Academic Writing

In qualitative research, transparent reporting is necessary to guarantee the findings' consistency, trustworthiness, and replicability. Researchers give a clear framework for evaluating the study's validity by adequately documenting each stage of the research process, beginning with data collection and continuing through the analysis and interpretation of the findings [122]. Through transparent reporting, other researchers can evaluate the reliability of the research, comprehend the circumstances under which the data was collected, and duplicate the study in various environments. Because of this, transparency improves ethical accountability, guaranteeing that the research process will respect the participants' experiences and remain free of prejudice on the researcher's side. Finally, transparency improves the overall quality of qualitative research by promoting a more trustworthy and reflective academic discourse. This is accomplished through the dissemination of information.

4.3.1 Adherence to Qualitative Research Reporting Standards

It is vital to adhere to established reporting formats to guarantee methodological transparency and boost the credibility of qualitative research. To facilitate the evaluation of research quality, these frameworks offer explicit rules for documenting the research process, guaranteeing consistency, and ensuring the process is documented. The following two reporting standards are well-recognized: A consolidated set of criteria for reporting qualitative research is called COREQ. This 32-item checklist focuses on fundamental features of qualitative research, such as the importance of reflexivity within the research team, the overall design of the study, and transparency in the interpretation of data [123]. Not only does it guarantee that researchers critically reflect on their roles and biases, but it also ensures that they provide descriptive explanations of their procedures. This framework, which consists of 21 items and is known as the Standards for Reporting Qualitative Research (SRQR), emphasizes methodological rigor, ethical issues, and clarity in data interpretation. Throughout the study, SRQR assists researchers in effectively communicating their analytical methodologies, ethical guidelines, and decision-making processes from beginning to end. For example, using COREQ promotes openness in energy efficiency research for buildings by providing specifics on collecting, analyzing, and interpreting input from participants regarding energy-saving devices [124]. This allows future researchers to reproduce or modify the study to accommodate various building types.

4.3.2 Presenting Qualitative Data in Academic Writing

An organized and coherent method is necessary for effectively presenting qualitative data in academic writing. This technique is essential to promote clarity and analytical depth. By utilizing the following methodologies, one may guarantee that the study's findings will be presented in an authoritative and significant manner: To present data logically and methodically, structured thematic narratives are utilized [125]. These narratives organize findings around clearly defined topics. With this technique's assistance, the reader can follow the study's evolution, highlighting essential patterns and correlations in the data. Literal Quotations: The use of actual quotations from participants not only lends an air of authenticity to the study but also offers a wealth of contextual insights into the experiences and opinions of the participants. The analysis is also grounded in real-world facts, which helps to make judgments more plausible. These quotations serve to ground the analysis. Visual Representations: Examples of visual representations include conceptual diagrams, coding matrices, and thematic maps that help simplify complicated data and illustrate the links between topics. The use of these visual aids helps readers comprehend the relationships and patterns that are included within the data, which improves both comprehension and analytical transparency in qualitative research. In zero-carbon (CO2) building research, visual representations such as energy flow diagrams may explain how different sustainable technologies interact, such as solar panels or heat recovery systems [126]. This helps to highlight energy efficiency patterns and methods for reducing carbon emissions.

4.3.3 Ethical Considerations in Publishing Qualitative Findings

Ethical when it comes to publishing the results of qualitative research, it is necessary to uphold ethical integrity throughout the whole process. The researchers are required to comply to specific ethical criteria to guarantee that the participants are treated with respect and that the findings of the research are credible:

- It is crucial to protect participants' identities, which is why participant anonymity is essential. Researchers should use pseudonyms and implement data-masking techniques to protect personal information and prevent unintentional identification [127]. This is especially important in sensitive research situations.
- Researchers must get the participants' explicit approval before incorporating any verbatim remarks in publications. This is known as "informed consent for direct quotations." By doing so, individuals are guaranteed to be fully informed of how their remarks will be portrayed and can select whether they will be directly quoted [128].
- Keeping from Overgeneralization: Researchers are responsible for using caution and avoiding extrapolating their findings beyond the unique setting of the study. An overgeneralization of the findings might result in a misunderstanding of the findings and brings into question the rigor of the methodology. In its place, results ought to be presented within the limitations of their respective contexts to uphold

ethical duty and guarantee the validity of conclusions obtained from the research [129].

In sustainable building research, for instance, it is of the utmost importance to guarantee the anonymity of participants while obtaining information from architects and engineers on the failures of green technology. This is because disclosing their names might affect their professional reputation or partnerships.

5 Conclusion

Qualitative data management and methodical collecting are essential to thorough research in many fields. With an emphasis on methodological alignment, ethical issues, and analytical rigor, this chapter has thoroughly examined various qualitative data-gathering techniques, such as focus groups, observations, interviews, and document analysis. Because qualitative research is interpretive, it is crucial to ensure that data collection methods and epistemological viewpoints are consistent to provide reliable and significant results. Structured, semi-structured, and unstructured interviews are the most popular ways to get participants' detailed viewpoints. Structured interviews are useful for large-scale investigations that need consistency because they make standardization and comparison easier. However, because of their strict structure, they could restrict exploratory ideas. Semi-structured interviews provide flexibility, allowing researchers to explore emerging topics while adhering to a guiding framework. Unstructured interviews, on the other hand, allow for impromptu discussion and in-depth examination of participants' experiences, yielding the richest qualitative data. The capacity of the researcher to build rapport, lessen response bias, and use reflexivity during the data-collecting process is crucial to the success of interviews. Focus groups are an effective technique for gathering group insights and understanding social interactions in a group context, in addition to interviews. They foster lively debates by capturing the complexity of participants' meaning-making and consensus-building. However, they also present unique challenges, such as the risk of certain voices dominating the conversation and ethical concerns regarding confidentiality in a group setting. The moderator's skills are crucial to minimize conformity bias, ensure balanced participation, and create an environment that encourages honest and open communication. Observational methods add another layer to qualitative research by allowing researchers to examine relationships, behaviors, and environmental factors in real-time. While participant observation, where the researcher actively engages in the environment, offers deep immersion and first-hand insights, it also raises ethical dilemmas about researcher bias and influence. In contrast, non-participant observation addresses these concerns by maintaining objectivity, though it may limit access to insider perspectives. Regardless of the method, ensuring the authenticity and integrity of observational data requires careful documentation through field notes and reflective journaling. Document and archive analysis enhances primary data collection by enabling researchers to study existing texts, historical documents, and institutional reports. This method allows for triangulating results without direct participant interaction, making it especially valuable for historical research, policy analysis, and discourse studies. It is essential to verify the reliability and validity of sources to avoid biases and misunderstandings that may arise from selective or inadequate documentation. In addition to data collection, qualitative data's effective organization and structuring significantly enhance methodological transparency and analytical coherence. The first step in processing qualitative data involves transcription, which transforms audio and video recordings into text formats for systematic examination. Automated transcribing technologies offer efficiency but require validation to correct potential errors, while manual transcription, though labor-intensive, boasts high accuracy. Organizing qualitative data into recognizable categories through coding—using open, axial, or selective techniques—facilitates identifying trends and themes. Advances in qualitative data analysis tools, such as NVivo, ATLAS.ti, and MAXQDA, have simplified data storage, retrieval, and theme analysis, thereby increasing the efficiency and repeatability of research. Ethical considerations remain crucial at every level of qualitative research, from data gathering to dissemination. Key ethical obligations include safeguarding participant privacy, obtaining informed consent, and adhering to institutional review board (IRB) guidelines. Techniques such as anonymization, encrypted storage, and restricted data access can mitigate the risk of data breaches and unauthorized disclosures. Additionally, ethical data-sharing agreements must balance participant privacy with transparency, especially regarding sensitive or private information. To establish trustworthiness in qualitative research, following Lincoln and Guba's (1985) four essential criteria—credibility, transferability, dependability, and confirmability—is essential. Triangulation, member checks, and prolonged engagement enhance credibility, ensuring that results accurately represent participants' realities. Detailed descriptions that allow readers to assess the relevance of findings to various contexts enhance transferability. Reflexivity and an open researcher posture ensure confirmability, while audit trails and consistent methodological documentation bolster dependability. Together, these criteria reinforce the dependability and rigor of qualitative research and distinguish it from subjective or anecdotal accounts. Addressing power dynamics and researcher bias is essential for conducting ethical qualitative research. Critical self-reflection regarding the researcher's influence on the study, or reflexivity, minimizes biases and enhances transparency. Techniques such as negative case analysis, peer debriefing, and bracketing in phenomenological research further reduce researcher-induced distortions. In cross-cultural research, ethical attention to linguistic, social, and political contexts is vital to avoid distortion and ensure culturally competent interpretations. Adhering to reporting standards like SRQR (Standards for Reporting Qualitative Research) and COREQ (Consolidated Criteria for Reporting Qualitative Research) is necessary for disseminating qualitative findings. By promoting methodological transparency, these frameworks ensure that qualitative research is thoroughly documented and reproducible. Verbatim participant quotes, well-organized topical narratives and visual aids such as conceptual diagrams and coding matrices are essential for effective qualitative reporting. Protecting participant identity, obtaining permission for direct quotes, and avoiding generalizations that extend beyond the study's context are critical ethical considerations in publication. Careful methodological rigor, ethical integrity, and systematic organization are vital for gathering and managing qualitative data. Throughout the research process, researchers must maintain reflexivity, ensure safe data handling procedures, and align data collection methods with the study's theoretical framework. While digital technologies have advanced qualitative data analysis, researchers must also weigh its ethical implications and limitations. By upholding transparency, methodological rigor, and ethical accountability, qualitative researchers contribute valuable insights that enrich scholarly debate, inform policy decisions, and enhance societal understanding. As qualitative methods evolve, technological advancements and multidisciplinary integration will further shape the field of qualitative inquiry, broadening its applicability and impact in current research.

References

1. Santiago-Delefosse M, Gavin A, Bruchez C, Roux P, Stephen SL (2016) Quality of qualitative research in the health sciences: analysis of the common criteria present in 58 assessment guidelines by expert users. Soc Sci Med 148:142–151
2. Levitt HM, Motulsky SL, Wertz FJ, Morrow SL, Ponterotto JG (2017) Recommendations for designing and reviewing qualitative research in psychology: promoting methodological integrity. Qual Psychol 4(1):2
3. Carroll LJ, Rothe JP (2010) Levels of reconstruction as complementarity in mixed methods research: a social theory-based conceptual framework for integrating qualitative and quantitative research. Int J Environ Res Public Health 7(9):3478–3488
4. Johnson JL, Adkins D, Chauvin S (2020) A review of the quality indicators of rigor in qualitative research. Am J Pharm Educ 84(1):7120
5. Ospina SM, Esteve M, Lee S (2018) Assessing qualitative studies in public administration research. Public Adm Rev 78(4):593–605
6. Asrifan A, Pratiwi WR, Bahri A, Jabu B, Rossydi A, Mulyadi M (2025) Qualitative methodology selection: key considerations for dissertation scholars. In: Qualitative research methods for dissertation research, pp 25–52. IGI Global Scientific Publishing
7. Hallberg LR (2006) The "core category" of grounded theory: making constant comparisons. Int J Qual Stud Health Well Being 1(3):141–148
8. Fusch PI, Fusch GE, Ness LR (2017) How to conduct a mini-ethnographic case study: a guide for novice researchers. Qual Rep 22(3):923–941
9. Kumar V, Bhosale MP, Nadgauda MTN, Lavanya C (2024) Research methodology-concepts & applications
10. Ruslin R, Mashuri S, Rasak MSA, Alhabsyi F, Syam H (2022) Semi-structured interview: a methodological reflection on the development of a qualitative research instrument in educational studies. IOSR J Res & Method Educ (IOSR-JRME) 12(1):22–29
11. Ellinor L, Girard G (2023) Dialogue: rediscover the transforming power of conversation. Crossroad Press
12. Kgosiemang TG (2024) Taking part in qualitative research: insights for researchers from participants (Master's thesis, University of Johannesburg (South Africa))
13. Carter SM, Shih P, Williams J, Degeling C, Mooney-Somers J (2021) Conducting qualitative research online: challenges and solutions. Patient-Patient-CentEd Outcomes Res 14(6):711–718

14. Drolet MJ, Rose-Derouin E, Leblanc JC, Ruest M, Williams-Jones B (2023) Ethical issues in research: perceptions of researchers, research ethics board members and research ethics experts. J Acad Ethics 21(2):269–292
15. Uwamusi CB, Ajisebiyawo A (2023) Participant observation as research methodology: assessing the defects of qualitative observational data as research tools. Asian J Soc Sci Manag Technol 5(3):19–32
16. Dicks B (2024) 14 qualitative research quality in multimodal research. The Sage Handbook of Qualitative Research Quality
17. Bryson A (2022) Recording lived experience: a toolkit for victims and survivors
18. Levitt HM, Morrill Z, Collins KM, Rizo JL (2021) The methodological integrity of critical qualitative research: principles to support design and research review. J Couns Psychol 68(3):357
19. Kiefer A (2024) Improving automatic transcription using natural language processing
20. Ntsobi MP, Costa AP, Kasperiuniene J, Brandão C, Ribeiro J (2024) Digital tools and techniques in qualitative research: digital skills and research optimisation. In: World conference on qualitative research, pp 1–25. Cham: Springer Nature Switzerland
21. Temara S (2025) Ethics for responsible data research: integrating cybersecurity perspectives in a digital era. In: IRB, human research protections, and data ethics for researchers, pp 103–144. IGI Global Scientific Publishing
22. Adler RH (2022) Trustworthiness in qualitative research. J Hum Lact 38(4):598–602
23. Ahmed SK (2024) The pillars of trustworthiness in qualitative research. J Med, Surg, Public Health 2:100051
24. Omodan BI (2022) A model for selecting theoretical framework through epistemology of research paradigms. Afr J Inter/Multidiscip Stud 4(1):275–285
25. Bryda G, Costa AP (2023) Qualitative research in digital era: innovations, methodologies and collaborations. Social Sciences 12(10):570
26. Gelo O, Brinkmann D, Benetka G (2008) Quantitative and qualitative research: Beyond the debate. Integr Psychol Behav Sci 42:266–290
27. Curry LA, Nembhard IM, Bradley EH (2009) Qualitative and mixed methods provide unique contributions to outcomes research. Circulation 119(10):1442–1452
28. Tracy SJ (2024) Qualitative research methods: collecting evidence, crafting analysis, communicating impact. John Wiley & Sons
29. Adeniran AO, Asifat JT, Familusi OB, Folorunso SS (2024) Critical analysis of research methods and misperceptions of statistical analysis. Int J Innov Manag, Econ Soc Sci 4(3):29–50
30. Bouncken RB, Qiu Y, Sinkovics N, Kürsten W (2021) Qualitative research: extending the range with flexible pattern matching. RMS 15(2):251–273
31. Khoa BT, Hung BP, Hejsalem-Brahmi M (2023) Qualitative research in social sciences: data collection, data analysis and report writing. Int J Public Sect Perform Manag 12(1–2):187–209
32. Almusaed A, Yitmen I, Almssad A (2024) Contemporary innovations and sustainable practices in the application of clay materials within architectural design and construction methodologies
33. Campion MA, Pursell ED, Brown BK (1988) Structured interviewing: raising the psychometric properties of the employment interview. Pers Psychol 41(1):25–42
34. Osborne N, Grant-Smith D (2021) In-depth interviewing. In Methods in urban analysis, pp 105–125. Singapore: Springer Singapore
35. Adeoye-Olatunde OA, Olenik NL (2021) Research and scholarly methods: semi-structured interviews. J Am CollE Clin Pharm 4(10):1358–1367
36. Ruslin R, Mashuri S, Rasak MSA, Alhabsyi F, Syam H (2022) Semi-structured interview: a methodological reflection on the development of a qualitative research instrument in educational studies. IOSR J Res & Method Educ (IOSR-JRME) 12(1):22–29
37. Cairns-Lee H, Lawley J, Tosey P (2022) Enhancing researcher reflexivity about the influence of leading questions in interviews. J Appl Behav Sci 58(1):164–188

38. Campion MA, Palmer DK, Campion JE (1997) A review of structure in the selection interview. Pers Psychol 50(3):655–702
39. Jianbin X Conducting semi-structured interviews in social service research: a practical guide
40. Jordan J, Clarke SO, Coates WC (2021) A practical guide for conducting qualitative research in medical education: part 1—how to interview. AEM Educ Train 5(3)
41. Karunarathna I, Gunasena P, Hapuarachchi T, Gunathilake S (2024). Comprehensive data collection: methods, challenges, and the importance of accuracy. Uva Clin Res 1–24
42. Chauhan RS (2022) Unstructured interviews: are they really all that bad? Hum Resour Dev Int 25(4):474–487
43. Knott E, Rao AH, Summers K, Teeger C (2022) Interviews in the social sciences. Nat Rev Methods Prim 2(1):73
44. Researchers can systematically gather high-quality qualitative data by following these criteria, which enhances their study's reliability and validity
45. Chae M, Kim B (2024) Reducing cultural barriers: a grounded theory approach to nursing student attitudes after multicultural education. Risk Manag Healthc Policy 2241–2253
46. Shaheen N, Shaheen A, Ramadan A, Hefnawy MT, Ramadan A, Ibrahim IA, Flouty O (2023) Appraising systematic reviews: a comprehensive guide to ensuring validity and reliability. Front Res MetrS Anal 8:1268045
47. Tian F, Liang P, Babar MA (2022) Relationships between software architecture and source code in practice: an exploratory survey and interview. Inf Softw Technol 141:106705
48. Akyıldız ST, Ahmed KH (2021) An overview of qualitative research and focus group discussion. Int J Acad Res Educ 7(1):1–15
49. Strasser S, Stauber C, Shrivastava R, Riley P, O'Quin K (2021) Collective insights of public-private partnership impacts and sustainability: a qualitative analysis. PLoS ONE 16(7):e0254495
50. Edwards-Fapohunda DMO (2024) The role of adult learning and education in community development: a case study of New York. Iconic Res Eng JS 8(1):437–454
51. Cole MB (2024) Group dynamics in occupational therapy: the theoretical basis and practice application of group intervention. Taylor & Francis
52. Tessema KA, Reilly S (2024) Managing peer groups in family business centers: directors' sensemaking practices. J Fam Bus Manag
53. Kunisch S, Denyer D, Bartunek JM, Menz M, Cardinal LB (2023) Review research as scientific inquiry. Organ Res Methods 26(1):3–45
54. Newman PA, Guta A, Black T (2021) Ethical considerations for qualitative research methods during the COVID-19 pandemic and other emergency situations: navigating the virtual field. Int J Qual Methods 20:16094069211047824
55. Rajaraman A, Austin JL, Gover HC, Cammilleri AP, Donnelly DR, Hanley GP (2022) Toward trauma-informed applications of behavior analysis. J Appl Behav Anal 55(1):40–61
56. Kang E, Hwang HJ (2021) Ethical conducts in qualitative research methodology: participant observation and interview process. J Res Publ Ethics 2(2):5–10
57. Maestre BO, Araujo HAA. Contributions to the application of non-participant observation as a research technique
58. Anjum G, Aziz M (2024) Advancing equity in cross-cultural psychology: embracing diverse epistemologies and fostering collaborative practices. Front Psychol 15:1368663
59. Jamie K, Rathbone AP (2022) Using theory and reflexivity to preserve methodological rigour of data collection in qualitative research. Res Methods Med & Health Sci 3(1):11–21
60. Altman J, Fibich P, Trotsiuk V, Altmanova N (2024) Global pattern of forest disturbances and its shift under climate change. Sci Total Environ 915:170117
61. Shin SYSSY, Miller SMS (2022) A review of the participant observation method in journalism: designing and reporting. Rev Commun Res 10
62. Dyar KL (2022) Qualitative inquiry in nursing: creating rigor. In: Nursing Forum (Vol 57, No 1, pp 187–200)
63. Bennett TD (2023) Interpretation is opinion: realigning the fact/opinion distinction in English defamation law. J Media Law 15(1):62–89

64. Pessanha F, Salah AA (2021) A computational look at oral history archives. ACM J Comput Cult Herit (JOCCH) 15(1):1–16
65. Kayesa NK, Shung-King M (2021) The role of document analysis in health policy analysis studies in low and middle-income countries: lessons for HPA researchers from a qualitative systematic review. Health Policy OPEN 2:100024
66. Kohtamäki M, Whittington R, Vaara E, Rabetino R (2022) Making connections: harnessing the diversity of strategy-as-practice research. Int J Manag Rev 24(2):210–232
67. Agyekum EB, Nutakor C, Khan T, Adegboye OR, Odoi-Yorke F, Okonkwo PC (2024) Analyzing the research trends in the direction of hydrogen storage–a look into the past, present and future for the various technologies. Int J Hydrogen Energy 74:259–275
68. Albahri AS, Duhaim AM, Fadhel MA, Alnoor A, Baqer NS, Alzubaidi L, Albahri OS, Alamoodi AH, Bai J, Salhi A, Santamaría J (2023) A systematic review of trustworthy and explainable artificial intelligence in healthcare: assessment of quality, bias risk, and data fusion. Inf Fusion 96:156–191
69. Nykyporets SS, Herasymenko NV, Chopliak VV (2024) Cognitive strategies impacting the structural composition of translated technical and scientific texts: an analysis of translation methodologies. Publishing House "Baltija Publishing"
70. Ulnicane I, Aden A (2023) Power and politics in framing bias in Artificial Intelligence policy. Rev Policy Res 40(5):665–687
71. Tempestini M (2025) Criticism goes social: exploring public architectural criticism through architecture awards. Archnet-IJAR: Int J Arch Res
72. Hasa K (2023) Examining the OECD's perspective on AI in education policy: a critical analysis of language and structure in the 'AI and the future of skills'(AIFS) document and its implications for the higher education (Doctoral dissertation, University of British Columbia)
73. Critical analysis and comparative policy evaluation examining language, rhetoric, and ideological influences on policies in various regions
74. Karunarathna I, Gunasena P, Hapuarachchi T, Gunathilake S (2024) Comprehensive data collection: methods, challenges, and the importance of accuracy. Uva Clin Res 1–24
75. Mettler T (2025) Computational text analysis for qualitative IS research: a methodological reflection. Commun Assoc Inf Syst 56(1):14
76. Craig SL, McInroy LB, Goulden A, Eaton AD (2021) Engaging the senses in qualitative research via multimodal coding: triangulating transcript, audio, and video data in a study with sexual and gender minority youth. Int J Qual Methods 20:16094069211013660
77. Eftekhari H (2024) Transcribing in the digital age: qualitative research practice utilizing intelligent speech recognition technology. Eur J Cardiovasc Nurs 23(5):553–560
78. Pandita K, Thakur PKS, Annamalai S (2023) Contextual transcription and summarization of audio using AI. In: Proceedings of the 5th international conference on information management & machine intelligence, pp 1–9
79. Sharma S, Dureja S, Saini D, Jose R, Pant R, Singh A (2025) Empowering impaired learners: technological advancements in higher education. Technol Disabil, 10554181251313711
80. Riazi M, Ghanbar H, Rezvani R (2023) Qualitative data coding and analysis: a systematic review of the papers published in the journal of second language writing. Iran J Lang Teach Res 11(1):25–47
81. Yıldız ET, Tam MS (2023) Using qualitative data analysis software (QDAS) in communication studies: a systematic review. Akdeniz Üniversitesi İletişim Fakültesi Dergisi 42:51–68
82. Kitchin R (2021) The data revolution: a critical analysis of big data, open data and data infrastructures
83. Javed H (2023) Challenges and requirements of heterogenous research data management in environmental sciences: a qualitative study (Master's thesis, H. Javed)
84. Paulus TM (2023) Using qualitative data analysis software to support digital research workflows. Hum Resour Dev Rev 22(1):139–148
85. Bazeley P (2021) Using NVivo for mixed methods research. In: The Routledge reviewer's guide to mixed methods analysis, pp 343–354. Routledge

86. Moncada M (2025) Should we use NVivo or excel for qualitative data analysis? Bull Sociol Methodol/Bull Méthodologie Sociol p 07591063251317084
87. Ahmad K, Maabreh M, Ghaly M, Khan K, Qadir J, Al-Fuqaha A (2022) Developing future human-centered smart cities: critical analysis of smart city security, data management, and ethical challenges. Comput Sci Rev 43:100452
88. Ramya R, Rajeswari G (2023) Cloud technologies in educational research. Education 5.0: revolutionizing learning for the future, p 90
89. Ahanger AS, Masoodi FS, Khanam A, Ashraf W (2024) Managing and securing information storage in the Internet of Things. In: Internet of Things vulnerabilities and recovery strategies, pp 102–151. Auerbach Publications
90. Gupta I, Singh AK, Lee CN, Buyya R (2022) Secure data storage and sharing techniques for data protection in cloud environments: a systematic review, analysis, and future directions. IEEE Access 10:71247–71277
91. Fischer DA, Goel K, Andrews R, van Dun CG, Wynn MT, Röglinger M (2022) Towards interactive event log forensics: detecting and quantifying timestamp imperfections. Inf Syst 109:102039
92. Kvale LH, Darch PT (2022) Privacy protection throughout the research data life cycle
93. Zuo Z, Watson M, Budgen D, Hall R, Kennelly C, Al Moubayed N (2021) Data anonymization for pervasive health care: systematic literature mapping study. JMIR Med Inform 9(10):e29871
94. Kangwa M (2023) Prevention of personally identifiable information leakage in ecommerce using offline data minimization and online pseudonymisation (Doctoral dissertation, The University of Zambia)
95. Mpamugo E, Ansa G (2024) Enhancing network security in mobile applications with role-based access control. J Inf Syst Inform 6(3):1872–1899
96. Serpico K (2024) Defining institutional review board application quality: critical research gaps and future opportunities. Research Ethics 20(1):19–35
97. Millum J, Bromwich D (2021) Informed consent: what must be disclosed and what must be understood? Am J Bioeth 21(5):46–58
98. Labadie C, Legner C (2023) Building data management capabilities to address data protection regulations: learnings from EU-GDPR. J Inf Technol 38(1):16–44
99. Wong J, Henderson T, Ball K (2022) Data protection for the common good: developing a framework for a data protection-focused data commons. Data Policy 4:e3
100. Jones KR, Gwynn EP, Teeter A (2019) Quantitative or qualitative: Selecting the right methodological approach for credible evidence. J Hum Sci Ext 7(2):5
101. Berkovich I, Grinshtain Y (2023) A review of rigor and ethics in qualitative educational administration, management, and leadership research articles published in 1999–2018. Leadersh Policy Sch 22(3):549–564
102. Franklin C, Ballan M (2001) Reliability and validity in qualitative research. Handb Soc Work Res Methods 4(273–292)
103. Nyirenda L, Kumar MB, Theobald S, Sarker M, Simwinga M, Kumwenda M, Johnson C, Hatzold K, Corbett EL, Sibanda E, Taegtmeyer M (2020) Using research networks to generate trustworthy qualitative public health research findings from multiple contexts. BMC Med Res Methodol 20:1–10
104. Golafshani N (2003) Understanding reliability and validity in qualitative research. Qual Rep 8(4):597–607
105. Loh J (2013) Inquiry into issues of trustworthiness and quality in narrative studies: a perspective. Qualitative Report 18(33)
106. Santos KDS, Ribeiro MC, Queiroga DEUD, Silva IAPD, Ferreira SMS (2020) The use of multiple triangulations as a validation strategy in a qualitative study. Ciencia saude coletiva 25:655–664
107. Edirisingha PA, Abarashi J, Ferguson S, Aitken R (2017) From "participant" to "friend": the role of Facebook engagement in ethnographic research. J Cetacean Res Manag 20(4):416–434
108. Finfgeld-Connett D (2010) Generalizability and transferability of meta-synthesis research findings. J Adv Nurs 66(2):246–254

109. Moss RH, Edmonds JA, Hibbard KA, Manning MR, Rose SK, Van Vuuren DP, Carter TR, Emori S, Kainuma M, Kram T, Meehl GA (2010) The next generation of scenarios for climate change research and assessment. Nature 463(7282):747–756
110. Onwuegbuzie AJ, Leech NL, Collins KM (2010) Innovative data collection strategies in qualitative research. Qualitative report 15(3):696–726
111. Sims DA (2023) Reimaging reflexivity through a critical theoretical framework: autoethnographic narratives on becoming a (de) colonised researcher. CritAl Stud Teach Learn 11(1):120–138
112. Donaldson SI, Christie CA, Mark MM (eds) (2014) Credible and actionable evidence: the foundation for rigorous and influential evaluations. Sage Publications
113. Rahman M, Deuchar R (2024) Ethics in qualitative criminological research: risks, boundaries, emotions, and reflexivity. Taylor & Francis
114. Weatherford J (2019) A Phenomenological study of how researchers practice reflexivity. University of Wyoming
115. Park YH, Lee DH, Lee JH (2024) A comprehensive review: robot-assisted treatments for gait rehabilitation in stroke patients. Medicina 60(4):620
116. Parkinson SE (2022) (Dis) courtesy bias:"Methodological cognates", data validity, and ethics in violence-adjacent research. Comp Pol Stud 55(3):420–450
117. Zhang Y, Segal A, Pompedda F, Haginoya S, Santtila P (2022) Confirmation bias in simulated CSA interviews: how abuse assumption influences interviewing and decision-making processes? Leg Criminol Psychol 27(2):314–328
118. Chen L, Tse HW, Wu D, Young MEDT (2021) Cross-cultural researchers' positionality in immigrant health research: Reflections on conducting research on Chinese immigrants' experiences in the United States. Int J Qual Methods 20:16094069211052190
119. Le HV (2024) Enhancing research insights: the impact of phenomenographic analysis in educational studies. Qual Res J
120. Almusaed A, Almssad A, Yitmen I, Wallhagen M, Yang YF, Iyer-Raniga U (2024) Integrative approaches in urban sustainability: architectural design, technological innovations and social dynamics in global contexts. BoD–Books on Demand
121. Almusaed A, Yitmen I, Almssad A (2024) Contemporary innovations and sustainable practices in the application of clay materials within architectural design and construction methodologies
122. Bingham AJ (2023) From data management to actionable findings: a five-phase process of qualitative data analysis. Int J Qual Methods 22:16094069231183620
123. Braun V, Clarke V (2024) How do you solve a problem like COREQ? A critique of consolidated criteria for reporting qualitative research. Methods Psychol 100155
124. Ahmadi H (2024) The impact of green walls on community well-being and environmental quality: a study of relevant professionals' perspectives in Vancouver (Master's thesis, Royal Roads University (Canada))
125. Naeem M, Ozuem W, Howell K, Ranfagni S (2023) A step-by-step process of thematic analysis to develop a conceptual model in qualitative research. Int J Qual Methods 22:16094069231205788
126. Omrany H, Ghaffarianhoseini A, Ghaffarianhoseini A, Dhawan K, Almhafdy A, Oteng D (2024) The use of city information modelling (CIM) for realizing zero energy community: a path towards carbon neutrality. In City Information Modelling (pp 215–247). Singapore: Springer Nature Singapore
127. Sampaio S, Sousa PR, Martins C, Ferreira A, Antunes L, Cruz-Correia R (2023) Collecting, processing and secondary using personal and (pseudo) anonymized data in smart cities. Appl Sci 13(6):3830
128. Kirk HR, Birhane A, Vidgen B, Derczynski L (2022) Handling and presenting harmful text in NLP research. arXiv:2204.14256
129. Guelmami N, Ben Ezzeddine L, Hatem G, Trabelsi O, Ben Saad H, Glenn JM, El Omri A, Chalghaf N, Taheri M, Bouassida A, Ben Aissa M (2024) The ethical compass: establishing ethical guidelines for research practices in sports medicine and exercise science. Int J Sport Stud Health 7(2):31–46

Data Analysis in Qualitative Research

Abstract Deep insight into human feelings, behaviors, and social structures is revealed via a potent instrument called qualitative data analysis. Unlike quantitative research, which depends on statistical computations, qualitative approaches capture the intricacy of actual interactions and provide rich, context-driven interpretations. This chapter shows how researchers could uncover significant trends and associations from unstructured data using methodical techniques to qualitative analysis, including theme coding, grounded theory, and discourse analysis. This chapter emphasizes best practices for guaranteeing credibility, validity, and openness in qualitative research by addressing methodological rigor, ethical concerns, and creative analytical tools. It also looks at how mixed-methods approaches, visualizing tools, and AI-driven analysis, which improve study clarity and impact, are changing their function. Whether looking at social behaviors, policy development, or sustainable design, qualitative research is still vital in helping to shape evidence-based choices and innovative ideas. Researchers, business leaders, and academics wishing to improve their qualitative research methods and uncover innovative ideas from challenging data should all read this chapter.

1 Introduction

An essential feature of academic research, qualitative data analysis, is the methodical study and interpretation of non-numerical data to provide critical understanding. England claims that quantitative research uses approaches meant to explain events using numerical data collecting, which is later subjected to mathematical analysis usually utilizing statistical methods [1]. In contrast, quantitative analysis stresses statistical correlations and numerical measures, qualitative analysis uses thorough descriptions and contextual knowledge to explore concepts, attitudes, and experiences. According to Carey, these methodologies allow scientists to examine cultural elements more successfully in a way consistent with public health viewpoints [2]. This analytical approach is much used in the humanities, social sciences, and other

A. Almusaed et al., *Practice of Research Methodology in Civil Engineering and Architecture*, https://doi.org/10.1007/978-3-031-97393-2_14

disciplines where the complexity of human experiences demands careful examination. Descriptive and exploratory, qualitative data spans interviews, focus groups, open-ended survey responses, field notes, and multimedia materials. Unlike numerical datasets for which statistical computation is available, qualitative data needs interpretative approaches to identify patterns, themes, and underlying meanings. Given the interpretive character of qualitative research, maintaining methodological rigor is essential to guaranteeing its validity and dependability. Singh notes partial application of techniques to improve rigor and credibility in qualitative research. Still, there are gaps in how quality standards are applied in important respects like credibility, transferability, reliability, confirmability, and ethical issues [3]. Above all, qualitative and quantitative research's varied methodologies and objectives define them distinctly. Whereas qualitative research helps to deepen knowledge of events through thorough narratives, quantitative research seeks to quantify variables and discover statistical correlations. Taherdoost contends that the great variety of research design choices makes it difficult for researchers to choose the most suitable technique for their study and separate methodological variations sufficiently [4]. While quantitative research is typically generalizable, qualitative research aims for depth and occasionally includes directly interacting with a smaller, carefully chosen group to investigate complex social and behavioral dynamics. Using the identification of patterns, meanings, and lived experiences, qualitative data analysis aims to deepen knowledge of social and behavioral events. When investigating subjective experiences, this method is beneficial as it lets scientists look at how people see, understand, and react to different facets of their surroundings. Methodologically, according to Shava and Shonhiwa, qualitative research often uses a case study design—is anchored in the interpretative paradigm. Usually, data collecting techniques consist of observations, interviews, and document analysis [5].

To find student participation and learning techniques, for instance, researchers may look at classroom interactions; to better know their relationships with engineering colleges, they might do patient interviews. Another crucial aspect is the ability of qualitative data analysis to probe social interactions, beliefs, and human behaviors. Unlike quantitative research, which isolates variables to test hypotheses, qualitative research explores the complexity of human occurrences inside their natural environments. This analytical technique helps identify social constructs, cultural influences, and personal stories impacting group and individual behavior. Using carefully qualitative data, researchers may identify emergent patterns and thematic trends, therefore enabling the construction of theoretical frameworks and models that enhance understanding of social occurrences [6]. Furthermore, qualitative data analysis is crucial for identifying trends, opportunities, and challenges in many sectors. In architectural engineering research, for instance, qualitative methods are used to explore user experiences in built environments, assess sustainability practices, and evaluate the impact of design decisions on occupant well-being. Researchers employ interviews, case studies, and observation to understand how architectural settings influence human behavior, comfort, and efficiency [7]. These qualitative methods inform evidence-based design solutions that meet functional and aesthetic criteria. While qualitative data analysis offers many advantages, it also presents significant methodological challenges, particularly regarding subjectivity and bias. Because qualitative research largely relies on interpretation, the researcher's perspectives, assumptions, and theoretical frameworks can inadvertently shape the outcomes. In contrast to quantitative research, which relies on numerical data for objective assessments, qualitative approaches require careful reflexivity to ensure validity and reliability in the study. Unlike quantitative research in which numerical data enables objective measurements, qualitative findings depend on human judgment and are therefore more sensitive to bias. Those who wish to reduce this anxiety have to be reflexive and consider their stance and its consequences on the course of the research. Moreover, triangulation by comparing various data sources, points of view, or techniques helps strengthen and validate qualitative findings. Another main challenge in qualitative data analysis is organizing enormous volumes of unstructured data. Unlike controlled quantitative statistics, qualitative data often consists of thorough textual or multimedia resources like field notes, interview transcripts, and video recordings. Qualitative data's sheer volume and complexity demand rigorous organization, coding, and classification to support critical research. Researchers often employ qualitative data analysis tools, such as NVivo, ATLAS.ti, or MAXQDA, to assist with coding and topic identification. Still, even with tools, qualitative research is an iterative, time-consuming endeavor requiring meticulous attention to detail and ongoing progress [8]. Maintaining credibility and reliability presents even another challenging aspect of qualitative research. Since qualitative research depends on other methods to establish the validity of results, instead of statistical validation. Member checking is a commonly utilized technique whereby researchers present their interpretations to study subjects to verify the validity and accuracy of the results. By integrating several points of view, peer debriefing—discussing analytical results with other researchers—enhances credibility. Keeping an audit trail that carefully notes methodological decisions and analytical results helps foster transparency and facilitates independent evaluation of the research process. Moreover, ethical questions are crucial in qualitative research, particularly about informed permission, confidentiality, and responsible participant representation. Qualitative research might include sensitive personal experiences, so researchers must closely ensure participant identification and data confidentiality protection. Moreover, the presentation of the results should be such that it respectfully and precisely reflects the participants' voices, avoiding misinterpretation or decontextualizing of their opinions [9]. Thorough investigation of challenging social and behavioral phenomena by researchers depends on qualitative data analysis. Emphasizing nuanced descriptions, interpretative meaning-making, subject study, and qualitative research provides insightful analysis that improves quantitative results. However, the method has flaws in research subjectivity, data management difficulty, and trustworthiness issues. These challenges demand rigorous methodological tools including reflexivity, triangulation, and member validation to raise the validity and dependability of qualitative findings. Although qualitative research continually evolves, its contributions to knowledge growth and pragmatic application are indispensable in many sectors [10].

2 Methods and Approaches in Qualitative Data Analysis

2.1 Content and Thematic Analysis for Building Engineering Research

In building engineering, knowledge of complicated events depends much on qualitative data analysis, which helps academics investigate industry methods, stakeholder viewpoints, and construction obstacles [11]. Content and topic analysis, among other qualitative research techniques, offer methodical ways to understand textual data. Particularly in qualitative research, theme analysis is extensively used to spot recurrent trends and main subjects. This approach is methodically based on data familiarization, coding, topic discovery, refining, and reporting. In engineering research, thematic analysis may be used to investigate professionals' encounters with sustainable building [12]. For example, interviews with engineers and building managers could reveal essential elements including material efficiency, regulatory difficulties, and energy performance. These revelations help academics better grasp how environmental ideas are included in building techniques and the obstacles to their acceptance. Examining qualitative data helps academics to gain a better understanding of how ecological factors affect building choices. Improving construction sector sustainability plans depends on this knowledge. By exposing real-world difficulties and possibilities, rigorous thematic analysis helps qualitative research contribute to industry breakthroughs, thereby directing policy creation, technical improvements, and more efficient building management methods. Conversely, content analysis methodically arranges textual material into preset categories, usually using quantitative metrics as word frequency counts. Content analysis offers a methodical approach to evaluate trends over several datasets, unlike theme study, which stresses interpretation [13]. Researchers examining project data from different construction companies, for example, could classify the material into essential topics like structural integrity, cost control, sustainability measures, and safety rule adherence. Comparing the frequency of these issues across studies helps scientists to spot industry-wide goals and new trends. Examining technical documents, regulatory texts, and past project records especially benefits from this approach. Two qualitative research techniques often employed in building engineering studies—that of topic and content analysis—are compared in Table 1. It offers valuable applications and field-based examples, emphasizing their main phases, benefits, and constraints. While content analysis methodically sorts material and provides qualitative and quantitative insights for building and engineering studies, theme analysis concentrates on spotting trends in textual data.

Table 1 Comparative study of content analysis and thematic approach in construction engineering research

Method	Description	Key steps	Advantages	Limitations
Thematic analysis [14]	A qualitative research method for identifying, analyzing, and interpreting patterns or themes within textual data. It helps researchers derive meaning from large volumes of textual information	• Familiarization with data • Coding • Theme identification • Refinement • Reporting • Verification	• Captures rich, in-depth insights • Flexible approach for different types of data • Can uncover hidden meanings and patterns	• Requires significant time and expertise • Interpretation can be subjective • Difficult to ensure reliability across multiple coders
Content analysis [15]	A systematic and objective approach to analyzing textual data by classifying content into predefined categories enables qualitative insights and quantitative assessment	• Defining categories • Systematic classification • Word frequency counts • Contextual analysis • Thematic interpretation	• Provides structured and systematic analysis • Enables quantitative insights through frequency analysis • Suitable for analyzing large datasets	• May overlook contextual nuances • Quantification may not fully capture meaning • Requires careful category definition

2.2 Construction Studies Grounded Theory and Narrative Analysis

Grounded theory is another often used qualitative method based on developing ideas from methodically obtained data. Unlike theme or content analysis, grounded theory uses an inductive coding technique rather than starting with established categories [16]. First, open coding is used to locate essential ideas; second, axial coding is used to create linkages between these ideas; and last, selective coding is used to improve the results by building a theoretical framework. Grounded theory uses building engineering to examine the acceptance of prefabricated construction methods. Interviews with engineers, project managers, and legislators help

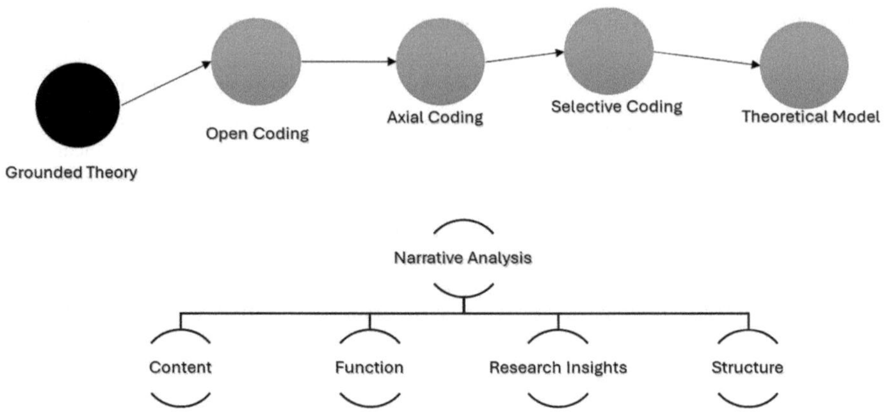

Fig. 1 Grounded theory and narrative analysis graphically represented. *Source* Authors

researchers to pinpoint essential elements affecting the acceptance of prefabrication, including design constraints, labor skill shortages, and supply chain inefficiencies: The resultant theoretical model clarifies for stakeholders the main forces behind prefabrication in building projects as well as the main obstacles [17]. Conversely, narrative analysis emphasizes how people create meaning using their experiences. This method is beneficial in knowing how builders see and deal with project difficulties. For example, examining human stories from engineers engaged in unsuccessful building projects can offer rich insights on risk management, communication failures, and decision-making mistakes. Researching these stories' structure, purpose, and substance helps one spot repeating elements causing project failures and suggest ways to better project management. Capturing historical viewpoints and enabling academics to record changes in engineering techniques across time, depending on firsthand experiences from seasoned experts, also makes narrative analysis quite helpful [18]. Figure 1 graphically shows the Grounded Theory and Narrative Analysis procedures in qualitative research. Grounded Theory develops a theoretical model by iteratively using the Open, Axial, and Selective Coding system. Structural, functional, and content analysis of narratives helps one to gain study ideas. This graphic emphasizes the building engineering study methodical technique to grasp patterns, themes, and meaning development.

2.3 Interpretative Strategies and Discourse Analysis in Construction Project Management

Another qualitative technique looking at language, power relationships, and social dynamics inside communication is discourse analysis [19]. This method is beneficial for researching how various stakeholders view problems connected to building in public debate. Scholars studying policy debates on smart cities, for instance, can examine how engineers, urban designers, and government officials present their viewpoints on sustainable infrastructure. Policymakers could stress economic development, while engineers would stress smart infrastructure's technological viability and long-term sustainability. Researchers can expose the underlying power dynamics and ideological forces influencing the evolution of smart cities and building policies using speech analysis, regulatory documentation, and media coverage [20]. Apart from these fundamental approaches, qualitative research in building engineering sometimes uses interpretive techniques to improve data analysis. Reflexivity, for example, is realizing how the researcher shapes the perception of the facts, guaranteeing the analytical process openness. Triangulation helps confirm results and raise dependability by combining several data sources or techniques. Contextualization helps place research results in their social, cultural, and historical settings, illuminating industry patterns and issues. Construction studies have used these interpretive techniques to examine the uptake of Building Information Modeling (BIM). Triangulating data from observational studies, project reports, and interviews helps academics fully grasp the advantages and drawbacks of BIM adoption and industry opposition to technological change [21]. Qualitative data analysis is necessary for investigating intricate building techniques, industry trends, and stakeholder relationships. Grounded theory helps create theoretical models from real-world data; themes and content analysis provide methodical means to find trends in textual data; narrative analysis captures personal and historical viewpoints; and discourse analysis investigates the part language plays in determining industry developments. Applying these approaches in building engineering research can help academics and professionals better grasp construction difficulties, new trends, and possible improvement in industry practice. Table 2 lists main interpretive techniques and discourse analysis approaches applied in qualitative research on building project management. Emphasizing their relevance in comprehending stakeholder viewpoints, industry trends, and linguistic dynamics, it underlines their descriptions, uses, and problems. To guarantee dependability in building studies, challenges like researcher bias, data discrepancies, and methodological complexity highlight the requirement of thorough investigation.

3 Techniques for Coding and Categorization

3.1 Qualitative Coding Stages

Analyzing textual, visual, or audio data in qualitative research involves systematically coding and categorizing patterns, themes, and relationships [28]. Coding is a fundamental analytical technique that aids researchers in structuring, understanding, and efficiently organizing vast volumes of qualitative data. Through thorough classification, researchers can extract significant insights, identify emerging trends, and develop theoretical frameworks that enhance knowledge of the studied phenomena

Table 2 Constructive project management interpretive strategies and discourse analysis

Aspect	Description	Challenges
Discourse analysis [19]	A qualitative technique that examines language, power relationships, and social dynamics in communication	It can be subjective and requires a deep contextual understanding of language and power structures
Application in construction research [22]	Used to analyze how stakeholders perceive construction-related issues in public debate, including sustainable infrastructure and smart cities	Difficulty in balancing perspectives of different stakeholders and avoiding bias in interpretation
Key interpretative strategies [23]	Techniques that enhance qualitative data analysis by ensuring validity, reliability, and contextual relevance	Time-consuming; requires expertise to ensure methodological rigor
Reflexivity [24]	Recognizing the researcher's influence on data interpretation ensures transparency in the analytical process	Can introduce researcher bias if not carefully managed
Triangulation [25]	Combining multiple data sources or methods to validate findings and improve reliability	Data inconsistencies across sources can affect reliability
Contextualization [26]	Situate research findings within social, cultural, and historical contexts to understand industry trends	Historical and cultural influences can be challenging to quantify
Example from construction studies [27]	Examining the adoption of Building Information Modeling (BIM) by triangulating data from project reports, interviews, and observational studies to assess benefits, challenges, and resistance to technological change	Resistance to change and data availability issues in construction studies

[29]. Qualitative coding labels data points based on their content, thus grouping material into themes or categories. This process can be deductive, where predefined codes guide the research, or inductive, where codes emerge from the data. By thoroughly analyzing these trends, researchers can create nuanced interpretations that reflect the complexity of human experiences, behaviors, and social relationships. Furthermore, coding ensures that results in qualitative research are grounded in the data, promoting rigor and transparency. It helps researchers understand diverse perspectives and contextual elements influencing the phenomenon. By transforming unstructured data into coherent, interpretable insights that support theoretical and practical advancements, systematic coding and classification ultimately enhance the credibility and depth of qualitative research [30].

3.1.1 Open Coding

The first phase in qualitative data analysis is open coding, where researchers break the data into smaller pieces and give descriptive codes highlighting essential trends [31]. This exploratory phase lets codes develop straight from the data rather than depending on predefined labels. Usually when reading transcripts closely, researchers label important parts with codes that reflect the significance of every segment. Open coding's main objective is to create an extensive collection of starting programs covering many facets of data [32]. Studies on sustainable building methods, for example, might produce labels like "energy-efficient materials," "construction waste management," or "renewable energy integration." These algorithms let researchers find several angles on the issue free from constraints based on predefined categories. This adaptability guarantees that the research stays receptive to new and data-driven ideas. Open coding, however, usually results in many codes, which call for grouping and refining them in the next phase [33]. This technique arranges similar codes into more general groups to preserve consistency and clarity in the analysis. As fresh ideas develop, researchers must evaluate, change, and improve the codes to guarantee that the results remain pertinent and thorough. In the end, open coding gives researchers a strong basis for more in-depth qualitative research so they may find significant trends and ideas in challenging data sets.

3.1.2 Axial Coding

In qualitative data analysis, axial coding is a crucial phase in which first open codes are methodically connected to create higher-order categories [34]. By identifying connections across codes, this technique enhances the analytical depth of the research, thereby promoting a more comprehensive understanding of the dataset. Researchers categorize and organize codes into overarching themes through axial coding, establishing conceptual links by analyzing relationships among context, interactions, and results. Axial coding aims to identify fundamental trends that define the dataset, thus facilitating the development of a coherent analytical framework [35]. In studies on green building certification, axial coding can uncover interactions between "material sustainability," "energy efficiency," and "long-term cost savings." Through these links, researchers may examine how industrial decision-making is influenced and how sustainability measures interact within building processes. By employing this approach, researchers transform scattered data points into a structured theoretical model that provides greater abstraction. Axial coding enhances interpretive coherence, clarifies conceptual boundaries, and aids in refining categories [36]. This method improves the validity and rigor of qualitative research by creating meaningful connections between codes, thus supporting the growth of theory and synthesizing findings. Lastly, axial coding transforms raw qualitative data into well-defined concepts, allowing for a more thorough investigation of complex social, organizational, and technological phenomena [33].

3.1.3 Selective Coding

The last level of qualitative data analysis, selective coding, allows researchers to find a core category by combining and synthesizing the whole dataset [37]. The main subject is this key category; all other categories work as supporting aspects adding context and depth. Selective coding's primary goal is to build a cohesive story or theoretical framework that reasonably clarifies the research results. During this phase, scientists create rational links between the subcategories found during axial coding and the central category. In studies of sustainable urban planning, for instance, the core category would be "resilient city design." Supporting topics concurrently may include "public transportation development," "climate adaptation strategies," and "community green spaces." Linking these components guarantees that the dataset is organized into a relevant and interpretable model using selective coding. At this point, one of the main difficulties is making sure the chosen core category comes from the data instead of reflecting researcher prejudice. Maintaining the validity of the results depends on maintaining methodological rigor. In essence, selective coding provides a disciplined model that emphasizes linkages in the dataset, therefore helping in theory construction [38]. Using this procedure, researchers can transcend descriptive analysis and reach theoretical conclusions, thereby fostering a better knowledge of the investigated phenomena.

3.2 Software-Assisted Versus Manual Coding

Both manual coding and computer-assisted qualitative data analysis tools (CAQDAS) offer different methodological advantages and limitations for qualitative data analysis [39]. Manual coding enables researchers to engage deeply with the data, fostering a complex and context-sensitive interpretation. This method is particularly beneficial for smaller datasets where thematic patterns can emerge through careful reading. However, it can be inherently time-consuming, susceptible to human error, and may lack consistency with larger datasets. In contrast, software-assisted coding tools like NVivo, ATLAS.ti, and MAXQDA enhance efficiency, repeatability, and systematic organization of large-scale qualitative data [40]. Through automated searches, hierarchical coding systems, and data visualization, these tools enable researchers to identify trends more accurately. Despite these advantages, software-assisted coding requires technical expertise and may lead to an over-reliance on automated classifications, which can diminish interpretive depth. The dataset's nature, the study's complexity, and the resources available will dictate whether manual or software-assisted coding is most suitable. A combination of both methods known as hybrid analysis can potentially enhance analytical rigor by balancing computational efficiency with qualitative richness [41].

3.2.1 Manual Coding

Without specific software, manual coding directly identifies, labels, and arranges qualitative data. Researchers classify and examine material by hand marking transcripts, using spreadsheets, or handling data on paper [42]. This approach guarantees complex interpretation by removing software-related expenses, giving more freedom in changing codes, and encouraging intimate interaction with the data. Manual coding is labor-intensive, particularly in processing big datasets, and prone to human mistakes and inconsistencies notwithstanding its advantages. It can be challenging to maintain dependability and repeatability over several phases of research, and it calls for exacting attention to detail. Nevertheless, manual coding is still valuable for small-scale qualitative research where researchers must closely investigate developing themes and subtle patterns. It is instrumental in exploratory research, when predefined coding structures are lacking, and researchers may dynamically change codes during the analytical process [43]. Although hand coding might not be feasible for big datasets, its interpretive depth and adaptability fit projects needing contextually rich, iterative analysis.

3.2.2 Software-Assisted Coding

Software for qualitative data analysis (CAQDAS) provides researchers with practical tools to manage, organize, and analyze substantial amounts of qualitative data [44]. Popular software packages such as NVivo, ATLAS.ti, and MAXQDA offer systematic coding, trend detection, and data visualization, enhancing analytical rigor and efficiency. NVivo supports survey integration, visual mapping, and advanced text searches, enabling many academics to work with multimodal datasets. It allows for hierarchical coding and analyzing various data forms, including text, audio, and video. ATLAS.ti provides network views that illustrate thematic connections, features a user-friendly drag-and-drop coding interface, and includes AI-powered text analysis tools that improve pattern detection. MAXQDA is particularly advantageous for mixed-methods research, integrating qualitative and quantitative data. It boasts robust visualization capabilities, such as code maps and heat maps, and facilitates cloud-based collaboration, allowing research teams to work seamlessly across different locations [45]. While software-assisted coding enhances productivity, maintains consistency, and accommodates large-scale qualitative research, it requires technical skills and may limit the researcher's direct engagement with the data. Conversely, manual coding enables greater flexibility and deeper interaction with data but is time-consuming and less scalable. The choice between manual and software-assisted coding depends on study objectives, dataset size, and available resources. Manual coding is suited for small-scale, exploratory studies; however, due to their systematic and reproducible nature, large and complex qualitative research projects greatly benefit from software-assisted methods [46]. Table 3 emphasizes the methodological approaches, benefits, limitations, and challenges of manual, software-assisted, and hybrid coding. While software-assisted coding enhances productivity, it requires technical knowledge. In contrast, manual coding ensures significant involvement but is labor-intensive. To address scalability issues, hybrid coding combines computational rigor with interpretive depth. The table aids researchers in selecting appropriate coding techniques based on dataset size and the complexity of the study.

3.3 Ensuring Reliability and Validity in Coding

In qualitative research, validity and reliability are crucial to ensure that coding procedures yield accurate, consistent, and meaningful findings [55]. While validity confirms that the identified themes genuinely reflect the data, establishing coding dependability reduces subjectivity. Inter-coder reliability tests, systematic creation of codebooks, and iterative improvements to the coding framework enhance coding accuracy and consistency in various ways. Reflexivity, peer debriefing, and triangulation minimize researcher bias, strengthening validity [58]. Structured coding techniques assist researchers in ensuring the repeatability and analytical rigor of qualitative results, thus enhancing the credibility and reliability of such findings [59]. In qualitative research, Fig. 2 shows triangulation to boost validity and trustworthiness. Three primary forms of data triangulation, investigator and methodological triangulation, help minimize bias, guarantee thorough analysis, and increase research dependability using several sources, points of view, and analytical strategies.

3.3.1 Intercoder Reliability

Intercoder reliability is the degree of agreement among several researchers using the same coding system over the same dataset [60]. High intercoder accuracy guarantees that coding is objective, consistent, and repeatable, lowering any biases in qualitative research. To get good intercoder reliability, compare their coding findings, and analyze variances, many researchers separately and independently code a part of the data. This procedure enhances analytical consistency using a consistent coding system and refined coding definitions. Two coders examining interviews, for instance, in research on the acceptance of renewable energy sources in metropolitan areas may often find themes like "public opinion," "technical feasibility," and "policy incentives." Consistent agreement on these issues reveals excellent intercoder dependability. Intercoder reliability promotes the validity of these results [55]. It increases qualitative research credibility by assuring uniform data interpretation among researchers, expanding the study's general analytical rigidity and dependability.

Table 3 Comparative analysis of qualitative coding methods

Feature	Manual coding	Software-assisted coding	Hybrid coding	Challenges
Methodological approach [47]	Hand-marking, labeling, and organizing data manually	Using CAQDAS to automate and structure qualitative analysis	Combining manual and software-assisted methods for a balanced approach	Maintaining consistency, avoiding research bias, managing large datasets
Advantages [48]	Greater interpretative depth, flexibility, cost-effectiveness	Efficient for large datasets, systematic organization, collaboration	Balances computational efficiency with qualitative richness	Technical learning curve, potential over-reliance on automation, software costs
Disadvantages [49]	Time-consuming, prone to human error, difficult to maintain consistency	Requires technical expertise, may reduce interpretative depth	Can be complex to implement, requires both manual effort and technical knowledge	Balancing manual effort with automation, integrating qualitative depth with computational efficiency
Best use cases [50]	Small-scale, exploratory research, nuanced interpretation	Large-scale qualitative research, mixed-methods studies, team-based projects	Studies requiring profound interpretation and systematic data organization	Ensuring replicability across research phases
Key tools/Techniques [51]	Spreadsheets, printed transcripts, hand annotations	NVivo, ATLAS.ti, MAXQDA, hierarchical coding, AI-assisted tools	Manual annotation combined with software for visualization and trend identification	Managing extensive coding manually without losing interpretative depth
Level of researcher engagement [52]	High—researchers deeply engage with data	Moderate—researchers interact through structured frameworks	High—researchers engage with both structured and unstructured coding	Software dependency and potential for misinterpretation of automated categorizations
Suitability for large datasets [53]	Low—inefficient for large datasets	High—designed for extensive data handling	High—software organizes data while researchers refine codes	Difficulties in scaling manual coding without losing detail
Suitability for small-scale studies [54]	High—ideal for small, nuanced qualitative studies	Moderate—can be used but may be excessive for small projects	High—researchers retain control over emerging themes	Integrating manual and automated approaches effectively

(continued)

Table 3 (continued)

Feature	Manual coding	Software-assisted coding	Hybrid coding	Challenges
Interpretative depth [55]	High—researchers maintain control over coding and meaning	Moderate—software categorizes data but may lose contextual richness	High—ensures contextual understanding with systematic organization	Risk of losing contextual meaning in automated analysis
Reproducibility and consistency [56]	Low—coding may vary between researchers over time	High—standardized coding ensures reproducibility	High—combines standardization with researcher-driven refinement	Ensuring thematic accuracy across different coding methods
Technical expertise required [57]	Minimal—no specialized software knowledge required	High—requires familiarity with software tools	Moderate—requires knowledge of both software and qualitative methodologies	Training requirements for both manual and software-based approaches

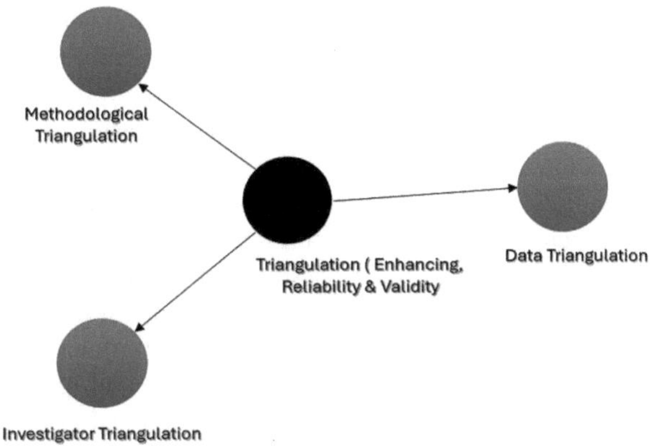

Fig. 2 Triangle model of qualitative research. *Source* Authors

3.3.2 Triangulation

By combining multiple data sources, researchers, or analytical techniques, triangulation, a systematic approach, validates study results, enhancing credibility and reducing researcher bias [61]. Triangulation improves the validity and reliability of qualitative research findings by cross-referencing various perspectives. There are several types of triangulations, each serving a different purpose:

- Data triangulation identifies consistent patterns using results from different sources, including expert interviews, policy evaluations, and field observations [62].
- In investigator triangulation, multiple researchers analyze the same dataset to compare interpretations, ensuring greater objectivity and helping to mitigate personal biases [61].
- Methodological triangulation employs qualitative and quantitative approaches such as interviews and statistical analysis to validate findings and enhance research rigor [63].

For example, in research on sustainable transportation, triangulation can be applied by incorporating evaluations of environmental impacts, traffic flow statistics, and stakeholder interviews. This approach ensures that results are not reliant on a single dataset or technique, thus strengthening the validity and credibility of the study. Triangulation provides a more comprehensive understanding of complex phenomena by integrating multiple viewpoints, ensuring that findings are more objective, reliable, and generalizable within qualitative research contexts.

3.3.3 Member Checking and Peer Debriefing

Member checking and peer debriefing are essential tools for verifying interpretations and enhancing credibility in qualitative research [64]. These techniques ensure that study results accurately reflect participant experiences, so coding judgments must remain impartial and rigorous. Returning preliminary results to research participants allows for evaluating whether the interpretations align with their experiences through member checking. This process ensures that the study properly represents individuals by enabling them to review and provide feedback, thereby improving categories, topics, and conclusions. By including direct participant validation, this approach minimizes misunderstandings and boosts the validity of the research. Peer debriefing involves the assessment of coding and analytical techniques by external researchers or professionals to identify potential biases or overlooked assumptions [65]. Introducing an outsider's perspective enhances coding structures and analytical outcomes, thus increasing the objectivity and reliability of qualitative research. It also fosters transparency by adding another layer of scrutiny, thereby reinforcing the methodological integrity of the study. Open, axial, and selective coding, among other methods, enable researchers to analyze qualitative data systematically. Triangulation, validation methods, and intercoder reliability assessments contribute to the credibility and dependability of qualitative research [66]. In sustainability studies, for instance, member checking may validate the authenticity of stakeholder perspectives on accepting renewable energy sources, while peer debriefing can refine coding judgments regarding policy frameworks. Choosing appropriate coding and validation methods ensures that research outcomes remain robust, meaningful, and aligned with study objectives.

Table 4 presents the main strategies to ensure dependability and validity in qualitative coding, emphasizing validity, intercoder reliability, triangulation, and member verification. These techniques reduce bias, enhance analytical consistency, and improve coding accuracy. While triangulation incorporates multiple perspectives, peer debriefing and member checking ensure honest and reliable results. Addressing coding challenges guarantees objective, systematic, reproducible interpretations of complex data, strengthening the rigor of qualitative research.

4 Interpreting and Presenting Qualitative Findings

An essential stage of research that calls for methodical data organization, analysis, and formal reporting is interpreting and presenting qualitative results [68]. This procedure guarantees that results are coherent, reliable, and fairly represent participant experiences. Large amounts of qualitative data must be meaningfully categorized, patterns and linkages must be established, and evidence-based interpretations must be used to bolster researchers' findings. Good qualitative research calls for a methodical methodology to guarantee dependability, openness, and clarity of data presentation. This entails spotting main ideas, combining direct data, and using transparent reporting systems. Moreover, ethical issues guarantee that participant points of view are politely and truthfully reflected [69]. The following sections describe key approaches to use empirical evidence to support interpretations, organizing research reports to improve readability and understanding, and synthesizing

Table 4 Ensuring reliability and validity in coding

Aspect	Description	Key techniques	Challenges
Validity and reliability [55]	Ensuring coding procedures yield accurate, consistent, and meaningful findings	Reflexivity, peer debriefing, triangulation, structured coding techniques	Ensuring coding consistency and reducing subjectivity in interpretation
Intercoder reliability [67]	Degree of agreement among multiple researchers using the same coding system	Comparing independently coded datasets, refining coding definitions, ensuring analytical consistency	Maintaining coder agreement over time and avoiding subjective bias
Triangulation [25]	Validating research findings by integrating multiple data sources, researchers, or methods	Data triangulation, investigator triangulation, methodological triangulation	Integrating different sources without contradictions and managing diverse datasets
Member checking and peer debriefing [65]	Confirming interpretations by involving participants and external reviewers to reduce bias	Returning preliminary findings to participants, external expert review, and refining coding structures	Ensuring participant engagement and minimizing external reviewer biases

qualitative data. Using these techniques helps qualitative researchers ensure their results are methodologically robust, insightful, and systematic.

4.1 Synthesizing Data for Insights

Organizing unstructured material into ordered categories, spotting essential themes, and creating linkages and patterns help to synthesize qualitative data [70]. This method turns uninteresting stories into insightful analysis that helps academics develop theoretical models and reach reasonable conclusions. Methodically organizing and labeling data allows researchers to find underlying linkages and repeating patterns across several points of view. This methodical technique improves data interpretation, therefore facilitating the recognition of causal linkages, trends, and correlations. In research on occupational stress, for example, topics including "job workload," "management support," and "work-life balance" could show up. Finding links among these issues helps scholars understand how organizational systems affect employee welfare [71]. This synthesis makes qualitative data more analyzable and actionable, enabling researchers to derive legitimate, evidence-based conclusions that add to academic knowledge and valuable applications.

4.1.1 Creating Themes and Categories

In qualitative data analysis, categorization comes first when researchers assign codes and themes to various data segments. This procedure is necessary to arrange several points of view into logical and significant conclusions [37].

- Open coding finds new concepts straight from the data.
- Axial coding polishes categories and creates links between themes.
- Selective coding arranges these groups into a primary story.

Research on workplace well-being may share common themes, such as "job satisfaction," "work-life balance," and "work-related stress" [72]. These themes enable researchers to logically and interpretably arrange their results and compare participant experiences. By methodically classifying data, researchers may examine trends, derive essential insights, and create theoretical models that advance their knowledge of the investigated topic.

4.1.2 Recognizing Relationships and Trends

Once the qualitative data has been organized, researchers must identify connections and recurring patterns to understand the topic comprehensively [73]. This process involves.

- comparing coded data across individuals or settings to uncover differences and similarities.
- identifying concurrent themes that reveal underlying trends in participants' experiences.
- establishing causal or correlational relationships between concepts, thereby enriching theoretical interpretations.

For example, a study on sustainability in architectural engineering research may reveal a trend in which energy-efficient building materials correlate with improved indoor environmental quality. Discovering such connections enables researchers to produce theoretically informed and data-driven conclusions. Qualitative analysis tools such as NVivo, ATLAS.ti, and MAXQDA assist in visualizing patterns through mind maps, topic clusters, and frequency matrices [74]. These tools enhance researchers' abilities to identify trends, refine analytical methods, and clearly and systematically present findings, thus improving the interpretability and depth of the research. Table 5 lists the main features of qualitative data synthesis and related difficulties such as data classification, coding systems, and pattern identification. Integrity, consistency, and meaningful interpretation are vital in qualitative research. Although analytical instruments like NVivo and MAXQDA enhance pattern recognition, researchers must navigate technological complexity, subjectivity, and bias. Addressing these challenges increases research validity, theoretical contributions, and data reliability.

Table 5 Challenges and strategies in synthesizing qualitative data

Aspect	Description	Challenges
Purpose of data synthesis [75]	Transforming unstructured narratives into insightful analysis to create theoretical models and derive conclusions	Ensure accuracy and avoid researcher bias in data interpretation
Key methods [76]	Organizing data into categories, spotting essential themes, and identifying relationships and trends	Managing large datasets while maintaining consistency in theme identification
Example themes [77]	Examples in workplace research include 'job workload,' 'management support,' 'work-life balance,' 'job satisfaction,' and 'work-related stress.'	Subjectivity in theme selection, requiring rigorous validation and peer review
Coding process [36]	Open coding (identifying concepts), Axial coding (refining categories, linking themes), Selective coding (developing a primary narrative)	Risk of over-coding or misinterpreting relationships between themes
Pattern recognition [78]	Comparing coded data across individuals/settings, identifying co-occurring themes, and establishing causal/correlational relationships	Difficulties in distinguishing correlation from causation in qualitative data
Analytical tools [79]	NVivo, ATLAS.ti, MAXQDA—Tools for visualizing patterns using mind maps, topic clusters, and frequency matrices	Technical learning curve for using software tools effectively and efficiently

4.2 Using Evidence to Support Findings

In qualitative research, honesty and openness rely on results firmly supported by empirical data. This requires anchoring interpretations in real-world participant experiences grounded in direct participant quotations, carefully selected case studies, and relevant data snippets [80]. Grounding findings in specific evidence helps researchers minimize overgeneralization and distortion, ensuring their interpretations remain close to the original facts. Providing comprehensive, verified evidence aids readers in understanding the clear connection between the facts and the interpretations, thereby enhancing the validity of the analysis [81]. This methodology demonstrates openness and trust, indicating that the research results are based on a thorough analysis of diverse, authentic voices. This transparent approach also helps to bolster confidence in the study results, thereby supporting the overall reliability and academic integrity of qualitative studies.

4.2.1 Integrating Direct Quotes, Excerpts, and Case Examples

Combining direct quotes, case studies, and observational passages anchors interpretations in participant experiences, thereby enhancing qualitative research [82]. These elements provide depth and authenticity to the context, supporting the validity and trustworthiness of the results:

- Direct quotes accurately capture participants' statements, preserving their feelings and perspectives. This approach emphasizes critical viewpoints and offers insight into personal experiences [82].
- Case studies provide comprehensive narratives highlighting human or contextual factors influencing actions or perceptions [83]. These stories illustrate how broader ideas manifest in specific settings, adding complexity and nuance.
- Observational data further enriches analysis by offering behavioral insights from real encounters, complementing thematic interpretations [84].

For example, a participant in urban sustainability research might say, "Public transport is unreliable, so people prefer driving despite traffic congestion."

This quotation highlights infrastructural inefficiencies and transit challenges in alignment with urban mobility and environmental impact concepts. When incorporating participant comments, researchers should:

- Select quotes that effectively underscore key concepts.
- Provide interpretative commentary to clarify the meaning of the excerpt.
- Anonymize participant identities to ensure ethical confidentiality.

Thoughtful integration of these components enables qualitative researchers to enhance the trustworthiness, reliability, and depth of their findings, facilitating ethical engagement with participants' voices and perspectives [85].

4.2.2 Preventing Misinterpretation and Overgeneralization

When restricted qualitative evidence is presented as always relevant, it leads to overgeneralization. This can result in erroneous conclusions that fail to represent the diversity of participant experiences accurately [86]. Researchers must adopt various measures to ensure ethical and precise reporting. First, one should recognize sample limitations and avoid making overly strong assertions. Since qualitative research is inherently context-dependent, results should be framed within the specific conditions of the study. To ensure that interpretations remain accurate and credible, researchers should differentiate between examples and broader patterns rather than making assumptions.

Moreover, caution should be taken with causal inferences. While proving causation can be challenging without additional empirical evidence, qualitative data can sometimes reveal trends and associations [87]. Researchers should carefully position their conclusions, emphasizing possible influencing factors instead of assuming a direct cause-and-effect relationship. For example, if five out of fifteen interview respondents express dissatisfaction with remote work, stating, "most employees dislike remote work" would be misleading. A more precise and nuanced interpretation would be, "One-third of participants reported challenges with remote work, particularly regarding communication and work-life balance." Clarifying the scope and context of study results ensures that findings are understood accurately and fairly. Maintaining transparency in data presentation enables researchers to enhance the validity of their qualitative findings by avoiding distortion and bias, thereby ensuring ethical and methodologically sound reporting [88].

Table 6 describes basic strategies for using evidence in qualitative research. It emphasizes direct quotes, case studies, and observational data to improve validity and transparency. It emphasizes essential difficulties such as avoiding prejudice, choosing sample passages, and stopping misunderstanding. Ethical issues like anonymity and appropriate data handling reinforce research integrity even more. Dealing with these issues improves data credibility and supports grounded, morally sound qualitative results.

Table 6 Ensuring credibility in qualitative evidence integration

Aspect	Description	Challenges
Purpose of evidence integration [89]	Ensuring qualitative findings are grounded in empirical data to enhance validity and transparency	Ensuring accurate representation without introducing bias in interpretation
Key methods [90]	Integrating direct quotes, case studies, and observational excerpts to reinforce research interpretations	Selecting representative excerpts that best illustrate key themes
Types of evidence [91]	Direct quotes (capturing participant experiences), case studies (comprehensive narratives), observational data (behavioral insights)	Maintaining balance between context-rich narratives and concise reporting
Preventing misinterpretation [92]	Recognizing sample limitations, distinguishing specific examples from broader patterns, and avoiding causal overgeneralization	Clarifying research scope to prevent overgeneralized conclusions
Ethical considerations [93]	Ensuring anonymity, selecting quotes thoughtfully, and maintaining transparency in data presentation	Ethically handling sensitive participant data while preserving authenticity

4.3 Structuring the Research Report

A well-organized qualitative research report must be morally sound, logically consistent, and aesthetically pleasing. Good organization guarantees that results are understandable, enabling readers to comprehend important conclusions and relevance [81]. Presenting themes, supporting data, and analytical interpretations make good research reporting appealing. Organizing the report logically, including an introduction, techniques, findings, and analysis, helps maintain clarity and interest. Credibility also depends on open reporting. Comprehensive methodological explanations will help researchers guarantee that results are reproducible and traceable. Using tables, diagrams, and idea maps improves understanding and facilitates access to difficult links between subjects [94]. Further adding to the quality and dependability of the report are ethical issues, including maintaining participant anonymity and guaranteeing fair portrayal of points of view. Researchers may clearly express perceptive, trustworthy, and insightful qualitative results using a methodologically sound, well-documented technique.

4.3.1 Writing Explicit and Coherent Results

A well-organized results section in qualitative research should clearly and logically present results so readers can easily trace the development from raw data to theme interpretation [95]. The presentation of results should maintain coherence, consistency, and clarity, ensuring that readers fully understand the relevance of the conclusions. Researchers should employ a methodical strategy to achieve this:

- Including Important Themes: Summarize the most significant results and highlight the study's main ideas [96].
- Offering Theme Breakdowns: Use direct quotations, case studies, and data snippets to emphasize present themes and support validity by demonstrating participant viewpoints [97].
- Understanding Results: Align outcomes with theoretical models and study objectives to clarify trends and connections [98].

For example, in a study on online learning experiences, a section on organized findings may include.

- Theme 1: Architects and engineers underlined difficulties like high starting prices, restricted market availability, and a lack of legislative incentives for using environmentally friendly products.
- Theme 2: Benefits of Integrating Sustainable Design Participants saw that passive design techniques and energy-efficient materials improved building performance and significantly lowered long-term operating costs.

In architecture and civil engineering, qualitative research can clearly and methodically convey helpful ideas and promote sustainable building techniques by presenting data organized. Logically arranging results enhances the credibility, clarity, and readability of researchers. Subheadings, tables, and conceptual frameworks can strengthen the report's structure, ensuring the results are well-documented and easily understood. A well-crafted findings section in qualitative research connects data and interpretation by providing insightful, evidence-based conclusions that uphold methodological rigor and transparency [99].

4.3.2 Qualitative Reporting Ethical Issues

Protecting participants' rights and ensuring the accurate and responsible presentation of results depend on maintaining ethical integrity in qualitative research. Ethical reporting encompasses several essential factors:

- Researchers have to safeguard participant identities using aliases, pseudonyms, or data anonymization methods to maintain confidentiality and anonymity [100]. Sensitive material should be removed or modified to prevent inadvertent publication.
- Participants must be well-informed about how their data will be used and provide explicit consent before their comments are included in study reports [101]. This ensures moral transparency and accountability.
- A well-documented study presents multiple perspectives rather than selecting stories that align with researchers' assumptions [102]. This enhances the study's credibility and validity.

In a study on sustainable building practices, researchers questioning architects and developers about obstacles to green construction must ensure that company names, project details, or proprietary techniques are anonymized to prevent the disclosure of private business information. Moreover, if a participant faces challenges related to financial constraints or regulatory approvals, their comments should be documented appropriately to avoid professional repercussions for their views. Prioritizing ethical responsibility helps qualitative researchers uphold academic integrity, respect participant confidentiality, and enhance study credibility. Ethical reporting ensures that conclusions are honest, respectful, and bias-free, promoting responsible information sharing in sustainable building research.

4.3.3 Concept Maps, Coding Frameworks, and Visual Representations

Improving the accessibility and clarity of qualitative research findings depends critically on visual tools. Integration of idea maps, coding frameworks, and structured diagrams helps researchers to clearly show linkages, patterns, and decision-making procedures, thereby improving the interpretability and appeal of complex data [103]. Commonly employed in qualitative research are several forms of visual representations:

- Concept maps provide links between ideas and topics, supporting their demonstration. In sustainable design, for instance, a concept map may explain how "thermal insulation" links to "energy efficiency" and "building performance."
- Coding frameworks generally use software-generated coding trees (e.g., NVivo) to show topic groupings and subcategories. Such models offer a disciplined way to show how qualitative data has been arranged and examined.
- Researchers may spot trends and connections among several subjects by using flow charts and matrices, which assist in visualizing relative insights and decision-making procedures.

A graphical model may show, for instance, in research on the acceptance of renewable energy in urban buildings:

- High expenses, ignorance, legal restrictions, and barriers to adoption.
- Government incentives, public awareness campaigns, technology development, all help adoption.

Graphs, coding diagrams, and mind maps help researchers improve qualitative results' readability, interpretability, and clarity [104]. These visual tools clarify complex relationships and enable stakeholders and legislators to understand important discoveries rapidly, increasing qualitative research's value and applicability. Emphasizing logical organization, ethical issues, and efficient data presentation, Table 7 summarizes main ideas for arranging qualitative research reports. It emphasizes difficulties such as juggling depth with readability, guaranteeing participant anonymity, and choosing suitable visual aids. In qualitative research, by tackling these difficulties researchers may improve clarity, credibility, and transparency, thereby ensuring that conclusions are well-documented, morally sound, and methodologically robust.

Presenting qualitative results requires methodical data synthesis, robust evidence backing, and honest reporting. Structured themes, strong links, and the integration of empirical data help researchers assure credibility and depth in their studies [108]. Ethical concerns and graphic representations improve transparency and accessibility further. A well-organized research report makes conclusions relevant and actionable by bridging qualitative insights with pragmatic uses.

5 Conclusion

Qualitative data analysis is fundamental to research methods focused on understanding human behavior, experiences, and social events. While quantitative research emphasizes numerical correlations and statistical validation, qualitative methods prioritize depth, context, and interpretive analysis. This chapter discusses the intricate procedures of qualitative data analysis alongside the necessity of methodological rigor, systematic coding, and ethical considerations to ensure the validity and reliability of research outcomes. In qualitative research, the ability to systematically

Table 7 Best practices and challenges in structuring qualitative research reports

Aspect	Description	Challenges
Purpose of structuring the research report [68]	Ensuring qualitative reports are logically organized, ethically sound, and visually clear for better interpretation	Ensuring clarity in complex data presentation and maintaining logical coherence
Key elements [105]	Presenting themes, supporting data, and analytical interpretations in a structured manner using subheadings, tables, and conceptual frameworks	Balancing detailed descriptions with concise reporting to maintain readability
Ethical considerations [106]	Maintaining participant anonymity, securing informed consent, and ensuring balanced representation of perspectives	Protecting participant identities while maintaining authenticity in data representation
Use of visual representations [107]	Using concept maps, coding frameworks, and structured diagrams to clarify relationships, patterns, and decision-making processes	Choosing the most effective visual tools without oversimplifying complex relationships

categorize and interpret large amounts of unstructured data is essential. Through open, axial, and selective coding, researchers can organize textual, visual, or auditory materials into relevant themes, thereby revealing patterns and connections that support theoretical advancements. Key tools for identifying trends in qualitative data include thematic and content analysis, which enable researchers to draw insightful conclusions about social behaviors, institutional processes, and industrial challenges. Merging grounded theory with discourse analysis further enhances the analytical process by providing a framework for understanding how language, power dynamics, and cultural factors influence human interactions. Although qualitative research offers a comprehensive exploration of complex issues, it also presents methodological challenges, particularly regarding researcher subjectivity and bias. Unlike quantitative research, qualitative analysis primarily relies on researcher interpretation, where statistical methods provide objective verification. This subjectivity necessitates using validation strategies such as triangulation, peer debriefing, and member checking to enhance the reliability and trustworthiness of results. Data management remains another significant challenge in qualitative research. Researchers must employ systematic coding techniques to effectively organize and analyze the large volumes of textual and multimedia data generated from observations, focus groups, and interviews. In this context, the choice between manual and software-assisted coding is critical. While manual coding is labor-intensive and susceptible to errors, it allows for a deeper engagement with the data. On the other hand, computer-assisted qualitative data analysis tools like NVivo, MAXQDA, and ATLAS.ti facilitate coding, boost productivity, and enhance the consistency of findings. However, software-assisted coding requires technical proficiency, and excessive reliance on automated classifications risks undermining the interpretive depth of qualitative results. Combining computational methods and manual coding provides the optimal blend that ensures analytical rigor and interpretive depth.

In qualitative research, ethical issues take top priority, particularly regarding confidentiality, informed consent, and responsible representation of participants' perspectives. When addressing sensitive subjects, researchers must take steps to protect participant confidentiality. Utilizing pseudonyms, anonymization techniques, and secure data storage ensures that participants' identities remain private. Another critical aspect is informed consent, which requires researchers to explain how participant data will be used clearly and to obtain explicit permission before including participant quotes in study publications. Ethical integrity also involves presenting multiple viewpoints, ensuring that study results do not merely reflect preconceived notions. In studies involving business leaders, such as those on sustainable architecture and urban development, anonymity is essential to prevent potential repercussions for anyone sharing negative opinions on company practices or legislative frameworks. In qualitative research, misinterpretation and overgeneralization pose significant risks. Unlike quantitative studies that aim for statistical generalizability, qualitative research is inherently context dependent. Results should be framed within the specific parameters of the research to avoid broad statements that extend beyond the sampled population. Researchers must distinguish specific examples from broader patterns, ensuring that empirical data informs conclusions rather than assumptions. Causal inferences should be cautiously approached, as qualitative data indicates correlations rather than definitive cause-and-effect relationships. Using precise language and well-contextualized remarks helps enhance the validity of qualitative results and prevent data distortion.

Results of qualitative research should be presented in a methodologically sound, cohesive, and organized manner. By maintaining a logical flow and analytical clarity, a well-structured research report allows readers to trace the development of results from raw data to thematic interpretation. Subheadings, tables, coding frameworks, and conceptual diagrams enhance readability and assist researchers in systematically presenting their findings. Visual aids like mind maps, thematic clusters, and frequency matrices highlight connections between topics and emphasize key trends, facilitating data analysis. These visual tools are particularly beneficial in studies examining industry trends, policy implications, and organizational behaviors, as they enable stakeholders and decision-makers to grasp the practical relevance of research results quickly. Direct participant quotes, case studies, and observational insights provide concrete evidence to support thematic interpretations and further validate qualitative research. Effective qualitative reporting also ensures that study objectives and theoretical frameworks align with research outcomes, thus bolstering the scholarly contribution of the work.

Many methodological innovations are changing the area as qualitative research develops. Large-scale qualitative research is becoming more effective as artificial intelligence and machine learning are integrated into qualitative analysis, improving pattern identification, text mining, and automated coding. Combining qualitative and quantitative approaches and mixed methods approaches allows statistical rigor to be merged with contextual richness, offering a more comprehensive understanding of study events. Ethical frameworks are also evolving to address growing concerns in qualitative research, particularly in digital ethnography and online data collection, where privacy and data security necessitate careful navigation. The increasing reliance on digital visualization tools and interactive data presentation transforms how qualitative research findings are conveyed, thus making research more accessible to various audiences, including legislators, business leaders, and community activists.

One cannot overemphasize the significance of qualitative data analysis in research. Its strength lies in its ability to reveal nuances, explore personal experiences, and provide a deeper understanding of social, cultural, and behavioral phenomena. Although methodological challenges such as data management, researcher bias, and ethical considerations must be cautiously addressed, applying rigorous analytical methods ensures the validity and reliability of qualitative findings. Through transparent reporting, triangulation, and systematic coding, researchers can enhance the dependability of their qualitative studies, thereby contributing valuable insights to academia, business, and policymaking. Ultimately, the role of qualitative research extends beyond theoretical exploration, offering practical solutions and guidance for evidence-based decision-making across numerous fields. Qualitative research will continue to be a powerful tool for understanding the complexities of human interactions and societal changes by consistently refining methodological techniques and embracing technological innovations.

References

1. England A (2021) Quantitative and qualitative research methods. Res Med Imaging Radiat Sci 71–96
2. Carey JW (1993) Linking qualitative and quantitative methods: Integrating cultural factors into public health. Qual Health Res 3(3):298–318
3. Singh N, Benmamoun M, Meyr E, Arikan RH (2021) Verifying rigor: analyzing qualitative research in international marketing. Int Mark Rev 38(6):1289–1307
4. Taherdoost H (2022) What are different research approaches? Comprehensive review of qualitative, quantitative, and mixed method research, their applications, types, and limitations. J Manag Sci & Eng Res 5(1):53–63
5. Muzari T, Shava GN, Shonhiwa S (2022) Qualitative research paradigm, a key research design for educational researchers, processes and procedures: a theoretical overview. Indiana J Humlties Soc Sci 3(1):14–20
6. Naeem M, Ozuem W, Howell K, Ranfagni S (2023) A step-by-step process of thematic analysis to develop a conceptual model in qualitative research. Int J Qual Methods 22:16094069231205788
7. Almusaed A, Yitmen I, Almssad A, Myhren JA (2024) Construction 5.0 and sustainable neuro-responsive habitats: Integrating the brain–computer interface and building information modeling in smart residential spaces. Sustainability 16(21):9393
8. Sutopo AH (2023) Unlocking the power of writing skills: insights and findings using Nvivo. Topazart

9. Bang TC (2024) Ensuring credibility and trustworthiness in qualitative inquiries. In applied linguistics and language education research methods: Fundamentals and innovations, pp 70–85. IGI Global
10. Taguchi N (2018) Description and explanation of pragmatic development: quantitative, qualitative, and mixed methods research. System 75:23–32
11. Shokri-Ghasabeh M, Chileshe N (2014) Knowledge management: barriers to capturing lessons learned from Australian construction contractors perspective. Constr Innov 14(1):108–134
12. Brewster DI (2022) A qualitative case study on achieving operational sustainability within a small business in the architectural and engineering industry (Doctoral dissertation, Trident University International)
13. Gheyle N, Jacobs T (2017) Content analysis: a short overview. Internal research note 10
14. Vaismoradi M, Turunen H, Bondas T (2013) Content analysis and thematic analysis: implications for conducting a qualitative descriptive study. Nurs Health Sci 15(3):398–405
15. Indulska M, Hovorka DS, Recker J (2012) Quantitative approaches to content analysis: identifying conceptual drift across publication outlets. Eur J Inf Syst 21(1):49–69
16. Thornberg R, Perhamus L, Charmaz K (2014) Grounded theory. Handb Res Methods Early Child Educ: Res Methodol 1:405–439
17. Wang Q, Shen C, Guo Z, Zhu K, Zhang J, Huang M (2023) Research on the barriers and strategies to promote prefabricated buildings in China. Buildings 13(5):1200
18. Killham JE (2014) Exploring the affordances of role in the online history education project "Place out of time:" A narrative analysis (Doctoral dissertation, University of Cincinnati)
19. Ullah I, Bibi M, Begam R, Khan T, Bibi N (2024) A critical discourse analysis of language patterns and power dynamics in social media discourse. J Appl Linguist TESOL (JALT) 7(4):914–925
20. Morozov E, Bria F (2018) Rethinking the smart city. Democratizing Urban Technology 2
21. Wang J (2023) Exploring digital collaboration across knowledge boundaries: a case study of the BIM-enabled construction project (Doctoral dissertation, University of Sheffield)
22. Brusselaers N, Mommens K, Macharis C (2021) Building bridges: a participatory stakeholder framework for sustainable urban construction logistics. Sustainability 13(5):2678
23. Rose J, Johnson CW (2020) Contextualizing reliability and validity in qualitative research: toward more rigorous and trustworthy qualitative social science in leisure research. J Leis Res 51(4):432–451
24. Engward H, Davis G (2015) Being reflexive in qualitative grounded theory: discussion and application of a model of reflexivity. J Adv Nurs 71(7):1530–1538
25. Santos KDS, Ribeiro MC, Queiroga DEUD, Silva IAPD, Ferreira SMS (2020) The use of multiple triangulations as a validation strategy in a qualitative study. Cienc Saude Coletiva 25:655–664
26. Welch C, Paavilainen-Mäntymäki E, Piekkari R, Plakoyiannaki E (2022) Reconciling theory and context: how the case study can set a new agenda for international business research. J Int Bus Stud 53(1):4–26
27. Alankarage S, Chileshe N, Samaraweera A, Rameezdeen R, Edwards DJ (2023) Guidelines for using a case study approach in construction culture research: application to BIM-enabled organizations. J Constr Eng Manag 149(9):05023009
28. Parameswaran UD, Ozawa-Kirk JL, Latendresse G (2020) To live (code) or to not: a new method for coding in qualitative research. Qual Soc Work 19(4):630–644
29. Shu X, Ye Y (2023) Knowledge discovery: methods from data mining and machine learning. Soc Sci Res 110:102817
30. Mettler T (2025) Computational text analysis for qualitative IS research: a methodological reflection. Commun Assoc Inf Syst 56(1):14
31. Mezmir EA (2020) Qualitative data analysis: an overview of data reduction, data display, and interpretation. Res Humanit Soc Sci 10(21):15–27
32. Deterding NM, Waters MC (2021) Flexible coding of in-depth interviews: a twenty-first-century approach. Sociol Methods Res 50(2):708–739

33. Skjott Linneberg M, Korsgaard S (2019) Coding qualitative data: a synthesis guiding the novice. Qual Res J 19(3):259–270
34. Vollstedt M, Rezat S (2019) An introduction to grounded theory with a special focus on axial coding and the coding paradigm. Compend Early Career Res Math Educ 13(1):81–100
35. Bryda G, Sadowski D (2024) January. from words to themes: AI-powered qualitative data coding and analysis. In: World conference on qualitative research, pp 309–345. Cham: Springer Nature Switzerland
36. Alhassan I, Sammon D, Daly M, Wibisono A, Kasraian L, Nagle T, Heavin C, Dennehy D, Zamani E, Qaffas A (2023) The use of open, axial and selective coding techniques: a literature analysis of IS research
37. Salmona M, Kaczynski D (2024) Qualitative data analysis strategies. In: How to conduct qualitative research in finance, pp 80–96. Edward Elgar Publishing
38. Amadi A (2023) Integration in a mixed-method case study of construction phenomena: from data to theory. Eng Constr Archit Manag 30(1):210–237
39. Ahmad A, Newman M (2010) Computer-assisted qualitative data analysis software: an illustration of limitations and advantages
40. Kyprianou A (2020) Strategies for hiring people with disabilities to prevent a skilled worker shortage (Doctoral dissertation, Walden University)
41. Sovacool BK, Axsen J, Sorrell S (2018) Promoting novelty, rigor, and style in energy social science: Towards codes of practice for appropriate methods and research design. Energy Res Soc Sci 45:12–42
42. Kuckartz U (2014) Qualitative text analysis: a guide to methods, practice and using software. Sage
43. Yamani S (2024) Exploration of techniques for working with sparse data when applying natural language processing to assist a qualitative data analysis of a COVID-19 open innovation community
44. Tian F, Liang P, Babar MA (2022) Relationships between software architecture and source code in practice: an exploratory survey and interview. Inf Softw Technol 141:106705
45. Keskin B, Salman B, Koseoglu O (2022) Architecting a BIM-based digital twin platform for airport asset management: a model-based system engineering with SysML approach. J Constr Eng Manag 148(5):04022020
46. Owusu-Mensah D, Naifei R, Brako L, Boateng P, Darkwah WK (2020) Analysis of production system management of Ghana's food and beverage industry: empirical evidence from spare parts inventory control, production quality and maintenance modeling. J Food Ind 4(1):1–43
47. Burrows S, Gurevych I, Stein B (2015) The eras and trends of automatic short answer grading. Int J Artif Intell Educ 25:60–117
48. Williams IP, Bryan S (2007) Cost-effectiveness analysis and formulary decision making in England: findings from research. Soc Sci Med 65(10):2116–2129
49. Chen JY, Barnes MJ (2014) Human–agent teaming for multirobot control: a review of human factors issues. IEEE Trans Hum-Mach Syst 44(1):13–29
50. Tight M (2017) Understanding case study research: Small-scale research with meaning. Sage
51. Peachey N (2018) Digital tools for teachers-trainers' edition V. 2 (Vol 186). PeacheyPublications Ltd
52. Vaughn LM, Whetstone C, Boards A, Busch MD, Magnusson M, Määttä S (2018) Partnering with insiders: a review of peer models across community-engaged research, education and social care. Health Soc Care Community 26(6):769–786
53. Boppiniti ST (2020) Big data meets machine learning: strategies for efficient data processing and analysis in large datasets. Int J Creat Res Comput Technol Des 2(2)
54. Vasileiou K, Barnett J, Thorpe S, Young T (2018) Characterising and justifying sample size sufficiency in interview-based studies: systematic analysis of qualitative health research over a 15-year period. BMC Med Res Methodol 18:1–18

55. O'Connor C, Joffe H (2020) Intercoder reliability in qualitative research: debates and practical guidelines. Int J Qual Methods 19:1609406919899220
56. Roberts K, Dowell A, Nie JB (2019) Attempting rigour and replicability in thematic analysis of qualitative research data; a case study of codebook development. BMC Med Res Methodol 19(1):1–8
57. Capilla R, Jansen A, Tang A, Avgeriou P, Babar MA (2016) 10 years of software architecture knowledge management: practice and future. J Syst Softw 116:191–205
58. Darawsheh W (2014) Reflexivity in research: promoting rigour, reliability and validity in qualitative research. Int J Ther Rehabil 21(12):560–568
59. Lemon LL, Hayes J (2020) Enhancing trustworthiness of qualitative findings: using Leximancer for qualitative data analysis triangulation. The qualitative report 25(3):604–614
60. Cheung KKC, Tai KW (2023) The use of intercoder reliability in qualitative interview data analysis in science education. Res Sci Technol Educ 41(3):1155–1175
61. Archibald MM (2016) Investigator triangulation: a collaborative strategy with potential for mixed methods research. J Mixed Methods Res 10(3):228–250
62. Natow RS (2020) The use of triangulation in qualitative studies employing elite interviews. Qual Res 20(2):160–173
63. Turner SF, Cardinal LB, Burton RM (2017) Research design for mixed methods: a triangulation-based framework and roadmap. Organ Res Methods 20(2):243–267
64. Soysal YILMAZ (2024) Reinterpreting the member checking validation strategy in qualitative research through the hermeneutics lens. Qual Inq Educ: Theory Pract 2(1)
65. Morse JM (2015) Critical analysis of strategies for determining rigor in qualitative inquiry. Qual Health Res 25(9):1212–1222
66. Coleman P (2022) Validity and reliability within qualitative research for the caring sciences. Int J Caring Sci 14(3):2041–2045
67. Campbell JL, Quincy C, Osserman J, Pedersen OK (2013) Coding in-depth semistructured interviews: problems of unitization and intercoder reliability and agreement. Sociol Methods Res 42(3):294–320
68. Tracy SJ (2024) Qualitative research methods: collecting evidence, crafting analysis, communicating impact. John Wiley & Sons
69. Miyazaki AD, Taylor KA (2008) Researcher interaction biases and business ethics research: respondent reactions to researcher characteristics. J Bus Ethics 81:779–795
70. Kumar V, Bhosale MP, Nadgauda MTN, Lavanya C (2024) Research methodology-concepts & applications
71. Zhang M, Di Fan D, Zhu CJ (2014) High-performance work systems, corporate social performance and employee outcomes: exploring the missing links. J Bus Ethics 120:423–435
72. Fuglestad S, Herje MVA (2023) The influence of work life balance on employee well-being: the mediating roles of job satisfaction and burnout/stress (Master's thesis, uis)
73. Ravindran V (2019) Data analysis in qualitative research. Indian J Contin Nurs Educ 20(1):40–45
74. Mortelmans D (2019) Analyzing qualitative data using NVivo. The Palgrave handbook of methods for media policy research, pp 435–450
75. Boldosova V, Luoto S (2020) Storytelling, business analytics and big data interpretation: literature review and theoretical propositions. Manag Res Rev 43(2):204–222
76. Gwyer R (2015) Identifying and exploring future trends impacting on academic libraries: a mixed methodology using journal content analysis, focus groups, and trend reports. New Rev Acad Librariansh 21(3):269–285
77. Kelly M, Soles R, Garcia E, Kundu I (2020) Job stress, burnout, work-life balance, well-being, and job satisfaction among pathology residents and fellows. Am J Clin Pathol 153(4):449–469
78. Vangala S (2018) Pattern recognition applied to continuous integration system
79. Noakes T, Harpur P, Uys C (2024) Noteworthy disparities with four CAQDAS tools: explorations in organising live Twitter data. Soc Sci Comput Rev 42(3):794–811
80. Paulus TM, Lester JN (2021) Doing qualitative research in a digital world. Sage Publications

81. Locke LF, Silverman SJ, Spirduso WW (2009) Reading and understanding research. Sage
82. Rockmann KW, Vough HC (2024) Using quotes to present claims: practices for the writing stages of qualitative research. Organ Res Methods 27(4):621–649
83. Paquet-Clouston M, García S (2023) On the dynamics behind profit-driven cybercrime: from contextual factors to perceived group structures, and the workforce at the periphery. Global Crime 24(2):122–144
84. Degeling C, Rock M (2020) Qualitative research for one health: from methodological principles to impactful applications. Front VetY Sci 7:70
85. Motulsky SL (2021) Is member checking the gold standard of quality in qualitative research? Qual Psychol 8(3):389
86. Yarkoni T (2022) The generalizability crisis. Behav Brain Sci 45:e1
87. Beach D, Pedersen RB (2016) Causal case study methods: foundations and guidelines for comparing, matching, and tracing. University of Michigan Press
88. Jacobs AM, Büthe T, Arjona A, Arriola LR, Bellin E, Bennett A, Björkman L, Bleich E, Elkins Z, Fairfield T, Gaikwad N (2021) The qualitative transparency deliberations: Insights and implications. Perspect Polit 19(1):171–208
89. McBrien B (2008) Evidence-based care: enhancing the rigour of a qualitative study. British Journal of Nursing 17(20):1286–1289
90. Jacobs A, Büthe T (2015) Qualitative transparency deliberations. In: Proposal to the APSA's QMMR organized section (approved by the QMMR Section Executive Committee
91. Merriam SB, Grenier RS (eds) (2019) Qualitative research in practice: examples for discussion and analysis. John Wiley & Sons. In: Tracy SJ (2024) Qualitative research methods: Collecting evidence, crafting analysis, communicating impact. John Wiley & Sons
92. Höfler M, Gloster AT, Hoyer J (2010) Causal effects in psychotherapy: counterfactuals counteract overgeneralization. Psychother Res 20(6):668–679
93. Buck AM, Ralston DF (2021) I didn't sign up for your research study: The ethics of using "public" data. Comput Compos 61:102655
94. Novak JD (2010) Learning, creating, and using knowledge: concept maps as facilitative tools in schools and corporations. Routledge
95. Saldana J (2011) Fundamentals of qualitative research. Oxford University Press
96. Thomas J, Harden A (2008) Methods for the thematic synthesis of qualitative research in systematic reviews. BMC Med Res Methodol 8:1–10
97. Hayes AS (2025) "Conversing" with qualitative data: enhancing qualitative research through large language models (LLMs). Int J Qual Methods 24:16094069251322346
98. Walkington C, Bernacki ML (2020) Appraising research on personalized learning: definitions, theoretical alignment, advancements, and future directions. J Res Technol Educ 52(3):235–252
99. Lal JP, Bedekar JS, Athawale BK (2023) Research methodology: theory. Academic Guru Publishing House, Practices and tools
100. Surmiak AD (2018) Confidentiality in qualitative research involving vulnerable participants: researchers' perspectives. In: Forum qualitative sozialforschung/forum: qualitative social research (Vol 19, No 3). DEU
101. Pietrzykowski T, Smilowska K (2021) The reality of informed consent: empirical studies on patient comprehension—systematic review. Trials 22:1–8
102. Bruce A, Beuthin R, Sheilds L, Molzahn A, Schick-Makaroff K (2016) Narrative research evolving: evolving through narrative research. Int J Qual Methods 15(1):1609406916659292
103. Tong A, Flemming K, McInnes E, Oliver S, Craig J (2012) Enhancing transparency in reporting the synthesis of qualitative research: ENTREQ. BMC Med Res Methodol 12:1–8
104. Marshall G, Freitas A, Jay C (2020) How researchers use diagrams in communicating neural network systems. arXiv:2008.12566
105. Van der Waldt G (2020) Constructing conceptual frameworks in social science research. TD: J Transdiscipl Res South Afr 16(1):1–9
106. Petrova E, Dewing J, Camilleri M (2016) Confidentiality in participatory research: challenges from one study. Nurs Ethics 23(4):442–454

107. Padilla LM, Creem-Regehr SH, Hegarty M, Stefanucci JK (2018) Decision making with visualizations: a cognitive framework across disciplines. Cogn Res: Princ Implic 3:1–25
108. Flynn BB, Sakakibara S, Schroeder RG, Bates KA, Flynn EJ (1990) Empirical research methods in operations management. J Oper Manag 9(2):250–284

Quantitative Research Methods

Systematic investigation in modern academic and professional fields depends on quantitative research techniques. Their power is in the capacity to record, measure, and interpret patterns using numerical data, enabling researchers to extract meaning from the visible world systematically, objectively, and repeatedly. As a methodological tool, quantitative research offers clarity and accuracy so academics may investigate correlations, validate ideas, and draw evidence-based conclusions. Whether in the social sciences, engineering, environmental studies, or education, converting events into measurable variables provides a strong lens to examine multifaceted problems with depth and precision. Beginning with a basic investigation of fundamental ideas of quantitative research, this parts defining qualities are technical tastes and philosophical commitments to openness, consistency, and analytical honesty; they include objectivity, replicability, and generalizability. Researchers using this practice are guided by organized systems that value design clarity, data-gathering consistency, and analytical responsibility. These components help the larger objective of creating information that holds up under examination and helps to create a common understanding. Central to this strategy is developing research questions that can be examined using empirical data. Often, these queries concern patterns, correlations, or causal mechanisms among well-defined variables. Quantitative research adapts to varied degrees of complexity and control using several study designs, including descriptive, correlational, and experimental formats. Every design provides a different perspective: descriptive techniques chart the shape of events, correlational studies find relationships, and experimental tactics try to isolate and evaluate causal factors. The fundamental dedication to methodical observation and critical interpretation stays constant regardless of the particular design. Making valid causal claims depends on employing experimental design. Deliberate structuring of situations and randomization, control, and blinding approaches help researchers draw surface-level relationships. Experimental thinking is more than technical execution; it requires conceptual clarity, in which counterfactual thinking and logical consistency support interpreting outcomes. This concept calls for attention to internal and external validity risks, challenges that, if ignored, could weaken the power of inferences. While maintaining the diversity of the research setting, techniques like analysis of covariance and propensity score matching help to protect the validity of results. Another foundation of quantitative research is survey techniques, which systematically grasp people's behaviors, views, and experiences on a broad scale. Surveys produce extensive datasets in research on built environments, social systems, or organizational settings that, when carefully designed, provide great insights. Survey instrument design has to balance thoroughness and clarity to guarantee that the generated data is relevant and understandable. Equally as crucial are the sampling techniques that support any survey-based inquiry. Researchers make important choices that influence the generalizability and reliability of their findings using careful inclusion selection. Strategies have to fit the study goals, limitations, and ethical issues, stressing the need for careful planning at every stage of the process. Statistical analysis connects raw data with relevant interpretation once the data are gathered. Researchers first grasp distributions and key tendencies using tools including descriptive statistics. Then, inferential methods broaden this knowledge to enable forecasts, comparisons, and the testing of theoretical assumptions. Statistical literacy is the capacity to select the right tests, grasp their assumptions, and read findings in light of the larger study setting; it combines technical knowledge with conceptual depth. Standard techniques like t-tests, ANOVA, correlation, and regression do not end but rather are components of a greater analytical story that links evidence to investigation. The analytical techniques supporting research issues must evolve as they get more complex. For those trying to model complicated linkages and latent structures, advanced methods in multivariate analysis provide a way ahead. Techniques such as structural equation modeling, factor analysis, and multiple regression allow one to explore data deeply and in more layers. These methods require a theoretical foundation and methodological accuracy since they rely on rigorous model specification, assumption testing, and interpretative caution. Though the researcher's critical judgment guarantees that the analysis is relevant and ethically correct, visualization tools and software platforms help to manage complexity. Emphasis is on the junction of conceptual knowledge and methodological application throughout this part. Quantitative research is a rigorous approach to viewing, probing, and interpreting the world, not just a collection of instruments. Every facet of the research process is permeated by ethical issues like data integrity, participant consent, and open reporting, hence underlining the duty of empirical investigation. Researchers interact with the lives, surroundings, and institutions the data represent as well as with the data. This brings a moral duty to do socially responsive, careful, respectful, and correct research. This part provides students and developing academics a road map to quantitative techniques. Whether they are analysts, designers, or interpreters, it provides them with the conceptual tools and critical awareness required to interact with data meaningfully. It also promotes a mindful attitude toward research as a human activity combining compassion and rigor. Doing this encourages readers to engage in a tradition of investigation honoring clarity, evidence, and the unending quest for knowledge.

Introduction to Quantitative Research

Abstract Emphasizing its fundamental ideas, methodological rigor, and relevance across various academic fields, this chapter provides a basic summary of quantitative research. A systematic study of numerical data and quantitative research aids in comprehending events, testing ideas, and establishing patterns or causal links. The chapter outlines essential aspects of quantitative research, including objectivity, generalizability, and replicability, supported by structured research designs and standardized tools. It also differentiates among descriptive, correlational, experimental, and quasi-experimental methods, each suited for different types of research. The roles of independent, dependent, and control variables are highlighted, along with the necessity of sampling techniques, validity, and reliability to ensure the accuracy and trustworthiness of conclusions. As a cornerstone of this approach, statistical analysis enables researchers to understand data and draw relevant conclusions. The book also addresses ethical issues like data integrity and informed consent. This chapter is a crucial roadmap for students and new researchers, equipping them with the conceptual tools and critical awareness needed to effectively plan, conduct, and evaluate quantitative investigations in academic and professional settings.

1 Introduction

The guiding structure for researchers navigating the complex process of investigation and knowledge creation is provided by research methodology. It entails a consistent and systematic approach to examining events, addressing research questions, and generating relevant insights [1, 2]. Quantitative research techniques are powerful tools in scientific investigation, primarily focusing on measuring and analyzing variables to yield objective, empirical outcomes. This methodology allows researchers to explore trends, validate hypotheses, and assess causal relationships by gathering and analyzing numerical data. Establishing itself as an essential strategy across various academic and professional fields, this methodological orientation emphasizes accuracy, objectivity, and replicability. According to Aliaga and Gunderson [3], quantitative research systematically investigates events through measurable data subjected

A. Almusaed et al., *Practice of Research Methodology in Civil Engineering and Architecture*, https://doi.org/10.1007/978-3-031-97393-2_15

to rigorous mathematical and statistical analysis. It aims to quantify events and variables using well-defined processes and organized tools. These procedures reflect carefully designed questions prioritizing data quality and minimizing human bias beyond mechanical processes. The ultimate objective is to achieve statistically valid results that inform theoretical advancement and sound decision-making [4]. Several key qualities distinguish quantitative research and enhance its credibility, with its reliance on numerical data paramount. This dependence allows statistical methods to thoroughly and accurately analyze outcomes [5]. Whether through surveys, tests, or observational methods, the systematic nature of data collection ensures the consistency, reliability, and validity of results. The guiding principle is standardization, whereby researchers adhere to procedures that maintain objectivity and reduce the likelihood of subjective influence. Generalizability is another distinguishing feature of quantitative research; researchers aim to extend results beyond the immediate study population to broader contexts by selecting appropriate samples [6]. This ability to generalize increases the applicability of research findings across diverse settings, populations, and timeframes. Various methods are employed for data collection, including surveys and questionnaires consisting of closed-ended questions, yielding measurable results through standardized instruments. These tools facilitate statistical aggregation and comparison, particularly for larger population studies. Conversely, experiments provide a controlled environment where researchers manipulate independent variables to observe their effects on dependent variables, thereby establishing causal relationships [7]. Observational methods effectively capture behaviors or events in natural settings, which can then be analyzed. Finally, secondary data analysis involves deriving new insights from existing datasets such as census data, organizational records, or previous surveys. Quantitative research offers numerous advantages, notably the objectivity it fosters, mitigating personal or interpretive biases through data-driven methods and standardized tools [8]. Larger sample sizes enhance statistical power and the generalizability of results, ensuring that findings can be reliably applied to wider populations. Additionally, the repeatability of quantitative studies bolsters their scientific rigor, enabling other researchers to replicate experiments and validate findings, a fundamental aspect of the scientific method. However, no research method is without flaws. A common critique of quantitative approaches is their potential lack of contextual depth; by focusing primarily on numerical indicators, this method may overlook the rich qualitative nuances associated with social interactions and human behavior [9]. The rigid framework of data collection tools can restrict the range of possible responses and hinder the exploration of complex or unexpected aspects of a phenomenon. Furthermore, although measurable, quantitative research often relies on self-reported data, which can be influenced by memory bias, social desirability, or misunderstandings of survey questions. Growing recognition of these limitations has led to increased interest in mixed-methods approaches that combine the strengths of both quantitative and qualitative research to provide a more comprehensive understanding of complex phenomena [10]. Nevertheless, quantitative research remains a cornerstone of academic inquiry, particularly in fields where measurement, prediction, and empirical validation are prioritized. For example, in the

Fig. 1 Core elements of quantitative research. *Source* Authors

construction sector, quantitative methods are invaluable for assessing project performance, ensuring structural safety, anticipating cost overruns, and analyzing labor productivity. By applying statistical models to large datasets, construction researchers and practitioners can develop data-driven strategies that enhance project outcomes and inform decision-making [11]. This chapter thoroughly examines quantitative research methods, exploring their defining characteristics, data collection techniques, statistical foundations, and applications in construction and other disciplines. Moreover, it presents a balanced evaluation of the advantages and disadvantages of this approach, providing readers with a nuanced perspective on the role of quantitative techniques in advancing professional and scientific research. Figure 1 illustrates the fundamental components of quantitative research and highlights three key aspects: main benefits, data collection techniques, and essential characteristics. The image demonstrates how quantitative research fosters impartiality, ensures replicability, and facilitates generalizable conclusions across various academic and professional fields by emphasizing numerical data, systematic processes, and empirical tools such as surveys and experiments.

2 Quantitative Research Methods

Quantitative research methods encompass a range of structured approaches designed to explore, measure, and explain phenomena using numerical data. These methodologies are commonly grouped into four main types: survey research, correlational research, causal-comparative research, and experimental research. Each approach has its distinct purpose and process, yet all are committed to objectivity, precision, and systematic inquiry. Survey research is perhaps the most widely used method in quantitative studies, allowing researchers to efficiently gather large volumes of data from a diverse population [12]. Through carefully designed questionnaires or interviews, survey research captures measurable responses to assess attitudes, behaviors, or characteristics. This method is particularly valuable for identifying trends or generalizing about a population. Correlational research examines relationships between two or more variables without manipulating them. It is instrumental in identifying patterns or associations, helping researchers understand whether and how variables move together, such as the relationship between academic performance and study habits. However, while correlation can indicate a connection, it does not imply causation. Causal-comparative research, or ex post facto research, seeks to identify cause and effect relationships by comparing groups that differ on a particular variable. Unlike experimental research, this method does not directly manipulate variables but instead relies on naturally occurring differences [13]. It is helpful in situations where experimental control is not feasible or ethical, such as when examining the impact of socioeconomic status on educational outcomes. Experimental research is the most rigorous quantitative method, manipulating one or more independent variables to observe their effects on a dependent variable. This approach allows researchers to establish causal relationships through controlled environments and random assignment. It is widely used in fields such as psychology, medicine, and engineering, where precision and control are critical. These four methods provide researchers with a comprehensive toolkit to investigate a broad range of questions [14]. Scholars can produce meaningful, evidence-based conclusions that contribute to theory, practice, and policy development across disciplines by choosing the most appropriate design based on their research goals. For example, in digital twin applications for buildings, experimental research may involve adjusting virtual HVAC parameters to measure effects on energy efficiency. Likewise, correlational research can explore the relationship between occupant density and real-time energy consumption, providing insights for more innovative building management and predictive maintenance strategies.

2.1 Survey Research

Surveys are beneficial for examining trends, attitudes, and behaviors in the construction sector. Survey research is one of the most commonly used fundamental techniques of quantitative research [12]. It collects data from a predetermined group of respondents using standardized instruments, such as online questionnaires, paper-based surveys, or structured interviews. These instruments are carefully designed to ensure the responses' clarity, consistency, and measurability. The main strength of survey research is its ability to gather data from a large and diverse population through random sampling techniques, thereby enhancing the generalizability and reliability of the results. Construction survey research can be broadly categorized into two types. Cross-sectional surveys provide a snapshot of industry practices or attitudes by collecting data concurrently. For instance, a survey could assess how construction workers perceive newly introduced workplace safety policies across various project sites [15]. Conversely, longitudinal surveys collect data from the

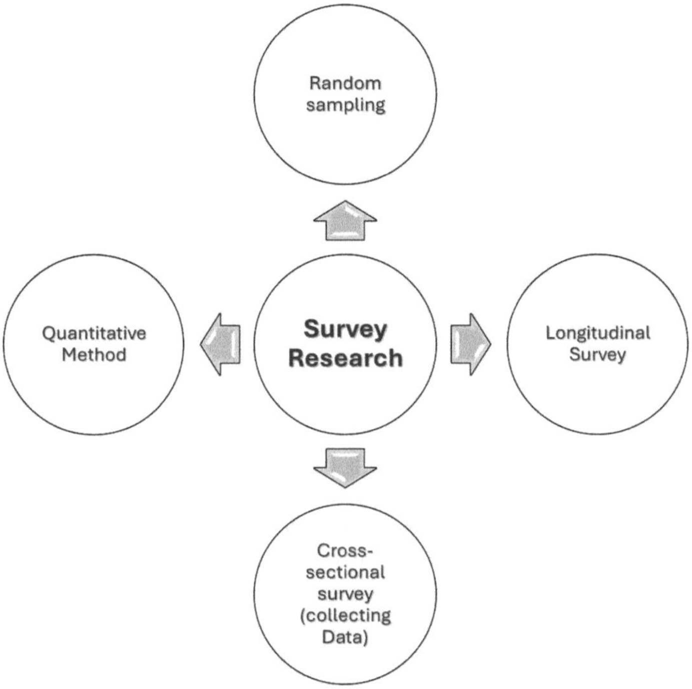

Fig. 2 Overview of survey research components in quantitative studies. *Source* Authors

same respondents over an extended period, allowing researchers to observe trends and changes. For example, over multiple construction cycles, longitudinal research could explore how the integration of Building Information Modeling (BIM) evolves in practice and its impact on reducing project costs and improving efficiency [16]. These surveys contribute to guiding industry standards, management decisions, and policies. Within a quantitative framework, Fig. 2 shows the fundamental elements of survey research. Presented in a circular configuration, the figure emphasizes key components, including data collection instruments, random sampling, and the two main survey types: cross-sectional and longitudinal, showing their related functions in organized data gathering and analysis.

2.2 *Correlational Research*

A key strategy in quantitative research, correlational studies aim to find and understand the links between two or more variables without actually changing them [17]. For example, this approach helps highlight patterns and connections that may not be immediately obvious but significantly impact project results, safety requirements, and overall efficiency in the building sector. Although it cannot prove causation, correlational studies offer insightful analysis that can guide decision-making, facilitate data-driven improvements in practice, and lead to further research. A correlational study, for instance, may investigate whether noise levels on building sites have a notable effect on worker output. By collecting information on performance measures and noise exposure, researchers could identify patterns indicating that increased noise levels are associated with decreased productivity and lower concentration [18]. Similarly, another study may examine how the longevity of buildings relates to the quality of building materials used. Identifying such a connection would help establish material standards and guide improved purchasing behavior. Ultimately, correlational studies are a vital initial step toward understanding complex relationships in the built environment and promoting the ongoing development of construction methods.

2.3 Causal-Comparative Research

Often referred to as ex post facto research, causal-comparative studies are a systematic approach for investigating cause-and-effect relationships by examining variations between groups where the independent variable has already occurred and cannot be altered [19]. This strategy benefits real-world building environments where experimental control is impractical or unethical. Instead of controlling factors, researchers observe naturally occurring variations and draw inferences about potential causal connections through comparative analysis. This approach, for example, effectively evaluates how specific practices or circumstances might influence outcomes in the building sector. For example, studies could compare projects managed with traditional project management strategies against those using Agile or Lean construction concepts. Researchers can determine how management styles impact project performance by analyzing completion timelines, communication efficiency, and budget compliance differences. A study might also compare the financial outcomes of buildings constructed using prefabrication methods with those built using traditional on-site construction. Although these factors have already been set in motion, causal-comparative research enables scholars and practitioners to gain insights from existing practices, highlight trends, and identify solutions that could enhance future building processes, bridging the gap between past experiences and future improvements. By highlighting its three fundamental elements, Fig. 3 graphically illustrates causal-comparative research. Centered around the key concept of "Causal-Comparative Research," (1) the Ex Post Facto Approach indicates that the independent variable has already occurred; (2) Cause-and-Effect Relationships aim to infer potential causal connections; and (3) Comparison of Groups demonstrates how differences among groups form the basis for analysis in this research design.

2.4 Experimental Research

Experimental research is the most rigorous and systematic of the various quantitative methods [20]. Its active manipulation of one or more independent variables to

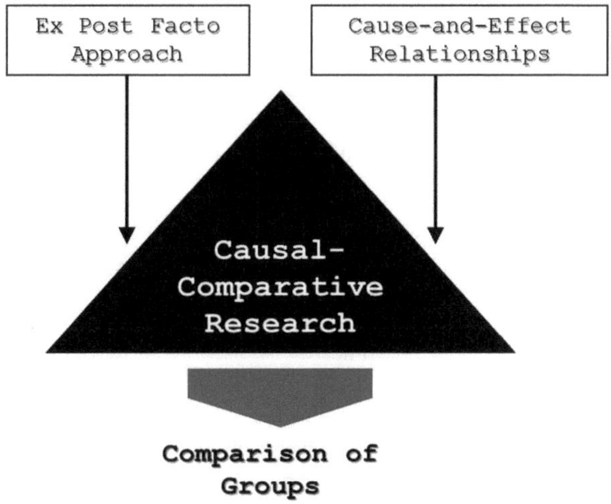

Fig. 3 Key elements of causal-comparative research. *Source* Authors

directly observe the consequences on a dependent variable distinguishes it, enabling researchers to establish clear and quantifiable cause-and-effect relationships. Typically grounded in scientific accuracy, this method employs control and experimental groups to differentiate the effects of the variable under study [21]. Experimental research provides a solid means to test new ideas and enhance performance in the construction sector [22]. For example, a researcher might investigate how AI-driven scheduling tools impact project productivity by randomly assigning construction teams to use conventional planning techniques or a new AI tool. The study can yield compelling evidence of the tool's effectiveness by comparing completion times, cost compliance, and productivity outcomes. Similarly, experimental studies could assess the energy-saving potential of newly developed sustainable building materials. Researchers can measure the benefits of innovation in sustainable design by constructing identical building models with different materials and monitoring their energy consumption. As illustrated in Table 1, this type of study connects scientific inquiry with practical application by offering a systematic and controlled approach to examining how technological and procedural interventions affect construction practices, fostering more efficient, sustainable, and informed decision-making in the built environment.

3 Characteristics of Quantitative Research

Fundamental qualities that define quantitative research ensure methodological rigor, empirical validity, and analytical precision [23]. These qualities form the foundation of quantitative research, allowing researchers to systematically study events, evaluate hypotheses, and obtain statistically validated findings. An organized and consistent methodological framework is one of the main features of quantitative

Table 1 Structured approach contextualizing quantitative research methods within the construction industry

Quantitative research method	Description	Examples in the construction industry	Key considerations
Survey research	Collects respondents' data using online surveys, polls, and questionnaires. Can be cross-sectional (single point in time) or longitudinal (over time)	– Assessing construction workers' safety perceptions on-site – Measuring client satisfaction with construction project delivery	– Random sampling improves generalizability – Cross-sectional surveys provide snapshots, while longitudinal surveys track changes over time
Correlational Research	Examine the relationship between two or more variables without manipulation. Determines the strength and direction of associations	– Relationship between construction site noise levels and worker productivity – Correlation between material quality and structural durability	– Correlation does not imply causation – Useful for identifying trends, but cannot determine cause-and-effect relationships
Causal-Comparative Research	Investigate cause-and-effect relationships by comparing groups based on existing differences. The independent variable is not manipulated	– Impact of different construction project management styles on project delays – Effect of prefabrication techniques on cost savings	– Cannot establish definitive causality – Useful when experimental manipulation is not possible
Experimental Research	Manipulates one or more independent variables to observe their effect on a dependent variable. Uses control and experimental groups for comparison	– Testing the impact of AI-driven project scheduling tools on construction efficiency – Evaluating the effectiveness of new sustainable building materials in reducing energy consumption	– Establishes causality due to controlled conditions – Requires random assignments to minimize bias

Source Author's compilation

research. This framework guides every stage of the research process, from developing research questions and hypotheses to collecting data, conducting statistical analysis, and interpreting results. Predefined processes ensure consistency, reduce researcher bias, and enhance the study's replicability, aiding in the development of reliable scientific knowledge [24]. The use of objective and measurable data is another key characteristic. Designed to yield numerical data suitable for statistical analysis, data collection techniques include controlled experiments, standardized tests, and structured surveys. While it provides a strong basis for empirical generalizations, this emphasis on quantification facilitates the identification of trends, correlations, and variations among variables. Furthermore, the focus on generalizability is remarkable. Results from quantitative studies can be extended to larger populations or different contexts using representative samples and probabilistic sampling methods [25]. This ability to generalize outcomes underscores the practical significance of the research and highlights its contributions to theoretical advancement, policy development, and industrial practices. Quantitative studies also require objectivity. Researchers maintain a neutral stance by employing tools and techniques that minimize personal influence and ensure data quality [26]. This commitment to objectivity enhances the study's validity and allows other researchers to verify findings. Additionally, quantitative research relies on statistical analysis. Statistical methods are employed to summarize data, evaluate the strength and direction of correlations among variables, and test hypotheses. These analytical methods enable researchers to draw meaningful inferences from data and assess the validity and relevance of their results. The defining characteristics of quantitative research, systematic methodology, objective measurement, generalizability, statistical rigor, and replicability, establish it as a cornerstone of empirical inquiry [27]. These qualities enhance the validity and applicability of research findings, necessitating quantitative approaches in engineering, health sciences, social sciences, economics, and the built environment.

3.1 Structured Data Collection Tools

Quantitative research is fundamentally driven by a commitment to gathering data consistently, dependably, and reproducibly [28]. Structured data collection tools such as surveys, polls, and questionnaires make this possible. These tools are helpful and essential for ensuring consistency and measurability of the data collected across settings and participants. Carefully crafted with closed-ended questions, rating scales, or predetermined response choices, structured tools enable researchers to quantify and analyze specific factors easily [29]. For example, a questionnaire sent to construction managers might include Likert scale questions regarding satisfaction with scheduling tools or multiple-choice questions on adherence to safety protocols. Each participant receives the same set of questions in the same manner, ensuring that the data collected remains constant and comparable. Both data collection and analysis benefit from this systematic approach, which reduces uncertainty and minimizes the influence of personal interpretation or researcher bias. A key advantage of this method is its facilitation of study replication with new groups or in different settings, which is crucial for generating trustworthy and generalizable results [30]. Structured data collection technologies provide the foundation for consistent quantitative findings by converting abstract research questions into tangible, analyzable data. For example, architectural engineering studies might utilize structured questionnaires to

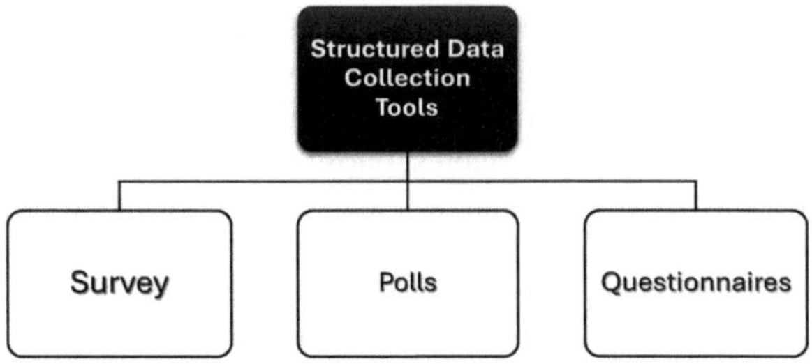

Fig. 4 Categories of structured data collection tools in quantitative research. *Source* Authors

assess indoor environmental quality satisfaction among building occupants. Respondents evaluate lighting, acoustics, and thermal comfort on a defined scale, enabling researchers to gather reliable data and identify trends that impact user experience and design effectiveness [31]. Figure 4 presents a simplified summary of key structured data collection technologies commonly utilized in quantitative studies. It categorizes the instruments into three main types: surveys, polls, and questionnaires, emphasizing their crucial role in enabling systematic, consistent, and repeatable data collection across various study environments and populations.

3.2 Representative Sample Size

One of the main advantages of quantitative research is its ability to generate results that are not only statistically valid but also significant beyond the specific population under study [32]. Careful use of a representative sample size, which includes participants accurately reflecting the larger population of interest, helps achieve this. Researchers aiming to understand patterns, behaviors, or outcomes must ensure their sample matches the essential characteristics of the larger population, such as age, occupation, education level, and geographic region [33]. Selecting the appropriate sampling method is crucial to this approach. Techniques like random sampling, where every individual in the population has an equal chance of being selected, help minimize bias and promote fairness in participant selection. Stratified sampling ensures that specific subgroups within a population, such as project managers, site engineers, and construction workers, are proportionally represented [34]. A suitably sized and well-selected sample enhances the credibility of the findings. It bolsters the study's external validity, enabling its conclusions to be reliably applied to larger populations. A representative sample thus assists researchers in transitioning from individual data points to broader concepts that inform future research, practice, and policy. The key components necessary for obtaining a representative sample in quantitative investigations are illustrated in Fig. 5. It emphasizes the importance of population alignment, appropriate sampling techniques, and relevant participant traits. These elements work together to enhance external validity and facilitate the generalizability of research findings to wider populations.

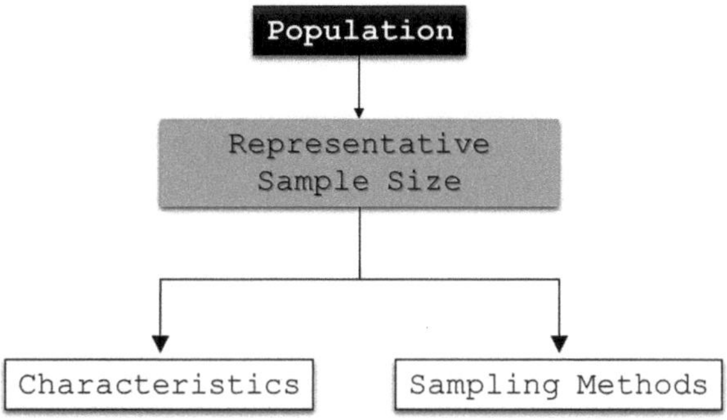

Fig. 5 Essential components of a representative sample in quantitative research. *Source* Authors

3.3 Closed-Ended Questions

A foundation of quantitative research, closed-ended questions provide a consistent and transparent way to collect data that aligns closely with the study's goals [35]. These questions allow respondents to choose the option that most accurately reflects their experience, opinion, or behavior from predetermined response choices like yes/no, multiple-choice, or Likert scales. Closed-ended questions transform complex human thoughts and experiences into quantitative, structured facts. They guide respondents toward specific answers by eliminating open interpretation, ensuring uniformity across all responses. This approach guarantees data consistency and facilitates the interpretation of results and statistical analysis [36]. For example, participants in a study assessing satisfaction with a new green building design would be asked to rate their comfort level with indoor temperature on a scale from one to five. Such clarity enables researchers to quickly compare answers, identify patterns, and draw evidence-based conclusions. Ultimately, closed-ended questions assist researchers in maintaining focus on their measurement objectives, resulting in understandable data that is useful for policy formation, decision-making, or design improvements in building and architectural engineering [37]. In biophilic architecture studies, for example, closed-ended questions could ask building occupants to evaluate their agreement with statements like "The presence of indoor plants enhances my well-being" on a Likert scale. Researchers employing this method can assess opinions on nature-based design elements and examine their psychological or physiological impacts. Figure 6 shows the main response categories in closed-ended inquiries, such as Yes/No, Multiple Choice, and Likert Scale, and emphasizes their main benefits. Essential instruments in structured quantitative research design, these formats guarantee consistency, direct respondents toward particular responses, and enable measurable data collection.

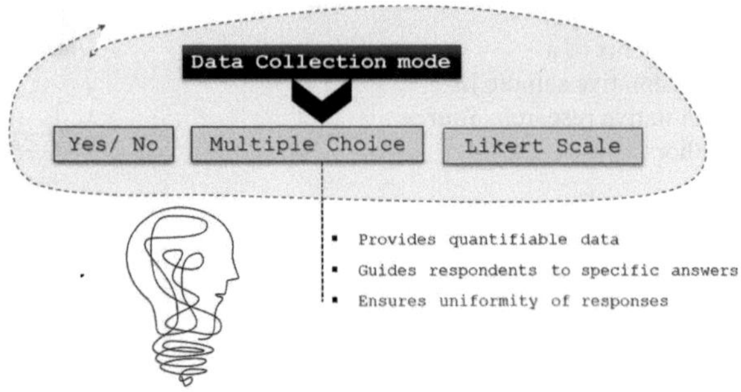

Fig. 6 Response formats and benefits of closed-ended questions in quantitative research. *Source* Authors

3.4 Review of Prior Studies

Reviewing previous studies is one of the most critical steps in a quantitative research study before entering the field or distributing any questionnaires [38]. This technique systematically investigates current literature, theoretical frameworks, empirical results, and methodological approaches relevant to the subject of interest. It's not just a box to tick; it's a vital opportunity to learn from the accomplishments and shortcomings of previous studies. Examining what has been researched helps researchers identify trends or knowledge gaps, understand how those factors have been assessed, and gain insightful analysis of essential variables [39]. This knowledge guides the formulation of research questions that are both relevant and significant and assists in creating well-informed hypotheses based on data rather than conjecture. Furthermore, examining previous studies enables researchers to build upon existing information instead of starting from ground zero. It ensures that new studies are linked to a broader scientific discussion and promotes academic continuity. Researchers can avoid duplication, add value, and position their work within an established theoretical and empirical context [30]. Ultimately, this process clarifies and guides the work's design, implementation, and interpretation by strengthening its foundation. Figure 7 illustrates the main functions of analyzing earlier studies, such as reviewing literature, identifying research gaps, formulating research questions, and generating hypotheses. This approach ensures that new studies contribute to existing knowledge, sustain theoretical continuity, and offer valuable, evidence-based insights to the academic field.

3.5 Quantitative Data Representation

Presenting data appropriately and effectively is just as crucial as gathering it. Data presentation is essential in quantitative research to enable researchers and their audiences to understand the meaning behind the figures [40]. Rather than letting conclusions lie buried in raw data sets, researchers utilize tables, charts, and graphs to bring their findings to life. These graphic tools serve several purposes. Tables are beneficial

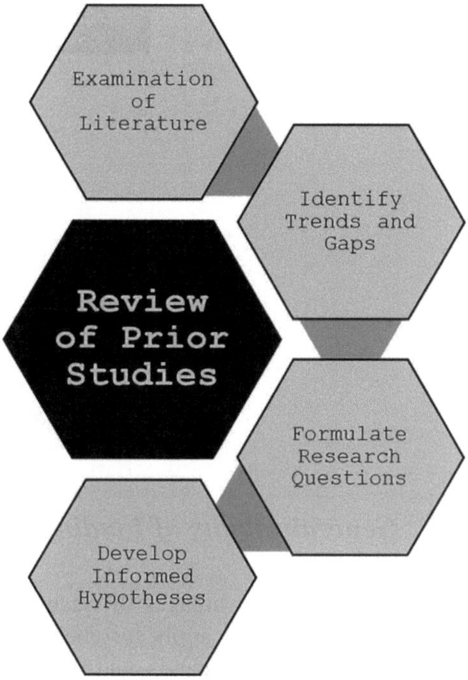

Fig. 7 Key functions of reviewing prior studies in quantitative research. *Source* Authors

for showing accurate numbers or different variables, allowing for precise, thorough comparisons. Bar graphs, pie charts, and line graphs, among other charts, illustrate trends and patterns, which aid in comparing outcomes across various populations, times, or settings. In particular, graphs are powerful tools for demonstrating relationships between variables, including correlations or changes in results [41]. More than cosmetic enhancements, these graphics improve the interpretation and accessibility of complex data for individuals without statistical expertise. They also promote transparency, allowing readers to view the evidence and verify the study's findings. Whether in engineering, public policy, or building design, their usage boosts the credibility and impact of the study, thereby facilitating wider dissemination of results and more confident use in practical decisions [42]. For example, bar charts comparing annual energy use across buildings with different insulation materials may illustrate quantitative data in energy-saving building research. Line graphs could display changes in energy use before and after smart thermostat installation. These visualizations allow architects, academics, and legislators to quickly understand performance variations, identify the most efficient solutions, and support data-driven decisions for creating structures that reduce energy use and promote environmental sustainability. Figure 8 shows a relationship diagram showing five fundamental traits of quantitative research: structured data collecting instruments, closed-ended questions, a representative sample size, a review of existing studies, and quantitative representation. The figure underlines how these components promote methodological rigor, consistency, and the legitimacy of research results.

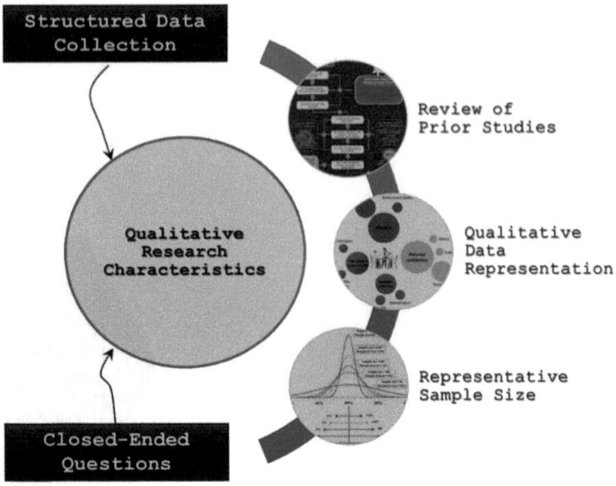

Fig. 8 Interconnected characteristics of quantitative research. *Source* Authors

3.6 Generalizability of Findings

Among the many advantages of quantitative research, its capacity to generalize results and confidently apply insights from a particular group of participants to a much larger population stands out [6]. In areas where large-scale informed decisions must be made, such as public policy, healthcare, education, or the built environment, this characteristic makes quantitative research particularly potent. However, methodological rigor, rather than chance, determines generalizability [43]. This necessitates consistent processes, explicit definitions of variables, and well-designed research tools. Equally crucial is the choice of a representative sample, wherein the studied individuals accurately reflect the variety and traits of the target population. Whether by random, stratified, or other probability-based methods, researchers ensure that their findings are not skewed or confined to a small population. This approach allows academics to offer evidence-based recommendations beyond their initial research [44]. For example, a study evaluating the performance of energy-efficient building designs in one area might guide the sustainability criteria for the entire nation. By doing this, quantitative research produces findings with relevance and influence beyond merely local responses; it drives change across sectors and communities. The logical development resulting in generalizable results in quantitative research is shown in Fig. 9. It underlines the need for scientific rigor and representative sampling as basic first steps. This obvious route shows how the research design increases the relevance of findings to more general populations, encouraging evidence-based practices and informed policy creation.

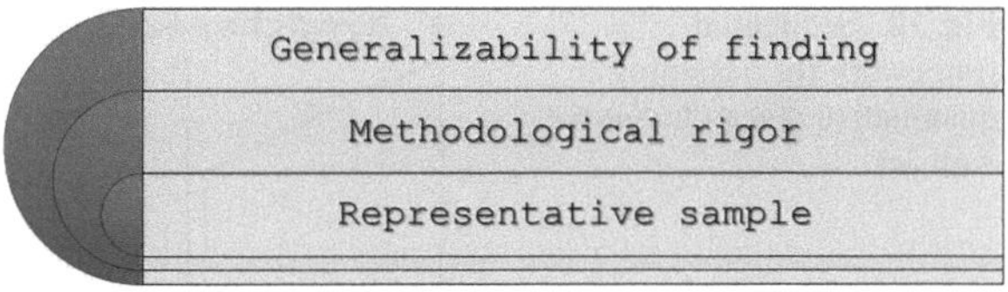

Fig. 9 Pathway to generalizability in quantitative research. *Source* Authors

4 Overview of Quantitative Research Process

A fundamental pillar of empirical scientific investigation, quantitative research provides a methodical and impartial approach to investigating, clarifying, and forecasting events using numerical data. Beginning with a theoretical framework, researchers develop specific, testable hypotheses. These hypotheses guide the selection of suitable methods and form the basis for a well-planned research project [45]. At this stage, a precise definition of the research issue is crucial, as it will determine the direction of the entire study. Once the study question and hypotheses are established, the focus shifts to identifying the target demographic and selecting a sample that accurately reflects it. This phase ensures that results can be generalized. To minimize bias and enhance the validity of the findings, researchers often employ probability-based sampling methods, such as random or stratified sampling. After obtaining a representative sample, researchers operationalize the variables of interest and select appropriate study strategies [5]. This involves choosing a study design that aligns with the research objectives and determining how abstract concepts will be measured, such as questionnaires, experiments, or observational techniques. Ensuring the reliability and comparability of the data hinges on clear and consistent measurement. Once data is collected, researchers apply statistical methods to analyze the findings. The data type and the hypotheses' arrangement dictate the selection of analytical tools [46]. These methods enable researchers to assess the significance of their results, explore correlations between variables, and identify patterns. The findings are interpreted within a broader theoretical and empirical context. This step highlights how the study contributes to ongoing academic discussions or practical applications by connecting new results with existing literature. This strategy empowers researchers to draw meaningful conclusions, evaluate implications, and formulate evidence-based recommendations. With a logically ordered and meticulously followed progression from theory and hypothesis development to data interpretation, the quantitative research process enables researchers to generate credible, generalizable, and actionable insights across various fields. A planned development in quantitative research design, shown in Fig. 10, emphasizes four fundamental phases: generating hypotheses, selecting a representative sample, determining a suitable study design, and evaluating results in theoretical and empirical settings. This step-by-step approach emphasizes the logical flow required to produce valid, generalizable, and contextually relevant results.

Fig. 10 Sequential framework for designing quantitative research. *Source* Authors

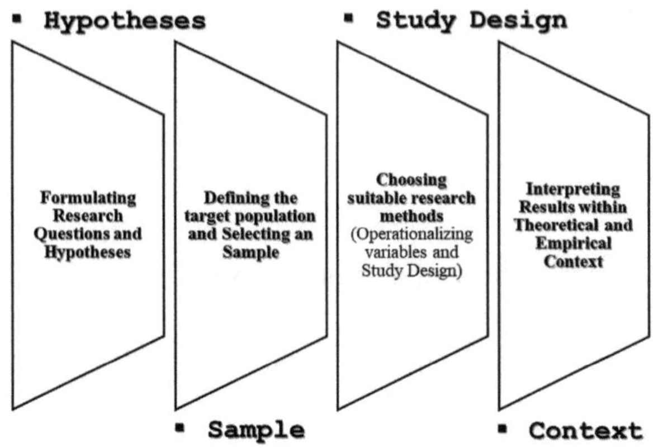

4.1 Formulating Research Questions and Hypotheses

One of the most critical steps in the quantitative research process is developing precise, well-defined research questions. It guides the whole investigation and affects every aspect, including the choice of approach and the reading of findings. A well-defined research question emphasizes what the researcher wants to investigate and helps guarantee the correctness and relevance of the study design, measurement equipment, and analytical techniques [47]. Executed correctly, this stage increases the general validity of the research results and reduces the possibility of methodological errors. Depending on the kind of research topic, quantitative research can take several shapes, including experimental, quasi-experimental, correlational, and descriptive designs, each serving a distinct goal and led by different kinds of inquiry [48]. Experimental studies let researchers find causal links by manipulating one or more independent factors and observing their influence on a dependent variable. Often regarded as the gold standard in this area, randomized controlled trials (RCTs) are frequently employed in sectors including medicine, education, and psychology to assess the effect of specific interventions [49] thoroughly. Not all quantitative studies, though, include manipulation; non-experimental research is fundamental in fields where direct intervention is unethical or impracticable. Researchers investigate naturally occurring associations using observational data or pre-existing databases in such situations. This strategy is widespread in the social sciences and epidemiology, where ethical or logistical considerations limit experimental manipulation [50]. Causal-comparative research within the non-experimental design category looks at the impact of an independent variable without directly changing it. By contrasting various groups, researchers evaluate how different exposures or situations could influence results. For example, a study could look at academic achievement between students taught via conventional instruction and those participating in online learning settings without designating participants to either group. Although this approach does not prove causation, it provides insightful analysis of possible patterns or effects [51]. Conversely, correlational studies seek to find statistical correlations between variables. Though it does not specify that one variable drives the other, it helps to address queries such as whether two variables are connected and how strong that link could be. Examining trends that call for more controlled research helps to highlight this approach. To show traits or patterns inside a particular community, descriptive research, including surveys and observational studies, aims to grasp what exists entirely rather than explain linkages or reasons. For instance, a national assessment of energy use patterns in residential structures could assist in highlighting demographic or geographic tendencies that could guide policy or focused actions [52]. These strategies emphasize the need to develop search questions that fit the study's goal. Whether one wants to verify a cause-and-effect link or merely describe a phenomenon, the question will guide the methodological road ahead and define the study's contribution to its field.

4.2 *Defining the Target Population and Selecting an Appropriate Sample*

Among the defining qualities of quantitative research is its ability to produce insights beyond the particular group of subjects being investigated. Careful definition of the target population and choice of a sample that precisely matches its traits makes this possible. Choosing a representative sample allows researchers to securely extrapolate their results to more significant populations, supporting the idea of external validity: the degree to which study results may be applied to settings beyond the sample itself [53]. Clearly defining the population of interest is the first stage in this procedure. This means picking particular qualities such as age, career, degree of education, geographic area, or other pertinent demographic or contextual elements. For example, in a research evaluating construction workers' use of sustainable materials, the target population would include professionals actively working in the construction sector, maybe divided by experience level or area. The selection of sampling method therefore becomes crucial once the population is specified. The aim is to make sure every minority in the population is fairly represented. Many well-known statistical methods help to do this. Stratified sampling guarantees proportional representation by randomly sampling from each group dividing the population into separate strata e.g., by role, area, or years of experience. When the population is geographically spread out, cluster sampling is appropriate since it lets researchers choose whole groups (clusters) instead of randomly selecting individuals. Especially when the population is properly organized, systematic sampling, where every nth person is chosen from a list, offers a straightforward yet powerful approach to get a representative sample [54]. Among these methods, random sampling is especially appreciated for improving the generalizability of results. Random sampling greatly lowers the danger of selection bias, a frequent danger to validity in research, by giving every population member an equal opportunity to be chosen. This impartial method guarantees that conscious or unconscious researcher decisions do not distort the sample and that the population's diversity is recorded as precisely as possible [55]. A statistical requirement and an ethical obligation are a well-chosen sample based on careful population definition and backed by a suitable sampling method. It guarantees that the study is inclusive, the results are believable, and the conclusions are relevant in practical environments. In the end, sampling is about reflecting the complexity of the population in a reasonable, analyzable shape that enhances the integrity and reach of quantitative research, not only about numbers. For instance, in a study on sustainable building materials, researchers can identify the target group as qualified green construction specialists all around Europe. The stratified sample would allow them to guarantee equitable representation from architects, engineers, and contractors, hence supporting generalizable findings on the adoption rates and preferences for eco-friendly building materials.

4.3 Choosing Suitable Research Methods, Including the Operationalization of Variables and Study Design

Selecting suitable research techniques and accurately describing how essential ideas will be measured, known as operationalization, is a fundamental step in quantitative research. This phase ensures that abstract theoretical ideas are converted into systematic, observable, analyzable, tangible, and quantifiable variables. Often, the outcome of a study depends on the clarity and consistency with which these variables are defined and quantified, as this directly influences the validity and reliability of the results. Any quantitative research classifies variables into three main categories: independent, dependent, and control. Independent variables are those that the researcher believes might affect or drive changes in other variables. Control variables are constant to avoid outside influences affecting the results, while dependent variables are the observed outcomes or phenomena. Careful management of these components helps researchers separate and evaluate the specific impacts of the independent variables more precisely [56]. Appropriate measuring scales must be used to quantify these variables if they are to be quantitatively examined. Nominal, ordinal, interval, and ratio scales have various uses depending on the data type. For categorical variables without natural order or numerical value, such as gender, employment, or type of building material, nominal scales are utilized. Although the precise distance between the ranks is not specified, ordinal scales rank-order data points such as satisfaction ratings or degrees of schooling. Interval scales like those measuring temperature in degrees Celsius or time on a clock offer equal distances between values but lack an actual zero point. One can compare variations, but not state one number as a multiple of another. Conversely, ratio scales possess all the characteristics of interval scales and a genuine zero, allowing the full spectrum of mathematical operations. Examples of ratio variables include height, weight, material cost, or energy use, which can be doubled, halved, or proportionally examined [57, 58]. The chosen scale of measurement greatly influences the statistical analysis

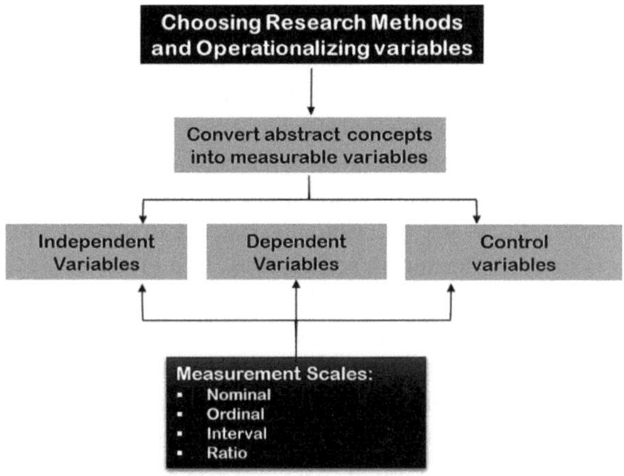

Fig. 11 Operationalizing variables and selecting research methods in quantitative research

of data. For example, although interval and ratio data allow for stronger parametric analyses like regression or analysis of variance (ANOVA), nominal and ordinal data may require non-parametric tests. Researchers, therefore, must match their operational definitions and measurement instruments with the analytical approaches they wish to apply and the nature of the variable. Furthermore, the study design must fit the type of data collection and the goals of the investigation. Testing causal linkages under controlled circumstances necessitates experimental designs. In contrast, non-experimental approaches such as correlational or descriptive research are more appropriate for investigating relationships or describing patterns in real-world environments. Choosing the proper design ensures that the study remains contextually relevant and methodologically valid. Rigorous quantitative research relies on the deliberate operationalization of variables and the strategic choice of study designs. These decisions guide the research, ensure the significance and analyzability of the data, and ultimately help provide consistent, generalizable, and meaningful results. The organized approach of transforming abstract ideas into quantifiable variables and classifying them as independent, dependent, or control variables is shown in Fig. 11. It also emphasizes the need of choosing suitable measurement scales nominal, ordinal, interval, and ratio for guaranteeing data accuracy, statistical relevance, and analytical clarity in quantitative research.

4.4 Selecting Appropriate Data Analysis Techniques

Quantitative research gathers measurable data using various systematic tools, including surveys, observational checklists, and standardized tests. Therefore, the worth of this information depends significantly on the authenticity and dependability of the tools employed. These two ideas guarantee that the study correctly reflects what it meant to measure and that outcomes are consistent and repeatable across several settings. Validity is the extent to which a tool assesses the desired idea. Quantitative research depends on several kinds of validity. Content validity assesses whether the components of the instrument reflect the assessed concept and fit with theoretical foundations [59]. Criterion validity evaluates how well one measure forecasts results depending on an accepted standard [60]. Often using convergent and discriminant validation methods, construct validity investigates whether the instrument reflects the theoretical construct it seeks to measure [61]. Often used in the creation of measuring scales, tools like factor analysis help to verify that individual items consistently reflect the underlying characteristics of a concept [62]. On the other hand, reliability is the consistency and stability of measurement across time or between raters. Standard methods of evaluating reliability are test–retest reliability, which looks at the instrument's stability over time; interrater reliability, which gauges consensus among several observers; and internal consistency, usually judged using Cronbach's alpha, which assesses how closely related a group of items within a scale [63]. Researchers guarantee their measurements are credible and relevant by addressing stringent validity and reliability concerns. This focus on methodological accuracy improves particular studies and helps to develop empirical knowledge more generally [45].

4.4.1 Research Designs and Analysis

Any good quantitative study starts with its research design. Whether experimental or quasi-experimental, the design determines how data is gathered, processed, and interpreted, finally determining the strength of the study's findings. Quantitative designs seek to create causal links or find significant correlations between variables. Three fundamental design components are often used to guarantee the validity of conclusions: pretests, control groups, and time-series analysis [45]. Pretests are a vital baseline that let researchers decide if seen changes following an intervention are caused by the treatment or were present previously. Lacking a pretest makes it challenging to tell whether the intervention caused improvement or whether it was caused by pre-existing issues [64]. From conventional tests to observational or behavioral baselines customized to fit the study goals, these pretests can take many shapes [65]. Equally as crucial in separating the consequences of the intervention are control groups. Researchers can more certainly ascribe variations to the intervention by contrasting results between a group subjected to the treatment and a group not. Particularly in practical environments, the difficulty is guaranteeing that the control and experimental groups are sufficiently similar, which requires meticulous planning and, where feasible, random assignment [48, 66]. Examining how results change across several time points helps time-series analysis to support conclusions. This method records long-term patterns instead of depending on one post-intervention observation, hence providing insight into whether an intervention's benefits are persistent, delayed, or transient. It benefits from longitudinal research on behavior modification, policy implementation, or performance tracking [67].

4.4.2 Choosing Analytical Tools

Selecting analytical tools becomes crucial to deriving credible and relevant findings once data has been gathered. Quantitative studies provide a variety of statistical methods; the choice of a suitable one relies on the kind of data and the study objective [51]. Using metrics including mean, standard deviation, frequencies, and percentages, descriptive analysis summarizes important characteristics, giving a picture of the dataset. Although frequency and proportion are suitable for categorical data, continuous variables are represented by central tendency and dispersion statistics [62]. When comparing groups, comparative analysis is applied; several methods assist in this depending on the number and kind of groups being researched. When comparing the means of two independent groups, t-tests are used to apply [68]. While Factorial ANOVA looks further by examining interactions among several independent variables, Analysis of Variance (ANOVA) compares several groups [69]. By including covariates to offset outside effects, Analysis of Covariance (ANCOVA) improves group comparisons [70]. Examining the impact of independent variables on several dependent variables at once, Multivariate Analysis of Variance NOVA) is appropriate [51]. The chi-squared test offers insight into distributional variations by comparing observed and expected frequencies in categorical data [71]. Producing coefficients from -1.0 to + 1.0, correlational analysis investigates the degree and direction of correlations between several variables. While Spearman's rho and point-bacterial correlation are utilized for ordinal and mixed-type data, Pearson's correlation is suitable for continuous variables [29]. One must remember that correlation alone does not indicate causality; outside confounding factors could affect relationships [72]. The chi-squared test is a useful parallel to correlation for categorical variables [71]. Regression analysis evaluates predictive connections, hence extending correlation. Multiple regression lets researchers investigate how a mix of independent variables affects a dependent variable [73]. Particularly when random assignment is impractical, more sophisticated methods like hierarchical regression and structural equation modeling (SEM) aid in revealing deeper causal routes [74]. At last, many quantitative research studies still demonstrate causation as their principal goal. Experimental designs using random assignment reduce bias and control for confounding factors, providing the most substantial proof of causation [51]. However, no experimental techniques, such as path analysis or SEM, offer substitute ways for modeling complicated causal links when such designs are impractical. However, to provide reliable outcomes, they need bigger samples and rigorous methodological criteria [56]. The selection of analytical tools is shown in Fig. 12.

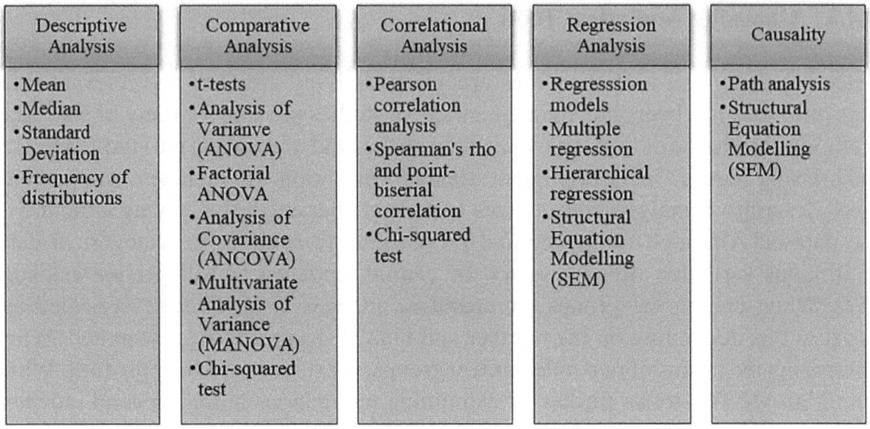

Fig. 12 Selection of analytical tools. *Source* Authors

4.5 Interpreting Results Within the Context of Theoretical and Empirical Literature

At its foundation, scientific research focuses on reducing the probability of arriving at false or misleading findings rather than establishing theories in an absolute sense. Statistical tools—especially p-values, which forecast the likelihood that the findings seen in a study might have happened by random chance—are crucial to this approach. Traditionally, a threshold of $p < 05$ has been commonly recognized to represent statistical significance, implying less than a 5% likelihood that the results are attributable to chance alone [75]. On the other hand, modern studies increasingly acknowledge that a result's relevance is not entirely represented just by statistical significance. Scholars have therefore begun to give more attention to effect size, for example, Cohen's d, which measures the size of the observed difference or correlation. By emphasizing the practical importance of the results [68, 76], this change promotes a more nuanced understanding of research outcomes beyond binary judgments of "significant" or "not significant." Furthermore, the integrity of a study's conclusions depends heavily on the rigor of its design and analytical technique. Researchers must carefully select approaches that fit their research topics and ensure that the statistical methods they use are appropriate for the kind and organization of their data. Doing so enhances the validity and dependability of their results and supports the study's contribution to evidence-based practice and policy. Advancements in artificial intelligence (AI) and machine learning (ML) provide exciting prospects to refine research design, enhance data analysis, and reveal intricate patterns that might not be discernible under conventional statistical methods as one looks forward. By opening the door for more complex and powerful scientific study, these developing technologies could significantly improve quantitative research methods' accuracy, flexibility, and scalability [77].

4.6 *Statistical Testing in Quantitative Construction Research: Process, Interpretation, and Application*

In quantitative construction engineering and management research, statistical tests are fundamental tools for hypothesis testing, inference, and decision-making. The application of these tests typically follows a structured and sequential process (Fig. 2), enabling researchers to derive valid conclusions from sample data. This process includes hypothesis formulation, data selection, test computation, and result interpretation through the p-value and significance level. The outcomes of these tests are usually presented in the results section, after descriptive statistics (e.g., means, standard deviations) and exploratory analyses. This procedure has been widely validated and standardized across recent empirical construction studies, particularly in domains such as project performance evaluation [78, 79], safety risk analysis [80, 81], and quality control optimization [82, 83]. Figure 13 shows Statistical Testing in Quantitative Construction Research: Process, Interpretation, and Application.

4.6.1 Hypothesis Formulation

Statistical inference begins with the articulation of two competing hypotheses:

- Null hypothesis (H_0): Assumes no significant difference or relationship exists between the studied variables.
- Alternative hypothesis (H_1): Proposes a significant difference or relationship.

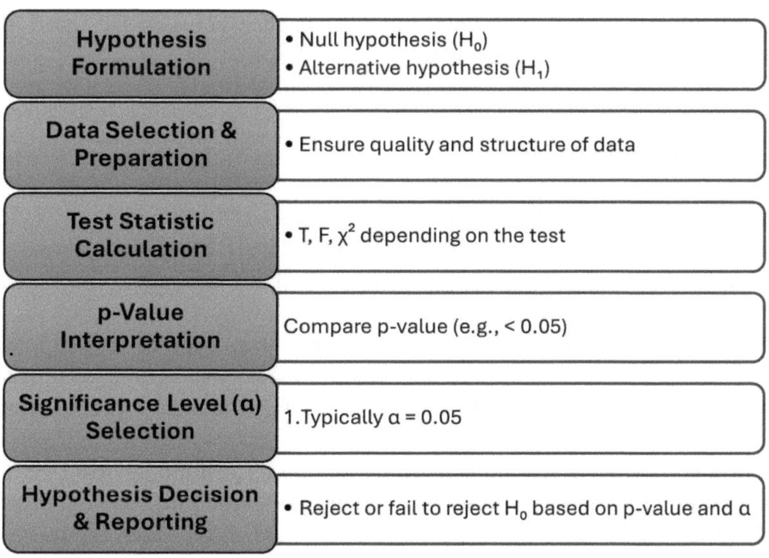

Fig. 13 Statistical testing in quantitative construction research: process, interpretation, and application. *Source* Authors

For example, in a construction context, researchers might investigate whether using prefabricated components significantly reduces project delays compared to traditional construction (H_1) or whether no such difference exists (H_0).

4.6.2 Data Selection and Preparation

The quality and organization of the data determine a statistical test's integrity. Researchers must enter raw data gathered from a suitably big and representative sample. Most building databases contain project-level measures, such as cost, length, safety incidents, or personnel-level data, such as skill level, labor hours. Data should be structured such that every row is an independent observation and every column is a variable. Construction projects use statistical analysis software like SPSS, R, Python, or Excel. For example, a research study on construction workers' non-compliance with personal protective equipment (PPE) can include hundreds of worker survey responses classified by project type or contractor experience [84].

4.6.3 Test Statistic Calculation

Modern statistical methods automatically calculate the test statistic the foundation for significance testing T, F, \u03c7\u00b2. These numbers under the null hypothesis measure the difference between predicted and observed results. For example, test statistics are commonly employed in construction research to evaluate performance indicators among groups; for airport project delivery, an ANOVA test might be performed to compare time and cost performance between Design-Build and Traditional Delivery approaches [85].

4.6.4 P-Value Interpretation

The p-value is a core outcome of most statistical tests, indicating the probability that the observed data would occur under the null hypothesis. In construction research, a small p-value (typically < 0.05) suggests that observed differences or relationships are unlikely to have happened by chance.

For example, suppose a researcher finds that the p-value for the difference in safety violations between unionized and non-unionized sites is 0.021. This provides statistically significant evidence to reject the null hypothesis and infer a meaningful difference between the groups.

Importantly, recent discussions in academic circles caution against overreliance on p-values due to their susceptibility to misinterpretation [86]. Some journals now encourage the inclusion of confidence intervals and effect sizes for more robust interpretation.

4.6.5 Significance Level (α) Selection

The significance level (α) defines the threshold below which results are statistically significant. Conventionally set at 0.05, researchers interpret p-values relative to this cut-off. Lower α levels, such as 0.01 or 0.001, are adopted in high-stakes studies, such as those analyzing structural failure rates or occupational hazards, to minimize Type I errors.

Construction Example: If a study on the compressive strength of concrete samples treated with a new admixture yields a p-value of 0.004, and α is set at 0.01, the result is considered significant, justifying further adoption of the treatment.

4.6.6 Hypothesis Decision and Reporting

The final step involves a decision to reject or fail to reject the null hypothesis based on the p-value and chosen α. This decision should be reported clearly in the results section of the research work, along with a concise summary of the tests used, the test statistics, and the exact p-value.

Example:

"An independent samples t-test indicated a significant difference in average schedule overruns between projects managed using Agile vs. Waterfall approaches ($t = 2.87$, $p = 0.007$). Thus, H_0 was rejected in favor of H_1." Table 2 summarizes some of the tests commonly used in construction quantitative research.

5 Research Designs and Statistical Analysis in the Construction Industry

The selection of research design, especially within experimental and quasi-experimental frameworks, is fundamental in construction research, as it determines how data is gathered, analyzed, and translated into practical findings. Establishing causal linkages, monitoring progress over time, and assessing intervention efficacy depend on a well-structured design. Three main elements—pretests, control groups, and time-series analysis—each enhancing the study's internal validity and the strength of its findings—are central to this strategy. These methodological approaches are particularly relevant in the construction sector, where studies often examine complicated, dynamic factors, including project efficiency, labor productivity, and safety outcomes. Pretests, for instance, can help define baseline conditions before implementing a new project management approach. By comparing results between employees who participated and those who did not, control groups allow researchers to isolate the impacts of a training program. Conversely, time-series analysis can demonstrate whether changes like reduced workplace accidents or increased productivity are sustained over time or are merely temporary fluctuations [64]. Construction researchers can better evaluate the effects of innovations, policies, and technologies

Table 2 Quantitative research structure applied to construction

Research focus	Examples in construction	Type of data	Statistical Test needed	Name of test
Association between two categorical variables	Is accident frequency associated with worker certification level (certified vs. non-certified)?	Categorical	A test for association between two categorical variables	Chi-squared (χ^2) test
Relationship between two scale variables	Is there a correlation between project duration and cost overrun percentage?	Scale (continuous)	A test to quantify a relationship (correlation/line fit)	Correlation and/or regression
Difference in a scale variable between two independent groups	Do subcontracts. Do performance scores differ between local and foreign contractors?	Scale: two independent groups	A test for differences between two unlinked groups	Independent T-test
Difference in a scale variable in the same group over time	Has construction worker productivity changed before and after the implementation of BIM tools?	Scale: The same group was measured twice	A test for differences in linked observations	Paired T-test
Difference in a scale variable across more than two groups	Does average safety compliance vary across residential, commercial, and infrastructure projects?	Scale: More than two groups	A test for group differences across multiple independent categories	ANOVA (Analysis of variance)
Identifying underlying factors influencing responses	Can safety perception survey items be grouped into key latent factors such as "Training", "Supervision", and "Personal Attitudes"?	Multiple variables	A statistical method to reduce and interpret variables	Factor analysis

Source Author's compilation

in real-world settings through appropriate statistical analysis and rigorous research designs. This evidence-based approach enhances the validity of results and aids in making more informed decisions across the building industry.

5.1 Experimental Design in Construction Research

By offering a methodical framework for testing hypotheses and defining cause-and-effect connections, experimental design is vital to building research. Through controlled experiments, researchers can alter specific factors such as construction techniques, materials, or training interventions to observe their direct effects on outcomes like project timeframes, safety performance, and cost efficiency. Experimental designs enhance the internal validity of results and reduce bias by incorporating elements such as random assignment, control groups, and pre-/and post-measures. This methodological discipline fosters the development of evidence-based methods, allowing the building sector to adopt more efficient and creative ideas grounded in consistent facts.

5.1.1 Assessing Change: The Role of Pretests

One of the key questions in research is whether an observed outcome represents a genuine change. In construction management, for instance, suppose a company implements a new safety training program and subsequently observes a decrease in workplace accidents. Without a pretest, whether the reduction was due to the training program or pre-existing safety improvements is unclear. A pretest is a baseline measure, enabling comparisons before and after an intervention. For example, a study by Yang and Fan [87] conducted pre- and post-test analyses with SPSS to evaluate safety knowledge and awareness changes, while a five-point Likert scale captures student perceptions of VR's motivational impact.

5.1.2 Establishing Causation Through the Use of Control Groups

Other factors may help explain the result even when a change is observed. For instance, a construction company adding new tools to increase efficiency might see greater output. On the other hand, if a simultaneous wage rise occurred, it would be unclear whether the enhanced efficiency stemmed from the equipment or the salary increase. The impact of an intervention is isolated using a control group. Developing a digital technology adoption model to assist potential adopters in assessing and choosing specific digital technologies can help them in construction research by enhancing their competitive edge. The model is a benchmarking tool for companies already utilizing digital technology to evaluate and measure their relative competitive position [88].

5.1.3 Evaluating Long-Term Effects: The Use of Time Series Analysis

Evaluating whether seen improvements endure over time is another key component of research design. Applying new work procedures could first show changes in construction, but it is essential to know if these advantages last. Time series analysis lets academics monitor performance over long durations. For instance, a study by Pan and Zhang [89] examined the state of knowledge by combining process mining and time series analysis into the virtual component using sophisticated data mining methods. The project sought deep insights from extensive IoT data collected from the physical side and kept in the cloud BIM.

5.2 Statistical Analysis in Construction Research

After a study plan is set, the vital stage in building research is choosing suitable statistical tools to examine the gathered data. Statistical analysis lets researchers find patterns, organize comparisons, assess correlations between variables, and develop evidence-based conclusions. Descriptive statistics, t-tests, ANOVA, correlation, and regression analysis are among the techniques that may be applied depending on the data type and study objectives. These methods guide decision-making, measure performance indicators, and help to determine how well building practices work. An appropriate statistical study guarantees that results are consistent and relevant to building situations.

5.2.1 Descriptive Analysis in Construction Workforce Studies

Descriptive statistics analyze the benefits and challenges of wearable safety devices in the construction sector. For instance, a survey on respondents' demographics may categorize respondents by gender, profession, employer type, and highest level of education [90]. These categorical variables are typically analyzed using frequencies and percentages. Additionally, continuous variables of wearable safety devices, such as struck-by objects, falling from a high level, slips, and trips, can be reported using means and standard deviations.

5.2.2 Group Comparisons Using ANOVA to Evaluate Construction Training Methods

Construction research depends on comparing several groups. For instance, studies might assess whether various approaches to training delivery provide different outcomes in construction safety knowledge and training [91]. The test scores (dependent variable) were normally distributed, and each participant's score was independent of the scores of other participants, so a one-sample t-test was run to answer the research question, "Are the overall scores of the population different from a hypothetical mean?". A One-Way ANOVA was used to answer the study question, "Do different education or training delivery modes affect the retention of safety knowledge conveyed in the training toolbox?". Given its ability to compare several levels or kinds of an independent variable, i.e., education/training delivery modes, about their influence on a dependent variable, i.e., test scores, this statistical test was considered suitable.

5.2.3 Predicting Pro-environmental Behavior in Construction Waste Management

Using the Pearson correlation coefficient as the evaluative criterion, Pearson's correlation study evaluates the strength of the relationship between two continuous variables [92]. The findings show a notable link between Pro-environmental Behavior (PEB) and its antecedents, including stress and personal and organizational elements [93]. Several linear regression studies examined Causal elements affecting PEB [94]. Key variables like the p-value were utilized to evaluate the dependability and validity of the regression model using the stepwise approach. The likelihood ratio test decided variable entrance and removal; inclusion thresholds were $p < 0.05$, and exclusion levels were $p > 0.10$. Statistically significant were p-values under 0.05. Model I designated co-worker PEB as the dependent variable; eight human and organizational elements were added as independent variables. Every incorporated element fulfilled the acceptance requirements described in [95, 96] by showing statistically significant p-values ($p < 0.05$), hence strengthening the validity and robustness of the model.

5.2.4 Explaining Causal Relationships with Structural Equation Modeling

The highest degree of quantitative research seeks to determine causal links to understand whether variables are connected and how and why one affects another. When addressing complex issues involving several interacting elements, this becomes crucial. Structural equation modeling (SEM), which allows researchers to test theoretical models and investigate direct and indirect interactions between variables, is one of the most powerful techniques for exploring these complex interactions. Studies where social, technological, and environmental aspects are intertwined especially benefit from SEM. It thoroughly analyzes how various factors affect one another, highlighting the deeper structures that underlie results. SEM provides a comprehensive approach to identifying pathways for innovation and improvement in contexts where human-centered technologies, resilience, and sustainability intersect, such as in building, infrastructure, or organizational development. Figure 14 illustrates how statistical analysis and quantitative research methodologies assist in addressing such

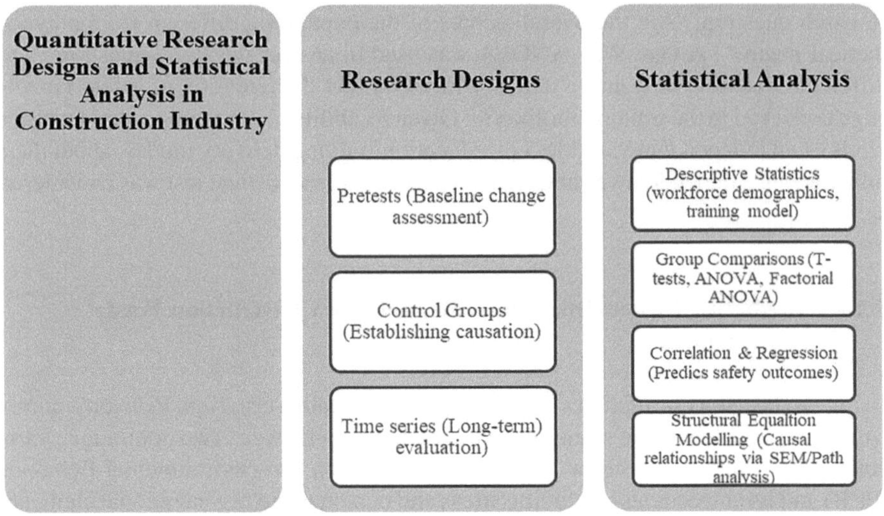

Fig. 14 Quantitative research designs and statistical analysis in the construction industry. *Source* Author's compilation

questions by providing the means to transition precisely and deeply from observation to explanation.

6 Advantages and Disadvantages of Quantitative Research

Through numbers, quantitative research offers a strong foundation for investigating and interpreting the world. Its power provides objective, methodical, and statistically rigorous results required for hypothesis testing, pattern identification, and evidence-based decision-making support. Researchers can use numerical data and organized approaches to give replicable, quantifiable, and generalizable outcomes for larger populations. Quantitative research is fundamental in social sciences, business, and engineering, where large-scale trends and interactions must be evaluated correctly [5, 45]. However, for these advantages, quantitative research has notable drawbacks as well. One significant difficulty is its failure to completely reflect the depth and complexity of human behavior and social environment. Emotional experiences, personal motivations, and subtle cultural or social impacts often exceed what can be measured using statistical models or organized tools. Many quantitative measures' rigorous structure could limit people's capacity to share different points of view or experiences, hence oversimplifying or omitting important insights [45]. Although scientific investigation is still founded on quantitative studies, its most significant value surfaces when employed carefully and, when appropriate, supplemented with qualitative approaches offering more depth and contextual knowledge. Quantitative techniques, for instance, may be used to evaluate the compressive strength of several concrete mixes in building materials research. Researchers can objectively identify which material performs best under particular circumstances using standardized tests and statistical data analysis, guaranteeing the findings' dependability and generalizability across building projects.

6.1 Advantages of Quantitative Research

Through the gathering and examination of numerical data, quantitative research provides a systematic and organized way to understand complex processes. Its main benefits include generating objective, repeatable, and generalizable results that can inform decisions across various sectors. Researchers can identify trends, validate theories, and accurately measure interactions using statistical methods and standardized tools. This approach is particularly valuable in large-scale investigations, as findings must apply to broader populations. Furthermore, quantitative research supports evidence-based practices by supplying consistent data that can be utilized to evaluate programs, policies, and theoretical models.

6.1.1 Reliability and Replicability

Among the most valued qualities of quantitative research is its consistency and repeatability. These characteristics provide lasting strength and credibility to scientific results. A reliable study indicates that the outcomes will be consistent if the same techniques are used under the same circumstances. Carefully planned research design, uniform data collection techniques, and clear variable definitions all contribute to this consistency, which is not accidental but rather a result of these factors ensuring the accuracy and repeatability of the measurements [97]. Replicability advances this concept. It indicates that other researchers can follow the same methods and reach equivalent conclusions regardless of the original study. Building scientific knowledge relies on this process, allowing results to be confirmed, questioned, or expanded upon. In practical sectors like engineering, healthcare, or construction, replicability ensures that, once validated, innovations can be successfully implemented and scaled in real-world environments. Ultimately, the reliability and replicability of quantitative research foster a culture of openness, accountability, and confidence in academic work, qualities vital for advancing theoretical knowledge and practical application.

6.1.2 Objective and Systematic Data Analysis

Especially when processing and analyzing data, a defining characteristic of quantitative research is its commitment to objectivity and systematic examination. Unlike more interpretative research methods, quantitative studies primarily rely on statistical techniques and mathematical models that follow established principles. This approach ensures that outcomes are not influenced by personal judgment, preconceptions, or expectations, but rather reflect clear trends in the data itself [46]. Researchers can openly and impartially evaluate correlations between variables by employing accepted statistical tests like regression analysis, ANOVA, or correlation coefficients. This reduces the potential for researcher bias, which can subtly affect the interpretation of findings in more subjective approaches. When utilized correctly, statistical methods offer an unbiased perspective that allows researchers to identify trends, validate theories, and draw more reliable conclusions. Atapattu et al. [98] provide a compelling example by developing a multiple regression analysis (MRA)-based model to predict the eventual cost of road construction projects at the pre-design stage. Their research utilized data from 43 completed New Zealand projects; the objective nature of the statistical method enabled them to identify key cost-driving factors and create a forecast model based on real data [98].

6.1.3 Perceived Credibility and Prestige

Often considered more scientific, the quantitative method has a solid basis in numerical data, statistical analysis, and computational techniques. In sectors where data-driven decision-making is vital, such as banking, engineering, and construction, this focus on measurable facts makes it more attractive [99] Reliable data is essential for process optimization, risk reduction, and outcome enhancement in various sectors. For example, in the construction industry, creative frameworks have been created to include advanced digital technologies, including Building Information Modeling (BIM), Digital Twins, and Extended Reality tools such as Virtual Reality, Augmented Reality, and Mixed Reality into daily operations [100, 101]. These solutions provide comprehensive, open, real-time tracking of construction site progress. Researchers often utilize structural equation modeling (SEM) to examine such systems, allowing them to judge the strength and validity of correlations between variables. This statistical study guarantees the practical and technical correctness of the tools presented. Empirical studies show time and time again how integrated digital technology may change more creative, efficient, and open building management systems.

6.1.4 Technological Integration

Today's fast-moving technology, particularly in big data analytics, artificial intelligence (AI), and machine learning, rapidly enables quantitative research. These technologies transform how researchers manage data, allowing them to swiftly and precisely examine large datasets. For example, recent developments have changed how construction projects are planned and executed. One interesting breakthrough is the integration of Deep Learning with Digital Twin technologies, supporting the broader movement toward Construction 4.0 [102]. This combination creates a more connected, open, and responsive building environment by facilitating real-time data flow from various sources, including sensors, models, and field reports. Combining simulation and optimization technologies helps professionals and researchers make better decisions during the planning and building stages, improving efficiency and reducing risk. These advancements enhance the analytical capacity of quantitative research and bring it closer to practical application in dynamic, data-rich environments.

6.2 Disadvantages of Quantitative Research

Although quantitative research provides clarity, accuracy, and organization to scientific investigation, it also has drawbacks, particularly regarding the complexity of human experiences. One of the key difficulties is that figures can convey only part of the narrative. Human ideas, feelings, and actions often possess subtleties that predefined questions or numerical scales cannot fully express. Consequently, this approach may occasionally overlook the fundamental significance or driving forces behind people's behavior. The inflexible nature of quantitative techniques is another limitation. There is limited opportunity for exploring unexpected insights or adapting to new developments during the study, since they rely on set instruments such as tests or surveys. Rich, multifaceted realities can also be oversimplified into measurable factors, sometimes distorting issues and making the results seem distant from actual situations. While quantitative studies excel at identifying trends and drawing conclusions, they may miss the human aspect of the research.

6.2.1 Overemphasis on Numerical Data

A frequent criticism of quantitative studies is their inclination to emphasize figures at the expense of the background in which those numbers reside. Although statistical trends and quantifiable results are helpful, they do not necessarily reflect the complexity of social systems or human behavior. Focusing primarily on numerical associations could lead researchers to miss significant social, cultural, or environmental elements that profoundly influence the data [103]. For example, while two initiatives could produce comparable performance measures, the underlying causes, such as labor culture, leadership dynamics, or community impact, could vary greatly. When the study focuses only on correlations or averages, these deeper insights can get missed. Lacking a thorough knowledge of context, one runs the danger of reaching practically incomplete and technically correct conclusions. Therefore, whereas quantitative studies provide clarity and comparability, they should be complemented with careful interpretation to prevent oversimplifying complicated reality [103].

6.2.2 Challenges in Research Model Development

Developing a strong and reliable research model in quantitative studies is not simple. From the beginning, it calls for clarity, accuracy, and careful planning in addition to mere statistics. The whole process depends on how well the hypotheses are crafted, how precisely the variables are characterized, and how carefully the samples are chosen. These actions set the groundwork for significant, data-driven insights when done correctly. But sometimes things don't go as planned. Poorly phrased survey questions, samples that don't reflect the population, or even subtle researcher preconceptions can all cause bias or mistakes that silently compromise the whole study. Although occasionally minor, these errors might provide false results that compromise the study's validity [45]. Creating a robust research model in quantitative work is a careful balancing act. Designing a survey that portrays the reality it seeks to comprehend requires both technological correctness and human understanding.

6.2.3 Potential for Misleading Conclusions

Though generally considered objective and scientific, quantitative research is not free of human impact. Every dataset and statistical test result is based on a researcher's decisions about what to measure, how to frame inquiries, and which techniques to use. Even accidental decisions can show previous ideas or expectations that influence the result of the research [104]. For instance, a survey question may be subtly biased or a variable defined too narrowly, which would cause the results to seem valid on the surface but distort the underlying reality. Likewise, overemphasizing statistical significance without regard to context might produce results that appear necessary but have little relevance. Thus, well-executed quantitative research can also lead to false conclusions if every stage, from design to interpretation, is not attended to. Understanding this possibility motivates academics to approach data with humility and diligence, always considering how their choices affect the narrative the figures tell [104].

6.2.4 Limited Ability to Capture Contextual Influences

One drawback of quantitative research is its emphasis on numbers and measurable factors, which frequently overlooks the human aspect of the narrative. Unlike qualitative approaches, which deeply explore people's experiences, ideas, and emotions, quantitative research usually depends on standardized instruments, such as tests or surveys, that do not provide much possibility for personal expression or unanticipated revelations [104]. Studying complicated, real-world environments where context matters, such as cultural norms, social dynamics, or environmental influences, can be particularly troublesome. Though hard to quantify with a multiple-choice question or a grading scale, these elements influence how people act and react. Consequently, essential subtleties could be missed, and the results could appear too unrelated or straightforward from the daily reality they want to mirror. Although quantitative studies offer clarity and organization, one should consider what they could omit and, when feasible, combine them with approaches that bring back the background [104].

Often driven by cutting-edge technology and statistical tools, quantitative research offers transparent, objective, and replicable insights. Helping academics find trends, validate hypotheses, and back data-driven choices is essential in engineering, business, and the social sciences. Though its methodical, numbers-based approach has many benefits, it can occasionally miss the depth of human experience. Real-world problems include managing construction projects, guaranteeing worker safety, or grasping organizational dynamics that typically have emotional, social, and cultural dimensions that numbers alone cannot completely clarify. That is why it is imperative to balance qualitative approaches with quantitative ones. Combining these strategies helps researchers investigate what is occurring, why, and how—this combination of depth and data results in a more thorough, human-centered awareness of our multifaceted problems.

7 Conclusion

A fundamental pillar of the empirical sciences, quantitative research provides a methodical, systematic, and objective way to examine events across several fields. Emphasizing its advantages in hypothesis testing, pattern discovery, and predictive modeling using numerical data and statistical analysis, this chapter critically examines the fundamental ideas and methods of quantitative research. Quantitative research enables evidence-based conclusions that significantly advance scientific debate by emphasizing measurement accuracy, replicability, and generalizability. Quantitative research under positivist and post-positivist paradigms is based on the belief that reality is measurable, observable, and exists apart from human perception. These philosophical underpinnings shape the creation of research questions, direct the hypothesis formulation, and define the choice of suitable methodological strategies. Working under this paradigm, researchers seek to remain impartial and minimize prejudice in order to generate repeatable results relevant to more general audiences. The chapter describes the defining characteristics of quantitative research: standardized data collecting tools, such as surveys, tests, and structured observations, alongside well-defined variables and statistical methods meant to examine trends, relationships, and causal links. Particularly for verifying theoretical assumptions, quantitative research is very successful by these methodological characteristics guaranteeing consistency and comparability between experiments. By ensuring that the sample correctly reflects the more significant population of interest, probability sampling methods help to promote the generalizability of results further. Various research designs are investigated, each with different goals and fit for specific study objectives. Descriptive designs record and summarize events as they happen; correlational designs look at the strength and direction of relationships between variables without suggesting causation; causal-comparative designs look at group differences when experimental manipulation is impossible; and experimental designs let you directly test causal relationships under controlled conditions. Research goals, the type of variables involved, and the degree of experimental control accessible to the researcher guide the choice of a suitable design. Fundamental to guaranteeing the trustworthiness of quantitative results, the chapter also discusses the key concepts of validity and reliability. Encompassing internal, external, and construct validity aspects, validity is the extent to which a tool or technique correctly assesses what it purports to measure. By contrast, reliability is the consistency and stability of measurement across time, settings, and observers. These standards, taken together, support the methodological discipline needed for strong scientific investigation. Among the strategies covered are instrument creation, pilot testing, and statistical verification as well as those for minimizing validity concerns and assuring dependability. The quantitative research process also includes ethical issues as another critical component. Researchers have to maintain participant protection, confidentiality, and informed consent. Maintaining academic integrity and building public confidence in research results depend on data collection, processing, and reporting openness. Notably, the chapter also emphasizes the need for careful data interpretation. Although statistical significance, e.g., $p \leq 0.05$, remains a conventional criterion, researchers are more and more pushed to evaluate effect sizes, e.g., Cohen's d, as a way to determine the practical importance of their results. This change shows a rising awareness that, in reality, statistically significant findings do not always correspond to substantial or influential conclusions. Looking ahead, developments in machine learning and artificial intelligence provide hope for more improvement of the design, analysis, and interpretation of quantitative data, hence providing advanced tools for modeling complicated systems and obtaining deeper insights. From construction to education to healthcare to environmental science, these developments are anticipated to increase the scope and accuracy of quantitative approaches. A strong and flexible tool for producing empirical information is quantitative research. Its strengths are in its methodological clarity, capability for generalization, and ability to generate replicable results. Its efficacy, however, lies in careful design, ethical purity, and data-critical involvement. A sophisticated awareness of quantitative research's ideas and techniques is still vital for researchers seeking theoretical knowledge and practical solutions as they face more challenging inquiries.

References

1. Leedy PD, Ormrod JE (2010) Practical research: planning and design, 9th edn. Pearson Educational International, Boston
2. Williams C (2007) Research methods. J Bus & Econ Res (JBER) 5(3). https://doi.org/10.19030/jber.v5i3.2532
3. Aliaga M, Gunderson B (2002) Interactive statistics. Sage Publications, Thousand Oaks
4. Goertzen MJ (2017) Introduction to quantitative research and data. In: Goertzen (ed) Applying quantitative methods to e-book collections (Vol 53, No 4). Library Technology Reports, Expert Guides to Library Systems and Services, Atlanta: Ala Techsource
5. Bryman A (2021) Social Research Methods. Oxford University Press

6. Lim WM (2024) What is qualitative research? an overview and guidelines. Australas Mark J 14413582241264619
7. Remler DK, Van Ryzin GG (2025) Control, exogeneity, and directness: understanding and designing quasi-and natural experiments. Am J Eval 46(1):62–89
8. Odiadi J (2024) Qualitative and quantitative research in business studies: managing. SSRN 4876332
9. Usman AC, Al-Hendawi M, Bulut S (2025) Approaches to qualitative research: a narrative literature review. Adv Med, Psychol, Public Health 2(2):81–95
10. Mrabti L, Alaoui ZB (2024) Balancing qualitative and quantitative research methods: insights and applications. In: Data collection and analysis in scientific qualitative research, pp 87–118. IGI Global
11. Pantović V, Vidojević D, Vujičić S, Sofijanić S, Jovanović-Milenković M (2024) Data-Driven decision making for sustainable IT project management excellence. Sustainability 16(7):3014
12. Kittur J (2023) Conducting quantitative research study: a step-by-step process. J Eng Educ Transform 36(4):100–112
13. Haslam SA, Steffens NK, Cruwys T, McGarty C (2024) Research methods and statistics in psychology
14. Jin J, Zhu Y, Dong G, Zhang Y, Yang X, Zhang C, Wen JR (2024) Flashrag: a modular toolkit for efficient retrieval-augmented generation research. arXiv:2405.13576
15. Sawacha EOO (1993) An investigation into safety attitudes and safety performance in the construction industry (Doctoral dissertation, Brunel University School of Engineering and Design PhD Theses)
16. Almusaed A, Yitmen I, Almssad A, Myhren JA (2024) Construction 5.0 and sustainable neuro-responsive habitats: Integrating the brain–computer interface and building information modeling in smart residential spaces. Sustainability 16(21):9393
17. Mohajan HK (2020) Quantitative research: a successful investigation in natural and social sciences. J Econ Dev, Environ People 9(4):50–79
18. Themann CL, Masterson EA (2019) Occupational noise exposure: a review of its effects, epidemiology, and impact with recommendations for reducing its burden. J Acoust Soc Am 146(5):3879–3905
19. Mosley MD (2022) Continual improvements in information and technology: a quantitative ex post facto (Causal Comparative) Study Design (Doctoral dissertation, University of Phoenix)
20. Hamilton D, McKechnie J, Edgerton E, Wilson C (2021) Immersive virtual reality as a pedagogical tool in education: a systematic literature review of quantitative learning outcomes and experimental design. J Comput Educ 8(1):1–32
21. Diener E, Northcott R, Zyphur MJ, West SG (2022) Beyond experiments. Perspect Psychol Sci 17(4):1101–1119
22. Abed J, Rayburg S, Rodwell J, Neave M (2022) A review of the performance and benefits of mass timber as an alternative to concrete and steel for improving the sustainability of structures. Sustainability 14(9):5570
23. Duckett LJ (2021) Quantitative research excellence: study design and reliable and valid measurement of variables. J Hum Lact 37(3):456–463
24. Ng JY, Lee MS, Liu JP, Steel A, Wieland LS, Witt CM, Cramer H (2024) How can meta-research be used to evaluate and improve the quality of research in the field of traditional, complementary, and integrative medicine? Integr Med Res 13(3):101068
25. Baltes S, Ralph P (2022) Sampling in software engineering research: a critical review and guidelines. Empir Softw Eng 27(4):94
26. Haq ZU, Rasheed R, Rashid A, Akhter S (2023) Criteria for assessing and ensuring the trustworthiness in qualitative research. Int J Bus ReflectS 4(2)
27. Agarwal SA, Chauhan NKS (2024) Basic concepts of research methodology & statistics. Academic Guru Publishing House
28. Minh NTT (2025) Exploring qualitative data collection methods for dissertation research. In: Qualitative research methods for dissertation research, pp 81–124. IGI Global Scientific Publishing

29. Bhalerao AK, Tiwari R (2024) Leveraging online platforms for effective data collection, surveys and visualization. Innov Digit Ext, 19.
30. Tuval-Mashiach R (2021) Is replication relevant for qualitative research? Qual Psychol 8(3):365
31. Song W, Calautit JK (2024) Inclusive comfort: a review of techniques for monitoring thermal comfort among individuals with the inability to provide accurate subjective feedback. Build Environ 111463
32. Taherdoost H (2022) What are different research approaches? comprehensive review of qualitative, quantitative, and mixed method research, their applications, types, and limitations. J Manag Sci & Eng Res 5(1):53–63
33. Mancilla S, Wences G, Hernández-López E, Cohen I (2025) Sub-spatial prediction of votes integrating socioeconomic, educational, and age strata with machine learning and topological data analysis. J Big Data 12(1):59
34. Butaki LC (2024) The impact of skilled construction workers' turnover on the performance of construction projects in Kenya: a case study of judiciary projects (Doctoral dissertation, University of Nairobi)
35. Karunarathna I, Gunasena P, Hapuarachchi T, Gunathilake S (2024) Data collection fundamentals: a guide to effective research methodologies and ethical practices. ResearchGate 65(65), 65. https://www.researchgate.net/publication/383155577_Data_Collection_Fundamentals_A_Guide_to_Effective_Research_Methodologies_and_Ethical_Practices
36. Karunarathna I, Gunasena P, Hapuarachchi T, Gunathilake S (2024) Comprehensive data collection: methods, challenges, and the importance of accuracy. Uva Clin Res 1–24
37. Hahn J, Heiler S, Kane MB, Park S, Jensch W (2022) The information gap in occupant-centric building operations: lessons learned from interviews with building operators in Germany. Front Built Environ 8:838859
38. Bihu R (2021) Questionnaire survey methodology in educational and social science studies. Int J Quant Qual Res Methods 9(3):40–60
39. Hongyun T, Sohu JM, Khan AU, Junejo I, Shaikh SN, Akhtar S, Bilal M (2025) Navigating the digital landscape: examining the interdependencies of digital transformation and big data in driving SMEs' innovation performance. Kybernetes 54(3):1797–1825
40. Khoa BT, Hung BP, Hejsalem-Brahmi M (2023) Qualitative research in social sciences: data collection, data analysis and report writing. Int J Public Sect Perform Manag 12(1–2):187–209
41. Wasserthal J, Breit HC, Meyer MT, Pradella M, Hinck D, Sauter AW, Segeroth M (2023) TotalSegmentator: robust segmentation of 104 anatomic structures in CT images. Radiol: Artif Intell 5(5):e230024
42. Kakolu S, Faheem MA (2024) Building trust with generative AI chatbots: exploring explainability, privacy, and user acceptance. Iconic Res Eng JS 8(3):823–834
43. Moeller J, Dietrich J, Neubauer AB, Brose A, Kühnel J, Dehne M, Pekrun R (2022) Generalizability crisis meets heterogeneity revolution: determining under which boundary conditions findings replicate and generalize
44. Green Y, Gully A, Roth Y, Roy A, Tucker JA, Wanless A (2023) Evidence-based misinformation interventions: challenges and opportunities for measurement and collaboration. Cargengie Endow Int Peace January, 9
45. Creswell JW, Creswell JD (2023) Research design: qualitative, quantitative, and mixed methods approaches. SAGE Publications
46. Babbie E (2020) The practice of social research. Cengage Learning
47. Yin RK (2022) Case study research and applications: design and methods. SAGE Publications
48. Shadish WR, Cook TD, Campbell DT (2001) Experimental and quasi-experimental designs for generalized causal inference. Houghton Mifflin
49. Hernán MA, Robins JM (2023) Causal inference: what if? Chapman & Hall
50. Maxwell JA (2021) Qualitative research design: an interactive approach. SAGE Publications
51. Tabachnick BG, Fidell LS (2019) Using multivariate statistics. Pearson
52. Fowler FJ (2020) Survey research methods. SAGE Publications
53. Cochran WG (2022) Sampling techniques. Wiley

54. Lohr S (2021) Sampling: design and analysis. Chapman & Hall
55. Groves RM et al (2020) Survey methodology. Wiley
56. Kline RB (2020) Principles and practice of structural equation modeling. Guilford Press
57. Stevens SS (2021) On the theory of scales of measurement. Science
58. DeVellis RF (2022) Scale development: theory and applications. SAGE Publications
59. Haynes SN, Richard DC, Kubany ES (2020) Content validity in psychological assessment. Psychological Assessment
60. Anastasi A, Urbina S (2021) Psychological testing. Pearson
61. Cronbach LJ, Meehl PE (2022) Construct validity in psychological tests. American Psychological Association
62. Field A (2021) Discovering statistics using SPSS. SAGE Publications
63. Nunnally JC, Bernstein IH (2021) Psychometric theory. McGraw-Hill
64. Campbell DT, Stanley JC (2015). Experimental and quasi-experimental designs for research. Ravenio Books
65. Dimitrov DM, Rumrill PD (2003) Pretest-posttest designs and measurement of change. Work 20(2):159–165
66. Brydges R (2019) The role of control groups in experimental research. J Res Methods 42(3):215–229
67. Kazdin AE (2017) Single-case research designs: methods for clinical and applied settings. Oxford University Press
68. Cohen J (1988) Statistical power analysis for the behavioral sciences. Routledge
69. Gelman A, Hill J, Vehtari A (2020) Regression and other stories. Cambridge University Press
70. Maxwell SE, Delaney HD (2004) Designing experiments and analyzing data: a model comparison perspective. Psychology Press
71. Greenwood C, Nikulin MS (1996) A guide to chi-squared testing. Wiley, New York
72. Pearl J (2009) Causal inference in statistics: an overview. Stat Surv 3:96–146
73. Hair JF, Hult GTM, Ringle CM, Sarstedt M (2019) A primer on partial least squares structural equation modeling (PLS-SEM). Sage
74. Byrne BM (2016) Structural equation modeling with AMOS: basic concepts, applications, and programming. Routledge
75. Wasserstein RL, Lazar NA (2016) The ASA's statement on p-values: context, process, and purpose. Am Stat 70(2):129–133
76. Lakens D (2013) Calculating and reporting effect sizes to facilitate cumulative science. PLoS ONE 8(6):e100129
77. Jordan MI, Mitchell TM (2015) Machine learning: Trends, perspectives, and prospects. Science 349(6245):255–260
78. Li CZ, Tam VW, Hu M, Zhou Y (2024) Lean construction management: a catalyst for evaluating and enhancing prefabricated building project performance in China. J Build Eng 94:109930
79. Nejatyan E, Sarvari H, Hosseini SA, Javanshir H (2023) Determining the factors influencing construction project management performance improvement through earned value-based value engineering strategy: a Delphi-based survey. Buildings 13(8):1964
80. Mohandes SR, Zhang X (2021) Developing a holistic occupational health and safety risk assessment model: an application to a case of sustainable construction project. J Clean Prod 291:125934
81. Li X, Wang C, Kassem MA, Alhajlah HH, Bimenyimana S (2022) Evaluation method for quality safety risks in prefabricated building construction using SEM–SDM approach. Int J Environ Res Public Health 19(9):5180
82. Guo K, Zhang L (2022) Multi-objective optimization for improved project management: current status and future directions. Autom Constr 139:104256
83. Wang X, Cheng C, Zhang J, Ma G, Li J, Jin J (2023) Real-time monitoring and quality assessment of subgrade compaction: key factors and ANN model. Acta Geotech 18(6):3349–3366

84. Al-Bayati AJ, Rener AT, Listello MP, Mohamed M (2023) PPE non-compliance among construction workers: an assessment of contributing factors utilizing fuzzy theory. J Safety Res 85:242–253
85. Khalef R, El-adaway IH (2023) Advancing airport project delivery: a comparison of design–build and traditional schedule and cost performance methods. J Manag Eng 39(6):04023041
86. Wasserstein RL, Schirm AL, Lazar NA (2019) Moving to a world beyond "p < 0.05". Am Stat 73(sup1):1–19
87. Yang YF, Fan CC (2025) Evaluating the effectiveness of Virtual Reality (VR) technology in safety management and educational training: an empirical study on the application and feasibility of digital training systems. Interact Learn Environ 1–29
88. Ling FY, Heng GTH, Chang-Richards A, Chen X, Yiu TW (2023) Impact of digital technology adoption on the comparative advantage of Singapore's architectural, engineering, and construction firms. J Constr Eng Manag 149(12):04023125
89. Pan Y, Zhang L (2021) A BIM-data mining integrated digital twin framework for advanced project management—automation. Construction 124:103564
90. Ibrahim K, Simpeh F, Adebowale OJ (2025) Benefits and challenges of wearable safety devices in the construction sector. Smart Sustain Built Environ 14(1):50–71
91. Hashem M, Mehany MS, Killingsworth J, Shah S (2019) An evaluation of training delivery methods' effects on construction safety training and knowledge retention—a foundational study. Int J Constr Educ Res 1–19
92. Cohen J, Cohen P, West SG, Aiken LS (2013) Applied multiple regression/correlation analysis for the behavioral sciences. Routledge
93. Hao JL, Yu S, Tang X, Wu W (2022) Determinants of workers' pro-environmental behaviour towards enhancing construction waste management: contributing to China's circular economy. J Clean Prod 369:133265
94. Denis DJ (2021) Applied univariate, bivariate, and multivariate statistics: Understanding statistics for social and natural scientists, with applications in SPSS and R. John Wiley & Sons
95. Keith TZ (2019) Multiple regression and beyond: an introduction to multiple regression and structural equation modeling. Routledge
96. Yitmen I, Almusaed A, Alizadehsalehi S (2024) Facilitating construction 5.0 for innovative, sustainable and resilient buildings: opportunities and challenges for implementation. Smart Sustain Built Environ
97. Cohen L, Manion L, Morrison K (2018) Research methods in education, 8th ed. Routledge
98. Atapattu CN, Domingo N, Sutrisna M (2024) Using multiple regression analysis, a conceptual cost estimation model for the pre-design stage of road projects. J Financ Manag Prop Constr 29(3):388–404
99. Alizadehsalehi S, Yitmen I (2023) Digital twin-based progress monitoring management model through reality capture to extended reality technologies (DRX). Smart Sustain Built Environ 12(1):200–236
100. Yitmen I, Almusaed A, Hussein M, Almssad A (2025) AI-driven digital twins for enhancing indoor environmental quality and energy efficiency in smart building systems. Buildings 15(7):1030
101. Almusaed A, Yitmen I, Almssad A (2023) Reviewing and integrating aec practices into industry 6.0: strategies for smart and sustainable future-built environments. Sustainability 15(18):13464
102. Kor M, Yitmen I, Alizadehsalehi S (2023) An investigation for integration of deep learning and digital twins towards construction 4.0. Smart Sustain Built Environ 12(3):461–487
103. Silverman D (2021) Doing qualitative research. 1–100. SAGE Publications
104. Lim WM (2024) What is quantitative research? An overview and guidelines. Australas Mark J 14413582241264622

Experimental Design and Causal Inference

Abstract This chapter presents a thorough, methodologically sound analysis of experimental design and causal inference, clarifying their key roles in empirical research across scientific fields. It meticulously explains the concepts that enable researchers to make legitimate causal claims, including counterfactual thinking, randomization, control conditions, replication, and blinding. Particular attention is given to threats to internal and external validity and the analytical techniques used to mitigate them, such as analysis of covariance, instrumental variables, and propensity score matching. The chapter links abstract causal frameworks with practical research design through theoretical explanation and applied examples, thus enhancing scientific investigation's accuracy, replicability, and ethical integrity.

1 Introduction

At its core, empirical scientific study involves making reasonable and methodical assumptions about causality. The development of "what if" questions arises quickly and continuously throughout the research and design process, as Chen (2022) emphasizes. Not only crucial for scientific research, this type of counterfactual thinking is also embedded in the professional judgment of architects and engineers. Whether designing new buildings, systems, or technologies, these professionals consistently encounter choices that require predictive assessment. They employ empirical knowledge, computer simulations, and data-driven analytical techniques to navigate ambiguity and evaluate potential outcomes [1]. Each design iteration serves as a hypothesis tested against physical models, historical facts, or predictive algorithms, reinforcing the intrinsic causal logic that underpins both design science and experimental research. Researchers commonly find statistical correlations across variables across disciplines—including the natural sciences, social sciences, and engineering as a foundation for producing explanatory insights. However, distinguishing whether these correlations indicate underlying causal mechanisms instead of false or confounding links calls for methodologically strong techniques. Enabling researchers

A. Almusaed et al., *Practice of Research Methodology in Civil Engineering and Architecture*, https://doi.org/10.1007/978-3-031-97393-2_16

to isolate and evaluate causal ideas under controlled circumstances depends on experimental design. Causal inference is especially vital in engineering fields, where industrial processes' optimization, stability, and control rely on an exact knowledge of cause-effect dynamics, as Rotari and Kulahci (2025) point out. Historically, the epistemological gold standard for determining causality has been randomized controlled trials. However, the growing interest in obtaining causal inferences from observational datasets is driven by the spread of high-resolution sensor technology and the accelerating expansion of data collecting capacity. Although such data-driven strategies present interesting opportunities, they also require careful methodological analysis to handle the natural constraints of non-randomized data when asserting causality [2]. Controlling for extraneous variability, a well-conceived experimental design provides the methodological basis for generating internally valid judgments about causal effects [3]. Fundamentally, an experiment is defined by systematically observing the consequent changes in one or more dependent variables brought on by manipulating one or more independent variables [4]. When included in a design that includes random assignment, control groups, and suitable procedural rigor, this manipulation offers the circumstances required to separate the causal effect of the treatment or intervention [5]. Unlike observational studies, where confounding factors could mix with the variables of interest, experimental designs let researchers control the research environment deliberately, lowering the likelihood of alternate explanations for reported results. Causal inference's conceptual basis is the counterfactual of what would have happened to the same unit of analysis in the absence of treatment. Experimental designs seek to simulate this counterfactual situation using comparison groups projected to be comparable at baseline, as it is difficult to see concurrently both the treated and untreated results for the same person [6]. Randomization is still the most dependable way to obtain baseline equivalency between experimental groups since it guarantees the impartial distribution of known and unknown confounding factors across treatment conditions. This methodological precaution lets the intervention itself appropriately account for any post-treatment variations. Advanced signal processing methods, as Ferwana and Varshney propose, can also help overcome causal inference difficulties by offering additional analytical tools for separating causal impacts [7]. The reasoning behind the experiments also includes using control conditions. Control groups provide researchers with reference points that help them separate the consequences of the therapy from natural variation or other temporal factors. While repeating results over several trials, populations, or settings helps to strengthen and generalize findings, including pretests, posttests, or both increases the capacity to see change over time. Moreover, methods like blocking and matching can lower variability linked to certain covariates, improving statistical efficiency and analytical accuracy [8]. However, even the most well-crafted experiment is vulnerable to several validity challenges, especially if the methodological assumptions underlying causal inference are not followed. Internal validity is the degree to which a study, free from confounding factors, shows a causal link between the independent and dependent variables [9]. If not properly expected and controlled, threats such as selection bias, maturation effects, historical events, testing reactivity, and attrition might undermine its validity. At the same time, especially when experiments are run under very controlled or artificial circumstances that may not reflect real-world complexity, external validity, the degree to which causal results can be extrapolated beyond the particular experimental setting, must also be considered [10]. Researchers may use quasi-experimental designs when random assignment is impractical or immoral; these designs preserve the logic of intervention and comparison but lack randomization [11]. Although such designs are naturally more susceptible to selection effects and confounding, their practical applicability in field situations is considerable. Therefore, quasi-experiments usually include pretest–posttest comparisons, interrupted time-series studies, or matched-group approaches to reduce validity concerns and support causal assertions. Furthermore, the field of causal inference has developed to encompass other statistical methods meant to mimic the characteristics of randomized trials in observational data. Furthermore, the field of causal inference has evolved to include additional statistical methods designed to replicate the features of randomized trials when analyzing observational data [12]. A graphical summary showing the fundamental ideas organizing a thorough experimental Design is shown in Fig. 1. "Experimental Design" is in the core; from there, five basic elements branch out: randomization, control conditions, replication, blocking and matching, and blinding techniques. By lowering bias, controlling variability, and improving interpretability, every principle adds distinctively to the validity and dependability of causal inference. The visual design provides a straightforward but thorough conceptual framework, stressing the interconnection of these ideas. This figure offers a starting point for researchers dedicated to methodological accuracy in empirical studies spanning scientific fields.

Guided by theoretical knowledge and backed by sensitivity analyses, methods such as propensity score matching, instrumental variable estimation, regression discontinuity designs, and difference-in-differences estimators allow researchers to estimate causal effects under explicit assumptions. Although these methods do not replace experimental control, they provide strong instruments for causal thinking when trials are unfeasible. Experimental research is also a philosophical and ethical quest, not only a technical one. It calls for openness about constraints, critical engagement with the assumptions ingrained in their designs, and methodological decisions that fit research goals, setting, and ethical criteria. Sound experimental thinking calls for both deductive clarity and inductive caution. Although causal theories have to be logically developed, their empirical validation has to consider the intricacies and contingencies natural in social systems and human behavior [13]. The two pillars of rigorous empirical research are experimental design and causal inference. Applied wisely, these frameworks help researchers to separate causation from correlation, evaluate the efficacy of treatments, and produce analytically accurate and practically relevant results. Answering particular research questions and progressing scientific knowledge in a methodical, reproducible, and ethically responsible way require, as this chapter will explain, combining design concepts with suitable inference techniques.

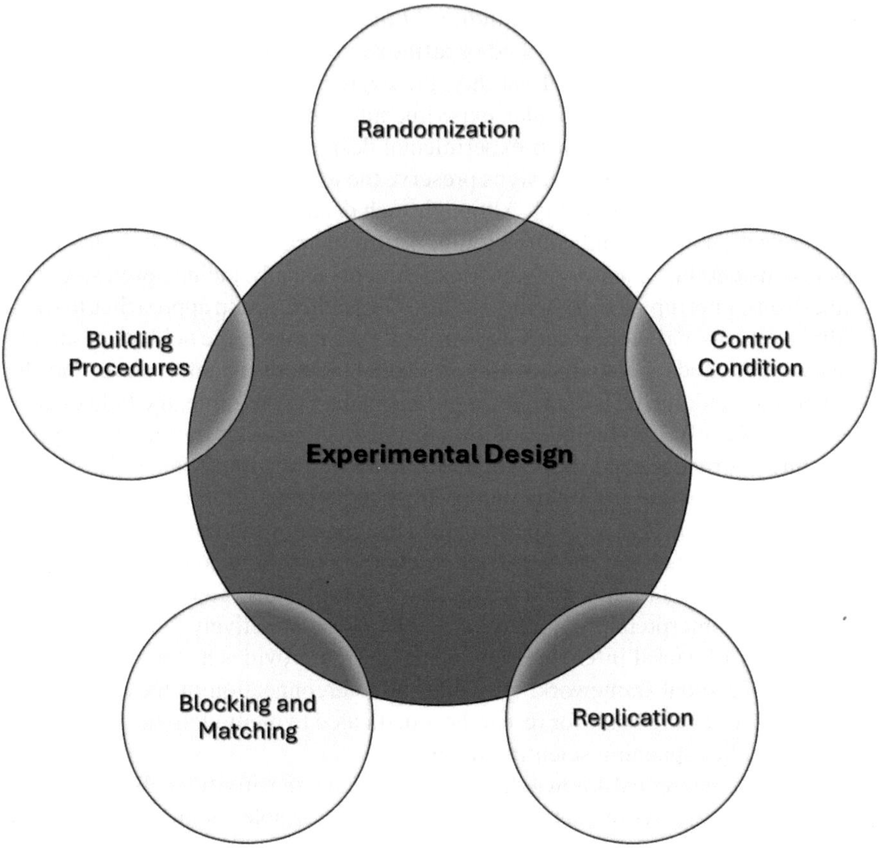

Fig. 1 Graphical abstract: foundations and principles of experimental design. *Source* Authors

2 Foundations and Principles of Experimental Design

Empirical study to prove causality is based on a thorough, well-organized experimental design [14]. Experimental design offers the framework for firmly validating causal assertions, manipulating variables, and testing hypotheses. Reflecting the scientific dedication to objectivity, replicability, and inferential robustness, the ideas supporting experimental design are procedural and philosophical [15]. Fundamentally, experimental design tries to mimic the counterfactual condition of what would have happened without the intervention using methodical manipulation and comparison. Researchers can identify the real impact of an intervention by controlling for possible sources of bias and variability, improving the study's internal validity [16]. Moreover, experimental design bridges theory with actual testing. It converts abstract theoretical ideas into operational research models that can be examined practically. Researchers can match methodological execution with conceptual frameworks by deliberately structuring treatment settings, control mechanisms, and measurement

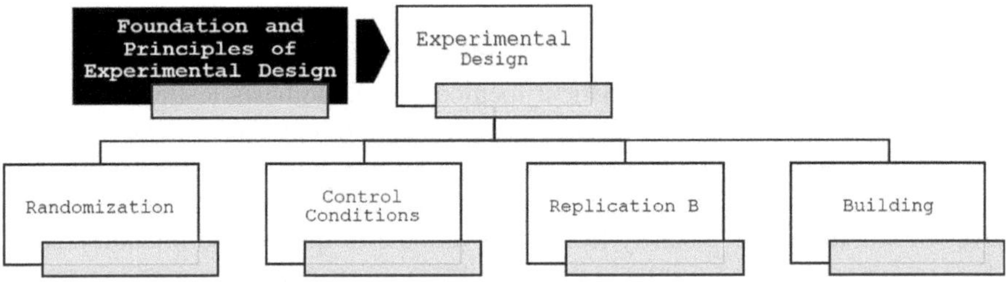

Fig. 2 Foundations and principles of experimental design. *Source* Authors

techniques [17]. This alignment improves empirical testing's fidelity and strengthens the validity and interpretation of the results. Ethical issues are also intimately related to the underpinnings of experimental design. The integrity of an experiment depends not just on technical rigor but also on following the values of fairness, openness, and respect for participants. Especially in medicine, education, and public policy, practical limits and ethical responsibilities must be weighed against randomization, control groups, and blindfolding [18]. The basic ideas of experimental design essential for isolating and quantifying causal effects, randomization, control conditions, replication, blocking and matching, and blinding procedures are extensively investigated in this part. Every concept strengthens the strength of empirical findings in laboratory and field environments by contributing to reducing bias, managing variability, and producing valid inferences [19]. The basic elements of experimental design randomization, control conditions, replication, and blinding are shown in Fig. 2. These concepts guarantee objectivity, lower prejudice, and improve internal validity. The framework connects theory and practice, allowing academics to test ideas while rigorously preserving ethical norms and methodological integrity.

2.1 *Randomization*

Often regarded as the strongest means of establishing causal links, randomization is a fundamental principle of experimental design [4]. It involves assigning experimental units, whether individuals, groups, or organizations, to treatment and control conditions based on a random method. This process, on average, ensures that both observed and unobserved confounding factors are evenly distributed across all experimental groups [20]. Therefore, randomization is a methodological tool against systematic bias and enhances the study's internal validity: probability theory and statistical inference support randomization. Without random assignment, the distribution of covariates between groups could be uneven, allowing external influences such as age, socioeconomic status, educational level, or unmeasured psychological factors to distort the relationship between the independent and dependent variables. Randomization mitigates this risk by disrupting potential connections between treatment assignments and confounding influences [21]. This ensures that treatment is the only systematic difference among groups, allowing for a more accurate attribution of any observed outcome variations to the intervention. Furthermore, randomization facilitates the application of statistical methods for hypothesis testing. Any post-intervention variations can be analyzed using formal inferential techniques such as t-tests, ANOVA, and regression modeling, as random assignment produces groups that are statistically equal in expectation [22]. Randomized experiments distinguish themselves in this interpretive clarity from observational or quasi-experimental studies, where treatment effects must be assessed under more stringent assumptions and are inherently more prone to bias. Various methods exist for implementing randomization, including simple, block, and stratified randomization. Decisions regarding these approaches are guided by sample size, the need to ensure balance among subgroups, and the study's logistical constraints [23]. For example, block randomization could be utilized in architectural engineering studies to evaluate the thermal performance of building envelope materials under controlled environmental conditions. Experimental units such as wall assemblies or facade systems can be classified by material type and then randomly assigned to test settings, ensuring that material heterogeneity does not bias findings. In addition to its traditional applications, randomization is increasingly integrated with advanced analytical and computational techniques. According to Ferwana and Varshney (2022), signal processing methods can enhance causal inference in high-dimensional or temporally dynamic experimental data. Such cross-disciplinary innovations expand the utility of randomization beyond conventional frameworks, permitting adaptation to new challenges in data-rich research environments [7]. Randomization also preserves the ethical integrity of experimental research. Eliminating systematic favoritism or selection bias in treatment distribution promotes the values of justice and transparency [24]. Ethical review boards may view randomization as a protective mechanism that fosters participant equity. However, it is essential to recognize that randomization does not guarantee accuracy. Thus, rigorous implementation must be paired with adequate sample sizes and precise outcome measurements. Poor adherence to procedures, attrition, or group contamination may undermine the validity of causal claims. Therefore, one should regard randomization as a necessary component of a broader methodological rigor rather than a guarantee of validity. The credibility of experimental research hinges on a dynamic and adaptive concept known as randomization [25]. It ensures group equivalence, enables statistical inference, enhances ethical fairness, and fosters methodological innovation. Its primary role in experimental design underscores its enduring importance across various fields and research methodologies. Emphasizing its scientific foundation, usual practices, ethical significance, and developing uses, Table 1 offers a methodical summary of the fundamental elements of randomization in experimental design. Every aspect has related difficulties that could show up during execution. This synthesis helps researchers expect constraints and choose suitable randomization techniques, supporting the strength and integrity of experimental causal inference.

Table 1 Key dimensions and challenges of randomization in experimental design

Aspect	Description	Challenges
Purpose [26]	To ensure equal distribution of confounding variables across groups	Requires strict adherence to protocol to prevent post-randomization imbalances
Methodological basis [27]	Grounded in probability theory and statistical inference	Assumes sufficient sample size for equivalence; imbalance may occur in small samples
Standard techniques [28]	Simple, block, and stratified randomization	Selecting the appropriate technique depends on study design complexity and logistical constraints
Ethical significance [29]	Promotes fairness and transparency in group allocation	Must address concerns about fairness in sensitive or vulnerable populations
Limitations	Does not address post-assignment biases (e.g., attrition, contamination)	Risk of dropouts or non-compliance undermining internal validity
Emerging applications [30]	Integration with computational models and signal processing methods	Requires technical expertise and interdiscip

2.2 *Control Conditions*

Control is a fundamental component of strong experimental design; control conditions allow scientists to isolate the causal influence of the independent variable [31]. For example, passive building research helps assess how specific architectural interventions, such as insulation techniques, window types, or ventilation strategies, impact performance outcomes, including thermal comfort, energy consumption, and indoor air quality. A control group is exposed to a standard or baseline condition and does not receive the experimental treatment. This establishes a baseline against which the performance of the intervention group can be meaningfully compared [32]. Utilizing control conditions facilitates counterfactual thinking, the idea that the control group reflects what would have occurred without the intervention. Drawing valid causal inferences depends on this approach, particularly in applied building science, where environmental factors interact. Moreover, well-designed control groups enhance transparency and interpretability by enabling researchers to attribute observed effects to specific design modifications instead of confounding variables. Figure 3 illustrates the basic function of control groups in experimental research. Researchers can separate the impact of the independent variable by comparing results between a control group (receiving no intervention) and an intervention group (undergoing the therapy). By reducing confounding variables and providing a precise evaluation of whether observed changes result from the intervention itself, this arrangement helps support legitimate causal inference.

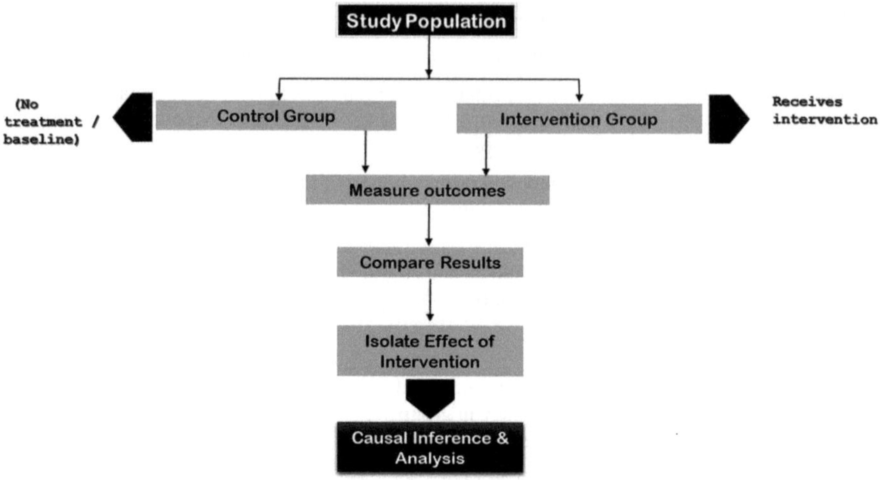

Fig. 3 Role of control groups in experimental design. *Source* Authors

Key Points

- Definition: Control conditions in experimental investigations serve as the baseline for comparison, receiving no intervention or a standard version.
- Control conditions enable counterfactual thinking to determine whether the experimental variable causes observed changes.
- Effect Decomposition: Multiple control groups can help untangle complex phenomena like placebo or Hawthorne effects.
- Advanced Designs: Control conditions in factorial studies assess interactions between various variables, including wall insulation and ventilation types.

2.3 Replication

A fundamental component of the scientific method, replication is essential to confirm the robustness and correctness of study results. It means rerunning a study with the same or a comparable approach to see whether the original outcomes can be consistently repeated [33]. Using replication, researchers can distinguish between results that indicate genuine effects and those that might have resulted from random mistakes, methodological constraints, or particular contextual elements. Replication helps the legitimacy and generalizability of scientific information by confirming or challenging first results [15]. This approach also aids in spotting trends, honing theories, and laying a more solid basis for future study. In disciplines struggling with reproducibility, where consistent evidence is vital for policy-making, healthcare procedures, or technological developments, replication is particularly critical. In the

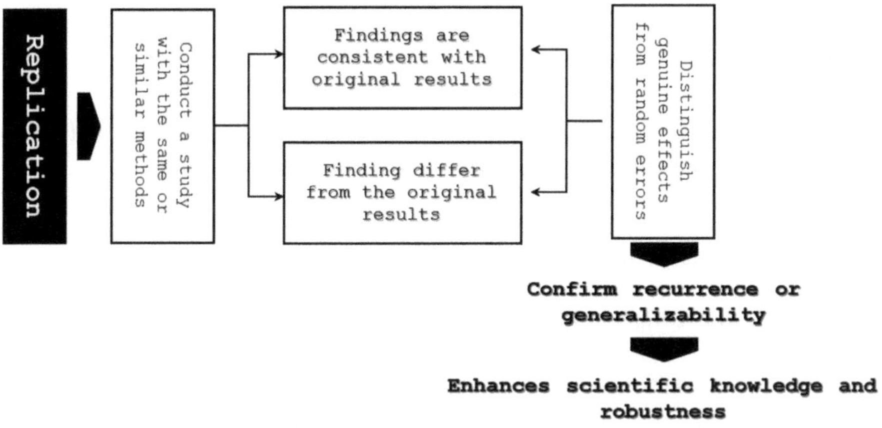

Fig. 4 Replication in scientific research

end, replication guarantees scientific integrity using consistent, methodical investigation across several research studies and environments, supporting results rather than only incidental or context-dependence. For example, the repeated testing of green roof systems across several climates illustrates replication in sustainable building research. Researchers confirm the environmental advantages of green roofs in different urban environments by repeating tests on thermal insulation and stormwater management, validating their consistent performance [34]. The logical flow of replication in scientific study is shown in Fig. 4. The study begins by employing comparable techniques and then compares them with the original results. Replication lets researchers identify real effects from random errors, whether outcomes agree or differ. This approach enhances scientific dependability by verifying correctness or generalizability. Replication finally strengthens scientific knowledge, especially in disciplines where evidence-based decision-making is essential for policy, healthcare, or technology.

2.3.1 Types of Replication

There are two main categories of replication: direct replication and conceptual replication.

- Direct replication refers to the precise duplication of a study using the same procedures, instruments, and sample characteristics as the original. Its purpose is to confirm the reliability of specific findings under identical conditions [35]. For example, in cognitive psychology, direct replications are often conducted to verify experimental effects related to memory recall or decision-making processes.
- Conceptual replication, on the other hand, tests the same underlying hypothesis but with different experimental designs, measurement tools, or participant populations [35]. This type of replication is particularly valuable for assessing the

generalizability of findings across contexts and populations. In medical research, conceptual replications might involve testing a treatment in different age groups or geographic regions to determine its broader applicability.

Both forms of replication play unique and complementary roles in strengthening causal inference and enhancing the credibility of scientific knowledge.

2.3.2 The Reproducibility Crisis

In many disciplines, including psychology, economics, machine learning, and the built environment, concerns regarding the reproducibility of scientific results have grown during the last twenty years [36]. Reproducibility problems in machine learning can stem from data leaking, overfitting, proprietary datasets, and insufficient model training and evaluation process documentation. The scientific community increasingly supports open-source code, data sharing, and uniform benchmarks to improve openness and reproducibility. Reproducibility is becoming more important in research building, especially with digital twin's virtual reproductions of actual structures employed for simulation, monitoring, and optimization. When researchers create digital twins to evaluate structural health or energy efficiency, results must be repeatable across several buildings and settings [37]. Studies, for instance, that use digital twins to model HVAC system performance under different occupancy circumstances call for constant replication to confirm their predictive accuracy and practical relevance. Not being able to replicate these simulations under comparable conditions could result in faulty policy recommendations in sustainable architecture or inconsistent design choices. Calls for open data environments, transparent reporting standards, and independent validation inside architectural and engineering research have risen with this rising knowledge to guarantee that results are scientifically sound and globally relevant rather than context-bound aberrations.

2.3.3 Institutional Responses

Journals, funding organizations, and universities have progressively embraced established procedures to promote scientific research openness, responsibility, and reproducibility in response to the reproducibility crisis [38]. These institutional shifts are especially relevant to sustainable building research, since results can significantly impact green technology investments, policy choices, and design plans on a large scale. The widespread promotion of pre-registration represents a significant advancement. Researchers who pre-register their study ideas, hypotheses, and analytical methods on platforms like the Open Science Framework provide a well-defined framework before data collection begins. In sustainable construction research, this practice ensures that performance claims related to energy efficiency, material durability, or lifespan assessments are not altered post hoc to align with preferred outcomes [39]. Another crucial policy is the encouragement or requirement of open data and code sharing. By publicly releasing datasets and simulation models, such as energy modeling or life cycle impact assessments, researchers enable others to replicate building performance studies [40]. For instance, when data inputs, building geometry models, and calculation methods are accessible for verification and reuse, research on the thermal performance of passive housing designs or green roofing systems gains credibility and relevance [41]. Furthermore, the use of registered reports is becoming increasingly popular. This publishing model involves peer review of the study's methodology and analysis plan before data collection [42]. As long as the approved method is adhered to, journals commit to publishing the study regardless of its outcomes. For example, this approach is beneficial in sustainable design for long-term field investigations, such as post-occupation evaluations or simulations of digital twins, where environmental, social, or behavioral outcomes may vary unpredictably over time [41]. These institutional measures help maintain individual studies' validity and enhance data's overall reliability in sustainable building research. Integrating reproducibility into the study design and dissemination process ensures that findings are trustworthy and applicable to real-world scenarios across diverse technical, cultural, and climatic contexts. Table 3 summarizes critical institutional policies designed to enhance research openness and replicability. It highlights the structural initiatives journals, funders, and academic institutions implement to address reproducibility issues and promote methodological integrity across scientific fields, ensuring each solution serves related challenges.

2.3.4 The Strategic and Analytical Role of Replication in Evidence Synthesis

Advancing methodological integrity and strategic knowledge-building in scientific research depends on replication studies being critical [47]. One of their most important contributions is their inclusion in meta-analyses and systematic reviews, where they assist in synthesizing results across a broad spectrum of research. Including successful and failed replications helps researchers to more precisely and generally estimate impact sizes, evaluate study heterogeneity, and find moderating factors like age, gender, cultural background, or research design. In areas like education, including replication studies into more comprehensive analyses has been critical in assessing the efficacy of instructional interventions across various systems, therefore offering a stronger empirical basis for policy-making [48]. Apart from their analytical value, replication deliberately encourages a reliable and cumulative scientific culture. Replicating initiatives helps discourage dubious research procedures and academic fraud through improved openness and methodological rigor. Moreover, replication is a road for theoretical improvement and a tool for validation [49]. It helps to specify the boundary conditions of events and can expose previously unobserved relationships or patterns that only become apparent in particular settings or populations. Therefore, replication can help generate new ideas and support theory building [50]. Although it is laborious, the importance of replication in improving

Table 3 Institutional responses to enhance reproducibility in research

Response	Description	Purpose	Challenges
Pre-registration [43]	Researchers register hypotheses, designs, and analysis plans in advance on platforms like OSF or ClinicalTrials.gov	Prevents selective reporting and p-hacking and enhances methodological transparency	It may limit flexibility during exploratory research and is underused in specific disciplines
Open data and code sharing [44]	Mandates sharing of datasets, simulation models, and code with the research community	Facilitates verification, replication, and cumulative knowledge building	Privacy concerns, proprietary data restrictions, and lack of standardization hinder effective sharing
Registered reports [45]	Peer review occurs before data collection; journals commit to publishing based on the study plan, not the results	Addresses publication bias; values replication studies regardless of outcome	Time-intensive review process; limited availability across journals and disciplines
Cultural shift in academia [46]	There is a broader emphasis on transparency, reproducibility, and openness in publishing and funding policies	Encourages Replication, reduces questionable practices, and improves research credibility	Resistance to change: incentives often still favor novel findings over replication or transparency

the dependability of scientific findings and their value to society cannot be emphasized enough. It guarantees that scientific information adds significantly to academic development and public decision-making and increases the basis for future studies. From this perspective, any evidence-based field dedicated to long-term influence and scientific integrity must replicate rather than luxury.

2.4 Blinding Procedures

By lowering prejudices arising from the expectations of participants, researchers, or both, blinding techniques, also known as masking, help improve empirical studies' credibility and integrity [51]. Blinding's primary goal is to keep data collecting and interpretation objective so that the measured results are due to the intervention under study rather than other psychological or behavioral effects. Research designs using single blinding let the participants remain ignorant of their group assignment, reducing the probability of placebo effects or behaviors influenced by their treatment assumptions [52]. Double blinding minimizes risks such experimenter bias, group differential treatment, or subtle cues that could influence participant behavior or responses by keeping the experimenters blind to group assignments, hence extending this concept [53]. In disciplines like medicine and psychology, where results typically rely on subjective reports or evaluations susceptible to expectation-related errors, the importance of blinding becomes especially clear. Though it has benefits, blindness is not always practically possible. Research on public policy implementation, behavioral therapies, or educational programs may naturally call for knowledge about the treatment circumstances among participants and facilitators [54]. Researchers have to use different techniques to offset prejudice in such situations. These could include using objective measurement instruments, third-party blind group allocation assessments, or sophisticated statistical methods to account for confounding variables. Using automated data collecting and processing, which reduces human intervention, technological developments have helped to strengthen research even more [55]. Implementing pre-analytical preparations and independent data audits has also grown more usual, so supporting openness and improving the dependability of results, especially in investigations when complete blinding cannot be preserved. For instance, in studies on innovative architecture assessing user happiness in adaptive lighting systems, people who are uninformed of the system's automation may offer more correct answers on comfort and well-being. Likewise, when the researchers are oblivious to the algorithm's active state, their user feedback readings stay impartial [56]. Such design increases the dependability of results in contexts where technological mediation and human perception interact.

3 Types of Experimental Design

Experimental research is a fundamental method for exploring causal relationships, testing hypotheses, and assessing interventions in the pursuit of scientific knowledge [57]. Well-structured experimental designs are essential for generating valid, reliable, and replicable results across various disciplines, from psychology to engineering sciences. An experiment's methodological integrity and final success depend not only on its underlying premise or conceptual framework but also on the accuracy with which its design is constructed. The rigor of outcome assessment, the methodical allocation and management of participants, and the control of extraneous variables all fall within this design framework [58]. Traditionally, experimental designs are classified according to several fundamental dimensions: the structural configuration of the study (e. g., group size, sequencing of interventions), the degree of control over confounding variables, the strategy used for participant assignment (randomized, non-randomized, or repeated measures), and the contextual appropriateness of the design to particular research settings, whether laboratory-based or field-oriented [59]. These categories are not random taxonomies but rather fundamental factors of a study's internal validity—the extent to which causal links can be confidently deduced—and its external validity, reflecting the generalizability of the results across populations and situations. Furthermore, these design elements directly relate to the ethical and practical viability of the study, highlighting their crucial importance in

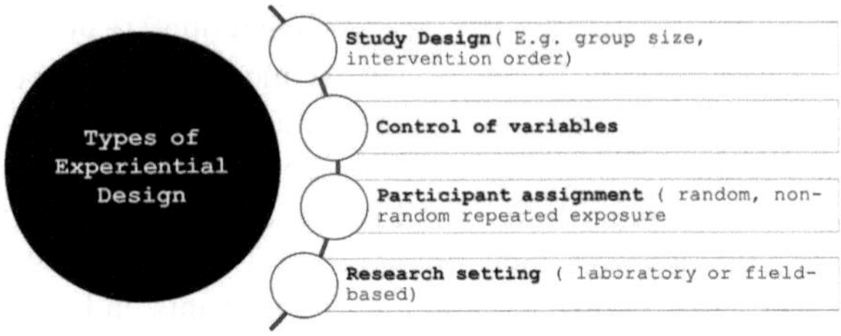

Fig. 5 Types of experimental design. *Source* Authors

generating scientifically strong and socially responsible knowledge. Internal validity refers to confidence in attributing observed effects to the changed variables rather than external influences [60]. External validity indicates how the findings can be generalized beyond the experimental environment. No single design suits everyone; the choice depends on the research objective, ethical considerations, available resources, and the nature of the examined phenomena. Significant types of experimental designs are outlined in this section, focusing on their defining characteristics, strengths and limitations, and relevance to various scientific settings [61]. Based on four fundamental dimensions of study structure, management of variables, participant assignment, and research setting, Fig. 5 classifies experimental designs. This categorization helps academics choose suitable designs that align with their goals, ensuring methodological rigor, ethical feasibility, and optimal internal and external validity across scientific fields.

3.1 *True Experimental Designs*

Authentic experimental designs are widely recognized as the gold standard in experimental research due to their robust methodological framework for establishing causal relationships [62]. These designs are characterized by the random assignment of participants to treatment and control groups, the manipulation of independent variables, and the inclusion of control conditions to isolate the effects of the intervention. Randomization is central to minimizing selection bias and controlling for confounding variables, ensuring that observed outcome differences can be confidently attributed to the experimental manipulation rather than extraneous influences [63]. This high level of internal validity—the degree to which a study can rule out alternative explanations for its findings—makes authentic experimental designs indispensable in disciplines such as psychology, medicine, engineering, and education, where precise attribution of effects is critical. Prominent authentic experimental designs include the pretest–posttest control group design, which measures change over time; the posttest-only control group design, which assesses outcomes after intervention without pre-measurement; and the Solomon four-group design, which combines both to control for testing effects. Despite their complexity, these designs

Fig. 6 Core structure of authentic experimental designs

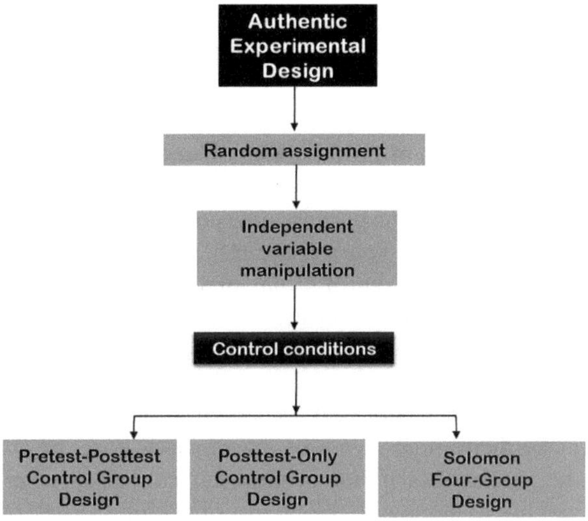

provide rigorous empirical evidence under controlled conditions and remain foundational for validating hypotheses and informing evidence-based practice in laboratory and applied settings [64]. For example, a genuine experimental design was used in building materials research to assess how nano-silica additions affected concrete strength. With the latter receiving different nano-silica concentrations, researchers randomly allocated concrete samples to control and treatment groups [65]. Both groups underwent compressive strength tests before and after the treatments. Results confirmed the additive's influence under controlled laboratory circumstances by showing statistically significant increases in strength for treated samples. Emphasizing random assignment, manipulation of the independent variable, and control conditions, Fig. 6 shows the basic building blocks of genuine experimental designs. It also emphasizes the three main design types pretest–posttest, posttest-only, and Solomon four-group utilized to guarantee high internal validity and consistent causal inference in controlled experimental research across practical fields.

3.1.1 Control Group Design Pretest–Posttest

At least two groups, experimental and control, with individuals randomly allocated to each, form this time-honored experimental design [66]. The dependent variable is the baseline set by both groups' pretests. While the control group receives no intervention, the experimental group gets the treatment. Both groups are post-tested after the intervention to gauge changes. This design allows scientists to identify between-group differences attributable to the therapy and within-group changes. However, it may cause testing effects, where knowledge of the assessment affects results; including a pretest helps regulate initial variability [67]. Particularly in controlled settings where internal validity is prioritized, this approach is often used for its strength in demonstrating causal links.

3.1.2 Design of Posttest-Only Control Group

Widely utilized, the posttest-only control group design is a precise experimental technique omitting a pretest phase [68]. Participants are randomly assigned to either the experimental or control group, and just a post-intervention measurement (posttest) is used to evaluate the effect of the treatment. This design ignores the pretest phase. It is especially beneficial in study environments where running a pretest is unrealistic because of time limits, logistical issues, or the possibility of pretesting affecting participant behavior [69]. In certain studies, exposing participants to a pretest might make them more aware of the research goals or anticipated results, undermining the genuineness of their posttest replies. The approach preserves the study's internal validity by removing the pretest, reducing the risk of such testing effects. Random assignment guarantees that pre-existing variations between groups are relatively dispersed on average despite lacking baseline measurements. The causal reading of post-test differences between groups rests on this premise of equivalence. The main benefits of the design are its methodological simplicity, lower participant load, and resistance to testing artifacts [70]. Its major drawback is the absence of early comparison data, making it impossible to examine within-group variation and verify that the experimental and control groups were equal initially. This could be a concern in research with limited sample numbers or when randomization does not adequately balance essential factors. Still, especially in applied environments where testing conditions must be simplified without compromising analytical rigor, the posttest-only control group design stays a strong and generally accepted method in experimental research when randomization is done correctly and sample sizes are sufficient.

3.1.3 Designs of Factorials

Representing a solid foundation in experimental research, factorial designs allow the study of two or more independent variables and their potential interacting effects on a dependent variable [71]. The 2×2 factorial design, which is frequently used, evaluates the significant effects of each factor and their interaction at two levels for every component. This approach helps determine whether the extent of one independent variable influences the impact of another, thereby enhancing the study's ecological validity and explanatory power [72]. Introducing more variables and levels increases the complexity of factorial designs; they require careful planning, additional resources, and larger sample sizes to maintain adequate statistical power. Commonly employed in controlled settings like hospitals or laboratories, these designs enable precise control of experimental conditions. While they are excellent for establishing causal relationships, the artificial nature of these settings may limit their generalizability to real-world situations. Conversely, within-subject designs, also known as repeated measures designs, involve subjecting the same participants to every experimental condition [73]. Each participant thus serves as their control, reducing the impact of individual differences on the observed outcomes. In this structure, fewer participants are required than in between-subject designs, increasing statistical power and efficiency. For example, an architectural engineering project assessed thermal efficiency in a controlled climate chamber under various wall orientations (north-facing vs. south-facing) and types of insulating materials (mineral wool vs. polyurethane). Guiding energy-efficient façade designs for colder regions, the 2×2 factorial study revealed a significant interaction between material and orientation [74]. The strength of this approach lies in its ability to detect minor treatment effects by reducing inter-individual variation. However, within-subject designs present specific methodological challenges. Among these are carry-over effects, where the influence of one experimental condition may persist and impact performance in the subsequent condition. Improper management can lead to fatigue, learning, or practice effects that skew results. Researchers sometimes mitigate such risks by systematically varying the sequence of conditions across participants [75]. Other issues include order effects, where task repetition boosts performance, and demand characteristics, where participants modify their behavior based on their perceptions of the study's intent. Despite these limitations, within-subject designs are frequently utilized in cognitive psychology, behavioral science, and therapeutic trials, particularly in settings that require multiple assessments of the same individuals [76]. Their methodological efficiency and heightened sensitivity to treatment effects ensure they remain vital tools in experimental research.

Three commonly used experimental designs—pretest–posttest, posttest-only, and factorial designs—are compared in Table 4. It outlines their defining qualities, benefits, drawbacks, and practical issues. While it runs the risk of testing effects, the pretest–posttest method is valuable for measuring change over time. Although it lacks baseline data, the post-test-only design circumvents these impacts. Despite requiring more complex planning and larger samples, the factorial design allows multiple variables to be tested simultaneously, enhancing explanatory power. This table illustrates how each design has unique benefits and trade-offs, assisting researchers in selecting the most suitable method for rigorous and context-relevant experimental study.

3.2 *Quasi-experimental Design*

Research settings where random assignment to experimental and control groups is impractical, unethical, or impossible typically utilize quasi-experimental techniques [77]. These methods are standard in applied sectors such as engineering, infrastructure development, environmental research, and urban planning, where interventions occur in real-world contexts that cannot be artificially controlled or modified. Although quasi-experimental designs do not possess the same degree of internal

Table 4 Comparative analysis of experimental designs: features, strengths, limitations, and methodological challenges

Design type	Key features	Strengths	Limitations	Challenges
Pretest–Posttest	Participants are randomly assigned to either experimental or control groups. Both groups complete a pretest and a posttest to assess changes	This design allows researchers to measure change over time and to control for initial group differences	The pretest may influence participants' responses in the posttest, introducing testing effects	It is challenging to prevent sensitization caused by repeated exposure to the same test instruments
Posttest-Only	Participants are randomly assigned, but only a posttest is administered after the intervention, without a pretest	The design is easy to implement and avoids the influence of pretesting on participants' behavior	It does not provide baseline data, making it harder to assess within-group change or verify group equivalence	The primary methodological concern is ensuring the two groups are equivalent at baseline without a pretest
Factorial design	Two or more independent variables are tested simultaneously, using within or between-subjects approaches	The design enables the analysis of both main and interaction effects, improving explanatory power	It requires a larger sample size and more complex planning to maintain validity and power	Managing multiple variables and experimental conditions increases logistical and analytical complexity

validity as authentic experimental designs, primarily due to the lack of randomization, they are essential for generating significant insights in environments where controlled research is unfeasible. Quasi-experiments depend on pre-existing groups, naturally occurring conditions, or time-based occurrences instead of random allocation, introducing potential biases and confounding variables [78]. Nevertheless, to address these limitations and draw reasonable inferences about causal relationships, researchers can employ various design approaches and statistical methods, such as matching, statistical controls, and longitudinal assessments. The strength of quasi-experimental designs lies in their external validity; they operate in environments that closely mirror real-life operational situations, making their results directly relevant to actual engineering decisions [79]. Frequently utilized in engineering research, quasi-experimental techniques evaluate the implementation of structural retrofitting solutions, new technologies, or regulatory changes. For example, by comparing traffic wear data, noise levels, or maintenance frequency with similar districts using traditional materials, researchers can assess the performance of a new pavement material deployed in selected urban areas. In structural engineering, a quasi-experiment could also explore how retrofitting public buildings with seismic dampers influences structural response and damage reports before and after implementation by using structures without retrofitting as a control group. By analyzing energy consumption trends before and after installation and contrasting these trends with buildings that did not experience any energy interventions, researchers in energy systems engineering could assess how solar panel installations affect administrative buildings or public schools [80]. Moreover, in large-scale infrastructure evaluations, when withholding a beneficial intervention from a comparison group would be challenging or potentially harmful, quasi-experiments are often the only ethical and practical option. For instance, research on post-earthquake rehabilitation would find it neither ethical nor feasible to randomize structural strengthening across buildings or areas. Similarly, municipal planners cannot randomize drainage improvements across districts when evaluating stormwater management systems. Yet, quasi-experimental designs compare flood occurrences before and after the intervention across treated and untreated areas [81]. Thus, quasi-experimental methods serve as a methodological bridge between the complexities of field-based research and the rigor of randomized controlled trials. They allow researchers to project the effects of changes while considering logistical constraints, ethical dilemmas, and the natural variations inherent in engineering contexts. As applied research becomes increasingly complex, particularly in multidisciplinary fields like environmental engineering, sustainable design, and urban infrastructure planning, the demand for thorough yet flexible evaluation techniques has also grown. Quasi-experimental techniques address this demand by providing a reliable approach to test hypotheses, evaluate technology or policy effects, and inform long-term planning in a scientifically sound and practically grounded manner [82]. Although they require meticulous interpretation, rigorous statistical modeling, and careful design, quasi-experimental designs remain crucial to engineering research methodologies, mainly when translating findings into policy, design standards, and innovation-driven solutions. Unethical or impossible designs typically utilize quasi-experimental techniques. Figure 7 shows how quasi-experimental methods are applied in engineering to assess interventions where randomization is not feasible. Emphasizing design techniques like longitudinal studies and statistical controls to guarantee external validity and plausible causal inference, the figure shows usual uses such as structural retrofitting, solar energy installations, and stormwater systems.

3.2.1 Characteristics and Rationale

Unlike genuine experimental designs, quasi-experiments lack random allocation. Instead of randomly assigning individuals or units to treatment and control groups, they rely on pre-existing groups or naturally occurring circumstances [83]. This structural variation jeopardizes internal validity since one cannot assume group equivalence. Thus, results may be influenced by selection bias, aging, regression to the mean, or concurrent historical events, complicating the formulation of clear causal inferences. Despite these limitations, quasi-experimental designs remain valuable and widely used, mainly supported by thorough planning, meticulous sampling, and

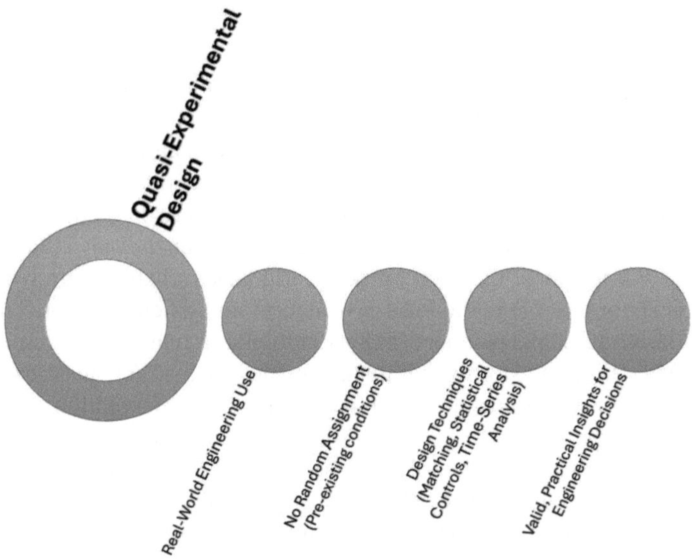

Fig. 7 Application of quasi-experimental designs in engineering research. *Source* Authors

appropriate statistical adjustments [84]. Frequently, the rationale for utilizing quasi-experimental methods lies in their adaptability to real-world contexts. Researchers may encounter ethical constraints that make random assignment either impractical or inappropriate, for instance, in situations where denying a control group access to beneficial treatments or resources would violate ethical standards. Practical challenges also arise from organizational, institutional, or community-based environments where structural barriers impede the implementation of randomization techniques [85]. Natural settings may offer research opportunities where the study process does not affect interventions or policy alterations but provides a foundation for comparison. Quasi-experimental designs are crucial in engineering fields, particularly urban planning, civil infrastructure, and architectural engineering. Evaluating the impact of newly enacted building codes, insulation standards, or seismic retrofitting laws may render randomization across buildings or communities unethical or unfeasible. Researchers can employ matched group comparisons, propensity score adjustments, time series analyses, and pre- and post-intervention data comparisons among comparable yet non-randomized groups to enhance causal interpretations [86]. Particularly for applied research where experimental control is inherently limited, these methods enable researchers to analyze large-scale, real-world phenomena while upholding scientific rigor. Ultimately, quasi-experimental designs represent a reasonable balance between scientific robustness and contextual relevance. They facilitate examining complex interventions within existing systems, yielding analytically valid results relevant for sustainable development planning,

Table 5 Characteristics, rationale, and challenges of quasi-experimental designs

Aspect	Description	Challenges
Definition	Quasi-experiments do not involve random assignment; they rely on pre-existing groups or naturally occurring conditions	Lack of group equivalence may bias outcomes
Limitations	Prone to internal validity threats such as selection bias, maturation, history effects, and regression to the mean	Difficult to isolate the actual effect of the intervention
Validity strategies	Use matched groups, statistical controls (e.g., ANCOVA, propensity scores), and longitudinal designs to support causal inference	Requires advanced statistical expertise and careful design
Ethical rationale	Enables research in contexts where withholding treatment or access to resources from a control group would be unethical	Ethical clarity may reduce control over experimental conditions
Practical rationale	Useful where systemic or institutional constraints make randomization infeasible	Limited researcher control over assignment and context
Natural setting use	Suitable for studying interventions or changes that occur independently of researcher involvement	Potential for unmeasured confounding variables
Research value	Balances scientific rigor with ecological validity and is appropriate for applied and real-world investigations where randomization is impossible	Generalizability can be strong, but causal claims require cautious interpretation

engineering advancements, and policy formulation [87]. Table 5 outlines quasi-experimental designs' defining characteristics, rationale, and methodological obstacles. It emphasizes their dependence on non-randomized settings, ethical and practical benefits, and internal validity concerns. The table underlines the significance of thorough design and analytical accuracy and the requirement of specific techniques to reduce prejudices.

3.2.2 Types of Quasi-experimental Designs

Several design patterns comprise quasi-experimental research, each suitable for specific study environments where randomization is unfeasible [88]. The Non-Equivalent Control Group Design (NECGD) and the Interrupted Time Series Design (ITSD) are the most frequently used, which offer systematic methods for evaluating interventions in naturalistic settings [89]. The Non-Equivalent Control Group Design contrasts a control group that does not receive the intervention with an experimental group that does. However, these groups are not formed through random assignment. Typically, both groups undergo pretests and posttests, allowing researchers to calculate treatment effects by factoring in specific baseline variances [90]. While it resembles randomized controlled trials, this design remains susceptible to selection bias. Statistical methods such as propensity score matching, analysis of covariance

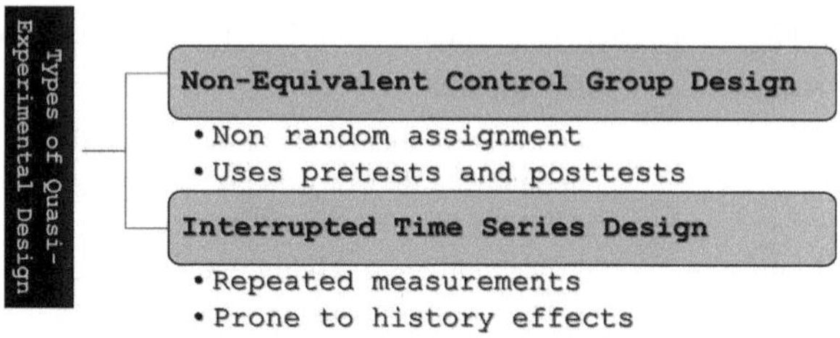

Fig. 8 Key types of quasi-experimental designs. *Source* Authors

(ANCOVA), and regression adjustments are employed to enhance group comparability and bolster the validity of causal inferences to mitigate these concerns. The Interrupted Time-Series Design requires collecting multiple measurements before and after intervention implementation. This enables researchers to assess whether the intervention significantly deviated from an established trend. Although highly effective in evaluating long-term effects, this approach may be influenced by historical factors and external events concurrent with the intervention, necessitating techniques like control series for stronger inferences [91]. For example, a study evaluating an energy-efficient construction code analyzed commercial building electricity usage using an interrupted time-series approach. Monthly energy data collected over five years indicated a significant reduction in usage following the code's adoption, suggesting the policy's effectiveness in light of seasonal fluctuations and long-term consumption patterns. Figure 8 illustrates two primary quasi-experimental designs: the non-equivalent Control Group and Interrupted Time Series. It highlights essential characteristics, including the absence of randomization, the use of pretests, and the implementation of multiple assessments. The figure facilitates comparison by emphasizing design logic, common validity, and inference limitations.

3.2.3 Ensuring Validity and Relevance in Quasi-experimental Research

One of the main difficulties in quasi-experimental studies is handling internal validity concerns arising from the lack of randomization [92]. Among the most prevalent dangers are selection bias, where pre-existing group disparities affect results; maturation, which refers to natural changes in participants over time; and historical effects, wherein unrelated external events influence the findings [93]. Other issues include testing impacts, which occur when exposure to a pretest influences posttest performance, and instrument changes throughout measures. Researchers employ techniques such as baseline measurements, creating non-equivalent comparison groups, and sophisticated statistical controls like regression correction, propensity score matching, or ANCOVA to offset these constraints [94]. Including qualitative data through interviews, observations, or case records can further enhance interpretation and support credibility through methodological triangulation. In applied disciplines, quasi-experimental approaches prove beneficial when randomized controlled trials are unfeasible or ethically questionable. This is particularly true in technical research, educational initiatives, and public policy assessments, where changes must be evaluated in real-world situations. Architectural engineering often utilizes such designs to assess the effects of post-occupancy building performance, acoustic regulations, energy retrofitting, or new building codes [95]. Researchers can isolate intervention effects with significant contextual relevance by comparing treated and untreated buildings before and after implementation. The need for methodological flexibility increases as research becomes more application-oriented. Quasi-experimental designs satisfy this need by providing systematic yet flexible frameworks capable of producing evidence under practical constraints [96]. Their ongoing connection with monitoring and evolving data analysis tools enhances their usefulness. Although they might not entirely supplant randomized trials regarding internal validity, quasi-experimental methods are crucial in generating practical insights in dynamic and complex systems.

3.3 Complicated and Hierarchical Designs

Modern studies often include data structured hierarchically or in layers [97]. These setups create statistical correlations among observations that contradict the independence assumption common in many traditional statistical models. Ignoring such dependencies could lead to skewed parameter estimations, underreported standard errors, and erroneous conclusions. Researchers use multilevel or hierarchical designs specifically created to address the structured nature of nested data to solve this complexity [98]. These models recognize that factors operating at several levels of analysis affect observations, making them not independent. Individual-level traits and broader environmental or structural circumstances linked with group membership may concurrently impact a given outcome. Hierarchical models work by partitioning the overall variance in an outcome variable into components that correspond to various levels within the data structure [99]. This method allows researchers to identify and measure variance attributable to individual units and higher-level groupings. It also facilitates estimating relationships within and between levels, enabling a more precise and comprehensive understanding of the observed data dynamics. Furthermore, these designs allow for the study of cross-level interactions, wherein variables defined at one level can influence the relationship between predictors and outcomes at another level. Random slopes and random intercepts allow group differences to affect baseline values and correlations' strength [100]. This capacity to simulate diversity across settings enhances the accuracy and interpretation of statistical inferences. Research aiming to reflect the complexity of real-world environments, where multilevel influences are the norm rather than the exception, is especially well-suited for hierarchical models. By aligning analytical methods with the layered nature of empirical phenomena, these models contribute to the creation of generalizable knowledge [101]. Complex and hierarchical designs offer a robust methodological framework for examining data with nested structures. These designs enhance empirical research's validity, reliability, and explanatory capacity across many scientific fields by employing complex modeling of variation and interaction while accounting for dependencies across levels.

3.3.1 Application and Hierarchical Structure

In large-scale experimental or observational investigations, data in engineering research often exhibit a hierarchical or nested structure [98]. For example, consider a study assessing how well a new composite material performs under stress. Level 1 might consist of individual samples, Level 2 could encompass several production lines or batches, and Level 3 might refer to manufacturing sites. Each level introduces new sources of variation that could influence the observed results. Neglecting this multilevel framework can lead to undesirable statistical effects, such as biased parameter estimates, reduced standard errors, and elevated Type I error rates [102]. Units embedded within the same context or group violate the assumption of independence among observations, contributing to this issue. For example, samples generated in the same batch or at the same site are likely more comparable to each other than those from different sources. Hierarchical designs utilize multilevel modeling (MLM) or hierarchical linear modeling (HLM) to address such complexity [103]. These models enable researchers to examine the effects of micro-level (specimen) and macro-level (batch or facility) factors by separating total variation into components attributed to each hierarchical level. This approach facilitates investigating how material characteristics vary under different manufacturing conditions and identifies the sources of systematic variation. In sustainable building studies, samples of hempcrete panels (Level 1) are collected from various curing chambers (Level 2) and distributed across different climate zones (Level 3). Multilevel modeling assists in determining whether variations in strength are due to local environmental factors, manufacturing techniques, or natural variability among individual panels [102]. For example, digital twin research collects sensor data (Level 1) from HVAC systems distributed across floors (Level 2) in various smart buildings (Level 3). A hierarchical approach helps ascertain whether anomalies in energy usage arise from local equipment issues, floor-level usage patterns, or building-wide operational inefficiencies [104]. Multilevel models encompass fixed effects, which reflect the average contribution of variables across all levels, and random effects, which capture the variability associated with higher-level groups [105]. This framework ensures that conclusions accurately represent the complexity of engineering systems, allowing for a more precise and consistent analysis. Especially in studies involving nested designs, multi-site experiments, or multi-phase production processes, properly accounting for hierarchical structures in engineering research enhances the validity and robustness of results.

3.3.2 Design Features and Flexibility

Well-suited to real-world research settings where data are organically stacked across several layers, hierarchical and complex experimental designs provide significant flexibility [106]. One of the unique qualities of such designs is their ability to mimic cross-level interactions, where the impact of a lower-level variable, such as individual behavior, depends on traits of a higher-level context, such as organizational policy or environmental conditions [107]. This allows researchers to investigate how broader background elements shape the influence of individual-level predictors, thereby enriching the understanding of outcomes. Furthermore, hierarchical designs enable modeling random slopes and intercepts, which helps predict treatment effect variance across various units or environments. Differences in infrastructure, climate, or management practices can cause a sustainability solution to affect different buildings or areas [108]. By considering such diversity, these designs offer a more nuanced understanding of the dynamics within multilevel systems. They have been widely employed across several fields. Hierarchical models assist in examining departmental or team performance and work habits in organizational studies. Environmental studies facilitate examining individual energy use within building-level features, including insulation or ventilation systems [109]. Their capacity to accommodate hierarchical data structures qualifies them to explore multilevel interactions in numerous practical settings. However, employing hierarchical designs also introduces significant methodological challenges. Their inherent complexity in design and analysis requires advanced statistical knowledge and specialized software capable of managing multilayer models. Additionally, sufficient statistical power relies on adequate data at each hierarchical level, which can be limited in many practical environments [110]. Despite these challenges, hierarchical designs provide a robust analytical framework that closely aligns with the layered nature of social, environmental, and institutional systems. They enhance external validity and generalizability by more accurately replicating the conditions under which real-world events occur. More broadly, the variety of available experimental designs empowers researchers to tailor their methodological approach to the specific needs of their research questions, contextual realities, and logistical constraints [111]. Although ethical, practical, or budgetary issues might render them unfeasible, genuine experimental designs offer the highest degree of control and support robust causal findings. Within-subject designs can enhance statistical efficiency and sensitivity but require careful management of internal hazards such as carryover effects. While they necessitate rigorous analytical techniques to substantiate causal claims, quasi-experimental designs provide valuable alternatives in field settings where random assignment is impractical [112]. Finally, complex and hierarchical designs can yield multilevel insights and model systems that more accurately reflect the diverse aspects of human behavior and structural conditions [113]. The appropriate selection and precise application of experimental designs directly affect empirical studies' credibility, reproducibility, and interdisciplinary relevance. Each type of design has its benefits and limitations; their strategic application is crucial for advancing scientific knowledge across various disciplines. Hierarchical designs, for example, in energy-efficient structures, help examine how individual occupant behavior interacts with building-level characteristics, such as insulation quality or HVAC systems. The effectiveness of energy-saving behaviors may depend on the design criteria of the building, thereby enabling researchers to uncover cross-level interactions between user behavior and structural energy efficiency policies.

4 Conclusion

One of empirical research's most persistent and fundamental goals is still the methodical quest of causality. Positioning experimental design and causal inference as essential frameworks for generating consistent and legitimate scientific information, this chapter has described the conceptual underpinnings, methodological structures, and analytical tools that control these processes. At the crossroads of the philosophy of science, statistical reasoning, and empirical research, experimental design offers the structural conditions for causal claims to be created, tested, and supported with methodological rigor. By means of methodical organization of the study environment, experimental design essentially focuses on controlling dangers to internal validity. Random assignment, defining control conditions, and using replication techniques are not only technical choices but also basic commitments to lowering prejudice and strengthening inference believability. These elements taken together create the framework wherein the researcher may separate the impact of an intervention while reducing the effect of confounding variables and extraneous factors. The causal conclusion drawn from such setups is not only observable but also based on well-substantiated counterfactual reasoning. Central to the reasoning of experimental design is the capacity to approach the counterfactual state the situation that would have occurred without the treatment. Experimental frameworks mimic this unobservable state by means of randomization and organized comparisons, hence allowing estimates of treatment effects with least systematic error. This logic guarantees that observed links reflect reasonable causal processes rather than false correlations or model misspecifications, hence penetrating the heart of inferential thinking and beyond statistical manipulation. The dependability and legitimacy of scientific assertions are founded on replication, an epistemic safeguard and methodological guideline. Reproducibility is a required prerequisite for cumulative knowledge and a measure of empirical robustness. Replication serves as a quality control tool in an academic environment more focused on openness and responsibility by means of which it disciplines inference, identifies false positives, and enhances the generalizability of results. The chapter underlines that a scientific culture that values methodological conservatism, epistemic humility, and long-term reliability over isolated or context-bound outcomes depends on replication. Quasi-experimental designs are a necessary substitute where rigorous randomization is unworkable, unethical, or impossible. Although they lack the random assignment mechanism, these designs use statistical techniques and strategic design components to approximate the assumptions of experimental control. In field-based or naturalistic settings, matching processes, longitudinal studies, and interrupted time-series techniques let researchers make conditional causal inferences. Though naturally more susceptible to internal validity concerns, quasi-experimental designs offer significant epistemological bridges between scientific purity and practical applicability. By reducing different kinds of expectation-induced bias, blinding techniques help to strengthen the validity of experimental inference even further. Whether single, double, or assessor-based, these processes help the objectivity of result measurement and interpretation. They shield scientific observations from psychological, behavioral, or institutional factors that could otherwise skew findings. Although different contexts may have different implementation limits, the philosophical justification for blinding is constant with the more general goals of inferential neutrality and epistemic discipline. Diverse research goals, settings, and limitations have guided the development of experimental designs. Among these are pretest–posttest systems, posttest-only frameworks, within-subject designs, factorial setups, and hierarchical models. Every design is characterized by its ability to govern variation, its sensitivity to particular validity concerns, and its fit with the epistemic objectives of the research question. Rigor in theoretical thinking, logistical issues, and ethical viability guide design types' choice, modification, and use. Furthermore, the growing complexity of empirical events calls for sophisticated design configurations acknowledging data's multilevel and nested character. Using hierarchical models and multilevel designs, researchers may include dependencies inside data structures, improving statistical accuracy and understanding results. These frameworks emphasize the need to match methodological structures with the ontological characteristics of the event under study, supporting the more general idea that research design must react to theoretical abstraction and empirical complexity. Ultimately, a larger epistemological undertaking has interdependent parts called experimental design and causal inference rather than separate methods. With analytical clarity and methodological openness, they allow researchers to pose and answer questions of causal importance, judge among rival hypotheses, and help to advance theory. In areas where policy, intervention, or innovation depends on inferential accuracy, they also act as normative standards for scientifically valid evidence. This chapter again emphasizes that good causal inference exists before the intentional, open, and theoretically informed use of experimental design ideas. Using these methodological commitments, scientists can progress from simple observation to explanation, from association to causality, and from isolated results to cumulative knowledge. The timeless importance of experimental design rests in its capacity to adapt, extend, and maintain the criteria of scientific validity, reproducibility, and epistemic integrity as the difficulties of empirical research get more multidisciplinary and complicated.

References

1. Chen X, Abualdenien J, Singh MM, Borrmann A, Geyer P (2022) Introducing causal inference in the energy-efficient building design process. Energy Build 277:112583
2. Rotari M, Kulahci M (2025) Correlation to causality. Qual Eng 37(1):162–172
3. Ahmed LKFN (2023) Political science-research methodology. Success Publications

4. Diener E, Northcott R, Zyphur MJ, West SG (2022) Beyond experiments. Perspect Psychol Sci 17(4):1101–1119
5. Em S (2024) Exploring experimental research: methodologies, designs, and applications across disciplines. Designs, and Applications Across Disciplines
6. Lynnc YL (2024) Investigating the impact of multilevel treatment options on statin patient outcomes: a counterfactual analysis (Doctoral dissertation, University of Minnesota)
7. Ferwana I, Varshney LR (2022) Optimal recovery for causal inference. arXiv: 2208.06729
8. Morris TP, Walker AS, Williamson EJ, White IR (2022) Planning a method for covariate adjustment in individually randomised trials: a practical guide. Trials 23(1):328
9. Matthay EC, Glymour MM (2022) Causal inference challenges and new directions for epidemiologic research on the health effects of social policies. Curr Epidemiol Rep 9(1):22–37
10. Burger HU, Gerlinger C, Harbron C, Koch A, Posch M, Rochon J, Schiel A (2021) The use of external controls: to what extent can it currently be recommended? Pharm Stat 20(6):1002–1016
11. Thomas CG, Thomas CG (2021) Experimental research
12. Reich BJ, Yang S, Guan Y, Giffin AB, Miller MJ, Rappold A (2021) A review of spatial causal inference methods for environmental and epidemiological applications. Int Stat Rev 89(3):605–634
13. Emadi J (2023) The development of a design theory for web based information systems. J Robot Spectr 1:013–023
14. Kamble S, Rana NP, Gupta S, Belhadi A, Sharma R, Kulkarni P (2023) An effectuation and causation perspective on the role of design thinking practices and digital capabilities in platform-based ventures. Technol Forecast Soc Chang 193:122646
15. Haig BD (2022) Understanding replication in a way that is true to science. Rev Gen Psychol 26(2):224–240
16. Steeger CM, Buckley PR, Pampel FC, Gust CJ, Hill KG (2021) Common methodological problems in randomized controlled trials of preventive interventions. Prev Sci 22(8):1159–1172
17. Muzahid AJM, Kamarulzaman SF, Rahman MA, Murad SA, Kamal MAS, Alenezi AH (2023) Multiple vehicle cooperation and collision avoidance in automated vehicles: survey and an AI-enabled conceptual framework. Sci Rep 13(1):603
18. Rad M, Rajab Dizavandi F, Rajab Dizavandi A (2023) The effect of education through telenursing on the caregiver burden among family caregivers of COVID-19 patients: a randomized clinical trial. Perspect Psychiatr Care 2023(1):2545639
19. Liu L, Jones BF, Uzzi B, Wang D (2023) Data, measurement and empirical methods in the science of science. Nat Hum Behav 7(7):1046–1058
20. Byrnes JE, Dee LE (2025) Causal inference with observational data and unobserved confounding variables. Ecol Lett 28(1):e70023
21. Sun Z, Jiang T, Zhang M, Li Y, Zhang J, Sun Y, Yu X (2024) Causal relationship between hypertension and epilepsy: a mendelian randomization study. Acta Epileptologica 6(1):8
22. Yu Z, Guindani M, Grieco SF, Chen L, Holmes TC, Xu X (2022) Beyond t test and ANOVA: applications of mixed-effects models for more rigorous statistical analysis in neuroscience research. Neuron 110(1):21–35
23. Althubaiti A (2023) Sample size determination: a practical guide for health researchers. J Gen Fam Med 24(2):72–78
24. Alabi M (2024) Ethical implications of AI: bias, fairness, and transparency
25. Guillermo JCL, Garcia YA, Hechavarria EL, Lozano WG, Maguiña WME, Yon GG, Cáceres YS Integrating quantitative and qualitative paradigms to improve transdisciplinary research and educational innovation: interpretation of the scientific method
26. Griffith DA, Plant RE (2022) Statistical analysis in the presence of spatial autocorrelation: selected sampling strategy effects. Stats 5(4):1334–1353
27. Vorland CJ, Brown AW, Dawson JA, Dickinson SL, Golzarri-Arroyo L, Hannon BA, Allison DB (2021) Errors in implementing, analyzing, and reporting randomization within obesity and nutrition research: a guide to their avoidance. Int J Obes 45(11):2335–2346

28. Berger VW, Bour LJ, Carter K, Chipman JJ, Everett CC, Heussen N, Randomization Innovative Design Scientific Working Group Robert A Beckman (2021) A roadmap to using randomization in clinical trials. BMC Med Res Methodol 21:1–24.
29. O'Sullivan L, Aldasoro E, O'Brien Á, Nolan M, McGovern C, Carroll Á (2022) Ethical values and principles to guide the fair allocation of resources in response to a pandemic: a rapid systematic review. BMC Med Ethics 23(1):70
30. Prochazka A, Vysata O, Marik V (2021) Integrating the role of computational intelligence and digital signal processing in education: emerging technologies and mathematical tools. IEEE Signal Process Mag 38(3):154–162
31. Robinson ZP, Helms ER, Trexler ET, Steele J, Hall ME, Huang CJ, Zourdos MC (2024) N of 1: optimizing methodology for the detection of individual response variation in resistance training. Sports Med 54(8):1979–1990
32. Dormal V, Vermeulen N, Mejias S (2021) Is heart rate variability biofeedback useful in children and adolescents? a systematic review. J Child Psychol Psychiatry 62(12):1379–1390
33. Nuijten MB (2022) Assessing and improving robustness of psychological research findings in four steps. Avoiding questionable research practices in applied psychology. Springer International Publishing, Cham, pp 379–400
34. Iffland R, Förster K, Westerholt D, Pesci MH, Lösken G (2021) Robust vegetation parameterization for green roofs in the EPA stormwater management model (SWMM). Hydrology 8(1):12
35. Derksen M, Morawski J (2022) Kinds of replication: Examining the meanings of "conceptual replication" and "direct replication." Perspect Psychol Sci 17(5):1490–1505
36. Nosek BA, Hardwicke TE, Moshontz H, Allard A, Corker KS, Dreber A, Vazire S (2022) Replicability, robustness, and reproducibility in psychological science. Annu Rev Psychol 73(1):719–748
37. Shen T, Li B (2024) Digital twins in additive manufacturing: a state-of-the-art review. Int J Adv Manuf Technol 131(1):63–92
38. Bausell RB (2021) The problem with science: the reproducibility crisis and what to do about it. Oxford University Press
39. Almusaed A, Almssad A, Yitmen I (2024) Sustainable built environment and its implications on real estate development: a comprehensive analysis. In: Integrative approaches in urban sustainability-architectural design, technological innovations and social dynamics in global contexts. IntechOpen.
40. Fnais A, Rezgui Y, Petri I, Beach T, Yeung J, Ghoroghi A, Kubicki S (2022) The application of life cycle assessment in buildings: challenges, and directions for future research. Int J Life Cycle Assess 27(5):627–654
41. Almusaed A, Yitmen I (2023) Architectural reply for smart building design concepts based on artificial intelligence simulation models and digital twins. Sustainability 15(6):4955
42. Waltman L, Kaltenbrunner W, Pinfield S, Woods HB (2023) How to improve scientific peer review: Four schools of thought. Learn Publ 36(3):334–347
43. Imai T, Toussaert S, Baillon A, Dreber A, Ertaç S, Johannesson M, Villeval MC (2025) Preregistration and pre-analysis plans in experimental economics (No 220). I4R Discussion Paper Series
44. Monks T, Harper A (2025) Computer model and code sharing practices in healthcare discrete-event simulation: a systematic scoping review. J Simul 19(1):108–123
45. Chambers CD, Tzavella L (2022) The past, present and future of registered reports. Nat Hum Behav 6(1):29–42
46. Chiarelli A, Loffreda L, Johnson R (2021) The art of publishing reproducible research outputs: supporting emerging practices through cultural and technological innovation
47. Adeoye MA (2024) Mastering the basics: a guide to research methodology for effective writing and publication. Chalim J Teach Learn 4(1):30–41
48. Buinwi U, Okatta CG, Johnson E, Buinwi JA, Tuboalabo A (2024) Enhancing trade policy education: a review of pedagogical approaches in public administration programs. Int J Appl Res Soc Sci 6(6):1253–1273

49. Marefat F, Hassanzadeh M, Hamidi F (2024) Incorporating replication in higher education: supervisors' perspectives and institutional pressures. Account Res 1–22
50. Flake JK, Davidson IJ, Wong O, Pek J (2022) Construct validity and the validity of replication studies: a systematic review. Am Psychol 77(4):576
51. Paolini S, Gibbs M, Sales B, Anderson D, McIntyre K (2024) Negativity bias in intergroup contact: Meta-analytical evidence that bad is stronger than good, especially when people have the opportunity and motivation to opt out of contact. Psychol Bull
52. Bhatia A, Appelbaum PS, Wisner KL (2021) Unblinding in randomized controlled trials: a research ethics case. Ethics Hum Res 43(2):28–34
53. Pleskac TJ, Kyung E, Chapman G, Urminsky O Single-or double-blind review? a field study comparing preference, reliability, bias, and validity
54. Connors EH, Prout J, Vivrette R, Padden J, Lever N (2021) Trauma-focused cognitive behavioral therapy in 13 urban public schools: Mixed methods results of barriers, facilitators, and implementation outcomes. Sch Ment Heal 13:772–790
55. Yarkoni T, Eckles D, Heathers JA, Levenstein M, Smaldino PE, Lane JI (2021) Enhancing and accelerating social science via automation: challenges and opportunities. Harv Data Sci Rev 3(2)
56. Stray J (2023) The AI learns to lie to please you: preventing biased feedback loops in machine-assisted intelligence analysis. Analytics 2(2):350–358
57. Remler DK, Van Ryzin GG (2021) Research methods in practice: Strategies for description and causation. Sage Publications
58. Stone JC, Glass K, Clark J, Ritskes-Hoitinga M, Munn Z, Tugwell P, Doi SA (2021) The MethodologicAl STandards for epidemiological research (MASTER) scale demonstrated a unified framework for bias assessment. J Clin Epidemiol 134:52–64
59. Erath TG, Pellegrino AJ, DiGennaro Reed FD, Ruby SA, Blackman AL, Novak MD (2021) Experimental research methodologies in organizational behavior management. J Organ Behav Manag 41(2):150–181
60. Ahsan MJ (2024) Unlocking sustainable success: exploring the impact of transformational leadership, organizational culture, and CSR performance on financial performance in the Italian manufacturing sector. Soc Responsib J 20(4):783–803
61. Taherdoost H (2022) What are different research approaches? Comprehensive review of qualitative, quantitative, and mixed method research, their applications, types, and limitations. J Manag Sci Eng Res 5(1):53–63
62. Butcher K, Yodsuwan C (2024) Building "causal realism" in experimental studies within hospitality and tourism. Int J Contemp Hosp Manag 36(1):259–273
63. DiFrancisco-Donoghue J, Varga P (2024) Experimental/intervention esports research methodology. In: Routledge handbook of esports (pp 148–160). Routledge
64. de Alencar JN, Oliveira MHDJ, Sampaio MCN, Rego MF, Nunes R (2024) A journey through philosophy and medicine: from Aristotle to evidence-based decisions. Philosophies 9(6):189
65. Ahmed HU, Mohammed AS, Faraj RH, Qaidi SM, Mohammed AA (2022) Compressive strength of geopolymer concrete modified with nano-silica: experimental and modeling investigations. Case Stud Constr Mater 16:e01036
66. Huang Y, Richter E, Kleickmann T, Richter D (2022) Class size affects preservice teachers' physiological and psychological stress reactions: an experiment in a virtual reality classroom. Comput Educ 184:104503
67. Pan SC, Rivers ML (2023) Metacognitive awareness of the pretesting effect improves with self-regulation support. Mem Cognit 51(6):1461–1480
68. Widely utilized, the posttest-only control group design is a precise experimental technique omitting a pretest phase
69. Höhler J, Barreiro-Hurlé J, Czajkowski M, Dessart FJ, Ferraro PJ, Li T, Rommel J (2024) Perspectives on stakeholder participation in the design of economic experiments for agricultural policymaking: Pros, cons, and twelve recommendations for researchers. Appl Econ Perspect Policy 46(1):338–359

70. D'Ettorre F, Banaei M, Ebrahimy R, Pourmousavi SA, Blomgren EMV, Kowalski J, Madsen H (2022) Exploiting demand-side flexibility: State-of-the-art, open issues and social perspective. Renew Sustain Energy Rev 165:112605
71. Anderson EM, Niska JR, Niska EE (2023) Factorial design. In: Translational radiation oncology (pp 327–330). Academic Press
72. Ramzan M, Razi U, Quddoos MU, Adebayo TS (2023) Do green innovation and financial globalization contribute to the ecological sustainability and energy transition in the United Kingdom? Policy insights from a bootstrap rolling window approach. Sustain Dev 31(1):393–414
73. Guloglu B, Caglar AE, Pata UK (2023) Analyzing the determinants of the load capacity factor in OECD countries: evidence from advanced quantile panel data methods. Gondwana Res 118:92–104
74. Hou J, Zhang L, Dewancker BJ, Meng X, Hou C (2022) Research on energy-saving factors adaptability of exterior envelopes of university teaching-office buildings under different climates (China) based on orthogonal design and EnergyPlus. Heliyon 8(8)
75. Renn O, Laubichler M, Lucas K, Kröger W, Schanze J, Scholz RW, Schweizer PJ (2022) Systemic risks from different perspectives. Risk Anal 42(9):1902–1920
76. Timm I, Giurgiu M, Ebner-Priemer U, Reichert M (2024) The within-subject association of physical behavior and affective well-being in everyday life: a systematic literature review. Sports Med 54(6):1667–1705
77. Perumal RV, Singh MD (2021). Experimental and quasiexperimental designs. In: Nursing research in Canada-E-Book: methods, critical appraisal, and utilization, 230
78. Gao Y, Li M, Sun S (2023) Field experiments in operations management. J Oper Manag 69(4):676–701
79. Frau G (2022) Unified power format modeling in modern electronic design: a quasi-experimental approach (Doctoral dissertation, Northcentral University)
80. Almusaed A, Yitmen I, Almssad A (2023) Reviewing and integrating aec practices into industry 6.0: strategies for smart and sustainable future-built environments. Sustainability 15(18):13464
81. Eger CG (2023) Causal effects of green infrastructure on stormwater hydrology and water quality (Doctoral dissertation, Syracuse University)
82. He X, Madigan D, Yu B, Wellner J (2025) Statistics at a crossroads; who is for the challenge? arXiv:2503.22945
83. Veltri GA (2021) Experiments. In: Research handbook on analytical sociology, pp 392–413. Edward Elgar Publishing
84. Gawali RB (2023) Research methodology and statistical methods. Academic Guru Publishing House
85. Goodkind JR, Amer S, Christian C, Hess JM, Bybee D, Isakson BL, Shantzek C (2017) Challenges and innovations in a community-based participatory randomized controlled trial. Health Educ Behav 44(1):123–130
86. Linden A, Adams JL (2011) Applying a propensity score-based weighting model to interrupted time series data: improving causal inference in programme evaluation. J Eval Clin Pract 17(6):1231–1238
87. Saeed K (2019) Towards sustainable development: essays on system analysis of national policy. Routledge
88. Ballance OJ (2024) Sampling and randomisation in experimental and quasi-experimental CALL studies: issues and recommendations for design, reporting, review, and interpretation. ReCALL 36(1):58–71
89. Cook TD (2014) Testing causal hypotheses using longitudinal survey data: a modest proposal for modest improvement. National Academy of Education, Washington, DC
90. Valente MJ, MacKinnon DP (2017) Comparing models of change to estimate the mediated effect in the pretest–posttest control group design. Struct Equ Modeling 24(3):428–450
91. Leeuw FL, Vaessen J (2009) Impact evaluations and development: NONIE guidance on impact evaluation

92. Maciejewski ML (2020) Quasi-experimental design. Biostat Epidemiol 4(1):38–47
93. Benge CL, Onwuegbuzie AJ, Robbins ME (2012) A model for presenting threats to legitimation at the planning and interpretation phases in the quantitative, qualitative, and mixed research components of a dissertation. Int J Educ 4(4):65
94. Whittaker TA (2020) The comparison of latent variable propensity score models to traditional propensity score models under conditions of covariate unreliability. Multivar Behav Res 55(4):625–646
95. Yitmen I, Al-Musaed A, Yücelgazi F (2021) ANP model for evaluating the performance of adaptive façade systems in complex commercial buildings. Eng Constr Archit Manag 29(1):431–455
96. Goldfarb A, Tucker C, Wang Y (2022) Conducting research in marketing with quasi-experiments. J Mark 86(3):1–20
97. Roth WJ, Gil B, Makowski W, Marszalek B, Eliášová P (2016) Layer like porous materials with hierarchical structure. Chem Soc Rev 45(12):3400–3438
98. Boyle MH, Willms JD (2001) Multilevel modelling of hierarchical data in developmental studies. J Child Psychol Psychiatry 42(1):141–162
99. Gelman A, Pardoe I (2006) Bayesian measures of explained variance and pooling in multilevel (hierarchical) models. Technometrics 48(2):241–251
100. Feingold A (2013) A regression framework for effect size assessments in longitudinal modeling of group differences. Rev Gen Psychol 17(1):111–121
101. Luciano MM, Mathieu JE, Park S, Tannenbaum SI (2018) A fitting approach to construct and measurement alignment: the role of big data in advancing dynamic theories. Organ Res Methods 21(3):592–632
102. Dedrick RF, Ferron JM, Hess MR, Hogarty KY, Kromrey JD, Lang TR, Lee RS (2009) Multilevel modeling: a review of methodological issues and applications. Rev Educ Res 79(1):69–102
103. Bash KL, Howell Smith MC, Trantham PS (2021) A systematic methodological review of hierarchical linear modeling in mixed methods research. J Mixed Methods Res 15(2):190–211
104. Jiang X (2010) A high-fidelity energy monitoring and feedback architecture for reducing electrical consumption in buildings. University of California, Berkeley
105. Leeuw JD, Meijer E (2008) Introduction to multilevel analysis. In: Handbook of multilevel analysis (pp 1–75). New York, NY: Springer New York
106. Cabrera C, Paleyes A, Thodoroff P, Lawrence ND (2023) Real-world machine learning systems: a survey from a data-oriented architecture perspective. arXiv:2302.04810
107. Kabisch N, Frantzeskaki N, Pauleit S, Naumann S, Davis M, Artmann M, Bonn A (2016) Nature-based solutions to climate change mitigation and adaptation in urban areas: perspectives on indicators, knowledge gaps, barriers, and opportunities for action. Ecol Soc 21(2)
108. Omer AM (2009) Energy use and environmental impacts: a general review. J Renew Sustain Energy 1(5)
109. Almusaed A, Almssad A (2012) Passive and low energy housing by optimization. Eff Therm Insulation—Oper Factor Passiv Build Model
110. Royle JA, Dorazio RM (2008) Hierarchical modeling and inference in ecology: the analysis of data from populations, metapopulations and communities. Elsevier
111. Lim WM (2024) What is qualitative research? An overview and guidelines. Australas Mark J. 14413582241264619
112. Pattison S, Gutwill J, Auster R, Cannady M (2019) Experimental and quasi-experimental designs in visitor studies: a critical reflection on three projects. Visitor Studies 22(1):43–66
113. Kozlowski SW, Chao GT, Grand JA, Braun MT, Kuljanin G (2013) Advancing multilevel research design: capturing the dynamics of emergence. Organ Res Methods 16(4):581–615

Survey Research and Sampling Techniques

Abstract Survey research is vital for understanding the complex construction industry and environmental phenomena. The population of interest defines a study's focus, ensuring relevance, validity, generalizability, and sampling accuracy. Sampling techniques, the cornerstone of survey research, determine the representativeness and reliability of findings. This chapter explores the categories of sampling techniques, offering insights into their selection and application in the context of construction and built environment studies. Emphasis is placed on aligning sampling strategies with research objectives, scope, and the availability of sampling frames rather than the inherent characteristics of the techniques. A survey questionnaire is a structured tool for collecting quantitative or mixed-methods data on preferences, attitudes, perceptions, and behaviors. Real-world examples from built environments illustrate the effective use of these methods.

1 Introduction

Offering a methodical and organized way to gather and examine numerical data, survey research is among the most frequently utilized techniques in quantitative research. Its primary goal is to detect trends, clarify connections, and generalize results to broader populations [1–3]. Fundamentally, this approach provides researchers with the means to investigate specific problems or large-scale social events using the power of numbers and statistics. Creswell [4] describes quantitative research as examining correlations between variables using mathematical methodologies, especially statistical tools, to produce objective, replicable outcomes. The systematic approach of survey research is its defining characteristic. Typically, it employs consistent, objective data collection methods such as surveys or questionnaires. Carefully crafted, these instruments gather reliable data from samples that reflect the larger population, enhancing the generalizability of the study results. The high dependability of survey research is one of its strengths: well-designed surveys can be replicated, and their findings can be confirmed in various settings

A. Almusaed et al., *Practice of Research Methodology in Civil Engineering and Architecture*, https://doi.org/10.1007/978-3-031-97393-2_17

or over time. Furthermore, survey research begins with well-defined research objectives. These queries are designed to elicit specific, quantifiable answers that can be statistically examined. Before data collection begins, the research design is established to ensure rigor and minimize bias [5]. Generally, numerical outcomes undergo several statistical tests that help interpret the data meaningfully. Eventually, when the design permits, this approach aids in identifying causal links, testing theories, or guiding future predictions. Tools like SPSS, R, and other computational software play a crucial role by enabling researchers to manage and evaluate complex data sets accurately [6] Foundational in many fields, including the social sciences, health studies, education, and engineering, survey research emphasizes accuracy and replicability. Policy creation, strategic planning, and applied decision-making rely on correct findings from structured data in these domains. A key component of survey research is sampling, which involves selecting a subset of people, groups, or organizations to represent a broader population [7]. Sampling is critical in ensuring that a study's findings are meaningful and manageable since surveying an entire population is often impractical or unattainable. Selecting an appropriate sampling method is one of the most significant aspects of the research process, as the quality of a sample directly influences the accuracy and generalizability of the results [3]. Researchers must align their sampling strategies with their research objectives, the scope of their investigation, and the available resources. This involves choosing among various sampling methods, which are sometimes categorized as probability or non-probability. By improving the study's representativeness, probability sampling, such as simple random, stratified, or cluster sampling, provides every member of a population with a known chance of being included [7]. On the other hand, exploratory research or situations where constraints limit the use of random procedures may rely on non-probability sampling methods such as convenience or purposive sampling. Survey research is essential in the construction sector and the broader built environment. These areas enable academics and professionals to evaluate methods, understand stakeholder perspectives, and support evidence-based decision-making across urban planning, project execution, and sustainability efforts. Surveys, for example, can be used to measure public satisfaction with infrastructure projects, assess the effectiveness of new green building technologies, or monitor safety practices among construction workers. As a quantitative method, survey research is thoroughly examined in this chapter. It begins by exploring the philosophical foundations of survey research and addresses target and accessible populations [8]. The chapter then delves into sampling strategies, emphasizing probability and non-probability approaches, and describes commonly employed questionnaire surveys. Finally, it evaluates the benefits of this approach using real-world examples from the construction sector and built environment, demonstrating how survey research contributes to informed decision-making and practical knowledge. Emphasizing organized instruments, representative sampling, and numerical data processing, Fig. 1 shows the principal elements and interactions inside survey research. Research frameworks and computer tools support the methodical, repeatable, and generalizable character of quantitative approaches employed throughout scientific fields.

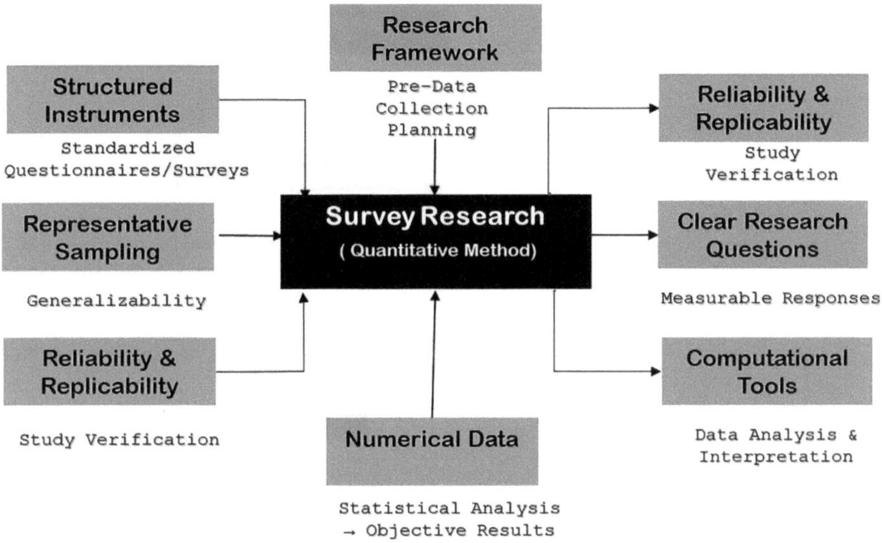

Fig. 1 Conceptual framework of survey research as a quantitative method

2 Survey Research

Survey A standard and powerful method in quantitative research is survey research [9]. A researcher collects data using this approach by showing a carefully chosen group of people a planned set of questions. This approach helps to investigate and grasp a big population's traits, views, actions, or attitudes [10]. Researchers can corroborate and question theoretical models using statistical analysis of the replies. Usually, surveys use standardized instruments like questionnaires or organized interview formats. These tools can have closed-ended questions that are simpler to measure and open-ended ones that let more complex answers come through [11]. The design of these instruments is essential; a well-constructed survey guarantees that the questions are clear, impartial, and consistent with the research goals. Surveys can be sent via websites, in-person interviews, phone calls, or mailed forms, allowing researchers to contact large and varied participant groups [10]. A basic idea in survey research is sampling. A representative sample is chosen as surveying a whole population is seldom practical. The aim is to guarantee that this subset accurately represents the essential traits of the population so that the results may be confidently generalized [12]. Survey research has more scalability than more qualitative approaches, including case studies. It may help to highlight trends across big data sets, hence improving the reliability and generalizability of the study [10]. A fundamental technique in quantitative research, survey research is illustrated in Fig. 2 as a simplified process flow. Starting with a planned set of questions, the researcher collects data from a selected sample using them. The diagram highlights three key elements: employing a representative sample, utilizing consistent tools, and understanding population

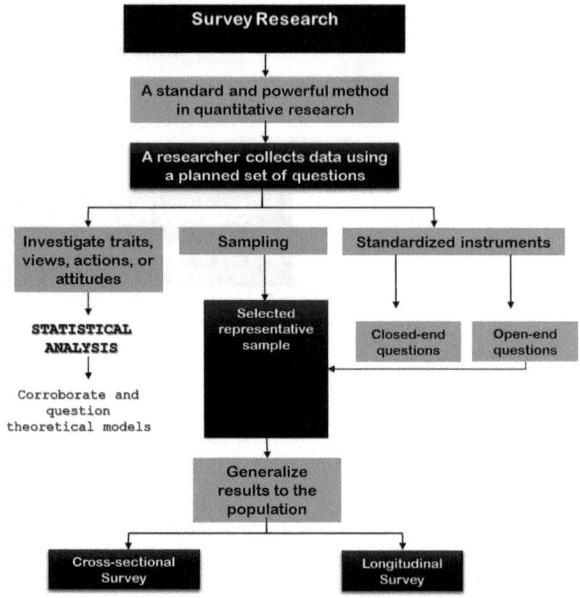

Fig. 2 Process flow of survey research in quantitative studies

characteristics and behaviors. These components, taken together, help gather consistent, quantifiable data. Statistical analysis then subjects the responses to it, allowing the researcher to confirm or question theoretical theories. Careful sample and instrument design enable a vital step: generalizing the findings to the larger population. Based on the data collection time frame, the graphic ultimately categorizes survey research into two main types: cross-sectional and longitudinal. This graphic depiction underscores the significance of survey research in producing empirical, generalizable outcomes by clearly illustrating its logical and methodological development.

2.1 *Cross-Sectional Surveys*

When the aim is to capture a population snapshot at a single point, cross-sectional surveys are among the most commonly used and practical methods in quantitative research [13]. Data is collected from a sample representing the larger target population in this observational study, allowing researchers to examine and evaluate several variables simultaneously [10]. Their efficiency helps explicitly cross-sectional surveys be quite helpful: researchers can compare various groups, including age, gender, occupation, or area, without having to wait for long periods or follow participants over time. All participants in a cross-sectional design are evaluated under the same conditions and time frame. Apart from the primary variables under study, they frequently share similar background characteristics. This method best identifies current trends, correlations, and potential group variations [14]. A cross-sectional survey, for instance, could investigate public attitudes on urban growth or examine how variation in demographic categories affects housing condition satisfaction. The main benefits of cross-sectional surveys are their time and cost efficiency [15]. Especially when resources are limited, these surveys are simpler to conduct and analyze since data is gathered once rather than over months or years. The approach has drawbacks: it cannot determine causation or track changes in attitudes or actions over time. Longitudinal studies are better suited for such discoveries. Still, cross-sectional surveys provide a strong and consistent foundation for evidence-based insights when the goal is to evaluate a vast population's present views, actions, or traits. They are instrumental in public health, education, social sciences, and the construction industry research, where fast, comparable, and generalizable data is often needed [10]. A cross-sectional poll, for example, may investigate how city residents view the application of digital twins in urban planning. Researchers can contrast opinions on data privacy, smart infrastructure, and general awareness by gathering responses from urban developers, city officials, and residents at one point [16]. This overview highlights deficiencies and directly informs improved communication and policy decisions on digital innovation in cities.

2.2 *Longitudinal Surveys*

Though they provide something more powerful—the capacity to see how things evolve—longitudinal surveys have some parallels with cross-sectional surveys, both observational [17]. Often, longitudinal studies track the same people or groups across months or years over several data collection points instead of recording a single snapshot. This method allows researchers to investigate possible causal connections, long-term trends, and gradual changes. In areas where understanding change is crucial, these surveys are beneficial. In medicine, for instance, they assist in monitoring patient health outcomes and the progression of chronic diseases. In the social sciences, they help examine educational growth, job patterns, or social mobility. Longitudinal data enables economists to track changes in income, consumer behavior, or housing stability. They are often used in market research in the corporate world to evaluate brand loyalty, customer satisfaction, or shifting consumer preferences over time [18]. One of the main advantages of longitudinal studies is their ability to let researchers investigate how personal experiences and behaviors evolve beyond a surface-level perspective. Data collected at several intervals helps researchers find patterns otherwise concealed in a one-time survey. For example, they may see how people's views change in response to changing legislation or how a digital product's popularity evolves. Unlike experimental investigations, longitudinal surveys don't involve interventions. They are correlational, meaning that researchers observe and assess various variables without attempting to influence the outcome. Because they occur spontaneously, they are particularly beneficial for examining changes in the real world. Cross-sectional surveys offer a snapshot, while longitudinal surveys provide a movie showing the flow of time, the rhythm of change, and the stories behind the

Table 1 Comparison between cross-sectional and longitudinal surveys

Aspect	Cross-sectional surveys	Longitudinal surveys
Timeframe	One point in time	Multiple points over time
Participants	Different individuals	Same individuals followed over time
Purpose	Snapshot of current views or traits	Track changes and trends
Data collection	Once	Repeated
Advantage	Quick and cost-effective	Shows development and long-term effects
Limitation	No causality or trend data	Time-consuming, possible dropouts

figures. These are effective means of grasping the complexity of human behavior and the evolution of society.

Two primary methods in observational quantitative research, cross-sectional and longitudinal surveys, are compared in Table 1. Because they conveniently capture present beliefs or behaviors, cross-sectional surveys offer a snapshot of a community at a particular time. On the other hand, longitudinal surveys let researchers track trends, changes, and possible causal links by collecting data from the same individuals over time. Although cross-sectional approaches are economical and quick, longitudinal research provides a more in-depth understanding of changes. This comparison helps researchers to choose the most appropriate approach by stressing the benefits and drawbacks of each technique.

3 Population of Interest and Target Population

3.1 Population of Interest

The population of interest forms the foundation of any well-designed study, as it precisely specifies who or what the research is about. Whether the focus is on people, families, businesses, or institutions, this category encompasses those whose qualities, experiences, or actions are central to the study's goals [19]. Essentially, this is the group to which the study's conclusions will apply, thus assigning significance and worth to them. Understanding the population of interest helps researchers maintain focus and ensures that the survey has relevant boundaries. Instead of seeking to investigate "everyone," which is rarely practical or scientifically valid, researchers can concentrate on a clearly defined segment of the larger population. Through this targeted approach, the validity and generalizability of the findings are improved, thereby avoiding the common pitfall of making overreaching conclusions that do not apply to the individuals examined [20]. Researchers must also establish specific limits driven by geographic, demographic, economic, and even political factors [21], thus refining the population of interest. These boundaries may include particular areas,

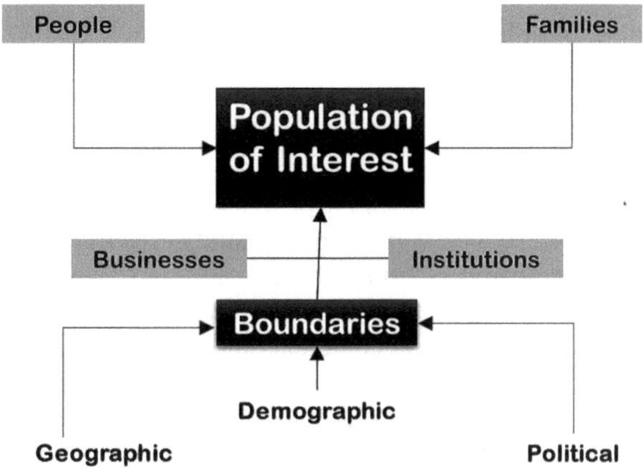

Fig. 3 Key components defining the population of interest

age groups, educational levels, occupations, or family structures. Legal and social categories may also come into play. The goal is to select a group that is both accessible and relevant. Researchers often outline specific inclusion and exclusion criteria within these parameters to determine who should be studied [22]. For example, a study on urban housing might exclude short-term visitors and include only those who have lived in the same area for at least one year. However, even with clear standards, practical constraints might lead to the unintentional exclusion of some ideal candidates or the inclusion of others. Ultimately, the demographic of interest should reflect those most affected by the study subject. Results can accurately convey the experiences and conditions that matter through a sample representing the real-world population under examination [19]. The main idea of the population of interest in the study is shown in Fig. 3. It emphasizes pertinent units of analysis, including people, families, businesses, and institutions. It demonstrates how geographic, demographic, economic, and political boundaries help define and refine the population pertinent to a particular study.

3.2 *Target Population*

The target population is the specific group of individuals, organizations, or entities a researcher intends to analyze directly [23]. The actual group from which data will be gathered is not merely a random subset of the larger population of interest, making it vital for the entire research process [24]. The target population must be carefully defined, as the study's findings rely on the responses from this group. This definition process unfolds gradually rather than all at once. Typically beginning with a broad concept, researchers subsequently refine it. The goal is to strike the right balance between enrolling enough individuals to collect significant data while excluding those unrelated to the research objectives. However, a definition that is too narrow might not provide adequate information, and one that is too broad risks distorting the results with irrelevant data. This refining technique minimizes mistakes and prevents erroneous findings by disregarding individuals not aligned with the study's aims [24]. Similar to the larger population of interest, the target population must be well defined and documented. This knowledge ensures that all parties involved in the study, including researchers, funders, and other stakeholders, understand who is being investigated and why. A clear understanding of the population's scope aids in planning the study effectively and utilizing resources efficiently [24]. Researchers establish inclusion and exclusion criteria to identify the target group better. These criteria determine who is eligible to take part and who is not. For instance, criteria could encompass age ranges, income levels, geographic locations, occupations, education, or social group membership. These factors ensure the study remains aligned with its objectives and gathers relevant and valid data [25]. Once the target population is defined, researchers create a sampling frame, a specific list or database of all qualified individuals who can be contacted and invited to participate. This stage connects actual data collection to the theoretical framework of the investigation. Maintaining clarity, focus, relevance throughout the process, and a well-defined target population backed by a reasonable sampling frame helps produce high-quality research [26]. It ensures that the outcomes are valid, reliable, and reflective of the group the investigation aims to understand. The main steps in specifying a target population within a research project are illustrated in Fig. 4. The graphic depicts four interconnected components: the sample frame, inclusion and exclusion criteria, identification, and refinement. These elements work together to ensure the chosen population is relevant, accessible, and aligned with the research objectives. Effective identification and refinement help narrow a large group into a reasonable and pertinent subset. While the sample frame ensures adequate access to participants, clearly defined inclusion and exclusion criteria maintain focus and integrity. This systematic approach enhances research reliability and methodological accuracy.

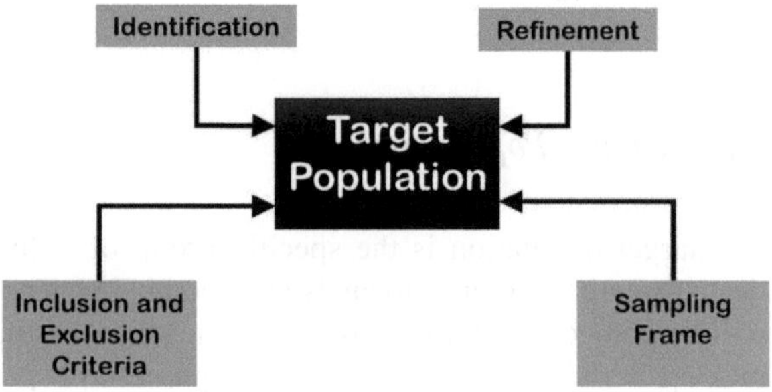

Fig. 4 Key stages in defining the target population

4 Sampling Framework

Any well-organized research endeavor depends on the sampling frame. It is the actual list or set of units, individuals, families, institutions, or companies from which a sample is taken [27]. Simply put, the sampling frame connects the target population to the study's sample. The target population specifies who should be included based on specific criteria; the sampling frame converts this ideal into a workable list of people or organizations the researcher can reach. Building a sampling frame requires careful consideration. The target population should be reflected as closely as possible to ensure relevant and representative outcomes. This means that researchers first need a well-defined target population. Once established, they can create a comprehensive list, such as a membership roster, customer database, employee directory, or voter list, comprising all who potentially qualify to participate based on the study's inclusion and exclusion criteria [27]. Even with the best intentions, differences could exist among the target population, the sampling frame, and the broader population of interest. For example, some individuals in the sampling frame may not qualify. Imagine a business that sends a survey to all staff members but wishes to examine only its customer-facing personnel. Consequently, even though they are outside the focus of the study, recently promoted managers or those on leave could be included in the dataset [28]. These discrepancies can create noise or bias in the findings, potentially leading to erroneous or misleading conclusions. The sampling frame can also be lacking or overly narrow. Imagine if the list lacks sufficient participants or omits significant subcategories. In such cases, the study might not have the statistical power to identify meaningful patterns or, in qualitative research, fail to reach saturation when no new insights are uncovered [29, 30]. A poorly constructed sampling frame could compromise even the most well-designed research. Defining and refining the sampling frame is not a one-time task. As the study progresses, it should be reviewed to ensure it aligns with the research objectives and participant criteria. This could involve adding contacts to a database to target underrepresented populations or removing ineligible individuals from a mailing list. A well-thought-out sampling frame ensures that the data collected is not only accessible but also accurate and genuinely representative of the group the researcher seeks to understand [31]. Its importance in ensuring the study findings' validity, reliability, and usefulness to inform real-world decisions cannot be overstated. The tiered structure of population selection in a study is shown in Fig. 5. From the general population of interest to the more particular target population, the operational sampling frame, and ultimately the chosen sample, each concentric circle denotes a narrower focus. This image shows how study goals guide methodical participant selection to guarantee relevance and accuracy.

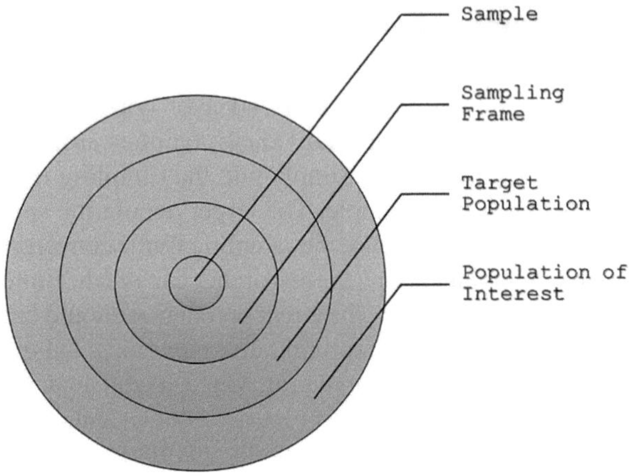

Fig. 5 Interconnections between population of interest, target population, sampling frame, and sample in research methodology (Adapted from Ref. [6])

5 Sampling Techniques for Quantitative Research

Designing research involves first establishing a suitable sample approach [4, 32, 33]. The possibly limitless number of responders makes gathering data from the whole target population in engineering research unrealistic. A census is not only impractical but also time-consuming and expensive. By contrast, when a quantitative study's unit of analysis is an organization, e.g., a company, or a single representative from each company, data collection from the whole population could be possible. Managing such large databases, nevertheless, presents major logistical issues. The particular study issues and goals being addressed will determine whether to sample the total population or a subgroup. While a sample is a portion of the population chosen for data collecting and analysis, the population is the whole set of components pertinent to the study [32, 33]. Sampling lets academics draw educated conclusions about the population without looking at every unit or person. The sample technique employed, therefore, determines much of the veracity of these conclusions. Two general sampling methods in quantitative research are probability sampling and non-probability sampling. Relying on random selection, probability sampling techniques include simple, stratified, cluster, and systematic random sampling. This guarantees that every population component has a known and non-zero probability of being included in the sample [34]. In quantitative research, such techniques are usually favored as they lower the likelihood of selection bias and enhance the dependability and generalizability of the findings. On the other hand, non-probability sampling methods ignore random selection. These consist of snowball, quota, purposive, and convenience sampling techniques. Although they might be simpler and cheaper, these techniques have more potential to skew the research and might not correctly reflect the population [35]. Research is a necessary and challenging process of choosing a representative sample from a community. Results could be skewed if the sample does not correctly represent the population's traits, restricting the generalizability of the conclusions [34–36]. Therefore, scientists must carefully develop their sampling

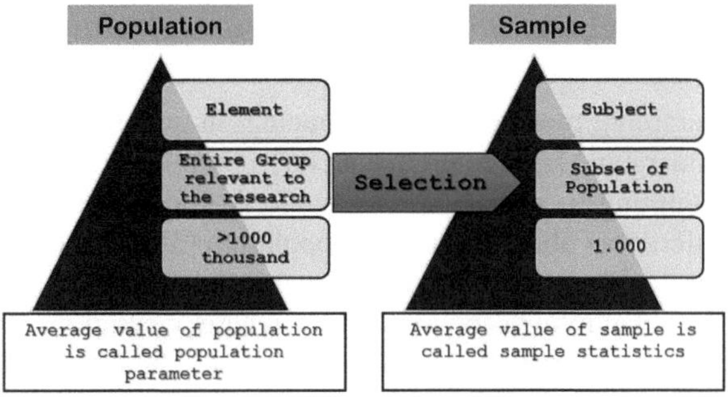

Fig. 6 Basic terminology in sampling techniques (Adapted from Ref. [37])

plans, ensuring the selected approach fits the research objectives, available resources, and the population under examination. Eventually, the outcome of a quantitative research project depends on how well the sample reflects the larger population and the data gathered. Sampling is a foundation of sound scientific research rather than just a technical process. Figure 6 shows the distinction between a population and a sample in a research setting. The population on the left is usually big and frequently too massive to examine; it is the whole group pertinent to the research. This population is used to draw a smaller group or sample through a process known as selection. The sample on the right is intended to represent the bigger group, even though it has fewer people. Efficient and relevant data analysis depends on researchers estimating population traits using the sample's average value.

Many sample strategies exist, from more rigid to more flexible methods, each with natural benefits and drawbacks. Several elements influence the selection of a suitable sampling technique: the research goals, the population's characteristics, the resource availability, and the analytical rigor called for. For example, less strict methods like snowball and quota sampling could be suitable in exploratory research or when the target population is hard to reach. In settings where representativeness is not the primary concern or probability-based procedures are impractical, these non-probability techniques let researchers quickly collect data. By contrast, investigations where lowering bias and guaranteeing the generalizability of results are vital call for more rigorous, statistically sound techniques, such as simple random, stratified, or cluster sampling. By ensuring that every population unit has a known and frequently equal likelihood of being included in the sample, these probability-based methods enhance the internal and external validity of the study [37]. No sampling technique is universally better. Researchers have to consider the particular setting of their work while making decisions. A well-chosen sample approach directly affects the credibility and application of the study's findings and increases data collection's viability and efficiency [37].

5.1 Importance of Sampling Techniques

Sampling methods are fundamental in quantitative studies, enabling researchers to understand significant, frequently complicated populations. Researchers depend on sampling to choose a reasonable fraction, or sample, from which they can draw inferences about the whole population [38] as gathering data from every person in a particular group is not often practical. This is particularly crucial in disciplines like the behavioral or social sciences, where populations might be large and varied. Trying to test or survey every individual would be logistically, monetarily, and practically impossible [39]. Instead, researchers want to choose a sample that closely reflects the traits of the larger population. When this is done correctly, the results obtained from the sample can be confidently extrapolated. On the other hand, a sample that is very different from the population it is meant to represent could produce skewed or deceptive results. Therefore, selecting the appropriate people or units to include in a study is not just a technical issue but also fundamental to guaranteeing the validity and reliability of the work [40]. Sampling is a strategic tool, not only one for saving time or skipping corners. Well-chosen sampling methods offer a systematic approach to guide decisions about whom or what to include in a study. Choosing participants could change depending on the research objectives, setting, and available resources. Some research asks for strict, statistically valid sampling techniques that give every population member an equal possibility of selection. Others, primarily exploratory or early-phase studies, might gain from more flexible methodologies where representativeness is less critical. Usually, sampling techniques fall into two broad categories: non-probability sampling and probability sampling. When the objective is to guarantee statistical correctness and reduce bias, probability sampling methods like random or stratified sampling are perfect. When time or participant access is constrained, non-probability techniques such as convenience or purposive sampling are frequently employed. Every strategy has advantages and disadvantages; no one is flawless. The efficacy of a sampling technique depends on how closely it fits the research goals and design. A careful sampling approach improves data quality and supports the general influence and validity of the research findings [37].

5.2 Types of Sampling Techniques in Quantitative Research

Generally, sampling methods fall into two basic categories: probability sampling and non-probability sampling [2, 13, 33, 34, 38]. Each has specific benefits and drawbacks that fit various study environments; these two groups reflect fundamentally distinct ways of gathering samples from a population. Usually, when the population is clearly defined and reachable, probability sampling is employed. This approach guarantees randomization and minimizes selection bias by ensuring every unit in the population has a known and non-zero probability of being chosen. This method allows researchers to use statistical theory to calculate sampling error and draw reasonable

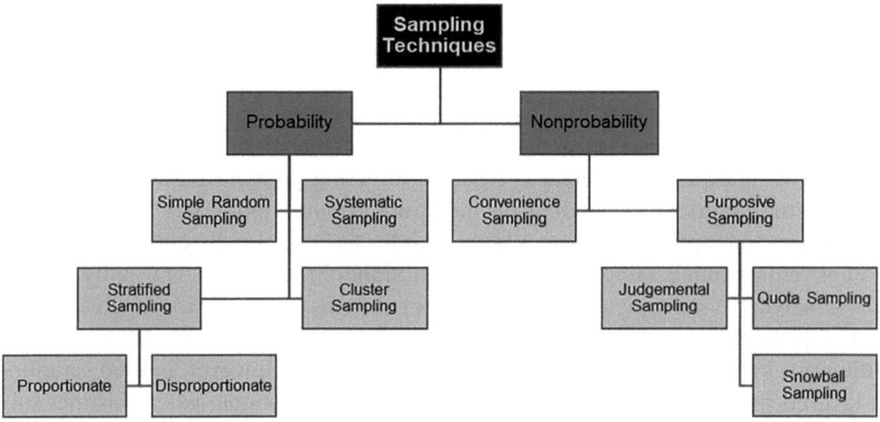

Fig. 7 Types of sampling techniques in quantitative research (Adapted from Ref. [37])

conclusions about the population. Consequently, for quantitative investigations where generalizability and rigor are essential, probability sampling is sometimes regarded as the gold standard [2, 33]. On the other hand, non-probability sampling is applied when a total population list is unavailable, inconvenient, or impossible to guarantee random selection. Dealing with hidden, hard-to-reach, or specific communities could lead to this. In such situations, studies depend on more practical techniques that do not ensure equal inclusion of every demographic factor [37]. Although non-probability techniques could create bias, particularly in exploratory phases or when contextual depth and flexibility precede broad generalizability, they might produce insightful results [13, 38]. Figure 7 shows that these two broad categories cover a range of subtypes, each designed for particular research requirements and settings. For instance, probability sampling comprises simple random, stratified, systematic, and cluster sampling, each providing unique ways to improve representativeness and lower bias in various situations [34]. Conversely, non-probability sampling techniques comprise snowball sampling, purposive sampling, quota sampling, and convenience sampling. Often, they are used when time limits, resource constraints, or access concerns preclude using random selection techniques. Though they lack statistical generalization like probability sampling, when utilized correctly and openly, these techniques can help draw significant findings [37]. A fuller investigation of these sampling subtypes, their main features, suitable uses, and influence on study validity and generalizability is provided in the next part.

5.2.1 Probability Sampling

A basic approach in quantitative research, probability sampling gives every population component a known and equal opportunity to be chosen. This randomization in selection increases the probability that the sample correctly reflects the population and reduces the risk of bias. Consequently, results from probability sampling are usually more statistically valid and generalizable than those from non-probability techniques [34]. Probability sampling's strength is in its objectivity and systematic design. Guided by probability theory, the selection process lets researchers estimate sampling mistakes and boldly draw conclusions about the larger population. This is especially crucial in research targeting great accuracy, repeatability, and external validity. Depending on the structure and complexity of the population under study, probability sampling methods can be generally classified into two groups. Unrestricted sampling, such as basic random sampling, is usually enough for somewhat homogeneous and simple populations. This approach presupposes that all people or units are equally available and relevant to the study [37] and needs little prior knowledge about the traits of the population. However, limited sampling is favored in more complicated or heterogeneous populations, where distinct subgroups may vary greatly in traits [41]. Stratified, systematic, and cluster sampling, among other methods, provide more order to the selection process, guaranteeing sufficient representation of important sub-groups. This extra framework increases the general dependability of the results [37] and helps researchers prevent over- or under-representation of specific segments. Every probability sampling method is appropriate for certain study objectives and serves a particular purpose. To give an example:

- Stratified sampling is appropriate when the population has subgroups that should be proportionally represented.
- For significant, geographically spread populations, cluster sampling is adequate.
- When working with ordered lists, systematic sampling provides a simple solution.

Table 2 offers a conceptual framework highlighting when and how these probability sampling methods are most applicable, summarizing them. The table is meant to help researchers choose the most suitable approach depending on study design and population complexity. The next part looks at each subtype, covering its fundamental characteristics, pragmatic benefits, and factors for application in several research settings.

- Simple Random (Unrestricted) Sampling

Random sampling is a technique in which each element within a population has a known and equal probability of being selected. This ensures an unbiased selection process, enhancing the sample's representativeness and increasing the research findings' generalizability [40]. The primary advantage of this method is its ability to minimize selection bias, making it one of the most representative sampling techniques. However, random sampling can become impractical when dealing with large populations due to the high costs and logistical challenges of maintaining an up-to-date and accessible population list. When random sampling is infeasible, alternative sampling techniques (as outlined in Table 2) may be employed to ensure a representative selection while addressing practical limitations [13, 37].

Table 2 Symbolic illustration of probability sampling techniques (Adapted from Ref. [34])

1. Simple random sampling					
A	B	C	D	E	Five random numbers were selected from a range of 1 to 25, yielding the sample elements **2, 4, 13, 20, and 22**. INotably,no elements from **Group B** were included in the selected sample
1	6	11	16	21	
2	7	12	17	**22**	
3	8	**13**	18	23	
4	9	14	19	24	
5	10	15	**20**	25	
2. Systematic sampling					
A	B	C	D	E	A random number between **1 and 5** was selected, yielding **4**. The resulting sample consists of the population elements **4, 7 (4 + 5 × 1), 14 (4 + 5 × 2), 19 (4 + 5 × 3), and 24 (4 + 5 × 4)**. Notably, all selected elements originate from a single row
1	6	11	16	21	
2	7	12	17	22	
3	8	13	18	23	
4	**9**	**14**	**19**	**24**	
5	10	15	20	25	
3. Stratified sampling					
A	B	C	D	E	A random number was selected from **1 to 5** within each stratum **A–E**. The resulting sample comprises the population elements **2, 6, 13, 20, and 21**. Notably, one element was selected from each column
1	**6**	11	16	**21**	
2	7	12	17	22	
3	8	**13**	18	23	
4	9	14	19	24	
5	10	15	**20**	25	
4. Cluster sampling					
A	B	C	D	E	Three clusters (**A, C, and E**) were randomly selected. Within each selected cluster, one or two elements were chosen randomly, resulting in the sample **5, 12, 15, 21, and 25**. Notably, no elements were selected from clusters **B and D**. Alternatively, the researcher may choose an entire cluster and collect **all elements** within it (e.g., **1, 2, 3, 4, 5**)
1	6	11	16	**21**	
2	7	**12**	17	22	
3	8	13	18	23	
4	9	14	19	24	
5	10	**15**	20	**25**	

- Systematic Sampling

Systematic sampling is a probability sampling technique in which subjects from a larger population are selected at fixed intervals. This method involves choosing a random starting point and selecting every nth element based on a predetermined interval. One of the primary advantages of systematic sampling is its simplicity and efficiency in implementation. However, it may introduce systematic bias if the population exhibits a hidden pattern that aligns with the sampling interval, potentially affecting the sample's representativeness. Consequently, careful consideration of the population structure is necessary to ensure the validity of the results [13, 37].

- Stratified Sampling

Stratified sampling is an appropriate technique when the population consists of distinct sub-groups with varying characteristics that necessitate separate analysis and comparison. This method involves dividing the population into meaningful strata based on specific attributes, ensuring that each subgroup is adequately represented in the sample. Compared to random and systematic sampling, stratified sampling requires additional time and effort to accurately define and classify the strata. Since data within each stratum are relatively homogeneous while differences exist across strata, the researcher must determine whether to collect data proportionately or disproportionately. In proportionate stratified sampling, the sample size from each stratum is selected based on its relative size within the population. In contrast, disproportionate stratified sampling allows for unequal representation when certain strata are of more significant analytical interest or require oversampling due to their smaller size. This approach enhances the precision of statistical comparisons and improves the generalizability of findings [13, 37].

- Cluster Sampling

Cluster sampling is a method in which the population is divided into distinct groups, or clusters, and a random sample of these clusters is selected for study. Each cluster is intended to be a microcosm of the entire population, containing a diverse mix of elements. Therefore, data within each cluster are heterogeneous, while the clusters are homogeneous relative to each other. Researchers can choose one or multiple clusters for data collection. Cluster sampling is often more cost-effective and logistically feasible than other methods, especially when dealing with large, dispersed populations. However, it may introduce a higher risk of sampling error and can lead to biased results if the selected clusters do not accurately represent the population [13, 37].

5.2.2 Non-probability Sampling

In non-probability sampling designs, the probability of each element in the population being selected as a subject is not equal. While this approach may limit the generalizability of findings, certain studies prioritize exploratory insights over broad applicability, making non-probability sampling appropriate [33]. This method is beneficial for obtaining preliminary results within constrained budgets and timeframes. Despite its limitations, non-probability sampling remains a valuable tool for achieving quick and cost-effective outcomes. The subsequent section will delve into various non-probability sampling techniques, providing a detailed overview of each method [13, 37]. Using a 5×5 matrix of grouped components (A–E), Table 3 shows four different probability sampling techniques: basic random, systematic, stratified, and cluster sampling. Every method has its selection strategy: simple random sampling picks all components from one group depending on convenience; systematic sampling selects particular items across chosen groups; stratified sampling guarantees one selected item represents each group; and cluster sampling randomly selects whole clusters and includes all or some elements inside them. This table shows how various sampling techniques influence distribution and representativeness within a dataset.

- Convenience Sampling

Convenience sampling is a non-probability sampling technique characterized by selecting readily accessible subjects. While this method offers advantages in speed, efficiency, and cost-effectiveness, it is essential to note that it may limit the generalizability of the findings due to potential sampling bias [13, 37].

Table 3 Symbolic illustration of non-probability sampling techniques (Adapted from Ref. [34])

1. Simple random sampling

A	B	C	D	E	
1	**6**	11	16	21	Group **C** was selected due to its **convenient time and location** for data collection. Consequently, all elements within this group were included in the sample, selecting elements **6, 7, 8, 9, and 10**. No elements were selected from groups **A, B, D, or E**
2	**7**	12	17	22	
3	**8**	13	18	23	
4	**9**	14	19	24	
5	**10**	15	20	25	

2. Systematic sampling

A	B	C	D	E	
1	6	11	16	21	The researcher identified groups **A, C, and D** as **representative and convenient** for the study. Within each selected group, **one or two elements** were chosen based on their **typicality and accessibility**. The resulting sample consists of elements **1, 3, 14, 18, and 20**. Notably, no elements were selected from groups **B and E**
2	7	12	17	22	
3	8	13	**18**	23	
4	9	**14**	19	24	
5	10	15	**20**	25	

3. Stratified sampling

A	B	C	D	E	
1	6	11	**16**	21	A **quota sampling approach** was implemented, requiring the selection of **one element from each group** (A–E). Each group's selection was based on **judgment or convenience**, resulting in the sample sizes of **4, 7, 14, 16, and 24**. Notably, **one element was chosen from each column group**
2	7	12	17	22	
3	8	13	18	**23**	
4	9	**14**	19	24	
5	10	15	20	25	

4. Cluster sampling

A	B	C	D	E	
1	**6**	11	16	21	Elements **3 and 6** were randomly selected from groups **A and B**. **Element 2** corresponds to **element 20**, while **element 6** corresponds to **elements 22 and 23**. As a result, the final sample comprises **3, 6, 20, 22, and 23**. Notably, no elements were selected from **group C**
2	7	12	17	**22**	
3	8	13	18	**23**	
4	9	14	19	24	
5	10	15	**20**	25	

- Purposive Sampling

In specific research scenarios, selecting participants based solely on convenience may compromise the study's objectives. For example, convenience sampling becomes inadequate when a study necessitates insights from individuals with specific expertise, such as financial managers. In such cases, purposive sampling is more appropriate, as it involves deliberately selecting individuals with the requisite knowledge or characteristics pertinent to the research. This method ensures that the data collected is relevant and informative, aligning with the study's goals [13, 37].

- Judgmental Sampling

Judgmental sampling, or purposive sampling, is a non-probability sampling technique in which the researcher selects participants based on their expertise or specific characteristics relevant to the study. This method ensures that the information gathered is reliable and pertinent to the research objectives. In certain studies, judgmental sampling is the most appropriate approach to obtaining meaningful and insightful data [13, 37].

- Quota Sampling

Quota sampling is a non-probability sampling method that ensures representation of specific subgroups within a population by setting quotas for each subset. Typically, the number of subjects selected from each subgroup reflects their proportion in the overall population. While quota sampling may limit the generalizability of findings, it is beneficial for ensuring that minority subgroups are adequately represented in a study [13, 37].

- Snowball Sampling

Snowball sampling, or chain-referral sampling, is a non-probability sampling technique employed to identify and recruit participants within specific networks or communities. The process begins by selecting initial subjects who meet the study's criteria; these participants are then asked to refer others within the same target population. This referral process continues until the desired sample size is achieved. Snowball sampling is particularly effective for accessing hard-to-reach or hidden populations, such as individuals engaged in illicit activities or members of marginalized communities, where traditional sampling methods may be inadequate. However, researchers should be mindful of potential biases, as the non-random nature of referrals can affect the sample's representativeness [13, 37].

6 Advantages and Disadvantages of Each Type of Sampling Technique

One of the most essential choices a researcher makes during the design stage of a study is selecting the appropriate sampling method [42]. Apart from the quality and validity of the data, this choice can affect the practical viability of the study as well. Different sampling techniques have advantages and drawbacks; educated decision-making depends on knowledge of these trade-offs. Especially in early-stage or exploratory research, certain studies may choose rapid, easy, and affordable sampling techniques [43]. For example, convenience or snowball sampling are simple and don't need a comprehensive population list. These techniques, therefore, frequently sacrifice representativeness and restrict the degree to which findings may be extended outside the sample. Conversely, probability-based techniques like simple random or stratified sampling provide greater control and rigor. These techniques guarantee that every population unit has a known selection probability, increasing the likelihood that the sample represents the larger population. Consequently, results are more dependable and generalizable. The trade-off, therefore, is that these strategies sometimes call for more time, work, and money [44]. Stratified sampling, for instance, is a method that improves accuracy but adds complexity by splitting the population into subgroups and sampling from every one. Cluster sampling strikes a compromise between coverage and efficiency. When researching significant, geographically spread populations, it is helpful. Rather than sampling people, researchers sample whole clusters such as schools or neighborhoods, cutting travel and logistical expenses. The drawback is that people inside the same cluster could be more similar than those in different clusters, lowering the data's variety. A middle-ground strategy is systematic sampling, in which every nth item is chosen from a list [45]. Its simplicity of use and randomization help. It can, however, be biased if there is an underlying pattern in the list that corresponds to the sampling interval. In the end, the context of the study determines the optimal sampling method. Researchers must consider statistical correctness and what is ethical and feasible in their unique environment. Table 4 thoroughly assesses the pros and cons of every sampling method to help researchers select the most suitable option for their study.

7 Survey Questionnaire

Collecting accurate and reliable data from field studies is essential for the success of survey research. Researchers must obtain respondents' opinions in a trustworthy and precise manner, which presents considerable methodological challenges. A well-designed questionnaire is widely recognized as an effective tool for data collection in survey research [46].

The questionnaire is a fundamental instrument in survey research for collecting primary data on attitudes, beliefs, and behaviors across diverse populations. It remains

Table 4 Summary of advantages and disadvantages of sampling techniques (Adapted from Ref. [34])

Sampling techniques	Description	Advantages	Disadvantages
Probability sampling			
Random sampling	All elements within the population are systematically **identified and enumerated** Each component has an **equal probability** of being selected as a subject in the study	Minimized bias and optimal generalizability of findings to the target population	It is high-cost and, in some cases, infeasible to obtain a comprehensive list of the entire population
Systematic sampling	A random starting point nnn is selected from the sampling frame, and every ninth element is systematically chosen as a subject	Straightforward and efficient when the sampling frame is readily available	Potential susceptibility to systematic bias
Stratified sampling	Initially, the population is stratified into relevant subgroups	Stratifying the population enables representative sampling within each subgroup, facilitating meaningful comparisons	Demands additional time
Cluster sampling	The population is divided into distinct clusters, such as geographic or area-based groups	Cost-effective data collection, particularly in area-based clustering	Lacks generalizability
Non-probability sampling			
Convenience sampling	The choice of elements is based on their ease and convenience	Affordable, quick, and easily accessible elements	It cannot be generalized to the entire population
Judgemental sampling	The choice of elements is based on their ability to provide information according to their experience	In certain studies, this is the sole practical and logical technique	It cannot be generalized to the entire population. Meaningful

(continued)

Table 4 (continued)

Sampling techniques	Description	Advantages	Disadvantages
Quota sampling	Each population subgroup is adequately represented by setting quotas proportional to its numbers	Quota sampling is efficient for including minority subgroups in research studies	Difficult to apply universally
Snowball sampling	Initial subjects recruit subsequent participants, expanding the sample through their networks	Cost-effective and helps researchers reach hard-to-access subjects	Referral-based sampling can introduce bias, compromising the ability to generalize findings to the entire population

a crucial tool for assessing public sentiment, including trust in scientific institutions. However, despite its widespread application, the questionnaire has methodological limitations. Since its inception in the nineteenth century, it has undergone continuous refinement to enhance its validity and reliability. Notably, Francis Galton, a British polymath of the late nineteenth century known for his contributions to statistical correlation analysis and controversial eugenics theories, acknowledged its shortcomings even as he employed it to support his hypotheses on the determinants of scientific excellence [47]. Barr et al. [48] defined a questionnaire as "a systematic compilation of questions submitted to a sample population to obtain desired information." Kuper and Kuper [49] describe it as "a structured and standardized set of questions, either administered to an individual, a small population, or, most commonly, respondents in a sample survey." Engel and Schutt [50] further elaborate on this definition, describing a questionnaire as "the instrument containing the questions in a self-administered survey."

From these definitions, it can be inferred that a survey questionnaire consists of a structured set of questions used in research, employing either a quantitative or mixed-methods approach. It is a fundamental data collection instrument, enabling researchers to align survey design with research objectives. These objectives may encompass the measurement of respondents' preferences, attitudes, perceptions, behaviors, and factual knowledge. The structured and standardized nature of the questionnaire allows respondents to select from predetermined options, express their views on various issues, and provide evaluative judgments [2, 46].

Key characteristics of a well-designed questionnaire survey are as follows:

- The questionnaire is structured and standardized to maintain respondent engagement and ensure completion.
- Questions are designed based on research objectives and may be closed-ended or open-ended. Closed-ended questions facilitate response comparison and statistical analysis, whereas open-ended questions allow respondents to provide detailed,

descriptive responses, leading to deeper insights. Each question is formulated to address a single idea, thereby minimizing ambiguity.
- The application of a survey questionnaire depends on factors such as research subject, available resources, geographical scope, time constraints, and sociocultural considerations.
- The survey questionnaire is primarily used in survey research, particularly within quantitative and mixed-methods research frameworks.

Adhering to these principles can help researchers enhance the validity and reliability of their survey instruments, ultimately improving the quality and applicability of their findings.

The credibility and success of quantitative and mixed-methods research depend on field-based survey investigations producing precise, trustworthy data. Theoretical and practical issues arise from validating and collecting respondents' ideas, attitudes, and behaviors in a trustworthy manner. Among the most often used instruments for negotiating these difficulties, the questionnaire has become a key tool in empirical research because to its adaptability, uniformity, and scalability [49].

7.1 Types of Survey Questionnaire

7.1.1 Classification and Administration of Survey Questionnaires

Survey questionnaires can be categorized based on their structure, response format, and mode of administration. The response format is typically classified into open-ended and closed-ended questions. At the same time, the administration methods include (i) self-administered mailed questionnaires, (ii) group-administered questionnaires, (iii) household drop-off survey questionnaires, (iv) online survey questionnaires, and (v) face-to-face interviews [50]. A detailed description of these methods is provided below.

- Self-Administered Mailed Questionnaire

A self-administered mailed questionnaire is distributed to respondents via postal mail for independent completion. The researcher is not present during the process, allowing respondents to answer the questions at their convenience and return the completed questionnaire in a prepaid envelope [51]. This method minimizes interviewer bias but is subject to low response rates due to potential participant disengagement.

- Group-Administered Questionnaire

A group-administered questionnaire involves respondents completing the survey in a structured group setting. Participants are gathered in a designated location and instructed to complete the questionnaire independently, without interaction with one another. This method ensures higher response rates while allowing researchers to provide clarifications if necessary [52].

- Household Drop-Off Survey Questionnaire

In this method, researchers personally deliver the questionnaire to respondents' homes or workplaces [53]. During the drop-off process, the researcher explains the study's objectives, emphasizes the importance of participation, and clarifies any ambiguities in the survey instrument. Respondents may return the completed questionnaire via postal or courier services, or a mutually agreed-upon collection point may be established. This approach enhances response rates by fostering rapport between the researcher and respondents, increasing engagement and willingness to participate [54].

- Online or Web-Based Survey Questionnaire

An online survey questionnaire is distributed electronically, typically via email, and respondents complete it using an interactive web-based form. According to Ritter and Sue [55], an effective online survey should include a secure login system, clear instructions, logically structured questions, and a concluding thank-you message acknowledging respondents' contributions.

Before utilizing an online survey, researchers must carefully consider the study's objectives and the target population. The selection of participants should be aligned with the research focus, distinguishing between public respondents and individuals with domain-specific expertise. Additionally, the researcher must determine the most appropriate question format, including multiple-choice, open-ended, or closed-ended responses, ensuring alignment with the study's analytical framework [25]. Although online surveys offer advantages such as cost-effectiveness, efficiency, and broad geographic reach, they also present certain limitations. Specifically, populations such as the elderly, individuals with low literacy levels, and those lacking internet access are inherently excluded from participation. Moreover, even internet-used respondents may encounter technical difficulties, potentially impacting response rates and data reliability. By selecting the appropriate questionnaire administration method, researchers can optimize data collection processes, enhance response quality, and ensure the validity and generalizability of survey findings [56].

7.1.2 Types of Questions Used in Survey Questionnaires

A survey questionnaire is a structured set of questions designed to collect data systematically. The length and complexity of a questionnaire are influenced by multiple factors, including the research objectives, available resources, the researcher's capacity, and time constraints. For instance, an extensive questionnaire may be necessary in a PhD-level study to address a significant knowledge gap and explore complex research questions in depth. Conversely, when the research scope is narrow, such as collecting feedback on products or services, or when funding and time are limited, a more concise questionnaire with a smaller sample size may be appropriate [46].

Survey questionnaires employ various types of questions, as identified by scholars such as Neuman [57], Grinnell and Unrau [58], and Trochim [53]. These include open-ended and closed-ended questions and more specific formats such as multiple-choice, dichotomous, nominal, ordinal, and interval-scale questions. Additionally, structured response formats like the Likert scale and contingency questions are commonly utilized.

Broadly, survey questions can be classified into two primary categories: open-ended and closed-ended questions. Open-ended questions allow respondents to provide unrestricted, detailed responses, offering qualitative insights. In contrast, closed-ended questions present predefined answer choices, facilitating more straightforward data quantification and statistical analysis. Most structured question types, including multiple-choice and Likert-scale items, fall under the closed-ended category due to their fixed response options determined by the researcher [59].

The choice of question type depends on the study's objectives, the desired depth of responses, and the analytical methods employed. A well-designed questionnaire strategically combines different formats to enhance data reliability, validity, and overall research effectiveness. This survey questionnaire is designed to illustrate various types of questions commonly used in research, as shown in Table 5.

7.2 Best Practices for Designing a Survey Questionnaire

Designing a well-structured survey questionnaire is essential to ensure data validity, reliability, and overall research effectiveness. A poorly designed questionnaire can introduce bias, reduce response rates, and compromise the quality of collected data. Below are the key best practices:

i. Clearly Define Research Objectives

   Before designing the questionnaire, researchers must establish clear research objectives. This ensures that every question aligns with the study's goals and contributes to meaningful data collection.

Example: If the objective is to assess public perception of AI-based digital twins in facility management, questions should focus on awareness, adoption, and perceived effectiveness.

ii. Choose the Right Question Types

   Survey questions are broadly categorized into formats of open-ended (qualitative insights) and closed-ended (quantifiable responses). A well-balanced questionnaire incorporates both to capture comprehensive insights.

Example:

- Open-ended: *"What challenges have you faced in using AI-based predictive maintenance systems in building facilities?"*

Table 5 Examples of the types of questions used in the survey questionnaire

Survey Questionnaire

Title: Assessing the Implementation of AI-based Digital Twins for Predictive Maintenance in Building Facilities: A Focus on Indoor Environment Quality

Introduction:

Dear Participant,

This survey aims to gather insights into the adoption and effectiveness of AI-based Digital Twins for predictive maintenance in building facilities, particularly in monitoring and enhancing **Indoor Environmental Quality (IEQ)**. Your responses will help improve the integration of digital twin technology in facility management.

The survey consists of three sections:

1. **General Information**
2. **Digital Twin Integration**
3. **AI-based Predictions and Performance**

Your participation is voluntary, and responses will remain confidential.

Section 1: General Information

1. **What is your role in building facility management?**
    - Facility Manager
    - Building Engineer
    - Energy Manager
    - Researcher
    - Other (please specify) __________
2. **What type of building do you manage or work with?**
    - Commercial
    - Residential
    - Industrial
    - Healthcare
    - Educational
    - Other (please specify) __________
3. **Have you implemented or considered implementing Digital Twins in your facility?**
    - Yes, currently implemented
    - Considering implementation
    - Not yet considered
    - No interest
4. **What are the main challenges you perceive in implementing Digital Twins for predictive maintenance?** *(Select all that apply)*
    - High cost
    - Lack of expertise
    - Data security concerns
    - Integration with existing systems
    - Resistance to adoption
    - Other (please specify) __________

Section 2: Digital Twin Integration

Please rate the following statements on a scale from 1 (Strongly Disagree) to 5 (Strongly Agree).

Statement	1 (SD)	2 (D)	3 (N)	4 (A)	5 (SA)
Our facility has a real-time data integration system for predictive maintenance.	☐	☐	☐	☐	☐

(continued)

Table 5 (continued)

The Digital Twin model accurately represents the building's physical and operational conditions.	☐	☐	☐	☐	☐
Implementing Digital Twins has improved equipment efficiency and reduced maintenance costs.	☐	☐	☐	☐	☐
Digital Twin integration has enhanced indoor environmental quality (IEQ) monitoring.	☐	☐	☐	☐	☐
The system provides actionable insights for maintenance planning and fault detection.	☐	☐	☐	☐	☐
We experience challenges in real-time data synchronization between physical assets and their digital representation.	☐	☐	☐	☐	☐

Section 3: AI-based Predictions and Performance
Please rate the following AI-based capabilities in Digital Twins using the Likert scale from 1 (Strongly Disagree) to 5 (Strongly Agree).

Statement	**1 (SD)**	**2 (D)**	**3 (N)**	**4 (A)**	**5 (SA)**
AI-driven Digital Twins improve fault detection and predictive maintenance.	☐	☐	☐	☐	☐
AI algorithms accurately predict failures and schedule maintenance before breakdowns occur.	☐	☐	☐	☐	☐
AI-based systems provide real-time insights for optimizing indoor air quality (IAQ).	☐	☐	☐	☐	☐
The integration of AI enhances energy efficiency and HVAC performance.	☐	☐	☐	☐	☐
AI-driven models improve occupant comfort and productivity through better climate control.	☐	☐	☐	☐	☐
AI-based digital twins can effectively detect and mitigate risks related to poor air quality (e.g., CO2 levels, humidity, temperature).	☐	☐	☐	☐	☐

Section 4: Future Perspectives and Adoption

1. **What level of priority would you assign to adopting AI-based Digital Twins in your facility?**
   - o High Priority
   - o Moderate Priority
   - o Low Priority
   - o No Priority
2. **What additional functionalities would you like AI-driven Digital Twins to have?**
   - o Advanced climate control optimization
   - o Automated energy consumption predictions
   - o Enhanced integration with IoT sensors
   - o Predictive analytics for asset performance
   - o Other (please specify) __________
3. **Would you be interested in receiving further training or knowledge about AI-based Digital Twins?**
   - o Yes
   - o No

(continued)

Table 5 (continued)

☐ **What are the primary challenges you have experienced in implementing AI-based Digital Twins for predictive maintenance in building facilities?**

…….

…….

…….

☐ **How has the integration of Digital Twins influenced the monitoring and management of Indoor Environment Quality (IEQ) in your facility?**

…….

…….

…….

☐ **In your opinion, what key improvements or additional features could enhance the effectiveness of Digital Twins for predictive maintenance and IEQ management?**

…….

…….

…….

☐ **Can you describe any specific examples of AI-driven Digital Twin technology successfully optimizing maintenance operations or improving indoor environmental conditions?**

…….

…….

…….

Conclusion:

Thank you for participating in this survey. Your insights will improve the implementation of AI-based Digital twins for predictive maintenance and Indoor Environment Quality. If you would like to receive the results of this study, please provide your contact details (optional).

- Closed-ended: *"On a scale of 1 to 5, how effective do you find AI-based digital twins in improving facility management?"*

iii. Ensure Question Clarity and Neutrality

Questions must be clear, concise, and free from bias to prevent misinterpretation and response distortion. Ambiguous or leading questions can influence responses, reducing the survey's validity.

Example:

- Biased: "*Do you agree that AI-based digital twins are the best solution for predictive maintenance in building facilities?*"

- Neutral: "*What are your thoughts on AI-based digital twins as a solution for predictive maintenance in building facilities?*"

iv. Maintain Logical Flow and Organization

A well-structured questionnaire should begin with general questions before progressing to specific ones, minimizing respondent fatigue and confusion.

Example:

- Start with introductory questions to ease respondents in.
- Organize questions thematically (e.g., awareness, adoption, challenges, improvements).
- Use skip logic to guide respondents through relevant sections.

v. Pretest and Pilot the Questionnaire

Conducting a pilot test before deploying a full-scale survey helps identify unclear questions, technical issues, and response biases.
Steps in Pilot Testing:

- Select a small group from the target population.
- Collect feedback on question clarity and difficulty.
- Revise based on responses.

vi Balance Length and Response Burden

A lengthy questionnaire may lead to survey fatigue and increased dropout rates. Studies suggest surveys exceeding 15–20 min significantly decrease response rates [60].

Example:

- Keep surveys concise (10–15 min max) unless in-depth responses are required.
- Use progress indicators in online surveys.
- Prioritize essential questions; avoid redundancy.

7.3 Strengths of the Survey Questionnaire Method

Questionnaires and scales efficiently collect data on individuals' characteristics, behaviors, and opinions. The census, for instance, gathers comprehensive information about a large population's traits, actions, and perspectives. Survey questionnaire methods enable researchers to obtain detailed data beyond mere demographic characteristics, facilitating the measurement of attitudes and opinions that are challenging to assess through observational techniques.

The strengths of survey questionnaire methods include:

- *Efficient Data Collection*: They allow for rapidly collecting extensive information from large populations.
- *Flexibility*: These techniques can be integrated with other methods to enhance the accuracy of results.
- *Reliability*: When the sample is representative, the data collection tool yields accurate and dependable data.
- *Cost-Effectiveness*: Quantitative surveys are generally more economical and quicker than alternative data collection methods.
- *Quantitative Assessment*: They enable researchers to numerically quantify individuals' attitudes numerically, facilitating statistical analysis.

These advantages underscore the value of quantitative surveys in research endeavors.

8 Conclusions

Survey research continues to be a robust and adaptable method in the toolbox of modern researchers, particularly within the realms of social science, construction, public health, education, and urban development. The strength of this approach lies in its structured nature, which allows for the systematic collection and analysis of data, uncovering patterns, testing theories, and guiding practical decisions. This chapter explores the many facets of survey research, from its conceptual foundation to its applied techniques, highlighting its versatility and methodological rigor. At the heart of survey research is a deep commitment to representation. Understanding who we are studying is as crucial as what we are learning. The concepts of the population of interest and the target population provide a structured lens through which researchers define and focus their studies. This clarity is essential for setting boundaries and ensuring that findings are relevant and applicable to the real world. Without this precision, even the most carefully designed studies risk producing results that are either misleading or too broad to be useful. Building upon this foundation is the critical process of sampling. Surveying entire populations is rarely feasible, so researchers must select a subset, ideally one that mirrors the broader group in meaningful ways. This is where sampling strategies come into play. Whether through probability methods such as simple random, stratified, and cluster sampling, or non-probability approaches like convenience and purposive sampling, the goal remains to ensure that the sample accurately reflects the population being studied. Each method has its strengths and trade-offs. Probability sampling, for example, offers higher levels of generalizability and less bias but often comes with logistical and financial challenges. Non-probability sampling, while quicker and more accessible, may sacrifice some accuracy. Understanding when and how to use these techniques is vital for achieving credible results. In some instances, hybrid strategies may be adopted to leverage the benefits of both approaches. Equally important is the concept of the sampling frame, which serves as the bridge between theoretical concepts and real-world execution. This frame must be carefully constructed and continually evaluated to ensure alignment with the research goals. A flawed or incomplete sampling frame can easily lead to skewed or invalid results, even when the rest of the research design is sound. For example, one of the most compelling applications of survey research is its ability to adapt across various domains. In construction and the built environment, for instance, surveys help planners assess public attitudes toward new projects, evaluate sustainability initiatives, and gather feedback on digital technologies such as AI-based digital twins. This makes survey research not just a data collection method, but a tool for communication, policy design, and participatory governance. Moreover, the survey questionnaire remains a cornerstone of this methodology. Whether administered online, in person, by mail, or through interviews, a well-crafted questionnaire can uncover rich insights. The' structure, clarity, and neutrality of questions determine the data quality collected. From broad public opinions to detailed professional feedback, questionnaires are adaptable and capable of producing both qualitative nuance and quantitative precision. Throughout this chapter, we've also seen the importance of maintaining methodological coherence. Each part of the survey must align with the overarching research objectives, from defining populations, building a sampling frame, selecting the proper sampling technique, and crafting the questionnaire. This coherence ensures that the results are statistically valid and contextually meaningful. Survey research and sampling techniques form the backbone of quantitative research efforts across disciplines. Researchers can produce credible, relevant, and impactful findings by grounding studies in well-defined populations, employing strategic sampling methods, and designing thoughtful instruments. The lessons from this chapter reinforce the importance of thoughtful planning, methodological precision, and ethical responsibility in conducting survey-based research. When applied carefully, these techniques illuminate the present and shape smarter decisions for the future.

References

1. Babbie ER (2010) The practice of social research, 12th edn. Wadsworth Cengage
2. Hartas D (2010) Quantitative research as a method of 4. In: Educational research and inquiry: Qualitative and quantitative approaches, vol 65
3. Nardi PM (2018) Doing survey research: A guide to quantitative methods. Routledge
4. Creswell JW (2014) Research design: qualitative, quantitative, and mixed methods approaches, 4th edn. SAGE Publications, Incorporated
5. Karunarathna I, Gunasena P, Hapuarachchi T, Gunathilake S (2024) Data collection fundamentals: a guide to effective research methodologies and ethical practices. ResearchGate 65(65):65. https://www.researchgate.net/publication/383155577_Data_Collection_Fundamentals_A_Guide_to_Effective_Research_Methodologies_and_Ethical_Practices
6. Ismail H (2024) Performance evaluation of AI-based healthcare systems in the United Arab Emirates: a comparative analysis of diagnostic accuracy and predictive capability using SPSS. Master's thesis, Rochester Institute of Technology
7. Ahmed SK (2024) How to choose a sampling technique and determine sample size for research: a simplified guide for researchers. Oral Oncol Rep 12:100662

8. Reddy D, Pulluru K (2024) Principles of statistics & research methodology. Academic Guru Publishing House
9. Ginevičius R (2024) Complex quantitative assessment of social research survey results. J Int Stud (2071–8330) 17(4)
10. Ghana A (2023) An overview of quantitative research methods. Int J Multidiscipl Res Anal 6(8)
11. Jensen D, Whiles BB, Mirza M (2025) Surveys and questionnaires. In: Translational urology. Academic Press, pp 189–194
12. Willie MM (2024) Population and target population in research methodology. Golden Ratio Soc Sci Educ 4(1):75–79
13. Hossan D, Dato'Mansor Z, Jaharuddin NS (2023) Research population and sampling in quantitative study. Int J Bus Technopreneurship (IJBT) 13(3):209–222
14. Dražanová L, Gonnot J, Heidland T, Krüger F (2024) Which individual-level factors explain public attitudes toward immigration? A meta-analysis. J Ethn Migr Stud 50(2):317–340
15. Lalani HS, Tessema FA, Kesselheim AS, Rome BN (2024) Availability and cost of expensive and common generic prescription drugs: a cross-sectional analysis of direct-to-consumer pharmacies. J Gen Intern Med 39(12):2187–2195
16. Almusaed A, Almssad A, Alasadi A, Al-Asadi F (2025) Sustainable reimagining of urban habitats: navigating Iraq's housing crisis through socio-technological and environmental approaches. Ain Shams Eng J 16(2):103280
17. Taubeneder R, Roehrich JK, Tyler BB, Squire B, Gnyawali DR (2024) Managing coopetition dynamics: a longitudinal study of a multiparty alliance formation in a large utilities project. J Oper Manag 70(6):875–903
18. Tahir AH, Adnan M, Saeed Z (2024) The impact of brand image on customer satisfaction and brand loyalty: a systematic literature review. Heliyon
19. Omair A (2025) Sample size estimation and sampling techniques for selecting a representative sample. J Health Special 2(4):142
20. Salkind NJ (2009) Exploring research, 7th edn. Pearson Education, New Jersey
21. Bakibinga P, Kabaria C, Kyobutungi C, Manyara A, Mbaya N, Mohammed S, Yeboah G (2019) A protocol for a multi-site, spatially-referenced household survey in slum settings: methods for access, sampling frame construction, sampling, and field data collection. BMC Med Res Methodol 19(2019):1–8
22. Okeke NI, Alabi OA, Igwe AN, Ofodile OC, Ewim CPM (2024) Customer journey mapping framework for SMEs: enhancing customer satisfaction and business growth. World J Adv Res Rev 24(1)
23. Baqi A, Abdeldayem M, Aldulaimi S (2024) Assessing the role of direct public engagement in shaping the UAE nuclear energy sustainability image. Int J Organ Anal 32(9):1669–1688
24. Sudman S (1976) Applied sampling. Academic Press
25. Kalleberg AL, Marsden PV, Aldrich HE, Cassell JW (1990) Comparing organisational sampling frames. Adm Sci Q 35(4):658–688
26. Ackerman B, Schmid I, Rudolph KE, Seamans MJ, Susukida R, Mojtabai R, Stuart EA (2019) Implementing statistical methods for generalising randomised trial findings to a target population. Addict Behav 94:124–132
27. Prandner D, Weichbold M (2019) Building a sampling frame for migrant populations via an onomastic approach: lesson learned from the Austrian immigrant survey 2016. In: Survey methods: insights from the field. pp 1–20
28. Tabachnick BG, Fidell LS (2018) Using multivariate statistics, 6th edition). Pearson.
29. Hackshaw A (2008) Small Studies: Strengths and Limitations. Eur Respir J 32(5):1141–1143
30. Fusch P, Ness L (2015) Are we there yet? Data saturation in qualitative research. Qualit Rep 20(9):1408–1416
31. Subbarayudu CBV, Singh HP, Karamthoti MB, Rao M (2024) A conceptual approach to research methodology and data analytics. Academic Guru Publishing House
32. Cooper DR, Schindler PS, Sun J (2006) Business research methods, 9th edn. McGraw-Hill Irwin

33. Sekaran U, Bougie R (2016) Research methods for business: a skill building approach, 7th edn. John Wiley & Sons
34. Malhotra NK, Das S (2010) Marketing research: an applied orientation, 6th edn. Pearson Education, Prentice
35. Bryman A, Bell E (2015) Business research methods, 4th edn. Oxford University Press
36. Collier JE, Bienstock CC (2007) An analysis of how nonresponse error is assessed in academic marketing research. Mark Theory 7(2):163–183
37. Sarker M, AL-Muaalemi MA (2022) Sampling techniques for quantitative research. In: Islam et al. (eds) Principles of social research methodology. Springer Nature Singapore, Singapore, pp 221–234
38. Marshall MN (1996) Sampling for qualitative research. Fam Pract 13(6):522–526
39. Eddon PB, Scheepers R (2012) Towards the improved treatment of generalization of knowledge claims in IS research: drawing general conclusions from samples. Eur J Inf Syst 21(1):6–21
40. Zikmund WG, Babin BJ, Carr JC, Griffin M (2013) Business research methods (9th **edn**). South-Western Cengage Learning, Hampshire, United Kingdom
41. Krishna VV, Veettil PC (2022) Gender, caste, and heterogeneous farmer preferences for wheat varietal traits in rural India. PLoS ONE 17(8):e0272126
42. Cash P, Isaksson O, Maier A, Summers J (2022) Sampling in design research: eight key considerations. Des Stud 78:101077
43. Loria F, Picciotto S, Adamo G, Zendrini A, Raccosta S, Manno M, Zarovni N (2024) A decision-making tool for navigating extracellular vesicle research and product development. J Extracell Vesicles 13(12):e70021
44. Lewandowski RA, Cirella GT (2023) Performance management systems: trade-off between implementation and strategy development. Oper Manag Res 16(1):280–295
45. Chiou R, Jefferies E, Duncan J, Humphreys GF, Lambon Ralph MA (2023) A middle ground where executive control meets semantics: the neural substrates of semantic control are topographically sandwiched between the multiple-demand and default-mode systems. Cereb Cortex 33(8):4512–4526
46. Chowdhury SR, Oakksa MA, Ahmmed F (2022) Survey questionnaire. In: Islam MR, Khan NA, Baikady R (eds) Principles of social research methodology. Springer Nature, Singapore, p. 131–156
47. Berenbaum MR, Thorp HH (2024) Publishing survey research. Proc Natl Acad Sci USA 121(18):e2406826121–e2406826121
48. Barr AS, Davis RA, Johnson PO (1953) Educational research and appraisal. Philadelphia 7: J.B Lippincott Company, pp viii+362
49. Kuper A, Kuper J (ed.) (1999) The social science encyclopaedia. Routledge
50. Engel RJ, Schutt RK (2009) Fundamentals of social work research. Sage Publications
51. Christmann P, Gummer T, Häring A, Kunz T, Oehrlein AS, Ruland M, Schmid L (2024) Concurrent, web-first, or web-only? How different mode sequences perform in recruiting participants for a self-administered mixed-mode panel study. J Surv Stat Methodol 12(3):532–557
52. Pham TT, Lingard H, Zhang RP (2023) Factors influencing construction workers' intention to transfer occupational health and safety training. Saf Sci 167:106288
53. Trochim WM (2007) Research methods. Biztantra
54. Li H, Han L, Ao Y, Wang Y, Wang T (2022) Influences of the built environment on rural school children's travel mode choice: the case of Chengdu. Int J Environ Res Public Health 19(15):9008
55. Ritter LA, Sue VM (2007) The survey questionnaire. N Dir Eval 115:37–45
56. Chelghoum A (2024) Unveiling respondents' collaboration in online questionnaire responses. ALTRALANG J 6(2):377–388
57. Neuman WL (2011) Social research methods: qualitative and quantitative approaches, 7th edn. Pearson
58. Grinnell RM, Unrau YA (2008) Social work research and evaluation: Foundations of evidence-based practice. Oxford University Press

59. Baburajan V, e Silva JDA, Pereira FC (2020) Open-ended versus closed-ended responses: a comparison study using topic modeling and factor analysis. IEEE Trans Intell Transp Syst 22(4):2123–2132
60. Farahani AV, Leinonen M, Ruotsalainen L, Jokisalo J, Kosonen R (2025) Predicting summer indoor temperatures in Nordic apartments considering heatwaves forecasts. Energy Build 115630

Statistical Analysis in Quantitative Research

Abstract In quantitative research, statistical analysis is essential; it is the foundation that supports our interpretation of data, testing of hypotheses, and drawing of significant findings. This chapter examines the fundamental statistical tools researchers use when dealing with numbers. Beginning with the fundamentals, such as descriptive statistics, which help to summarize data using averages, spreads, and patterns, it unpacks them. It then turns to inferential statistical techniques, letting us draw informed conclusions about a bigger population from a smaller sample. Illustrations of how they are utilized in research will reveal justifications for commonly used statistical tests, including *t*-tests, ANOVA, correlation, and regression. The chapter underlines the significance of data quality and dependability and covers probability theory, a vital component of compelling statistical findings. Statistical software is necessary in today's research context; we quickly examine how these tools improve accuracy and simplify analysis. Using pragmatic ideas and unambiguous direction, this chapter seeks to enable researchers to conduct robust statistical studies, firmly understand their results, and steer clear of typical errors. It emphasizes the need to be statistically literate to generate a credible, evidence-based analysis.

1 Introduction

Quantitative research is essential for helping us grasp patterns, correlations, and trends across a broad spectrum of sectors, from healthcare and social sciences to engineering and business [1]. Statistical analysis, a toolkit enabling researchers to understand statistics, evaluate hypotheses, and reach conclusions beyond the particular sample being researched, sits at the core of this strategy [2]. Statistical techniques help researchers clarify and organize their studies to assess data fairly and arrive at well-founded conclusions. The methods and instruments of statistical analysis have significantly evolved with time. Historically, scholars largely depended on traditional methods such as *t*-tests, ANOVA, and linear regression. However, powerful computing methods make statistical analysis more precise and insightful [3]. The emergence of machine learning, which combines data-driven algorithms with

A. Almusaed et al., *Practice of Research Methodology in Civil Engineering and Architecture*, https://doi.org/10.1007/978-3-031-97393-2_18

conventional statistics to enable sophisticated pattern detection and prediction, is one notable development [4]. Another change is the increasing use of Bayesian statistics, which provides a flexible approach to update our knowledge as new information becomes available. This makes it perfect for uncertain or changing research contexts [5]. Statistical analysis is not only about number crunching. It enables academics to summarize vast data sets, validate theories, and create models that forecast or clarify behavior. Standard methods for investigating how various variables interact and whether particular assumptions hold are regression analysis, factor analysis, and structural equation modeling (SEM) [6]. These techniques direct decision-making and reveal concealed links. By lowering prejudice and simplifying results to duplicate, they also significantly increase the validity of studies [7]. Statistical techniques have been pushed even more by the explosion of big data and artificial intelligence (AI) in recent years. Massive dataset have enabled academics to depend on machine learning models to manage complicated issues, including hundreds or perhaps thousands of variables [8]. These models reveal patterns that would be undetectable using conventional methods by themselves. Resampling techniques such as bootstrapping and permutation tests have grown more significantly concurrently. Especially when dealing with tiny samples or skewed distributions, these methods allow researchers to get more dependable findings without assuming much about the data. They offer a flexible and straightforward approach to measure variability and validate hypotheses. Combining hierarchical models with Bayesian inference is another developing trend. This method lets researchers examine data organized in layers, such as pupils in classrooms or hospital patients, while delicately considering ambiguity [9]. It's more flexible than conventional approaches and is becoming increasingly popular in sectors where data is complicated and diverse. The growth of available statistical tools such as R, Python, SPSS, and others has also facilitated advanced research for all kinds of academics [10]. These tools provide pre-built functions, simple interfaces, and strong visualizations, enabling users to concentrate more on interpretation and less on manual computations. By linking basic approaches, new technologies, and advanced tools, Fig. 1 shows how statistical analysis underpins quantitative research. Emphasizing how big data and contemporary technologies improve insight generation, it reveals the evolution from conventional exams to machine learning and AI-driven solutions. The chart helps academics grasp the changing terrain of quantitative analysis and how to use it successfully across sectors by stressing these connections.

Four main facets of statistical analysis in quantitative research are covered in this chapter [11]. First, it examines how researchers perform statistical analysis, selecting the appropriate techniques and using them correctly to produce reliable outcomes [12]. Second, it addresses how to analyze those findings, knowing what the figures indicate and how they connect to the study topic. Third, it investigates how well-organized reports, charts, and visual tools help one to convey results. It finally presents Generative AI (GenAI), a new technology that can assist data analysis by automating process components, enabling researchers to produce insights more quickly and efficiently [13]. These components taken together guarantee that statistical analysis

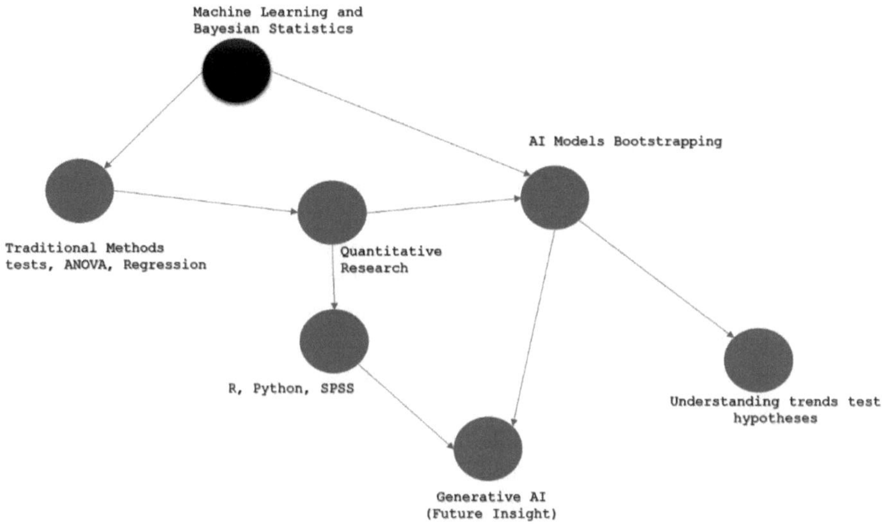

Fig. 1 Mapping the evolution of statistical analysis in quantitative research

is not only technically correct but also relevant and accessible, thereby enabling researchers to convert data into essential knowledge.

2 Significance of Statistical Analysis in Quantitative Research

More than just numbers, statistics is a potent instrument enabling people in many different sectors to make better choices, enhance services, create superior goods, and guarantee quality [14]. Its significance is much more evident in quantitative studies. Statistical analysis enables researchers in healthcare, engineering, education, or business to find trends, challenge assumptions, and derive significant insights from data. It converts raw data into insights that can propel actual changes. Building on a broad bibliographic study demonstrating how commonly statistics are utilized in professional practice and research, this part emphasizes how statistical techniques provide a more exact and organized knowledge of complicated research issues and help technical correctness [15]. By using statistical methods, researchers can investigate correlations between variables, test hypotheses, and assess how well their data supports particular ideas or suggestions. Powerful software and computational tools have greatly increased the function of statistics in the digital era. These developments have changed how research is conducted, hence accelerating, improving accuracy, and increasing accessibility. Today's Researchers can use sophisticated models, automate complex computations, and manage bigger data sets. Consequently, overcoming research challenges in both natural and social sciences now heavily relies on statistical analysis [16]. Several fundamental processes define statistics: data gathering, data organization, data analysis, and result interpretation in a logical fashion. Especially when dealing with a bigger population sample, it's utilized to build wise choices on evidence. Without these procedures, spotting variations, comprehending trends, or consistently making forecasts would be challenging [17]. A significant component is quantitative analysis, which emphasizes examining numerical data gathered from surveys, polls, or pre-existing dataset. Researchers can better grasp particular problems or draw generalizations about a population using computer-based tools and mathematics [18]. A healthcare researcher, for instance, could utilize statistical data to find disease risk factors; an education researcher could look at student achievement across several institutions. Many disciplines—science, engineering, business, and the social sciences use quantitative data analysis since it enables strong, evidence-based choices [19]. However, knowing each variable's measurement level determines whether one can properly perform this study. This refers to how data is categorized, such as categories, ranks, or numerical values; it then decides which statistical techniques are suitable. Selecting the incorrect approach could produce false findings. A statistical test that doesn't fit the examined data could invalidate a researcher's results. This can compromise the general quality of the study and result in false conclusions. That is the reason [20] why knowing how to fit measurement levels with the appropriate statistical techniques is so crucial. Statistical analysis, in brief, is the engine driving high-quality quantitative research. It enables academics to get from data gathering to practical discoveries. Statistical analysis clarifies what is happening, why it matters, and how we can react by combining well-chosen techniques with contemporary tools and explicit knowledge. Anyone trying to get significant and reliable outcomes must know how to utilize statistics properly as research gets increasingly data-driven. Table 1 provides a clear picture of how statistical analysis benefits academics in healthcare, industry, and education. It demonstrates how statistics may transform raw data into actual insights and guide wise choices. It also highlights typical problems simultaneously, such as employing incorrect techniques or misinterpreting data that could compromise consistent outcomes. Doing a thorough study requires knowledge of both sides.

3 Statistical Analysis Techniques for Quantitative Study

Statistics is the use of several mathematical tools and methods to enable researchers gather, arrange, characterize, and analyze numerical data. These statistical techniques become absolutely necessary in research, notably in quantitative investigations [21]. They provide the basis for forming findings and guiding decisions as well as the measurement and organization of data. Statistical analysis is fundamental to grasping intricate patterns and relationships across many disciplines, including design, engineering, and construction, as well as the data produced by empirical research. Importantly, statistical methods go beyond number management [22]. Used correctly, they enable researchers to draw significant conclusions, assess the strength

Table 1 What statistical analysis brings to quantitative research and the challenges that come with it

Aspect	Description	Challenges
Purpose of statistical analysis	Supports decision-making, improves services and products, ensures quality	Misinterpretation of results due to lack of statistical literacy
Application areas	Widely used in healthcare, engineering, education, and business	Sector-specific data complexities and inconsistent data formats
Contribution to research	Helps detect patterns, test assumptions, and generate insights	Risk of using unsuitable models or drawing causal conclusions from correlational data
Role in digital era	Enhanced by computational tools; enables automation and handling large dataset	Over-reliance on software without understanding underlying assumptions
Key processes	Involves data collection, organization, analysis, and interpretation	Poor data quality or bias in collection can compromise analysis
Quantitative analysis	Analyzes numerical data from surveys and dataset to explain phenomena	Limitations in sample representativeness or low response rates in surveys
Importance of measurement levels	Correct measurement levels ensure valid statistical methods and results	Misclassification of data types can lead to invalid conclusions

of correlations, spot trends, and challenge hypotheses. Often closely related to the study objectives and hypotheses, these insights enable academics to go from raw data to informed conclusions supporting or disputing their theoretical frameworks. Usually, statistical techniques fall into two broad categories: inferential statistics and descriptive statistics [23, 24]. Every category in quantitative research has a unique purpose; knowing the distinction between them is vital for choosing the suitable analytical method. Descriptive statistics offer a means of summarizing and structuring data so that researchers may readily grasp and convey essential characteristics of the dataset. These methods comprise variability, including range, variance, and standard deviation, and measurements of central tendency, including the mean, median, and mode. These techniques allow researchers to show a detailed data picture without overgeneralizing outside the investigated group. On the other hand, inferential statistics let researchers draw informed generalizations or forecasts about a broader group based on data collected from a smaller portion. This results in one of the most important statistical differences: the distinction between populations and samples. Valid and dependable statistical analysis depends on this difference [25]. The whole group of people, events, or things a researcher wants to investigate or draw conclusions about constitutes a population. It reflects the larger group the researcher wants to generalize the results to. In architectural engineering, for example, this could cover all building projects using a given material or design approach, or all professional architectural engineers operating under a particular sector. Researchers usually utilize a sample, a smaller, carefully chosen fraction that reflects the traits of the whole, since studying an entire community is sometimes unrealistic. This sample has to be representative to guarantee the validity of the findings. Should it not be, the study results could be erroneous or deceptive. Choosing this subset, also known as sampling, is especially important since prejudice could undermine the study [26]. Researchers can generalize to the whole population by using inferential techniques applied to the data gathered from the sample. Architectural engineers, for instance, might examine thermal performance data from a sample of green buildings to draw more general conclusions regarding energy efficiency policies. They could also look at structural load reactions in prototype models to project full-scale performance. Descriptive and inferential statistical methods let architecture engineers efficiently compile performance measures, evaluate design ideas, and offer proof supporting improved planning, material utilization, or structural behavior. Statistical analysis is fundamental to generating strong and consistent results in architectural engineering research, whether averaging indoor air quality measures, examining how façade design affects heat gain, or simulating structural stress courses. Figure 2 provides a straightforward and obvious picture of how statistical analysis operates in quantitative research. Starting with data gathering and organization, descriptive and inferential statistics help interpret the figures. These technologies are vital in sectors like design, engineering, and building. Particularly for architectural engineers, they enable the transformation of raw data into valuable insights that direct better design and decision-making.

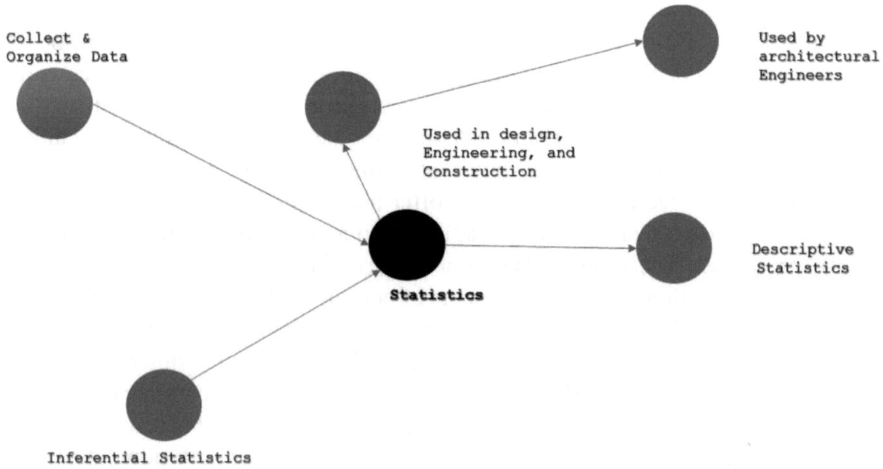

Fig. 2 Simplified overview of statistical analysis in quantitative research (*Source* authors)

3.1 Descriptive Statistics

Descriptive statistics form the backbone of any quantitative research effort. They offer a structured way to make sense of raw data by summarizing it into meaningful figures and patterns that are easy to understand and interpret. Rather than aiming to make predictions or generalizations about a larger population, descriptive statistics are focused on clearly describing what the data shows within the scope of a study [27]. This makes them particularly valuable in the early stages of research, helping to uncover patterns, detect anomalies, and lay the groundwork for more advanced statistical analyses. At the heart of descriptive statistics are two core dimensions: central tendency measures and variability. Measures of central tendency are designed to capture what could be considered "typical" in a dataset. The mean, median, and mode are the most commonly used [28]. The mean gives the average value by dividing the sum of all values by the number of observations. The median represents the middle value when all the data is ordered, offering a less sensitive measure to extreme values. The mode identifies the most frequently occurring value in the dataset, which can help recognize common responses or repeated measurements. However, knowing the central value is not enough; researchers also need to understand how spread out the data is. This is where measures of variability come in [29]. These include the standard deviation, which tells us how much the data points tend to deviate from the mean; the range, which is simply the difference between the highest and lowest values; skewness, which describes whether the distribution is symmetrical or slanted to one side; and kurtosis, which indicates whether the distribution is more flat or peaked compared to a normal distribution. These indicators help researchers gauge the consistency or diversity within the data, and whether there are outliers or unusual patterns warrant further attention. Descriptive statistics are about producing numbers and making data understandable [30]. When used thoughtfully, descriptive statistics enable researchers to quickly grasp the shape and structure of their dataset and present these findings clearly and compellingly to others. This clarity is essential when communicating results to decision-makers, practitioners, or interdisciplinary teams who may not be familiar with statistical jargon. In addition to summarizing individual variables, descriptive analysis can include examining relationships between variables, particularly through visual methods such as scatter plots or cross tabulations [31]. Visualizations such as histograms, bar charts, and frequency polygons offer intuitive insights that are sometimes missed in tables or summary statistics alone. These tools help bring the data to life and make it easier to communicate findings effectively. It is also essential to distinguish between parameters and statistics. A parameter is a measure that describes a characteristic of an entire population, such as the actual average height of all architectural engineers in a country, while a statistic refers to a summary based on a sample taken from that population [32]. Because it's usually impractical or impossible to measure an entire population, researchers rely on well-structured samples and use descriptive statistics to estimate population characteristics cautiously. Another essential task in descriptive analysis is the classification and grouping of raw data. Before meaningful insights can be drawn, data must often be sorted and structured. This includes determining the range of the dataset and deciding how to group values, such as dividing test scores into intervals or categorizing responses into themes. Proper classification improves the clarity of the results and sets the stage for building charts, graphs, and summary tables. According to Agresti and Finlay [33], some of the most widely used techniques in descriptive statistics include: frequency distribution analysis, which shows how often each value or category appears; testing for normality, which assesses whether the data follows a bell-shaped curve using measures like skewness and kurtosis; and percentile calculations, which rank data points within the dataset to show how a particular value compares to others [34]. Also included are the standard measures of central tendency and dispersion and tools for evaluating correlations between variables, helping to identify whether two factors move together, either positively or negatively. Graphical data representation through charts and plots remains one of the most effective ways to summarize and communicate data patterns clearly and intuitively. Descriptive statistics provide researchers with a robust toolkit for understanding and summarizing their data [35]. They are often the first step in any quantitative study, offering the essential foundation needed before moving into more complex inferential techniques. When used well, descriptive statistics transform raw numbers into compelling stories, helping researchers explain what is happening in their data and why it matters. The most often used tools to characterize and interpret data are shown in Table 2. It covers techniques to locate normal values, such as the average or most common number, to observe data dispersion, and to verify whether the data follows a normal pattern. It also includes graphic charts, correlations, and percentiles, among other methods. These instruments, taken together, enable academics to simplify and clarify raw data.

3.1.1 Example: Statistical Analysis in Built Environment Research

Research Topic:

- The impact of building height on indoor air quality (IAQ) in urban residential areas.

Objective:

- To analyze how building height affects indoor air quality (measured by PM2.5 concentration) in urban residential buildings.

1. Frequency Distribution Analysis

A survey was conducted in 50 residential buildings of varying heights (low-rise, mid-rise, and high-rise), measuring PM2.5 concentrations ($\mu g/m^3$) inside each building (Table 3).

Table 2 Key tools for summarizing data with descriptive statistics

Category	Technique	Description
Measures of central tendency	Mean	Average of all values
Measures of central tendency	Mode	Most common value
Measures of central tendency	Median	The middle value is when sorted
Measures of variability	Standard deviation	Shows how much values vary from the average
Measures of variability	Range	Difference between the highest and lowest values
Measures of variability	Skewness	Shows if the data is symmetrical or lopsided
Measures of variability	Kurtosis	Shows if the data is flat or peaked
Core descriptive techniques	Frequency distribution	Shows how often values appear
Core descriptive techniques	Normality test	Checks if data follows a regular pattern
Core descriptive techniques	Percentiles	Shows how data points rank compared to others
Core descriptive techniques	Central tendency summary	Combines mean, median, and mode
Core descriptive techniques	Dispersion summary	Combines range, SD, and quartiles
Core descriptive techniques	Correlation measures	Measures relationship between two variables
Core descriptive techniques	Graphical representation	Uses charts like histograms and line graphs

Table 3 Frequency distribution analysis

PM2.5 range ($\mu g/m^3$)	Frequency	Percentage (%)
0–10	5	10
11–20	8	16
21–30	15	30
31–40	12	24
41–50	10	20

The highest frequency of PM2.5 values falls between 21 and 30 $\mu g/m^3$, indicating that most residential buildings experience moderate pollution levels.

2. Testing for Normality

- To check whether PM2.5 concentrations follow a normal distribution, the Shapiro–Wilk test and skewness and kurtosis values were calculated:
- Shapiro–Wilk Test: $p = 0.08$ ($p > 0.05$, normal distribution assumption holds).

- Skewness: 0.23 (close to 0, indicating symmetry).
- Kurtosis: 2.8 (within the normal range of 2–3).
- Conclusion: PM2.5 levels in buildings follow a roughly normal distribution.

3. Percentile Calculations

- Percentiles help understand how PM2.5 concentrations vary across different buildings.
- 25th Percentile (Q1) = 18 $\mu g/m^3$ (25% of buildings have PM2.5 levels below this value).
- 50th Percentile (Median, Q2) = 27 $\mu g/m^3$ (Half of the buildings have PM2.5 levels below this value).
- 75th Percentile (Q3) = 38 $\mu g/m^3$ (75% of buildings have PM2.5 levels below this value).
- This indicates that most buildings have moderate to high pollution levels.

4. Measures of Central Tendency

- The following measures summarize PM2.5 concentrations across all buildings:
- Mean (Average): 28.6 $\mu g/m^3$.
- Median (Q2): 27 $\mu g/m^3$.
- Mode: 25 $\mu g/m^3$.
- Interpretation: The mean and median are close, indicating a symmetrical distribution of air quality values.

5. Measures of Dispersion

- To understand the spread of PM2.5 levels, the following statistics were calculated:
- Range: 45–5 $\mu g/m^3 = 40\ \mu g/m^3$.
- Standard Deviation: 9.3 $\mu g/m^3$ (variation of PM2.5 across different buildings).
- Interquartile Range (IQR): Q3–Q1 = 38–18 = 20 $\mu g/m^3$ (spread of middle 50% of data).
- Interpretation:
- A higher standard deviation suggests significant variability in indoor air pollution across buildings.
- The IQR of 20 $\mu g/m^3$ shows that PM2.5 concentrations vary considerably among different residential buildings.

6. Correlation and Relationship Measures

- The Pearson correlation coefficient (r) between building height and PM2.5 concentrations was calculated:
- $r = -0.67$ (Moderate negative correlation).
- $p = 0.002$ (Statistically significant at $p < 0.05$).
- Interpretation:

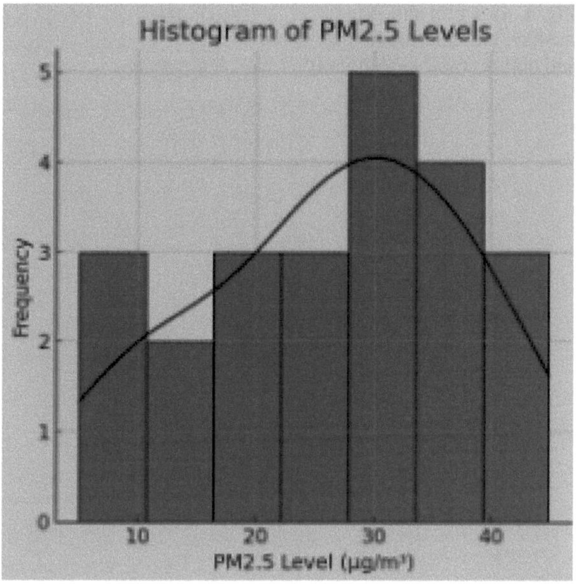

Fig. 3 Histogram showing the frequency of PM2.5 values, with a standard curve overlay (*Source* Author)

- As building height increases, PM2.5 levels decrease (*higher floors have better air quality*).
- The negative correlation suggests that low-rise buildings are more affected by urban pollution than high-rise buildings.

7 Graphical Data Representation

(a) Histogram of PM2.5 Levels.

- A histogram displays the distribution of PM2.5 levels across the sampled buildings (Fig. 3).

(b) Box Plot of PM2.5 by Building Height Category.

- A box plot illustrates how PM2.5 concentrations vary among low-rise, mid-rise, and high-rise buildings (Fig. 4).
- Low-rise buildings: Higher median PM2.5 and more variability.
- Mid-rise buildings: Moderate PM2.5 levels with some outliers.
- High-rise buildings: Lower PM2.5, less variation.

(c) Scatter Plot: Building Height versus PM2.5

- A scatter plot visualizes the relationship between building height and PM2.5 concentrations (Fig. 5).

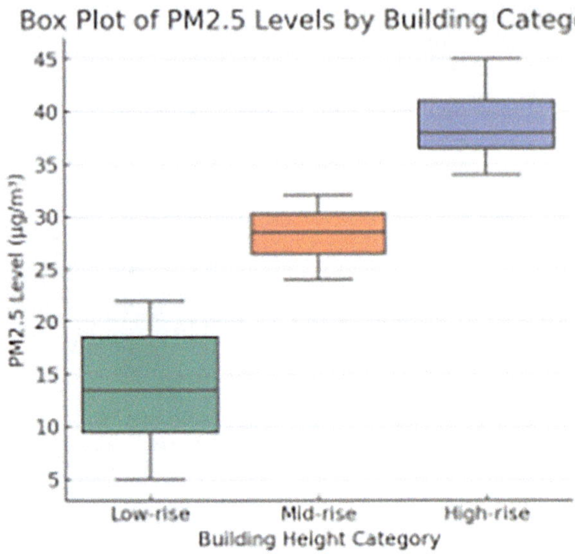

Fig. 4 Box plot showing median, interquartile range, and outliers (*Source* Authors)

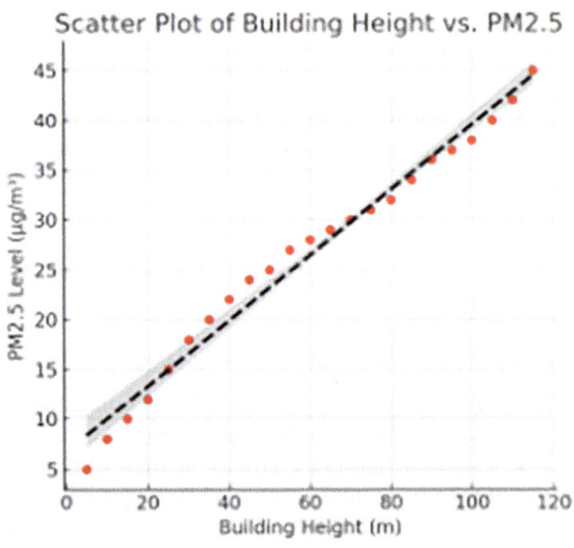

Fig. 5 Scatter plot with a negative trend line indicating decreasing PM2.5 levels with increasing height (*Source* Authors)

3.2 *Inferential Statistics*

Inferential Researchers using inferential statistics may go beyond merely describing the data they have gathered. Rather than restricting analysis to what has been directly seen, inferential techniques let one make conclusions about a broader group, the population, based on data from a smaller, representative portion of the sample [35].

Sampling starts the process. Instead of gathering data from every person or case in the population, researchers operate with a fraction that reflects its essential traits. The statistics derived from this sample, known as statistics, are then applied to forecast unknown parameters of the larger population. These projections are not done blindly; probability theory offers the foundation for researchers to measure their confidence in their results [36]. This mathematical basis guarantees that the insights obtained from sample data are statistically valid and reliable, supporting inferential statistics. According to Howell [37], a statistic is a numerical figure that summarizes some characteristic of the sample and, utilizing well-crafted techniques, supports inferences about the broader population from which the sample was collected. Kothari [38] underlines even more that inferential statistics are about evaluating the dependability of findings rather than only about drawing them. Rather than random fluctuation in the sample, this is accomplished by estimating how likely the findings represent actual population patterns using probability-based methods. Usually, inferential techniques are split into two main categories:

- Estimation is the process of ascertaining the estimated value of a population parameter. This can be shown as a range of values (confidence interval) that probably contains the actual population value or as a single best guess (point estimate) [39]. Confidence levels help to highlight how exact the forecast is and how much uncertainty is included.
- Theory: Based on the sample data, testing is performed to assess if a particular assumption about a population holds [40]. Setting up a null hypothesis, a default assumption that no impact or difference exists, starts this process. Researchers then use statistical tests to decide if the data provides sufficient evidence to overturn this assumption in favor of an alternate explanation.

Considering uncertainty, these two domains provide the basis of inferential statistical practice, enabling researchers to draw data-driven conclusions and logical decisions [41]. Instead of just stating what the data reveals, inferential statistics ask, "What does this tell us about the larger picture?" How certain can we be, then?

Across fields, inferential statistics help make essential choices and create thorough, evidence-based studies [21]. Always honest about the boundaries of what can be asserted, they ensure that results are pertinent to the particular study population and have significant insights that may apply more generally.

3.2.1 Confidence Intervals (Interval Estimation)

A Fundamental to inferential statistics, a confidence interval (CI) offers a range of values within which an unknown population characteristic, such as a mean or proportion, is likely to fall [42]. A confidence interval shows an upper and lower boundary rather than a single estimate, providing researchers a better understanding of the accuracy and uncertainty of their estimates. Data from a sample helps to compute this interval. Tied to a particular confidence level, it indicates how probable the interval is to hold the actual population value. Common confidence levels include 90, 95, and 99%. A 95% confidence interval, for example, indicates that about 95% of the intervals produced would include the actual population parameter if the same study were run several times with different random samples [43]. Confidence intervals are a sensible approach given the variability from operating with samples instead of complete populations. The obtained statistics could vary somewhat from one study to another since no two samples are precisely alike. The confidence interval provides a statistically based range indicating the possible error in researchers' estimates, compensating for this sample variability [44]. Several elements, including the sample size, the data variability, and the selected confidence level, determine the breadth of a confidence interval. Larger sample sizes usually result in tighter intervals, suggesting more exact projections. On the other hand, high data variability or a higher confidence level might lead to broader intervals, therefore indicating more uncertainty. A confidence interval does not indicate a 95% probability that the actual parameter falls within any specific range [45]. Instead, it shows a long-term frequency interpretation: 95% of all similarly built intervals will have the precise value. This difference emphasizes the probabilistic character of statistical inference and advises researchers to read confidence ranges carefully. Confidence intervals are beneficial because they provide more information than a straightforward point estimate. Although a single estimate provides the best guess, the interval helps readers know how much that estimate might fairly change. In academic research and pragmatic decision-making, such as policy assessment, clinical trials, engineering performance analysis, and quality control, CIs are therefore quite effective [46]. Confidence intervals connect sample data to population-level conclusions. Researchers who offer a clear and understandable estimate of uncertainty can report results with a fair perspective of confidence and caution.

3.2.2 Hypothesis Testing (Tests of Significance)

Hypothesis Among the most often applied techniques in inferential statistics is hypothesis testing. Using data gathered from a sample, hypothesis testing offers a methodical approach for researchers to examine assumptions or assertions about a population [47]. Hypothesis testing employs a clear and logical method to decide if the evidence supports or contradicts a specific idea rather than depending on guessing or subjective judgment. The process starts with two opposing claims defined:

a. The Null Hypothesis (H_0) is the beginning assumption. It usually implies no effect, difference, or relationship in the population [48]. It is the default stance researchers want to test against.
b. The Alternative Hypothesis (H\u2081) is what the researcher believes or expects to show that there is an effect, difference, or relationship. Hypothesis testing assesses whether the data offer strong enough evidence to reject the null hypothesis in favor of the alternative.

Once the hypotheses are articulated, researchers choose and use a suitable statistical test, such as a *t*-test, chi-square test, or ANOVA, depending on the kind of data and research topic [49]. These tests investigate whether the patterns observed in the sample data would probably have occurred by random chance if the null hypothesis were true. Assuming the null hypothesis is accurate, the *p*-value, which indicates the likelihood of seeing outcomes at least as extreme as those discovered in the sample, helps us to comprehend this process. The *p*-value answers the question: How rare is this outcome if nothing truly happens in the population?

A *p*-value below 0.05 indicates statistically significant findings and rejection of the null hypothesis. This implies that the observed pattern is unlikely to be caused by mere chance [50]. Should the *p*-value be 0.05 or above, we keep assuming the null hypothesis could still be true since there is insufficient evidence to reject it boldly. But, one should keep in mind that hypothesis testing merely enables researchers to make a reasoned conclusion depending on the available data; it does not show whether a hypothesis is true or incorrect. Every statistical result is uncertain; hypothesis testing admits this by applying a significance level (\u03b1) usually set at 0.05 or 0.01 [51]. This score indicates the likelihood of a Type I error, or rejecting the null hypothesis when it is true, resulting in a false positive. Hypothesis testing is, in reality, a decision-making framework rather than just a statistical instrument. It lets researchers from all fields, including doctors, engineers, scientists, and others, assess data and make educated decisions about their jobs. Hypothesis testing offers a consistent approach to distinguishing between actual effects and random variation, whether evaluating a new design method, comparing treatments, or studying behavioral trends. Ultimately, hypothesis testing transforms uncertainty into an organized assessment. By following a logical sequence from articulating the hypotheses to evaluating the data and interpreting the findings, researchers can reach conclusions that are not only data-driven but also statistically sound.

3.2.3 Methods of Statistical Inference

Inferential Researchers draw a bigger population via statistical inference based on data gathered from a sample. It lets us, all the while recognizing the natural uncertainty of operating with incomplete knowledge, make evidence-based decisions, informed forecasts, and logical conclusions [49]. Point estimation, interval estimation, and hypothesis testing are three basic approaches on which inferential statistics depends to perform this correctly. Point Estimation determines a single, best-guess value for an unknown population parameter. A researcher, for instance, might project the population's average value using the sample mean, or average value from a sample [52]. Although this offers a simple and obvious outcome, it does not reflect the degree of uncertainty around that projection. Point estimates are therefore frequently accompanied by confidence intervals. Interval estimating extends point estimating by adding a range of reasonable values into which the population parameter will probably fall. This is called a confidence interval. Interval estimation conveys confidence, such as 95%, that the population parameter resides within the computed range rather than recommending a single fixed value. This approach increases the estimate's interpretability and dependability and offers a more accurate view of uncertainty. Hypothesis Testing is a methodical approach to assess whether observed data supports or contradicts a particular claim or assumption about the population. Researchers do statistical tests to evaluate if the evidence is strong enough to reject the null hypothesis by establishing a null hypothesis (usually saying no effect or difference exists) and an alternative hypothesis (showing a significant impact or difference). This approach helps confirm research results or make decisions. These three approaches combined constitute the foundation of statistical inference. Each has a unique but complementary function: point estimation provides a precise estimate, interval estimation expresses the confidence around that estimate, and hypothesis testing assesses if a specific assertion withstands examination. Their basis in probability theory links all these methods. Probability helps us measure how probable it is that our judgments, based on limited evidence, reflect actual population tendencies. It lets researchers avoid overgeneralization and consider random variance. Grounding inference in probability helps statisticians guarantee that the insights obtained from sample data are mathematically valid and generalizable with obviously specified margins of error and confidence [53]. These inferential methods are crucial for thorough, transparent, and dependable research across disciplines, engineering, health sciences, social research, environmental studies, and beyond. Statistical inference techniques offer a strong basis for transforming raw data into meaningful conclusions that can guide theory and practice, whether one is estimating average outcomes, identifying feasible ranges, or assessing the validity of a hypothesis.

3.2.4 Example: Statistical Analysis in Built Environment Research

Research Topic:

- *The impact of green roofs on indoor temperature reduction in urban residential buildings*

Objective:

- Using confidence intervals, hypothesis testing, and statistical inference techniques, to investigate whether green roofs significantly lower indoor temperature in urban residential structures relative to conventional roofs.

i. *Confidence Interval (Interval Estimation)*

  - Forty residential buildings were studied, twenty with green roofs and twenty with conventional roofs. Recorded across summer months, the average indoor temperature was as follows: data displayed in Table 4.
  - To determine the correct mean indoor temperature for each group [54], we compute a 95% confidence interval (CI) using the formula below:

Table 4 The average indoor temperature over summer months

Roof type	Sample size (n)	Mean indoor temperature (°C)	Standard deviation (°C)
Green roofs	20	24.8	1.5
Conventional roofs	20	27.2	1.8

$$CI = \overline{X} \pm Z_{\alpha/2} \times \frac{\sigma}{\sqrt{n}}$$

where:

$\overline{X}$ = Sample Mean

$Z_{\alpha/2}$ = Critical Z-value for the desired confidence level (e.g., 1.96 for 95% confidence)

σ = Standard Deviation

n = Sample Size

Confidence Interval Calculations

For Green Roofs:

$$CI = 24.8 \pm (1.96 \times \frac{1.5}{\sqrt{20}})$$

$$CI = 24.8 \pm (1.96 \times 0.335)$$

$$CI = 24.8 \pm 0.66$$

$$95\% CI = [24.14, 25.46]$$

For Conventional Roofs:

$$CI = 27.2 \pm (1.96 \times \frac{1.8}{\sqrt{20}})$$

$$CI = 27.2 \pm (1.96 \times 0.402)$$

$$CI = 27.2 \pm 0.79$$

$$95\% CI = [26.41, 27.99]$$

Interpretation:

- Buildings with green roofs have an estimated mean inside temperature between 24.14 and 25.46 °C [55].
- Buildings with conventional roofs have an estimated true mean inside temperature between 26.41 and 27.99 °C [55].
- The non-overlapping confidence intervals suggest a probable significant difference between the two groups.

ii. *Hypothesis Testing (Tests of Significance)*

We use a two-sample *t*-test to determine whether the difference in mean indoor temperature between green roofs and conventional roofs is statistically significant.

Hypothesis

- Null Hypothesis (H_a): There is no significant change in interior temperature between green roofs and conventional roofs ($\mu_1 = \mu_2$).
- Green roofs, according to the alternative hypothesis (H_1), have considerably lower indoor temperature as compared to conventional roofs ($\mu_1 < \mu_2$) [56].

The test statistic (t) is computed using a two-tailed independent *t*-test as follows:

$$t = \frac{\overline{X}_1 - \overline{X}_2}{\sqrt{\frac{s_1^2}{n_1} + \frac{s_2^2}{n_2}}}$$

where:

$\overline{X}_1$ = Mean for Green Roofs

$\overline{X}_2$ = Mean for Conventional Roofs

s_1, s_2 = Standard Deviations

n_1, n_2 = Sample Sizes

T-Test Calculation

$$t = \frac{24.8 - 27.2}{\sqrt{\frac{1.5^2}{20} + \frac{1.8^2}{20}}}$$

$$t = \frac{-2.4}{\sqrt{\frac{2.25}{20} + \frac{3.24}{20}}}$$

$$t = \frac{-2.4}{\sqrt{0.1125 + 0.162}}$$

$$t = \frac{-2.4}{\sqrt{0.2745}}$$

$$t = \frac{-2.4}{0.524}$$

$$t = -4.58$$

Using a t-distribution table with 38 degrees of freedom ($n_1 + n_2 - 2 = 40 - 2$) and a significance level of 0.05, the critical t-value is $\pm$ 2.02.

Since $|t| = 4.58 > 2.02$, we reject the null hypothesis (H_0) and conclude that green roofs significantly reduce indoor temperature.

P-Value Interpretation

The p-value associated with $t = -4.58$ is $p < 0.001$, meaning the likelihood of obtaining these results by chance is less than 0.1%.

Conclusion: Green roofs have a statistically significant impact in reducing indoor temperatures.

iii. *Methods of Statistical Inference*

This study demonstrates the three key methods of inferential statistics:

i. Point Estimation

- The sample means (24.8 °C for green roofs and 27.2 °C for conventional roofs) are used as point estimates of the actual population means.

ii. Interval Estimation (Confidence Intervals)

- The 95% confidence intervals for each roof type were computed, showing that indoor temperatures in buildings with green roofs are significantly lower.

iii. Hypothesis Testing

- A two-sample t-test confirmed that the temperature reduction in buildings with green roofs is statistically significant, rejecting the null hypothesis.

This study backs up the finding that green roofs considerably lower indoor temperatures in urban residential structures. With confidence intervals and hypothesis testing verifying that green roofs have a quantifiable cooling impact, the results are statistically significant. These findings could guide energy-efficient building design, sustainable architecture, and urban planning.

3.3 Analysis of Variance (ANOVA): Concept, Assumptions, and Applications

3.3.1 Concept of ANOVA

Analysis of Variance (ANOVA) is a widely used statistical method for analyzing quantitative data, particularly for comparing differences among multiple groups. Professor Fisher first introduced the concept of variance, developing the theoretical foundations of ANOVA and demonstrating its practical applications [57]. Subsequent contributions from Professor Snedecor and other statisticians further refined the technique.

ANOVA is primarily used to test the homogeneity of means across different groups by partitioning total variation in the data into two components:

i. Variation due to chance (within-group variability)
ii. Variation due to specific causes (between-group variability) [58].

This method allows researchers to determine whether observed differences among group means are statistically significant or result from random variation. One of the most commonly applied forms is one-way ANOVA, which decomposes variance to assess the effect of a single independent variable on a dependent variable [59].

3.3.2 Assumptions of ANOVA

Several assumptions must be met for ANOVA results to be statistically valid. While ANOVA is considered robust to mild violations of these assumptions, strict adherence improves the reliability of conclusions [58–60].

Key Assumptions:

i. Random Sampling: Subjects must be selected through a simple random sampling process to ensure a representative sample [61].
ii. Normality: The dependent variable should be normally distributed within each group. This can be verified using standard quantile plots, where data points should align closely to a straight line.
iii. Homogeneity of Variance: The variance of the dependent variable should be approximately equal across all groups. This can be tested by calculating the largest to smallest sample standard deviation ratio, which should not exceed 2:1.
iv. Independence of Observations: Observations within and across groups must be independent.

Violations of these assumptions can lead to inaccurate statistical inferences, requiring adjustments such as data transformation or nonparametric alternatives.

3.3.3 Applications of ANOVA

ANOVA is applicable in experimental research where more than two groups are compared, extending beyond the ***t*-test**, which is limited to two-group comparisons.

One-Way ANOVA

One-way ANOVA is used when there is one independent variable (categorical) with three or more groups and a dependent variable (continuous, interval, or ratio level). The test produces an F-ratio, determining whether differences among means are statistically significant [62].

Two-Way ANOVA

Two-way ANOVA extends the analysis by including two independent variables, allowing researchers to examine:

- The main effect of each independent variable.
- The interaction effect between the two independent variables.

Assumptions for Two-Way ANOVA:

- The population distribution should be approximately normal.
- The samples must be independent.
- Population variances should be equal across groups.
- Sample sizes should be equal to maintain statistical accuracy.

Post Hoc Testing

While ANOVA determines whether a significant difference exists, it does not specify which groups differ. Post hoc tests such as Tukey's HSD or Bonferroni correction are used to identify pairwise differences between group means.

3.3.4 Example: Application of ANOVA in Built Environment Research

Research Topic:

- *The effect of building materials on indoor temperature in residential housing*

Objective:

- To analyze whether different building materials (Concrete, Brick, and Wood) significantly influence indoor temperature regulation in residential buildings using one-way ANOVA, two-way ANOVA, and post hoc testing.

Table 5 Normality test

Building material	Shapiro–Wilk *p*-value	Normality decision
Concrete	0.08	Normal ($p > 0.05$)
Brick	0.12	Normal ($p > 0.05$)
Wood	0.06	Normal ($p > 0.05$)

Table 6 Levene's test for equality of variances

Levene's test *p*-value	Decision
0.18	Homogeneous ($p > 0.05$)

Data Collection: Random Sampling

A random sample of 45 residential buildings was selected, with 15 houses per building material type (Concrete, Brick, and Wood).

- The houses were randomly selected from three urban areas, ensuring a representative sample.
- Indoor temperature was measured during peak summer hours (12:00–16:00) over two weeks.

Checking ANOVA Assumptions

- Before conducting ANOVA, we verify that the key statistical assumptions are met.

Normality Test

- To check whether indoor temperature follows a normal distribution, the Shapiro–Wilk test and a normality plot were used as shown in Table 5.

Since *p*-values are all greater than 0.05, the normality assumption holds for all groups.

Homogeneity of Variance Test

To verify that the variance of indoor temperature is similar across materials, we use Levene's test for equality of variances as shown in Table 6.

H_0: All groups have equal variances

H_1: At least one group has a different variance

Since $p > 0.05$, we fail to reject H_0 meaning the homogeneity of variance assumption holds.

Table 7 Data summary (mean indoor temperature in °C)

Building material	Mean temperature (°C)	Standard deviation (°C)	Sample size (n)
Concrete	28.5	1.2	15
Brick	27.3	1.5	15
Wood	25.8	1.1	15

Table 8 ANOVA calculation

Source of variation	Sum of squares (SS)	Degrees of freedom (df)	Mean square (MS)	F-ratio	*p*-value
Between groups	34.2	2	17.1	8.92	0.0012
Within groups	57.1	42	1.36	–	–
Total	91.3	44	–	–	–

Independence of Observations

- Each house was measured independently, ensuring no overlap in measurements.
- Randomization of sample selection minimizes bias and maintains independence.

One-Way ANOVA: Testing the Effect of Building Material on Indoor Temperature

Hypotheses

- Null Hypothesis (H_0): *There is no significant difference in indoor temperature among the three building materials.*
- Alternative Hypothesis (H_1): *At least one building material leads to a significantly different indoor temperature.*

Data Summary (Mean Indoor Temperature in °C)

Data summary for mean indoor temperature in °C is shown in Table 7.

ANOVA Calculation

$$F = \frac{\text{Between} - \text{group variability}}{\text{Within} - \text{group variability}}$$

ANOVA calculation is shown in Table 8 [62].

Since $p = 0.0012$ (<0.05), we reject H_0 and conclude that building material significantly affects indoor temperature.

Table 9 Tukey's HSD test

Comparison	Mean difference (°C)	p-value	Significance
Concrete versus brick	1.2	0.08	Not significant
Concrete versus wood	2.7	0.002	Significant
Brick versus wood	1.5	0.04	Significant

Post Hoc Testing: Identifying Differences Among Groups

Since ANOVA indicates a significant effect, we conduct Tukey's HSD test to determine which building materials differ, as shown in Table 9

Conclusion:

Concrete and Wood show a significant difference in indoor temperature.

Brick and Wood also show a significant difference.

Concrete and Brick do not differ significantly.

Two-Way ANOVA: Effect of Building Material and Roof Type on Indoor Temperature

A Two-Way ANOVA is performed to analyze the combined effect of building material and roof type (Flat vs. Sloped).

Hypotheses

Main Effect 1 (Building material): *Building material significantly affects indoor temperature.*

Main Effect 2 (Roof type): *Roof type significantly affects indoor temperature.*

Interaction Effect: *Building material and roof type interact to influence indoor temperature.*

Two-Way ANOVA Results

Two-way ANOVA results are shown in Table 10.

Interpretation

Building material significantly affects indoor temperature ($p = 0.001$).

Roof type also significantly affects indoor temperature ($p = 0.004$).

There is no significant interaction between building material and roof type ($p = 0.11$).

Table 10 Two-way ANOVA results

Source	Sum of squares (SS)	Df	Mean square (MS)	F-ratio	*p*-value
Building Material	30.2	2	15.1	9.10	0.001
Roof Type	18.4	1	18.4	8.50	0.004
Interaction	6.2	2	3.1	2.15	0.11
Error	50.1	40	1.25	–	–

Conclusion

Key Findings:

1. One-way ANOVA: Building material significantly affects indoor temperature.
2. Post Hoc Testing:
    - Wood reduces indoor temperature more than Concrete or Brick.
    - No significant difference between Concrete and Brick.
3. Two-way ANOVA:
    - Both building material and roof type independently influence indoor temperature.
    - No significant interaction between the two variables.

Implications for Built Environment Research:

- Wood-based structures improve thermal comfort compared to concrete or brick buildings.
- Sloped roofs help in better heat dissipation during summer.
- Future research should explore the long-term energy efficiency of different materials and roof types.
- This example illustrates the practical application of ANOVA in built environment research, demonstrating its role in evaluating material efficiency, optimizing building design, and informing sustainable construction practices.

3.4 Regression Analysis: Concepts, Assumptions, and Applications

3.4.1 Introduction to Regression Analysis

Regression analysis is a statistical technique used to identify relationships between variables and predict outcomes based on observed data. It provides a mathematical equation that models these relationships, enabling researchers to make data-driven predictions [63]. In addition to predicting values, regression analysis generates various statistical measures, including the p-value (which tests statistical significance) and the correlation coefficient (which quantifies the strength of the relationship).

Basic regression techniques include scatter plots and linear regression. In contrast, more advanced methods, such as multiple regression, are used when analyzing the effect of numerous predictor variables on a single outcome [61].

Definition and Purpose

Regression analysis investigates the predictive power of independent variables on a dependent variable. It allows researchers to quantify how much changes in an independent variable influence the dependent variable, assuming all other factors remain constant [64]. This technique is fundamental in quantitative research, as it identifies which independent variables significantly affect the dependent variable and explores the nature of these relationships, which may suggest causal inferences.

3.4.2 Types of Regression Analysis

Simple Regression

Simple regression involves one independent variable (predictor) and one dependent variable (outcome). The relationship is modeled using the linear equation [64]:

$$Y = a + bX + \epsilon$$

where:

Y = Dependent variable

X = Independent variable

a = Intercept

b = Regression coefficient (slope)

ϵ = Error term

Multiple Regression

Multiple Regression involves two or more independent variables predicting a single dependent variable [64].

The general form of the regression equation is:

$$Y = a + b_1X_1 + b_2X_2 + \ldots + b_nX_n + \epsilon$$

where each X represents an independent variable and b_n represents its corresponding coefficient.

Multiple regression assesses the combined effect of several predictors, controlling for confounding factors and increasing the model's explanatory power [58].

Assumptions of Regression Analysis

For regression analysis to yield valid and reliable results, several assumptions must be met [59]:

Linearity

The relationship between independent variables (IVs) and the dependent variable (DV) should be linear.

Checked using scatter plots to ensure data points form a straight-line pattern.

No Multicollinearity

Multicollinearity occurs when independent variables are highly correlated, leading to unreliable coefficient estimates.

Tested using the Variance Inflation Factor (VIF); a VIF above 10 suggests problematic multicollinearity.

Independence of Residuals

The residuals (errors) should be independent of each other.

Tested using the Durbin-Watson statistics, where values between 1.5 and 2.5 indicate acceptable independence.

Homoscedasticity (Constant Variance of Residuals)

The variance of residuals should be consistent across all levels of the independent variable.

Checked by plotting standardized predicted values against standardized residuals.

If residuals fan out or form a pattern, the assumption is violated.

Normality of Residuals

Residuals should follow a normal distribution, which ensures reliable confidence intervals and hypothesis tests.

Assessed using histograms, Q-Q plots, and the Shapiro–Wilk test.

No Influential Outliers

Outliers or influential cases should not unduly affect the regression model.

Detected using Cook's Distance, where values above 1 indicate problematic influence.

Failure to meet these assumptions may lead to biased or inefficient estimates, requiring data transformations or alternative regression methods.

Applications of Regression Analysis

Regression analysis is widely used in various scientific fields to model relationships and predict future outcomes.

Example: Built Environment Research

Research Topic: The Impact of Building Height and Window-to-Wall Ratio (WWR) on Indoor Temperature.

Objective:

To determine how building height (m) and window-to-wall ratio (%) influence indoor temperature (°C) in urban residential buildings.

Data Collection (Random Sampling)

A random sample of 50 residential buildings was selected, measuring:

- Building height (m)—Independent Variable (IV1)
- Window-to-wall ratio (%)—Independent Variable (IV2)
- Indoor temperature (°C)—Dependent Variable (DV).

Regression Model Implementation

A multiple linear regression model was constructed:

$$\textit{Indoor Temperature} = a + b_1(\textit{Building Height}) + b_2(\textit{WWR}) + \epsilon$$

The regression output is shown in Table 11.

Interpretation:

Building Height: For every additional meter in building height, indoor temperature decreases by 0.08 °C ($p = 0.001$).

WWR: For every 1% increase in window-to-wall ratio, indoor temperature increases by 0.15 °C ($p = 0.007$).

Table 11 Regression output

Variable	Coefficient (b)	Standard error	p-value	Significance
Intercept (aa)	30.5	1.2	0.002	Significant
Building height	−0.08	0.02	0.001	Significant
Window-to-wall ratio	0.15	0.04	0.007	Significant

Table 12 Evaluation of assumptions for regression model

Assumption	Test applied	Result	Decision
Linearity	Scatter plot	Linear relationship	Assumption holds
No multicollinearity	VIF	VIF < 2.0 for all IVs	Assumption holds
Independence of residuals	Durbin-Watson test	1.87 (acceptable)	Assumption holds
Homoscedasticity	Residuals versus. predicted	No fan-shaped pattern	Assumption holds
Normality of residuals	Q-Q plot and Shapiro–Wilk	$p = 0.12$ (normal)	Assumption holds
No influential outliers	Cook's distance	Max value < 1.0	Assumption holds

Both predictors are statistically significant ($p < 0.05$), confirming their impact on indoor temperature regulation.

Evaluating Assumptions for Regression Model

Evaluation of assumptions for regression model is shown in Table 12.

All assumptions are satisfied, confirming the validity of the regression model.

Conclusion and Implications

Building height negatively correlates with indoor temperature, indicating that taller buildings experience lower indoor temperatures.

Higher window-to-wall ratios increase indoor temperatures, suggesting a trade-off between natural lighting and thermal performance. The regression model effectively predicts indoor temperature variations, aiding architects and urban planners in climate-responsive building design.

3.5 *Correlation Analysis: Concepts, Assumptions, and Applications*

3.5.1 Definition and Purpose of Correlation Analysis

Correlation analysis is a statistical technique used to measure the strength and direction of an association between two variables. Unlike regression, correlation does not establish cause-and-effect relationships but identifies the degree to which two variables co-vary [61]. The correlation coefficient quantifies this relationship on a scale from -1 to $+1$, where:

$+1$ indicates a perfect positive correlation (both variables increase together).
0 indicates no correlation (no linear relationship between variables).
-1 indicates a perfect negative correlation (one variable increases while the other decreases).

Correlation analysis is widely applied in research to determine whether and how strongly two variables are related [61].

3.5.2 Types of Correlation Coefficients

Several correlation coefficients exist, with the two most commonly used being.

Pearson's Product-Moment Correlation Coefficient (R)

Measures the strength and direction of a linear relationship between two continuous variables (interval or ratio scale).

Assumes data normality, linearity, and homoscedasticity [58].

Formula for Pearson's *r*:

$$r = \frac{\sum(X_i - \overline{X})(Y_i - \overline{Y})}{\sqrt{\sum(X_i - \overline{X})^2}.\sqrt{\sum(Y_i - \overline{Y})^2}.}$$

where X and Y are the two variables, and $\overline{X}$ and $\overline{Y}$ There are their respective means.

Spearman's Rank Correlation Coefficient (P or Rho)

A non-parametric test that measures the strength and direction of a monotonic relationship between two variables.

Used when data are ordinal or do not meet the assumptions of normality and linearity [26].

Formula for Spearman's ρ:

$$\rho = \frac{6\sum d_i^2}{n(n^2 - 1)}$$

d_i is the difference between paired ranks, and n is the number of observations.

3.5.3 Assumptions of Correlation Analysis

Assumptions for Pearson's Correlation

i. Linearity: The relationship between variables must be linear, meaning data points should form a straight-line pattern in a scatter plot.
ii. Normality: Both variables should be normally distributed (assessed using histograms or the Shapiro–Wilk test).
iii. Homoscedasticity: The variance of residuals should remain constant across all independent variable values.

Assumptions for Spearman's Correlation

Ordinal or Continuous Data: At least one variable must be ordinal, interval, or ratio level. Monotonic Relationship: Variables must move consistently increasing or decreasing, even if not linearly. No Normality Requirement: Unlike Pearson's r, Spearman's ρ does not assume normality and is more robust to outliers.

3.5.4 Interpretation of Correlation Coefficients

Interpretation of correlation coefficients is shown in Table 13.

Positive versus Negative Correlation

Positive correlation ($+r$ or $+\rho$): As one variable increases, the other also increases.

Negative correlation ($-r$ or $-\rho$): As one variable increases, the other decreases (inverse relationship).

Table 13 Interpretation of correlation coefficients

Correlation coefficient (r or ρ)	Strength of relationship
±0.00–0.10	No or very weak correlation
±0.11–0.30	Weak correlation
±0.31–0.50	Moderate correlation
±0.51–0.70	Strong correlation
±0.71–1.00	Robust correlation

3.5.5 Statistical Significance of Correlation

The statistical significance of a correlation coefficient is tested using a *p*-value:

$p < 0.05 \rightarrow$ The correlation is statistically significant, unlikely to have occurred by chance.
$p \geq 0.05 \rightarrow$ The correlation is insignificant, meaning the relationship could be due to random variation.

The standard error of a correlation coefficient is used to calculate confidence intervals, ensuring that the observed correlation is not an anomaly.

3.5.6 Example: Correlation Analysis in Built Environment Research

Research Topic:

The Relationship Between Building Height and Wind Speed in Urban Environments

Objective:

To determine whether building height (m) is correlated with average wind speed (m/s) in an urban area.

Data Collection:

- A sample of 30 buildings was randomly selected, measuring:
- Building Height (m)—Continuous independent variable
- Wind Speed at Rooftop Level (m/s)—Continuous dependent variable.

3.5.7 Pearson's Correlation Analysis

Hypotheses

Null Hypothesis (H_0): *There is no significant correlation between building height and wind speed.*

Alternative Hypothesis (H_1): *There is a significant correlation between building height and wind speed.*

Pearson's Correlation Output

Pearson's correlation output is shown in Table 14.

Table 14 Pearson's correlation output

Variable	Mean	Standard deviation
Building height (m)	50	12.5
Wind speed (m/s)	4.2	1.8
Pearson's r	0.76	–
p-value	0.002	–

Table 15 Spearman's rank correlation output

Spearman's ρ	p-value
0.72	0.004

Interpretation

The correlation coefficient $r = 0.76$ indicates a strong positive correlation.

The correlation is statistically significant since $p = 0.002$ (<0.05).

This suggests that taller buildings experience higher wind speeds at rooftop levels.

3.5.8 Spearman's Correlation Analysis

If the dataset contains ordinal building height categories (e.g., Low-Rise, Mid-Rise, High-Rise), Spearman's rank correlation would be more appropriate.

Spearman's Rank Correlation Output

Spearman's rank correlation output is shown in Table 15.

Since $\rho = 0.72$ and $\rho < 0.05$, a statistically significant positive correlation exists between building height category and wind speed.

3.5.9 Conclusion and Implications

Key Findings: Pearson's correlation analysis showed a strong, significant positive correlation between building height and wind speed ($r = 0.76$, $\rho = 0.002$). Spearman's correlation analysis confirmed that the correlation holds even when height is categorized ($\rho = 0.72, \rho = 0.004$).

i. These findings suggest urban planners should consider wind effects when designing tall buildings to minimize potential wind-related issues.

Practical Implications:

- Architects should integrate aerodynamic building designs to mitigate high rooftop wind speeds.
- Urban designers can implement wind barriers or green infrastructure to regulate airflow around tall buildings.

3.6 Multiple Correlation and Regression: Concepts, Assumptions, and Applications

3.6.1 Definition and Purpose of Multiple Correlation and Regression

Multiple Correlation

Multiple correlation measures the strength of association between one dependent variable and two or more independent variables. Unlike simple correlation, multiple correlation considers multiple predictors simultaneously to assess their combined effect on the dependent variable.

Multiple Regression

Multiple regression extends simple regression by incorporating two or more independent variables to predict a single dependent variable. The multiple regression equation takes the form [64]:

$$Y = a + b_1X_1 + b_2X_2 + \dots + b_mX_{nM} + \epsilon$$

where:

Y = Dependent variable (outcome)

$X_1, X_2, \dots, X_m$ = Independent variables (predictors)

a = Intercept (constant)

$b_1, b_2, \dots, b_m$ = Regression coefficients

ϵ = Error term

Regression analysis is not purely associative but also predictive, allowing researchers to estimate outcomes based on given values of independent variables [64].

3.6.2 Assumptions of Multiple Regression Analysis

For multiple regression to produce reliable results, certain assumptions must be met [65].

Sample Size

A minimum sample size is required for reliable regression estimates.

A commonly used formula:

104 + m104 + m (where mm is the number of independent variables)
50 + 8m50 + 8m (alternative formula)

Example: If 4 independent variables are used, the required sample size is

$$\max(104 + 4, 50 + 8 \times 4) = \max(108, 82) = 108.$$

Multicollinearity of Independent Variables

Multicollinearity occurs when two or more highly correlated independent variables ($r > 0.9$).

This can distort regression coefficients, making it difficult to assess the unique contribution of each predictor.

Tolerance values < 0.1 indicate problematic multicollinearity and should be addressed by removing highly correlated predictors.

Linearity

Multiple regression assumes a linear relationship between independent and dependent variables.

Verified using scatter plots; nonlinear relationships may require data transformation or polynomial regression.

Absence of Outliers

Outliers can skew regression estimates and reduce the model's accuracy.

Identified using:

- Scatter plots
- Leverage statistics
- Cook's Distance (values > 1 indicate influential points).

Homoscedasticity

The variance of residuals (errors) should be constant across all values of independent variables.

If variance increases or decreases systematically, heteroscedasticity is present, which can be identified using residual plots.

Normality

Both independent and dependent variables should be normally distributed.

Tested using:

- Histograms
- Q-Q plots
- Shapiro–Wilk test.

Violations of these assumptions may require data transformation or robust regression techniques.

3.6.3 Types of Multiple Regression

Multiple regression can be conducted using different approaches depending on the research objective:

Standard Multiple Regression

All independent variables are entered simultaneously into the model.

Used when there is no theoretical basis for prioritizing variables.

Hierarchical Multiple Regression

Independent variables are entered in a specific order, based on theoretical justifications.

Allows researchers to evaluate the incremental effect of adding new predictors.

Stepwise Multiple Regression

The statistical software selects independent variables automatically based on statistical significance.

This approach is data-driven but may lead to overfitting if not used carefully.

Table 16 Regression output and interpretation

Variable	Coefficient (b)	Standard error	p-value	Significance
Intercept (a)	30.5	1.2	0.002	Significant
Building height (X1)	−0.08	0.02	0.001	Significant
Window-to-wall ratio (X2)	0.15	0.04	0.007	Significant

Example: Multiple Regression in Built Environment Research

Research Topic:

The impact of building height and window-to-wall ratio (WWR) on indoor air temperature in urban residential buildings

Objective:

- To determine how building height (m) and window-to-wall ratio (%) influence indoor air temperature (°C).

Data Collection (Random Sample of 50 Buildings)

- Building height (X_1, in meters)
- Window-to-wall ratio (X_2, in percentage)
- Indoor temperature (Y, in °C).

Regression Model:

$$\text{Indoor Temperature} = a + b_1(\text{Building Height}) + b_1(\text{WWR}) + \epsilon$$

Regression Output and Interpretation

Regression Output and Interpretation are shown in Table 16.

Interpretation:

- Building Height (X_1): *For every additional meter, indoor temperature decreases by 0.08 °C ($p = 0.001$).*
- Window-to-Wall Ratio (X_2): *For every 1% increase in WWR, indoor temperature increases by 0.15 °C ($p = 0.007$).*
- Both predictors are statistically significant ($p < 0.05$).

Assumption Testing for Regression Model

Assumption testing for the regression model is shown in Table 17.

Since all assumptions are satisfied, the regression model is valid and reliable.

Table 17 Assumption testing for regression model

Assumption	Test applied	Result	Decision
Sample size	Tabachnick and fidell rule	$104 + 2 = 106$	Assumption met
Multicollinearity	VIF < 2.0	No high correlations	Assumption met
Linearity	Scatter plots	Linear relationship	Assumption met
Absence of outliers	Cook's distance	Max value < 1.0	Assumption met
Homoscedasticity	Residual plots	Evenly spread residuals	Assumption met
Normality	Q-Q plot and Shapiro–wIlk	$p = 0.12$ (normal)	Assumption met

Conclusion and Implications

Key Findings:

1. Building height negatively affects indoor temperature (taller buildings are cooler).
2. Higher WWR increases indoor temperature, suggesting a trade-off between natural lighting and thermal performance.
3. Multiple regression confirms that both predictors significantly influence indoor temperature.

Practical Implications:

- Architects should optimize window-to-wall ratios for improved thermal efficiency.
- Urban planners should consider building height regulations to mitigate heat accumulation.
- Sustainable building designs should balance daylighting with thermal comfort.

3.7 Applications of Correlation and Regression Analysis

Correlation and regression analysis serve three primary purposes in scientific research: testing hypotheses, analyzing associations between variables, and making predictions.

3.7.1 Testing Hypotheses About Cause-and-Effect Relationships

Regression analysis can be used to examine causal relationships between variables.

The researcher manipulates the independent variable (X) and measures its effect on the dependent variable (Y).

Example: Different dosages of an admixture (X) are administered, and its effect on concrete strength (Y) is measured.

3.7.2 Identifying Associations Between Variables

Correlation analysis assesses whether two variables are associated, without establishing causality.

Neither variable is directly manipulated; both are naturally occurring.

If an association is found, three possibilities exist:

- X influences Y
- Y influences X
- A third variable influences both X and Y
  - Example: Examining the relationship between CO_2 levels and indoor air quality.

3.7.3 Predicting Values of One Variable Based on Another

- Regression models can estimate the expected value of a dependent variable (Y) given a known value of an independent variable (X).
- This is particularly useful in forecasting and trend analysis.
- Example: Predicting housing prices based on square footage and location factors.

Researchers can examine relationships, test hypotheses, and develop predictive models by applying correlation and regression analysis, enhancing data-driven decision-making across various scientific disciplines.

4 Quantitative Data Analysis with GenAI

Integrating Generative AI (GenAI) tools into quantitative research methodologies can transform data analysis, interpretation, and visualization. However, compared to qualitative analysis, the application of GenAI in quantitative research has received relatively less attention, partly due to concerns regarding these algorithms' opaque, "black-box" nature. Recent advancements in AI transparency, model interpretability, and trustworthiness have increased confidence in these tools, paving the way for their adoption in handling large, complex dataset and enhancing analytical efficiency [66]. This section explores key areas where GenAI can support quantitative research, particularly visual analytics, statistical integration, and natural language interactions.

4.1 Visual Analytics and Pattern Identification

A significant contribution of GenAI to quantitative research lies in visual analytics and pattern recognition. These AI-driven applications can process large dataset rapidly and generate dynamic visualizations, helping researchers identify trends, anomalies, and relationships that may not be immediately apparent through conventional statistical methods [67].

One key advantage of GenAI-powered visualization is its ability to generate data representations through natural language prompts. For instance, a researcher could input a dataset and request, "*Show the correlation between economic indicators and stock market performance over the past decade.*" The AI system would produce a relevant visualization, potentially revealing insights overlooked in traditional analyses. This capability improves efficiency and facilitates exploratory data analysis, allowing researchers to examine data from multiple perspectives, leading to new hypotheses and research directions [67].

Recent advancements in GenAI visualization tools have further expanded their utility. For example, in 2024, Anthropic introduced "Artifacts" to Claude, enabling the tool to interpret dataset and generate visual representations, including custom charts, graphs, and infographics [66]. These capabilities are particularly valuable in disciplines dealing with big data, such as genomics, climate science, and social network analysis, where conventional visualization methods often struggle to capture the complexity of multidimensional data. However, while AI-generated visualizations enhance data exploration, their interpretation and contextualization still require domain expertise to ensure accuracy and alignment with research objectives.

4.2 Integration with Statistical Software and Computational Analysis

GenAI tools are increasingly integrated with traditional statistical software to create hybrid analytical systems combining AI-driven insights with conventional statistical techniques. Widely used programming environments such as Python and R, which form the backbone of statistical research, are now being embedded into GenAI platforms, enabling researchers to leverage AI-generated insights and established statistical methodologies [69]. For example, in a longitudinal study, a GenAI tool could assist in:

- Preprocessing and cleaning data by identifying missing values, outliers, or inconsistencies.
- Suggesting appropriate statistical models based on data characteristics.
- Generating preliminary insights before applying traditional statistical techniques for hypothesis testing.

By integrating GenAI tools into statistical workflows, researchers can enhance analytical efficiency, reduce errors, and facilitate large-scale data processing. This synergy bridges the gap between AI-powered automation and traditional statistical rigor, allowing researchers to adopt GenAI capabilities without disrupting established methodologies. Furthermore, such integration addresses concerns regarding reproducibility, as AI-generated insights can be validated using conventional statistical techniques [67].

4.3 Natural Language Interactions for Statistical Analysis

Another is its capacity to do sophisticated statistical analyses via natural language directions, among the most thrilling achievements in applying Generative AI (GenAI) in quantitative research. Researchers no longer have to write code or traverse complex statistical software menus; they can instead only request the AI to carry out activities using simple, daily language. This change might make sophisticated statistical techniques available to a larger population, including those lacking a formal statistical background [66]. Picture being able to type a request like:

- Run a multiple regression using these variables and describe the results.
- Perform a factor analysis and highlight the key components.
- From the SPSS file, analyze the ANOVA and emphasize essential results.

GenAI makes this kind of interaction feasible. This method facilitates statistical analysis by allowing researchers from many disciplines—including social science, education, public health, and design to use quantitative methods without knowing complicated tools or terminology. However, this increased accessibility has significant obligations as well. GenAI may make analysis more natural and quick, but it does not eliminate the necessity of statistical knowledge. Relying entirely on artificial intelligence without a precise understanding of the methodologies employed could lead to misinterpretation of results or incorrect application of the approach. Even with strong artificial intelligence capabilities, a reasonable basis in statistical literacy is still absolutely vital. Furthermore, researchers must be educated on utilizing GenAI and evaluating and validating the results critically as they become more incorporated into statistical education and research processes. While AI can help with interpretation, the researcher must ensure the study design, research topic, and data constraints guide the analysis. GenAI-driven natural language interfaces are changing our interaction with data. Especially for people outside of conventional quantitative fields, they simplify data exploration and understanding. However, with this simplicity comes the requirement of critical thought, rigorous validation, and continuous education. Statistical analysis in the future will not supplant human judgment; rather, it will be about improving it using innovative technologies cooperating with knowledgeable researchers.

5 Conclusions

The foundation of quantitative research is statistical analysis, which offers researchers the means to read, evaluate, and convey ideas inspired by data. From descriptive statistics that summarize data traits to inferential statistics that support hypothesis testing and population generalizations, we have investigated the fundamental statistical methods used in quantitative research throughout this chapter. Choosing suitable techniques that fit the study goals and data traits is a basic statistical analysis component. Descriptive statistics, including measures of central tendency (mean, median, and mode) and measures of dispersion (range, variance, and standard deviation), offer initial insights into data distributions. Conversely, inferential statistics let researchers evaluate hypotheses using t-tests, ANOVA, chi-square tests, and regression analysis, enabling them to make significant conclusions about their study populations. Hypothesis testing, a methodical way to confirm or reject assumptions based on sample data, is one of the essential elements of quantitative research. Hypothesis testing's use of probability, significance levels (*p*-values), and confidence intervals guarantees that research results are statistically strong and not only the result of chance. The chapter has also underlined the need for effect sizes, power analysis, and sample size determination to improve the validity and dependability of statistical findings. Moreover, modern quantitative research depends on statistical software to guarantee accuracy and simplify data analysis. Software tools like SPSS, R, and Python help researchers to run complicated statistical tests and display data swiftly. However, applying statistical methods properly requires precise knowledge of underlying assumptions, possible prejudices, and ethical issues in data analysis. One important point from this chapter is that research design includes statistical analysis, which influences the validity of results rather than just a mechanical process. Researchers must be painstaking in data collecting, cleaning, and analysis to prevent typical traps like misinterpretation of results, over-reliance on *p*-values, and neglect of statistical assumptions. Strictly using statistical analysis increases the validity of research findings, supporting evidence-based decision-making in many fields. Researchers must be flexible and continually improve their analytical abilities to preserve the quality and influence of quantitative research as statistical techniques change with computer and data science developments.

References

1. Dehalwar K, Sharma SN (2023) Fundamentals of research writing and uses of research methodologies. Edupedia Publications Pvt Ltd.
2. Creswell JW, Creswell JD (2023) Research design: qualitative, quantitative, and mixed methods approaches. SAGE Publications
3. Bryman A (2022) Social research methods. Oxford University Press
4. James G, Witten D, Hastie T, Tibshirani R (2021) An introduction to statistical learning with applications in R. Springer
5. Gelman A, Hill J, Vehtari A (2023) Regression and other stories. Cambridge University Press

6. Cohen J, Cohen P, West SG, Aiken LS (2021) Applied multiple regression/correlation analysis for the behavioral sciences. Routledge
7. Field A (2022) Discovering statistics using IBM SPSS statistics. SAGE
8. Shmueli G, Bruce PC, Gedeck P, Patel NR (2022) Data mining for business analytics: concepts, techniques, and applications in R. Wiley
9. Hesterberg T (2021) What teachers should know about the bootstrap: resampling in the undergraduate statistics curriculum. Am Stat 75(3):236–248
10. Wickham H, Grolemund G (2021) R for data science: import, tidy, transform, visualize, and model data. O'Reilly Media
11. Lopez JC, Cuba MB, Aquize RMV, Mitta BEC (2022) Relevance of statistical analysis in quantitative research. Universidad Ciencia y Tecnología 26(113):49–56
12. Upton G, Cook I (2014) A dictionary of statistics (3rd edn). Oxford University Press
13. Cano-Marin E (2024) The transformative potential of Generative Artificial Intelligence (GenAI) in business: a text mining analysis on innovation data sources. ESIC Market 55(2):e333–e333
14. Javaid M, Haleem A, Singh RP, Sinha AK (2024) Digital economy to improve the culture of industry 4.0: a study on features, implementation, and challenges. Green Technol Sustain 100083
15. Megahed FM, Chen YJ, Ferris JA, Knoth S, Jones-Farmer LA (2024) How can generative AI models such as ChatGPT be (mis) used in SPC practice, education, and research? An exploratory study. Qual Eng 36(2):287–315
16. La Trobe (2020) Statistics. LibGuides at La Trobe University
17. Niaz M, Nwagwu U (2023) Managing healthcare product demand effectively in the post-COVID-19 environment: navigating demand variability and forecasting complexities. Am J Econ Manage Bus (AJEMB) 2(8):316–330
18. Muijs D (2010) Doing quantitative research in education with SPSS. SAGE
19. Hair JF, Black WC, Babin BJ, Anderson RE (2019) Multivariate data analysis, 8th edn. Pearson Education Limited
20. Mweshi GK, Muhyila M (2024) Determining a statistical analysis for the quantitative study. Adv Soc Sci Res J 11(7):187–231
21. Kumar R, Lal K (2025) Basic research and statistics in special education. Perfect Writer Publishing
22. Almusaed A, Yitmen I, Almssad A, Myhren JA (2024). Construction 5.0 and sustainable neuro-responsive habitats: integrating the brain–computer interface and building information modeling in bright residential spaces. Sustainability 16(21):9393
23. Alem DD (2020) An overview of data analysis and interpretations in research. Int J Acad Res Educ Rev 8(1):1–27
24. Li S (2022) Quantitative research methods in ISLA. Instructed second language acquisition research methods. John Benjamins Publishing Company, pp 31–54
25. Ghanad A (2023) An overview of quantitative research methods. Int J Multidiscip Res Anal 6(08):3794–3803
26. Weiss NA (1999) Introductory statistics. Addison Wesley
27. Hinkle DE, Wiersma W, Jurs SG (2003) Applied statistics for the behavioral sciences, 5th edn. Houghton Mifflin Company, Boston, MA
28. Malakar IM (2023) Conceptualizing central tendency and dispersion in applied statistics. Cognition 5(1):50–62
29. Fein EC, Gilmour J, Machin T, Hendry L (2022) Statistics for research students. University of Southern Queensland, Darling Heights, Australia
30. Black K (2023) Business statistics: for contemporary decision making. John Wiley & Sons
31. Hashmi SAM. SPSS-driven analysis for sustainable development research
32. Almusaed A, Yitmen I (2023) Architectural reply for innovative building design concepts based on artificial intelligence simulation models and digital twins. Sustainability 15(6):4955
33. Devore JL, Berk KN, Carlton MA (2021) Modern mathematical statistics with applications. Springer Nature

34. Iacobucci D, Román S, Moon S, Rouziès D (2025) A tutorial on what to do with skewness, kurtosis, and outliers: new insights to help scholars conduct and defend their research. Psychol Market
35. Allen M, Poggiali D, Whitaker K, Marshall TR, van Langen J, Kievit RA (2021) Raincloud plots: a multi-platform tool for robust data visualization. Wellcome Open Res 4:63
36. Agresti A, Finlay B (1997) Statistical methods for the social sciences (3rd edn. Prentice Hall
37. Howell DC (2002) Statistical methods for psychology, 5th edn. Duxbury, Pacific Grove, CA. Retrieved from https://www.sheffield.ac.uk/polopoly_fs/1.43991!/file/Tutorial-14-correlation.pdf
38. Kothari CR (2004) Research methodology, 2nd edn. New Age International (p) Limited, publisher, New Delhi
39. Gödel M, Bode N, Köster G, Bungartz HJ (2022) Bayesian inference methods to calibrate crowd dynamics models for safety applications. Saf Sci 147:105586
40. Hennink M, Kaiser BN (2022) Sample sizes for saturation in qualitative research: a systematic review of empirical tests. Soc Sci Med 292:114523
41. Desai J, Watson D, Wang V, Taddeo M, Floridi L (2022) The epistemological foundations of data science: a critical review. Synthese 200(6):469
42. Rovetta A, Mansournia MA, Vitale A (2024) For a proper use of frequentist inferential statistics in public health. Global Epidemiol 8:100151
43. Nanda A, Mohapatra BB, Mahapatra APK, Mahapatra APK, Mahapatra APK (2021) Multiple comparison test by Tukey's honestly significant difference (HSD): does the confidence level control the type I error? Int J Stat Appl Math 6(1):59–65
44. Pustejovsky JE, Tipton E (2022) Meta-analysis with robust variance estimation: expanding the range of working models. Prev Sci 23(3):425–438
45. Coral DE, Smit F, Farzaneh A, Gieswinkel A, Tajes JF, Sparsø T, Franks PW (2025) Subclassification of obesity for precision prediction of cardiometabolic diseases. Nat Med 31(2):534–543
46. Gandhi H, Pakanati D (2024) Developing policy violation detection systems using CIS standards
47. Adeniran AO, Asifat JT, Familusi OB, Folorunso SS (2024) Critical analysis of research methods and misperceptions of statistical analysis. Int J Innov Manage Econ Soc Sci 4(3):29–50
48. Dementyev A (2022) T-test and hypothesis testing (explained)
49. Reddy D, Pulluru K (2024) Principles of statistics & research methodology. Academic Guru Publishing House
50. Margarian A (2022) Beyond p-value-obsession: when are statistical hypothesis tests required and appropriate? German J Agric Econ (GJAE) 71(4):213–226
51. Yan M, Shi W, Gong P, Xie Y, Zhang K, Li X, Wei H (2025) A U-shaped non-linear association between serum uric acid levels and the risk of Hashimoto's thyroiditis: a cross-sectional study. Front Endocrinol 16:1514857
52. Lakens D (2022) Sample size justification. Collabra: Psychol 8(1):33267
53. Miller C (2024) Investigating novice statistics students' reasoning about and with sampling distributions. Doctoral dissertation, University of Georgia
54. Freedman D, Pisani R, Purves R (2007) Statistics. WW Norton & Company, London
55. Koushan S (2025) Studying the energy performance of double-skin facade configurations for low-rise residential buildings in Buffalo, New York. Master's thesis, State University of New York at Buffalo
56. Welch BL (1947) The generalization of 'STUDENT'S' problem when several different population variances are involved. Biometrika 34(1–2):28–35
57. Bryman A, Cramer D (2005) Quantitative data analysis with SPSS 12 and 13: a guide for social scientists. Routledge, London
58. Neuman WL (2007) Basics of social research, 2nd edn. Pearson Education Inc
59. Armstrong RA, Eperjesi F, Gilmartin B (2002) The application of analysis of variance (ANOVA) to different experimental designs in optometry. Ophthalmic Physiol Opt 22(3):248–256

60. Huck S (2015) Statistical misconceptions, classic. Routledge
61. Singh D, Rao PV (2007) A surface roughness prediction model for arduous turning process. Int J Adv Manuf Technol 32:1115–1124
62. Gogtay NJ, Deshpande S, Thatte UM (2017) Principles of correlation analysis. J Assoc Phy Ind 65:78–81
63. O'Brien D, Sharkey Scott P (2012) Correlation and regression. In: Chen H (ed) Approaches to quantitative research—a guide for dissertation students. Oak Tree Press
64. Quirk TJ, Rhiney E (2016) Multiple correlations and multiple regressions. Excel for statistics. Springer, Cham
65. Tabachnick BG, Fidell LS (2007) Experimental designs using ANOVA, vol 724. Thomson/Brooks/Cole, Belmont, CA
66. Kamalov F, Santandreu Calonge D, Gurrib I (2023) New era of artificial intelligence in education: towards a sustainable multifaceted revolution. Sustainability 15(16):Article 16
67. Perkins M, Roe J (2024) Generative AI tools in academic research: Applications and implications for qualitative and quantitative research methodologies. arXiv preprint arXiv:2408.06872
68. Anthropic (2024) What are artifacts and how do I use them? | Anthropic Help Center. https://support.anthropic.com/en/articles/9487310-what-are-artifacts-and-how-do-i-use-them
69. Machlis S (2023) 8 ChatGPT tools for R programming. InfoWorld. https://www.infoworld.com/article/2338386/8-chatgpt-tools-for-r-programming.html

Advanced Data Analysis in Quantitative Research

Abstract Bridging theory, methodology, and application, this chapter offers a thorough and pragmatic investigation of advanced data analysis methods in quantitative research. Traditional statistical approaches can fail to capture the dynamic interaction of factors in real-world data as modern research issues become more complicated. The chapter presents and discusses several multivariate techniques including Multiple Linear Regression, MANOVA, Exploratory and Confirmatory Factor Analysis, and Structural Equation Modeling with their conceptual underpinnings, assumptions, software implementation, and interpretative depth in mind to meet this difficulty. Ensuring that researchers use methods correctly and preserve methodological rigor and theoretical clarity, emphasis is on model specification, assumption testing, sample size adequacy, and open reporting. Visual frameworks and practical examples help to show even more how these methods expose hidden patterns, model causal links, and guide theory-driven research. Ultimately, the chapter promotes a thoughtful, data-literate, and ethically based approach to quantitative analysis, enabling researchers to generate strong, generalizable, and significant results. This tool is vital for academics hoping to improve their analytical accuracy and significantly support evidence-based policies in education, sustainability, psychology, and social sciences.

1 Introduction

Quantitative research is the foundation for empirical social, natural, and applied scientific studies [1]. At its core, numerical data evaluates hypotheses, identifies patterns, and draws generalizable conclusions [2]. The relevant analytical procedures must evolve as research topics become more complex and data collection techniques become more sophisticated to ensure accurate and pertinent interpretations. This chapter introduces advanced data analysis in quantitative research, setting the stage for a deeper exploration of methods that extend beyond fundamental statistical procedures. Advanced data analysis involves the application of

A. Almusaed et al., *Practice of Research Methodology in Civil Engineering and Architecture*, https://doi.org/10.1007/978-3-031-97393-2_19

sophisticated statistical models and a comprehensive understanding of data structure, measurement theory, and the assumptions underpinning statistical methods [3]. Maintaining clarity and rigor throughout the analytical process necessitates a strategic approach to selecting the most suitable tools for a research question. The goal is to uncover statistically robust and substantively meaningful insights rather than merely conducting studies. In this context, advanced analysis goes beyond descriptive statistics and basic inferential tests, such as *t*-tests or ANOVAs, to explore areas including multivariate modeling, structural equation modeling, multilevel modeling, and longitudinal analysis [4]. Recognizing that data rarely exists in perfect form is fundamental to advanced analysis. Researchers often face missing data, outliers, non-normal distributions, or measurement errors. Addressing these challenges requires a combination of critical judgment and methodological expertise. Ensuring data quality and analytical integrity involves techniques such as multiple imputations, robust estimates, and variable transformation. Furthermore, the increasing availability of large dataset, often called "big data," compels researchers to adopt various data cleaning, management, and analysis strategies, frequently aided by programming environments like R or Python or specialized tools such as SPSS, Stata, or SAS [5]. Advanced data analysis also relies on the model selection and evaluation process. Unlike basic research approaches that test straightforward hypotheses, advanced quantitative research typically involves comparing multiple competing models. The most appropriate model is primarily determined by cross-validation, information criteria (AIC, BIC), and goodness-of-fit indices [6]. This technical and conceptual approach requires researchers to align their statistical choices with theoretical expectations and the characteristics of their data. Moreover, contemporary data analysis emphasizes interpreting interactions, indirect effects, and conditional correlations, allowing researchers to grasp the nuances and complexities of reality. For instance, mediation analysis explores the underlying mechanisms by which one variable influences another. At the same time, moderation research examines how the relationship between two variables varies at different levels of a third variable. Structural equation modeling offers a robust framework for integrating these dynamics into a cohesive analytical approach, encompassing factor and path analysis within a single model [7]. An increasing reliance on longitudinal and multilevel data represents another significant trend in modern quantitative research. Longitudinal data enables researchers to observe changes and infer causal relationships more clearly. Time series analysis and growth curve modeling are designed to manage such data efficiently. Conversely, multilevel modeling accommodates hierarchical data structures, such as patients nested within hospitals or students nested within classrooms. These methods facilitate the examination of cross-level interactions and help mitigate statistical biases that may arise from overlooking data dependencies [8]. As advanced data analysis evolves, predictive analytics and machine learning are becoming increasingly essential. Predictive methods prioritize accuracy in forecasting outcomes, whereas traditional inferential statistics focus on hypothesis testing and parameter estimation. Research processes are progressively merging with techniques such as neural networks, support vector machines, random forests, and decision trees. Although these approaches may sacrifice understanding for predictive power, they provide valuable tools in contexts where prediction accuracy and handling large dataset are critical. The ethical responsibilities of researchers expand along with data complexity [9]. Principles of transparency, replicability, and a commitment to avoid p-hacking or data dredging should guide advanced data analysis. Researchers should thoroughly describe their methods and findings, make their data and code available whenever feasible, and remain vigilant against biases that could distort interpretations. The credibility of research findings relies heavily on the rigor and integrity of the analytical process, especially when trust in science is paramount [10]. Consequently, advanced data analysis is as much an art as a science. While statistical software and algorithms provide the technical tools, the true value of analysis lies in the researcher's ability to derive insightful conclusions and narrate compelling stories based on data. This process requires technological literacy and critical thinking, encompassing questioning, reflection, and iteration. Mastering advanced data analysis entails understanding methodologies, recognizing their appropriate application, acknowledging their limitations, and considering their implications for theory, practice, and policy. A hierarchical chart showing the link between quantitative research techniques and sophisticated data analysis is shown in Fig. 1. It emphasizes three fundamental elements of advanced analysis: hypothesis assessment, complex statistical techniques, and the management of analytical difficulties. This graphic underlines the main part advanced technologies play in increasing the use of quantitative research and clarifies the conceptual framework of the chapter.

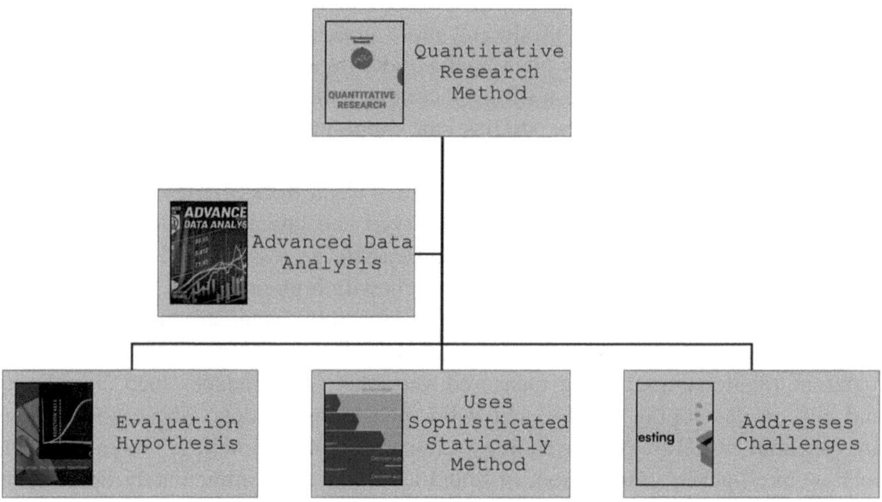

Fig. 1 Structure of quantitative research and advanced data analysis

2 Multivariate Statistical Techniques

2.1 Multivariate Techniques Unveiled

A key component of sophisticated quantitative research methods, multivariate statistical approaches allow researchers to explore the interrelationships among multiple variables simultaneously [11]. Unlike univariate methods, which focus on a single variable, or bivariate methods, which examine correlations between two variables, multivariate approaches align with the natural complexity of real-world data by integrating several dimensions into a unified analytical framework. This multidimensionality enhances accuracy and complexity, enabling researchers to forecast, explain, and characterize events [12]. In empirical studies, it is rare for a single element to influence a result in isolation. Instead, dynamic, interdependent systems necessitate the control of variables. For instance, environmental conditions, psychological well-being, age, gender, educational level, and socio-economic status may all influence personal health outcomes or academic success. Multivariate methods provide the necessary tools to capture this complexity, offering insights that more straightforward analytical techniques cannot deliver [13].

One of the primary advantages of multivariate analysis is its capability to account for confounding factors while evaluating the specific contributions of each predictor. By isolating the relevant effects from background noise, this functionality enhances the internal validity of the results [14]. Consequently, multivariate techniques often improve external validity by producing more generalizable outcomes that reflect the multifactorial nature of the phenomena under investigation. Furthermore, multivariate methods facilitate the testing of complex theoretical models. Rather than analyzing separate hypotheses, studies can simultaneously assess direct, indirect, and interaction effects across multiple variables [15]. This capability is critical in the social and behavioral sciences, where theoretical frameworks often suggest networks of causal relationships involving both observed and latent constructs. Techniques such as structural equation modeling, multivariate regression, and factor analysis enable the empirical validation of such models with high precision. Another crucial aspect of multivariate methods is their ability to reveal latent structures within data sets. For example, factor analysis identifies underlying dimensions that explain apparent relationships among measured variables. This reduces data complexity and aids in developing reliable and valid measurement tools. These applications are essential in fields like psychology, education, and marketing, where concepts such as intelligence, satisfaction, or brand loyalty are not immediately observable. Moreover, employing multivariate methods enhances the transparency and rigor of quantitative studies. They promote comprehensive analytical approaches that are empirically grounded, philosophically informed, and methodologically robust [16]. However, utilizing these strategies requires a solid foundation in statistical reasoning and proficiency with relevant software tools. Model specification, assumption testing, and result interpretation also need critical thinking [17]. Multivariate statistical methods are not merely advanced data analysis tools; they represent a paradigm

Table 1 Summary of core multivariate statistical techniques and their applications

Technique	Purpose	Variables	Use case
MLR [19]	Predict outcome from multiple predictors	Continuous DV, mix of IVs	Health risk modeling
EFA/CFA [20]	Identify/test latent factors	Continuous	Scale development
MANOVA [21]	Compare groups on several DVs	Categorical IVs, multiple DVs	Education, intervention studies
SEM [22]	Test complex models (latent + observed)	Mixed	Theory testing
PCA [23]	Reduce dimensions, retain variance	Continuous	Data simplification
Cluster analysis [24]	Group similar cases	Any	Market segmentation
Discriminant analysis [25]	Predict group membership	Categorical DV, continuous IVs	Classification tasks
Canonical correlation [26]	Explore two variable sets	Continuous	Cross-domain variable linkage

of thought that embraces empirical phenomena's complexity, interconnectedness, and contextual richness. Mastery of these methods enables researchers to gain more profound, more consistent insights and contributes to developing rigorous, evidence-based knowledge across diverse scientific disciplines [18]. A summary of important multivariate statistical methods, Table 1 emphasizes their primary goals, types of variables, and typical uses. This summary helps to compare and choose suitable techniques rapidly, depending on study goals. The table is a valuable tool for researchers conducting advanced quantitative studies in the social, behavioral, and applied sciences by grouping methods, including multiple regression, factor analysis, and structural modeling.

2.2 *Core Techniques and Their Applications*

A diverse array of multivariate statistical methods has been developed to manage contemporary research's increasing complexity effectively; each technique is tailored to meet specific analytical needs and accommodate various data types [27]. These methods enable researchers to analyze multiple factors simultaneously, uncover intricate patterns, and assess theoretical models that extend beyond simple cause-and-effect relationships. By incorporating various predictors and outcomes within a single analytical framework, multivariate techniques provide a more comprehensive understanding of the phenomena under investigation [28]. They hold particular importance in psychology, education, public health, economics, and social sciences, where numerous interconnected variables often affect observed results. Furthermore, multivariate methods enhance statistical analyses' explanatory and predictive power by

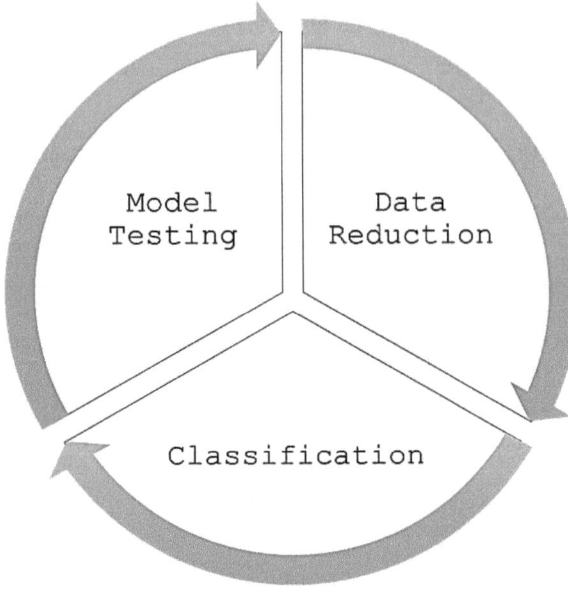

Fig. 2 Core techniques and applications in multivariate analysis

controlling for confounding variables, estimating interaction effects, and addressing measurement error [29]. While some techniques prioritize model testing, classification, or exploring latent variables, others focus on data reduction to clarify complex data sets. As the analytical complexity of quantitative research increases, proficiency with these tools becomes essential for conducting sound, reliable, and insightful studies [30]. The following section outlines key multivariate methods, highlighting their primary objectives, appropriate data types, and practical application areas. This summary helps determine the most suitable approach based on research design, data characteristics, and analytical goals. The main areas of multivariate analysis model testing, data reduction, and classification are shown in Fig. 2. Every approach in the study helps to provide better explanations and predictions. Emphasizing application domains and decision criteria, the picture briefly summarizes method choice in sophisticated quantitative research.

2.2.1 Multiple Linear Regression (MLR)

Among the most often used basic methods of multivariate statistical analysis, Multiple Linear Regression (MLR) ranks high [31]. It is meant to model the linear relationship between two or more independent (predictor) variables and a single continuous dependent one. MLR's primary goal is to forecast the degree to which each predictor variable affects the variance in the dependent variable after considering the influence of other predictors in the model [32]. MLR is a strong tool in both explanatory and predictive research since it allows one to isolate and investigate the particular impact of every variable. Assuming all other predictors stay constant, each predictor in an MLR model is given a regression coefficient indicating the expected change in the dependent variable linked with a one-unit change in that predictor [33]. These coefficients let researchers evaluate the relative significance of several variables and offer information on the direction and strength of associations. Moreover, measures like R-squared, modified R-squared, and standard error of the estimate let one assess the general fit of the model, thereby showing how well the predictors together account for the variation in the result. In practical disciplines, including education, health sciences, economics, business, and psychology, MLR is incredibly beneficial. Widely applied across fields, Multiple Linear Regression (MLR) is a strong statistical tool for investigating the concurrent impact of many predictors on a single outcome variable [34]. For example, in sustainable building research, MLR can be used to examine how annual energy consumption or indoor thermal comfort levels are affected by building orientation, insulation quality, material sustainability index, occupancy behavior, and HVAC efficiency. The technique is also widely used in consumer behavior research, labor market analysis, and risk assessment models where many independent variables are believed to affect a dependent outcome jointly. MLR's main advantage is its capacity to compensate for confounding variables, allowing researchers to separate the particular influence of one predictor while statistically considering the presence of others [35]. This function guarantees more exact and nuanced conclusions and significantly improves internal validity. MLR, on the other hand, produces reliable findings depending on many statistical assumptions. Among these are linear correlations between the dependent variable and predictors, independence of observations, homoscedasticity (constant variance of residuals), regular distribution of residuals, and lack of multicollinearity among predictors. Ignoring these assumptions could compromise the model's integrity and produce inaccurate or ineffective estimations. Researchers often utilize statistical software packages like SPSS, R, Stata, or Python to apply MLR, which offer diagnostic tools for assumption testing and visualisation features [36]. A fundamental and flexible analytical tool in quantitative research, MLR lets researchers investigate complicated variable interactions, adjust for confounding factors, and produce theoretically based and practically valuable ideas. Figure 3 briefly depicts Multiple Linear Regression (MLR), stressing its fundamental goal of modeling the linear relationship between several predictors and a single outcome variable. MLR's main components are assumptions, model fit indicators, and regression coefficients. The graph also underlines MLR's practical uses in sustainable building design, economics, and psychology, highlighting its value in explanatory and predictive studies with complicated variable interactions.

2.2.2 Exploratory and Confirmatory Factor Analysis

Typically employed in the early stages of research, factor analysis is utilized when there is little or no prior knowledge regarding the underlying structure of the data [37]. For example, Exploratory Factor Analysis may be used in passive building studies to uncover the underlying elements that influence indoor thermal comfort.

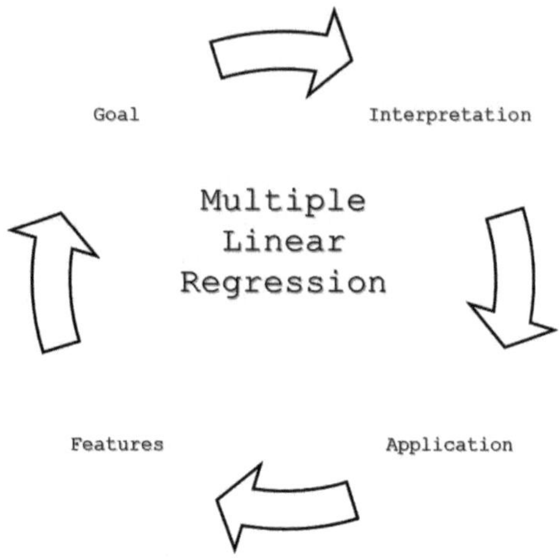

Fig. 3 Overview of multiple linear regression (MLR)

Common trends are identified by examining room temperature, humidity, window size, insulation level, and air exchange rate. Grouping these related factors assists researchers in recognizing latent constructs like "envelope efficiency" or "natural ventilation effectiveness," which aid in the design and assessment of passive building performance during the initial stages of study. Its purpose is to explore potential factor structures by collecting variables that display strong intercorrelations, suggesting the same latent construct may influence them. Key steps in EFA include:

- Examining the factorability of data with tools such as the Kaiser—Meyer—Olkin (KMO) test and Bartlett's test of sphericity guarantees the dataset is appropriate for factor analysis [38]. For example, researchers might collect data on occupant comfort and indoor climatic conditions in passive building studies to evaluate whether the factors are appropriately connected. This ensures that significant latent structures can be grouped from signs such as window use, CO_2 levels, and air temperature.
- Common approaches are primary axis factoring and maximum likelihood, which assist in extracting the first factors for shared variation across variables [37]. These techniques, for instance, can examine passive building performance and reveal hidden dimensions like thermal envelope quality or daylight efficiency. This stage transforms a massive set of comfort and energy indicators into a small number of understandable and controllable variables for more research [39].
- Criteria, including eigenvalues greater than 1, screen plots, and parallel analysis, decide how many elements to keep in the model. For example, in passive building research, this approach could suggest that three main components, thermal comfort, ventilation, and daylighting, capture most of the variation in the dataset. This helps to define separate performance areas that direct building design changes and policy recommendations.
- Techniques such as varimax (orthogonal) and oblimin (oblique) improve interpretability by defining the link between variables and factors and simplifying

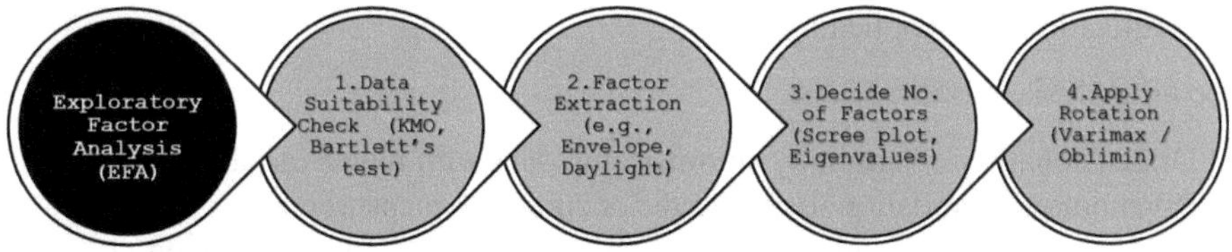

Fig. 4 Process of exploratory factor analysis in passive building (*source* authors)

factor loadings [38]. For example, when assessing passive buildings, rotation might help to identify which factors most accurately reflect ideas like indoor environmental quality or energy efficiency. The resulting building allows more exact findings on how specific design aspects support general sustainability performance factors [40].

Figure 4 illustrates the simplified Exploratory Factor Analysis (EFA) process used in passive building research. This approach involves evaluating data appropriateness, extracting latent variables, determining the number of relevant components, and applying rotation algorithms. This technique can uncover hidden correlations among factors such as indoor temperature, humidity, and ventilation, thereby informing the development of performance-based design concepts for passive structures.

EFA is especially valuable when designing new measurement instruments or scales, such as questionnaires assessing new psychological or behavioral constructs, where the factor structure is unknown [41].

I. Confirmatory Factor Analysis (CFA)

Conversely, confirmatory factor analysis (CFA) is a hypothesis-driven method employed when the researcher has an a priori model based on theory or previous exploratory factor analysis (EFA) findings [42]. Typically utilizing structural equation modeling (SEM), CFA tests whether the data fits a specific factor structure. In CFA, the researcher specifies [43]:

- The number of factors,
- Which observable variables load onto each factor, and
- The relationships between factors (e.g., whether they are correlated or independent).

The model's fit to the observed data is evaluated using goodness-of-fit measures such as the Comparative Fit Index (CFI), Root Mean Square Error of Approximation (RMSEA), and Standardized Root Mean Square Residual (SRMR) [44]. Using CFA, theoretical models and measurement scales are often validated in energy-efficient building research. For instance, CFA helps verify whether observed variables such as insulation quality, HVAC system efficiency, window glazing, and energy consumption accurately reflect theoretical dimensions like thermal performance, system optimization, and envelope integrity when designing a scale to evaluate energy efficiency performance. This validation ensures that the scale consistently quantifies the desired energy efficiency constructs.

II. Utilization and Applications

Factor analysis aids in:

- Dimensionality reduction by summarizing data with fewer variables (factors).
- Psychometric validation to ensure the accuracy of measurement instruments.
- Theory building through the identification and verification of hidden construct structures.

Research across cultures: CFA is essential for assessing measuring group invariance. The theoretical foundation, rigorous statistical testing, and thoughtful interpretation define the appropriate use of EFA and CFA. Misapplication could lead to inaccurate judgments about the dependability and correctness of constructs [45].

2.2.3 Multivariate Analysis of Variance (MANOVA)

An extension of univariate ANOVA, MANOVA lets researchers simultaneously investigate the effect of one or more independent factors on two or more connected dependent variables [46]. Unlike ANOVA, which looks at one dependent variable at a time, MANOVA evaluates the group impact of the independent variable(s) on the multivariate result, considering any relationships among dependent variables. This integrated strategy helps to lower the possibility of Type I errors, which could result from running several ANOVAs independently on each dependent variable, by providing a more complete knowledge of the data [47]. When researchers wish to investigate how a grouping variable, intervention, or experimental condition influences a series of related results, MANOVA is especially useful. For example, in educational research, a study could evaluate how a new teaching approach affects students' performance in several areas, including academic knowledge, critical thinking, and communication skills. MANOVA lets these results be examined simultaneously rather than conducting three distinct ANOVAs, therefore providing a more complete picture of the efficacy of the intervention and determining if group differences are consistent across all assessed outcomes. For example, in the framework of environmental behavior or sustainable building practices, MANOVA could assess how demographic variables or awareness campaigns affect a mix of energy-saving behaviors, ecological attitudes, and building usage patterns. Researchers get a greater understanding of the more significant consequences of interventions or traits using a combined analysis of these results, hence correcting for the interdependence among dependent variables. Table 2 provides an overview of the key features and practical issues related to MANOVA. It highlights its definition, benefits, appropriate applications, and interpretive focus. The table also outlines common MANOVA-related concerns, such as statistical assumptions, interpretive complexity, and sensitivity to sample factors. This structured overview emphasizes MANOVA's applications and limitations in multivariate research contexts.

Table 2 Key features and challenges of multivariate analysis of variance (MANOVA)

Aspect	Description	Challenges
Definition [48]	An extension of ANOVA that assesses the impact of one or more independent variables on two or more related dependent variables simultaneously	Requires multivariate normality and homogeneity of variance–covariance matrices
Key advantage [49]	Addresses correlations between dependent variables and minimizes the risk of Type I errors that arise from conducting multiple separate ANOVAs	Interpreting results can be complex, especially when significant multivariate effects are present
When to use [50]	When analyzing multiple related outcomes influenced by a grouping variable, intervention, or experimental condition	Sensitive to outliers and unequal sample sizes between groups
Focus [51]	Evaluates the overall multivariate effect instead of the individual variable impact in isolation	Requires larger sample sizes to ensure statistical power across various dependent variables
Benefit [52]	Offers a more comprehensive understanding of effects across multiple dimensions and addresses interdependent variables in a unified analysis	It is challenging to apply when dependent variables are weakly correlated or violate statistical assumptions

2.2.4 Structural Equation Modeling (SEM)

Structural Equation Modeling (SEM) is a comprehensive and versatile statistical technique that combines elements of factor analysis and multiple regression analysis into a unified framework [53]. It allows researchers to examine complex relationships among observed variables and unobserved latent constructs [54]. Unlike traditional regression models, which are limited to assessing direct effects between independent and dependent variables, SEM accommodates both direct and indirect (mediated) effects, thereby capturing the more nuanced pathways through which variables influence one another. SEM is particularly valuable when testing theoretical models that involve multiple interrelated constructs, as it enables the specification and evaluation of an entire conceptual framework rather than focusing on isolated relationships. This holistic perspective supports a deeper understanding of the structural interdependencies among constructs. It can incorporate feedback loops, hierarchical variable arrangements, and reciprocal causation [55]. Moreover, SEM provides tools for evaluating model fit, allowing researchers to assess how well the proposed model aligns with the observed data. Another essential feature of SEM is its ability to handle measurement error by including latent variables, representing theoretical constructs inferred from multiple indicators. This leads to more accurate and reliable estimation of relationships than models that rely solely on observed variables [56]. Furthermore, SEM can be applied across different populations or groups, making it suitable for testing measurement invariance and comparing models between groups. SEM offers a flexible and robust approach to investigating complex theories and advancing empirical research across various disciplines [57]. For example, in the

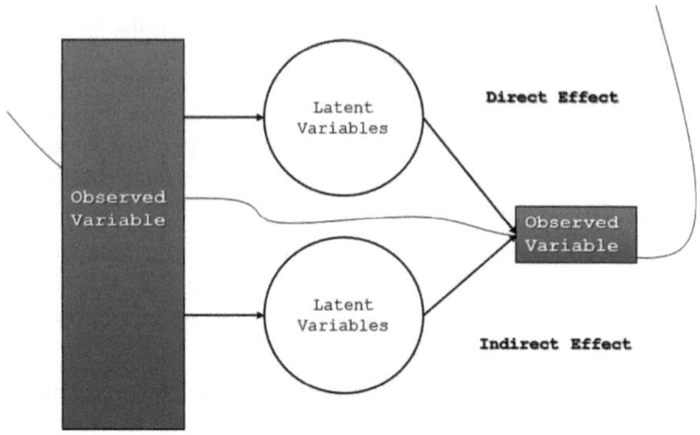

Fig. 5 Structural equation modeling framework with observed and latent variables

study of ecological construction materials, SEM can be used to explore how material characteristics such as thermal performance, recyclability, and cost impact users' acceptability, which is shaped by perceived environmental advantages. This helps researchers grasp the intricate interplay between technical characteristics and human perceptions of sustainable building materials. Figure 5 illustrates the foundational framework of Structural Equation Modeling (SEM), emphasizing the connections between observable variables and latent components. Demonstrating SEM's capacity to describe complex, mediated effects, arrows indicate both direct and indirect pathways. The graphic highlights SEM's dual focus: measuring latent variables from observed indicators and the structural relationships among these constructs within a theoretical framework.

3 Methodological and Practical Considerations

Although multivariate methods provide great analytical power and the capacity to reveal intricate correlations inside data, they also present a variety of methodological and practical issues that researchers have to handle to guarantee valid, reliable, and interpretable results [58]. First, the assumptions underlying these methods, such as multivariate normality, linearity, homogeneity of variance–covariance matrices, and the lack of multicollinearity, must be vigorously examined. These violations of assumptions could result in erroneous inferences, higher error rates, or skewed parameter estimates. Applying any multivariate model calls for extensive preliminary data screening and diagnosis first. Second, multivariate analysis is mainly influenced by the quality of measurement [59]. Poorly defined conceptions can destabilize models, badly designed survey questions, or untrustworthy data, especially when dealing with latent variables like factor analysis or structural equation modeling [60]. Therefore, preliminary actions directly affecting the quality of the analysis are instrument creation and validation. Third, multivariate analysis has more stringent sample size criteria than univariate techniques. For instance, techniques such as Structural Equation Modeling or MANOVA generally call for high sample sizes to provide consistent parameter estimation and adequate statistical power [61]. Small samples can lead to overfitting, non-convergence, or questionable model fit indices. The interpretation of outcomes in multivariate analysis is more complex and calls for a strong theoretical basis [62]. Researchers should be careful in their findings, especially when investigating indirect effects, interaction terms, or multi-level structures. Transparent reporting and appropriate model validation increase the credibility and replicability of results.

3.1 Evaluation of Statistical Assumptions

An essential first step in any multivariate analytic technique is evaluating statistical assumptions [27]. Whether the underlying assumptions of the statistical methods have been satisfied determines much of the credibility of the results, the validity of the conclusions, and the general resilience of the study findings. All multivariate methods, including MANOVA, multiple regression, factor analysis, and Structural Equation Modeling (SEM), depend on certain mathematical assumptions that must be checked and verified before analysis [63]. Among the most important is the assumption of multivariate normality, which calls for a normal distribution across all model variables and their combinations. If this premise is violated, the estimators used in maximum likelihood estimation may become skewed, leading to false p-values or misleading confidence intervals. Initial insights come from tests for univariate normality, such as Shapiro–Wilk or Kolmogorov–Smirnov [64]. However, multivariate-specific tests like Mardia's skewness and kurtosis measurements should be utilized to evaluate multivariate normality. Visual tools such as Q-Q plots and histograms also aid in identifying anomalies. Linearity requires that the relationships between the independent and dependent variables be linear. Scatterplots and partial regression plots are commonly employed to test this [65]. Nonlinearity can lead linear models to under- or overestimate effects if they cannot adequately capture curved or complex data patterns. Accurate prediction of standard errors relies on homoscedasticity, or the equal variance of residuals across all levels of the independent variables. Violations can lead to ineffective parameter estimates and adversely affect hypothesis testing. Residual plots and statistical tests such as Levene's and Breusch-Pagan tests can be used to evaluate this assumption [66]. The premise of independence of observations suggests that data points should not influence one another. This is typically true for between-subjects designs but not for repeated measures or clustered data. If independence is compromised, conventional techniques may produce overly optimistic significance values. Methods like Generalized Estimating Equations (GEE) or multilevel modeling (hierarchical linear modeling) may be necessary in such cases [67]. Another critical issue in multivariate analysis is multicollinearity, which arises from strong correlations among independent variables, resulting in unstable regression coefficients and complicating result interpretation. Variance Inflation Factors (VIF), condition indices, or inspection of correlation matrices are used to test this assumption. If multicollinearity is detected, researchers can consider eliminating redundant variables, combining them, or utilizing methods such as Principal Component Analysis (PCA) to reduce dimensionality. It is also essential to examine outliers and influential cases. Even a small number of extreme data points can skew parameter estimates, affecting the generalizability of findings. Techniques like Mahalanobis distance, Cook's distance, and leverage values help identify problematic findings. In reality, breaches of these assumptions call for appropriate remedies rather than invalidating the study. Researchers might opt for statistical techniques that are less sensitive to assumption violations, use robust estimation methods, resample data through bootstrapping, or apply data transformations (e.g., log, square root) [68]. Proper documentation of assumption testing and corrective measures enhances the study's validity and improves its transparency and reproducibility. Scientific judgment relies on assumption verification, not mechanical processes. Including multivariate normality, linearity, homoscedasticity, data independence, multicollinearity, and outlier influence, Table 3 thoroughly summarizes the main statistical assumptions supporting multivariate analysis methods. The table specifies the definition of each assumption, acceptable diagnostic tools, possible repercussions of violations, and suitable correction measures. This methodical overview helps researchers to methodically check assumptions before analysis, guaranteeing the validity and dependability of findings. Researchers might improve their multivariate results' openness, accuracy, and reproducibility using data transformation, strong modeling, or dimensionality reduction addressing assumption violations.

All these elements are essential for generating reliable and significant study results, demonstrating the researcher's attention to detail, understanding of the data, and commitment to methodological rigor [69]. For example, evaluating statistical assumptions in studies on recycling building materials is crucial when assessing how material origin, processing technique, and application environment affect structural performance. To ensure accurate interpretation, researchers must verify assumptions such as normality, linearity, and homoscedasticity before employing multivariate approaches. Failing to meet these criteria could distort relationships between factors like compressive strength or lifespan and hinder efforts to utilize recycled materials for building purposes.

3.2 Sample Size and Power

An essential first step in any multivariate analytic technique is the evaluation of statistical assumptions [70]. Whether the underlying assumptions of the statistical methods have been satisfied determines much of the credibility of the results, the validity of the conclusions, and the general resilience of the study findings. Specific mathematical assumptions must be evaluated and confirmed before analysis drives multivariate methods, including MANOVA, multiple regression, factor analysis, and Structural Equation Modeling (SEM) [48]. Among the most important is the

Table 3 Summary of statistical assumptions, diagnostics, and remedies in multivariate analysis

Assumption	Description	Diagnostic tools	Consequences if violated	Remedies
Multivariate normality	All variables and combinations must adhere to a normal distribution	Shapiro–Wilk, Kolmogorov–Smirnov, Mardia's test, Q-Q plots, histograms	Biased estimates, incorrect *p*-values, misleading confidence intervals	Data transformation, robust estimation, bootstrapping
Linearity	Linear relationships between independent and dependent variables	Scatterplots, partial regression plots	Under-/overestimation/overestimation of effects, misfitting models	Polynomial terms, data transformation, nonparametric models
Homoscedasticity	Equal variance of residuals across levels of independent variables	Residual plots, Levene's test, Breusch-Pagan test	Inefficient estimates, unreliable standard errors, biased significance tests	Weighted least squares, data transformation
Independence of observations	Observations should not influence each other	Study design review, time/order checks	Inflated Type I error, incorrect standard errors	Generalized Estimating Equations (GEE), multilevel modeling
Multicollinearity	Independent variables should not be highly correlated	Variance Inflation Factor (VIF), correlation matrices, condition indices	Unstable coefficients, reduced interpretability	Remove or combine variables, Principal Component Analysis (PCA)
Outliers and influential cases	Extreme data points can distort results	Mahalanobis distance, Cook's distance, leverage values	Skewed parameter estimates, poor model generalizability	Remove, winsorize, or robustify data; conduct sensitivity analyses

assumption of multivariate normality. This calls for a normal distribution across all model variables and their combinations. Should this premise be broken, the estimators employed in maximum likelihood estimation could become skewed, and the statistical tests could produce false *p*-values or deceptive confidence ranges. While multivariate normality has to be evaluated using multivariate-specific tests like Mardia's skewness and kurtosis measurements, tests for univariate normality (such as Shapiro–Wilk or Kolmogorov–Smirnov) offer preliminary insights. Visual tools such as Q-Q plots and histograms also help to find anomalies [71]. Linearity is the need for the relationships between the independent and dependent variables to be linear. Scatterplots and partial regression plots are commonly used to test this. Nonlinearity can cause linear models to underestimate or overestimate effects if they cannot represent curved or complicated data patterns. Accurate standard error estimation depends on homoscedasticity, or equal variance of residuals across all degrees of the independent variables [72]. Violations might have an impact on hypothesis testing and result in ineffective parameter estimates. Residual plots and statistical tests such as Levene's or Breusch-Pagan's can help to evaluate this assumption. The premise of independence of observations suggests that the data points are not affected by one another. This is usually true in between-subject's designs but not in repeated measurements or clustered data. Should independence be compromised, conventional techniques could generate unreasonably hopeful significance values. In such situations, methods like Generalized Estimating Equations (GEE) or multilevel modeling (hierarchical linear modeling) can be used. Another critical issue in multivariate analysis is multicollinearity [68]. It results from a strong correlation among independent variables, which causes unstable regression coefficients and makes result interpretation more difficult. Variance Inflation Factors (VIF), condition indices, or correlation matrix inspection are used to test this assumption. Should multicollinearity be found, researchers can consider eliminating redundant variables, merging them, or using methods such as Principal Component Analysis (PCA) to lower dimensionality. Furthermore, influential cases and outliers must be examined since even a small number of extreme data points might skew parameter estimations and compromise the generalizability of findings [73]. Problematic observations are identified by methods including Mahalanobis distance, Cook's distance, and leverage values. Practically, breaches of these assumptions do not automatically invalidate the study but instead call for suitable corrections. Researchers might pick other statistical approaches less vulnerable to assumption violations, utilize robust estimate techniques, resample the data via bootstrapping, or perform data transformations such as log or square root [74]. Corrective actions and assumption testing are well documented, which helps to increase the study's validity, openness, and repeatability. Scientific judgment depends on assumption verification, which is not a mechanical chore. All necessary for generating reliable and significant study results show the researcher's attention to detail, data knowledge, and methodological rigor dedication. Examining how material origin, processing techniques, and application environment affect structural performance calls for assessing statistical assumptions in studies on recycling construction materials [75]. The fundamental statistical assumptions of multivariate analysis methods are summarized in Table 4. From multivariate normality and linearity to multicollinearity and model robustness, it highlights seven important areas with their explanations, diagnostic methods, repercussions if broken, and suitable treatments. Using this thorough framework, researchers may methodically assess their data before analysis, improving the validity and dependability of their findings. The table promotes methodological rigor and openness in multivariate research across psychology, education, sustainability, and engineering by combining assumption testing with corrective techniques.

Table 4 Key assumptions and diagnostic strategies in multivariate analysis

Assumption/Issue	Description	Diagnostic tools	Consequences if violated	Remedies
Multivariate normality	Variables and their combinations should follow a normal distribution	Mardia's test, Shapiro–Wilk test, Q-Q plots, and histograms	Biased estimates and invalid significance tests	Data transformation, robust estimation, and bootstrapping
Linearity	Assumes a straight-line relationship between predictors and outcomes	Scatterplots, residual plots, partial regression plots	Misrepresentation of relationships; model underfit	Polynomial terms, data transformation, non-linear modeling
Homoscedasticity	Residual variance should be equal across all predictor values	Levene's test, Breusch-Pagan test, residual plots	Inefficient estimates, inflated standard errors	Weighted least squares, transformation, robust standard errors
Independence of observations	Data points should not influence one another	Design assessment, autocorrelation checks	Overstated significance, underestimated variance	Multilevel modeling, Generalized Estimating Equations (GEE)
Multicollinearity	Predictors should not be highly correlated	Variance Inflation Factor (VIF), condition index, correlation matrix	Unstable coefficients, loss of interpretability	Remove variables, combine predictors, use PCA or ridge regression
Outliers and influential cases	Extreme values can distort analysis	Mahalanobis distance, Cook's distance, and leverage values	Skewed results, poor generalizability	Winsorize, remove outliers, robust regression, sensitivity analysis
Model robustness and transparency	Overall model validity depends on assumption checks and corrections	Combined use of diagnostics and transparent documentation	Cumulative bias, non-replicable results	Clear reporting, simulation checks, robustness testing

To guarantee correct interpretation, researchers must verify assumptions such as normality, linearity, and homoscedasticity before using multivariate approaches [76]. Not meeting these criteria could distort correlations between factors like compressive strength or durability and impede attempts to use recycled resources for building usage. Moreover, employing Structural Equation Modeling to investigate correlations between building envelope quality, HVAC efficiency, and user energy behavior requires a sufficient sample size in energy efficiency building research [77]. A tiny sample could lead to an erroneous model fit or unstable parameter estimations, which would complicate drawing legitimate conclusions. Especially when assessing direct and indirect impacts on general energy performance within residential or commercial building settings, a sample size of at least 300 participants may be required to guarantee consistent findings if the study includes several latent constructs and observed indicators.

3.3 Model Specification and Fit

Multivariate statistical methods depend on model formulation and fit evaluation, especially in advanced techniques like Structural Equation Modeling (SEM) [78]. Based on theory, prior research, or conceptual reasoning, the model definition explicitly outlines the hypothesized connections between observable and latent variables. This covers deciding whether measurement models relationships between indicators and latent constructs are included, selecting which variables are predictors or outcomes, and defining the direction and type of associations e.g., causal, correlational, mediated [79]. Testing theoretical assumptions in a statistically relevant manner depends on a well-specified model. It should reflect the conceptual framework supporting the research and the data structure. While under-specification (omitting important routes or constructs) could provide skewed estimates and erroneous conclusions, over-specification (adding too many parameters) can cause overfitting and a loss of generalizability [80]. Thus, model specification has to be based on robust theoretical rationale and empirical data. Once the model is defined, its fit has to be assessed to see how well the suggested structure matches the data. A good model fit indicates that the empirical data confirms the theoretical assumptions. On the other hand, a lousy model fit could point to difficulties in the expected correlations, false beliefs, or concerns like missing variables or measurement inaccuracies. Model fit is assessed using a combination of statistical indices, which include:

- **Chi-square (χ^2) test**: Assesses overall fit but is sensitive to sample size [81].
- **Comparative Fit Index (CFI) and Tucker-Lewis Index (TLI)**: Values close to or above 0.95 indicate a good fit [82].
- **Root Mean Square Error of Approximation (RMSEA)**: Values below 0.06 suggest an excellent fitv [83].
- **Standardized Root Mean Square Residual (SRMR)**: Values under 0.08 indicate an acceptable fit [84].

These indices provide different points of view and should be read collectively instead of separately. Should a model show a lousy fit, researchers might modify it. This approach guarantees that adjustments are defensible based on theory rather than data-driven investigation alone, revisiting theoretical assumptions and looking at modification indices for possible improvements. Notably, re-specification should be performed judiciously to prevent capitalizing on chance, compromising the model's validity and replicability. Model fit assessment measures theoretical adequacy, not just statistical [85]. It helps to find whether the proposed relationship structure holds in practice and whether the model sufficiently reflects the topic under study. Researchers improve their multivariate studies' validity, interpretability, and scientific contribution using meticulous model specification and rigorous fit testing. Model specification in SEM, for instance, can specify how variables, including material origin, processing technique, and lifetime data, affect perceived environmental value and reuse potential in material passport studies on recycled construction materials [86]. A lousy model fit can suggest that the expected correlations among these variables are out of line and need modification. Correct specification guarantees that the model correctly shows how recycled materials support sustainable building design and circular construction plans.

3.4 *Interpretation of Multivariate Results*

Interpreting results from multivariate analyses requires more than a surface-level examination of statistical outputs such as coefficients and *p*-values [87]. It is a theoretically informed and methodologically rigorous process that evaluates how multiple variables interact, how they collectively contribute to outcomes, and what these relationships imply for the underlying theoretical framework. Multivariate techniques, primarily Structural Equation Modeling (SEM), MANOVA, and multiple regression, generate complex interdependencies among variables that must be understood in the research context and the hypothesis's structure [88]. This process of interpretation is central to drawing valid, replicable, and meaningful conclusions from multivariate data.

3.4.1 Understanding Variable Interactions and Multivariate Contributions

A key objective in multivariate analysis is to assess the significance, magnitude, direction, and contextual meaning of relationships between variables. Standardized coefficients (e.g., beta weights in regression or standardized path coefficients in SEM) enable researchers to evaluate the relative influence of predictors on outcome variables. [89] Simultaneously, indices such as R^2 provide insight into the proportion of variance explained by the model. Theoretical expectations must guide interpreting these relationships, considering both direct and indirect pathways and the potential for measurement error. In analyses such as MANOVA, the multivariate significance of predictors across multiple dependent variables must be distinguished from univariate outcomes [48]. This distinction is crucial for understanding how a single factor may have different effects across dimensions of the dependent construct. Similarly, in SEM, researchers must differentiate between the measurement model, which validates the relationship between observed indicators and latent constructs, and the structural model, which captures the hypothesized relationships among latent variables. Evaluating multivariate outputs should also consider multicollinearity, suppression effects, or redundancy among predictors, which can distort the actual effect size or statistical inference [90]. Moreover, significance levels alone may not reflect substantive importance, so it is necessary to supplement statistical interpretation with theoretical justification and effect size measures.

3.4.2 Mediating and Moderating Mechanisms

Beyond assessing direct associations, multivariate techniques allow for testing mediating and moderating variables, providing insight into the processes and conditions under which relationships occur. This aspect of interpretation enhances the explanatory power of multivariate models and contributes to a more nuanced understanding of causal mechanisms. A mediator is a variable that explains the relationship between an independent and a dependent variable [91]. For example, in environmental psychology or sustainable design research, a researcher may hypothesize that environmental knowledge mediates the relationship between green building exposure and pro-environmental behavior. Mediation analysis, especially within SEM, often involves decomposing total effects into direct and indirect effects, the latter being tested through bootstrapping to establish statistical significance with greater precision and fewer distributional assumptions. Conversely, a moderator is a variable that affects the strength or direction of the relationship between two different variables [92]. For example, socio-economic status might moderate the effect of energy-saving awareness on behavioral change, strengthening the relationship among individuals with higher income levels. Moderation is typically assessed through interaction terms, and the interpretation requires probing and visualizing the interaction to understand the nature of the conditional effects. Both mediating and moderating relationships must be theoretically grounded to avoid post-hoc explanations. Their inclusion should be based on conceptual models or empirical precedent, not solely on statistical convenience [93]. Correctly interpreting these mechanisms strengthens the explanatory coherence of multivariate models and ensures that conclusions are methodologically sound and theoretically relevant. Interpreting multivariate results is a multidimensional task requiring a synthesis of statistical insight, theoretical grounding, and methodological precision. It is not merely a mechanical step in analysis but a critical stage in the research process determining empirical findings' validity, depth, and impact [94]. Figure 6 illustrates the key elements used to interpret multivariate analysis findings. The circular structure emphasizes the interrelatedness

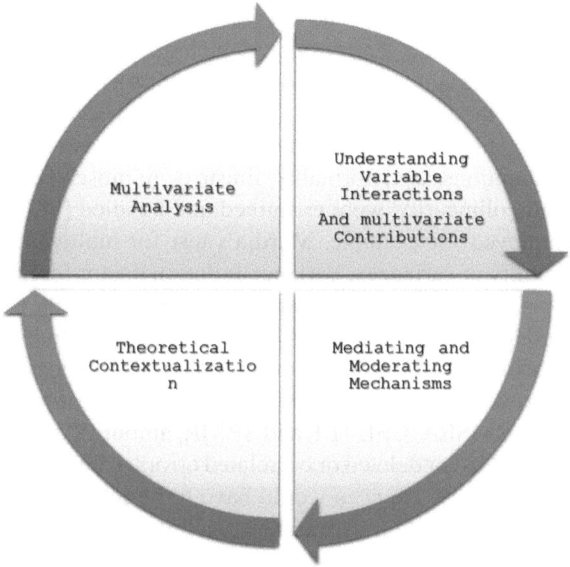

Fig. 6 Representation of key components in the interpretation of multivariate results (*Source* authors)

of understanding variable interactions, evaluating model contributions, and identifying mediating and moderating effects. This approach represents the iterative and integrative process necessary in multivariate interpretation, where theoretical frameworks provide context for statistical results. Particularly in complex studies such as Structural Equation Modeling or MANOVA, the diagram serves as a conceptual roadmap, demonstrating that valid interpretation necessitates statistical rigor, methodological introspection, and theoretical consistency.

3.5 Transparency and Reporting

Especially when using sophisticated multivariate methods, quantitative research's legitimacy, reproducibility, and scientific integrity are built on openness and thorough reporting [95]. The more complicated analytical models get, the more need for clarity in their construction, validation, and interpretation. Transparent reporting guarantees that practitioners, reviewers, and other researchers can grasp, assess, reproduce, and maybe expand upon the analytical framework and results of the study [96]. This fundamental element depends on advancing cumulative knowledge and preserving confidence in empirical evidence. Transparency starts first and foremost with a clear expression of model construction. This comprises stating the theoretical justification for the model, the variables involved, the type and direction of hypothesized associations, and the selected statistical method. For example, in Structural Equation Modeling (SEM), researchers have to differentiate between the measurement model, which specifies how observed variables reflect latent constructs, and the structural model, which describes causal pathways among those constructs. Every choice, including choosing indicators, adding or removing control variables, or applying interaction terms, should be supported theoretically and methodologically [97]. Second, it is necessary to record methods of assumption testing. Researchers must clarify how advanced statistical models' underlying assumptions, e.g., normality, linearity, homoscedasticity, independence, and lack of multicollinearity, were examined and handled [98]. Describing the diagnostic tests employed for example, Mardia's test for multivariate normality, Levene's test for equality of variances, Variance Inflation Factor for multicollinearity, and how infractions were fixed, for example, by transformation, robust estimation, or alternative modeling techniques helps others to evaluate the integrity of the analytical process critically. Third, scholars must explain the choices taken during model review and improvement. This covers how changes were justified and how model fit was assessed using RMSEA, CFI, TLI, and SRMR, among other indices. For example, if pathways were added or deleted or correlated errors were generated depending on modification indices, these changes would have to be justified using statistical advice and theoretical justification. Arbitrary model fitting based only on statistical criteria without theoretical foundation compromises the scientific validity of the findings. Presenting findings to improve clarity and accessibility is fundamental to open reporting [99]. Multivariate studies benefit significantly from visual representations such as structural models, factor loading tables, and path diagrams. These graphic tools let readers understand how constructs are operationalized and interrelated, compare relative effect sizes, and grasp intricate interactions between variables. For instance, a well-designed path diagram in SEM shows latent variables, observable indicators, standardized path coefficients, and error terms in an integrated schematic, enabling interpretation and critique. Furthermore, to give a more complete image of the significance of the findings, researchers should publish statistical significance, effect sizes, and confidence intervals [100]. Apart from stating the last model, it is helpful to highlight any rival tested models and justify the choice of one model over others. This kind of comparison modeling improves openness and helps to provide more complex knowledge of the events under investigation. Transparency also includes the accessibility of data and code where ethical and practical issues allow. Sharing data sets, analytical scripts, and codebooks helps to foster open science practices and enables independent verification and secondary studies. At the very least, researchers should offer thorough documentation of variables, coding schemes, and processing steps where complete data sharing is not feasible, such as for privacy reasons. One should admit limits and doubts. Every analysis includes limitations connected to sample size, measurement accuracy, missing data, or the generalizability of results; no model is flawless [101]. Openly addressing these shortcomings increases the credibility of a study and indicates the researcher's dedication to methodological rigor and intellectual integrity; it does not diminish the study. The essential elements for guaranteeing openness in multivariate research are summarized in Table 5. It explains why each component is vital and specifies what researchers should report, including model construction, assumption testing, and fit evaluation. A good tool for researchers using sophisticated quantitative methods across sectors, the table boosts

Table 5 Main elements of transparent reporting in multivariate analysis

Aspect	What to report	Why it matters
Model design	Variables, paths, theory, and methods used	Ensures clarity and theoretical grounding
Assumption testing	Tests used and how violations were addressed	Validates the accuracy of results
Model fit	Fit indices (e.g., RMSEA, CFI) and changes made	Shows model quality and justification of modifications
Visual outputs	Diagrams, tables, and effect sizes	Enhances understanding and communication
Alternative models	Compare models and reasons for final choice	Supports robustness and transparency
Data and code access	Dataset or coding documentation	Promotes reproducibility and open science
Limitations	Sample size, measurement issues, generalizability	Demonstrates honesty and scientific integrity

reproducibility, improves interpretability, and supports methodological rigor using unambiguous reporting norms derived from complicated processes.

Detailed reporting and openness are ethical and epistemic imperatives, not just procedural criteria. They guarantee that academic, policy, and professional environments may responsibly apply, correctly reproduce, and critically assess quantitative research. Transparency is a crucial bridge between complicated statistical methods and the more general scientific community in the context of advanced multivariate analysis, where the subtleties of modeling might hide important choices and assumptions [102].

3.6 *Software Proficiency and Analytical Expertise*

Implementing multivariate statistical methods requires knowledge of software and analytical skills in advanced quantitative research [27]. The complexity of modeling tasks demands a thorough understanding of statistical theory and the technical ability to execute analyses accurately within specialized software environments, as these techniques often involve the simultaneous analysis of several variables, latent constructs, and hierarchical relationships. Modern multivariate methods, including Confirmatory Factor Analysis (CFA), Structural Equation Modeling (SEM), Multivariate Analysis of Variance (MANOVA), and Latent Class Analysis (LCA), primarily depend on software systems capable of performing sophisticated mathematical calculations and estimation algorithms [103]. Commonly used software like SPSS, Stata, R, SAS, AMOS, Mplus, and LISREL offer specific features for different studies. SPSS, for example, is appropriate for those conducting conventional multivariate processes, while Mplus and AMOS are more specialized for

SEM and latent variable modeling [104]. While it requires more excellent programming knowledge, R offers unmatched versatility and extensibility for exploratory and confirmatory modeling through packages like lavaan, semTools, and psych. Despite their power, these software tools are not infallible and require significant expertise to operate correctly. Errors in coding, faulty data preparation, improper variable selection, or misinterpreting output can compromise study outcomes. Misspecifying a structural model reversing causal paths, incorrectly fixing or releasing parameters, or ignoring associated error terms can lead to misleading conclusions, even if model fit indices show an acceptable match. Similarly, overlooking missing data, non-normal distributions, or outliers could skew parameter estimations and compromise the generalizability of results. Therefore, researchers must be knowledgeable about the statistical techniques and skilled at navigating the software interface. This includes understanding the theoretical implications of pathways, loadings, and residuals; model diagnostics; fit indices (e.g., RMSEA, CFI, TLI, SRMR); and knowledge of estimation techniques (e.g., Maximum Likelihood, Weighted Least Squares) [105]. Without this knowledge, scholars risk perceiving statistical modeling as a mechanical process disconnected from its theoretical foundations and empirical implications. Researchers should undergo rigorous statistical theory and software implementation training to mitigate these risks. This training should encompass applicable education on data management, assumption testing, model construction, and analyzing complex output beyond initial courses. Combining courses, seminars, guided tutorials, and practical research experience fosters methodical fluency, enabling researchers to transition from mechanical application to critical engagement with their analytical tools [106]. Engaging a methodologist or statistics consultant in team research environments also ensures that analytical choices are methodologically sound and provides valuable oversight. Researchers using open-source platforms like R or Python should use extensive documentation, community forums, and peer-reviewed codebases that promote proper usage and continuous learning [107]. In addition to technical expertise, academics must adopt a thoughtful and critical approach to the analytical process. They should frequently ask whether the models are appropriate, explore different specifications, and consider the theoretical implications of statistical results. Analytical expertise encompasses the ability to conduct multivariate analysis alongside the judgment to determine when specific models are unsuitable, when results may be unstable, or when findings are likely to be misinterpreted. Conducting rigorous multivariate studies necessitates software knowledge and analytical ability as fundamental qualifications [108]. These skills enable researchers to implement complex analytical techniques reliably, assess their models' validity, and attain statistically and theoretically sound results. The skill sets of scientists committed to methodological rigor, transparency, and scientific integrity must evolve as the demands of data-intensive research change [109]. Table 7 emphasizes the elements of software competency and analytical knowledge required for conducting a thorough multivariate analysis. It underscores the importance of selecting suitable software, understanding statistical methods, avoiding common mistakes, and engaging in ongoing methodological training. The table also highlights the significance of teamwork and community resources in enhancing accuracy, transparency, and the quality of complex quantitative studies.

Multivariate statistical methods let academics interact more actively with the complex nature of real-world data by allowing them to go beyond basic pair-wise comparisons [110]. Unlike univariate or bivariate techniques, which frequently oversimplify relationships by isolating variables, multivariate methodologies enable simultaneous analysis of several connected variables, offering a more holistic and realistic depiction of the phenomena under study. This ability is beneficial in disciplines where conceptions are complex and cannot be captured by single-variable models, such as education, psychology, public health, environmental science, and engineering. By allowing interdependencies across variables, multivariate techniques provide a more sophisticated investigation of theoretical frameworks. Researchers may examine latent structures underneath observable results, moderating circumstances, indirect routes, and direct links [111]. Advanced possibilities for modeling complicated causal structures, quantifying the influence of confounding factors, and accounting for measurement error are provided by techniques including multiple

Table 7 Software proficiency and analytical expertise in multivariate analysis

Key aspect	Description	Purpose/importance
Software selection	Choosing appropriate tools like SPSS, R, Stata, SAS, AMOS, Mplus, or LISREL	Ensures the chosen software matches the complexity and goals of the analysis
Understanding capabilities	Knowing which methods each software supports (e.g., SEM in AMOS, CFA in Mplus, flexibility in R)	Enables efficient implementation of specific multivariate techniques
Technical proficiency	Ability to navigate software interfaces, write code, and manage data correctly	Prevents errors in data entry, coding, and model execution
Statistical knowledge	Understanding estimation methods, diagnostics, and model fit indices (e.g., RMSEA, CFI, TLI, SRMR)	Supports accurate model building, evaluation, and interpretation
Error avoidance	Preventing issues like mis-specified models, overlooked assumptions, or misinterpreted outputs	Maintains reliability and validity of findings
Training and development	Continuous learning through courses, workshops, tutorials, and research practice	Builds methodological fluency and critical analytical thinking
Collaborative expertise	Consulting statisticians or using community resources (e.g., R forums, peer-reviewed scripts)	Enhances quality control and fosters accurate, reproducible, and well-informed research decisions

regression, MANOVA, factor analysis, and structural equation modeling (SEM) [112]. Consequently, findings from multivariate studies usually tend to be more accurate, reliable, and generalizable. Furthermore, using multivariate methods helps create data-driven insights that are both theoretically guided and practically important [113]. Appropriately used with close regard to assumptions, model design, and open reporting, these techniques enable researchers to find patterns, validate theories, and produce knowledge that significantly advances science [114]. Thus, multivariate analysis is a thorough method of empirical investigation rather than just a statistical instrument; it allows precise, repeatable, and perceptive research across many fields.

4 Conclusion

In several fields, empirical research has increasingly benefited from advanced data analysis in quantitative studies. The demand for strong, versatile, and interpretable analytical tools becomes more pressing as research topics become complex and dataset become more multidimensional. Multivariate statistical techniques meet this need by allowing the simultaneous study of related variables and constructs and facilitating a deeper and more holistic understanding of underlying patterns and mechanisms. This chapter has demonstrated that advanced quantitative techniques embody technical accuracy, conceptual complexity, and methodological integrity. Far more than merely mathematical tools, techniques like Multiple Linear Regression (MLR), Exploratory and Confirmatory Factor Analysis (EFA/CFA), Multivariate Analysis of Variance (MANOVA), and Structural Equation Modeling (SEM) serve as vehicles for theory testing, hypothesis refinement, and practical insight. Each approach offers specific advantages depending on the study goals, the data type gathered, and the model's assumptions. Advanced data analysis hinges on the understanding that no model exists outside of context. Multivariate techniques best capture the multifaceted nature of real-world events. Researchers navigate a complex web of interacting variables whether they are assessing educational interventions, studying sustainable construction systems, understanding consumer behavior, or evaluating policy outcomes. Multivariate methods acknowledge this complexity by providing a coherent framework for evaluating direct, indirect, and moderating interactions while addressing measurement error, confounding variables, and sample diversity. However, this analytical skill entails responsibility. Advanced techniques require careful consideration of assumptions such as multivariate normality, linearity, homoscedasticity, data independence, and multicollinearity absence. Violations of these assumptions can significantly undermine model validity, leading to erroneous findings and distorted estimates. Researchers should, therefore, regard data screening and model diagnostics as essential first steps in the analytical process, not optional ones. Equally critical is the issue of sample size and statistical power. Insufficient sample sizes not only compromise the accuracy of parameter estimates but also limit the ability to identify significant effects, particularly in models involving multiple indicators, pathways, or levels of analysis. A priori power analyses and expectations of data quality issues ensure that the results of multivariate analyses are both credible and reproducible. Sound multivariate analysis also relies on model specification and fit assessment. Particularly in SEM and CFA, where complex theoretical frameworks are evaluated, rigorous model development must precede analysis. The scientific relevance of the model can be compromised by overfitting, under-specification, or data-driven adjustments lacking a theoretical basis. Statistical modeling must, therefore, be grounded in a solid theoretical rationale supported by empirical data and published transparently. Interpreting multivariate outcomes is a multifaceted endeavor. Researchers must go beyond statistical significance to assess the results' practical utility, theoretical implications, and contextual interpretation. Understanding the roles of mediating and moderating mechanisms enriches the causation narrative and allows for a more nuanced exploration of how and under what circumstances relationships emerge. In doing so, scholars create a more dynamic and integrated body of knowledge. Furthermore, the increasing integration of multivariate statistics with computational tools such as R, Python, and machine learning environments has expanded the possibilities for quantitative research. Multivariate techniques are essential to data-intensive discovery and innovation, from large-scale educational evaluations to real-time environmental monitoring. Their methodological diversity and enduring relevance are underscored by their adaptation to various fields, from psychology and sociology to engineering and sustainability. Nevertheless, advanced data analysis embodies scientific thinking rather than merely a technological exercise. It necessitates openness to methodological critique, a discerning eye on limitations, and integrity in reporting. The integrity of the research process influences the credibility of results just as much as the accuracy of mathematical calculations. Upholding the reliability of scientific work hinges on ethical considerations, including data privacy, reproducibility, and preventing p-hacking. Advanced data analysis in quantitative research provides the statistical framework for understanding the complexities of the world in measurable and significant ways. When applied with precision, theoretical clarity, and ethical rigor, these techniques yield statistically sound, socially and scientifically relevant findings. They foster a culture of evidence-based research, enhance the dialogue between theory and data, and enable scholars to address genuinely relevant issues. The pillars of academic excellence will always be the principles of methodological robustness, transparency, and critical interpretation as the landscape of data analysis evolves.

References

1. Mohajan HK (2020) Quantitative research: a successful investigation in natural and social sciences. J Econ Develop Environ People 9(4):50–79
2. Tracy SJ (2024) Qualitative research methods: collecting evidence, crafting analysis, communicating impact. John Wiley & Sons
3. Foster JJ, Yavorsky C, Barkus E (2005) Understanding and using advanced statistics: a practical guide for students

4. Keith TZ (2019) Multiple regression and beyond: an introduction to multiple regression and structural equation modeling. Routledge
5. Kabacoff R (2022) R in action: data analysis and graphics with R and Tidyverse. Simon and Schuster
6. Ding J, Tarokh V, Yang Y (2017) Bridging AIC and BIC: a new criterion for autoregression. IEEE Trans Inf Theory 64(6):4024–4043
7. Staphorst L (2024) Technology forecasting for national research and education networks using structural equation modelling-based context-sensitive data fusion. Doctoral dissertation, University of Pretoria (South Africa)
8. Aguinis H, Gottfredson RK, Culpepper SA (2013) Best-practice recommendations for estimating cross-level interaction effects using multilevel modeling. J Manag 39(6):1490–1528
9. Hesse A, Glenna L, Hinrichs C, Chiles R, Sachs C (2019) Qualitative research ethics in the significant data era. Am Behav Sci 63(5):560–583
10. Kretser A, Murphy D, Bertuzzi S, Abraham T, Allison DB, Boor KJ, Yada R (2019) Scientific integrity principles and best practices: recommendations from a scientific integrity consortium. Sci Eng Ethics 25:327–355
11. González-Díaz RR, Bustamante-Cabrera GI (2021) Predictive sequential research design to study complex social phenomena. Entropy 23(5):627
12. Zhao L (2021) Event prediction in the significant data era: a systematic survey. ACM Comput Surv (CSUR) 54(5):1–37
13. Denis DJ (2021) Applied univariate, bivariate, and multivariate statistics using python: a beginner's guide to advanced data analysis. John Wiley & Sons
14. Neri M, Politis A, Krause DA, Carli M, Vtanen T (2024) Speaker distance estimation in enclosures from single-channel audio. IEEE/ACM Trans Audio, Speech, Lang Process 32:2242–2254
15. Band N, Kadmon R, Mandel M, DeMalach N (2022) Assessing the roles of nitrogen, biomass, and niche dimensionality as drivers of species loss in grassland communities. Proc Natl Acad Sci 119(10):e2112010119
16. Themelis C, Sime JA, Thornberg R (2023) Informed grounded theory: a symbiosis of philosophy, methodology, and art. Scand J Educ Res 67(7):1086–1099
17. Sulistyanto H, Prayitno HJ, Sutama S, Narimo S, Sutopo A (2023) The effectiveness of hybrid learning-based adaptive media to empower student's critical thinking skills: Is it really for VARK learning style? Asian J Univ Educ 19(1):95–107
18. Karunarathna I, De Alvis K, Gunasena P, Jayawardana A (2024) Creating value through literature reviews: techniques for identifying research gaps
19. Xu KL, Guo J (2024) A new test for multiple predictive regression. J Financ Economet 22(1):119–156
20. Baharum H, Ismail A, Awang Z, McKenna L, Ibrahim R, Mohamed Z, Hassan NH (2023) The study adapted instruments based on confirmatory factor analysis (CFA) to validate measurement models of latent constructs. Int J Environ Res Public Health 20(4):2860
21. Dugard P, Todman J, Staines H (2022) Multivariate analysis of variance (MANOVA). In: Approaching multivariate analysis, 2nd edn. Routledge, pp 55–82
22. Li CH (2021) Statistical estimation of structural equation models with a mixture of continuous and categorical observed variables. Behav Res Methods 53(5):2191–2213
23. Migenda N, Möller R, Schenck W (2021) Adaptive dimensionality reduction for neural network-based online principal component analysis. PLoS ONE 16(3):e0248896
24. Dalmaijer ES, Nord CL, Astle DE (2022) Statistical power for cluster analysis. BMC Bioinformatics 23(1):205
25. Backhaus K, Erichson B, Gensler S, Weiber R, Weiber T (2023) Discriminant analysis. Multivariate analysis: an application-oriented introduction. Springer Fachmedien Wiesbaden, Wiesbaden, pp 203–263
26. Helmer M, Warrington S, Mohammadi-Nejad AR, Ji JL, Howell A, Rosand B, Murray JD (2024) On the stability of canonical correlation analysis and partial least squares with application to brain-behavior associations. Commun Biol 7(1):217

27. Mertler CA, Vannatta RA, LaVenia KN (2021) Advanced and multivariate statistical methods: practical application and interpretation. Routledge.
28. Borsboom D, Deserno MK, Rhemtulla M, Epskamp S, Fried EI, McNally RJ, Waldorp LJ (2021) Network analysis of multivariate data in psychological science. Nat Rev Methods Primers 1(1):58
29. Yu L, Liu W, Wang X, Ye Z, Tan Q, Qiu W, Chen W (2022) A review of practical statistical methods used in epidemiological studies to estimate the health effects of multi-pollutant mixture. Environ Pollut 306:119356
30. Amini Farsani M, Babaii E, Beikmohammadi M, Babaii Farsani M (2022) Mixed-methods research proficiency for applied linguists: a PLS-path modelling approach. Qual Quantity 1–26
31. Valentini M, dos Santos GB, Muller Vieira B (2021) Multiple linear regression analysis (MLR) applied for modeling a new WQI equation for monitoring the water quality of Mirim Lagoon, in the state of Rio Grande do Sul—Brazil. SN Appl Sci 3:1–11
32. Shams SR, Jahani A, Kalantary S, Moeinaddini M, Khorasani N (2021) The evaluation on artificial neural networks (ANN) and multiple linear regressions (MLR) models for predicting SO_2 concentration. Urban Climate 37:100837
33. Weisburd D, Wilson DB, Wooditch A, Britt C, Weisburd D, Wilson DB, Britt C (2022) Multiple regression. Adv Stat Criminol Criminal Justice 15–72
34. Dargahi P, Nasseri S, Hadi M, Nodehi RN, Mahvi AH (2023) Prediction models for groundwater quality parameters using a multiple linear regression (MLR): a case study of Kermanshah, Iran. J Environ Health Sci Eng 21(1):63–71
35. MacKinnon DP, Lamp SJ (2021) A unification of mediator, confounder, and collider effects. Prev Sci 22(8):1185–1193
36. Mehmetoglu M, Venturini S (2021) Structural equation modelling with partial least squares using Stata and R. Chapman and Hall/CRC
37. Schreiber JB (2021) Issues and recommendations for exploratory factor analysis and principal component analysis. Res Social Adm Pharm 17(5):1004–1011
38. Govindasamy P, Isa NJM, Mohamed NF, Noor AM, Ma L, Olmos A, Green K (2024) A systematic review of exploratory factor analysis packages in R software. Wiley Interdiscipl Rev: Comput Stat 16(1):e1630
39. Almusaed A, Almssad A, Yitmen I, Wallhagen M, Yang YF, Iyer-Raniga U (2024) Integrative approaches in urban sustainability: architectural design, technological innovations and social dynamics in global contexts. BoD–Books on Demand
40. Fnais A, Rezgui Y, Petri I, Beach T, Yeung J, Ghoroghi A, Kubicki S (2022) The application of life cycle assessment in buildings: challenges, and directions for future research. Int J Life Cycle Assess 27(5):627–654
41. Ike OO, Ugwu LE, Enwereuzor IK, Eze IC, Omeje O, Okonkwo E (2023) Expanded-multidimensional turnover intentions: scale development and validation. BMC Psychol 11(1):271
42. Nájera P, Abad FJ, Sorrel MA (2023) Is exploratory factor analysis always to be preferred? A systematic comparison of factor analytic techniques throughout the confirmatory–exploratory continuum. Psychol Methods
43. Kyriazos T, Poga-Kyriazou M (2023) Applied psychometrics: estimator considerations in commonly encountered conditions in CFA, SEM, and EFA practice. Psychology 14(5):799–828
44. Goretzko D, Siemund K, Sterner P (2024) Evaluating model fit of measurement models in confirmatory factor analysis. Educ Psychol Measur 84(1):123–144
45. Bashkov BM, Coggeshall WS (2021) Using a measurement framework to examine item quality/functioning. Basic elements of survey research in education: addressing the problems your advisor never told you about, pp 237–265
46. Din IU, Hayat Y (2021) ANOVA or MANOVA for correlated traits in agricultural experiments. Sarhad J Agric 37(4):1250–1259

47. Chen WH, Uribe MC, Kwon EE, Lin KYA, Park YK, Ding L, Saw LH (2022) A comprehensive review of thermoelectric generation optimization by statistical approach: Taguchi method, analysis of variance (ANOVA), and response surface methodology (RSM). Renew Sustain Energy Rev 169:112917
48. Ntumi S (2021) Reporting and interpreting multivariate analysis of variance (MANOVA): Adopting the best practices in educational research. J Res Educ Sci (JRES) 12(14):48–57
49. Shahat MA, Emam MM, Alhinai M, Omara EM, Alhabsi N, Alhosni K, Al-Balushi SM (2025) Enhancing middle school science education: Evaluating a competency-based STEM training program for teachers. Soc Sci Human Open 11:101457
50. Rahnenführer J, De Bin R, Benner A, Ambrogi F, Lusa L, Boulesteix AL, Topic group "High-dimensional data" (TG9) of the STRATOS initiative (2023) Statistical analysis of high-dimensional biomedical data: a gentle introduction to analytical goals, common approaches and challenges. BMC Med 21(1):182
51. Giner-Sorolla R, Montoya AK, Reifman A, Carpenter T, Lewis NA Jr, Aberson CL, Soderberg C (2024) Power to detect what? Considerations for planning and evaluating sample size. Pers Soc Psychol Rev 28(3):276–301
52. Maulidi A, Girindratama MW, Soeherman B, Andono FA (2024) Organizational culture, power distance and corruption: the mediating role of religiosity. Indonesian Account Rev 14(2):203–221
53. Han X, Li Z, Chen H, Yu M, Shi Y (2025) Structural equation model in landscape performance research: dimensions, methodologies, and recommendations. Land 14(3):646
54. Goller M, Hilkenmeier F (2022) PLS-based structural equation modelling: an alternative approach to estimating complex relationships between unobserved constructs. Methods for researching professional learning and development: challenges, applications and empirical illustrations. Springer International Publishing, Cham, pp 269–292
55. Reumers L, Hameleers N, Hilderink H, Bekker M, Jansen M, Ruwaard D (2024) Quantifying reciprocal relationships between poverty and health: combining a causal loop diagram with longitudinal structural equation modelling. Int J Equity Health 23(1):87
56. Baptista ML, Goebel K, Henriques EM (2022) Relation between prognostics predictor evaluation metrics and local interpretability SHAP values. Artif Intell 306:103667
57. Russo D, Stol KJ (2021) PLS-SEM for software engineering research: an introduction and survey. ACM Comput Surv (CSUR) 54(4):1–38
58. Kyriazos T, Poga M (2023) Dealing with multicollinearity in factor analysis: the problem, detections, and solutions. Open J Stat 13(3):404–424
59. Ullmann T, Heinze G, Hafermann L, Schilhart-Wallisch C, Dunkler D, TG2 of the STRATOS initiative (2024) Evaluating variable selection methods for multivariable regression models: a simulation study protocol. Plos One 19(8):e0308543
60. Hammond P, Opoku MO (2023) The mediating effect of going concern and corporate reporting in the relationship between corporate governance and investor confidence in financial institutions. Heliyon 9(10)
61. Ching-Hong Li J, Nesca M, Michael Waisman R, Cheng Y, Man Chung Tze V (2021) A robust effect size measure Aw for MANOVA with non-normal and non-homogenous data. Methodol Innov 14(3):20597991211055949
62. Depaoli S, Winter SD, Liu H (2024) Under-fitting and over-fitting: the performance of Bayesian model selection and fit indices in SEM. Struct Equ Model 31(4):604–625
63. Maiti J (2022) Multivariate statistical modeling in engineering and management. CRC Press
64. Konopatov AV, Shidlovsky YV, Shtil AA, Bylino OV (2024) Kolmogorov-Smirnov and Shapiro-Wilk single distribution analysis methods in examining sample structure, strength of genetic background, and effects of interventions in lifespan and aging studies. bioRxiv, 2024–12
65. Kamel AR, Abonazel MR (2023) A simple introduction to regression modeling using R. Comput J Math Stat Sci 2(1):52–79
66. Osemeke RF, Igabari JN, Christian ND (2024) Detection and correction of violations of linear model assumptions by means of residuals. J Sci Innov Technol Res

67. Austin PC, Kapral MK, Vyas MV, Fang J, Yu AYX (2024) Using multilevel models and generalized estimating equation models to account for clustering in neurology clinical research. Neurology 103(9):e209947
68. Dey D, Haque MS, Islam MM, Aishi UI, Shammy SS, Mayen MSA, Uddin MJ (2025) The proper application of logistic regression model in complex survey data: a systematic review. BMC Med Res Methodol 25(1):15
69. Lim WM (2024) What is qualitative research? An overview and guidelines. Aust Market J 14413582241264619
70. Backhaus K, Erichson B, Gensler S, Weiber R, Weiber T (2021) Multivariate analysis. Springer Books 10(1):973–978
71. Antony Seba P, Bibal Benifa JV (2022) Perceptive analysis of chronic kidney disease data through conceptual visualization. In: International conference on computational intelligence in pattern recognition. Springer Nature Singapore, Singapore, pp 110–122
72. Đalić I, Terzić S (2021) Violation of the assumption of homoscedasticity and detection of heteroscedasticity. Decis Making: Appl Manage Eng 4(1):1–18
73. Ododo F, Addotey N (2025) Understanding the influence of outliers on machine learning model interpretability. Int J Afr Sustain Develop Res
74. Wilcox RR (2023) A guide to robust statistical methods, vol 746. Springer, New York, p 37
75. Almusaed A, Yitmen I, Myhren JA, Almssad A (2024) Assessing the impact of recycled building materials on environmental sustainability and energy efficiency: a comprehensive framework for reducing greenhouse gas emissions. Buildings 14(6):1566
76. Knief U, Forstmeier W (2021) Violating the normality assumption may be the lesser of two evils. Behav Res Methods 53(6):2576–2590
77. Tawakoni E, Nikkhah A, Zomorodian ZS, Tahsildoost M, Hoonejani MR (2022) Estimating the impact of occupants' behaviour on energy consumption by Pls-SEM: a case study of Pakdel Residential Complex in Isfahan, Iran. Front Sustain Cities 4:700090
78. Albahri AS, Alnoor A, Zaidan AA, Albahri OS, Hameed H, Zaidan BB, Yass AA (2021) Based on the multi-assessment model: towards a new context of combining the artificial neural network and structural equation modelling: a review. Chaos, Solitons Fractals 153:111445
79. Devezer B, Buzbas EO (2023) Rigorous exploration in a model-centric science via epistemic iteration
80. Neely ML, Pieper CF, Gu B, Dmitrieva NO, Pendergast JF (2023) Exploration of model misspecification in latent class methods for longitudinal data: correlation structure matters. Stat Med 42(14):2420–2438
81. Sathyanarayana S, Mohanasundaram T (2024) Fit indices in structural equation modeling and confirmatory factor analysis: reporting guidelines. Asian J Econ Bus Account 24(7):561–577
82. Montoya AK, Edwards MC (2021) The poor fit of model fit for selecting number of factors in exploratory factor analysis for scale evaluation. Educ Psychol Measur 81(3):413–440
83. Afework T, Wondimagegnehu A, Alemayehu N, Kantelhardt EJ, Addissie A (2021) Validity and reliability of the Amharic version of supportive care needs survey-short form 34 among cancer patients in Ethiopia. BMC Health Serv Res 21(1):484
84. Pavlov G, Maydeu-Olivares A, Shi D (2021) Using the standardized root mean squared residual (SRMR) to assess exact fit in structural equation models. Educ Psychol Measur 81(1):110–130
85. West SG, Wu W, McNeish D, Savord A (2023) Model fit in structural equation modeling. Handb Struct Equ Model 2:184–205
86. Jemal KM, Kabzhassarova M, Shaimkhanov R, Dikhanbayeva D, Turkyilmaz A, Durdyev S, Karaca F (2023) Facilitating circular economy strategies using digital construction tools: framework development. Sustainability 15(1):877
87. Seßler K, Fürstenberg M, Bühler B, Kasneci E (2025) Can AI grade your essays? A comparative analysis of large language models and teacher ratings in multidimensional essay scoring. In: Proceedings of the 15th international learning analytics and knowledge conference, pp 462–472

88. Bagozzi RP (2022) Structural equation models in consumer research: exploring intuitions and deeper meanings of SEMs
89. Bhakta N, Lechner CM (2021) To score or not to score? A simulation study on the performance of test scores, plausible values, and SEM, in regression with socio-emotional skill or personality scales as predictors. Front Psychol 12:679481
90. Zheng Z, Liu K, Zhou Y, Debliquy M, Bittencourt C, Zhang C (2025) Response to letter to the Editor from Y. Takefuji on "Beyond principal component analysis: enhancing feature reduction in electronic noses through robust statistical methods". Trends Food Sci Technol 104918
91. Abu-Bader S, Jones TV (2021) Statistical mediation analysis using the sobel test and hayes SPSS process macro. Int J Quant Qual Res Methods
92. Ali AAA, AlZgool M, Alzoraiki M, Milhem M, Al-Absy MSM (2023) Moderating effect of strategic planning on the relationship between career path planning and job performance. Sustainability 15(11):8490
93. Pires PB, Prisco M, Delgado C, Santos JD (2024) A conceptual approach to understanding the customer experience in E-Commerce: an empirical study. J Theor Appl Electron Commer Res 19(3):1943–1983
94. Kumar A, Praveenakumar SG (2025) Research methodology. Authors Click Publishing
95. Razlan NM, Samsuddin SF, Abrizah A (2024) Embracing transparency and openness in scholarly publishing: insights from the Malaysian social sciences researchers. Malays J Libr Inf Sci 29(3):19–45
96. Braun V, Clarke V (2025) Reporting guidelines for qualitative research: a values-based approach. Qual Res Psychol 22(2):399–438
97. Kohler U, Class F, Sawert T (2024) Control variable selection in applied quantitative sociology: a critical review. Eur Sociol Rev 40(1):173–186
98. Marisetty N (2024) Mitigating assumption violations in regression analysis: insights from 20 years of currency pair data. Available at SSRN 5068940
99. Murari A, Rossi R, Spolladore L, Lungaroni M, Gaudio P, Gelfusa M (2023) A practical utility-based but objective approach to model selection for regression in scientific applications. Artif Intell Rev 56(Suppl 2):2825–2859
100. Lovakov A, Agadullina ER (2021) Empirically derived guidelines for effect size interpretation in social psychology. Eur J Soc Psychol 51(3):485–504
101. Lakens D (2022) Sample size justification. Collabra: Psychol 8(1):33267
102. Alghazzawi D, Ullah H, Tabassum N, Badri SK, Asghar MZ (2025) Explainable AI-based suicidal and non-suicidal ideations detection from social media text with enhanced ensemble technique. Sci Rep 15(1):1111
103. Dong Q, Chen X, Dong S, Ni F (2021) Data analysis in pavement engineering: an overview. IEEE Trans Intell Transp Syst 23(11):22020–22039
104. Whittaker TA, Schumacker RE (2022) A beginner's guide to structural equation modeling. Routledge
105. Zulkifli NR, Aimran N, Deni SM. The effect of regularized unweighted least squares on CFI, TLI, RMSEA and TLI in structural equation modeling
106. Gumbo M (2024) Development of a comprehensive design handbook for enhancing mechanical engineering practices for a company
107. Guo P (2021) Ten million users and ten years later: Python tutor's design guidelines for building scalable and sustainable research software in academia. In: The 34th annual ACM symposium on user interface software and technology, pp 1235–1251
108. Dehalwar K, Sharma SN (2023) Fundamentals of research writing and uses of research methodologies. Edupedia Publications Pvt Ltd
109. Umbach G (2024) Open Science and the impact of Open Access, Open Data, and FAIR publishing principles on data-driven academic research: Towards ever more transparent, accessible, and reproducible academic output? Stat J IAOS 40(1):59–70
110. Burckhardt P, Nugent R, Genovese CR (2021) Teaching statistical concepts and modern data analysis with a computing-integrated learning environment. J Stat Data Sci Educ 29(sup1):S61–S73

111. Cheung GW, Cooper-Thomas HD, Lau RS, Wang LC (2021) Testing moderation in business and psychological studies with latent moderated structural equations. J Bus Psychol 36:1009–1033
112. Kline RB (2023) Principles and practice of structural equation modeling. Guilford publications
113. Melo A, Câmara MM, Pinto JC (2024) Data-driven process monitoring and fault diagnosis: a comprehensive survey. Processes 12(2):251
114. Fellows RF, Liu AM (2021) Research methods for construction. John Wiley & Sons

Data Interpretation and Academic Writing

A key change arises as the research process evolves from empirical observation and analysis into academic communication; this change shapes the influence and credibility of the entire effort by defining the clarity of interpretation and the accuracy of writing. Emphasizing the transformation of data into meaning and the conversion of meaning into knowledge, which is spread and preserved through academic writing, Part V is the climax of this journey. This part addresses how researchers understand and express their results and position their work within the broader scientific discourse. The emphasis is therefore twofold: writing as structured distribution and interpretation as intellectual synthesis. Interpreting research results requires deep engagement with the data, a reflective attitude toward the original research questions, and a careful analysis of how the results support, challenge, or extend current theoretical frameworks; it goes far beyond merely presenting results in tables or graphs. Interpretation involves intellectual honesty, acknowledging what the data reveals and what it does not. Publicly debating limitations, welcoming unexpected outcomes, and clearly defining the boundaries of generalizability are essential. By situating results within a credible academic context, these elements enhance rather than undermine the validity of a study. Additionally, interpretation is inherently dialogical; it is not solely a singular cognitive process. It lays the groundwork for future research, engages with academic controversies, and responds to previous literature. Discussing results bridges the gap between empirical discovery and theoretical advancement and connects practical significance with scholarly dialog. Thus, effectively analyzing findings yields noticeable implications for practical engineering, education, and policy applications and enhances collective knowledge. This interpretive process cannot be isolated; it must be grounded in a well-constructed and critically engaged literature review. Synthesizing previous studies is a fundamental activity that shapes every aspect of a study's design, analysis, and interpretation rather than serving as a retrospective exercise. A thorough literature review identifies key ideas, theoretical conflicts, and methodological gaps. It situates the contribution that the current study aims to make and establishes the framework for the research topic. Without this framework, even the most intriguing findings risk straying from the academic discussions they intend to inform. Therefore, the literature review is historical and forward-looking; it assesses past achievements and indicates where future scholarship should be directed. This process guides researchers toward appropriate methods, relevant discussions, and unresolved questions warrant further investigation. This function becomes increasingly important as multidisciplinary approaches gain traction, necessitating scholars to draw from various disciplines and traditions. A well-structured review is an intellectual roadmap that assists the reader and enhances the author's academic credibility. Once results are comprehended and literature is synthesized, the task shifts to organizing the study into a coherent, legible, and academically rigorous report. The structure of a research report serves as a narrative framework connecting empirical observation with theoretical reflection, rather than merely a formula to follow. From the introduction to the conclusion, every part of the report serves a specific purpose that advances the argument and ensures reader comprehension. A cohesive framework enhances the research's contribution and clarity by helping translate complex data and concepts into accessible conclusions. Effective organization ensures the communication of ethical considerations, methodological decisions, and analytical interpretations. Transparency in this regard is vital as it facilitates replication, fosters trust, and maintains the integrity of academic work. Furthermore, a well-structured report can demonstrate the social relevance of the study by illustrating how comprehensive academic research addresses practical challenges and informs actionable solutions. In this way, the organization connects academic rigor with social impact. Beyond structure lies style, the medium for conveying scholarly discussions. Academic writing style is a methodological necessity, not merely an aesthetic choice. Fundamental attributes that lend strength and precision to academic communication include formality, clarity, and objectivity. Proper citations, consistent terminology, and avoiding logical fallacies or unsubstantiated claims are also critical. Mastering academic writing requires meticulous attention to detail and familiarity with broader rhetorical conventions across disciplines. Rigor in writing reflects rigor in thinking. A scholar's integrity becomes evident through the deliberate articulation of arguments, thorough presentation of data, and transparent acknowledgment of limitations. This part underlines the importance of writing as a crucial component of the research process, not just the concluding stage. It determines how knowledge is generated, validated, and shared. Developing one's academic voice involves both style and a commitment to one's discipline, the subjects of inquiry, and the broader society that research aims to serve. These interpretation, synthesis, and scholarly presentation principles are particularly impactful in crafting academic papers and theses. Every academic output requires a tailored approach to complexity, creativity, and methodological transparency, regardless of whether at the undergraduate, master's, or doctoral level. Typically, undergraduate coursework emphasizes fundamental research design and basic comprehension. Master's theses advance to methodological sophistication and critical synthesis, while doctoral dissertations demand creativity, theoretical contributions, and high academic independence. Understanding these distinctions enables researchers to cultivate academic maturity and produce work aligned with their educational and professional aspirations. Ultimately, dissemination concludes the journey from writing to impact.

Research achieves its purpose when it reaches its intended audiences: academic peers, practitioners, policymakers, or the general public. Dissemination pertains to communication, not merely publication. It necessitates a strategic awareness of how to adapt messages for different stakeholders, ensure accessibility without compromising rigor, and responsibly share knowledge in inclusive and culturally sensitive ways. Essential elements of this effort include dissemination strategies, knowledge translation, stakeholder engagement, and public scholarship. This final component completes the research journey, transitioning from the individual task of interpretation and writing to the collective responsibility of disseminating and applying knowledge. Its thoroughness, clarity, visibility, value, and impact assurances empower scholars to contribute meaningfully to their fields and society through their work.

Interpretation of Research Findings and In-Depth Discussion

Abstract This chapter thoroughly examines and interprets research findings, highlighting their importance within the broader academic and professional landscape. It rigorously evaluates the extent to which the findings corroborate or contest established hypotheses, enhancing the existing corpus of knowledge. The discourse transcends a simple data report, offering a comprehensive analysis highlighting the results' practical and theoretical ramifications. The chapter also discusses the study's intrinsic limitations, recognizing elements that may influence the validity and generalizability of the findings. By openly addressing these limitations, the analysis enhances the research's credibility and proposes avenues for future investigations to rectify recognized deficiencies. Additionally, it delineates prospective pathways for further inquiry, emphasizing chances for practical implementation in disciplines like architecture and engineering. This section situates the research within a comprehensive theoretical framework. It examines its practical implications, therefore highlighting the study's academic significance, outlining its contributions to the field, and establishing a basis for future investigations to expand upon the insights produced. This analytical method reinforces the research's contribution to enhancing academic dialogue and practical comprehension within the field.

1 Introduction

An in-depth explanation and evaluation of the study's findings are provided in the discussion chapter, which is one of the most important portions of any dissertation [1]. The major purpose of this section is to assess the aims of the study and to contextualize those findings within the larger academic and professional milieu. This chapter aims to explain the significance of the results by demonstrating how they contribute to the current body of knowledge and whether they support or contradict the theories and concepts that have been developed. The commentary should not just consist of a summary of the facts but should include a thoughtful analysis emphasizing the results' relevance. The researcher must explain how the findings answer the research

A. Almusaed et al., *Practice of Research Methodology in Civil Engineering and Architecture*, https://doi.org/10.1007/978-3-031-97393-2_20

questions or hypotheses, contributing to the field of study [2]. To do this, it is necessary to discuss the results and acknowledge any limitations or biases that may have been present in the research. Recognizing these limitations enables the reader to evaluate the findings more properly and ensures that the study continues to be transparent and adheres to rigorous scientific standards. Additionally, the conversation frequently suggests new avenues for study in the future, which offers the possibility of addressing problems that have not yet been answered or overcoming restrictions that have been highlighted [3]. An organized and systematic approach is taken in the construction of the conversation. In the beginning, it provides an overview of the most important findings and interprets how they connect to the aims of the overall research project. Following this, a comparison and contrast of the findings with previously published research is carried out, during which parallels, differences, and the distinctiveness of the present investigation are identified. Following this, the chapter delves into the constraints placed on the research, analyzing how these constraints influenced the validity and generalizability of the findings. The discussion concludes by looking to the future and pointing out areas that may benefit from more research to either expand upon the contributions made by the study or fill in gaps in the related field [4]. The reader should have a clear sense of the academic significance of the study as well as its possible influence on future research and practice by the time the conversation is over. This chapter is a turning point in the dissertation since it establishes a connection between the acquired empirical data and the theoretical framework presented in prior chapters. It offers a critical perspective through which to examine the findings, highlighting the relevance of such discoveries in the academic sphere and the realm of practical applications [5]. This approach is critical in sectors like architecture and engineering because the interpretation of findings must consider the consequences of new technologies, materials, or techniques in the actual world. As an illustration, while assessing environmentally friendly building materials, it is vital to go beyond the technical performance of the materials. An in-depth examination has to consider how these materials can potentially lessen the negative effects on the environment, such as carbon emissions, and how they might improve the long-term resilience of buildings [6]. Therefore, the conversation has to go beyond the immediate data points to evaluate the broader implications, aligning the research findings with bigger societal, environmental, and technological trends. The critical discussion after the interpretation of the results is necessary to engage with the facts in a relevant manner. This part provides a comprehensive analysis of the study's contributions to the field, noting areas in which it either confirms or contradicts previous research. It may require contrasting conventional and innovative approaches to building, evaluating the effectiveness of new materials, and considering the consequences of these approaches for the economy, the environment, and society [7]. In addition to emphasizing the study's breakthroughs, comparisons of this kind assist in placing research within the context of the current body of knowledge. Within the context of the debate, addressing the study's limits is an essential component. To preserve the integrity of scientific research, it is necessary to acknowledge these limits. These may result from decisions made during the design phase of the research project, such as the size of the sample, the limitations of the methodology, or external circumstances beyond the researcher's control [8]. When it comes to research in architecture and engineering, for instance, some of the limits that may be encountered include the difficulty of evaluating the long-term endurance of creative building approaches or the difficulty of scaling new materials across diverse climates. In recognizing these limitations, the researcher can give a more detailed analysis, enabling subsequent researchers to either enhance the study or investigate alternate methods. In addition to the academic analysis, the discussion should consider the wider implications of the results for professional practice, policy, and more research [9]. This is especially true in dynamic industries such as architecture and engineering, where new findings can drastically impact industry norms and public regulations. For instance, research on energy-efficient construction approaches might lead to changes in regulatory policies or establish new standards for environmentally responsible architecture to follow. The researcher ensures that the effect of the study resonates with both scholarly and professional audiences by placing the findings within these larger settings and introducing them into the discussion [10]. As a result, the discussion chapter has a dual purpose: it summarizes the study's findings and opens a critical debate with the existing literature and practice. Demonstrating how the study adds to the progress of knowledge in the field is essential, and this synthesis is an essential part of that demonstration. The discussion is more than just an interpretation of the facts; it also establishes the framework for future innovation and inquiry, giving a steppingstone upon which academics and practitioners may build [11]. The discussion chapter is an essential component of the dissertation because it establishes a connection between the outcomes of empirical research and the wider academic and professional framework [12]. Not only does it include a critical evaluation of the findings, but it also involves acknowledging the study's limits and making ideas for further research. In doing so, the discussion confirms the study's findings and assures that the findings are relevant and can be applied in both theoretical and practical situations. The research contributes to the advancement of the area using this meticulous analysis, encouraging further discovery and innovation [13].

2 Developing a Dissertation Discussion

Developing a well-structured research discussion is an essential component of the research process [14]. This is because the discussion serves as the basis for understanding and evaluating the findings of the research effort. Within this part, the researcher can comprehensively analyze how the findings correspond to the initial research questions, hypotheses, or objectives. It does this by bridging the gap between the findings of the research and the larger academic context, demonstrating how the study contributes to the existing body of knowledge. The researcher can highlight new insights, trends, or unexpected outcomes by interpreting the data considering the current literature. Furthermore, the conversation offers a forum to confront the study's limits beyond interpretation. Recognizing these constraints does not lessen the significance of the study; instead, it reflects the honesty and transparency of the academic community [15]. Through an in-depth analysis of the limitations or shortcomings of the study, the researcher makes it possible for subsequent researchers to improve or broaden the scope of the work. As an additional point of interest, the comment section allows you to propose possible directions for further research. Additionally, this may entail advocating additional research to be conducted on factors, suggesting different research approaches, or recognizing new research topics that surfaced throughout the study [10]. A comprehensive and perceptive discussion not only improves the general clarity and coherence of the dissertation but also draws attention to the intellectual contributions, analytical abilities, and capacity of the researcher to interact critically with the research subject [16]. Therefore, doing so strengthens the academic value of the work as well as the more enormous ramifications it has within its area.

2.1 Analysis and Interpretation of the Information Obtained from Research

This chapter is devoted to presenting the findings and providing an in-depth explanation. When the findings are first offered, they are presented in a way that is both clear and concise, which ensures that the data is transferred efficiently [17]. Following this, a thorough analysis is carried out, emphasizing the need for a comprehensive approach to conclude the relevance of the data concerning the study questions or hypotheses. For example, a study investigating the reuse of construction materials may demonstrate that recycled concrete aggregates exhibit strength equivalent to standard concrete in specific applications. The findings of this investigation will be given in terms of the quantitative data on compressive strength. Then, an analysis will be conducted to investigate the ramifications of this result [18]. To provide an interpretation, analyzing how this result lends credence to the concept that reusing construction materials might preserve structural integrity while also encouraging sustainability would be necessary. Furthermore, the study would consider any trends, such as the cost-effectiveness of recycling materials or the environmental advantages, and it would also address constraints, such as the variability in material quality or the circumstances in which reuse is most viable [19]. This research not only substantiates the possibility of reusing materials but also significantly contributes to the larger conversation about environmentally responsible building methods. By critically interacting with the findings, the study not only highlights the relevance of these results for increasing sustainable building design but also underscores its role in promoting environmental efficiency in the construction sector, which was one of the keys aims of the study [20].

2.1.1 The Communication of the Results of Research

To effectively distribute the findings of research, it is necessary to use supplementary tools such as graphs, tables, and other visual aids [21]. This set of tools is essential since they help improve the clarity of complicated material, making it easier for the audience to understand. By graphically depicting complex data sets, they assist readers in gaining a better understanding of the most critical results more expediently and efficiently, hence underlining the overall value of the findings. For example, in research centered on energy use in environmentally friendly buildings, incorporating tables or graphs that compare the levels of energy efficiency of various construction materials may greatly assist comprehension. The reader can rapidly grasp which items offer the best efficiency through visual aids, eliminating the need for the reader to comb through considerable amounts of text. This technique makes the key results more understandable and highlights the significance of the findings about environmentally responsible building practices. By employing these tools, researchers can guarantee that their data is conveyed clearly and concisely [22]. This makes it simpler for readers to connect with the study and obtain an interpretation of the findings.

2.1.2 Interpreting the Results

Following the presentation of the results, the subsequent crucial stage is interpreting those results. In this step, the data are reviewed once more, and an analysis is performed to see whether the findings are consistent with the primary hypotheses or assumptions developed at the beginning of the research [23]. The interpretation must have a solid connection to the theoretical framework formed earlier in the study and the empirical information obtained throughout the research process. Let's say, for instance, that the data indicates that using an environmentally friendly material dramatically reduces energy usage. In such a scenario, it is essential to investigate how this result is consistent with current ideas and earlier research conducted in the field. In addition, this analysis should determine if the data supports the hypotheses or whether they question them, which will provide a complete comprehension of the outcomes [24]. In addition, it is essential to consider the broader academic and practical implications of these discoveries, particularly about the design of architectural structures and sustainability [25]. This allows the analysis to contribute to the ongoing study and place the findings within the context of the broader discourse. It also provides insights that could influence future research and applications in the applicable subject.

2.2 Tackling the Research Problem

When interpreting the research outcomes, making a clear connection between those findings and the initial study subject and aims is necessary. This stage is vital for demonstrating how the data answers the research questions or hypotheses at the beginning of the study. It is essential to examine the methodology and explain how the research design, data collection, and analytic procedures aligned with the study's goals to successfully relate the findings to the research issue [26]. This makes it possible to link the findings to the research issue effectively. Suppose the research was conducted to investigate the influence that environmentally friendly materials have on the energy efficiency of buildings. In that case, the findings must be evaluated considering this question. The researcher can clearly and concisely demonstrate how the results directly react to the study question if they provide precise details on the technique utilized. On top of that, establishing a connection between the findings and the theoretical framework produced earlier in the research helps establish the interpretation within a more comprehensive framework. It reassures the researcher of the robustness of their interpretation. Using this strategy, you may achieve two outcomes: clarifying how the research helps address the initial issue and strengthening the study's overall validity [27]. By openly addressing the research topic via this lens, the interpretation becomes more robust, guaranteeing that the results are well-supported by both the data and the research method.

2.2.1 Making Connections Between the Results and the Research Questions

Establishing a strong and logical relationship between the study's findings and the questions asked at the beginning of the research is of the utmost importance to guarantee academic coherence and rigor. A systematic relationship must be established between each data point and the primary research topic, and the researcher must communicate these links clearly and concisely [28]. By doing so, the relevance of the results is elevated, and the general cohesion of the dissertation is enhanced. For example, suppose the research question intends to investigate the influence of different building materials on the thermal insulation of urban areas [29]. In that case, the section on the outcomes must immediately interact with the investigation. It is essential that the interpretation thoroughly elucidates how the data leads to a more in-depth understanding of the usefulness of the material in improving thermal performance in certain conditions. It is of the utmost importance to provide evidence that the results are not necessarily different from the study aims but rather intimately intertwined with them. The researcher strengthens the study's analytical framework by rigorously matching the findings with the starting hypothesis. Enhancing the intellectual and structural integrity of the dissertation is accomplished via this methodology, which guarantees that the findings are pertinent and contribute significantly to the progression of knowledge in the subject. This methodology guarantees that the findings are relevant and contribute significantly [30].

2.2.2 Placing the Results Within the Context of the Theoretical Framework

The research findings must be positioned within the theoretical framework created in previous parts of the study to guarantee that the research is rigorous and scholarly. The researcher's capacity to relate empirical evidence to pre-existing theoretical conceptions is highlighted by this alignment, which demonstrates a profound comprehension of academic discourse. By incorporating the findings into the theoretical framework, the researcher demonstrates how the study contributes to existing theoretical discussions in the academic community [31]. By comparing the findings to previously known ideas, this method verifies the findings and allows one to question or build upon previously accepted paradigms by utilizing fresh data. For example, if the research investigates environmentally responsible building techniques, the findings should be regarded as pertinent ideas of ecological design or sustainability [32]. By doing so, the study is placed within the larger academic framework, and the results are brought to light regarding how they either support or reject the already available information. The researcher establishes the study's significance and ability to enhance academic thinking by connecting the investigation's findings and the theoretical framework. In addition to ensuring that the dissertation maintains conceptual coherence overall, this emphasis on maintaining conceptual coherence can reassure the audience about the quality and reliability of the research. This makes a significant contribution to the area.

2.3 Recognizing the Limitations Constraints

In academic research, acknowledging the limitations of a study is an essential component that demonstrates intellectual rigor and maturity [33]. It is important to recognize that any research endeavor, regardless of how it was designed or carried out, is subject to certain limits. Openly addressing these limitations ensures that the research process is transparent and credible. In the following part, we will attempt to shed light on any possible biases, methodological flaws, or external influences that may have had a role in influencing the findings. For example, sample size, the methodology of data collection, or the study's focus are all potential constraints sources. There is a possibility that the results cannot be generalized due to the small sample size, and some data collection approaches may cause biases, mainly when subjective responses are involved. Additionally, the study may be limited by time or resources, affecting the data analysis's depth or the findings' comprehensiveness. In addition, it is of the utmost importance to consider any assumptions about the study since they might impact on how the results are interpreted. In addition to providing a well-rounded interpretation of the findings, the researcher exhibits a full awareness of the bounds of the study by identifying and expressing these restrictions [34]. Because of this openness, the reader can critically evaluate the findings inside their appropriate context, which contributes to an overall improvement in the dependability and integrity of the study.

2.3.1 Restrictions Imposed by Methodology

To properly evaluate the research limits, it is necessary to consider the methodological restrictions that may have impacted the study results [35]. These restrictions may result from factors such as the size of the sample, the accuracy of the measurements, or uncontrollable external variables. To maintain transparency and guarantee the validity of the study findings, it is essential to acknowledge the difficulties encountered on the methodological front. For example, in research that assesses the performance of construction materials across various climate zones, the uniformity of the results may be affected by the diverse weather conditions [36]. Such external influences, which are beyond the researcher's control, have the potential to add data variability, hence restricting the consistency of the conclusions. Like the previous point, the generalizability of the results may be jeopardized if the sample size or selection criteria are too limited. The study's credibility is improved by acknowledging these methodological constraints, which also stimulates more research to be conducted to solve these issues. By demonstrating transparency regarding the possible limitations of the study, the researcher contributes to a more comprehensive comprehension of the findings and enables subsequent research to build upon the work that has already been done using more advanced approaches [37].

2.3.2 Allowed Potential Biases

Not only is it necessary to acknowledge the limits of the methodology, but it is also crucial to acknowledge the potential biases that may have influenced the study [38]. Many different parts of the research process might give rise to biases. Some of these factors include the design, the selection of participants, and the data-collecting techniques. To display a careful and critical attitude when reviewing the study outcomes, it is recommended that these biases be acknowledged. An example would be biases in selecting participants, which might occur if the sample does not appropriately reflect the population studied, leading to skewed conclusions. In a similar vein, methods of data collecting, such as self-reported surveys or interviews, have the potential to create bias if participants offer comments that are subjective or intended to gain social approval. Even the research design, such as how questions are framed or the chosen variables, might accidentally lead to biased conclusions [39]. By addressing these possible biases openly and honestly, the researcher promotes openness and indicates that they have a balanced view of the data. This disclosure strengthens the study's credibility, and readers can critically evaluate the results considering these potential impacts. This, in turn, encourages more research to be conducted to eliminate biases in future studies.

2.4 *Contrasting the Results with the Previously Published Literature*

For a comprehensive examination of research findings, it is necessary to compare them with previously published material to highlight the contribution that they have made to the academic area [40]. This allows the researcher to highlight the uniqueness of the work and its position within the larger scholarly discourse by comparing the results of the study with those of previous research. It is crucial to make this comparison to determine if the findings agree with current beliefs and practices, whether they question them, or whether they go beyond them. Suppose the research investigates the utilization of digital twins in the administration of building projects, for instance. In that case, the findings need to be examined, considering the previous research that has been conducted on this technology. Suppose the findings are consistent with previously known information, such as the capacity of digital twins to improve the efficiency of projects and lower costs [41]. In that case, this comparison will reinforce the validity of the current study and the work done in the past. If, on the other hand, the results are different, indicating limitations in the use of digital twins that were not anticipated, this gives a chance to evaluate or revise existing hypotheses and suggest new research prospects. Critical engagement with the current body of literature may better determine the study's contribution to advancing knowledge on the topic. Whether the findings support or contradict the findings of a previous study, this comparison contributes to a better understanding of digital twins and the potential they provide while also providing significant insights for research that will be conducted in the future.

2.4.1 Contrasting the Results with Previously Published Literature

It's essential to connect your findings with research like yours. This will demonstrate whether the results are consistent or divergent. Using this method helps you place your research into the context of the existing body of knowledge, providing a more precise understanding of how your work adds to the area. Validating the dependability of your findings and reiterating their relevance within the larger academic context may be accomplished by connecting your findings with other research that has been conducted previously [42]. For example, if your study analyzes the efficacy of smart building technologies in enhancing energy efficiency, comparing your findings to those of other studies that investigated comparable technologies might shed light on recurring patterns or distinctions. When the findings of your study are consistent with those of previous research, they lend credence to the legitimacy of both your work and the studies that came before it [42]. Alternatively, could your findings go in different directions? In that case, this difference may point to new directions that need to be investigated further or bring to light gaps in the existing body of research. Regardless of the circumstances, when you correlate your results with previous studies, you increase the depth of your analysis. It bolsters the overall argument and ensures that your research is meaningfully integrated into the ongoing discourse that is taking place in the academic community.

2.4.2 Observing and Demonstrating Originality

Even though it is crucial to connect research findings with current literature, it is equally essential to demonstrate the study's originality [43]. The study should highlight its unique contributions to the field, whether through pioneering discoveries, innovative methodology, or new theoretical insights. This section should underscore these contributions. Emphasizing the work's uniqueness is critical to establishing its relevance in academic discourse. For example, if the research introduces a novel method for assessing the thermal properties of new construction materials, it should be presented as a significant contribution. Clearly define how the study differs from or expands upon prior research, advancing the existing understanding of the subject. Reiterate the importance of the work to establish it as a crucial addition to the field [44]. To emphasize its value, articulate how the findings differ from, build upon, or challenge previous research. Demonstrating the study's potential to influence future research and contribute to ongoing scholarly discussions enhances its academic value.

2.5 Make Suggestions for the Direction of Future Research

The capacity to lay the groundwork for further investigation is a defining characteristic of a well-structured academic discussion [45]. This section should describe areas requiring additional exploration to address the study's limitations or investigate new questions that have emerged from the results. Proposing future research initiatives is crucial for highlighting the work's broader significance and its potential to contribute to the ongoing expansion of knowledge in the field. For example, if the study revealed methodological shortcomings or external factors that might have influenced the findings, recommending research to address these limitations would be valuable. Additionally, if unexpected discoveries or new lines of inquiry emerge, they should be proposed for further exploration. Specifying the subsequent stages of research extends the study's relevance and establishes a foundation for future researchers to build. This approach ensures the work catalyzes continued scholarly progress, enhancing its contribution.

2.5.1 Addressing the Limitations of Individual Studies in Subsequent Research

Addressing limitations identified in previous research is crucial for suggesting solutions that can be applied in future studies [43]. By tackling these constraints, the credibility of research and its contribution to the broader academic discourse can be significantly enhanced. For instance, if a small sample size limits a study, recommending more comprehensive and diverse samples in future research would be advisable. This approach would improve the generalizability and robustness of the findings, leading to a more comprehensive understanding of the subject. Additionally, if methodological limitations, such as reliance on self-reported data or cross-sectional designs, are present, future studies could adopt more rigorous methodologies, including objective measurements or longitudinal approaches, to strengthen the validity and reliability of the results. Implementing these methodological improvements would lead to more accurate and reliable outcomes. Acknowledging the limitations of a study and offering carefully considered suggestions to address them demonstrates a commitment to advancing academic research [46]. This proactive approach enables subsequent research to build on existing findings, resulting in deeper insights and a more nuanced understanding of the field.

2.5.2 Exploring Unconventional Research Routes and Methods

Identifying new research directions is essential for deepening the understanding of study findings and ensuring a focused approach for future research endeavors [47]. This involves addressing any existing constraints and exploring unique questions that emerge from the study. Such investigations might include examining alternative resources, applying the study to different contexts, or assessing the long-term impact of the variables considered. For example, if research concentrated on the short-term benefits of environmentally friendly construction materials, future studies could explore the performance of these materials over extended periods or in diverse climatic conditions. Highlighting these new lines of inquiry demonstrates an awareness of the study's broader implications and potential impact on future academic work [44]. This approach also encourages researchers to build upon existing findings and delve into previously unexplored areas within the field, thus contributing to the continuous growth of knowledge. Outlining these potential avenues for future research ensures that the study is viewed not as isolation but as part of a more extensive academic dialogue. This approach emphasizes the importance of study and the ability to foster further understanding of the subject matter.

2.6 *Expanding the Context: Impact on the Discipline*

The concluding section should emphasize the broader academic significance of the study, highlighting its contribution to advancing the field [44]. This part thoroughly analyzes the results' relevance and explores their potential implications for current and future research. This analysis demonstrates its importance and potential impact on ongoing research by positioning the work within the discipline's larger theoretical and empirical frameworks. For example, suppose a study examines recent advancements in environmentally friendly building materials. In that case, clarifying how these findings contribute to developing more efficient construction methods and enriching the existing knowledge on sustainability is crucial [48]. It is also essential to identify how the study might inspire future research, whether by opening new avenues of inquiry or refining existing theoretical frameworks. Drawing connections between the results and the broader academic discourse helps illustrate the uniqueness of the contribution and its potential to shape future research directions. This comprehensive evaluation reaffirms the study's significance within the field, highlighting its role in fostering academic progress and sparking new scholarly discussions.

2.6.1 Assessing the Implications of the Study

It is vital to provide a precise definition of the relevance of the results to comprehend the contribution that they may have made to the study area [49]. This part emphasizes the study's broader applicability, whether through advancing theoretical frameworks, providing practical consequences for industry, or challenging established paradigms. When the effect of the study is highlighted, it illustrates the significance of the research in both academic and practical contexts. Suppose the research presents novel understandings of environmentally friendly building materials. In that case, it is necessary to explain how these discoveries assist in developing more environmentally friendly building practices. Describing how the study enhances or refines pre-existing theories is essential, mainly if the research offers a fresh theoretical approach [50]. On the other hand, highlighting how they present fresh opportunities for academic discussion or more research is paramount if the findings contradict the commonly held beliefs. It is essential to define the study's effect since doing so highlights its contribution to the academic field and increases its relevance for practitioners in the industry. With this technique, the study will undoubtedly be acknowledged as a significant contribution to the existing academic literature and practical applications, reiterating the research's value in driving future breakthroughs [51].

2.6.2 Participating in the Development of Emerging Trends

Research must be positioned within developing trends to remain relevant as academic fields continue to build [52]. To do this, it is necessary to emphasize the possible impact of the study's findings on future scholarly debate and their contribution to ongoing advancements within the subject. The study's importance extends beyond academics, affecting future governmental choices or spurring innovation, notably in environmentally friendly building practices and sustainable building materials. For example, suppose the study investigates developments in environmentally friendly building techniques. The findings may determine future sustainability trends, impacting academic research and corporate practices [53]. The significance of the study in terms of its contribution to broader social concerns, such as environmental sustainability, is emphasized by the fact that it is aligned with existing and expected forms of development. Involving oneself in these developing tendencies ensures that study will continue to be relevant and shows its ability to stimulate more research and innovation. Using this strategy, the study is positioned to become crucial in defining future policy and industrial orientations, guaranteeing that its effect will reverberate inside the academic community and beyond academic circles. In the discussion section of the research, the researcher offers a critical analysis, interpretation, and contextualization of their results. This is the most essential component of the dissertation. This chapter discusses important aspects, such as an in-depth analysis of the findings, an analysis of how the findings relate to the research objectives, and a discussion of any limits found throughout the study. In addition, it examines the findings of the current body of literature, determining whether there are any similarities or differences, and provides recommendations for the future of the study. In addition, the commentary emphasizes the broader ramifications of the study, thereby demonstrating the contribution that it has made to the field and providing a justification for its relevance [44]. The section explains how research advances knowledge and retains relevance for current and future investigations. This ensures that the research will influence the academic community. This is accomplished by setting up the study within the larger educational environment. Essential parts of a dissertation's discussion chapter are highlighted in Table 1, which includes critical analysis, research relevance, and literature comparison, among other things. To ensure the discussion portion is thorough and effective, it lists typical problems and ways to fix them.

3 Strategies and Structure for Crafting a High-Quality Discussion in Scientific Writing

3.1 The Process of Writing the "Discussion" Section of a Scientific Article

A scientific publication's "Discussion" portion is essential for understanding study findings and situating them within the larger framework of current knowledge [61]. This part requires a comprehensive examination and a deep comprehension of the conclusions and the pertinent literature. In research examining sustainable recycling materials in construction, for example, the "Discussion" must elucidate how

Table 1 Key components of the discussion chapter, associated challenges, and strategic solutions

Key element	Description	Common challenges	Resolution strategies
Critical interpretation of results [54]	It provides a deep, nuanced analysis of the research findings, explaining their significance, relevance, and potential implications	Avoiding misinterpretation, maintaining objectivity, and resisting the temptation to overstate the findings' importance	Use multiple frameworks or theories to cross-check interpretations and remain aligned with the research scope
Linking findings to objectives [55]	Examine how the findings directly answer the research questions and achieve the objectives	Ensuring that every conclusion is firmly rooted in the research objectives without making unsupported claims	Use structured comparisons and mapping techniques to connect findings to objectives visually
Acknowledgment of study limitations [56]	Identify and explain the limitations, such as methodological constraints, data quality issues, or external factors impacting results	Striking the right balance between transparency and undermining the study's credibility	Frame limitations as opportunities for further research and explain how they were managed
Comparative analysis with literature [57]	Discusses how the study's findings align or contrast with previous research, highlighting significant similarities, differences, or anomalies	Locating the most relevant, recent, and high-quality literature for valid comparison	Conduct a thorough literature review and critically evaluate the validity of the studies used for comparison
Recommendations for future research [58]	Suggest areas where further research is needed to address the remaining gaps or explore new questions raised by the current findings	Ensuring the recommendations are actionable, realistic, and directly tied to the study's scope and findings	Based recommendations on observed limitations and identified research gaps, maintaining practical and feasible suggestions
Highlighting broader contributions [44]	Demonstrates the study's impact on advancing knowledge and its broader implications for the field or industry	Convincingly demonstrating the significance without overstating the results' impact or novelty	Situate findings within both theoretical and practical contexts to show their real-world application and scholarly value

(continued)

Table 1 (continued)

Key element	Description	Common challenges	Resolution strategies
Justifying study's relevance [59]	Explains how the research remains valuable for current academic and practical applications and future research directions	Ensuring long-term relevance, particularly in fast-evolving fields, while avoiding speculative claims	Emphasize the study's connection to enduring challenges and emerging trends in the field
Academic contextualization [60]	Places the study within the broader academic discourse, showing how it contributes to ongoing debates or unresolved questions in the field	Successfully integrating the research into existing debates without being repetitive or redundant	Regularly engage with the latest scholarly publications and demonstrate how the research bridges gaps or introduces new ideas

the findings enhance the field, such as demonstrating that recycled materials like concrete aggregate improve structural integrity, mitigate environmental impact, or provide economic advantages. This section has to juxtapose current data with prior investigations, elucidating consistencies or differences in findings. If the study identified improved durability in recycled materials, the discussion might connect this to other current research while examining why previous studies may have had divergent results [62]. Maintaining a clear and ordered framework is crucial for coherently presenting data and emphasizing relevance. Additionally, the discourse should suggest future research avenues, including evaluating recycled materials' performance under diverse environmental circumstances and assessing their long-term sustainability advantages within the building sector. The objective is to illustrate how the study outcomes enhance understanding and contribute to continuing discourse within the discipline. Figure 1 shows the process of authoring the "Discussion" part, which is about environmentally friendly building materials that have been recycled. Examining study results, comparing them to previous research, describing discrepancies, emphasizing contributions, suggesting areas for future research, and keeping the flow organized are all essential steps. The arrows show the connections between these actions that make sense.

3.2 *The Structure of the Conversation*

A well-organized framework is necessary when producing an excellent "Discussion" part for a scientific publication. Not only does systematic organization improve clarity, but it also guarantees that the research findings are contextualized appropriately within the area. Because of this, the reader can better understand how the findings relate to the more significant body of scientific knowledge. The following

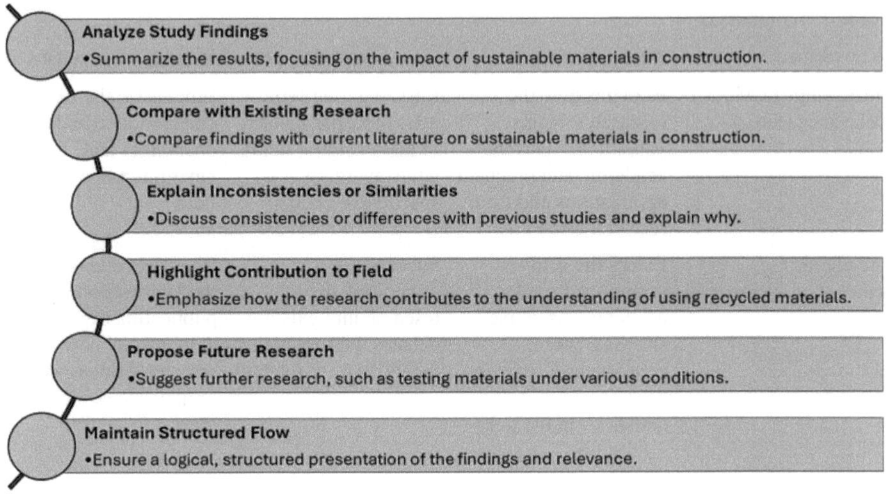

Fig. 1 Process of writing the "discussion" section of scientific work (*Source* Authors)

is an example of a structure generally known for arranging the "Discussion" section. This format strikes a balance between coherence and thoroughness.

3.2.1 The Introduction Paragraph

In the first paragraph of the "Discussion" section, you should briefly describe the most noteworthy findings from the study [63]. It is essential to establish a direct connection between these findings and the research question included in the "Introduction." This section aims to provide a seamless transition between the earlier results and the more in-depth analysis in the following paragraphs. When the reader is presented with a concise and straightforward introduction, such as "The primary findings of this study suggest…", they are given a brief overview of the research findings before delving into the more in-depth implications. For instance, in research that investigates the application of digital twins in environmentally responsible building practices, the opening paragraph may include a summary of the most important findings, such as how the utilization of digital twin technology enhances the efficiency of projects and decreases waste. Similarly, research focusing on daylight efficiency in buildings might synthesize data indicating how optimum daylight integration can reduce energy consumption and promote indoor comfort, directly addressing sustainability goals in contemporary architectural practices. Figure 2 depicts the links between the most essential academic writing topics. A focus is placed on the way the "Research Question" leads to the "Key Findings" and the link between those findings and the "Previous Research," "Discussion," and "Implications." Additionally, it illustrates how the conversation contributes to "Future Research" and, eventually, how it leads to "Conclusions."

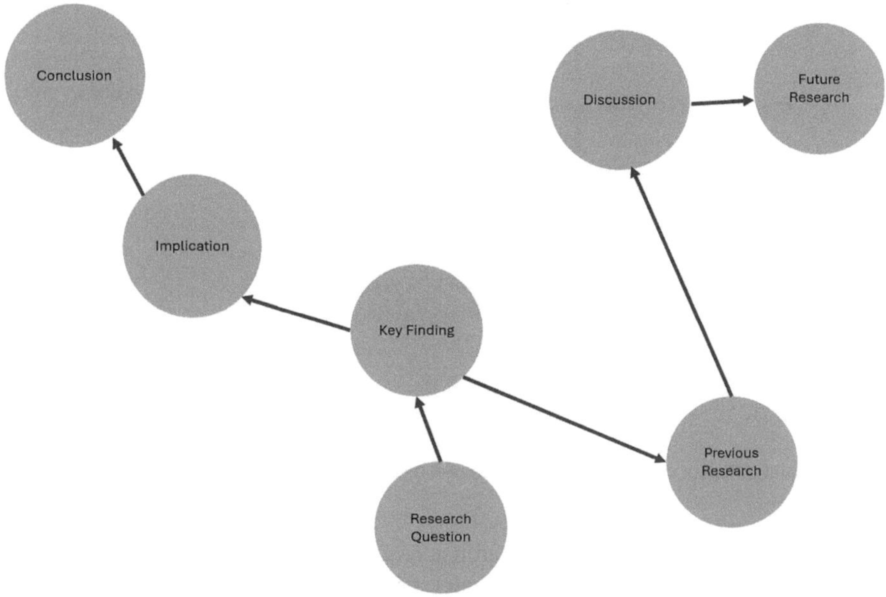

Fig. 2 Clear relationship diagram between key concepts in academic writing (*Source* Authors)

3.2.2 Body Paragraphs

Each succeeding paragraph should focus on putting these findings within the context of previous research, which should concentrate on specific outcomes or topic areas. The body paragraphs should methodically compare the results of the current study with those of earlier research, stressing the similarities and differences between the two sets of findings. For example, in a study that investigates the use of digital twins in environmentally responsible building practices, one paragraph can examine how the study's findings correspond with prior studies on the potential of technology to minimize the number of resources consumed. One such statement might be read as follows: "The findings indicate that the implementation of digital twin technology resulted in a 15% reduction in material waste, which is consistent with the findings of studies conducted by Y and Z" [64, 65]. An additional comparison might be made to address how the current study's findings extend on earlier research by indicating that digital twins can also improve energy use throughout the operating phase of buildings, which is a less researched element in literature. Similarly, a paragraph could be written on how the current research coincides with studies that have proven a reduction in energy usage by maximizing natural light. The body paragraphs should also mention a feasible explanation for any discrepancies found in earlier research. This might include variable meteorological circumstances or variances in computer modeling technologies. In addition, they can suggest potential future study paths, such as evaluating the effectiveness of digital twin technology in various construction settings or examining the effectiveness of daylighting in multiple building types and

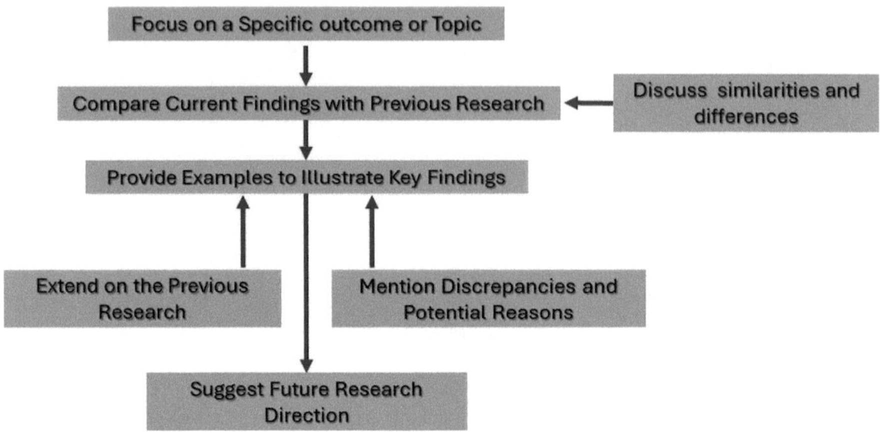

Fig. 3 Structured process for writing the discussion section in a scientific article (*Source* Authors)

climates. For example, "Further research is required to assess the long-term sustainability impacts of digital twins in complex urban construction projects" or "Future studies should explore how daylight efficiency can be maximized in varying latitudes to reduce energy consumption further" are examples of statements that could be made. A scientific publication's "Discussion" section might be written systematically, as seen in Fig. 3, which exemplifies this method. The procedure starts by concentrating on specific outcomes or subjects and comparing the present findings with those of earlier research. In addition to correcting disparities, presenting instances, explaining parallels and contrasts, and expanding past research, it emphasizes sharing examples. At the end of the process, suggestions are made on potential future study areas. This stage ensures that the debate is exhaustive and appropriately contextualized within the current body of literature.

3.2.3 Considering the Limitations

One of the most critical aspects of the "Discussion" part is recognizing the constraints placed on the research [66]. Incorporating observations throughout the discussion and using a dedicated paragraph are viable options for accomplishing this goal. The study's credibility is improved when the research limits are addressed openly and honestly. These limitations may pertain to the findings' capacity to be generalized, the size of the sample, or the methodological constraints. In the case of research on digital twins in construction, for instance, a paragraph under "Limitations" may mention that the simulation environment utilized in the study might not ideally mimic the circumstances seen in the real world. This may restrict the extent to which the findings can be generalized. The following might be included in the paragraph: "The limitations of this study include the reliance on a controlled digital simulation environment, which may not fully capture the complexities of real-world construction projects." In the same vein, the limits of the research on daylight efficiency might be related to the fact that the findings were exclusive to a particular geographic region. Because the patterns of daylight in tropical climates are notably different from those in the temperate zone locations that were investigated in this study, it is possible that the findings cannot be fully generalized to structures in tropical climates. By addressing these limits openly and honestly, the researcher indicates that they are aware of the scope and bounds of the study while preserving the credibility of the results [67]. The tactics for identifying the limits of research are outlined in Table 2. These strategies include embedding observations throughout the discussion, utilizing dedicated paragraphs, and expressing honest recognition. It brings to light problems such as ensuring consistent discussions of limits, avoiding an unduly narrow perspective of the study's scope while retaining credibility, and striking a balance between sustaining research value and keeping transparency.

Figure 4 shows how several measures might be taken to resolve research restrictions in academic studies. This begins with the awareness of the study's limits, followed by strategies such as integrating them throughout the discussion, employing a dedicated paragraph, or acknowledging them. Every strategy has further repercussions, such as enhancing the study's legitimacy, drawing attention to the limitations imposed by geography, and demonstrating a grasp of the extent of the research. The controlled example technique emphasizes geographical limits, highlighting the importance of transparency while addressing these limitations.

3.2.4 The Final Exposition Paragraph

The last paragraph of the "Discussion" section encapsulates the principal findings and arguments presented throughout the discourse [74]. This section aims to highlight the alignment of the findings with the initial research question posed at the outset of the inquiry. This closing explanation must reiterate the key results, providing a clear and concise summary that underscores the importance of these findings and the study aims. It is essential to leave the concepts and ideas as they are when writing this paragraph. In conclusion, one should briefly summarize the most critical points derived from the analysis and interpretation of the data. The reiteration technique enhances the discussion's internal coherence, ensuring the reader understands how the study's findings address the research issue. The conclusion would reiterate this contribution, highlighting its significance in fostering improvements in sustainable construction, mainly if the survey uncovered progress in sustainable building practices through innovative materials. Consequently, the ending paragraph functions as a last assertion that underscores the study's contribution, offering the reader a distinct understanding of the extensive ramifications and impact of the research. For example, the final paragraph may state: "In conclusion, this study illustrates that the incorporation of twin digital technology markedly improves resource efficiency and diminishes waste, rendering it a feasible approach for advancing sustainability in construction initiatives." This exemplifies the application of digital twins in sustainable building. A comparable conclusion may be articulated for a study on

Table 2 Strategies and challenges in acknowledging research limitations

Section	Subsection	Description	Challenges
Recognizing research limitations		The central aspect of the discussion section is to identify and acknowledge the research's limitations. This process ensures transparency and maintains study credibility	Identifying all potential limitations without undermining the perceived value of the research
Incorporating observations [68]	Throughout discussion	Limitations can be woven into the analysis and interpretation of results. Readers remain aware of constraints that may affect the interpretation of specific findings	Ensuring consistent mention of limitations without overwhelming the discussion with negatives
Generalizability issues [69]	Acknowledge aspects that limit findings' applicability to other contexts or populations	Balancing transparency with maintaining the broader applicability of results	
Dedicated paragraph [70]	Controlled study example	Discusses limitations in a focused paragraph, addressing specific constraints	Crafting a detailed explanation without diverting focus from the study's strengths
Geographic limitations [71]	Discusses how geographic factors can impact study applicability in different regions	Addressing the limitation without limiting the perceived scope of the study's results	

(continued)

Table 2 (continued)

Section	Subsection	Description	Challenges
Honest acknowledgment [72]	Improve Study Credibility	Open discussion of limitations enhances the reliability of the study, showing awareness of constraints and biases	Avoiding the perception of being overly critical or undermining the study's contributions
Shows awareness of scope [73]	Transparency about study boundaries helps readers evaluate the conclusions by showing what the study can and cannot address	Clearly defining the study's scope without appearing too restrictive	

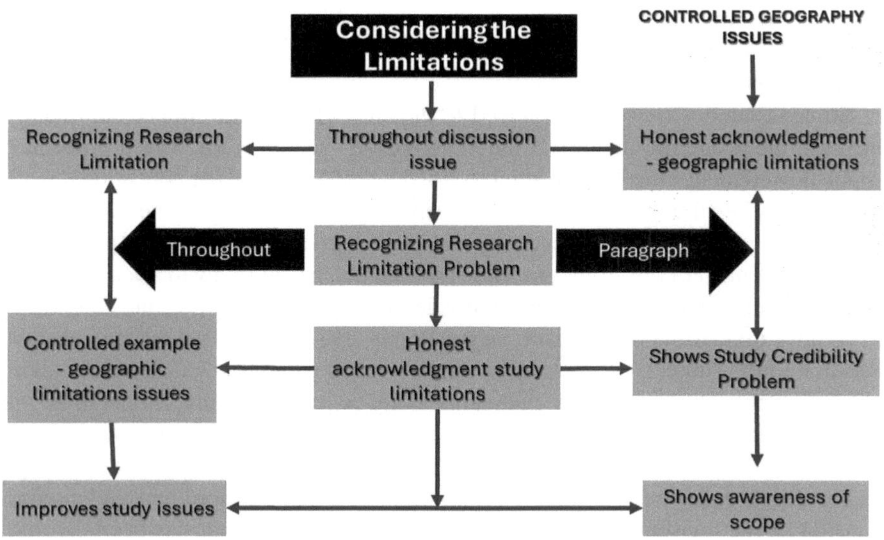

Fig. 4 Approaches to addressing research limitations in academic studies (*Source* Authors)

daylight efficiency: "This study substantiates that enhancing daylight utilization in architectural design diminishes dependence on artificial lighting and fosters energy efficiency, thus advancing overarching sustainability goals in architectural practice." [75] This conclusion guarantees the reader a comprehensive understanding of how the research findings contribute to the area and their direct relation to the first stated research problem. The iterative process of improving research findings to align with research objectives is depicted in Fig. 5. The emphasis is placed on summarizing the most critical findings, ensuring that the findings are aligned with the research

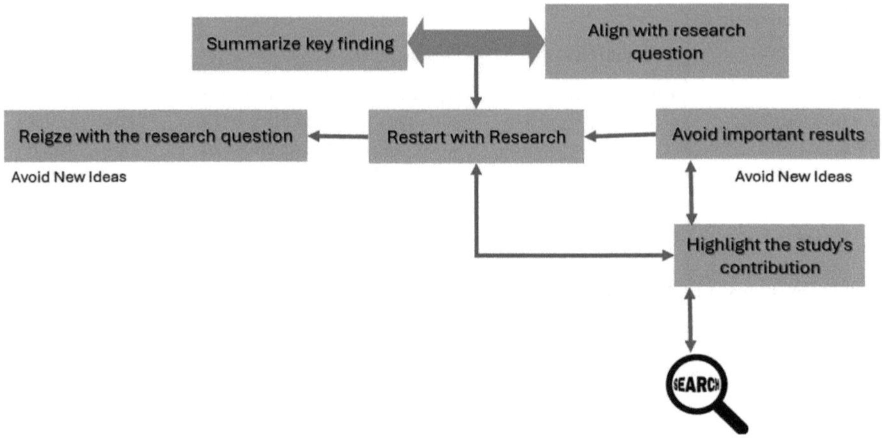

Fig. 5 Iterative research alignment and contribution mapping process (*Source* Authors)

question, avoiding neglecting essential results, and highlighting the study's contributions. If required, the process is restarted with more research to improve or realign the emphasis of the study.

Strategic tactics must be employed to enhance clarity and intellectual depth to increase the academic quality of the "Discussion" section [76]. The utilization of sophisticated tactics has the potential to considerably increase the educational worth of this part, even though it is essential to keep a regular framework. The "Discussion" section of a scientific study is necessary since it interprets the findings and places them within the context of the larger academic community. By adhering to a defined framework, adopting methodologies that assure depth and analytical rigor, and committing to iterative improvement, scholars can generate a "Discussion" that makes a valuable addition to the subject. It is helpful to employ comprehensive tactics such as taking a holistic perspective, critically evaluating using "why" questions, cooperating with colleagues to obtain varied opinions, and extensively interacting with current literature to reach better academic rigor [76]. These approaches encourage more in-depth examination, going beyond simple interpretations to give a more nuanced comprehension of the consequences of the study. A "Discussion" that has been carefully constructed provides insightful interpretations of the results and indicates opportunities for future research, contributing to the advancement of academic discussion and the growth of the field.

4 Interpretation of Key Findings

4.1 Significant Findings

Interpreting major findings is essential for extracting significant insights from research since it facilitates a deeper comprehension of the underlying dynamics inside a study. For example, the key findings of a study illuminate patterns and connections that elucidate critical facets of sustainable practices in building, especially with the utilization of reclaimed wood and material documentation. By meticulously analyzing these results, researchers may better understand the interrelationships among factors and their temporal interactions [77]. This study's principal conclusion is the robust positive link between using material passports and the recyclability of wooden building materials. Material passport's comprehensive records that monitor the composition and lifetime of construction materials are vital in advancing sustainable practices. The study's statistics indicate that structures with thoroughly recorded material passports are more prone to attain elevated recycling rates. This correlation demonstrates that material passports enhance the identification of recyclable components and promote sustainable construction methods. Utilizing material passports in reclaimed wood projects enables stakeholders to monitor the wood's provenance, age, and treatment history, assuring compliance with sustainability standards [78]. The report also shows significant trends in using recycled wood in buildings. Older structures with material passports have better material recycling rates than modern builds. This might be ascribed to detailed documentation of materials in ancient structures, encompassing their origin and any treatments received. Such records can facilitate reusing or repurposing resources such as recycled wood. Accessing comprehensive information on a material's history mitigates uncertainties in the recycling process, reducing the evaluation of the quality and safety of recycled wood for application in new building projects. This discovery underscores the need to preserve precise material documentation over a building's lifecycle since it can substantially improve recycling results. Furthermore, the research indicates that material passports might affect decision-making in sustainable building. In projects utilizing recycled wood, a material passport offers detailed information on the wood's condition, including its durability and prior uses. This information may facilitate informed decisions on the optimal integration of reclaimed wood into new structures, ensuring its application maximizes its lifetime and minimizes waste. Material passports provide a comprehensive understanding of material qualities, allowing project managers to integrate recovered wood with sustainability objectives, including reducing the carbon footprint in new structures. Comprehending these trends is essential for formulating focused solutions that tackle significant issues in sustainable building [79]. For example, acknowledging the need of thorough material documentation may result in policy proposals that promote or require the implementation of material passports for all new building endeavors, especially those utilizing recycled materials. This would guarantee the maintenance of comprehensive records from the beginning, enhancing the long-term recyclability of construction elements. The study underscores the significance of recycled wood in sustainable building practices, endorsing measures emphasizing utilizing recovered materials in new projects and fostering a circular economy within the construction sector. This study's notable findings underscore the importance of material passports in improving the recyclability of construction materials, mainly recycled wood [80]. The findings demonstrate that documentation is essential for attaining sustainable construction practices and facilitating improved material management throughout a building's lifespan. These insights establish a basis for formulating successful strategies and regulations that enable the transition to more sustainable and resource-efficient construction practices.

4.2 A Discussion of Unanticipated Final Results

Even though the study may have successfully accomplished its primary objectives, the unexpected results may provide valuable insights that warrant additional investigation. These kinds of results frequently bring to light intricacies in the data that were not anticipated during the study design phase, aiding in developing a more nuanced comprehension of the subject matter of the investigation [81]. An example would be if the research revealed that although painstakingly recorded through material passports, high-tech materials displayed lower-than-anticipated recyclability rates. This would be a challenge to the initial premise. Although these data imply that additional factors may have a more relevant role, the expectation may have been that comprehensive documentation would intrinsically improve the recyclability of materials. To understand these unexpected outcomes properly, thoroughly investigating the potential underlying reasons is vital [82]. One possible explanation for the decreased recyclability of high-tech materials is that their chemical compositions are more complex than those of other materials. Despite the availability of extensive data, such intricacy can be a barrier to recycling operations, making it more challenging to separate the elements and find new uses for them. Furthermore, even if detailed information on the composition of these compounds is available through material passports, it is possible that industrial procedures have not yet sufficiently evolved to handle these materials efficiently. Consequently, there is a gap between theoretical frameworks and the use of such frameworks in industry, as seen by the mismatch between the availability of documentation and practical recyclability. It is of the utmost importance to address these unexpected discoveries since doing so enables a more comprehensive evaluation of the study issue. The research can find areas where the existing methodology or industrial practices are lacking by conducting an in-depth analysis of the factors responsible for the lower recyclability rates. It is abundantly evident that merely enhancing documentation could not be sufficient if other systemic problems, such as limits imposed by technology or economic constraints, are not simultaneously addressed. This conclusion calls for more research to design recycling techniques to manage the complexity of high-tech materials [83]. Furthermore, these discoveries present a chance to refine different ideas and theories that have already been established. Researchers are prompted to evaluate and expand upon the existing body of information due to the challenge to the initial premise that material documentation directly corresponds with better recyclability. This reevaluation is critical because it allows for the design of more rigorous research questions that consider the varied character of material recycling. In addition, it emphasizes the significance of researching how industry norms and practices change to keep up with the growing utilization of sophisticated materials in the building and manufacturing industries. The fact that the study considered unexpected findings is another factor that leads to a more well-rounded discussion on the subject. The research highlights the dynamic nature of scientific investigation, characterized by the fact that the outcomes do not always meet initial expectations. This is accomplished by conceding that certain assumptions were not made in practice. A more honest evaluation of the existing state of knowledge is made possible due to this transparency in resolving disparities, highlighting the necessity of ongoing learning and adaptation within the profession. Examining these unanticipated findings enhances the study's conclusions by drawing attention to gaps in the currently utilized procedures and identifying areas requiring more exploration. It sheds light on the inherent challenges associated with recycling high-tech materials. It indicates that progress in documentation has to be supported by corresponding breakthroughs in the capacities of their respective industries. This method not only makes the study findings more relevant but also paves the way for more research that will be conducted in the future to bridge the gap between theoretical potential and actual execution [84]. According to Table 3, the relevance of unexpected discoveries in research is investigated, emphasizing the function these results play in enhancing knowledge and locating new study routes. In doing so, it exposes issues such as interpreting data without prejudice, adopting industrial processes, and bridging gaps between theoretical frameworks and practical implementation, highlighting the significance of doing extensive research.

5 Contextualizing Findings Within Literature

5.1 Comparison with Previous Studies

It is essential to place the research findings within the context of the current body of literature to explain how fresh insights add to the academic discipline [95]. Not only does this procedure put the present investigation within the context of the larger discourse, but it also assists in determining the relevance and originality of the study's findings. Scholars can confirm existing ideas, criticize prevalent assumptions, or give new views by comparing their findings with those of earlier studies. These findings highlight that material passports' impact varies with the materials' complexity and refine and nuance this general understanding. For instance, if previous research has consistently suggested that implementing material passports significantly improves recycling rates across various building types, then the current study's findings serve to refine and nuance this general understanding. To be more specific,

Table 3 Discussion of unanticipated final results: insights and challenges

Aspect	Description	Challenges
Unexpected results [85]	Unexpected outcomes can reveal unforeseen complexities in the initial study design, offering new insights	Interpreting unexpected outcomes accurately without preconceived biases
Relevance of findings [86]	Unexpected findings can help refine understanding of the study's subject matter, going beyond primary objectives	Integrating unexpected findings into the broader context of the research
Potential explanations [87]	Factors like complex chemical compositions or industry practices may influence lower recyclability	Understanding the complexity of factors affecting recyclability, such as material properties
Impact on initial hypothesis [88]	Challenges initial premises, showing that other factors may be more significant than initially believed	Revising hypotheses without undermining the overall study design or validity
Role of documentation [89]	Detailed records may not always guarantee better outcomes if other influencing factors exist	Balancing the importance of documentation with other influential factors
Industrial practices [90]	Even with detailed data, industry practices might not be advanced enough to process materials effectively	Adapting industry processes to handle new materials with varying properties
Theoretical versus practical gap [91]	This highlights a disconnect between theoretical frameworks and their practical application in the industry	Bridging the gap between theory and practice, especially when practical limitations arise
Importance of addressing unexpected results [92]	Critical for comprehensively evaluating study findings and identifying gaps in methodology or practice	Addressing unexpected results without overstating their significance in the study
Opportunities for further research [93]	Unexpected results provide direction for new research avenues, including refining existing theories	Formulating new research questions that align with emerging insights from unexpected results
Implications for industry [94]	This indicates the industry must adapt and evolve to manage complex materials effectively	Encouraging industry to adapt in ways that are both feasible and aligned with research findings

although material passports may generally improve the possibility of recyclability, the effectiveness of these passports may be greatly dependent on the kind and complexity of the materials involved, which calls into question the concept of a universal solution. This kind of comparison study is necessary to comprehend the areas where new research coincides with, diverges from, or expands the already available information. In constantly changing sectors, such as sustainable construction methods, evaluating how new research interacts with existing literature is of utmost significance. Researchers can highlight the unique contributions that their study has made while also acknowledging the work that has been done in the past thanks to this method. Researchers can identify sections of the literature that their study addresses or areas in which their findings pose new issues when thoroughly engaging with the existing body of research. For instance, if previous research had stressed the significance of material passports in improving recycling efficiency without considering material-specific characteristics, the current study's emphasis on material complexity provides a further layer of depth, contributing to the advancement of the debate within the area [96]. Additionally, researchers can explore their findings' broader implications when contextualizing their findings within literature. This is accomplished by placing the study in dialogue with previous research, which makes it possible to evaluate whether the findings of the study support previously established theories or suggest new avenues for further investigation. This conversation makes the study more rigorous academically and lays the groundwork for a more in-depth discussion and interpretation of the findings. It helps develop a narrative that places the study within the context of an ongoing scholarly debate, recognizing the continuities and the disruptions that it contributes to the field through its application. In sustainable building practices, where new technologies and techniques are routinely produced, such debates are essential for determining how fresh insights could affect future industry standards and practices. This is because we are constantly developing new technologies and processes. Using this technique also makes it easier to conduct an in-depth analysis of the possibilities and constraints of the study. When researchers compare their findings to those of past studies, they can determine the limits of their study's limits and the amount to which their findings may be generalized. For instance, if the current study's findings differ from those of past studies that generally endorsed material passports as a universal tool for enhancing recyclability, then it is vital to explore the causes that might be responsible for these variations. These disparities might be attributed to various factors, including variances in study methods, differences in geographic focus, and differences in the magnitude of the investigations. Acknowledging these variances increases the quality of the results drawn from the study and informs future research efforts by indicating areas that deserve more exploration. Contextualizing results within the body of literature is not just about matching new research with previously acquired information; it also involves actively contributing to the debate that is taking place within the academic community [97]. By contrasting and comparing the results of the study with those of earlier research, researchers can discover areas in which their findings match with, question, or enhance the already available information. Table 4 delineates the fundamental procedures for contextualizing study findings within existing literature, emphasizing their descriptions and related problems. It underscores the necessity of matching novel discoveries with the broader academic framework, recognizing concordances or divergences, and delineating contributions while confronting potential challenges.

Table 4 Key steps in contextualizing research findings and their challenges

Step	Description	Challenges
Contextualizing findings [98]	Integrating the study's findings within the broader scholarly discourse to demonstrate relevance and significance	Ensuring that the study's contributions are positioned accurately within the broader body of knowledge
Comparing with previous studies [99]	Evaluating the extent to which the study's outcomes corroborate or diverge from existing research	Identifying and synthesizing relevant studies to facilitate rigorous and meaningful comparisons
Identifying alignments/ divergences	Elucidating areas of concordance or discrepancy between the current study and prior research	Differentiating between minor variations and substantial theoretical or empirical contradictions
Implications and contributions [100]	Articulating how the study advances the field, introduces novel perspectives, or informs future research trajectories	Communicating the study's contributions without overstating its scope or impact on the existing literature

Through this critical engagement with literature, it is possible to arrive at a more nuanced understanding of the contributions made by the study. Additionally, it guarantees that new research is positioned as a component of a dynamic and growing academic debate. In addition, it makes it possible to have a more nuanced perspective on the topic, which acknowledges the complexity and particularities responsible for shaping the study's results. Ultimately, this procedure highlights the significance of placing new research within a broader academic framework to ensure the findings are relevant and meaningful within the continuing conversation [101]. Figure 6 depicts incorporating new research findings into the established body of literature. It comprises four essential steps: contextualizing findings, comparing them with prior studies, detecting alignments or divergences, and addressing ramifications and contributions. This systematic method improves comprehension of the study's significance and influence.

5.2 *Gaps Addressed by the Study*

One of the most important contributions that a study can make is the ability to recognize and fill in the gaps that exist within the existing body of knowledge [4]. This helps to ensure that research contributes to the advancement of the field rather than just repeating the findings of previous studies. It is of the utmost importance to address these knowledge gaps since doing so enables academics to improve upon previously established ideas, broaden conceptual frameworks, and present novel points of view. Within the scope of this investigation, the key gap being addressed is the

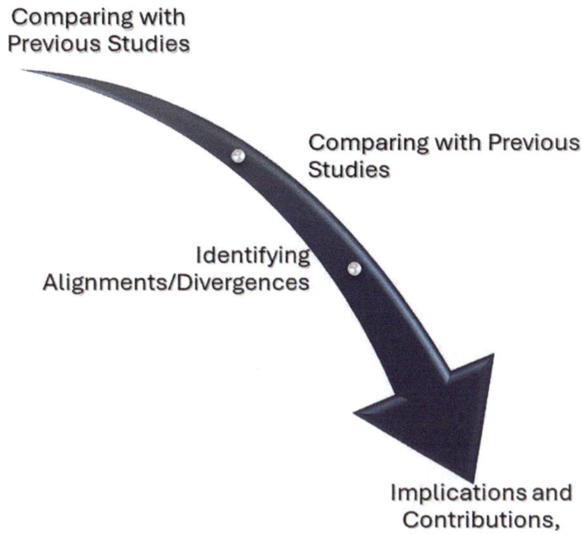

Fig. 6 Contextualizing research findings within literature (*Resource* Authors)

restricted evaluation of the influence that material passports have on the management of high-tech and complicated building materials. As a result of the fact that previous studies have mostly focused on traditional materials like concrete and steel, there is a lack of thorough comprehension of how material documenting procedures influence more complex and sophisticated materials. The focus of this research on recording high-tech materials using material passports allows for the provision of essential insights that fill a significant gap in the existing body of literature. Even though past studies have demonstrated that material passports have the potential to improve transparency, traceability, and recycling efficiency for conventional materials, the extent to which these advantages may be applied to more complex materials has remained largely unexplored [102]. The study can provide a more nuanced understanding by addressing this gap, which acknowledges that advanced materials present specific issues distinct from those connected with ordinary construction materials. According to the findings, although material passports provide significant benefits for tracking and managing standard building materials, deploying these passports for more complicated materials requires extra adaptations that are particular to the business. For example, the inherent complexity and variety of advanced materials demand the implementation of specialized documentation procedures distinct from those appropriate for more straightforward materials such as concrete. Based on these findings, it appears that a standardized method would not be adequate for all sorts of materials, highlighting the necessity of a more adaptable framework within the building industry. This enhanced approach addresses the hitherto overlooked aspect of adapting sustainable documentation techniques to developing building technology [103]. The study helps close a knowledge gap and lays the groundwork for future research and innovation in business. The research stimulates additional inquiry into how the construction industry may modify its tactics to manage complex materials more efficiently. This is accomplished by recognizing the constraints and opportunities connected with existing practices. Developing customized documentation protocols or adaptable frameworks with one-of-a-kind characteristics and lifespan requirements for advanced materials might fall under this category. To ensure that sustainability initiatives in the construction sector are inclusive, embracing a wider range of materials than those that have traditionally been addressed, developments of this kind are essential. As a result, the study provides a more in-depth comprehension of material passports' function in the environmentally responsible administration of various materials within the construction sector. It contradicts the notion that a universal method can be effective across all materials. Instead, it advocates for a strategy focused on the unique environment in which it is being implemented. The academic discourse is enriched because of this emphasis on differentiation and adaptation, which encourages stakeholders to realize the specific requirements of diverse materials while planning and implementing sustainable practices. Additionally, the findings gained by this study have major practical consequences for professionals working in the sector, politicians, and academics. Practitioners can build more successful methods for implementing material passports if they are better aware of the unique problems involved with the documentation of high-tech materials. Policymakers may utilize these findings to formulate policies that facilitate the seamless incorporation of innovative materials into environmentally responsible design and construction procedures [104]. To ensure that the transition toward sustainability continues to be both inventive and thorough, academics might utilize this study as a springboard for additional research, researching how documentation requirements could be altered to suit an increasing range of materials. As a result, this research not only fills a significant need in the existing body of knowledge but also prepares the way for a material management strategy that is more inclusive and flexible within the context of sustainable building.

6 Suggestions for Subsequent Investigations

This section delineates suggestions for future research derived from the gaps, limitations, and findings reported in the present study. These suggestions direct future research endeavors, offering a framework to enhance comprehension and practical applications of the study's topic. Addressing the gaps identified in the research enables the following studies to elucidate the knowledge areas that remain unexamined or inadequately articulated [105]. For example, if certain factors or linkages were inadequately explored owing to limitations in sample size or time, subsequent studies should seek to analyze these elements more comprehensively. This will assist in corroborating the findings across many settings and enhance the overall comprehension of the subject. Moreover, recommendations for methodological improvements are crucial for enhancing study quality. Enhancing study designs by utilizing bigger, more varied samples, integrating mixed-methods approaches, or applying sophisticated data analysis techniques can provide more robust and generalizable results. Future research may investigate longitudinal designs to comprehend temporal changes, offering a more dynamic perspective beyond the present study's limitations [106]. Proposing novel research directions is another essential component of this segment. Researchers might leverage the study's findings to investigate similar subjects or assess the ramifications of novel technology, cultural settings, or rising trends. These new routes can provide novel viewpoints and insights that enhance the area, fostering creativity and discovery. These suggestions will enable future research to build upon the foundation established by this study, advancing understanding of the subject. This guarantees that forthcoming studies stay pertinent, significant, and adaptable to the changing panorama of research issues.

6.1 Addressing Outstanding Discrepancies

Academic research is conducted primarily to locate and fill up knowledge gaps within a particular field of study [30]. Specific concerns may continue to be unsolved, which will require more research to be conducted despite the meticulous efforts of the current study. To thoroughly comprehend the subject matter, it is vital to address any unresolved disparities. There are frequently limits in the design or scope of the study that are the source of unresolved questions. As an illustration, if the research was centered on using material passports in a particular building type or demography, it could only consider some variances in different building settings or user groups. The scope of this study might be expanded in subsequent research to encompass a broader range of buildings or areas, assuring that the findings are relevant in various settings. Furthermore, the generalizability and application of the study would be improved because of this technique, which would help validate the results across a broader range of situations. Furthermore, unanticipated discoveries or developing tendencies can open new research methods that were not initially planned. Suppose the research on material passports uncovered an unexpected connection between the lifecycle of certain materials and their capacity to be recycled, for instance. In that case, subsequent research may dive more deeply into discovering the nature of this connection. The underlying mechanisms that make some materials more suitable for reuse would be understood better, ultimately improving environmentally responsible building practices. By addressing these emergent concerns, subsequent researchers will be able to build upon the fundamental discoveries, resulting in deeper insights and an expansion of the scope of the initial study. The impact that time has on the study's conclusions is yet another important topic that should be investigated. The current study was carried out over a very short period, which may have resulted in the researchers disregarding long-term differences in the usefulness of material passports in various settings. Through the recording of changes over time, longitudinal research has the potential to provide a more nuanced understanding. For example, it may be used to investigate how the implementation of material passports affects sustainable construction practices over many years. Using this methodology, it is possible to uncover patterns, fluctuations, and potential causal linkages that a cross-sectional study could overlook. This strategy guarantees that future research will capture the entire richness of the subject matter. The following table (Table 5) overviews the most essential parts of correcting unsolved differences in academic research. It highlights the significance of broadening the scope of the study, investigating new research routes, and gaining knowledge of the mechanisms at play. Additionally, it sheds insight into the difficulties often connected with each component, such as limitations on available resources, time commitments, and the complexity of data analysis [107].

Table 5 Addressing outstanding discrepancies in research

Aspect	Description	Challenges
Purpose of research [30]	Academic research seeks to identify and address deficiencies in knowledge within a specific domain	Identifying the most pertinent knowledge gaps requires extensive literature review and critical analysis
Unresolved discrepancies [108]	Specific issues may still need to be solved due to inherent study design or scope limitations	Prioritizing unresolved questions for further investigation is challenging, especially under constraints of time and resources
Expanding scope [109]	Subsequent research should aim to extend the study's scope to encompass a broader range of contexts, enhancing the generalizability of the findings	Expanding the scope may necessitate additional resources, larger sample sizes, and a more comprehensive data collection strategy
New research Avenues [110]	Unanticipated findings can create opportunities for exploring novel research directions beyond the initial scope of the study	Formulating new research questions or hypotheses based on emergent findings can be complex and require substantial conceptual development
Understanding mechanisms [111]	Investigating newly identified relationships is crucial for elucidating the underlying mechanisms, contributing to theoretical and practical advancements	Analyzing these mechanisms may require sophisticated analytical methods and interdisciplinary expertise
Impact of time [112]	Short-term studies may fail to capture the temporal dynamics of certain phenomena, while longitudinal research can provide a more nuanced understanding over time	Conducting longitudinal studies demands significant time investment and sustained commitment to rigorous data collection protocols
Potential insights [113]	Longitudinal research can uncover trends, variations, and causal linkages that cross-sectional studies might overlook, offering a more holistic understanding	Managing, processing, and analyzing extensive dataset over prolonged periods can be resource-intensive and technically demanding

6.2 Proposed Methodological Enhancements

To guarantee future research precision, dependability, and robustness, evaluating the procedures utilized in the studies and finding prospective enhancements is essential [114]. Developing more refined research methodologies can assist future studies in better addressing the complexity of the studied area, which will ultimately result in more nuanced and reliable conclusions. Within the context of methodological enhancement, one of the most important recommendations is to expand the sample size in the following investigations. The current study may have been restricted in its potential to generalize its findings because the sample size was either small or homogenous. It would be possible for future researchers to confirm the relevance of findings across various circumstances and demographics if they had access to a bigger and more varied sample. Fresh insights not obvious in the first study might be revealed by including individuals from various age groups, cultural backgrounds, or professional sectors in their participation. If this technique were taken, the findings would be less likely to be biased and more applicable to a wider range of situations. It would also ensure that the findings are not unduly unique to a single demographic. In addition, it is advised that mixed-methods study designs be utilized, which combine qualitative and quantitative approaches to research. Even though the present investigation may have concentrated on a single strategy, combining qualitative approaches with quantitative analysis, such as interviews and focus groups, can provide a more comprehensive perspective on the research subject. In contrast to quantitative analysis, which gives statistical support to the observed patterns, qualitative data offers profound insights into the experiences, views, and motives of the people participating in the study. Therefore, the results' reliability and validity are improved due to this mixed-methods approach, which increases the overall knowledge of the study subject by capturing both numerical patterns and the underlying context. To derive more relevant insights from complicated dataset, future research could potentially benefit from applying more advanced data analysis tools. Methods such as machine learning, structural equation modeling (SEM), and time-series analysis might assist researchers in exploring complex correlations between variables, which more conventional approaches may miss [115]. A deeper layer of analysis may be achieved using machine learning algorithms, which can examine enormous dataset to recognize patterns, forecast outcomes, or categorize data points with a high degree of accuracy. Structural equation modeling enables researchers to investigate the causal linkages between variables, offering a more in-depth comprehension of how various elements interact. On the other hand, time-series analysis can monitor changes and trends over time, providing insights into the temporal dynamics of a specific event. More complex and actionable findings may be obtained through these new methodologies, enabling researchers to arrive at more precise and pertinent conclusions to the study context. In addition, it is necessary for future research to improve the repeatability and transparency of research techniques when conducting a study [116]. It is important to guarantee that future researchers can reproduce the study and confirm its findings by clearly documenting the protocols for data collecting, analysis methodologies, and decision-making processes. To enhance credibility and facilitate the capacity of the wider academic community to build upon previously conducted research, transparency in technique is essential. In the future, research should try to record everything as thoroughly as possible, including providing extensive descriptions of sampling techniques, data management processes, and statistical calculations. This method enhances the study's dependability and contributes to developing an academic community that values openness and participation in collaborative endeavors. When doing future research, it is essential to consider longitudinal designs, another significant methodological development area. Many studies rely on cross-sectional data, which gives a picture of a certain time. On the other hand, longitudinal research monitors change over longer periods. This design is beneficial for understanding trends or the impacts of a certain variable over an extended period. Specifically, longitudinal studies have the potential to offer valuable insights into how attitudes or behaviors evolve because of treatments or changes in the external environment. This method provides a more comprehensive knowledge of causation and development, which is sometimes overlooked in studies conducted over shorter periods. Through longitudinal designs, researchers can unearth more profound patterns and offer a more all-encompassing perspective on the dynamics of the subject matter. As an additional recommendation for improving the rigor of the methodology, the adoption of strong experimental designs is also advocated. Randomized controlled trials (RCTs) or quasi-experimental designs might be utilized in future research projects, if they are viable, to create more robust causal conclusions. Because they eliminate bias via random assignment, randomized controlled trials (RCTs) are considered the gold standard for evaluating the efficacy of therapies. Researchers can directly compare the outcomes of the control group with those of the experimental group using this methodology, which provides clearer proof of the causal linkages between the two groups. It is possible to employ quasi-experimental designs where randomized controlled trials (RCTs) are not feasible [117]. This provides an alternate approach to analyzing cause-and-effect correlations in real-world scenarios. When these designs are used in subsequent research, it can dramatically increase the validity of the investigation findings. In addition, it is recommended that future studies adopt a more stringent approach to data validation and dependability assessment. To get findings that can be relied upon, it is essential to ensure that the instruments used for data collection are trustworthy and that the measurements are consistent across various settings. Reliability analysis, test–retest processes, and inter-rater reliability checks are some of the methods that may be utilized to confirm the data's consistency over time. In addition, using validity measures, such as construct and content validity, guarantees that the study properly captures the notions it seeks to measure. Using these methods, future research can provide more trustworthy, dependable results and be generally accepted within the academic community. The development of research methodology is an essential component in enhancing the quality and rigor of research. Expanding sample sizes, adopting mixed-methods techniques, leveraging new analytical tools, and applying longitudinal and experimental designs are all factors that can dramatically increase the depth and scope of future research. Furthermore, ensuring that succeeding studies build upon a sound methodological foundation is accomplished by emphasizing transparency, reproducibility, and rigorous data validation processes within the research process. Implementing these upgrades will not only result in findings that are more trustworthy and applicable to a wider range of situations, but they will also contribute to the continued development and improvement of research procedures in the area [118].

7 A Comprehensive Understanding of the Differences Between the "Results" and "Discussion" Sections

Researchers frequently encounter difficulties when attempting to differentiate between the components of a study effort titled "Results" and "Discussion" [119]. To successfully explain the findings of the research and the consequences of those findings, it is essential to clarify this distinction. The data gathered from the research are presented in the section under "Results," which does not include any interpretations. In addition to displaying the data using tables, figures, and text, it provides a concise and impartial summary of the findings. This part aims to provide straightforward and objective responses to the research questions by presenting the findings obtained from the research. The Section on Discussion, on the other hand, provides an interpretation of the relevance of these discoveries. It connects the findings and the study's aims, investigates the causes that lie under the surface, and analyses how the results correspond with the existing body of literature. The Discussion presents opportunities to analyze patterns, consider constraints, and suggest future study implications. The relevance of the findings and the prospective applications are explained, and the larger context of the findings is emphasized. Understanding what each part should contain in research writing is essential to ensuring clarity and coherence. To give the reader a holistic picture of the study's contribution to the field, the foundation is provided by a well-structured Results section [120]. In contrast, a thoughtful Discussion section adds depth to the overall presentation. When it comes to writing a research article that is both clear and compelling, this division is essential.

7.1 Principal Distinctions Between Results and Discussion Sections

The Results and Discussion sections fulfill separate functions in a research paper, each focusing on different facets of the subject [120]. The Results section emphasizes the objective presentation of study findings, including figures, tables, and descriptive statistics to summarize observations without interpretation. It addresses the inquiry, "What did we discover?" by presenting clear, concise evidence substantiating the study's aims. The Discussion section, in contrast, evaluates these findings, providing

Table 6 Key differences between results and discussion sections

Results section	Discussion section	Challenges
Answers the question "What?" – What was found in the research?	Answers the question, "So what? What do these findings mean?	Balancing detail without overwhelming the reader
Focuses on describing the experiments and outcomes	Focuses on interpreting the significance of the main findings	Avoiding excessive overlap between sections
States findings without interpreting them	Analyzes and interprets the results without repeating them	Maintaining objectivity while providing interpretation
Includes only data relevant to the study's objectives	It does not introduce new results; it supports statements with existing data	Ensuring consistency between results and interpretations
Use past tense to describe what was found	It uses past tense for the study's findings and present tense for interpretations	Deciding on the appropriate tense for each section
It can include tables, figures, and other visual data	Primarily text-based but can refer to figures and tables when interpreting findings	Effectively using visuals to complement the text
Results section	Discussion Section	Challenges

explanations and perspectives on their relevance. It addresses the question, "What do these results signify?" by examining patterns, comparing findings to prior research, and discussing any limits or implications. The Results part is factual and data-oriented, whereas the Discussion portion is analytical and interpretative, highlighting the results' contribution to the existing body of knowledge [121]. Comprehending these distinctions guarantees precision and efficacy in conveying the research findings. In studies, the Results and Discussion parts differ primarily in what is shown in Table 6. It emphasizes each segment's distinct functions, the information each should have, and the typical difficulties encountered when writing them. Accurate and clear presentation of study findings and interpretations is possible with an understanding of these distinctions.

7.2 *Checklist for Effectively Presenting Research Findings*

A research work results section is essential for presenting the study's findings in an understandable and organized manner [122]. It needs to concentrate on solving the query "What?" by providing a detailed account of the findings uncovered throughout the investigation. Objective reporting is essential, which means that the findings should be provided without any subjective interpretations or remarks, and the analysis should be reserved for the part that discusses the findings. For clarity, data should be shown using tables or figures; nevertheless, it is essential to prevent redundancies,

Table 7 Key considerations and challenges in writing the discussion section

Task	Have you done this?	Next steps	Challenges
Have you interpreted key results?	Yes/No	Deepen analysis of main findings	Ensuring interpretations are rooted in evidence without over-speculation
Did you discuss the significance of your findings?	Yes/No	Link findings to the research objectives	Effectively linking results to broader research objectives and contributions
Did you mention study limitations and future directions?	Yes/No	Add a section addressing limitations	Balancing honesty about limitations with maintaining the study's credibility
Is your argument logical and structured?	Yes/No	Refine for clearer flow and reasoning	Structuring arguments for clarity while accommodating complex data
Have you avoided overgeneralizing or overstating results?	Yes/No	Adjust language to ensure balanced interpretation	Avoiding overstated claims while emphasizing the study's significance
Task	Have you done this?	Next steps	Challenges
Have you interpreted key results?	Yes/No	Deepen analysis of main findings	Ensuring interpretations are rooted in evidence without over-speculation

which means that data should be displayed using either a table or a graph, but not both. In addition, the section must have a brief overview of the most important findings to facilitate a seamless transition into the upcoming discussion. This summary emphasizes the most relevant findings from the study, laying the groundwork for providing a more in-depth interpretation and analysis of the findings in the subsequent parts [10]. The most important duties for writing the Discussion section are outlined in Table 7, with the primary focus being on interpreting the data, highlighting the importance of the results, and addressing the study's limitations. In addition to highlighting problems such as assuring interpretations supported by evidence, tying findings to broader aims, and retaining clarity without overstating the study's contributions, it further directs the subsequent measures that should be taken.

8 Conclusion

After this chapter, a critical synthesis of the research findings is presented. This synthesis integrates empirical data with larger theoretical frameworks that apply to the domains of architecture and engineering. The importance of this part cannot be overstated when it comes to establishing the study's place within the broader body of knowledge and comprehending the significance of the study for both academic research and practical implementation. This conclusion aims to develop a complete understanding of the study's outcomes and their relevance in furthering knowledge within the field. A systematic interpretation of the data accomplishes this. The investigation goes beyond merely presenting the findings; instead, it delves thoroughly into the ramifications of the findings to bridge the gap between theory and practice. It emphasizes how the findings either validate or contradict previously established hypotheses, so providing fresh viewpoints that improve the current academic conversation. For example, when analyzing environmentally friendly construction materials, this study demonstrates how certain materials contribute to sustainability beyond their technical performance by addressing more enormous environmental implications. These impacts include lowering carbon emissions and boosting long-term resilience. All of these are examples of how specific materials contribute to sustainability. This nuanced method highlights the significance of going beyond isolated data points to examine the more considerable repercussions of research, matching the findings with broader societal, environmental, and technological trends. Such an approach stresses the requirement of looking beyond isolated data points. The acknowledgment of the limits of the study is an essential component of this conversation. This is because it strengthens the credibility of the research by addressing any biases, methodological constraints, and external influences that influence conclusions. Because of this transparency, readers can evaluate the validity and generalizability of the findings critically. This provides a balanced perspective that acknowledges both the strengths and places for development within the study. The maintenance of scientific integrity and the provision of a firm basis upon which future research may be built are both dependent on the existence of such openness. Additionally, this conclusion highlights intriguing possibilities for subsequent research, which encourages the examination of topics that were not fully addressed in the study that was being presented here. This section presents a road map for future researchers by offering prospective pathways for deeper exploration. Some examples of these potential avenues include the examination of innovative approaches or underexplored areas of sustainable building and construction practices. The study will not only add to the current academic scene, but it will also contribute to the continuous innovation and inquiry that is important for the growth of the discipline. This forward-looking viewpoint assures that the study will contribute to these things.

Furthermore, the practical ramifications of the findings are brought to light, with a special emphasis placed on these discoveries' significance to engineering and architecture. The findings of the study can influence the practices of the industry, to create modifications to legislation, and to set new standards in environmentally responsible building architecture. The findings that pertain to energy-efficient materials, for instance, have the potential to influence adjustments in construction rules, which would ultimately lead to more environmentally responsible practices being implemented throughout the sector. To ensure that study has a meaningful influence beyond academia, one of the most important strengths is the capacity to transform theoretical ideas into practical applications. This analysis of the research findings is both critical and reflective, providing a comprehensive view of the study's contributions to the field in which it was conducted. By ensuring that the findings are consistent with previously established theories, addressing the study's limits, and recommending avenues for further research, this analysis guarantees that the impact of the study will be both significant and long lasting. By acting as a link between actual data and theoretical ideas, it contributes to the enhancement of academic discourse in the fields of architecture and engineering, as well as to the development of a culture that encourages ongoing research and innovation. Therefore, study constitutes an essential addition to the progress of knowledge and will have a long-lasting impact on academic endeavors and professional practice.

References

1. Dehalwar K (ed) (2024) Basics of research methodology-writing and publication. EduPedia Publications Pvt Ltd
2. Karunarathna I, Gunasena P, Hapuarachchi T, Gunathilake S (2024) The role of scientific hypotheses in shaping modern research and innovation
3. Yeomans M, Boland FK, Collins HK, Abi-Esber N, Brooks AW (2023) A practical guide to conversation research: How to study what people say to each other. Adv Methods Pract Psychol Sci 6(4):25152459231183920
4. Karunarathna I, De Alvis K, Gunasena P, Jayawardana A (2024) Creating value through literature reviews: techniques for identifying research gaps
5. Robson C (2024) Real world research. John Wiley & Sons
6. Ghobadi M, Sepasgozar SM (2023) Circular economy strategies in modern timber construction as a potential response to climate change. J Build Eng 107229
7. Hernández H, Ossio F, Silva M (2023) Assessment of sustainability and efficiency metrics in modern methods of construction: a case study using a life cycle assessment approach. Sustain 15(7):6267
8. De Oliveira B (2023) Participatory action research as a research approach: advantages, limitations and criticisms. Qual Res J 23(3):287–297
9. Dwivedi YK, Jeyaraj A, Hughes L, Davies GH, Ahuja M, Albashrawi MA, Al-Busaidi AS, Al-Sharhan S, Al-Sulaiti KI, Altinay L, Amalaya S (2024) "Real impact": challenges and opportunities in bridging the gap between research and practice–making a difference in industry, policy, and society. Int J Inf Manag 102750
10. Lim WM (2024) What is qualitative research? an overview and guidelines. Australas Mark J 14413582241264619
11. Mpuangnan KN, Gugulethu BPZ (2024) Transforming research and pedagogy in higher education through scientific inquiry steps. Edelweiss Appl Sci Technol 8(5):74–87
12. Zhou H, Jiang FK (2023) 'The study has clear limitations': presentation of limitations in conclusion sections of PhD dissertations and research articles in applied linguistics. Engl Specif Purp 71:34–47
13. Kumar P, Thakur N, Kumar K, Kumar S, Dutt A, Thakur VK, Gutiérrez-Rodelo C, Thakur P, Navarrete A, Thakur N (2024) Catalyzing innovation: exploring iron oxide nanoparticles-Origins, advancements, and future application horizons. Coord Chem Rev 507:215750
14. Homod RZ, Yaseen ZM, Hussein AK, Almusaed A, Alawi OA, Falah MW, Abdelrazek AH, Ahmed W, Eltaweel M (2023) Deep clustering of cooperative multi-agent reinforcement learning to optimize multi chiller HVAC systems for smart buildings energy management. J Build Eng 65:105689

15. Alvesson M, Sandberg J (2024) The art of phenomena construction: a framework for coming up with research phenomena beyond 'the usual suspects.' J Manage Stud 61(5):1737–1765
16. Ibrahim MY, Jusoh MH (2024) 10 effective steps to write good literature review for an academic dissertation. Multidiscip Rev 7(1):2024009–2024009
17. Logeshwaran J, Shanmugasundaram N, Lloret J (2023) Energy-efficient resource allocation model for device-to-device communication in 5G wireless personal area networks. Int J Commun Syst 36(13):e5524
18. Guo J, Gao S, Liu A, Wang H, Guo X, Xing F, Zhang H, Qin Z, Ji Y (2023) Experimental study on failure mechanism of recycled coarse aggregate concrete under uniaxial compression. J Build Eng 63:105548
19. Ottenhaus LM, Yan Z, Brandner R, Leardini P, Fink G, Jockwer R (2023) Design for adaptability, disassembly and reuse–a review of reversible timber connection systems. Constr Build Mater 400:132823
20. Scrucca F, Ingrao C, Barberio G, Matarazzo A, Lagioia G (2023) On the role of sustainable buildings in achieving the 2030 UN sustainable development goals. Environ Impact Assess Rev 100:107069
21. Eberhard K (2023) The effects of visualization on judgment and decision-making: a systematic literature review. Manag Rev Q 73(1):167–214
22. Khoa BT, Hung BP, Hejsalem-Brahmi M (2023) Qualitative research in social sciences: data collection, data analysis and report writing. Int J Public Sect Perform Manag 12(1–2):187–209
23. Donegan S, Williamson P, D'Alessandro U, Tudur Smith C (2013) Assessing key assumptions of network meta-analysis: a review of methods. Res Synth Methods 4(4):291–323
24. Almusaed A, Alasadi A, Almssad A (2022) A research on the biophilic concept upon school's design from hot climate: a case study from Iraq. Adv Mater Sci Eng 2022(1):7994999
25. Almusaed A, Almssad A, Yitmen I, Wallhagen M, Yang YF, Iyer-Raniga U (2024) Integrative approaches in urban sustainability: architectural design, technological innovations and social dynamics in global contexts. BoD–Books on Demand
26. Peel KL (2020) A beginner's guide to applied educational research using thematic analysis. Pract Assess Res Eval 25(1)
27. Quintão C, Andrade P, Almeida F (2020) How to improve the validity and reliability of a case study approach? J Interdiscip Stud Educ 9(2):264–275
28. Purssell E, McCrae N (2020) How to perform a systematic literature review: a guide for healthcare researchers, practitioners and students. Springer Nature
29. Almusaed A, Alasadi A, Almssad A (2022) Research article a research on the Biophilic concept upon school's design from hot climate: a case study from Iraq
30. Maiya AK, Aithal PS (2023) A review-based research topic identification on how to improve the quality services of higher education institutions in academic, administrative, and research areas. Maiya AK, Aithal PS (2023) A review based research topic identification on how to improve the quality services of higher education institutions in academic, administrative, and research areas. Int J Manag, Technol, Soc Sci (IJMTS) 8(3):103–153
31. Smadi O, Chamberlain D, Shifaza F, Hamiduzzaman M (2021) Factors affecting the adoption of the community of inquiry framework in Australian online nursing education: a transition theory perspective. Nurse Educ Pract 55:103166
32. Almusaed A, Almssad A, Yitmen I (2024) Sustainable built environment and its implications on real estate development: a comprehensive analysis
33. Wine O, Spiers J, Burns KK, van Manen M, Vargas AO (2022) A case study unpacking the collaborative research process: eight essential components. Environ Sci Policy 131:209–220
34. Casteel A, Bridier NL (2021) Describing populations and samples in doctoral student research. Int J Dr Stud 16(1)
35. Taherdoost H (2022) What are different research approaches? comprehensive review of qualitative, quantitative, and mixed method research, their applications, types, and limitations. J Manag Sci & Eng Res 5(1):53–63
36. Almusaed A, Yitmen I, Myhren JA, Almssad A (2024) Assessing the impact of recycled building materials on environmental sustainability and energy efficiency: a comprehensive framework for reducing greenhouse gas emissions. Buildings 14(6):1566

37. Tomaszewski LE, Zarestky J, Gonzalez E (2020) Planning qualitative research: design and decision making for new researchers. Int J Qual Methods 19:1609406920967174
38. Bergen N, Labonté R (2020) "Everything is perfect, and we have no problems": detecting and limiting social desirability bias in qualitative research. Qual Health Res 30(5):783–792
39. Stantcheva S (2023) How to run surveys: a guide to creating your own identifying variation and revealing the invisible. Annual Review of Economics 15(1):205–234
40. Mason S, Merga MK, Canché MSG, Roni SM (2021) The internationality of published higher education scholarship: how do the 'top'journals compare? J Informet 15(2):101155
41. Shahzad M, Shafiq MT, Douglas D, Kassem M (2022) Digital twins in built environments: an investigation of the characteristics, applications, and challenges. Buildings 12(2):120
42. Pratt MG, Sonenshein S, Feldman MS (2022) Moving beyond templates: a bricolage approach to conducting trustworthy qualitative research. Organ Res Methods 25(2):211–238
43. Haddaway NR, Bethel A, Dicks LV, Koricheva J, Macura B, Petrokofsky G, Pullin AS, Savilaakso S, Stewart GB (2020) Eight problems with literature reviews and how to fix them. Nature Ecology & Evolution 4(12):1582–1589
44. Wickert C, Post C, Doh JP, Prescott JE, Prencipe A (2021) Management research that makes a difference: broadening the meaning of impact. J Manage Stud 58(2):297–320
45. Rowland DR (2023) Two frameworks to guide discussions around levels of acceptable use of generative AI in student academic research and writing. J Acad Lang Learn 17(1):T31–T69
46. Ali MA, Kamraju M (2023) Effective strategies for crafting research proposals in higher education. Int J Bus Manag Res 11(4):107–120
47. Kar AK, Dwivedi YK (2020) Theory building with big data-driven research–moving away from the "What" towards the "Why." Int J Inf Manage 54:102205
48. Li L, Luan H, Yin X, Dou Y, Yuan M, Li Z (2022) Understanding sustainability in off-site construction management: State of the art and future directions. J Constr Eng Manag 148(11):03122008
49. Lim WM, Kumar S, Ali F (2022) Advancing knowledge through literature reviews:'what', 'why', and 'how to contribute.' Serv Ind J 42(7–8):481–513
50. Brodie RJ, Peters LD (2020) New directions for service research: refreshing the process of theorizing to increase contribution. J Serv Mark 34(3):415–428
51. Dabić M, Marzi G, Vlačić B, Daim TU, Vanhaverbeke W (2021) 40 years of excellence: an overview of technovation and a roadmap for future research. Techno 106:102303
52. Deroncele-Acosta A, Jiménez-Chumacero RV, Gamarra-Mendoza S, Brito-Garcías JG, Flores-Valdivieso HG, Velázquez-Tejeda ME, Goñi-Cruz FF (2023) Trends in educational research for sustainable development in postgraduate education programs at a university in peru. Sustain 15(6):5449
53. Bakos J, Siu M, Orengo A, Kasiri N (2020) An analysis of environmental sustainability in small & medium-sized enterprises: patterns and trends. Bus Strateg Environ 29(3):1285–1296
54. Dodgson JE (2021) Critical analysis: the often-missing step in conducting literature review research. J Hum Lact 37(1):27–32
55. Johnson JL, Adkins D, Chauvin S (2020) A review of the quality indicators of rigor in qualitative research. Am J Pharm Educ 84(1):7120
56. Rowlands J (2021) Interviewee transcript review as a tool to improve data quality and participant confidence in sensitive research. Int J Qual Methods 20:16094069211066170
57. Tao Y, Both A, Silveira RI, Buchin K, Sijben S, Purves RS, Laube P, Peng D, Toohey K, Duckham M (2021) A comparative analysis of trajectory similarity measures. GIScience & Remote Sens 58(5):643–669
58. Ashraf MA, Yang M, Zhang Y, Denden M, Tlili A, Liu J, Huang R, Burgos D (2021) A systematic review of systematic reviews on blended learning: trends, gaps and future directions. Psychol Res Behav Manag 1525–1541
59. Mthuli SA, Ruffin F, Singh N (2022) 'Define, explain, justify, apply' (DEJA): an analytic tool for guiding qualitative research sample size. Int J Soc Res Methodol 25(6):809–821
60. Taylor LD (2022) Toil and trouble: contextualizing student success work at research universities. J High Educ 93(4):622–650

61. Luft JA, Jeong S, Idsardi R, Gardner G (2022) Literature reviews, theoretical frameworks, and conceptual frameworks: an introduction for new biology education researchers. CBE—Life Sci Educ 21(3):rm33
62. Almusaed A, Almssad A, Alasadi A, Yitmen I, Al-Samaraee S (2023) Assessing the role and efficiency of thermal insulation by the "Bio-Green Panel" in enhancing sustainability in a built environment. Sustain 15(13):10418
63. Popenoe R, Langius-Eklöf A, Stenwall E, Jervaeus A (2021) A practical guide to data analysis in general literature reviews. Nord J Nurs Res 41(4):175–186
64. Meng X, Das S, Meng J (2023) Integration of digital twin and circular economy in the construction industry. Sustain 15(17):13186
65. Rocca R, Rosa P, Sassanelli C, Fumagalli L, Terzi S (2020) Integrating virtual reality and digital twin in circular economy practices: a laboratory application case. Sustain 12(6):2286
66. Smith JJ, Amershi S, Barocas S, Wallach H, Wortman Vaughan J (2022) Real ml: recognizing, exploring, and articulating limitations of machine learning research. In: Proceedings of the 2022 ACM conference on fairness, accountability, and transparency. pp 587–597
67. Zina O (2021) The essential guide to doing your research project. Sage
68. Gurl T (2011) A model for incorporating lesson study into the student teaching placement: what worked and what did not? Educ Stud 37(5):523–528
69. Smith B (2018) Generalizability in qualitative research: Misunderstandings, opportunities and recommendations for the sport and exercise sciences. Qual Res Sport, Exerc Health 10(1):137–149
70. Mokhamar NW (2016) The impact of integrating reading and writing skills on Palestine technical college students' paragraph writing and attitudes. The impact of integrating reading and writing skills on Palestine technical college students' paragraph writing and attitudes
71. Lin H, Chen M, Lu G, Zhu Q, Gong J, You X, Wen Y, Xu B, Hu M (2013) Virtual geographic environments (VGEs): a new generation of geographic analysis tool. Earth Sci Rev 126:74–84
72. Patton MQ (1999) Enhancing the quality and credibility of qualitative analysis. Health Serv Res 34(5 Pt 2):1189
73. Aguinis H, Ramani RS, Alabduljader N (2018) What you see is what you get? enhancing methodological transparency in management research. Acad Manag Ann 12(1):83–110
74. Kelly GJ, Bazerman C (2003) How students argue scientific claims: a rhetorical-semantic analysis. Appl Linguis 24(1):28–55
75. Almusaed A, Yitmen I, Almssad A (2024) Application of clay materials within architectural design and construction methodologies. Developments in Clay Science and Construction Techniques. p 3
76. Duszak A (1994) Academic discourse and intellectual styles. J Pragmat 21(3):291–313
77. Casarrubea M, Jonsson GK, Faulisi F, Sorbera F, Di Giovanni G, Benigno A, Crescimanno G, Magnusson MS (2015) T-pattern analysis for the study of temporal structure of animal and human behavior: a comprehensive review. J Neurosci Methods 239:34–46
78. Almusaed A, Yitmen I, Almsaad A, Akiner İ, Akiner ME (2021) Coherent investigation on a smart kinetic wooden façade based on material passport concepts and environmental profile inquiry. Materials 14(14):3771
79. Almusaed A, Almssad A (eds) (2023) Sustainable smart cities: a vision for tomorrow. BoD–Books on Demand
80. Maraqa MJ, Spatari S (2022) BIM material passport to support building deconstruction and a circular economy. In: Construction in the 21st century 12th international conference (CITC 12). p 148
81. Drew CJ, Hardman ML, Hosp JL (2007) Designing and conducting education. Sage Publications
82. Reason J (2016) Managing the risks of organizational accidents. Routledge
83. Izatt RM, Izatt SR, Bruening RL, Izatt NE, Moyer BA (2014) Challenges to achievement of metal sustainability in our high-tech society. Chem Soc Rev 43(8):2451–2475
84. Baldassarre B, Konietzko J, Brown P, Calabretta G, Bocken N, Karpen IO, Hultink EJ (2020) Addressing the design-implementation gap of sustainable business models by prototyping: a tool for planning and executing small-scale pilots. J Clean Prod 255:120295

85. Weick KE, Sutcliffe KM (2015) Managing the unexpected: sustained performance in a complex world. John Wiley & Sons
86. Hancock DR, Algozzine B, Lim JH (2021) Doing case study research: a practical guide for beginning researchers
87. Sandin G, Peters GM (2018) Environmental impact of textile reuse and recycling–a review. J Clean Prod 184:353–365
88. Klayman J, Ha YW (1987) Confirmation, disconfirmation, and information in hypothesis testing. Psychol Rev 94(2):211
89. Curkovic S, Sroufe R, Melnyk S (2005) Identifying the factors that affect the decision to attain ISO 14000. Energy 30(8):1387–1407
90. Camp RC (2024) Benchmarking: the search for industry's best practices that lead to superior performance. CRC Press
91. Theoretical versus practical gap, this highlights a disconnect between theoretical frameworks and their practical application in the industry
92. Grimshaw JM, Eccles MP, Lavis JN, Hill SJ, Squires JE (2012) Knowledge translation of research findings. Implement Sci 7:1–17
93. Alvesson M, Kärreman D (2011) Qualitative research and theory development: mystery as method
94. Li G, Yang H, Sun L, Ji P, Feng L (2010) The evolutionary complexity of complex adaptive supply networks: a simulation and case study. Int J Prod Econ 124(2):310–330
95. Fang Z (2023) Demystifying academic reading: a disciplinary literacy approach to reading across content areas. Taylor & Francis
96. Lei G, Docherty R, Cooper SJ (2024) Materials science in the era of large language models: a perspective. Digital Discovery
97. Kervin L, Howard S, Jones R, McKnight A (2023) Research for educators 3e. Cengage AU
98. Halme M, Piekkari R, Matos S, Wierenga M, Hall J (2024) Rigour versus reality: contextualizing qualitative research in the low-income settings in emerging markets. Br J Manag 35(1):36–51
99. Botelho TL (2024) From audience to evaluator: when visibility into prior evaluations leads to convergence or divergence in subsequent evaluations among professionals. Organization Science
100. Swider BW, Yang J, Wang M (2024) The use of trajectories in management research: a review and insights for future research. J Manag 50(6):2012–2045
101. Healey MP, Leroy H, Post C, Potočnik K (2023) Changing the scholarly conversation: what it means, why it matters, and how to approach it in micro research. J Manage Stud 60(6):1633–1656
102. Zante G, Elgar CE, Hartley J, Mukherjee R, Kettle JP, Horsfall L, Walton A, Harper G, Abbott AP (2024) A toolbox for improved recycling of critical metals and materials in low-carbon technologies. RSC Sustain
103. Sánchez-Garrido AJ, Navarro IJ, García J, Yepes V (2023) A systematic literature review on modern construction methods in building: an integrated approach using machine learning. J Build Eng 73:106725
104. Khan M, McNally C (2023) A holistic review on the contribution of civil engineers for driving sustainable concrete construction in the built environment. Dev Built Environ 100273
105. Birkstedt T, Minkkinen M, Tandon A, Mäntymäki M (2023) AI governance: themes, knowledge gaps and future agendas. Internet Res 33(7):133–167
106. Grant M, Meissel K, Exeter D (2023) Promoting temporal investigations of development in context: a systematic review of longitudinal research linking childhood circumstances and learning-related outcomes. Educ Psychol Rev 35(1):19
107. Thompson G, Creagh S, Stacey M, Hogan A, Mockler N (2023) Researching teachers' time use: complexity, challenges and a possible way forward. The Australian Educational Researcher, pp 1–24
108. Chigbu UE, Atiku SO, Du Plessis CC (2023) The science of literature reviews: searching, identifying, selecting, and synthesising. Publications 11(1):2

109. Urminsky O, Dietvorst BJ (2024) Taking the full measure: integrating replication into research practice to assess generalizability. J Consum Res 51(1):157–168
110. Blagoev B, Hernes T, Kunisch S, Schultz M (2024) Time as a research lens: a conceptual review and research agenda. J Manag 50(6):2152–2196
111. Ma T, Cheng Y, Guan Z, Li B, Hou F, Lim ETK (2024) Theorising moderation in the configurational approach: a guide for identifying and interpreting moderating influences in QCA. Inf Syst J 34(3):762–787
112. Vogl S (2023) Mixed methods longitudinal research. In: Forum qualitative sozialforschung/forum: qualitative social research, vol 24, no 1
113. Chi B, Lee JB (2024) TOD effects on travel behavior: a synthesis of evidence from cross-sectional and longitudinal studies. J Transp Land Use 17(1):269–297
114. Shaheen N, Shaheen A, Ramadan A, Hefnawy MT, Ramadan A, Ibrahim IA, Hassanein ME, Ashour ME, Flouty O (2023) Appraising systematic reviews: a comprehensive guide to ensuring validity and reliability. Front Res MetrS Anal 8:1268045
115. Ye A, Bollen KA (2024) Path and direction discovery in individual dynamic factor models: a regularized hybrid unified structural equation modeling with latent variable. Multivar Behav Res 1–24
116. Wilson E, Ramage FJ, Wever KE, Sena ES, Macleod MR, Currie GL (2023) Designing, conducting, and reporting reproducible animal experiments. J Endocrinol 258(1)
117. Biddle N, Curchin K Randomised controlled trials and quasi-experimental designs. In: The Routledge handbook of human research ethics and Integrity in Australia. Routledge, pp 187–198
118. Enworo OC (2023) Application of Guba and Lincoln's parallel criteria to assess trustworthiness of qualitative research on indigenous social protection systems. Qual Res J 23(4):372–384
119. Casal JE, Kessler M (2023) Can linguists distinguish between ChatGPT/AI and human writing?: a study of research ethics and academic publishing. Res Methods Appl Linguist 2(3):100068
120. Karunarathna I, Gunasena P, De Alvis K, Jayawardana A (2024) Structured reviews: organizing, synthesizing, and analyzing scientific literature
121. Hvalshagen M, Lukyanenko R, Samuel BM (2023) Empowering users with narratives: examining the efficacy of narratives for understanding data-oriented conceptual models. Inf Syst Res 34(3):890–909
122. Lim WM, Kumar S (2024) Guidelines for interpreting the results of bibliometric analysis: a sensemaking approach. Glob Bus Organ Excell 43(2):17–26

Writing the Literature Review

Abstract A literature review is an essential element of scholarly research that systematically analyzes and synthesizes existing academic contributions on a specific topic. The main goal is to offer a thorough summary of the existing knowledge, allowing for identifying relevant theories, methodologies, and areas where further research is needed. The literature review plays a proactive role in identifying areas for further research, thereby contributing to advancing academic knowledge. It thoroughly analyzes peer-reviewed journals, books, conference proceedings, and other authoritative sources to assess the credibility and significance of the chosen research. This process provides a concise overview of previous discoveries and a thorough evaluation of the methods, theoretical frameworks, and consistency of research patterns. The literature review aids in situating the researcher's work within the broader academic conversation by identifying thematic patterns, significant debates, and unanswered inquiries. Moreover, it enables the generation of research inquiries and the choice of suitable methodologies. The literature review offers a coherent framework that takes the reader through the current information body, organized chronologically, topically, or conceptually. The primary purpose of the literature review is to discover any areas of research that have not been adequately addressed, establish the significance of the researcher's contribution, and establish a basis for furthering academic knowledge on the topic.

1 Introduction

1.1 The Critical Role of Literature Reviews in Advancing Business Research

The generation of knowledge in business research is fast rising, but it is also becoming more fragmented and multidisciplinary [1]. This makes it increasingly challenging to keep up with the latest advancements and evaluate the available evidence in specific

A. Almusaed et al., *Practice of Research Methodology in Civil Engineering and Architecture*, https://doi.org/10.1007/978-3-031-97393-2_21

areas. Within this setting, the literature review is a vital and transformational component in academic research, providing a methodical and thorough assessment of pre-existing scholarly works [2]. Concerns emerge regarding the quality and trustworthiness of conventional literature reviews due to their limitations, such as the need for more methodological rigor and the ad hoc nature of their conduction. As a result, the literature review has become more crucial as a research tool, offering a systematic and reliable strategy for combining research [2]. A literature review presents a thorough and inclusive summary of the existing information, enabling the recognition of pertinent ideas, approaches, and research areas that may contribute to future investigation. The literature review examines and evaluates the combined results of previous research, identifying essential contributions in the subject and areas that require additional inquiry. Consequently, it has a crucial impact on determining the course of new investigations and guiding researchers in formulating their own study inquiries, hypotheses, and theoretical structures. The literature review process systematically explores and assesses scholarly sources, such as peer-reviewed journals, books, conference proceedings, and other reputable publications. This thorough approach guarantees that researchers actively interact with the most relevant and current studies in their respective sectors [3]. Through an assessment of the merits, constraints, and approaches employed by different sources, the literature review goes beyond a simple summary, providing a discerning synthesis that reveals patterns, contradictions, and possible deficiencies in the current knowledge base. This analytical analysis enables researchers to determine the interconnectedness of different studies, whether via agreement, disagreement, or divergence. It also assists in placing their study into the broader academic conversation.

A well-conducted literature review is a cognitive foundation for future study, allowing researchers to find suitable theoretical frameworks, research techniques, and analytical approaches [4]. It establishes a basis for further study, guaranteeing that future investigations are influenced by and add to the current knowledge base. In addition, a thorough literature review showcases the researcher's comprehension of the topic and places their work within the context of ongoing academic discussions. Researchers prove the relevance and originality of their contributions to the subject by actively participating in ongoing discussions and carefully assessing previous findings [5]. Furthermore, a literature evaluation enables the recognition of research gaps and discrepancies, which are crucial for the progression of knowledge. These gaps frequently occur when the existing literature does not cover critical parts of a topic, lacks coherence, or uses approaches that are no longer applicable. By identifying these shortcomings, researchers may portray their work as tackling unanswered inquiries, therefore making a valuable contribution to the advancement of their area. Identifying these gaps in knowledge also aids in refining research questions and hypotheses, ensuring that new study not only builds upon previous work but also brings out fresh and original ideas. Equally important, a literature review plays a key role in refining research methodologies, ensuring that researchers adopt or enhance current procedures to answer their research inquiries more effectively [6]. The most important functions that a literature review serves in academic research are depicted in Fig. 1. The five critical roles covered in this article are a summary of

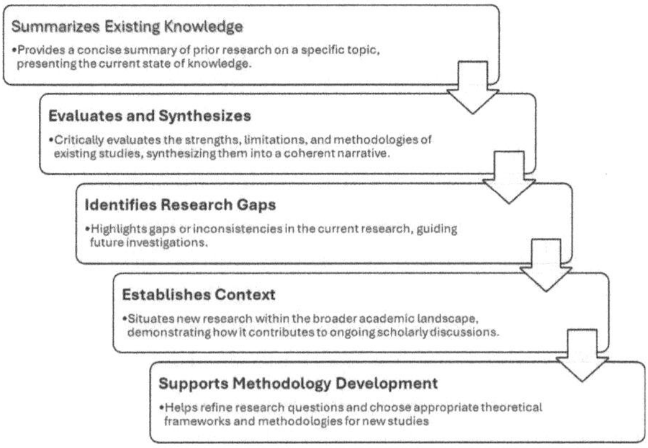

Fig. 1 Key functions of a literature review in academic research (*Source* Authors)

the current body of knowledge, evaluation and synthesis of research, identification of research gaps, the establishment of context, and assistance in creating methodology. Combining these components results in a comprehensive framework that may incorporate previous research and offer direction for new inquiries within a particular topic.

A literature review must be meticulously structured to accurately represent the issue's thematic, chronological, or theoretical extent [7]. This framework directs the reader in analyzing and combining the critical literature and emphasizes the connections and interactions between different researches. A theme structure entails a concentration on facets of the study issue, whereas a chronological review traces the progression of ideas throughout time [8]. A theoretical approach enables academics to analyze different conceptual frameworks and models that have influenced the discussion on a particular topic. A literature review identifies gaps and critically evaluates the methodological techniques employed in earlier investigations. This rigorous assessment guarantees that researchers may adopt or enhance current procedures to answer their research inquiries more effectively. Through careful examination of the constraints of previous approaches, researchers may guarantee that their current study design is more resilient and tackles the deficiencies highlighted in earlier studies [9]. Essentially, literature review is a fundamental aspect of academic research. It involves systematically and critically analyzing current information to produce new research. It is not only a passive representation of previous work but rather an active involvement in academic conversation, allowing scholars to recognize deficiencies, improve methodology, and contribute to the progress of their area. A well-executed literature review guarantees that new research is based on known knowledge and provides new insights that advance academic inquiry. This is achieved by carefully examining and synthesizing prior studies [10].

1.2 Categorization of Literature Review Functions and Roles in Academic Research

A literature review plays a crucial role in academic research, providing a structured evaluation of existing studies on a given topic. It serves multiple purposes, including summarizing knowledge, synthesizing findings, identifying research gaps, and guiding future investigations [11]. This categorization highlights its essential functions in supporting rigorous, informed, and meaningful research. The most crucial roles of a literature review are outlined in Table 1, which includes summarizing, synthesizing, and critically evaluating the current body of information. While simultaneously situating the current study within the context of the academic community, it emphasizes the significance of finding research gaps, reviewing theoretical frameworks, and providing direction for future research.

The organized method used in the literature review guarantees that the study will be original in its contributions and firmly rooted in previously acquired information. It is critical in determining the path that future investigations will take since it identifies gaps and synthesizes results from various sources. It is valuable because it contributes to the improvement of academic discourse and ensures that significant progress is made in the discipline.

2 Key Steps for Conducting a Literature Review

Creating a comprehensive and academic literature study entails many critical tasks. The following explains and expands on the five significant steps of producing a literature review.

2.1 Search for Relevant Material

The first thing that must be done is to perform a comprehensive and systematic search for academic literature pertinent to the research subject [14]. To complete this procedure, you must consult various trustworthy sources, including academic databases, journals, conference proceedings, books, and other publications. It is essential to use suitable keywords, search phrases, and Boolean operators to obtain relevant research easily and quickly. For instance, if the subject of the research is environmentally friendly building materials, the researcher might search databases such as ScienceDirect or Scopus using keywords such as "sustainable materials," "green building," and "environmentally friendly construction." The results of this search would include books, conference papers, and pertinent articles. Similarly, suppose the researcher is interested in a subject such as renewable energy technology. In that case, they might search databases such as IEEE Xplore or Engineering Village using phrases such as

Table 1 Overview of the functions, goals, and research role of the literature review

Section	Subsection	Details
Purpose of the literature review [12]	Summarization of existing knowledge	– Comprehensively evaluates previous writings and studies – Provides a concise summary of prior findings – Assesses completeness and accuracy – Identifies known and unknown aspects of the topic – Critically evaluates scholarly sources
	Synthesis and understanding	– Synthesizes existing research to clarify the state of knowledge – Contextualizes work by integrating diverse research – Highlights gaps for further investigation
Functionality of the literature review [12]	Critical evaluation	– Identifies, evaluates, and interprets prior research – Critique strengths, limitations, and methodologies of past studies – Assesses reliability and accuracy
	Identifying research gaps	– Identifies gaps or unresolved questions in current research – Exposes limitations or conflicting evidence – Assesses inconsistencies in findings
	Theoretical frameworks	– Reviews theoretical frameworks from prior studies – Establishes a foundation for future research – Provides a detailed analysis of research problems
Role in research process [13]	Contextualization and positioning	– Positions research within the broader scholarly conversation – Avoids duplication by highlighting prior work – Guides meaningful inquiry and ensures relevance

(continued)

Table 1 (continued)

Section	Subsection	Details
	Development of methodology	– Provides a systematic approach to understanding a field – Evaluate methodological rigor and critical themes – Bases research questions on informed understanding
	Guidance for future research	– Defines the scope and focus of future investigations – Traces development of ideas and concepts – Identifies influential studies and assesses their relevance

"renewable energy," "solar power," and "wind turbines." This inquiry would produce scholarly publications, conference papers, and technical reports on renewable energy technology.

2.2 Evaluate Sources

After gathering a sizable number of possible sources, the next critical step is thoroughly assessing each one's quality, trustworthiness, and relevancy [15]. This multistep procedure is crucial since it guarantees that the research is supported by reliable, high-caliber sources, enhancing the study's objectivity and scholarly value. The authorship, publishing venue, research methodology, and general academic quality of each source are carefully considered to ensure the literature review is built on solid foundations [16].

- **Authorship**: Researchers must evaluate the writers' qualifications, specializations, and associations [17]. An essay on Building Information Modeling (BIM) and sustainability, written by a specialist connected to a top university or research center for sustainable building, will probably provide more trustworthy information than one written by an unidentified author. A material passport specialist affiliated with a reputable organization will probably offer more thorough and precise research on circular economy applications in building and design.
- **Publication Venue**: Reputable academic publishers and peer-reviewed journals frequently provide credible information [18]. For example, an essay on BIM sustainability that appears in the Journal of Cleaner Production or Automation in Construction would be considered more credible than one that discusses the same subject in non-peer-reviewed journals. Sustainability research benefits from the strict standards and contributions of journals like Materials and Structures and

Sustainable Cities and Society. Publications published in these journals are more reliable and frequently referenced in related subjects.

- **Research Methodology**: An essential factor in determining the trustworthiness of a source is how rigorously it was conducted [19]. A solid and reproducible methodology is provided by research examining the influence of material passports on circular construction practices. The study uses big datasets from many case studies and incorporates life cycle analysis (LCA) to quantify environmental effects. Conversely, research that exclusively depends on restricted or non-generalizable data would be deemed less trustworthy. Comparably, studies on BIM's contribution to more sustainable buildings that use advanced tools such as Autodesk Revit or Green Building Studio for simulations and quantitative data analysis will likely produce more valuable results than studies that rely on theoretical talks.

The source's relevance to the study's subject is another factor that needs to be considered. The sources should explicitly address the research questions or hypotheses. Then, the sources should contribute fresh insights or critical evaluations of the subject matter [20]. An article that is well respected in the field of renewable energy could only be helpful if it concentrates on technology that has become obsolete or geographical regions that are relevant to the research that is now being conducted. Because of this, the source's content must be to the study objectives. For instance, when researching sustainable building materials, a researcher would evaluate the credibility and quality of sources from reputable journals such as Building and Environment or Sustainable Cities and Society. This evaluation would consider the author's background, the journal's reputation, and the methodology rigor level. Similarly, the dependability of a source would be evaluated in the field of renewable energy by considering the author's experience, the impact factor of journals such as Renewable and Sustainable Energy Reviews or IEEE Transactions on Sustainable Energy, and the robustness of the study strategy [21]. Evaluating sources is complex and involves several aspects, including careful examination of authorship, publication venue, study methods, and relevance. Using a comprehensive assessment of these characteristics, the researchers guarantee that their literature evaluation is founded on trustworthy and high-quality scholarship [22]. This, in turn, contributes to a more robust and relevant study conclusion.

2.3 *Identify Significant Themes, Critical Debates, and Existing Gaps in the Body of Knowledge*

After completing a comprehensive and methodical study of the chosen body of literature, an in-depth analysis is carried out to identify major themes, critical disputes, and current gaps in the body of knowledge [23]. A significant pattern that shapes the continuing debate surrounding the research issue may be identified through this analytical method, which also permits the discovery of repeating themes, viewpoints that disagree with one another, and the formation of significant patterns. Obtaining a more nuanced grasp of the present study state is accomplished via synthesizing these ideas. It is of the utmost importance to recognize such topics and arguments since doing so contributes to a more profound understanding of the fundamental ideas and areas of disagreement that are present within the subject. A further benefit of identifying gaps and limitations in the existing body of literature is that it assists in identifying areas that require further investigation [24]. When it comes to sustainable construction, for instance, the study can disclose a significant emphasis placed on issues such as the usage of recycled materials, the application of energy-efficient design techniques, and life-cycle assessments of building materials. On the other hand, there may be gaps in the literature about the long-term performance of sustainable materials or the effect of regional climate circumstances on the efficiency of these materials. This deficiency in comprehensive data may indicate a potential for more empirical research to analyze these factors over more extended periods [25]. Similarly, the field of renewable energy may see the emergence of significant issues such as photovoltaic solar energy systems, wind energy generation, and bioenergy technologies as the most prominent themes. The comparative effectiveness of various solar panel technologies or large-scale wind farms' environmental and social implications may be the subject of critical arguments within the body of published research in academic journals [26]. The study that investigates the integration of different renewable energy sources or the development of improved energy storage technologies, both of which are important to tackle the intermittency difficulties connected with renewable energy, may, nevertheless, be found to have gaps in its scope. Addressing these gaps would benefit the academic community and industry practitioners since it would yield valuable insights. For researchers to make a meaningful contribution to the ongoing conversation and highlight areas in which their study might fill existing holes or question prevalent beliefs, they must systematically identify these topics, disputes, and gaps [27].

2.4 *Outline the Structure*

A well-organized literature review is necessary when directing the reader through the study landscape. The following stage is to provide a detailed sketch of the structure, which will assist in presenting the information in a manner that is logical and consistent [28]. A specified structure acts as a roadmap, assisting in organizing essential parts, subsections, and transitions that make navigating the review easier [29]. This choice of structure is determined by the type of data being synthesized and the emphasis of the inquiry. The classification of the literature can be done in several ways, including chronologically, thematically, or by employing a combination of the two methodologies. Each approach offers several benefits, which vary according to the degree of difficulty and breadth of the study issue. In the case of a literature review on "digital twins and smart built environments," for instance, a theme framework may be utilized as a method of organization. Different features of digital twins, such as data integration, real-time monitoring, and predictive analytics, might be the subject of sections dedicated to examining them. The study might investigate breakthroughs, applications in the industry, and case studies that demonstrate the benefits and problems of integrating digital twin technology in intelligent buildings. These could be included within each area of the review [30]. These subjects might be investigated further in subsequent sections, such as the difficulties associated with data integration in Internet of Things devices or case studies that show practical application in urban planning. On the other hand, a historical framework could be appropriate for tracing the development of digital twin technologies over time. The review may develop into the architectural, engineering, and construction (AEC) sectors, beginning with its roots in the manufacturing industry [31]. Each section might concentrate on significant technical achievements, with subsections delving into developments such as the incorporation of artificial intelligence in digital twins or the utilization of blockchain technology to ensure the safety of data transfers. Readers would be able to comprehend the historical progression of the technology as well as its potential future tendencies if this framework were utilized. It is possible to present the research understandably and excitingly to improve the reader's comprehension of the subject matter. This may be accomplished by selecting the most suitable structure for the literature review, which might be chronological or topical [32].

2.5 Structured Approach to Writing the Literature Review

The introduction is crucial to the literature review, as it sets the stage for the entire document. It establishes the background of the research topic and specifies the review's objectives clearly and concisely. Furthermore, it outlines the organizational structure to guide the reader through the literature. This part aims to illustrate the relevance of the subject matter and explain how the literature review fits into the enormous scope of the investigations being conducted [33].

The literature review should provide a comprehensive overview of the present state of research. This is crucial as it helps the reader understand the current knowledge in the field. The critical interaction with the sources occurs in the literature review's body. The purpose of this section is not to describe individual research but rather to integrate data from various sources using analysis, comparison, and synthesis [34]. To do this, it is necessary to recognize recurring topics, areas of agreement or disagreement, and gaps in the current body of knowledge. During the literature presentation, it is of the utmost importance to establish connections between concepts derived from various sources, illustrating how these concepts complement or contrast. Additionally, this part should provide insights into how your work contributes to or diverges from the existing state of research. The review has to weave together many lines of thinking to offer a coherent narrative beyond merely describing the subject matter [35]. The objective is to develop a cohesive argument or viewpoint that provides support for the research question or hypothesis, as well as to synthesize the essential insights that have been gleaned from the existing body of literature. This definition entails drawing attention to significant theories, techniques, or discoveries that are particularly pertinent to your study and explaining the implications of those theories and procedures in a larger context. The review must conclude with a brief conclusion summarizing the most important results from the literature, drawing attention to significant gaps or limitations in the existing research, and suggesting areas that may be the focus of future study [36]. It is essential to emphasize the literature review's relevance to the study's objectives and explain how it has contributed to the decision-making process on the research direction [37]. For instance, a researcher doing a literature evaluation on environmentally friendly building materials could investigate various sources in architecture. They would conduct an in-depth analysis of the opportunities and constraints presented by multiple materials, investigate efficient design solutions regarding energy consumption, and evaluate the approaches utilized in lifetime evaluations [38]. By using this approach, not only would individual research be summarized, but it would also be synthesized to offer a thorough picture of the current state of knowledge in sustainable architecture. Similarly, a researcher in the engineering field may investigate the historical development of devices that utilize renewable energy sources. To discover technological achievements, problems, and current controversies, they would investigate the developments in the many categories of renewable energy products [39]. These findings would be incorporated into the literature review, highlighting gaps in the existing research and making suggestions for potential future avenues for the area. Figure 2 provides an overview of the actions required for a systematic literature review. It shows the process of looking for appropriate academic sources, assessing the quality of those sources, identifying significant topics and research gaps, arranging the review, and creating a complete and cohesive analysis.

Fig. 2 Key steps for conducting a literature review (*Source* Authors)

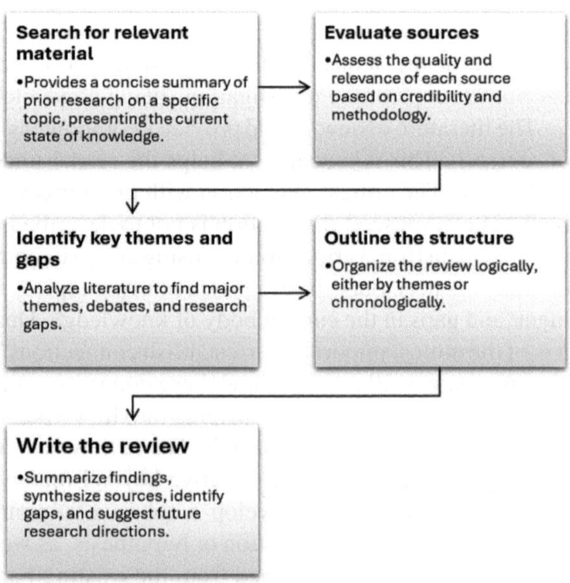

3 The Purpose of a Literature Review

A literature review is essential to academic research to present an organized synthesis of the current body of knowledge on a particular topic [40]. It serves various objectives, including exhibiting the researcher's grasp of the area, developing a theoretical framework, setting the study within a larger academic context, resolving gaps in the literature, and critically evaluating the present state of research. All these functions are accomplished using this document. These roles are vital to ensure that the study is pertinent, methodologically sound, and adds significantly to the field.

3.1 Displaying Proficiency in the Subject Matter and Academic Environment

The primary objective of a literature review is to demonstrate the researcher's comprehensive comprehension of the existing information about the study topic. This task requires providing a concise overview of previous research and a thorough analysis of its strengths and weaknesses, demonstrating the researcher's proficiency in understanding and navigating the scholarly terrain of their discipline [7]. By recognizing and addressing essential ideas, concepts, and methodological methods, the researcher shows their knowledge of the issue and ability to contribute to ongoing academic discussions. In addition, this activity emphasizes the intellectual context in which the new research is located, enabling the researcher to explain how their work aligns with, extends, or questions the current body of literature [41]. In the field of energy efficiency in architecture and construction, a thorough literature analysis would analyze the development of design methods that attempt to decrease energy usage in buildings. The researcher might assess studies examining passive and active design solutions, such as passive solar heating, natural ventilation, and incorporating renewable energy sources like solar panels or wind turbines [42]. By thoroughly analyzing these studies, the researcher would showcase their comprehension of how these strategies have been utilized in different situations, their comparative effectiveness, and the patterns developing in energy-efficient design. In addition, the assessment will emphasize the shift toward more intelligent energy management systems that are integrated using Building Information Modeling (BIM) [41]. These systems utilize energy simulations and real-time data to maximize energy efficiency across the whole lifespan of a building. This examination showcases the researcher's ability to combine previous and present advancements, situating their work within a historical and state-of-the-art research framework. In another example, the literature analysis on recycled building materials would specifically examine the growing emphasis in construction and architecture on using sustainable materials [42]. This trend aligns with the more significant effort to minimize the environmental impact of buildings. The researcher will thoroughly assess studies that examine the advantages and difficulties of integrating recycled materials, such as recovered wood, recycled concrete, or reused steel, into contemporary building endeavors. By examining the advantages and drawbacks of these materials, the researcher will showcase their comprehension of how these materials aid in reducing waste, conserving resources, and promoting the circular economy. In addition, they might analyze the technical efficacy of these materials in terms of their durability, structural soundness, and long-term viability. An extensive evaluation might also examine the progress made in using recycled materials in construction through developing technologies like bio-based or composite materials, demonstrating the researcher's knowledge of the most recent advancements in sustainable building methods [43]. The literature study will examine research investigating how architectural and engineering methods integrate ecological concepts to reduce the environmental effect of buildings. This may involve exploring ideas like biophilic design, which incorporates natural components into constructed spaces to enhance the well-being of occupants and decrease the need for artificial systems for heating, cooling, and lighting [44]. A critical evaluation may assess studies examining the efficacy of green roofs, living walls, or rainwater harvesting systems. This demonstrates the researcher's profound comprehension of how environmental design principles contribute to urban areas' sustainability and livability. In addition, the researcher might investigate the utilization of BIM for modeling and simulating environmental performance. This would enable designers to evaluate the possible effects of various environmental design techniques on energy consumption, water usage, and overall sustainability measurements [45]. By carefully analyzing literature on energy efficiency, recycled building materials, and environmental design, the researcher demonstrates their deep knowledge and positions their work within the broader academic and professional conversation. This interaction showcases the researcher's aptitude for amalgamating many viewpoints and pinpointing significant patterns and deficiencies in the discipline, ultimately establishing the groundwork for their investigation [46]. For example, they may find a discrepancy in the performance of recycled materials when used with energy-efficient technology. Alternatively, they may realize that current research on environmental design has not correctly included new digital tools such as BIM to maximize sustainability results. These insights enable the researcher to provide fresh information, tackle unanswered topics, and expand the limits of the discipline. A straightforward visual depiction of the goal and format of a literature review is shown in Figure 3. The diagram highlights a literature review's main goals and processes, including showcasing thorough subject-matter expertise, assessing methodological approaches, identifying essential theories and concepts, assessing prior research, and situating the study within the larger intellectual framework. The figure also illustrates how a literature study helps integrate recent and historical developments, aligning, extending, or challenging existing knowledge. This format serves as an example of the vital function that a literature review plays in providing a solid scholarly framework for new research.

Overall, the literature review showcases the researcher's proficiency, capacity to evaluate the current pool of information discerningly, and comprehension of the intricate intellectual context in which their work is situated [47]. Through exploring essential subjects such as energy efficiency, the use of recycled building materials,

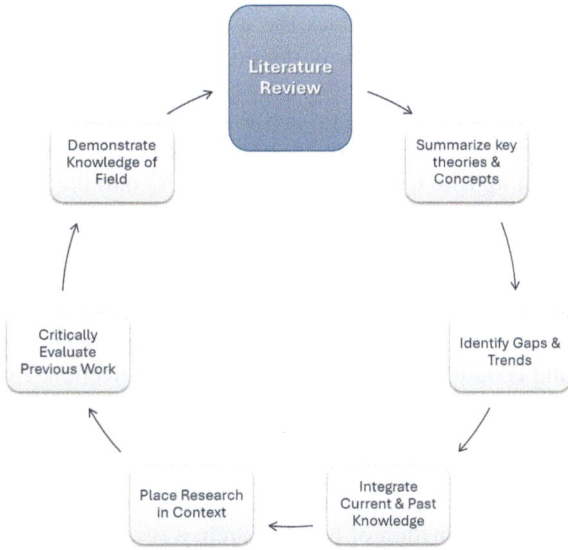

Fig. 3 Key objectives and structure of a literature review in academic research (*Source* Authors)

and environmental design, the researcher provides a solid groundwork for their study, demonstrating the significance and possible influence of their contribution to the area.

3.2 Creating a Theoretical Framework and Research Methodology

Constructing a theoretical framework and research technique is essential to every research project since it establishes a logical basis for the study [48]. A thorough and methodical literature review is crucial in this process, as it enables researchers to discover, assess, and incorporate well-established conceptual models, ideas, and procedures from prior studies. This procedure educates the researcher's theoretical stance and directs the methodological choices, ensuring that the study aligns with current knowledge while filling gaps and providing novel insights. The theoretical framework provides the conceptual foundation that supports the study, providing a framework through which research questions and hypotheses are developed [49]. Anchoring the study inside a well-established body of knowledge guarantees that the research questions are pertinent and that the study adds to the existing academic conversation. In the context of research on energy efficiency in buildings, a theoretical framework may utilize the Energy Efficiency First concept. This theory asserts that prioritizing energy-saving techniques is more important than developing new energy-producing technologies. A researcher might pinpoint deficiencies by examining the theoretical underpinnings of this concept about residential and commercial structures, such as its restricted implementation in specific climatic regions or building types [50]. This enables the alteration or enlargement of the framework to cater to the study situation. In researching sustainable building materials, the theoretical framework can be based on established models like the life-cycle assessment (LCA) approach [51]. This approach assesses the environmental impact of materials from the beginning of their production to their disposal at the end of their useful life. By incorporating Life Cycle Assessment (LCA) into the theoretical framework, researchers may thoroughly evaluate the environmental performance of materials such as bamboo or recycled steel. This evaluation involves comparing their performance to that of traditional materials. An extensive examination of the existing literature on LCA approaches can empower researchers to modify or expand these models to investigate novel materials or delve into less-studied elements of sustainability, such as the social and economic implications of material selection. Literature review plays a crucial role in determining the suitable research methods and instruments by identifying the most successful methodologies for studying comparable research issues. Simulation-based modeling and case study analysis are commonly used approaches in energy efficiency to evaluate the energy performance of buildings [52]. Through a thorough examination of past research, a researcher may discover that dynamic energy modeling tools such as Energy Plus or TRNSYS have been extensively employed to assess the thermal efficiency of insulating materials or energy retrofits. This evaluation would allow the researcher to include or adjust these simulation approaches in a manner that aligns with the unique goals of their study, therefore improving the dependability and accuracy of the research results. Regarding sustainable building materials, empirical approaches like experimental testing or field research are often used to assess novel materials' performance and environmental effects in real-world situations. For example, research on cross-laminated timber (CLT) frequently utilizes field experiments to evaluate the structural and thermal capabilities of traditional materials such as steel and concrete [53]. By examining the methodology utilized in these investigations, researchers may discern the most effective approaches and modify experimental procedures to assess the properties of developing materials such as hempcrete or mycelium-based composites. This guarantees that the research adds to the existing knowledge on sustainable construction while upholding scientific integrity. Furthermore, including well-established procedures in a study enhances the overall validity and trustworthiness of the research. Researchers can assure the reliability and applicability of their findings by adopting or adapting established methods from existing literature. Comparative case studies may be utilized in energy efficiency research to examine the energy performance of buildings in various climatic zones while ensuring consistency with previous studies. Similarly, in the field of sustainable materials, utilizing established methodology from earlier research on biodegradable or recycled materials increases the credibility of the research. It allows for meaningful comparisons across various studies [54]. A thorough and methodical literature study is required to develop a theoretical framework and research technique. Researchers can provide a strong intellectual and methodological basis for their work by critically examining and incorporating current ideas and procedures. In energy efficiency and sustainable building materials, creative research is crucial in advancing environmentally responsible practices in the architectural and construction industries [55].

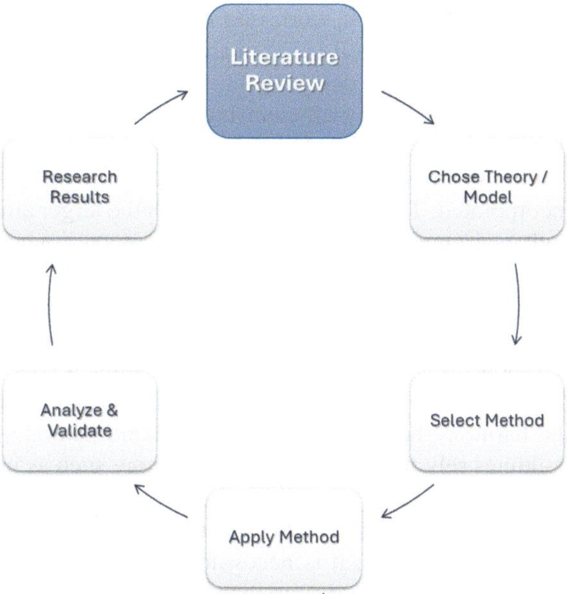

Fig. 4 Circular process of theoretical framework and research methodology development

A theoretical framework and research methods are developed through an iterative process, as depicted in Figure 4. An analysis of the existing literature, the selection of a suitable theory or model, the selection of a research technique, the application of that approach, the analysis and validation of findings, and the generation of research results are the steps.

3.3 Situating Research Within a Broader Academic Context

One of the most essential functions of the literature review is to situate research within the greater academic landscape. This is in addition to the fact that it is responsible for synthesizing pertinent studies [56]. To do this, it is necessary to draw a connection between the research and the ongoing debates, developing trends, and existing ideas that characterize the subject. As part of this process, researchers investigating biophilic architecture and energy-efficient structures must tie their findings to fundamental theoretical concepts and the most recent developments in sustainable design [57]. As an illustration, pioneers in biophilic architecture, such as Amjad Almusaed, have played a significant role in establishing a connection between the human desire for nature and the design of buildings. Using Almusaed theories as a foundation, a researcher may investigate how incorporating natural elements into buildings, such as green roofs, water features, or interior gardens, can increase the well-being of people and the energy efficiency [58]. One example is that green roofs may provide the function of insulation, lowering the energy needed for heating and cooling. Putting their work into context within the expanding conversation about sustainability and well-being in architecture, the researcher places their work within the context of such design techniques. One more illustration comes from research conducted on energy-efficient buildings. Natural ventilation, sun orientation, and daylighting are passive design methods increasingly included in mainstream architectural practices [59]. Researchers may investigate how these strategies are being incorporated. They link their study with the larger aim of lowering energy use in the built environment by citing these advancements, which aligns their research with the larger goals. After that, the researcher may add a new layer by investigating how these passive tactics interact with biophilic components, such as natural materials or open areas filled with light, to improve energy performance. As a component of this more comprehensive process of situating, it is essential to participate in ongoing discussions in the area [60]. One example is the prevalence of conversations on the harmony between technology and the natural world. Should energy-efficient buildings rely more on modern technology such as intelligent systems and artificial intelligence, or can they achieve equivalent efficiency by embracing nature-inspired solutions such as wind-driven ventilation or the strategic positioning of trees for shading? Therefore, it is necessary for researchers working in this field to define their stance in these discussions and demonstrate how their work either supports or challenges the existing way of thinking. Moreover, locating gaps in previously conducted research is frequently necessary in situating research within this framework [61]. For instance, several studies on energy-efficient buildings concentrate on the technical performance of systems such as heating, ventilation, air conditioning (HVAC), and insulation. This debate might be expanded upon by a researcher who investigates how biophilic designs, such as green walls or biomimetic materials, give environmentally friendly and energy-efficient benefits. This topic is less frequently explored in conventional energy efficiency studies. They contribute to filling a vacuum in the literature by doing so and highlighting the developing nature of sustainable construction techniques. To demonstrate how their work contributes to the expansion of the discipline, the researcher places their study within this more comprehensive academic framework. This may be accomplished by refining existing concepts, such as enhancing passive design methods by including natural features, or by adding new viewpoints, such as investigating the symbiotic link between biophilic architecture and energy efficiency. Both approaches are viable options. In addition to expanding knowledge and impacting practice, their contributions help shape the future research direction in sustainable architecture [62]. Within an academic framework, the fundamental components that contribute to forming the "Research Focus" are presented in Figure 5. According to the picture, four fundamental components are highlighted: "Literature Review," "Ongoing Debates," "Emerging Trends," and "Existing Ideas." The research subject is positioned within a more comprehensive academic framework through the interconnectedness of these components, which work together to accomplish this. Through synthesizing past studies, engaging with contemporary discourses, and identifying gaps or trends, each component plays an essential purpose, eventually refining and justifying the relevance of the research topic.

Fig. 5 Crucial elements determining the research topic in an academic setting

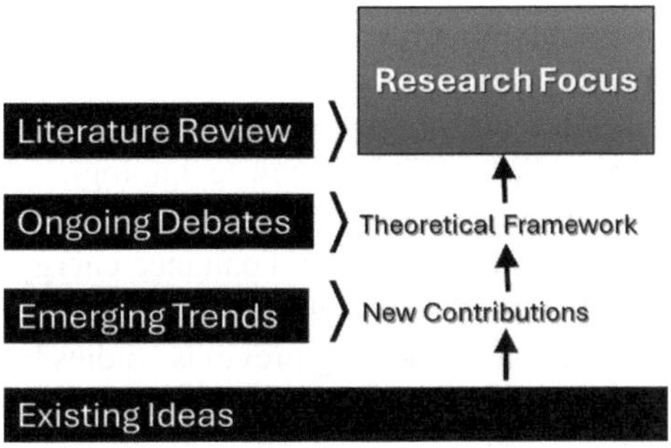

3.4 *Resolving Deficiencies or Making Contributions to Ongoing Discussions*

A literature review's primary goal is to discover gaps in knowledge or unsolved issues within the current body of scholarly work the research aims to tackle [63]. By doing a systematic and analytical assessment of the literature, researchers can identify specific areas where previous studies have not met anticipated outcomes, exposed discrepancies, or neglected significant unresolved problems. This procedure not only elucidates the distinct contribution of the research but also provides a rationale for the necessity of more investigation. By identifying and filling these gaps, the researcher demonstrates the novelty and significance of their study. In sustainable building, there is a lack of integration between energy-efficient materials and technologies and digital twins for improved performance tracking and lifecycle management [64]. Digital twins, which create digital replicas of real-world building systems, have the potential to improve sustainability by enabling real-time monitoring and predictive maintenance, hence minimizing energy waste. Nevertheless, there is a lack of available literature in this area, which creates a valuable chance to contribute substantially to both subjects.

In addition, a well-constructed literature review can enhance existing scholarly discussions and shape future research trajectories in this field. Regarding digital twins in architecture, current discussions mainly focus on their incorporation into sustainable practices. Through rigorous analysis and active involvement in these discussions, the researcher can offer fresh insights or other methodologies, thereby progressing the conversation and shaping the future trajectory of research in this field. Moreover, by finding areas of ambiguity, discord, or unexplored topics within the existing body of literature, the researcher's work is positioned as an expansion or reaction to these deficiencies. Many studies primarily examine sustainable construction materials, but only some explore the possibility of utilizing artificial intelligence and digital twins to simulate the environmental effects of various materials over a building's lifespan [65]. This gap allows academics to suggest creative solutions that improve sustainability by utilizing AI-powered tools in selecting materials and energy efficiency. A literature review helps discover gaps and significantly evaluates current research on the issue [66]. In addition to providing a concise overview of previous research, it entails assessing the investigations' caliber, methodological rigor, and theoretical advancements. Through a rigorous analysis of strengths and shortcomings, the researcher can discern recurring patterns in treating certain subjects, such as sustainable building approaches or digital twin technologies. Many studies support sustainable design principles, but there is frequently a need for more precise data on how digital twins can accurately assess and enhance energy efficiency during the whole lifespan of a structure [67]. Through a careful analysis of the literature, researchers show that they are not just duplicating previous studies but are providing new and original perspectives on the area. This approach has two purposes: firstly, it guides the research by suggesting topics that need more investigation, and secondly, it ensures that the study contributes significantly to the field. Therefore, a meticulously conducted literature evaluation clarifies research inquiries, develops theoretical frameworks, and propels scholarly dialogues. Each researcher contributes to advancing knowledge in areas like sustainable building and digital twin integration. The iterative process of doing a literature review is depicted in Fig. 6, which also demonstrates the important roles involved in moving the research forward. Following the completion of the literature evaluation, researchers identify knowledge gaps, which serve as a compass for determining the path of the research initiative. Contributing to academic discussions, finding potential solutions, and evaluating the strength of the existing literature are all subsequent steps in the process. This cycle strategy guarantees that the study will remain dynamic and receptive to the changes in scholarly discourse.

Finally, Table 2 summarizes the literature review's many roles in advanced research contexts. It highlights the literature review's flexibility in performing

Fig. 6 The literature review process and its functions

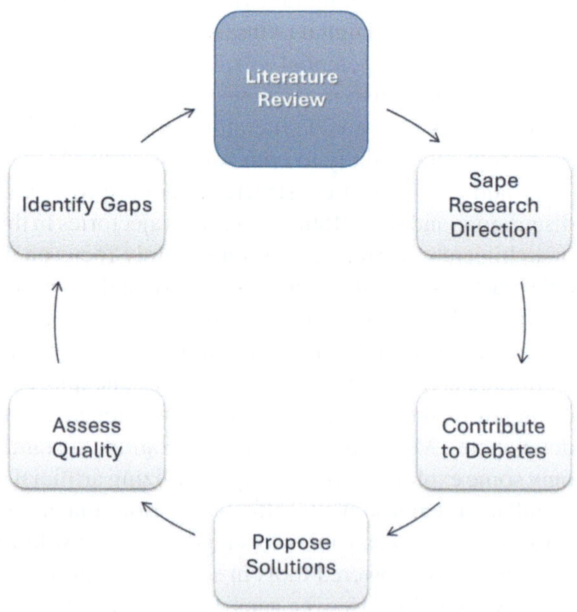

various academic tasks, such as showcasing thorough knowledge or pointing out new paradigms. It covers important issues, including creating research methods for Cyber-Physical Systems and charting the development of Building Information Modeling (BIM) in sustainable design. The table serves as another evidence of the value of literature reviews in developing intricate theoretical frameworks, enhancing scholarly discourse, and the identification of research gaps. Furthermore, it emphasizes how important it is for them to support research within frameworks that look forward and to facilitate multidisciplinary integration, especially when it comes to the convergence of artificial intelligence (AI) and building information modeling (BIM) in creating smart cities [20].

4 Comprehensive Literature Review: Key Processes and Critical Analysis

Any academic inquiry must include a thorough literature evaluation, necessitating the researcher to adhere to several vital procedures to guarantee depth, relevance, and rigor. Here is a thorough academic explanation of the essential factors to consider and the stages required, along with examples to help make things clear:

- **Establishing Boundaries and Objectives**: Clearly defining the parameters of the topic being studied is the first stage in creating a solid literature study. This unambiguously states the main research topic or goal [76]. A literature search on biophilic architecture and its impact on workplace productivity, for example, should only include research that examines the psychological impacts of architecture, sustainability, and green building trends. Establishing limits helps the researcher focus on the central problem and prevents them from deviating into other subjects. Moreover, the precise definition of objectives will direct the process of analyzing and synthesizing the literature.
- **Conducting an Extensive Search**: The next stage is thoroughly searching the academic literature after determining the study subject [77]. This entails searching through scholarly journals, conference papers, and academic books using a variety of academic databases (including Google Scholar, Scopus, and Web of Science) and applying filters. For instance, the search engine should use a mix of terms such as "sustainable urban planning," "energy-efficient buildings," and "smart city technologies" if the issue is energy efficiency in urban planning. Furthermore, to ensure that both fundamental and developing studies are represented, it is essential to use general terms and particular specialty keywords. A comprehensive search might provide many irrelevant results, while a minimal search can overlook pertinent multidisciplinary research.
- **Choosing Reputable Sources**: Selecting reliable sources requires assessing their relevance and reliability after obtaining a wide range of sources. Prioritizing peer-reviewed journals above non-academic sources or general publications is advisable [78]. To guarantee that the evaluation is based on reliable, current sources,

Table 2 Advanced and multifunctional purposes of a literature review with exemplary applications in complex research contexts

Purpose	Description	Exemplary applications
Demonstrating comprehensive knowledge of the field [68]	It showcases a deep and exhaustive understanding of the research domain by synthesizing critical theories, methodologies, historical trends, and landmark studies	In sustainable architecture, a literature review could trace the evolution of BIM from an essential design tool to an advanced system used for comprehensive environmental impact simulations while integrating its interaction with green certifications like LEED
Establishing a multilayered theoretical framework [69]	Integrates multiple theoretical perspectives and conceptual models, creating a nuanced framework that addresses complex, multidimensional research questions	In the study of Cognitive Digital Twins (CDTs) for Construction 5.0, a researcher could combine theories from AI-driven decision-making, human–machine interaction, and construction process optimization to create a multi-tiered theoretical framework
–Designing an advanced research methodology [20]	Develops a sophisticated methodological approach by drawing from diverse disciplines and prior empirical studies, ensuring methodological precision and innovation	For Cyber-Physical Systems (CPS) in construction, it is integrating methodologies from real-time data analytics, IoT systems, and predictive maintenance models to form a comprehensive research strategy that adapts to dynamic on-site conditions
Contextualizing Research within a Broader Academic and Industrial Framework [70]	Embeds the research within academic and industry-specific contexts, linking theoretical contributions to practical applications and societal needs	For BIM in sustainable architecture, I was situating the research within broader discussions on global sustainability goals, circular economy practices, and industry-driven innovations in energy-efficient building design and resource management
Identifying multidimensional gaps and research deficiencies [60]	Uncovers complex, layered research gaps by critically examining the intersections of different research fields, exposing areas that have been underexplored or overlooked	A review on BIM for energy-efficient buildings might reveal gaps in the literature regarding integrating renewable energy systems with real-time data analysis in post-construction phases, highlighting a multifaceted gap in design and operations

(continued)

Table 2 (continued)

Purpose	Description	Exemplary applications
Contributing to and refining complex scholarly debates [71]	It enriches and extends ongoing debates by introducing new variables, theoretical nuances, or empirical findings that challenge, refine, or expand existing scholarly discussions	In Construction 5.0, they contribute to debates on AI-human collaboration by offering empirical evidence on how Cognitive Digital Twins can foster adaptive project management strategies, influencing debates on human creativity vs. AI automation in construction
Critically evaluating the state of research with multivariate analysis [72]	Performs a comprehensive and critical evaluation of the current research landscape by examining methodological, theoretical, and empirical dimensions across various studies	Evaluating the efficacy of BIM in sustainable design might involve a multi-method comparison of energy simulations, life cycle assessments (LCA), and post-occupancy evaluations, identifying inconsistencies and limitations in current modeling approaches
Justifying research within a high-impact, future-oriented framework [73]	Establishes a strong justification for the research by situating it within future-oriented trends and high-impact areas, emphasizing its potential contribution to pressing global challenges	In Cyber-Physical Systems, justifying research on real-time safety monitoring could be contextualized within the increasing complexity of urban mega-projects and the growing demand for automation in risk management for sustainable and resilient construction

(continued)

Table 2 (continued)

Purpose	Description	Exemplary applications
Identifying and analyzing emerging paradigms and technological disruptions [74]	Detects new paradigms, disruptive technologies, and shifts in academic thought, providing insights into their implications for the research field and potential future developments	A literature review could explore the rising integration of AI with BIM in smart cities, highlighting the disruption brought by machine learning and data-driven decision-making in urban planning and intelligent infrastructure design
Fostering interdisciplinary integration and knowledge transfer [75]	Facilitates synthesizing ideas from multiple disciplines, encouraging cross-disciplinary innovation and fostering knowledge transfer across fields, domains, and sectors	In Cyber-Physical Systems, a literature review might integrate concepts from data science, robotics, and structural engineering, advancing CPS applications in construction by borrowing analytical methods from disciplines like artificial intelligence and automation

use recent papers from publications like Energy and Buildings or The Journal of Urban Planning and Development. Each study's methodological soundness, frequency of citations, and the reputation of the journal in which it is published must also be considered. This stage ensures the literature review is based on the best available research, avoiding using out-of-date or unreviewed sources.

- **Structuring the Review**: A critical component of drafting a literature review is logically arranging the gathered material [33]. The literature may be arranged methodologically, chronologically, or thematically, depending on the study type. For instance, the literature might be categorized under themes like machine learning in design optimization, AI-driven robots in construction, and the integration of AI with BIM in a study on advances in AI applications in the construction industry. The review's logical structure and accessibility are guaranteed by the theme approach, which also makes it easier for readers to follow the concepts and advancements as they evolve in the area.
- **Examining Literature**: This stage entails critically analyzing and summarizing earlier studies [34]. For example, it is crucial to take note of methodological strategies (such as qualitative surveys and quantitative analysis of productivity measures) and any contradictory findings while examining research on biophilic design. Finding recurrent themes, contemporary trends, and gaps in literature is another aspect of analyzing the body of research. For example, despite several studies attesting to the psychological advantages of the design approach, a dearth of empirical evidence on the long-term economic benefits of biophilic design exists. This may represent a research gap that needs to be addressed.
- **Offering a Balanced Critique**: A thorough literature review weighs the advantages of earlier research while critically analyzing its drawbacks [79]. For example, possible bias induced by dependence on simulation-based research instead of real-world implementation data should be considered when analyzing studies on AI applications in urban planning. Furthermore, examining the methods employed in different studies (e.g., small sample sizes and absence of control groups) improves the review's academic rigor. It demonstrates a thorough grasp of the subtleties of research.
- **Finding Research Gaps**: A crucial objective of the literature review is finding gaps in the corpus of current research [80]. For example, a study of energy-efficient building materials may point out that, although materials for temperate climates have been the subject of much research, studies on energy efficiency in severe weather—such as high humidity or desert regions—remain relatively unexplored. Finding these gaps enables the researcher to explain the study's applicability by framing their effort as filling a vacuum in the literature.
- **Composing with Coherence and Clarity**: A precise and rationally organized academic literature review is essential [81]. The wording should be professional yet succinct, and each part should flow naturally into the next. Technical phrases should be clarified and should not be used in unnecessary jargon. Academic standards should also be adhered to while using citations, guaranteeing that each assertion or concept is correctly ascribed to its source. This rigorous approach places the study inside the larger academic conversation and legitimizes the effort.

- **Revision and Refinement**: Revision is necessary to guarantee that the review is correct and clearly expressed [82]. Rereading the review, fixing grammar or clarity errors, and getting input from mentors or coworkers is essential. For example, input from an environmental studies specialist guarantees that the most current and pertinent information is included in a review of sustainable practices in architecture.
- **Keeping Up to Date**: Since the academic subject is constantly changing, it is critical to incorporate the most recent findings into the literature review [83]. For instance, the most recent developments in artificial intelligence (AI) and Internet of Things (IoT) technology should be considered while evaluating research on smart cities. Updating the review regularly contributes to keeping it scholarly and relevant.

A literature review involves a critical engagement with the state of knowledge in the topic rather than just a synopsis of previous research. The review offers synthesis, critique, and gap identification, laying the groundwork for future study. Figure 7 outlines the essential steps involved in creating a comprehensive literature review. It includes defining research boundaries, conducting exhaustive searches, selecting credible sources, structuring the review, analyzing literature critically, identifying research gaps, and ensuring clarity through revision and regular updates.

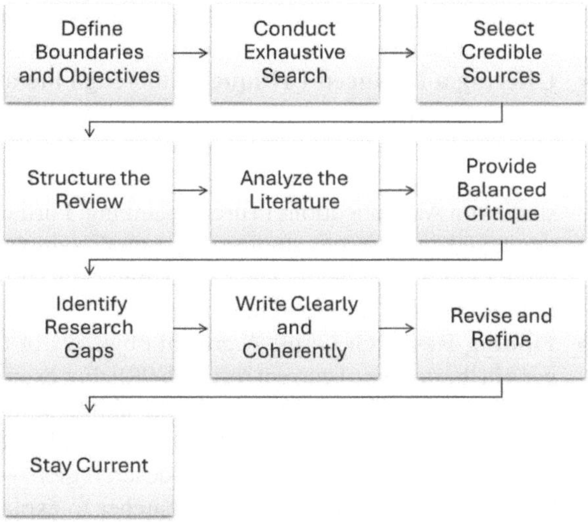

Fig. 7 Key steps for conducting a comprehensive literature review

5 Conclusion

A literature review is an essential part of academic research because it provides a methodical and analytical analysis of the previous academic work that has been done on a specific topic. Its primary objective is to synthesize the existing body of knowledge, detect pertinent ideas and approaches, and draw attention to gaps in the existing body of literature that will serve as a foundation for future study. An approach to the well-structured literature review guarantees that it goes beyond merely providing a summary of past work and provides a complete analysis that places the researcher's study within the context of the larger academic community. This helps to guarantee that the study is relevant, trustworthy, and rooted in the current body of knowledge. It establishes a sound intellectual basis for future investigations, which helps ensure the research is relevant. A literature study that has been meticulously carried out serves numerous vital roles. The first benefit is that it assists in identifying important themes, continuing debates, and unanswered research issues that provide a path for further investigation. This is accomplished by conducting an in-depth analysis of various sources, including academic texts, journals subjected to peer review, and other authoritative references. In the process of doing this study, scholars can obtain a better understanding of the present state of knowledge and are better prepared to participate in academic conversation. Additionally, the literature review allows researchers to assess the benefits and drawbacks of the procedures and theoretical frameworks already in use, providing them with a well-informed foundation for refining their study design. Researchers can give fresh insights, build on sound foundations, and address areas that have been missed or poorly researched if they acknowledge the strengths and gaps that exist within the discipline. A significant contribution to the advancement of academic discourse is made by the literature review, which, in addition to summarizing and assessing the current body of knowledge, makes it possible to identify areas that require additional inquiry. This approach is vital for generating research questions that address substantial gaps in the area. Selecting proper procedures to discover and produce relevant results is also essential. A simple mechanical technique is not all that is involved in the detection of these gaps. Despite this, it requires a high level of engagement with the existing body of literature, during which the researcher identifies patterns, discrepancies, and unexplored regions that require further investigation. A literature study, for instance, can reveal a requirement for further empirical research on incorporating biophilic design components in energy-efficient buildings. This is the case in domains such as architecture. Through observations of this nature, the groundwork is laid for a study that makes theoretical and practical contributions to the field. In addition, the literature review helps to cultivate intellectual rigor by ensuring that references to trustworthy and high-quality sources inform the study. It allows researchers to build on previously acquired information, guaranteeing their work is new and valuable. As new research builds on previous studies and contributes to existing scholarly debates, this process makes it easier for academics to converse with one another. To demonstrate how their study contributes to the advancement of knowledge in a meaningful way, researchers situate their work within the more considerable academic discussion by critically interacting with the existing body of literature. It is not enough to summarize the research findings; the literature review is essential to academic research. It provides a critical foundation for the creation of new research, identifies gaps in the subject, and synthesizes both the existing knowledge and the gaps that have been identified. When this is done, a strong basis is laid for future investigations, and it also contributes to the continuous evolution and development of the academic field. When researchers make knowledgeable, critical, and inventive contributions, they may ensure that their work contributes significantly and meaningfully to advancing the scientific discipline.

References

1. Sheng J, Amankwah-Amoah J, Wang X (2017) A multidisciplinary perspective of big data in management research. Int J Prod Econ 191:97–112
2. Snyder H (2019) Literature review as a research methodology: an overview and guidelines. J Bus Res 104:333–339
3. Tennant JP, Dugan JM, Graziotin D, Jacques DC, Waldner F, Mietchen D, Elkhatib Y, Collister LB, Pikas CK, Crick T, Masuzzo P (2017) A multi-disciplinary perspective on emergent and future innovations in peer review. F1000Research 6
4. Dehalwar K (ed.) (2024) Basics of research methodology-writing and publication. EduPedia Publications Pvt Ltd
5. Leong FT, Austin JT (eds.) (2023) The psychology research handbook: a guide for graduate students and research assistants. SAGE publications
6. Karunarathna I, Bandara S, Jayawardana A, De Alvis K, Gunasena P, Hapuarachchi T, Ekanayake U, Rajapaksha S, Gunawardana K, Aluthge P, Gunathilake S Bridging research gaps: how to write a focused and critical literature review
7. Antons D, Breidbach CF, Joshi AM, Salge TO (2023) Computational literature reviews: method, algorithms, and roadmap. Organ Res Methods 26(1):107–138
8. Dong J, Dong S, Buckingham L (2023) How does a research topic evolve into a research field? A bibliometric analysis of metadiscourse research. ibérica (45):163–189
9. Edelman NL (2023) Trauma and resilience informed research principles and practice: a framework to improve the inclusion and experience of disadvantaged populations in health and social care research. J Health Serv Res Policy 28(1):66–75
10. van Riel A, Snyder H (2024) Enhancing the impact of literature reviews: guidelines for making meaningful contributions. Spanish J Market-ESIC 28(3):250–265
11. Maiya AK, Aithal PS (2023) A review-based research topic identification on how to improve the quality services of higher education institutions in academic, administrative, and research areas. Int J Manag Technol Soc Sci (IJMTS) 8(3):103–153
12. Chigbu UE, Atiku SO, Du Plessis CC (2023) The science of literature reviews: searching, identifying, selecting, and synthesising. Publications 11(1):2
13. Cooksey RW (2024) Systemic contextualisation: situating and positioning. In: Unity from diversity: pluralist systemic thinking for social and behavioural research. Springer Nature Singapore, Singapore, pp 165–222
14. Booth A, James MS, Clowes M, Sutton A (2021) Systematic approaches to a successful literature review
15. Johnson JL, Adkins D, Chauvin S (2020) A review of the quality indicators of rigor in qualitative research. Am J Pharm Educ 84(1):7120

16. Mauer B, Venecek J (2022) Writing the literature review. Sin: trategies for Conducting Literary Research, 2e
17. Kuperman VY, Sokol GH (2024) On the causes and ramifications of multi-authorship in science. Scientometrics 1–21
18. Publication Venue: Reputable academic publishers and peer-reviewed journals frequently provide credible information
19. Giddings LS, Grant BM (2009) From rigour to trustworthiness: validating mixed methods. Mixed Methods Res Nurs Health Sci 13(1):119–134
20. Sovacool BK, Axsen J, Sorrell S (2018) It promotes novelty, rigor, and style in energy social science: towards codes of practice for appropriate methods and research design. Energy Res Soc Sci 45:12–42
21. Zhu Y, Liu J, Hu Y, Xie Y, Zeng D, Li R (2024) Distributionally robust optimization model considering deep peak shaving and uncertainty of renewable energy. Energy 288:129935
22. Belcher BM, Rasmussen KE, Kemshaw MR, Zornes DA (2016) Defining and assessing research quality in a transdisciplinary context. Res Evaluat 25(1):1–17
23. Schick-Makaroff K, MacDonald M, Plummer M, Burgess J, Neander W (2016) What synthesis methodology should I use? A review and analysis of approaches to research synthesis. AIMS Public Health 3(1):172
24. Eerkes-Medrano D, Thompson RC, Aldridge DC (2015) Microplastics in freshwater systems: a review of the emerging threats, identification of knowledge gaps and prioritization of research needs. Water Res 75:63–82
25. Wixom BH, Watson HJ (2001) An empirical investigation of the factors affecting data warehousing success. MIS Quart 17–41
26. McKenna R, Pfenninger S, Heinrichs H, Schmidt J, Staffell I, Bauer C, Gruber K, Hahmann AN, Jansen M, Klingler M, Landwehr N (2022) High-resolution large-scale onshore wind energy assessments: a review of potential definitions, methodologies and future research needs. Renew Energy 182:659–684
27. Lapan SD, Quartaroli MT, Riemer FJ (eds) (2011) Qualitative research: an introduction to methods and designs. John Wiley & Sons
28. Simsek Z, Fox B, Heavey C (2023) Systematicity in organizational research literature reviews: a framework and assessment. Organ Res Methods 26(2):292–321
29. Véras LGD, Medeiros FL, Guimarães LN (2019) Systematic literature review of sampling process in rapidly-exploring random trees. IEEE Access 7:50933–50953
30. Ghansah FA, Lu W (2024) Major opportunities of digital twins for smart buildings: a scientometric and content analysis. Smart Sustain Built Environ 13(1):63–84
31. Hatoum MB, Nassereddine H (2024) Determining decision-making factors for technology adoption in the construction industry
32. Many JE, Fyfe R, Lewis G, Mitchell E (1996) Traversing the topical landscape: exploring students' self-directed reading-writing-research processes. Read Res Q 31(1):12–35
33. Galvan JL, Galvan MC (2017) Writing literature reviews: a guide for social and behavioral sciences students. Routledge
34. Randolph J (2019) A guide to writing the dissertation literature review. Pract Assess Res Eval 14(1):13
35. Hemmings C (2011) Why stories matter: the political grammar of feminist theory. Duke University Press
36. Fink A (2019) Conducting research literature reviews: from the internet to paper. Sage publications
37. Govindan K, Rajendran S, Sarkis J, Murugesan P (2015) Multi-criteria decision-making approaches for green supplier evaluation and selection: a literature review. J Clean Prod 98:66–83
38. da Cruz TT, Las-Casas B, Dias IKR, Arantes V (2024) Nanocelluloses as sustainable emerging technologies: state of the art and future challenges based on life cycle assessment. Sustain Mater Technol e01010

39. Kadenic MD, Ballantyne AG, Olsen TB, Enevoldsen P, Gross A (2024) Proactive understanding of the multi-level stakeholder acceptance of a novel renewable energy technology: chemical storage of solar energy. Energy Res Soc Sci 108:103370
40. Newman J (2024) Promoting interdisciplinary research collaboration: a systematic review, a critical literature review, and a pathway forward. Soc Epistemol 38(2):135–151
41. Almusaed A, Yitmen I, Almssad A (2023) Reviewing and integrating aec practices into industry 6.0: Strategies for smart and sustainable future-built environments. Sustainability 15(18):13464
42. Almusaed A, Yitmen I, Myhren JA, Almssad A (2024) Assessing the impact of recycled building materials on environmental sustainability and energy efficiency: a comprehensive framework for reducing greenhouse gas emissions. Buildings 14(6):1566
43. Rosa Latapie S, Abou-Chakra A, Sabathier V (2023) Bibliometric analysis of bio-and earth-based building materials: current and future trends. Constr Mater 3(4):474–508
44. Almusaed A, Yitmen I, Almssad A (2024) Contemporary innovations and sustainable practices in applying clay materials within architectural design and construction methodologies
45. Feng H, Kassem M, Greenwood D, Doukari O (2023) Whole building life cycle assessment at the design stage: a BIM-based framework using environmental product declaration. Int J Build Pathol Adapt 41(1):109–142
46. Creamer EG (2024) Visual displays in qualitative and mixed method research: a comprehensive guide. Taylor & Francis
47. Alejandra Martinez-Daza M, Valencia Quecano LI, Guzmán Rincón A (2024) Conceptual model for the assessment of academic productivity in research seedbeds from a systematic review. Eur J Educ Res 13(2)
48. Alvesson M, Sandberg J (2024) The art of phenomena construction: a framework for coming up with research phenomena beyond 'the usual suspects.' J Manage Stud 61(5):1737–1765
49. Partelow S (2023) What is a framework? Understanding their purpose, value, development and use. J Environ Stud Sci 13(3):510–519
50. Almusaed A, Almssad A, Yitmen I (2024) Sustainable built environment and its implications on real estate development: a comprehensive analysis
51. Pan Y, Zhu M, Lv Y, Yang Y, Liang Y, Yin R, Yang Y, Jia X, Wang X, Zeng F, Huang S (2023) Building energy simulation and its application for building performance optimization: a review of methods, tools, and case studies. Adv Appl Energy 10:100135
52. Gan VJ, Lo IM, Ma J, Tse KT, Cheng JC, Chan CM (2020) Simulation optimisation towards energy efficient green buildings: current status and future trends. J Clean Prod 254:120012
53. De Araujo V, Aguiar F, Jardim P, Mascarenhas F, Marini L, Aquino V, Santos H, Panzera T, Lahr F, Christoforo A (2023) Is cross-laminated timber (clt) a wood panel, a building, or a construction system? A systematic review on its functions, characteristics, performances, and applications. Forests 14(2):264
54. Moshood TD, Nawanir G, Mahmud F, Mohamad F, Ahmad MH, AbdulGhani A, Kumar S (2022) Green product innovation: a means towards achieving global sustainable product within biodegradable plastic industry. J Clean Prod 363:132506
55. Umoh AA, Adefemi A, Ibewe KI, Etukudoh EA, Ilojianya VI, Nwokediegwu ZQS (2024) Green architecture and energy efficiency: a review of innovative design and construction techniques. Eng Sci Technol J 5(1):185–200
56. Urquhart C, Cheuk B, Lam L, Snowden D (2024) Sense-making, sensemaking and sense making—a systematic review and meta-synthesis of literature in information science and education: an annual review of information science and technology (ARIST) paper. J Assoc Inform Sci Technol
57. Almusaed A, Almssad A, Yitmen I, Wallhagen M, Yang YF, Iyer-Raniga U (2024) Integrative approaches in urban sustainability: architectural design, technological innovations and social dynamics in global contexts. BoD–Books on Demand
58. Almusaed A, Almusaed A (2011) Biophilic architecture hypothesis. Biophilic and bioclimatic architecture: analytical therapy for the next generation of passive sustainable architecture. 39–46

59. Almusaed A, Yitmen I, Almsaad A, Akiner İ, Akiner ME (2021) Coherent investigation on a smart kinetic wooden façade based on material passport concepts and environmental profile inquiry. Materials 14(14):3771
60. Jones SR, Torres V, Arminio J (2021) Negotiating the complexities of qualitative research in higher education: essential elements and issues. Routledge
61. Alvesson M, Sandberg J (2013) Has management studies lost its way? Ideas for more imaginative and innovative research. J Manage Stud 50(1):128–152
62. Bibri SE, Krogstie J (2017) Smart sustainable cities of the future: an extensive interdisciplinary literature review. Sustain Cities Soc 31:183–212
63. Torraco RJ (2016) Writing integrative literature reviews: using the past and present to explore the future—human resource development review. 15(4):404–428
64. Tahmasebinia F, Lin L, Wu S, Kang Y, Sepasgozar S (2023) Exploring the benefits and limitations of digital twin technology in building energy. Appl Sci 13(15):8814
65. Pan Y, Zhang L (2021) Roles of artificial intelligence in construction engineering and management: a critical review and future trends. Autom Constr 122:103517
66. Lim WM, Kumar S, Ali F (2022) Advancing knowledge through literature reviews:'what', 'why', and 'how to contribute.' Serv Ind J 42(7–8):481–513
67. Teng SY, Touš M, Leong WD, How BS, Lam HL, Máša V (2021) Recent advances on industrial data-driven energy savings: digital twins and infrastructures. Renew Sustain Energy Rev 135:110208
68. Lim WM, Kumar S (2024) Guidelines for interpreting the results of bibliometric analysis: a sensemaking approach. Glob Bus Organ Excell 43(2):17–26
69. Walther J, Sochacka NW, Benson LC, Bumbaco AE, Kellam N, Pawley AL, Phillips CM (2017) Qualitative research quality: a collaborative inquiry across multiple methodological perspectives. J Eng Educ 106(3):398–430
70. Dwivedi YK, Jeyaraj A, Hughes L, Davies GH, Ahuja M, Albashrawi MA, Al-Busaidi AS, Al-Sharhan S, Al-Sulaiti KI, Altinay L, Amalaya S (2024) "Real impact": Challenges and opportunities in bridging the gap between research and practice–Making a difference in industry, policy, and society. Int J Inform Manag 102750
71. Reio TG Jr (2021) The ten research questions: an analytic tool for critiquing empirical studies and teaching research rigor. Hum Resour Dev Rev 20(3):374–390
72. Marsh HW, Bornmann L, Mutz R, Daniel HD, O'Mara A (2009) Gender effects in the peer reviews of grant proposals: a comprehensive meta-analysis comparing traditional and multilevel approaches. Rev Educ Res 79(3):1290–1326
73. Stauffer M, Seifert K, Ammann N, Snoeij JP (2021) Policymaking for the long-term future: improving institutional fit. Working paper at the Simon Institute for Longterm Governance. Online. https://www.researchgate.net/publication/351050249_Policymaking_for_the_Long-term_Future_Improving_Institutional_Fit. Accessed 17 Sep 2022
74. Cagnin C, Havas A, Saritas O (2013) Future-oriented technology analysis: ITs potential to address disruptive transformations. Technol Forecast Soc Change 80(3):379–385
75. Klein JT (2022) Building capacity for transformative learning: lessons from crossdisciplinary and cross-sector education and research. Environ Dev Sustain 1–14
76. Hancock DR, Algozzine B, Lim JH (2021) Doing case study research: a practical guide for beginning researchers
77. Varsha PS, Chakraborty A, Kar AK (2024) How to undertake an impactful literature review: understanding review approaches and guidelines for high-impact systematic literature reviews. South Asian J Bus Manag Cases 13(1):18–35
78. Pooja K, Upadhyaya P (2024) What makes an online review credible? A systematic review of the literature and future research directions. Manag Rev Quart 74(2):627–659
79. Booth A, Martyn-St James M, Clowes M, Sutton A (2021) Systematic approaches to a successful literature review. 1–100
80. Haque AB, Arifuzzaman BM, Siddik SAN, Kalam A, Shahjahan TS, Saleena TS, Alam M, Islam MR, Ahmmed F, Hossain MJ (2022) Semantic web in healthcare: a systematic literature review of application, research gap, and future research avenues. Int J Clin Pract 2022(1):6807484

81. Golden-Biddle K, Locke K (2006) Composing qualitative research. Sage Publications
82. Gabbay DM, Rodrigues OT, Russo A (2010) Revision, acceptability and context: theoretical and algorithmic aspects. Springer Science & Business Media
83. Pencheon D, Wright J (2002) Keeping up to date through lifelong learning. The evidence base for diabetes care pp 727–752

Structuring the Research Report

Abstract An ordered research report enhances theoretical understanding and practical applications. The careful organization of its elements, from the introduction to the conclusion, is crucial for connecting unique theoretical contributions with valuable insights. This approach fills significant gaps in current knowledge and links research outcomes with urgent real-world issues. The introduction contextualizes the study within a broader academic, sociological, or practical framework, elucidating its relevance, delineating its extent, and providing a cohesive basis for its aims and research questions. The essay emphasizes the significance of interdisciplinary methods, which integrate theoretical frameworks with empirical evidence to tackle complex issues. Supported by strong mixed-methods frameworks, transparent reporting, and ethical compliance, methodological rigor guarantees valid, trustworthy, and generalizable results. Moral issues, including informed permission and participant confidentiality, are essential for maintaining the integrity and credibility of study findings. The results of a methodically organized research report have significant significance for academics, policymaking, and practice. Researchers reveal deficiencies and stimulate novel avenues of investigation, whereas practitioners and policymakers provide empirical suggestions that guide decision-making and foster innovation. This strategy improves academic rigor and societal relevance by tackling issues like sample diversity and contextual restrictions while promoting iterative research procedures. Ultimately, it creates a framework that guarantees research significantly advances knowledge and has quantifiable effects on intricate social challenges.

1 Introduction

To establish the background of the study, define its scope, and articulate its relevance and significance, the introduction is an essential component of every research report [1]. It provides direction to the reader by situating the research within the context of

A. Almusaed et al., *Practice of Research Methodology in Civil Engineering and Architecture*, https://doi.org/10.1007/978-3-031-97393-2_22

its more considerable academic, practical, or societal environment, therefore illustrating why the study is essential and beneficial. The introduction lays the groundwork for a logical and significant study by referring to the history, the explanation of the problem, the aims, the research questions, and the investigation's relevance. The foundation of an engaging introduction is the Background and Context section, which provides a thorough and organized summary of the research issue and establishes the study's framework [2]. This section gives readers the background information they need to understand the study's significance and applicability by summarizing essential developments, trends, and difficulties in the subject. As a guide for the research project, the Introduction, on the other hand, describes the goals and parameters of the investigation [1]. The Background presents pertinent information and evidence that puts the research topic in perspective while critically analyzing current literature. It articulates the problem statement, identifies shortcomings, gaps, or problems within the present body of knowledge, and substantiates these concerns with specific evidence [3]. This part is crucial for drawing readers in and building credibility because it is their first substantial interaction. Although the format and length of background sections might differ from manuscript to manuscript, excellent ones have one thing in common: they offer a thorough, accurate, and well-written examination of the study issue, demonstrating the author's experience and successfully capturing the audience's attention. The subject is placed within its larger philosophical or practical context in the Background and Context section [4]. An example would be examining the historical evolution of building technology and sustainable design practices in the context of research on energy efficiency and sustainable buildings. It has the potential to bring attention to the changing energy requirements in metropolitan areas and the urgent requirement to reduce carbon emissions by implementing new solutions. Using this methodology, the study is contextualized by demonstrating its relevance to contemporary concerns, such as attaining energy efficiency and mitigating the implications of climate change, while simultaneously exposing gaps in the existing knowledge and practices [5]. By establishing the framework for a thorough and knowledgeable debate, this part sets the scene for the study and directs readers toward the significance and direction of the investigation. The rationale further demonstrates the importance of the study from the researcher's viewpoint, highlighting the need for the inquiry and its potential to progress in the field. In addition to convincing readers of the importance of the research, a well-written background lays a strong basis for the objectives and research questions of the study [6]. A research work background is usually the first part that aims to clearly state the central issues or queries the study tries to answer. Pointing out unsolved problems or underexplored regions highlights the necessity for more research. Figure 1 delineates the logical evolution inside a research framework. The Background section starts by offering an overview of the topic and emphasizing critical problems. The Rationale subsequently elucidates the significance and necessity of the investigation. The Research Questions delineate the precise inquiries the study intends to explore to finalize the research's conceptual framework.

Considering the surrounding circumstances, the problem statement precisely defines the issue or knowledge gap the research seeks to answer. This component

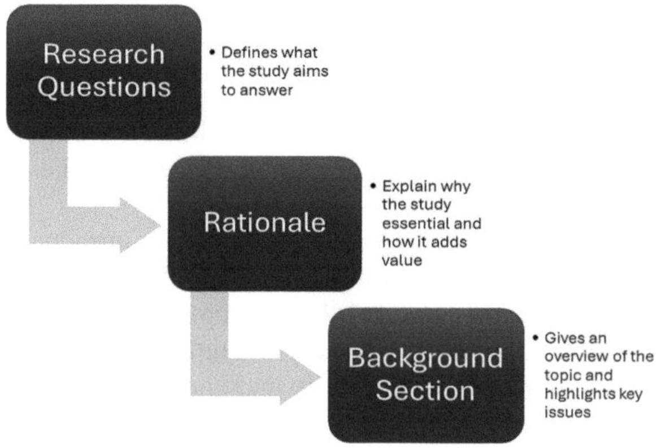

Fig. 1 Flow of research structure: background, rationale, and research questions (*Source* Authors)

shifts from a broad overview of the area to an accurate emphasis on the study possible, which is of absolute importance. In the context of circular economy buildings, for instance, the problem statement might highlight the limited integration of circular design principles in current construction practices, such as the reuse and recycling of materials, or the lack of scalable frameworks for achieving material circularity in urban development [7]. These examples are examples of how the problem statement could be used. When a problem statement is well-articulated, it not only reveals the limits or flaws of previous research but also emphasizes the urgency and importance of filling the identified gap. For example, there is a need for sustainable ways to decrease waste in construction and increase resource efficiency. The research objectives and questions in sustainable building materials provide a precise framework for the study by articulating its aims and the specific issues it intends to explore. The purpose is to create a quantifiable framework to examine the difficulties of using sustainable materials in construction. This structure guarantees that the study stays concentrated and consistent with its objective. The study may encompass finding significant economic, technological, and regulatory obstacles to the broad use of sustainable building materials and assessing their environmental and cost-saving advantages compared to conventional materials. The research questions direct the methodological approach, refining the focus of the investigation. Such inquiries may include: "What are the principal economic constraints impeding the utilization of sustainable building materials in construction endeavors?" What is the perception of building experts about the long-term advantages of sustainable materials compared to traditional alternatives? The introduction delineates specific targets and issues, so establishing a systematic framework for the inquiry ensures the study's scope is exact, and its significance is evident [8].

In the introduction, the study's relevance is an essential component that addresses the significance of the research and the contributions it intends to make to either academic understanding or practical applications. In addition to justifying the importance of the study, this part also emphasizes the potential of the research to contribute to the advancement of theoretical knowledge, to inform professional practice, or to affect the formation of policy [9]. By emphasizing these contributions, the introduction highlights the broader influence that the study has had and successfully convinces readers of the relevance of the study. By demonstrating how the study fills in existing gaps, offers novel insights, and contributes to creating solutions to critical difficulties within the area, this component indicates how the research represents a significant contribution.

An essential component of an academic introduction is demonstrating the study's originality. This often involves a brief discussion of relevant literature to establish how the current research builds upon or differs from previous studies [3]. Highlighting the study's originality strengthens its value and assures readers that it offers new approaches or insights on the subject under investigation. The introduction typically includes a concise summary of the methodological approach, giving readers an overview of how the study will achieve its objectives. This section might outline the study design, data collection methods, and analytical techniques, ensuring they align with the research questions and goals. While not exhaustive, this preview helps readers grasp the scope of the study and the methodology used to reach its conclusions. Its clarity and logical progression characterize a successful introduction. It smoothly transitions from general context to specific details, guiding readers through the research's background, problem statement aims, and significance [10]. By using clear and consistent language, the introduction ensures accessibility to a broad audience, including those who may not be experts in the field. The introduction is the initial entry point into a research report by delivering a detailed summary of the study's setting, problems, aims, and importance. The introduction frames the survey as meaningful and valuable by skillfully integrating these parts. This aligns the research with the more considerable academic or practical debate that is taking place. In addition to capturing readers' attention, a well-written introduction establishes the framework for the logical evolution of the study [6]. It ensures that the research is perceived as being extraordinarily rigorous and significant. For instance, research on zero carbon dioxide emissions buildings that align with Agenda 2030 may concentrate on overcoming the obstacles that keep people from adopting environmentally responsible building methods. Even though there have been breakthroughs in low-energy technology, problems such as high costs, limited incentives, and public ignorance still need to be overcome. The research may seek to analyze the efficacy of policies, determine the most effective methods for incorporating renewable energy into urban areas, and examine the feasibility of retrofitting facilities for existing structures. Through the provision of insights into socio-economic and regulatory variables, the study has the potential to help Sustainable Development Goals 11 and 13, providing recommendations that governments and developers may put into action to scale sustainable practices and speed the transition to carbon–neutral buildings [11]. Figure 2 demonstrates the iterative structure of a typical research process. Starting with the Abstract, the study flows into the Introduction, which gives context and background. The Background & Context further defines the structure, leading to the Problem Statement that outlines the research gap. Next, Objectives & Questions explain the emphasis, followed by Methodology, which describes the technique employed. The Results & Discussion section examines results, culminating in

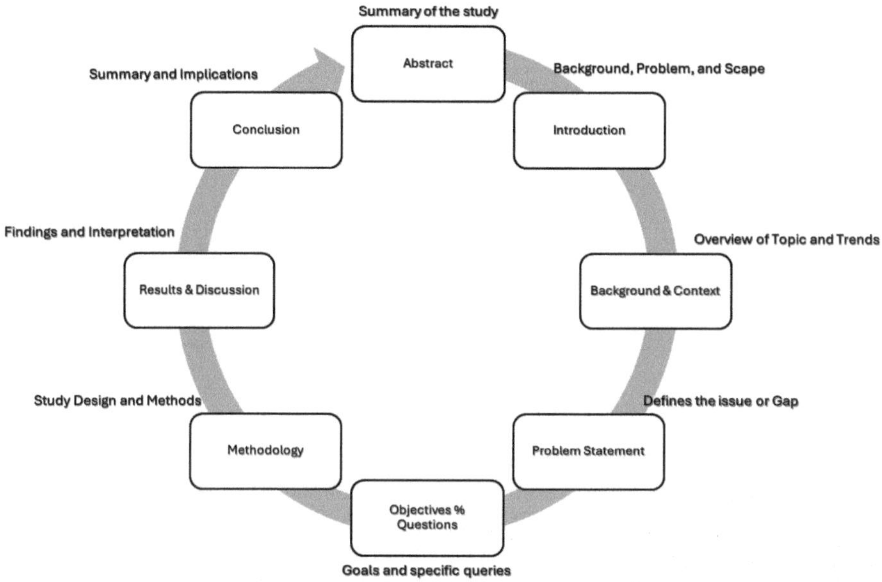

Fig. 2 Framework of a research process (*Source* Authors)

the Conclusion synthesizing ideas. The circular flow stresses the cyclical nature of research, where results may inspire improvements in earlier phases, creating a strong study framework.

2 A Comprehensive Methodology for Reliable and Valid Research: A Rigorous Mixed-Methods Approach

The rigorous technique utilized in this research study is described in this part precisely and methodically [12]. This section overviews the tools, methodologies, and processes to guarantee the study's repeatability, dependability, and validity. The research topic needs to be investigated using a mixed-methods approach, combining qualitative data for contextual depth and quantitative data for quantifiable insights. This technique ensured that the research problem was studied in its entirety. Additionally, the robustness of the findings was improved because of this dual method, which made it easier to analyze complicated events from various angles. Both clarity and accuracy are used to present the chosen approaches to encourage replicability and inspire meaningful application in comparable circumstances [13]. When it came to protecting the rights of the participants, ethical concerns such as informed consent and confidentiality were given the highest priority. The methodology demonstrates the study's dedication to dependability and relevance by adhering to the criteria that have been established and by keeping transparency throughout the process [14].

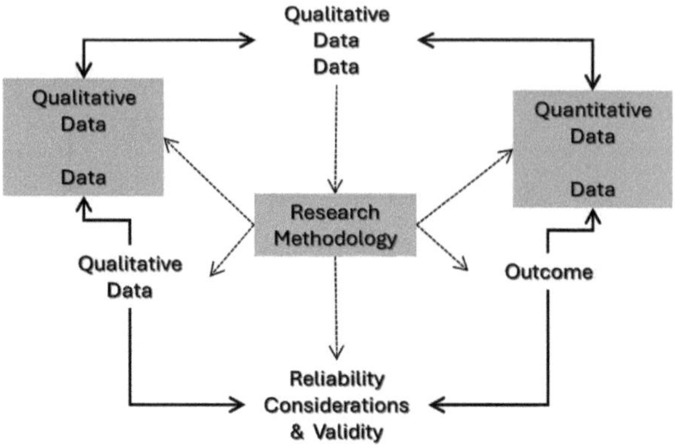

Fig. 3 Minimalistic research methodology framework (*Source* Authors)

Not only does this comprehensive method bolster the findings, but it also offers a significant reference for research that will be conducted in the future. Figure 3 delineates a reduced research technique framework comprising six fundamental aspects. The uppermost row includes "Qualitative Data," "Quantitative Data," and "Ethical Considerations," signifying the essential components of the investigation. The elements are linked to the bottom row, which displays the principal outcomes: "Reproducibility," "Reliability," and "Validity." This pristine and simple design underscores the organized progression from techniques to outcomes, accentuating the methodology's clarity and rigor.

2.1 *Objective*

A fundamental framework aimed at guaranteeing that the research process is transparent, systematic, and well-organized is achieved via the methodology [15]. In this part, the reason for the selected study design is explained, the procedures utilized for data collection are outlined, and the analytical approaches used to interpret and validate the findings are presented. The credibility, reproducibility, and general dependability of the study are all improved significantly due to the methodology's specific definition of these components. This methodological approach has as one of its primary goals the maintenance of the highest professional standards generally acknowledged within the scientific community. By following these standards, one can ensure that the findings are reliable, genuine, and relevant to advancing knowledge within the subject. It is not just that this dedication to rigor helps to strengthen the study's credibility. Still, it also makes it easier to apply the findings meaningfully in situations connected to the research [16]. In addition, the technique emphasizes clarity and coherence, which makes it possible for other researchers to reproduce or modify the strategy in subsequent studies that are comparable. This work contributes substantially to the larger body of knowledge by offering a comprehensive overview of the research design, the processes associated with data collecting, and the analytical methodologies utilized. Ultimately, this extensive and systematic approach improves the study's importance and relevance among the academic and professional communities, fostering continuing growth and innovation in the subject [17].

2.2 Research Design

This section employed a mixed-methods research methodology to facilitate a thorough and detailed analysis of the study topics. Ngulube [18] asserts that mixed-methods research amalgamates qualitative and quantitative approaches to encapsulate the intricacies of diverse realities. Upon identifying the research topic, it was concluded that successful resolution necessitated systematically gathering qualitative and quantitative data concurrently or sequentially. Thus, the study adopted a methodologically rigorous approach, employing basic or advanced mixed methods designs to guarantee the analysis's depth, reliability, and validity [18]. Integrating quantitative and qualitative approaches allowed for a holistic exploration of complex phenomena [19]. Quantitative data offered measurable insights into general trends, while qualitative data provided a deeper understanding of the context and intricacies underlying these trends. A sequential strategy was adopted to maximize the strengths of both methods. Initially, quantitative data were collected to identify broad patterns and establish a foundational understanding of the research problem. This step provided statistical evidence and measurable variables critical for shaping the subsequent phase of the research. The next phase involved qualitative research to uncover more decadent, nuanced interpretations of the quantitative findings. Techniques such as interviews or thematic analysis were employed to explore key themes, contextual factors, and subjective experiences that could not be captured through quantitative methods alone [20]. The integration of these approaches ensured robust and comprehensive analysis. Quantitative findings offered a clear, objective framework, while qualitative insights enriched the interpretation by revealing underlying complexities. This dual approach enhanced the study's reliability and validity, enabling a thorough and multidimensional understanding of the research problem. The mixed-methods design supported credible and actionable conclusions [21].

2.3 A General Overview of Data Collection and Analysis

It is possible to split the process into two main phases: quantitative and qualitative data gathering. Each stage has unique procedures to increase data quality, ethical compliance, and research integrity. This way, the process may be separated into two main stages to ensure a systematic and reliable data collection and analysis approach [22]. Quantitative data collection is accomplished through the utilization of structured online questionnaires that are intended to collect numerical data according to the goals of the study. A stratified random sample procedure is advised to guarantee complete demographic representation, reduce the likelihood of bias, and enhance the generalizability of the findings. By using this strategy, it is guaranteed that the sample will sufficiently reflect all of the subgroups that are contained within the population. Before participating in the study, participants must obtain ethical permission from an Institutional Review Board (IRB) or another appropriate ethics body [23]. Additionally, participants are required to submit their informed consent electronically. To ensure that the data continues to be accurate and reliable, it is necessary to rectify any missing information by employing procedures known as robust imputation [24]. On the other hand, qualitative data collection is centered on collecting in-depth insights through semi-structured interviews. It is recommended that these interviews be carried out with a purposeful sample of participants, who are chosen based on their capacity to supply a wealth of information and their relevance to the study objectives. Accessibility and flexibility are two benefits that may be gained by conducting interviews through videoconferencing, with each session lasting roughly forty-five minutes. During data management, it is of the utmost importance to safeguard the confidentiality of the participants by anonymizing any identifying information. It is recommended that quantitative data analysis be carried out using statistical tools, integrating both descriptive and inferential approaches. Inferential methods, like regression analysis and hypothesis testing, explore the relationships between variables and test research hypotheses [25]. Descriptive statistics, on the other hand, give an overview of variable distributions and provide a summary of the characteristics of the participants. Data must be resolved methodically to ensure the validity and reliability of the results are maintained. A full breakdown of the steps of data gathering and analysis is presented in Table 1. In the first phase, quantitative data was collected through structured questionnaires and stratified random sampling. This was done to guarantee that demographic populations were adequately represented and to prevent bias. The second phase of the research project focused on qualitative data by conducting semi-structured interviews to collect various opinions. Statistical methods such as regression and hypothesis testing were utilized for quantitative analysis. On the other hand, qualitative analysis utilized theme approaches, which ensured reliability using independent coding and stringent review procedures.

2.4 Comprehensive Explanation of Methodologies

A successful research approach that integrates qualitative and quantitative techniques is essential to meet the complex nature of research aims. Taherdoost [30] asserts that research can be examined through three primary paradigms: quantitative methods, emphasizing numerical data and statistical analysis; qualitative approaches, prioritizing descriptive and interpretive insights; and mixed methods, which integrate both to offer a comprehensive perspective [30]. Although these frameworks are essential for academic research, scholars frequently encounter difficulties in choosing the

Table 1 Data collection and analysis phases

Phase	Activity	Details
Phase 1: quantitative data collection [26]	**Instrument**: structured online questionnaires **Sampling**: stratified random sampling **Ethics**: institutional review board (IRB) approval obtained **Consent**: participants provided informed electronic consent	– Numerical data collected on factors in the study framework – Stratified random sampling ensured demographic representation, reduced bias, and enhanced generalizability
Phase 2: qualitative data collection [27]	**Instrument**: Semi-structured interviews **Sampling**: purposive sampling of survey respondents **Ethics**: anonymity is ensured by removing identifying information	– Interviews captured diverse perspectives – Participants' confidentiality is guaranteed through anonymization – Focused on a variety of opinions to enhance depth and representativeness
Quantitative data analysis [28]	**Techniques**: descriptive and inferential statistics **Methods**: regression analysis, hypothesis testing **Tools**: statistical software **Missing data**: addressed using imputation methods	– Descriptive statistics characterized sample features and variable distributions – Regression analysis and hypothesis testing explored relationships between variables – Ensured objectivity through robust imputation techniques for missing data
Qualitative data analysis [29]	**Methodology**: thematic analysis **Process**: familiarization, coding, theme development, refinement **Tools**: qualitative analysis software **Reliability**: independent coding by secondary researcher	– Repeated readings for data immersion – Coding segmented relevant data – Codes organized into broader themes aligned with study objectives – Reliability strengthened through independent coding and resolution of discrepancies

most suitable design for their investigations and understanding their specific applications. The mixed-methods technique provides a strategic advantage by improving the validity and reliability of research through triangulation, utilizing several data sources to corroborate conclusions. This methodological integration enhances the study's rigor and guarantees a more nuanced and thorough knowledge of complicated occurrences. This study employs an integrated methodology, merging qualitative and quantitative methods for a **comprehensive** and rigorous examination. The qualitative component provides detailed, contextual insights that enhance the interpretative depth of the study, whereas the quantitative phase methodically finds overarching themes and statistical patterns. The quantitative phase utilizes a rigorously organized data-gathering technique, including methodically developed questionnaires corresponding to the study's primary variables to guarantee methodological accuracy and relevance. Lim [31] emphasizes credibility, dependability, confirmability, and transferability in qualitative research to uphold rigorous academic standards [31]. The quantitative data collected from these structured surveys are evaluated to identify trends, examine correlations across variables, and experimentally confirm theoretical frameworks, thereby improving the study's validity and reliability. This study employs a comprehensive methodological approach, integrating qualitative and quantitative methodologies for a thorough analysis. The qualitative component provides detailed, contextual insights that enhance the interpretative depth of the study, whereas the quantitative phase methodically finds overarching themes and statistical patterns. The quantitative phase utilizes a structured data-gathering approach, including properly developed questionnaires linked to the study's primary variables to ensure methodological accuracy and relevance. Lim [31] emphasizes the importance of creating credibility, dependability, confirmability, and transferability in qualitative research to uphold rigorous academic standards [31]. The quantitative data collected from these structured surveys are evaluated to identify trends, examine correlations across variables, and experimentally confirm theoretical frameworks, thereby improving the study's validity and reliability. These interviews offer flexibility, enabling participants to express complex opinions and producing contextually rich data. Recurring themes in the qualitative data are systematically found, categorized, and interpreted through thematic analysis. This approach guarantees agreement with the study's goals while making obtaining significant findings easier [32]. The findings' interpretive quality is further enhanced by including semi-structured interviews, which provide insightful background information that deepens the study. Ethical concerns are essential to the research process to safeguard participants and ensure strict adherence to standards. Precise information on the study's goals, procedures, and participants' rights, including the option to discontinue participation at any time, is given to them. To ensure participant anonymity and confidentiality, informed permission is acquired online, and all identifying information is eliminated. An Institutional Review Board (IRB) grants ethical approval, guaranteeing that all research practices follow accepted moral norms and criteria. Sampling techniques are meticulously planned to ensure representativeness and relevance [22]. Stratified random sampling is used in the quantitative phase to ensure that critical demographic groups are represented proportionately. This method reduces sampling bias and improves the findings' applicability to a larger population. Purposive sampling chooses participants for the qualitative phase who can offer rich, in-depth information pertinent to the study's goals. This approach guarantees that a wide variety of viewpoints are represented in qualitative analysis, which advances a thorough comprehension of the phenomenon being studied. When conducting a comprehensive and objective inquiry, a mixed-methods approach is essential. Mixed-methods research (MMR) is a principled and integrative framework, as stated by Dawadi et al. [33]. This framework helps supplement traditional quantitative and qualitative approaches by bridging the gap between their respective paradigms to capitalize on the unique strengths of each of these methodologies [33]. MMR provides a solid basis for tackling complex research topics. This is accomplished by combining quantitative findings' generalizability and statistical rigor with qualitative data's contextual depth and richness. This combination of methodologies makes it possible to simultaneously investigate broad patterns and specific discoveries, guaranteeing that general tendencies do not obscure more nuanced points of view. Triangulation, a process that systematically cross-verifies results across various methods and data sources, is an additional method that helps strengthen the findings' validity and dependability. Using this procedure, the study's methodological rigor is improved, and the findings are guaranteed to be consistent and coherent. The implementation of MMR is a research strategy that is rigorous, balanced, and morally sound. It offers both theoretical and practical contributions to the field. This technique provides accurate, dependable, and relevant insights into the principal phenomena of the research by combining depth with breadth and rigor with flexibility. As a result, it contributes to advancing academic understanding and supports evidence-based applications in practice. An illustration of the mixed-methods architecture utilized to satisfy challenging research objectives may be seen in Fig. 4. Both the quantitative phase, which consists of systematic data collection and statistical analysis, and the qualitative phase, which includes semi-structured interviews and thematic analysis, are incorporated into the framework. A core component that ensures compliance with standards incorporates ethical concerns, including informed consent, participant confidentiality, and permission from the Institutional Review Board (IRB). There is a connection between these components through triangulation, which increases the overall validity and dependability of the findings. This technique strikes a compromise between the generalizability of broad trends and the contextual richness of qualitative insights, which ultimately results in a methodology that is robust, thorough, and responsible from an ethical standpoint.

3 A Rigorous Framework for Presenting Research Results

An accurate, well-organized, and objective presentation of the findings is provided in the portion of a study report devoted to the results [34]. This section is the cornerstone of academic communication. This component is vital to the integrity and rigor of the study because it converts raw data into a story that is easily understandable and rigorously adheres to factual reporting. The primary purpose of the results section is to briefly explain the outcomes, backed by statistical analysis and visual aids. This is done to ensure that the findings are communicated in a manner that is both clear and objective. One of the most critical aspects of this part is the implementation of proper statistical procedures to display the data methodically. Descriptive statistics are utilized to provide a concise summary of the data. These statistics highlight important metrics along the lines of measurements of central tendency (mean, median, mode) and dispersion (standard deviation, variance, and range) [35]. Readers can get an instant and intuitive comprehension of the dataset through these summaries, which enable them to comprehend the overall features of the variables

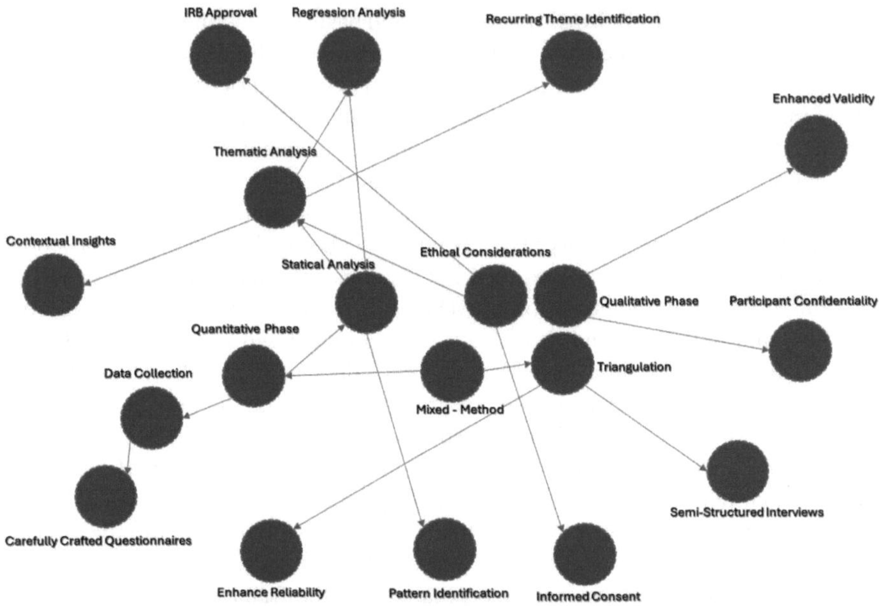

Fig. 4 Mixed-methods framework for research methodology (*Source* Authors)

studied. Inferential statistics are utilized when arriving at more comprehensive generalizations is necessary. It is possible to extrapolate findings outside the sample using hypothesis testing, correlation analysis, or regression modeling [36]. This facilitates confirming that the results are relevant within a more comprehensive setting. The accessibility and interpretability of the data are significantly improved via visual representations, which play a crucial role in this regard. Graphs, charts, and tables are deliberately used in the presentation to provide a concise illustration of trends, patterns, and correlations. Using bar charts or line graphs to illustrate changes over time is usual practice. Scatter plots, on the other hand, are an effective method for representing correlations between variables. Tables offer a comprehensive numerical breakdown of the data, enabling values to be compared and referenced quickly [37]. The results section can establish a balance between thorough reporting and reader engagement using these visual aids. This helps to reduce the cognitive burden that is involved with the processing of complicated statistical information. An additional essential component of this part is the focus on the most important discoveries. To ensure that readers can immediately identify the elements of the study that are most pertinent to them, it is essential to highlight noteworthy trends, patterns, or correlations discovered in the data. These results are presented clearly and concisely if the study reveals a strong positive connection between two variables or displays a statistically significant difference between groups [38]. On the other hand, no interpretations, guesses, or subjective analyses are presented to preserve this section's academic rigor and impartiality. Instead, the results are stated the same way they were observed, establishing an evidence-based foundation for later parts, such as the discussion and the conclusions. In contrast to the other sections of the study report, the section on the outcomes is distinguished by its emphasis on objectivity and impartiality among its contents. Because this part is purposefully devoid of any bias, it is concerned with the direct results of the study without making any inferences regarding causality, addressing ramifications, or providing any theoretical interpretations. It is via this kind of constraint that the results are presented in a manner that is objectively reflective of the facts, which enables the readers to form their judgments based on the evidence that is supplied. Not only is this approach necessary for preserving the honesty of the study, but it is also necessary for establishing confidence and credibility within the academic audience. The results part is further improved in clarity and efficacy thanks to the logical structuring of the section [39]. It is common practice to present the findings sequentially, that is, by studying objectives or hypotheses. This strategy guarantees the existence of a logical narrative, directing the reader through the findings in an organized fashion. An overview of the study aims and methodologies is often included at the beginning of the section. This serves to provide the context for the findings. After that, it systematically presents the findings, frequently classifying them according to the relevant topics or factors. This logical sequence guarantees that the data are accessible and understandable, making it easier to understand the study's conclusions [1].

In the research that investigates the influence of physical activity on mental health, for instance, the section under "Results" describes the participants' demographic characteristics and baseline measurements of the critical variables [40]. This would be followed by descriptive statistics summarizing the average exercise frequency and mental health ratings and inferential statistics demonstrating the variables' connections. Examples of such statistics are correlations and regressions. It would be beneficial to augment the statistical analysis with visual aids such as scatter plots or histograms, which would provide a representation of the data that is both clear and easy to understand. It is intended that the conclusion of the section would include a summary of the key findings, emphasizing outcomes that are statistically significant or unexpected observations. When reporting discoveries that are unexpected or out of the ordinary, transparency is also quite important [41]. It is possible that such results may not directly address the research questions; nonetheless, they do contribute to a thorough study account and give significant insights into research that will be conducted in the future. By the underlying commitment to neutrality that this section has made, these results are provided in a factual manner and without any interpretation. Furthermore, the results department consistently maintains academic excellence by meeting specified reporting requirements. As a result of the meticulous way the statistics are presented, the possibility of misunderstanding or ambiguity is significantly reduced. The presentation of statistical findings provides measures like p-values, confidence intervals, and effect sizes. This is done to provide a comprehensive and nuanced understanding of the findings [42]. Including this degree of specificity guarantees that the research can be replicated and enables colleagues to examine the findings' validity critically. It is essential to include a section on the results in a scholarly report since it is the portion that serves as the definitive record of the findings of the investigation. This organization's responsibility is to convey the study's findings unambiguously, concisely, and objectively while keeping to the highest standards of academic rigor. By utilizing statistical analysis and visual aids, emphasizing the most important findings, and ensuring that impartiality is maintained, the results section explains the facts that are both transparent and reliable [43]. This rigorous and structured methodology contributes to the study's credibility and creates the framework for meaningful debate and interpretation in the following parts. The section on the outcomes serves as the basis for the validity and significance of the research. Table 2 delineates a systematic framework for presenting research findings, underscoring objectivity, rigor, and clarity. Each facet is meticulously examined, pinpointing intrinsic obstacles and emphasizing prospects to improve academic integrity and influence. The framework underscores the necessity for utilizing rigorous statistical techniques, efficient visual representations, and compliance with reporting requirements to guarantee precise and unbiased dissemination of results. It also underscores logical structure, openness, and accessibility to accommodate varied audiences. This paradigm enhances study credibility by addressing problems and utilizing possibilities, facilitating meaningful interpretation and debate in succeeding parts.

4 A Thorough Examination of the Results Reveals Theoretical Ramifications, Practical Relevance, and Future Directions

An in-depth review of the findings sheds light on the theoretical implications, practical impact, and prospective future research possibilities associated with using these findings [53]. It is the discussion part that acts as the foundation of academic research. It offers a critical interpretation of the findings by situating them within the context of the study's aims and the larger body of already available literature. In this part, the more significant implications of the findings are discussed, their significance to theoretical frameworks and practical applications is evaluated, the study's limits are acknowledged, and recommendations for further research are provided. By synthesizing data-driven findings with theoretical and practical views, the discussion guarantees that the study contributes significantly to the discourse of the academic community and provides insights into practices implemented in the real world [54]. In addition, it facilitates the contextualization of the findings within the area, increasing their application and promoting information transfer between other domains. The integrated method used in this study underlines the academic and practical significance of the study, which in turn fosters intellectual growth and paves the path for future research into topics that still need to be answered and developing areas of interest.

Table 2 Challenges and opportunities in the critical framework for presenting research results

Aspect	Description	Challenges	Opportunities
Objective [44]	Present findings accurately, systematically, and impartially	Avoiding subjective interpretations or biases	Build trust and ensures credibility in research presentations
Data presentation [45]	Use descriptive and inferential statistics to summarize data effectively	Ensuring statistical methods are appropriate and accurate	Provides a foundation for rigorous, evidence-based conclusions
Visual representation [46]	Incorporate tables, graphs, and charts to display trends and relationships	Selecting suitable visuals that avoid oversimplification	Enhances understanding and engagement with intuitive data presentation
Key findings [47]	Highlight significant trends, patterns, and relationships without analysis	Avoiding excessive detail or interpretation that compromises objectivity	Draws attention to the most critical results, guiding discussions effectively
Objectivity [48]	Report outcomes without bias, speculation, or interpretation	Maintaining neutrality, especially for unexpected findings	Demonstrates impartiality, ensuring reliability and academic rigor
Organization [34]	Present findings in a logical order aligned with research objectives	Structuring data clearly to avoid confusion or fragmentation	Creates a coherent narrative, making the results easy to follow and interpret
Transparency [49]	Accurately report unexpected or abnormal findings	Identifying and reporting anomalies without overinterpretation	Adds depth and completeness to the research, encouraging further exploration
Reporting standards [50]	Include key metrics (p-values, confidence intervals, effect sizes) for clarity and rigor	Ensuring metrics are detailed yet interpretable for diverse audiences	Enhances credibility and adherence to scholarly standards, fostering reproducibility
Clarity and accessibility [51]	Use precise language and visuals to simplify complex findings	Avoiding overly technical jargon while maintaining rigor	It makes the research accessible to experts and non-experts, broadening its impact
Foundation for analysis [52]	Provide a solid basis for interpretation, discussion, and conclusions	Ensuring results are comprehensive and robust enough to support further analysis	Lays a strong groundwork for meaningful discussion and critical insights

4.1 *Results of the Investigation*

When subjected to a thorough examination, the findings disclose whether they agree or disagree with previously established ideas and studies. This provides a crucial basis for placing the research within the larger academic framework [55]. Through an analysis of the degree to which the study's findings are consistent with pre-existing scholarly frameworks, this comparative examination highlights the study's contribution to the advancement of understanding. The findings are consistent with the existing body of literature and strengthen the validity and robustness of the prevalent theoretical models, hence bolstering the empirical foundation on which they are based. On the other hand, findings that contradict the findings of previous research call into question the assumptions that have been made and call for a critical reevaluation of the paradigms that have been established. The existence of such differences opens possibilities for theoretical improvement. It encourages the development of alternative models that can fill in the gaps or solve the constraints found [56]. Using this methodical comparison, the research not only sheds light on its significance and influence but also encourages critical engagement with the topic, which contributes to the advancement of academic discourse and paves the way for novel inquiries that contribute to the growth of the discipline. For example, using solar panels and energy-efficient materials in residential structures has demonstrated a 30% reduction in energy usage, aligning with previous research on sustainable construction [57]. These findings underscore the economic feasibility of renewable energy and contest antiquated construction methods, facilitating the adoption of more sustainable building standards worldwide.

4.2 *The Repercussions*

Not only are the conclusions of this study significant from a theoretical standpoint, but they also have significant implications for the academic, professional, and policy-making communities [58]. The study fills essential gaps in the existing academic literature within the educational community by providing empirical data that either supports or refines preexisting theoretical frameworks. The basic understanding of the area is strengthened due to this contribution, which inspires future scholarly investigation. In a professional context, the findings provide insights that may be implemented and used to inform the best practices, which can drive innovation and efficiency in relevant industries. Regarding policymaking, the findings offer a solid foundation of evidence that can guide decision-making, which in turn helps promote the creation of informed and influential policies [59]. By combining academic research with practical application, this synthesis highlights the transformational potential of the results to generate significant societal change. As a result of the more enormous ramifications, the need for multidisciplinary collaboration is brought to light. This highlights the necessity of translating academic ideas into real-world solutions that handle complicated issues, eventually contributing to society's growth and improvement [60].

4.3 The Constraints

To provide a comprehensive analysis and contextualization of the findings of this investigation, it would be essential to acknowledge the study's limitations [61]. There are a few limitations that should be taken into mind, even though the research provides a substantial addition to the area. The findings cannot be generalized to more extensive or varied groups due to the small sample size, even though the sample size was acceptable for the aim of the study. Efforts were taken to ensure that the sample was diverse; nonetheless, there are inherent limits in capturing all demographic or contextual data, which may introduce minor biases and impair the results' ability to be representative [62]. Rather than weakening the study's legitimacy, these limits explain its scope and bounds, providing a platform for critical examination and future inquiries. When these limits are discussed openly, it strengthens the academic rigor and transparency of the study, which in turn encourages constructive examination and helps to advance the evolution of knowledge [63]. By addressing these issues, the research promotes future investigations to improve its insights and widen the scope of its usefulness.

4.4 Toward Future Research Directions

Several avenues for future research are proposed to build on the findings and address the limitations identified [64]. Expanding the sample size and ensuring greater demographic and contextual diversity would significantly enhance the generalizability and robustness of subsequent studies. Conducting comparative research across varied cultural, geographical, or organizational contexts could deepen the understanding of the phenomena and provide valuable cross-contextual insights. Methodologically, adopting longitudinal designs or experimental approaches would enable examining causal relationships and dynamic processes over time, offering a more comprehensive understanding of the underlying mechanisms. Additionally, integrating qualitative methods such as in-depth interviews or ethnographic research could uncover nuanced perspectives and better contextualize participants' lived experiences, further enriching the findings [65]. This study also identified several unexplored questions and unexpected findings, which offer promising opportunities for further investigation. Future research should prioritize translational approaches that bridge academic inquiry with practical applications. Applied studies, testing interventions, or strategies informed by the findings would enhance their real-world relevance and societal impact by providing actionable solutions to pressing challenges. The discussion section situates the findings within a broader academic and practical framework, critically analyzing their alignment with existing literature, theoretical contributions, and practical implications [66]. The study reinforces its commitment to academic rigor and societal relevance by openly addressing limitations and providing meticulously formulated recommendations. This integrative approach underscores the research's dual contribution: advancing theoretical frameworks while offering solutions that address real-world complexities, demonstrating its enduring significance to academia and society [67]. Figure 5 provides an overview of the complete evaluation of the research findings, highlighting the theoretical implications, practical applicability, and future research prospects of the findings discussed. Specifically, it highlights critical themes such as alignment with current frameworks, findings that professionals and politicians can implement, and new ways to resolve limits. This integrated framework bridges the gap between academic rigor and real-world application, encouraging intellectual progress, practical effect, and a platform for developing research in emerging areas of interest.

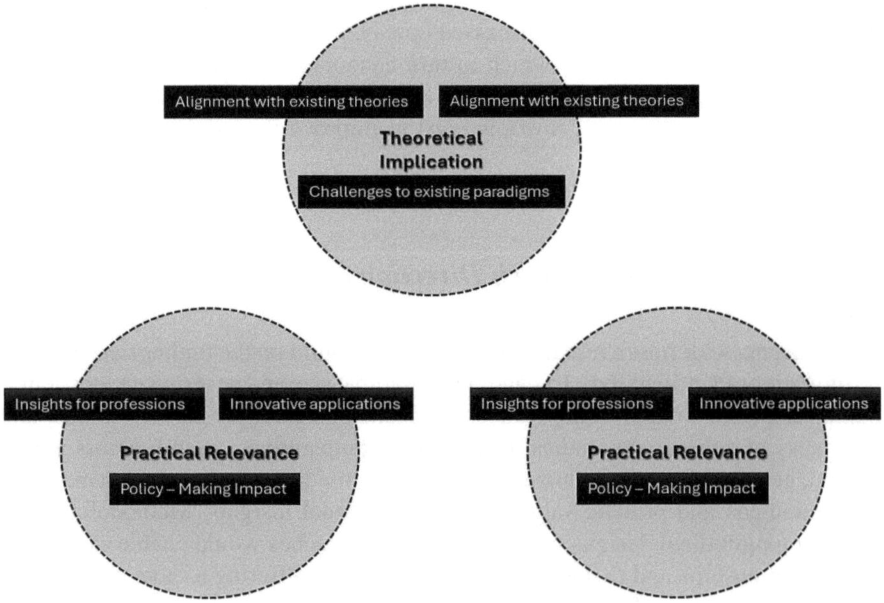

Fig. 5 A thorough examination of the results (*Source* Authors)

5 Conclusion and Recommendations

In addition, it provides an analytical comment on the more far-reaching implications of the study's contributions to the subject [68]. When tackling complex and varied situations, the research findings highlight the vital role collaboration should play across other disciplines. Working together helps stimulate the development of original, robust, and culturally appropriate solutions. This is accomplished by incorporating a variety of viewpoints and skills. In addition, the research sheds light on the crucial significance of evidence-based approaches in fostering scientific innovation, policy formulation, and social growth. Because of these findings, it is suggested that stakeholders put strategic frameworks into place to enable multidisciplinary interaction. This will ensure that information sharing transcends the borders of different disciplines. Policymakers should design initiatives encouraging collaborative research and development, and those policymakers should support them [69]. Researchers are strongly urged to use integrative techniques, which integrate theoretical insights with empirical applications. On the other hand, practitioners should prioritize the contextual adaptation of evidence-based tools and practices. This alignment is essential for generating sustainable, impactful, and equitable outcomes. For example, in sustainable building design, architects, engineers, environmental scientists, and urban planners must work together to achieve energy-efficient structures that can fulfill society's and the environment's requirements. Approaches that involve several disciplines guarantee that buildings not only reduce their carbon footprints but also improve the well-being of their occupants and their ability to withstand the effects of climate change [70]. This alignment is essential for generating sustainable, impactful, and equitable outcomes.

5.1 Recommendations

5.1.1 For Practitioners

Implementing approaches that are supported by evidence should be a top priority for practitioners to guarantee the successful incorporation of robust research findings into applications that are used in the real world [71]. These techniques improve the results' reliability, validity, and efficiency across various practical contexts. Moreover, promoting collaboration across academic lines is essential to efficiently solving complicated and varied problems. Through collaboration with specialists hailing from a wide range of fields, practitioners can jointly develop solutions that are both comprehensive and innovative, therefore addressing urgent challenges holistically [72]. These alliances provide synergistic effects, which in turn have the effect of multiplying the impact and breadth of efforts that are deployed. Furthermore, it is essential to scale up and improve proven tactics systematically supported by data to accomplish wider socioeconomic and environmental goals [73]. The efforts must be strategically connected with modern priorities and global concerns to guarantee that they will remain relevant and adaptable. Practitioners contribute to creating sustainable and impactful interventions that produce long-term value across sectors and assist the progress of equitable and resilient systems by continually refining and contextualizing approaches that have been demonstrated to be effective [74].

5.1.2 Relevance to Policymakers

Establishing initiatives and laws founded on the empirical facts offered in this chapter should be a priority for policymakers [75]. This will ensure that policies are both evidence-based and practicable. Adoption rates could be significantly increased by introducing focused interventions to address identified implementation gaps. In addition, formulating inclusive policies is of utmost importance since they guarantee fair results by addressing gaps associated with underrepresentation in terms of demographic classification or contextual factors. To better promote applied research activities, particularly those that bridge academic results with real-world applications, allocating funding explicitly for research and development is possible [76]. This technique makes it easier to translate research findings into policies and programs that can be put into action and successfully implemented within institutional structures. Policymakers can improve their programs' relevance, efficacy, and equality by using evidence-driven and inclusive policies [77]. This will ultimately result in the development of long-term benefits for society.

5.1.3 Implications for Researchers

Researchers must embrace a more comprehensive methodological framework, prioritizing mixed-method designs to capture quantitative patterns and qualitative subtleties [78]. This strategy makes it possible to gain a more in-depth comprehension of complicated phenomena and increases the level of rigor in the findings. In the future, one of the primary goals of research should be to find ways to overcome the constraints discovered in this study. One of the most critical areas for development is expanding the sample's diversity and investigating underrepresented scenarios. This will ensure that the results are more thorough and generalizable. Furthermore, the advancement of research into developing concerns and unanticipated results brought to light by this study has the potential to provide a solid platform for such investigations [79]. Research that bridges the gap between theoretical developments and practical applications should also be a focus. Translational research serves to bridge this gap. Researchers can guarantee that their work has a concrete influence on society by concentrating on creating actionable insights. This will allow them to drive innovation and address difficulties that are encountered in the real world. The contribution of research to both the academic world and society is anticipated to be significantly strengthened due to this simultaneous emphasis on methodological improvement and translational relevance [80].

5.2 Concluding Remarks and Reflections

This study contributes beyond the immediate findings, providing a solid platform for advancing theoretical frameworks and practical implementations [81]. This research highlights the dual value of the findings to both the academic community and society by critically examining the alignment of the results with the existing literature, questioning accepted assumptions, and suggesting insights that may be put into real-world practice. Integrating theoretical ideas with practical solutions is a particularly illustrative example of the transformational potential of interdisciplinary research. This integration demonstrates how academic inquiry may stimulate innovation and drive social advancement [82]. This study aims to demonstrate how rigorous research may significantly generate knowledge and practical application by addressing complex problems and suggesting answers supported by evidence. In looking to the future, the findings highlight the need to do ongoing research into the subject matter and urge for evidence-driven, collaborative programs involving various organizations and individuals. These activities are essential for progressing intellectual knowledge, resolving unresolved issues, and enhancing theory and practice [83]. Additionally, the study highlights the importance of translational research, bridging the gap between academic findings and their influence in the real world. Future attempts need to build on this work to guarantee that the contributions made by research will continue to be relevant and valuable. Through the promotion of continued cooperation and innovation, this research provides the framework for efforts that produce long-term value across various fields and industries. It is a stepping stone for ongoing attempts to link academic rigor with social demands [84]. This helps to ensure that contributions to both areas are sustainable and beneficial throughout a more extended period.

6 Conclusion

A well-structured research report is one of the most critical factors in enhancing academic comprehension and practical application. The rigorous structure of each component, beginning with the introduction and ending with the conclusion, guarantees that the study will ultimately accomplish its dual goals of contributing to theoretical frameworks and delivering insights that can be put into practice. Researchers can put their work within a larger academic and societal framework when they have a defined structure to operate within. This helps bridge the gap between theoretical novelty and practical usefulness. This integration highlights multidisciplinary cooperation because it enables research to handle complicated challenges by integrating theoretical constructions with real-world applications. This integration illustrates the relevance of interdisciplinary collaboration. When this is accomplished, research findings will contribute to improving intellectual debate and advancing society practically. By addressing gaps in the existing literature and the design of research questions that correspond with urgent concerns, the introduction of a research report emphasizes the study's relevance. This contextual base guarantees the content is relevant to academic audiences and practical stakeholders, such as policymakers and practitioners. A clear and repeatable framework is provided by a structured report, which further contributes to the enhancement of academic rigor. The methodological section is the most important part of this report because it allows for thorough explanations of the data-collecting procedures and analysis that align with the study's aims. Mixed-method approaches are particularly successful in providing a thorough knowledge of complex events. These approaches capture both quantitative trends and qualitative insights. The robustness and generalizability of the findings are strengthened when constraints, such as the variety of the sample or contextual biases, are effectively addressed. Incorporating ethical issues is another essential component of a high-quality study report. Not only does adherence to principles such as informed consent, confidentiality, and institutional norms secure the safety of participants, but it also strengthens the credibility and dependability of the research findings. Ethical rigor is essential to preserve trust and guarantee that the research findings may be utilized responsibly. The revolutionary potential of these findings is readily apparent regarding the contributions that research reports make to both the academic and practical spheres. These papers provide academics with a basis for future research by pointing out areas of concern and suggesting possibilities for further investigation. They provide practitioners and policymakers insights that can be put into action, which will help guide decision-making and give information that will support the implementation of evidence-based solutions. For instance, research on sustainable building practices may influence the creation of laws that promote energy-efficient construction and the incorporation of renewable resources. This, in turn, can contribute directly to achieving global sustainability objectives such as those established in Agenda 2030. Despite their merits, research reports are required to handle inherent problems. These issues include ensuring that all relevant information is presented, investigating possibilities that have not been thoroughly investigated, and preserving neutrality in the interpretation of data. By being transparent about the constraints of the reporting process, one may encourage constructive criticism and create the road for future changes. While the iterative character of research, in which discoveries motivate more exploration, is essential for advancing academic knowledge and practical applications, it is also indispensable. Researchers can improve their methodology and broaden the scope of subsequent studies when they openly confront the constraints of their work. The potential of translational research to bridge the gap between academic discoveries and practical application indicates the significance of transdisciplinary research. When theoretical ideas are aligned with the requirements of the actual world, academic contributions are sure to be intellectually rigorous and socially useful. For instance, research on incorporating renewable energy sources into urban planning can serve as a reference for formulating plans for the sustainable development of cities. In addition, future research should be built based on structured reports by increasing the variety of the samples, investigating cross-cultural situations, and utilizing longitudinal study methodologies. Innovative analysis methods and digital technologies can potentially improve the accuracy and significance of findings. At the same time, interdisciplinary collaboration continues to be essential in tackling global concerns. To produce comprehensive solutions that balance economic, social, and environmental goals, it is necessary to combine the experience of professionals from various sectors. It is impossible to exaggerate the significance of a well-structured research report in terms of its contribution to advancing academic and social goals. These studies offer a framework for addressing difficult problems and generating durable progress because they combine academic rigor with practical relevance. This methodical strategy guarantees that research makes a significant contribution to the expansion of knowledge and the improvement of society. It also encourages continuous cooperation and creativity while setting the groundwork for future undertakings that link academic inquiry with its influence in the real world.

References

1. Hancock DR, Algozzine B, Lim JH (2021) Doing case study research: a practical guide for beginning researchers
2. Tracy SJ (2024) Qualitative research methods: collecting evidence, crafting analysis, communicating impact. John Wiley & Sons
3. Chigbu UE, Atiku SO, Du Plessis CC (2023) The science of literature reviews: searching, identifying, selecting, and synthesising. Publications 11(1):2
4. Welch C, Paavilainen-Mäntymäki E, Piekkari R, Plakoyiannaki E (2022) Reconciling theory and context: how the case study can set a new agenda for international business research. J Int Bus Stud 53(1):4–26
5. Kurniawan TA, Meidiana C, Goh HH, Zhang D, Othman MHD, Aziz F, Anouzla A, Sarangi PK, Pasaribu B, Ali I (2024) Unlocking synergies between waste management and climate change mitigation to accelerate decarbonization through circular-economy digitalization in Indonesia. Sustain Prod Consum 46:522–542
6. Yadav R (2023) How to write a research paper: a guide book for neophyte students: a guide book for neophyte students. Shashwat Publication
7. Almusaed A, Yitmen I, Almssad A (2024) Contemporary innovations and sustainable practices in the application of clay materials within architectural design and construction methodologies
8. Kunisch S, Denyer D, Bartunek JM, Menz M, Cardinal LB (2023) Review research as scientific inquiry. Organ Res Methods 26(1):3–45
9. Dimitrova R, Wiium N (2021) Handbook of positive youth development: advancing the next generation of research, policy and practice in global contexts. Springer International Publishing, pp 3–16
10. O'Sullivan TA, Jefferson CG (2020) A review of strategies for enhancing clarity and reader accessibility of qualitative research results. Am J Pharm Educ 84(1):7124
11. Zhu W, Zhang J, Wang D, Ma C, Zhang J, Chen P (2023) Study on the critical factors influencing high-quality development of green buildings for carbon peaking and carbon neutrality goals of China. Sustain 15(6):5035
12. Li D, Xu N, Zhao Y, Zhou C, Zhang LP, Wu LZ, Zhang T (2022) A reliable and precise protocol for urea quantification in photo/electrocatalysis. Small Methods 6(9):2200561
13. Fabrigar LR, Wegener DT, Petty RE (2020) A validity-based framework for understanding replication in psychology. Pers Soc Psychol Rev 24(4):316–344
14. Carcary M (2020) The research audit trail: methodological guidance for application in practice. Electron J Bus Res Methods 18(2):166–177
15. Mehta A, Alaiashy O, Kumar P, Tamilinian V, Besong S, Balpande S, Verma S, Dhingra K (2024) Advancing model-based systems engineering in biomedical and aerospace research: a

comprehensive review and future directions. J Knowl Learn Sci Technol 3(4):133–147. ISSN: 2959–6386 (online)
16. Helmer S, Blumenthal DB, Paschen K (2020) What is meaningful research and how should we measure it? Scientometrics 125(1):153–169
17. Rossoni AL, de Vasconcellos EPG, de Castilho Rossoni RL (2024) Barriers and facilitators of university-industry collaboration for research, development and innovation: a systematic review. Manag Rev Q 74(3):1841–1877
18. Ngulube P (2022) Using simple and complex mixed methods research designs to understand research in information science. In: Handbook of research on mixed methods research in information science. IGI Global, pp 20–46
19. Amadi A (2023) Integration in a mixed-method case study of construction phenomena: from data to theory. Eng Constr Archit Manag 30(1):210–237
20. Spencer L, Radcliffe L, Spence R, King N (2021) Thematic trajectory analysis: a temporal method for analysing dynamic qualitative data. J Occup Organ Psychol 94(3):531–567
21. Molina-Azorin JF, Fetters MD (2023) Towards the future legitimacy of mixed methods designs: responsible mixed methods research for tackling grand challenges for the betterment of society. In: The Sage handbook of mixed methods research design. pp 485–495
22. Mweshi GK, Sakyi K (2020) Application of sampling methods for the research design. Arch Bus Rev 8(11):180–193
23. Guillén-Gámez FD, Ramos M (2021) Competency profile on the use of ICT resources by Spanish music teachers: descriptive and inferential analyses with logistic regression to detect significant predictors. Technol Pedagog Educ 30(4):511–523
24. Adhikari D, Jiang W, Zhan J, He Z, Rawat DB, Aickelin U, Khorshidi HA (2022) A comprehensive survey on imputation of missing data in internet of things. ACM Comput Surv 55(7):1–38
25. Mukasa ES, Christospher W, Ivan B, Kizito M (2021) The effects of parametric, non-parametric tests and processes in inferential statistics for business decision making—a case of 7 selected small business enterprises in Uganda. Open J Bus Manag 9(3):1510–1526
26. Campbell CM, Taylor-Clark T, Loan AL (2022) Protecting nurse survey participants: ethical considerations for conducting survey research among nurses. Clinical Ethics 17(4):391−408
27. Adeoye-Olatunde OA, Olenik NL (2021) Research and scholarly methods: semi-structured interviews. J Am CollE Clin Pharm 4(10):1358–1367
28. Cook RM (2021) Addressing missing data in quantitative counseling research. Couns Outcome Res Eval 12(1):43–53
29. Naeem M, Ozuem W, Howell K, Ranfagni S (2023) A step-by-step process of thematic analysis to develop a conceptual model in qualitative research. Int J Qual Methods 22:16094069231205788
30. Taherdoost H (2022) What are different research approaches? comprehensive review of qualitative, quantitative, and mixed method research, their applications, types, and limitations. J Manag Sci & Eng Res 5(1):53–63
31. Lim WM (2024) What is qualitative research? an overview and guidelines. Australas Mark J 14413582241264619
32. Christou PA (2022) How to use thematic analysis in qualitative research. J Qual Res Tour 3(2):79–95
33. Dawadi S, Shrestha S, Giri RA (2021) Mixed-methods research: a discussion on its types, challenges, and criticisms. J Pract Stud Educ 2(2):25–36
34. Karunarathna I, Gunasena P, De Alvis K, Jayawardana A (2024) Structured reviews: organizing, synthesizing, and analyzing scientific literature
35. Alabi O, Bukola T (2023) Introduction to descriptive statistics. In: Recent advances in biostatistics. IntechOpen
36. Shen X, Meinshausen N (2024) Engression: extrapolation through the lens of distributional regression. J R Stat Soc Ser B: Stat Methodol qkae108
37. Seo J, Xia Y, Lee B, McCurry S, Yam YJ (2024) MAIDR: making statistical visualizations accessible with multimodal data representation. In: Proceedings of the CHI conference on human factors in computing systems. pp 1–22

38. Paudel R, Kunwar V, Ahmed MF, AYedgarian V (2024) Work-life equilibrium: key to enhancing employee job satisfaction. Educ Adm: Theory Pract 30(7):1146–1159
39. Smith JD, Li DH, Rafferty MR (2020) The implementation research logic model: a method for planning, executing, reporting, and synthesizing implementation projects. Implement Sci 15:1–12
40. Eather N, Wade L, Pankowiak A, Eime R (2023) The impact of sports participation on mental health and social outcomes in adults: a systematic review and the 'Mental Health through Sport'conceptual model. Syst Rev 12(1):102
41. Quattrone P (2021) Seeking transparency makes one blind: how to rethink disclosure, account for nature and make corporations sustainable. Account, Audit & Account J 35(2):547–566
42. Hider U (2024) Beyond P-values: unpacking the clinical relevance of statistical significance, no 12266. EasyChair
43. Karunarathna I, Gunasena P, Hapuarachchi T, Gunathilake S (2024) Comprehensive data collection: methods, challenges, and the importance of accuracy
44. Papaspyridakos P, Vazouras K, Chen YW, Kotina E, Natto Z, Kang K, Chochlidakis K (2020) Digital versus conventional implant impressions: a systematic review and meta-analysis. J Prosthodont 29(8):660–678
45. Cooksey RW, Cooksey RW (2020) Descriptive statistics for summarising data. In: Illustrating statistical procedures: finding meaning in quantitative data. pp 61–139
46. Guetterman TC, Fàbregues S, Sakakibara R (2021) Visuals in joint displays to represent integration in mixed methods research: a methodological review. Methods Psychol 5:100080
47. Zhang Y, Huang Z (2024) Unveiling hidden patterns: a computational analysis of less commonly labeled ransomware families. Authorea Preprints
48. Westwood SJ, Parlatini V, Rubia K, Cortese S, Sonuga-Barke EJ (2023) Computerized cognitive training in attention-deficit/hyperactivity disorder (ADHD): a meta-analysis of randomized controlled trials with blinded and objective outcomes. Mol Psychiatry 28(4):1402–1414
49. Oakden-Rayner L, Gale W, Bonham TA, Lungren MP, Carneiro G, Bradley AP, Palmer LJ (2022) Validation and algorithmic audit of a deep learning system for detecting proximal femoral fractures in patients in the emergency department: a diagnostic accuracy study. Lancet Digit Health 4(5):e351–e358
50. Holmberg C (2024) Why we need to discuss statistical significance and p-values (again): addressing the underlying issue of different probability interpretations and actionable recommendations. Nord J Nurs Res 44:20571585241253176
51. Martínez Silvagnoli L, Shepherd C, Pritchett J, Gardner J (2022) Optimizing readability and format of plain language summaries for medical research articles: cross-sectional survey study. J Med Internet Res 24(1):e22122
52. Hanelt A, Bohnsack R, Marz D, Antunes Marante C (2021) A systematic review of the literature on digital transformation: insights and implications for strategy and organizational change. J Manage Stud 58(5):1159–1197
53. Vrontis D, Makrides A, Christofi M, Thrassou A (2021) Social media influencer marketing: a systematic review, integrative framework and future research agenda. Int J Consum Stud 45(4):617–644
54. Bibri SE (2021) Data-driven smart sustainable cities of the future: an evidence synthesis approach to a comprehensive state-of-the-art literature review. Sustain Futur 3:100047
55. Berman J, Smyth R (2015) Conceptual frameworks in the doctoral research process: a pedagogical model. Innov Educ Teach Int 52(2):125–136
56. Ma Y, Liu Y, Appolloni A, Liu J (2021) Does green public procurement encourage firm's environmental certification practice? The mediation role of top management support. Corp Soc Responsib Environ Manag 28(3):1002–1017
57. Umoh AA, Adefemi A, Ibewe KI, Etukudoh EA, Ilojianya VI, Nwokediegwu ZQS (2024) Green architecture and energy efficiency: a review of innovative design and construction techniques. Eng Sci & Technol J 5(1):185–200
58. Edler J, Karaulova M, Barker K (2022) Understanding conceptual impact of scientific knowledge on policy: he role of policymaking conditions. Minerva 60(2):209–233

59. Gabbay J, Le May A, Pope C, Brangan E, Cameron A, Klein JH, Wye L (2020) Uncovering the processes of knowledge transformation: the example of local evidence-informed policy-making in United Kingdom healthcare. Health Res Policy Syst 18:1–15
60. Osaba E, Villar-Rodriguez E, Del Ser J, Nebro AJ, Molina D, LaTorre A, Suganthan PN, Coello CAC, Herrera F (2021) A tutorial on the design, experimentation and application of metaheuristic algorithms to real-world optimization problems. Swarm Evol Comput 64:100888
61. Cascella M, Montomoli J, Bellini V, Bignami E (2023) Evaluating the feasibility of ChatGPT in healthcare: an analysis of multiple clinical and research scenarios. J Med Syst 47(1):33
62. Andrus M, Villeneuve S (2022) Demographic-reliant algorithmic fairness: characterizing the risks of demographic data collection in the pursuit of fairness. In: Proceedings of the 2022 ACM conference on fairness, accountability, and transparency. pp 1709–1721
63. Mellor D (2021) Improving norms in research culture to incentivize transparency and rigor. Educ Psychol 56(2):122–131
64. Post C, Sarala R, Gatrell C, Prescott JE (2020) Advancing theory with review articles. J Manage Stud 57(2):351–376
65. Meydan CH, Akkaş H (2024) The role of triangulation in qualitative research: converging perspectives. In: Principles of conducting qualitative research in multicultural settings. IGI Global, pp 98–129
66. Luft JA, Jeong S, Idsardi R, Gardner G (2022) Literature reviews, theoretical frameworks, and conceptual frameworks: an introduction for new biology education researchers. Educ Psychol 21(3):rm33
67. Aithal PS, Mishra N (2024) Integrated framework for experiential learning: approaches & impacts. Int J Case Stud Bus, IT Educ (IJCSBE) 8(1):145–173
68. Taylor AK, Kingstone T, Briggs TA, O'Donnell CA, Atherton H, Blane DN, Chew-Graham CA (2021) Reluctant pioneer': a qualitative study of doctors' experiences as patients with long COVID. Health Expect 24(3):833–842
69. Penuel WR, Riedy R, Barber MS, Peurach DJ, LeBouef WA, Clark T (2020) Principles of collaborative education research with stakeholders: toward requirements for a new research and development infrastructure. Rev Educ Res 90(5):627–674
70. Almusaed A, Almssad A, Yitmen I (2024) Sustainable built environment and its implications on real estate development: a comprehensive analysis
71. Moullin JC, Dickson KS, Stadnick NA, Albers B, Nilsen P, Broder-Fingert S, Mukasa B, Aarons GA (2020) Ten recommendations for using implementation frameworks in research and practice. Implement Sci Commun 1:1–12
72. Rush H, Marshall N, Bessant J, Ramalingam B (2021) Applying an ecosystems approach to humanitarian innovation. Technol Forecast Soc Chang 165:120529
73. Bachmann N, Tripathi S, Brunner M, Jodlbauer H (2022) The contribution of data-driven technologies in achieving the sustainable development goals. Sustain 14(5):2497
74. Dushkova D, Ivlieva O (2024) Empowering communities to act for a change: a review of the community empowerment programs towards sustainability and resilience. Sustain 16(19):8700
75. Bogenschneider K, Corbett T (2021) Evidence-based policymaking: envisioning a new era of theory, research, and practice. Routledge
76. Haeussler C, Assmus A (2021) Bridging the gap between invention and innovation: increasing success rates in publicly and industry-funded clinical trials. Res Policy 50(2):104155
77. Miao F, Holmes W, Huang R, Zhang H (2021) AI and education: a guidance for policymakers. Unesco Publishing
78. Levitt HM, Morrill Z, Collins KM, Rizo JL (2021) The methodological integrity of critical qualitative research: principles to support design and research review. J Couns Psychol 68(3):357
79. Dwivedi YK, Hughes L, Baabdullah AM, Ribeiro-Navarrete S, Giannakis M, Al-Debei MM, Dennehy D, Metri B, Buhalis D, Cheung CM, Conboy K (2022) Metaverse beyond the hype: multidisciplinary perspectives on emerging challenges, opportunities, and agenda for research, practice and policy. Int J Inf Manage 66:102542

80. Bardosh KL, de Vries DH, Abramowitz S, Thorlie A, Cremers L, Kinsman J, Stellmach D (2020) Integrating the social sciences in epidemic preparedness and response: a strategic framework to strengthen capacities and improve global health security. Glob Health 16:1–18
81. Wickert C, Post C, Doh JP, Prescott JE, Prencipe A (2021) Management research that makes a difference: broadening the meaning of impact. J Manage Stud 58(2):297–320
82. Tula OA, Daraojimba C, Eyo-Udo NL, Egbokhaebho BA, Ofonagoro KA, Ogunjobi OA, Gidiagba JO, Banso AA (2023) Analyzing global evolution of materials research funding and its influence on innovation landscape: a case study of us investment strategies. Eng Sci & Technol J 4(3):120–139
83. Li T, Miller EA, Krajcik JS (2023) Theory into practice: supporting knowledge-in-use through project-based learning. In: Fostering science teaching and learning for the fourth industrial revolution and beyond. IGI Global, pp 1–35
84. Brunet L, Fochler M, Müller R, Sigl L (2024) Modes of relevance in research. In: Towards understanding the promises and possibilities of doing relevance. Minerva

Academic Writing Style and Rigor

Abstract This chapter deconstructs the two foundations that shape high-impact academic work—style and rigor—in an era when knowledge is universal and academic integrity is unquestionable. It examines how formality, objectivity, and accuracy underpin credible research through clear and precise definitions; rigor ensures methodological integrity, critical depth, and ethical responsibility. Readers will acquire the essential skills to navigate citation systems (APA, MLA, Chicago, Harvard), avoid pitfalls such as plagiarism and logical fallacies, and enhance validity, reliability, and analytical depth. By equipping academics with valuable tools to elevate their academic voice, the chapter also addresses contemporary issues in writing, open-access publishing, and evolving peer review criteria. Whether you are a student, researcher, or educator, this book empowers you to write with honesty, critical thinking, and a significant contribution to your profession. A must-read for anyone serious about academic writing in the twenty-first century, it serves as a call to cultivate scholarly excellence beyond grammar and organization.

1 Introduction

A fundamental component of intellectual communication is academic writing. Within and across fields, it is the primary means of creating, assessing, and disseminating knowledge [1]. Far from merely a technical ability, academic writing reflects the academic community's epistemological foundations, methodological commitment, and ethical standards. Its core is characterized by a rigorous adherence to standards that promote transparency and accountability, logical coherence, clarity, and accuracy. These qualities distinguish academic writing from other forms of writing and qualify it for intellectual debate and research contribution. The significance of academic writing lies in the transfer of knowledge and in maintaining the quality, rigor, and legitimacy of scholarly work [2]. Formalized frameworks and consistent terminology facilitate critical engagement with complex ideas, the expression of evidence-based arguments, and the respectful acknowledgment of previous contributions. Thus, academic writing aids in organizing knowledge within fields and

A. Almusaed et al., *Practice of Research Methodology in Civil Engineering and Architecture*, https://doi.org/10.1007/978-3-031-97393-2_23

contributes to the broader advancement of research and education. The strength of academic writing is defined by two interrelated facets: style and rigor [3]. Style encompasses the linguistic and structural elements of academic writing, including formal language, logical organization, objective tone, and appropriate citation practices. These features help convey concepts, reduce ambiguity, and enhance readability. Academic style is not monolithic; it varies across fields and is shaped by specific methodological and rhetorical traditions. Nonetheless, shared values such as formality, objectivity, and accuracy form the foundation for transparent and ethical academic communication. In contrast, rigor pertains to the intellectual and methodological integrity of the research process and its written representation. It involves the ethical handling of data and sources, the careful design and implementation of research methods, and the systematic application of critical thinking. Rigor ensures that findings are credible, interpretations are justifiable, and contributions remain within ongoing academic discourse. From formulating research questions to reporting results, every stage of the research lifecycle necessitates reflexivity, receptivity to critique, and a commitment to maintaining high standards. Academic writing must also demonstrate ethical responsibility. Proper referencing and citation reflect academic integrity and respect for the intellectual efforts of others while also serving as tools for avoiding plagiarism [4]. Authors who document the origins of ideas and evidence promote transparency, allow verification, and situate their work within a network of academic dialogue. Standardized citation formats such as APA, MLA, Chicago, and Harvard ensure consistency and accuracy in citation standards that align writing with disciplinary expectations. Upholding rigor in academic writing extends beyond mere technical precision [5]. It involves fostering a deep understanding of the audience, purpose, and context of academic communication. Writers must engage critically with the literature, adhere to academic expectations, present fair arguments, and explicitly define the scope and limitations of their work. Such practices bolster the reliability of academic writing and enhance the credibility of the individual researcher and the broader research community. The demands placed on academic writing have increased in an increasingly complex and multidisciplinary research environment. Technological advancements, open-access publishing, global collaboration, and the incorporation of digital technologies have transformed how research is conceived, evaluated, and disseminated. These developments require a nuanced understanding of writing, not only as a cognitive and linguistic activity but also as a cultural and ethical practice. Writers must be prepared to navigate various requirements of academic publishing, address moral concerns such as authorship, data integrity, and peer review, and adopt a reflective approach toward the evolving standards of academic communication [6]. This work, therefore, highlights the essential aspects of objectivity, formality, and accuracy by emphasizing the primary characteristics that distinguish high-quality academic writing. It also explores the ethical significance of citation and referencing and their practical application in academic documentation. Additionally, it examines strategies to enhance academic rigor by improving validity, reliability, and analytical depth and identifies common pitfalls that compromise the quality of research-based writing. Collectively, these elements create a comprehensive framework for understanding and engaging with academic writing in a methodologically sound, ethically responsible, and stylistically appropriate manner. Writers prioritizing academic communication's stylistic and substantive facets will be better equipped to make meaningful contributions to their fields and the larger academic community [7]. From this perspective, academic writing is a practice to be nurtured rather than merely a skill to be acquired; it embodies the values of integrity, critical inquiry, and the collective pursuit of knowledge.

2 Defining the Characteristics of Academic Writing

Formal criteria guiding academic writing guarantee clarity, correctness, neutrality, and credibility, the foundations of intellectual conversation. Among the fifteen important ethical issues Guelmami et al. [8] found in this field were plagiarism, data falsification, AI misuse in authorship, result exaggeration, strategic self-citation, duplicate publication, undisclosed conflicts of interest, image manipulation, peer review abuse, ghost or gift authorship, poor data retention, IRB fabrication, lack of informed consent, and unethical human or animal experimentation [8]. As a means of communication within the academic community, it seeks to educate and significantly enhance continuous intellectual dialogue. The foundation of academic writing lies in the clear expression of ideas, a logical structure of content, and the careful use of evidence from reliable sources. Objectivity stands out as a distinctive characteristic of this writing style. Writers must remain objective, avoiding personal biases, emotional language, and unproven beliefs [9]. Using impersonal constructs, careful phrasing (hedging), and reliance on empirical or theoretical support for every assertion help maintain this objectivity. Distancing the writer from the argument emphasizes the concepts and their logical consistency within the academic context. Academic writing is also formal, requiring the avoidance of informal language and adherence to conventional syntax, punctuation, and discipline-specific terms. Typically structured around an introduction, methodology, analysis, discussion, and conclusion, academic writing is organized. This structure enhances the readability of complex content and aids in the logical development of ideas. Both language and arguments necessitate accuracy. Academic writing calls for the unambiguous expression of concepts, the definition of words, and the removal of uncertainty. Ramzan et al. [10] underline the importance of a thorough and multifarious strategy to properly handle the issues by stating that the linguistic aspects of academic papers are essential for their clarity and quality [10]. Paragraphs are organized to support a single, consistent point related to the overall thesis, while sentences are crafted to convey precise meaning. Furthermore, academic writing relies on proper citation and referencing, allowing readers to verify sources and explore the research background, thus acknowledging the work of other scholars. Failing to cite sources correctly may lead to academic fraud and undermine the quality of the work. A disciplined approach that combines clarity, objectivity, and analytical rigor enables academic writing to serve as a vehicle for knowledge building and distribution within and across fields. Adhering to these conventions of objectivity, formality, accuracy, organized thought, and proper attribution upholds its significance as a foundation of academic integrity and research [11].

2.1 *Objectivity, Formality, and Precision*

Academic writing relies on principles that promote openness, intellectual rigor, and academic reputation [12]. Among these conventions, the fundamental ones are objectivity, formality, and accuracy. These principles not only influence the language of scholarly works but also reflect the values of the academic community: commitment to objective investigation, rigorous thinking, and conceptual clarity. Objectivity in academic writing involves intentionally avoiding personal bias, emotional language, and unverified claims [13]. Scholarly work is expected to present logically organized, evidence-based arguments that engage critically with current literature. An impersonal tone often achieved through third-person constructions, passive voice, and careful modality (e.g., "it is suggested," "the data indicate") allows the focus to remain on the topic rather than the author [14]. This rhetorical stance emphasizes that creating academic knowledge must be grounded in verifiable data and methodologically rigorous analysis rather than personal belief or anecdotal observation. Thus, objectivity is not merely a stylistic choice but an ethical necessity underpinning academic discourse's credibility and reliability.

Formality, an essential aspect of objectivity in academic writing, reflects scholarly communication's disciplined, rule-driven nature [15]. Academic writing demands a register that favors discipline-specific vocabulary and a syntactically complex yet cohesive structure over colloquialisms, contractions, and informal idioms. Formality enhances the power of the text and signifies the writer's adherence to the norms and expectations of the academic community. Furthermore, it contributes to communication accuracy, which is crucial for presenting specialized knowledge. A formal tone ensures that intricate arguments are expressed with the seriousness, clarity, and respect required for academic engagement across various disciplines and cultural contexts [16]. The third pillar of effective academic writing is accuracy. Academic discourse's complexity and abstract nature necessitate precise language, well-defined ideas, and logically structured arguments. Consistent terminology, elimination of ambiguity, and meticulous calibration of Claims are essential for defining accuracy [17]. Scholarly standards reject vague generalizations and rhetorical exaggerations, as they distort meaning and undermine analytical integrity. Conversely, academic authors must articulate their conclusions clearly and accurately, define their terms, and constrain their arguments. Precision is a methodological imperative that supports replicability, critique, and the advancement of knowledge; it transcends mere language choice. For example, using recycled aggregate concrete (RAC) in structural applications has demonstrated similar compressive strength to traditional concrete, given suitable mix designs and processing methods [18]. This result emphasizes the potential of reusing building debris as a sustainable substitute in contemporary civil engineering practice. Along with their usual challenges, Table 1 shows

Table 1 Core academic writing principles and their associated challenges

Academic principle	Description	Common challenges
Objectivity	Upholding neutrality through evidence-based reasoning, detachment from personal opinions, and logical consistency	– Maintain a consistently impersonal tone without becoming vague or mechanical – Avoid unverified claims and emotional appeals – Balance critique with neutrality in literature engagement
Formality	Use discipline-specific language and structured syntax, and avoid an informal or conversational tone	– Avoid contractions, slang, and idiomatic expressions – Maintaining complexity while ensuring clarity – Tailoring tone to meet cross-disciplinary and intercultural academic standards
Precision (accuracy)	Ensuring precision in terminology, argument structure, and claim boundaries to support replicability and critique	– Clearly define key terms – Avoid vague generalizations and exaggerations – Ensuring conclusions are methodologically sound

three fundamental principles of academic writing: objectivity, formality, and accuracy. These values uphold the intellectual rigor of academic discourse. Objectivity demands impartiality and evidence-based thinking. Formality requires disciplined, organized language specific to a field. Precision ensures consistency in definitions and clarity. While adhering to academic norms across various disciplines, authors sometimes struggle to balance clarity with complexity, maintain impartiality, and avoid casual language.

Academic writing fundamentally embodies the integration of impartiality, formality, and accuracy. These values are interconnected elements rather than isolated traits that uphold the integrity of scholarly communication. By consistently applying these criteria, academic authors not only enhance the clarity and persuasiveness of their work but also uphold community standards of research quality, disciplinary coherence, and epistemic accountability.

2.2 *Citation and Referencing Standards in Academic Writing*

2.2.1 Importance of Citation

Academic writing depends significantly on citations, which ensure proper recognition of sources, ideas, and arguments [19]. Acknowledging the intellectual contributions of others and distinguishing original thoughts from borrowed material upholds academic integrity. This practice reflects both academic professionalism and an ethical duty. Using accurate citations, authors demonstrate their engagement with current research, respect for the academic community, and participation in an ongoing dialogue about knowledge creation. Moreover, citations allow readers to verify information, explore essential texts, and assess the validity of claims based on their original context. Incorrect citations, intentional or unintentional, can lead to serious academic dishonesty, including allegations of plagiarism [20]. Such infractions seriously undermine the academic credibility of the author and call into question the dependability, openness, and general academic value of the work. Maintaining academic rigor and ethical responsibility calls for rigorous following of accepted citation practices inside educational institutions and academic publishing. Proper reference gives intellectual ownership and allows the academic community to critically engage, replicate, and validate the research. For example, in architectural and engineering research, it is usual to cite empirical studies on thermal insulation, sun orientation, and natural ventilation systems as basic components in design rationale. Researchers show methodological soundness and help to the cumulative development of sustainable building practices by placing their arguments within the larger body of literature on energy-efficient materials and climate-responsive design. Ignoring appropriate citation practices not only increases questions of academic dishonesty but also separates the work from scholarly conversation. On the other hand, careful interaction with previous studies shows academic integrity, strengthens the legitimacy of new contributions, and supports the cooperative and iterative character of knowledge generation in the scientific and professional spheres [21]. The primary functions of academic writing are illustrated in Fig. 1. It highlights how proper referencing supports academic integrity, recognizes intellectual contributions, and enables readers to verify facts. The figure also emphasizes how citation fosters ethical responsibility, enhances transparency, and strengthens intellectual discourse. These elements demonstrate that academic communities' legitimacy, professionalism, and knowledge growth rely on accurate citation.

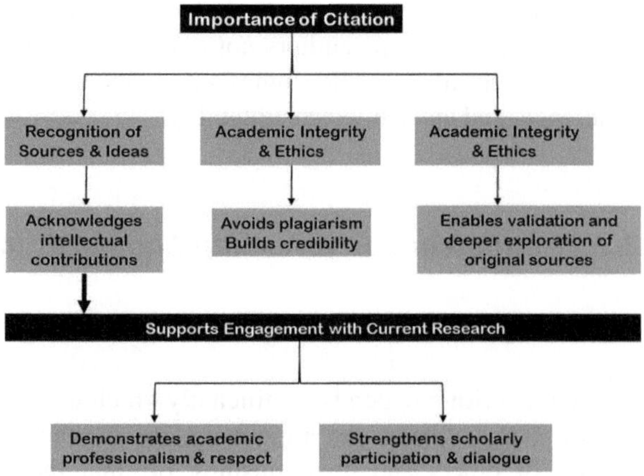

Fig. 1 Key functions of citation in academic writing (*Source* Authors)

2.2.2 Ensuring Traceability and Transparency

Citations allow readers to trace the origins of concepts, data, and arguments. This traceability enables others to verify claims, examine the original context, and explore related studies, ensuring that academic discourse remains open and accountable. Maintaining the integrity of scholarly communication relies on promoting academic openness, as it facilitates systematic knowledge accumulation and fosters critical engagement within academic communities. Zamiri and Esmaeili [22] argue that enhancing the quality and impact of educational outcomes in an era characterized by rapid technological advancement and an increasing emphasis on collaborative learning requires a comprehensive understanding of practical knowledge-sharing tools and strategies [22]. Through citations, researchers can identify theoretical frameworks, track the evolution of ideas, and situate their work within a broader intellectual tradition. This enhances the credibility of current research and facilitates interdisciplinary dialogue and collaboration. Moreover, traceable citations support the reproducibility of studies by allowing future scholars to replicate methodologies or challenge assumptions, thereby advancing the reliability and robustness of academic inquiry [23]. For example, authors in digital twin studies frequently mention foundational research on system interoperability, simulation accuracy, and real-time data integration. By citing earlier work on the virtual modeling of smart infrastructure, researchers validate their methods, enable method replication, and demonstrate how their contributions augment and extend current knowledge in cyber-physical systems and intelligent asset management.

2.2.3 Standard Referencing Systems

Standardized referencing systems provide organized and methodical frameworks for recording sources, guaranteeing clarity, consistency, and intellectual rigor across many academic fields. Reflecting an evolving methodological scene that values both efficiency and depth in data analysis, Kowalski et al. [24] note that the use of fast qualitative techniques has increased dramatically over the last decade, especially in the areas of quality improvement and health services research [24]. These systems guarantee that references are presented consistently and transparently. For instance, studies on thermal insulation, solar gain optimization, and energy modeling are documented in passive building research using consistent referencing techniques. This systematic citation approach achieves three objectives: ensuring methodological openness, facilitating the replication of results, and reinforcing the academic rigor necessary for advancing sustainable and climate-responsive architectural design. These systems promote academic integrity and enhance intellectual discourse by enabling readers to locate and verify sources. The APA style, widely used in the social sciences, is among the most common referencing systems and emphasizes an author-date citation method that prioritizes the recency of sources to reflect ongoing research developments [25]. In contrast, the MLA style, prevalent in the humanities, employs an author-page format suitable for fields requiring thorough textual

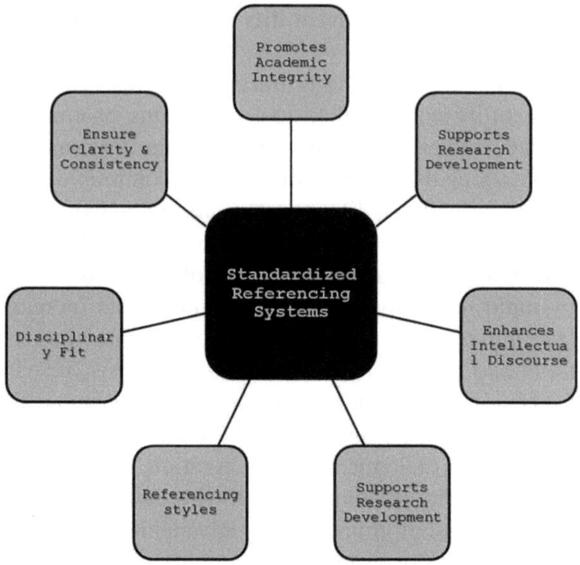

Fig. 2 Functions and disciplinary applications of referencing systems in academic writing (*Source* Authors)

analysis. Often adopted in the sciences, the Chicago Manual of Style supports the Author-Date system, while the Notes-Bibliography method is preferred in history and the humanities. This dual framework offers flexibility based on disciplinary needs. The Harvard style, widely recognized across various disciplines, is appreciated for its straightforward structure and is another respected author-date system. Each referencing system ensures accurate and consistent academic communication by providing precise rules for crediting diverse sources, such as books, journal articles, and digital media [26]. The fundamental functions and structural components of standardized referencing systems are illustrated in Fig. 2. It highlights their role in methodological transparency, integrity, and academic clarity. The figure also depicts how citation styles, including APA, MLA, Chicago, and Harvard, align with academic conventions. Through source verification and consistency across academic disciplines, these systems ensure accurate documentation, promote research development, and enhance intellectual discourse.

2.2.4 Disciplinary Differences

Citation Reflecting different priorities in how information is created, assessed, and transmitted, citation practices are firmly ingrained in academic fields' methodological and rhetorical traditions [27]. These variations are based on the epistemological foundations guiding academic research in every discipline, hence transcending style. Citation methods in the humanities, especially in history, philosophy, literature, and theology, stress interpretive depth, contextual richness, and close interaction with primary and secondary sources. Footnotes and endnotes are shared in these fields, enabling writers to offer detailed comments, source provenance, and more debate without breaking the main story [28]. For example, the Chicago Notes—Bibliography style accommodates this depth and provides a flexible and thorough format appropriate for complex historical and literary study. By contrast, citation methods in the social and natural sciences, including psychology, sociology, education, biology, and public health, emphasize brevity, recency, and replicability. The author-date systems used in the APA and Harvard styles emphasize the publication date, allowing readers to evaluate the relevance and currency of cited works rapidly [29]. This is especially important in fast-changing sectors where academic legitimacy and knowledge development depend on current results and empirical validation. Following discipline-specific citation rules means more than just formatting; it guarantees consistency with field-specific standards and reader expectations. This commitment strengthens the consistency and argument of academic writing, helps the peer review process, and increases the academic authority of the writer. Moreover, citation uniformity helps standardize academic discourse, promotes cross-disciplinary knowledge and research cooperation, and maintains the methodological integrity of every discipline [30]. Ultimately, the difference in citation methods across fields emphasizes how academic societies value, display, and assess evidence. Effective communication and preservation of the intellectual legacy and scholarly standards that characterize every academic field depend on respect of these conventions. Table 2 illustrates how citation practices differ across academic fields, reflecting various epistemological foundations and methodological goals. The social and natural sciences emphasize brevity, regency, and replicability through author-date systems, whereas the humanities focus on depth and contextual analysis using footnotes and narrative integration. These differences enhance academic rigor and support discipline-specific criteria through methodological consistency, thereby improving scholarly communication and fostering cross-disciplinary collaboration.

2.2.5 Common Challenges in Citation and Referencing

Managing citations and references can be difficult for academic authors in numerous ways, many of which can seriously compromise the quality and integrity of their work [31]. Correctly citing complicated or non-conventional sources is one frequent challenge. Datasets, personal interviews, social media information, or other digital media are among the materials that often fall outside conventional citation formats, generating uncertainty in the correct credit and formatting. Furthermore, authors sometimes find it challenging to keep citation style consistent throughout the work. When combining references from other sources or working on writing assignments, this problem becomes incredibly noticeable [32]. Even minor discrepancies in punctuation, capitalization, or date formatting, for example, can compromise the professional look of the writing and suggest a lack of attention to detail. Inadvertent plagiarism is another significant difficulty, usually resulting from ambiguous rules on when and how to cite. Unintentional academic fraud might result from improper paraphrasing, failure to differentiate original analysis from sourced material, or misunderstandings

Table 2 Disciplinary variations in citation practices

Disciplinary domain	Citation priorities	Common methods and styles	Rationale
Humanities	Interpretive depth, contextual richness, engagement with sources	Footnotes, endnotes; Chicago notes-bibliography	Allows detailed commentary and source provenance without interrupting narrative flow
Social sciences	Brevity, recency, replicability	Author-date systems; APA, Harvard	Facilitates quick assessment of relevance and supports empirical evaluation
Natural sciences	Precision, recency, replicability	Author-date systems; APA, Harvard	Emphasizes current findings and methodological transparency
Purpose of citation	Varies by epistemological traditions and disciplinary standards	Varies accordingly	Reflects methodological values and rhetorical norms within each academic tradition
Academic function	Enhance scholarly authority, aid peer review, ensure clarity	Discipline-specific application	Strengthens consistency, supports credibility, and aligns with reader expectations

about citation standards for common knowledge [33]. Novice researchers or students who are not entirely aware of the subtle standards for academic integrity find this problem particularly common. Moreover, the need to fulfill deadlines or word limits could drive some authors to rely too much on source material without sufficient critical involvement or creativity, raising the possibility of citation mistakes. Overcoming these obstacles requires a comprehensive knowledge of referencing rules, a commitment to academic integrity, and meticulous attention to detail [34]. These mistakes can be significantly reduced using citation management systems, referring to style guidelines (e.g., APA, MLA, Chicago), participating in peer review, or asking for supervisor comments. In the end, showing both the academic legitimacy of the writer and their appreciation for the intellectual contributions of others, understanding citation techniques is a vital component of academic writing. Figure 3 shows three fundamental issues commonly encountered in academic citation and referencing: citing complex sources, maintaining consistency, and preventing unintentional plagiarism. These problems highlight essential obstacles that compromise the professional quality of scientific writing and academic integrity. Addressing these issues requires knowledge of citation standards, careful attention to detail, and the use of appropriate referencing tools and institutional policies.

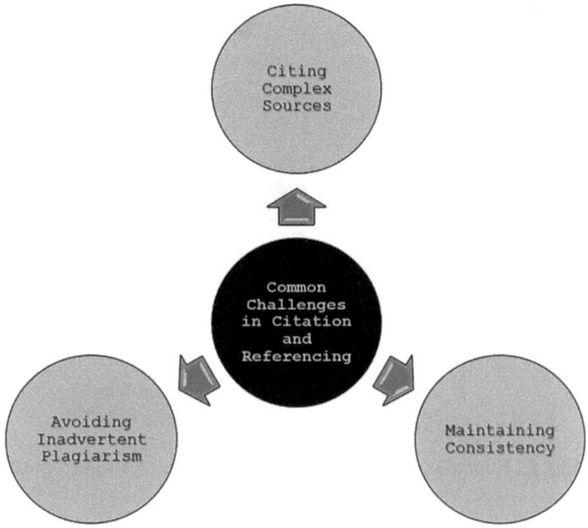

Fig. 3 Key challenges in citation and referencing practices (*Source* Authors)

2.2.6 Ethical Considerations

Citations reflect a profound ethical aspect fundamental to the quality of academic work, serving not only as a technical or stylistic necessity [35]. Proper citation methods ensure that intellectual contributions are recognized appropriately and that the academic record presents a fair, accurate, balanced literary account. Incorrect use of citations, whether intentional or accidental, can significantly undermine this integrity. Excessive self-citation, citation stacking among networks of authors or journals, and the deliberate exclusion of opposing or foundational sources all contribute to the distortion of the intellectual landscape and the manipulation of academic measures such as impact factors and h-indexes, thereby misleading readers, reviewers, and assessors [36]. Such practices may result in reputational harm, retractions, or broader repercussions on academic trust by undermining the neutrality of scientific communication. Ethical referencing urges academics to engage thoroughly with relevant literature, to acknowledge opposing viewpoints, and to avoid exaggerating the significance of their work or that of close colleagues. It promotes transparency in distinguishing between original thoughts and borrowed ideas and strict adherence to disciplinary standards on attribution [37]. Upholding ethical norms in citation fosters academic responsibility and cultivates a culture of respect, openness, and critical engagement. Ensuring that credit is duly given and that academic contributions are evaluated based on merit rather than manipulation facilitates the advancement of knowledge [38]. In this context, citation stands as a symbol of academic integrity, professional ethics, and commitment to the shared ideals of the research community rather than merely serving as a documentation tool. For example, incorporating recycled crushed concrete as an aggregate in fresh concrete mixes has been shown to reduce environmental impact and minimize construction waste

Fig. 4 Ethical considerations in citation practices (*Source* Authors)

while maintaining sufficient structural performance for non-load-bearing elements. This method fosters the cyclical use of materials in the construction industry and promotes sustainable building practices. Key ethical aspects of citation procedures in academic writing are illustrated in Fig. 4. It highlights the necessity of proper citation to acknowledge intellectual contributions and warns against misuse, such as excessive self-citation and citation stacking. These unethical practices can undermine the credibility of research and distort academic metrics. The emphasis illustrates that promoting academic integrity requires fairness, transparency, and accountability in citation practices across all scholarly disciplines.

3 Ensuring Rigor in Academic Research and Writing

A fundamental pillar of an academic work's reputation, relevance, and scholarly value is academic rigor [39]. Far above superficial criteria like grammatical accuracy or syntactic grace, rigor engages with the core of academic research: its intellectual coherence, methodological strength, theoretical underpinning, and ethical purity. It encompasses the techniques and values that ensure epistemologically sound, methodically designed, critically informed, well-articulated, and comprehensible research. Rigor exists at the intersection of form and substance; it is as much about the procedural correctness of research techniques as it is about the depth of engagement with academic discussions and the ability to contribute actively to ongoing intellectual conversations. In its broadest sense, academic rigor is consistently dedicated to literary quality throughout all phases of the research and writing process [40]. Rigor is reflected in the clarity of conceptual frameworks, the appropriateness and transparency of methodological choices, the accuracy of data interpretation, and the logical structure of academic argumentation from the initial formulation of research questions to the final presentation of results. It also requires situating one's work within the existing body of knowledge, demonstrating a comprehensive understanding of the topic, and the capability to question and expand its boundaries critically [41]. This necessitates intellectual reflexivity and technical expertise to continually assess one's assumptions, address limitations, and remain receptive to alternative perspectives. Moreover, academic rigor is not a static or uniform concept; various disciplinary paradigms, knowledge orientations, and evolving criteria for research quality shape it. What qualifies as rigor in a positivist empirical study may differ somewhat from the expectations in interpretive or constructivist research approaches. In the sciences, rigor often emphasizes reproducibility, statistical validity, and experimental control; in the humanities, it may be expressed through theoretical richness, argumentative coherence, and nuanced textual analysis [42]. Despite these contextual variations, a common thread persists: serious research adheres to the highest standards of analytical accuracy, methodological openness, and ethical responsibility. Rigor must be viewed as a continuous and iterative process rather than a fixed endpoint [43]. It demands ongoing involvement throughout the research lifecycle, encompassing problem identification, literature evaluation, research design, data collection and analysis, and dissemination of results. Each stage requires intentional, informed decisions that align with the study's issues and the standards of the academic community. Thus, rigor serves as both a process and a principle, functioning as a procedural guide and a benchmark for assessing the scholarly value of academic work [44]. This section examines the conceptual framework of academic rigor, offering a theoretical perspective to understand its various aspects. It also provides valuable strategies to enhance validity, reliability, and analytical depth in scholarly writing while addressing persistent challenges that threaten research rigor, such as logical fallacies, methodological inconsistencies, and unfounded claims. Pursuing academic rigor is not merely a legal obligation; it represents an intellectual and ethical commitment to producing knowledge that is transparent, responsible, and capable of withstanding rigorous scrutiny within and outside one's disciplinary community. For instance, sustainable architecture employs energy-efficient materials, green roofs, and passive solar design to reduce environmental impacts and improve building efficiency. Architects who prioritize resource conservation and ecological balance design structures that decrease carbon emissions and promote occupant health, long-term durability, and alignment with sustainable development objectives [45]. The necessary dimensions that make up academic rigor—including methodological integrity, intellectual reflexivity, ethical responsibility, and disciplinary specificity—are shown in Table 3. It stresses rigor as both a technique and a value underpinning academic legitimacy throughout all phases of research. The table also highlights typical dangers to rigor. It illustrates how practical applications, such as sustainable architecture, represent rigorous academic standards by integrating theory, evidence, and real-world influence.

Table 3 Dimensions and attributes of academic rigor in research and writing

Dimension	Description	Key attributes	Example/application
Definition and core focus	Rigor is foundational to research's credibility, relevance, and scholarly value	Intellectual coherence, methodological robustness, ethical integrity	Core principle in all high-quality academic work
Form and substance	Rigor exists at the intersection of research methods and intellectual engagement	Procedural accuracy, conceptual clarity, deep scholarly engagement	Balanced use of empirical data and theoretical analysis
Process orientation	Rigor is iterative and continuous throughout the research lifecycle	Evident from research question formulation to dissemination	Maintained through all stages: design, execution, analysis, reporting
Conceptual frameworks	Clear, logical frameworks guide scholarly inquiry	Transparent methodology, structured argumentation, relevant theory	Framing sustainable architecture within ecological design theory
Disciplinary variation	Disciplinary norms and paradigms shape rigor	Empirical reproducibility (sciences); theoretical depth (humanities)	Sciences: statistical controls; Humanities: critical textual analysis
Intellectual reflexivity	Scholars must critically assess their assumptions and remain open to critique	Recognition of limitations, responsiveness to counterarguments	Acknowledging trade-offs in green building certifications in architecture
Ethical responsibility	Rigor demands ethical practices in data collection, analysis, and citation	Academic honesty, transparency, avoidance of bias	Proper attribution in interdisciplinary literature reviews
Benchmarking and standards	Rigor functions as a principle and evaluative benchmark	Consistency with disciplinary standards, scholarly merit	Peer-reviewed publication as a measure of research rigor
Common threats to rigor	Pitfalls that compromise research quality must be avoided	Logical fallacies, methodological inconsistencies, unsupported claims	Overgeneralizing sustainability outcomes without empirical backing
Practical illustration	Real-world applications reflect rigorous academic principles	Integration of theory and practice, problem-driven research	Sustainable architecture using green roofs and passive solar design

3.1 Conceptualizing Rigor in Scholarly Contexts

A foundation of significant and reliable research is academic rigor. From developing research questions to sharing results, it reflects a mindset that controls the entire research process, not just a methodological need [46]. Rigor is a methodical, exact, and consistent attitude toward investigation that guarantees intellectual honesty, openness, and credibility. It should not be mistaken for complication or difficulty. Instead, rigor emphasizes all the qualities required to produce consistent and legitimate knowledge: clarity, depth, and methodological consistency. A well-defined research goal initiates a rigorous academic approach. Academics must clearly state their goals and demonstrate their connection to broader theoretical and practical contexts [47]. This requires a careful reading of current literature and a critical evaluation of established work. By doing so, scholars can identify gaps, conflicts, or opportunities for further study by situating their work within an ongoing academic dialogue. Critical thinking is at the core of rigor. This intellectual attitude urges academics to scrutinize the weight of evidence supporting various claims, question assumptions, and challenge conventional wisdom. Thorough research delves deeper to uncover underlying mechanisms, reconcile opposing viewpoints, and present well-substantiated arguments rather than accepting superficial justifications [48]. Such inquiries demand intellectual discipline and an openness to contradiction, complexity, and change. Another important aspect of academic excellence is methodological transparency. Researchers must articulate and justify their methodological choices concerning study topics and settings. Every stage, from sampling and data collection to analysis and interpretation, should be clearly articulated and logically connected to the overall study design, whether employing qualitative, quantitative, or mixed methodologies. This transparency enables other researchers to evaluate the validity of the results, replicate the study, or build upon its contributions. Rigor also entails ethical and procedural responsibility. Researchers are responsible for ensuring their work meets academic integrity standards, including accurate data representation, avoidance of bias, and respect for research subjects [49]. Furthermore, they must present their results in a way that allows others to verify and access them. This commitment to transparency enhances the study's validity and fosters a collaborative scholarly environment. By upholding high research standards, academic rigor collectively advances knowledge. Scholars bolster the credibility of academic work by being transparent in their methodologies, supporting their arguments with logical reasoning, and robustly defending their claims. Rigor is a fundamental principle underlying academic integrity and quality, serving as more than just a procedural necessity [50]. It embodies a commitment to depth, accuracy, and accountability qualities that enhance individual studies and contribute to the academic community. Researchers who embrace rigor uphold their fields' integrity and aid in advancing knowledge. Emphasizing their goal and the usual difficulties connected with each, Table 4 outlines the basic elements of academic rigor. It emphasizes the need of ethical responsibility, logical consistency, methodological openness, critical thinking, and well defined research goals. Presenting these components together helps the table to

Table 4 Fundamental components, goals, and issues of academic rigor

Element	Explanation	Purpose	Common challenges
Clear research purpose	Defining specific and pertinent research questions	Guides the study and connects it to broader academic goals	Risk of vague or overly broad questions
Critical thinking	Questioning assumptions and evaluating evidence	Ensures depth and quality in analysis	Biases or lack of reflection can weaken conclusions
Transparent methods	Clearly explain the process for collecting and analyzing data	Allows others to understand, verify, or replicate the study	Poor documentation or unclear procedures
Logical coherence	Ensuring coherence among objectives, methods, and conclusions	Builds credibility and strengthens arguments	Inconsistencies can confuse or mislead readers
Ethical accountability	Maintain honesty, fairness, and respect during the research process	Maintain integrity and trust in the research	Overlooking ethical issues or poor data handling practices

be a useful summary for knowing how rigor enables credible and relevant academic work while recognizing shared challenges.

3.2 *Strategies for Enhancing Validity, Reliability, and Depth*

Academic rigor is the result of intentional, informed, and sustained efforts that support the integrity and substance of scholarly work; it is not an unintentional byproduct of study [51]. From the first question development to the last analysis and distribution stages, it shows a dedication to quality throughout the study process. Rigor is ingrained in how researchers think about issues, interact with the literature, choose and use techniques, analyze data, and report results. It calls for intellectual discipline, methodological honesty, and critical self-reflection. Validity, depth, and dependability are essential aspects of academic rigor [52]. These aspects are complementary and together guarantee that research results are not only correct and consistent but also intellectually relevant and analytically rich. Validity concerns the degree to which a study measures or reflects what it purports to investigate. Reliability is the consistency and stability of such measures or interpretations across time, context, or observer [53]. Depth is the analytical complexity of the piece and its capacity to interact critically with ideas, data, and consequences beyond surface-level observations [54]. These three components are quality markers and protections against bias, misperception, and superficiality. Their presence indicates that the study can endure examination and positively contribute to the academic community. However, being rigorous is not guaranteed. Improving these aspects calls for methodologically sound, culturally relevant, and reflective of discipline norms focused approaches. Though different fields define validity, reliability, or deep differently, the fundamental dedication to openness, coherence, and critical involvement is universal [55]. Researchers must be deliberate in their methodological decisions, defend their approaches with theoretical and empirical justification, and keep a consistent analytical perspective throughout their work. The quest for rigor has to be deliberate and ingrained in every stage of academic investigation, whether using qualitative, quantitative, or mixed-method techniques [56]. Thus, academics create excellent work and help build the legitimacy, development, and ethical basis of academic knowledge shared by many others. Ensuring academic rigor, for example, in sustainable architectural engineering could include verifying energy simulation models against actual building performance data, using consistent material sustainability testing, and carefully assessing the long-term environmental effect of design decisions using life cycle analysis to guarantee dependability, validity, and depth in the research results.

3.2.1 Enhancing Validity

Validity refers to the degree to which a research project accurately reflects the phenomena it aims to explore [57]. Regardless of methodological orientation, it is a crucial criterion in assessing the' soundness, credibility, and usefulness of research results. A valid study demonstrates that the research tools, techniques, and interpretations effectively represent the constructs or real-world situations under investigation. Enhancing validity requires methodical planning and methodological consistency from the initial research design to data processing and interpretation [58]. Validity in quantitative studies is typically categorized into three types: construct, internal, and external. Construct validity assesses whether the tools employed, such as surveys, tests, or experimental procedures, accurately represent the theoretical constructs being investigated. Researchers should utilize well-defined variables to enhance construct validity, choose or develop psychometrically validated tools, and align measurement scales with established theoretical frameworks [59]. Focusing on the causal relationships between variables, internal validity necessitates rigorous control of confounding factors through randomization, control groups, and statistical control techniques. External validity pertains to the generalizability of results, which can be improved through appropriate sampling, replication studies, and inclusion of diverse participants. Qualitative research frames validity regarding credibility, authenticity, and trustworthiness, as it typically aims to reflect participants' experiences and contextual realities accurately rather than generalize [60]. Ensuring credibility involves triangulation using multiple sources, investigators, or techniques, member checking where participants validate the researcher's interpretations, and sustained field participation to gain an in-depth understanding of the setting. Maintaining an audit trail, which records decisions, assumptions, and analytical procedures throughout the study, enhances transparency and allows for external assessment of the research process and rationale. Strengthening validity depends on ensuring alignment between study questions, methodologies, and data interpretation [61]. Researchers must confirm that their methodological choices correspond with their study questions and that their findings are firmly rooted in the collected data. This entails avoiding overgeneralization, maintaining coherence between theoretical frameworks and empirical data, and delineating the scope and limitations of the investigation. Pretesting research tools, seeking expert advice, and providing rich contextual details yield a more valid inquiry across both paradigms. A thorough review of existing literature also clarifies conceptual concepts and guides the development of legitimate analytical frameworks [55]. Ultimately, enhancing validity is not a one-time task but an ongoing process that must be reevaluated at every stage of the research cycle to ensure the study yields accurate, contextually grounded, and methodologically sound results. Figure 5 illustrates a conceptual framework outlining the multidimensional tactics necessary to enhance validity in quantitative and qualitative research paradigms. At the center of the figure is the key concept of validity, which is essential for establishing research findings' credibility, accuracy, and trustworthiness. The graph depicts how validity divides into two distinct yet complementary branches: quantitative and qualitative research, each equipped with various tools. This approach highlights construct, internal, and external validity as crucial components in quantitative research. Methodological tools such as psychometrically sound instruments, statistical controls, randomization, and appropriate sampling strategies reinforce these characteristics. Meanwhile, validity in the qualitative pathway is associated with credibility, which is achieved through triangulation, member checking, prolonged engagement, and maintaining an audit trail. These strategies ensure that findings are transparent and verifiable, with interpretations firmly grounded in the participants' reality.

3.2.2 Enhancing Reliability

An essential element of high-quality research is its dependability, which means consistency, stability, and replicability of study results [52]. It guarantees that the outcomes represent actual, repeated observations rather than those of random error, measurement inaccuracy, or subjective bias. A dependable study confirms the veracity of its conclusions by showing consistent results when run under comparable circumstances. Improving dependability starts with methodically standardizing research techniques. This calls for creating and using unambiguous rules controlling every phase of the research process. Supported by thorough instructions to reduce variability in execution, quantitative research includes measuring equipment running consistently across time and samples [62]. Structured interview guides, coding systems, and pre-defined analytical processes guarantee consistent data interpretation and help strengthen dependability in qualitative research. Pilot testing is vital since it lets researchers evaluate research tools' clarity, usability, and dependability before they are employed at scale. Running a pilot study helps to find and fix unclear items, technological problems, or procedural errors that could compromise consistency [63]. This proactive improvement guarantees consistent instrument operation and stable data generation. Establishing inter-rater reliability is vital in

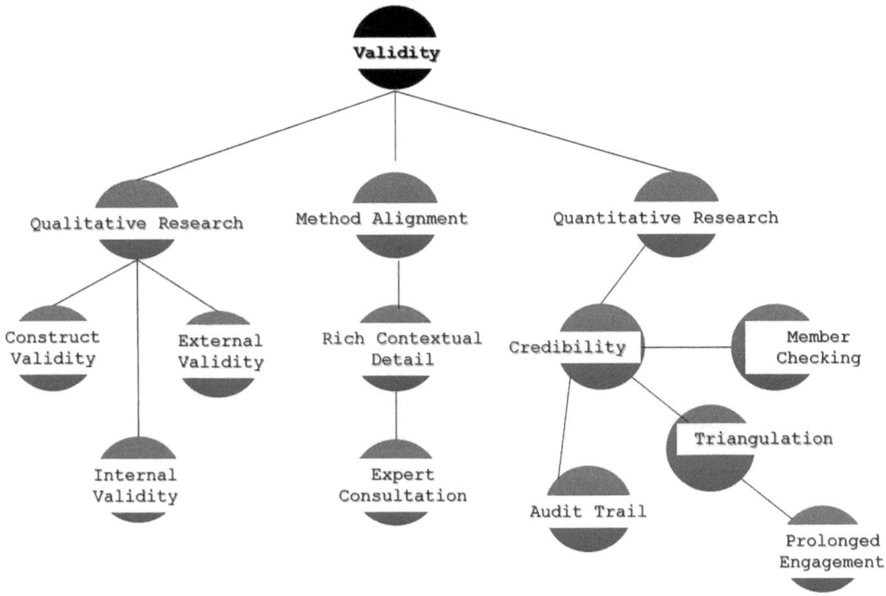

Fig. 5 Conceptual framework for enhancing validity in research (*Source* Authors)

studies with many observers or coders [64]. This approach trains people to use the same coding or rating criteria and use statistical techniques to evaluate their level of agreement. High inter-rater reliability guarantees consistency in the subjective or qualitative data analysis by suggesting that interpretations are not overly reliant on individual points of view. Reliability is also shown in the academic writing's presentation of research. Consistent language, systematic reporting of procedures, clear, coherent, logically constructed arguments, and systematic reporting of procedures all improve the perceived and absolute trustworthiness of academic work [52]. Researchers should keep terminological clarity and avoid contradictions that could cause interpretation uncertainty or misunderstanding. Moreover, keeping thorough records of every study step including instrument construction, data administration, coding selections, and analytical choices helps to increase dependability even more. Such openness lets other studies assess or duplicate the methodological soundness of the work. For example, studies on sustainable and smart buildings show that consistent energy consumption measurements and sensor-based monitoring techniques increase dependability [65]. For example, accurate comparisons across several smart buildings using calibrated sensors lets one measure indoor temperature and CO2 levels. Researchers guarantee the replicability of energy efficiency and occupant comfort results in intelligent building systems by using consistent data collecting techniques and repeating tests under comparable environmental circumstances. Improving dependability requires a thorough approach, including methodological accuracy, procedural uniformity, inter-rater consistency, and academic clarity. These techniques together help provide strong, reliable results that further rigorous and

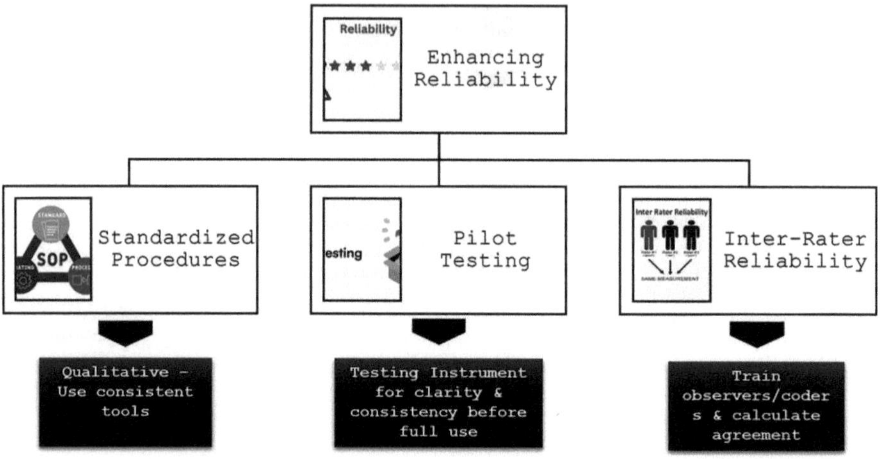

Fig. 6 Key strategies for enhancing research reliability (*Source* Authors)

repeatable scientific knowledge. The fundamental methods for enhancing research dependability shown in Fig. 6 include standardized processes, pilot testing, inter-rater reliability, and open academic reporting. These interrelated elements guarantee consistent, repeatable outcomes and support empirical investigations' validity across quantitative and qualitative research approaches, including uses in smart building systems.

3.2.3 Enhancing Depth

A distinguishing characteristic of serious study is depth in academic writing. It demonstrates the writer's ability to engage with complex ideas, examine relationships between concepts, and generate significant insights beyond superficial observations and simple explanations [66]. Depth reflects the scholar's critical and introspective attitude, emphasizing their effort to advance understanding in a specific area rather than merely transmit knowledge. Academic depth is established through comprehensive and dynamic interactions with the literature [67]. A well-developed literature review transcends a basic summary of current research by revealing knowledge gaps, identifying discrepancies, addressing theoretical and methodological controversies, and emphasizing underlying trends. Guided by analytical rather than descriptive goals, a thorough study draws from foundational and recent works. Situating a study within a broader academic context helps the researcher frame their work in response to ongoing issues or emerging trends, thereby demonstrating its relevance and uniqueness [68]. Engaging with complex and nuanced debates further enhances depth. Academic writing should avoid simplistic cause and effect thinking and binary oppositions. Instead, it should embrace complexity by exploring multiple perspectives, including theoretical angles and anticipating counterarguments. A comprehensive study connects specific findings to broader theoretical or social implications by linking micro-level details to macro-level interpretations [69]. This multi-layered approach increases the clarity and complexity of the study's narrative. Moreover, depth involves synthesizing rather than merely describing or cataloging sources. Academic writing should illustrate how different theories, discoveries, or concepts interact, conflict, or converge [70]. This requires an evaluative stance that assesses sources' strengths, weaknesses, and interrelationships. Synthesizing data demonstrates the author's analytical abilities and shows how their work expands upon and contributes to existing knowledge. Notably, depth is associated with critical thinking. Scholars must continually question their assumptions, methodological choices, and interpretations [71]. This reflexive approach enhances the study's validity and promotes intellectual honesty. It ensures that academic writing progresses through ongoing inquiry, adaptation, and engagement with complex realities rather than remaining static or formulaic. Deepening transforms academic writing from a passive retelling of others' work into an active, reflective, and creative contribution to scholarly discourse [66]. This enhances the impact, relevance, and intellectual value of academic research. Research on digital twins reveals substantial insights in construction studies. These tools integrate predictive analytics, BIM models, and real-time sensor data to simulate building performance over time. Rather than merely outlining the technology, researchers examine its impact on lifecycle management, highlight interoperability challenges, and explore its role in sustainability. This comprehensive analysis demonstrates significant engagement by connecting technological implementation with strategic decision-making and broader industrial transformation. Figure 7 illustrates a circular structure that emphasizes the iterative process of achieving depth in academic writing. From conducting a literature review and identifying gaps to synthesizing ideas and engaging in critical reflection, the paradigm underscores continuous involvement. This cyclical framework highlights that depth is not a linear progression but evolves through dynamic interactions among theory, evidence, and reflection, ultimately leading to original and significant academic contributions.

3.2.4 Supporting Rigor Through Language and Citation

Fundamental instruments for articulating and maintaining academic rigor are language and citation. Effective academic communication depends on clarity, accuracy, and appropriate expression [72]. These linguistic elements strengthen the logic of arguments, the coherence of academic papers, and their readability. Rigor in language use goes beyond grammatical accuracy; it includes carefully choosing words, consistently defining essential concepts, and constructing logically sequenced and thematically cohesive paragraphs [73]. Avoiding unnecessary language, mainly when not required for disciplinary discussions, enhances accessibility without sacrificing intellectual rigor. Likewise, a neutral tone and formal register contribute to

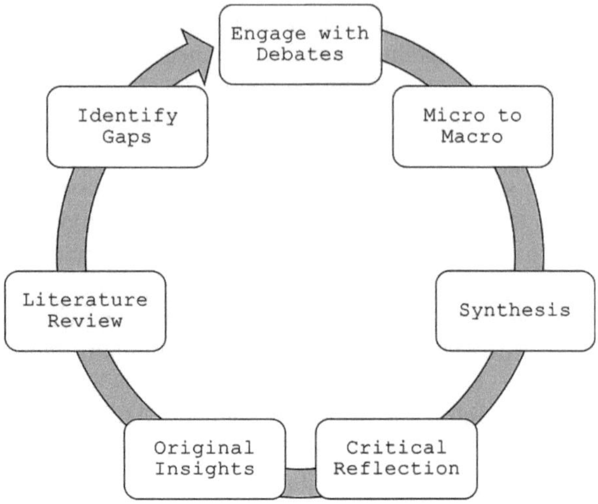

Fig. 7 Circular model for enhancing depth in academic writing (*Source* Authors)

the trustworthiness of the writing and align with academic standards. Equally important is the ethical and proper use of citations. Referencing demonstrates academic integrity, serving not simply as a formality [74]. It fosters transparency, acknowledges creative contributions, and establishes ownership of ideas. Through appropriate citation, scholars enable others to trace the sources of ideas, assess the basis of assertions, and situate new work within ongoing academic discourse. Citations enhance the credibility of the presented work and help establish responsibility through ethical markers and navigational tools [75]. However, supporting rigor entails much more than merely adding citations. Scholars must actively engage with the literature, critically evaluating, synthesizing, and responding to contemporary ideas. Citing becomes a means to advance knowledge in this dialogic interaction with academic sources rather than merely a legal obligation [76]. The analytical depth of a study improves through engagement with differing perspectives, articulating original interpretations, and highlighting theoretical and methodological conflicts, which also reflects the researcher's academic growth. Enhancing the foundational aspects of academic rigor, validity, reliability, and depth requires precise wording and careful citation. Each element supports logical conclusions and strengthens the argument; methodological consistency boosts trustworthiness while critical engagement sharpens analytical insight [61]. These practices foster a stringent academic culture characterized by consistency, transparency, and reflective inquiry when woven into the research and writing process. By doing so, academics uphold disciplinary standards and contribute to creating ethically grounded, significant, and credible information [77]. Figure 8 illustrates a dynamic, cyclical model where precise language, organized writing, and ethical citation collaborate to foster academic rigor. Each stage contributes to developing legitimate, reliable, and analytically rich research. The directed arrows emphasize this process's continuous and reinforcing nature, highlighting that diligent

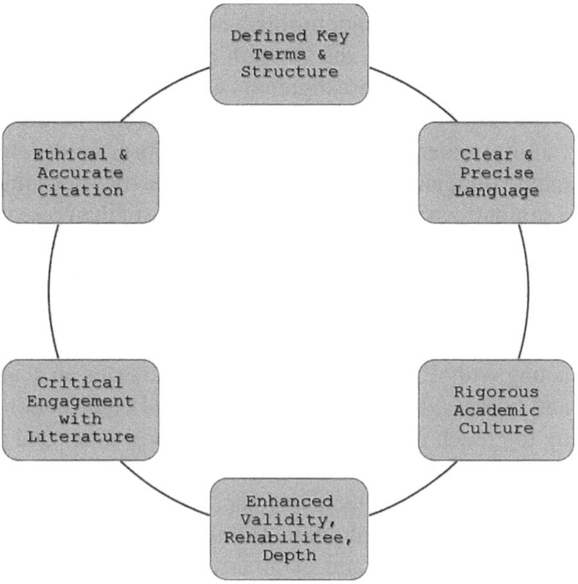

Fig. 8 Circular flow of academic rigor through language and citation (*Source* Authors)

research arises from an ongoing cycle of clarity, engagement, and accountability in academic communication.

3.3 *Identifying and Avoiding Common Pitfalls*

Although academic rigor is the foundation of high-quality research, academics of all experience levels often face difficulties that could undermine their work's strength, reliability, and integrity [78]. The first stage in avoiding these frequent traps and maintaining the criteria anticipated in academic communication is acknowledging them. Logical fallacies are among the most common problems. Often stemming from a lack of critical examination, these logical fallacies undermine the validity of arguments. Overgeneralizations, applying small or unrepresentative sample results to an entire population, lead to false conclusions [79]. Presenting false dichotomies ignores diverse viewpoints and reduces complicated events to two opposed results. Appeals to authority can erode credibility if they are not backed by critical interaction with the referenced source. Causal assertions without supporting data are the most troubling. Academic rigor requires an evidence-based, theoretically consistent, methodologically sound, unambiguous, logical reasoning framework. A second important trap is relying on unproven assertions or anecdotal data. Although personal experiences could improve contextual awareness, they should not replace empirical facts [80]. Academic writing has to be founded on critically examined, methodically gathered data. Arguments based on untested assumptions or isolated examples fall short of academic validity and undermine the reader's trust in the conclusions offered. Another major problem is the lack of methodological openness. Readers find it difficult to assess the strength of the results when researchers fail to sufficiently explain how their investigations were carried out, including sample techniques, data collecting tools, and analytical approaches. Scientific research depends on replicability; other researchers cannot confirm or develop on the work without exact methodological information [81]. Academic rigor thus calls for researchers to offer a straightforward and thorough narrative of their research method. Confirmation bias, the inclination to selectively include data supporting one's premise while omitting opposing data, is another frequent issue. This compromises objectivity and skews the story of the study. Academic writing's rigor includes thinking about other points of view, responding to opposing views, and using various data sources. Critically interacting with different points of view increases an academic paper's analytical depth and fairness. Ultimately, the intellectual contribution of a study can be constrained by too much dependence on secondary sources and a dearth of creative interpretation [82]. Academic writing should include critical synthesis, contextual awareness, and theoretical application rather than just a recounting of others' views. Creating an anchored original voice, formed by careful investigation and expression, shows academic maturity and maintains the criteria of academic greatness [83]. Scholars may improve the legitimacy, depth, and general influence of their academic output by knowing and correcting these typical traps. Including logical fallacies, anecdotal evidence, and methodological opacity, Table 5 offers a concise summary of typical traps eroding academic rigor. Every pitfall has a summary, common problems, and practical avoidance techniques. Researchers wishing to improve the credibility, objectivity, and depth of their academic writing using critical awareness and methodical refining will find this helpful table.

4 Enhancing Academic Writing Through Iteration and Ethical Practice

Reflecting, criticism, and ethical obligation shape academic writing's dynamic, repetitive process rather than a one-time act of expression [84]. Producing a high-quality academic text calls for several rounds of drafting, editing, and polishing to satisfy the criteria of clarity, coherence, and intellectual integrity. The power of academic writing is in both what is said and how properly and carefully it is built [85]. Producing rigorous and trustworthy academic work calls for ethical considerations, revision techniques, and iterative drafting, all discussed in this part.

Table 5 Common pitfalls in academic writing: descriptions, challenges, and solutions

Common pitfall	Description	Challenges	How to avoid
Logical fallacies	Use of flawed reasoning such as overgeneralizations, false dichotomies, unsupported causal claims	Difficulty in identifying invalid arguments; risk of uncritical acceptance of flawed logic	Ensure arguments are logically sound, evidence-based, and theoretically grounded
Anecdotal evidence	Relying on personal experiences or unverified claims instead of systematic evidence	Temptation to generalize from personal experience; lack of empirical support	Use empirical data and validate claims through systematic research
Methodological opacity	Failure to clearly explain research design, data collection, or analysis methods	Readers cannot assess validity or replicate the study; perceived lack of credibility	Provide a transparent account of methodology, including tools and procedures
Confirmation bias	Presenting only evidence that supports one's views while ignoring counterarguments	Skewed interpretation of data; undermines objectivity and academic fairness	Engage with multiple perspectives and include contradictory evidence where relevant
Overreliance on secondary sources	Summarizing existing literature without providing critical analysis or original insights	Lack of originality; perceived as derivative or superficial research	Critically analyze sources and develop an independent scholarly voice

4.1 Peer Review and Iterative Drafting's Function

Engaging in several revisions is one basic approach to improving scholarly writing. Often, first drafts are exploratory places where ideas are created and preliminary arguments are tested [86]. However, to guarantee that the result satisfies the standards of academic quality, these early versions usually need considerable polishing. Iterative drafting helps authors to find contradictions, clarify hazy concepts, and arrange material more rationally. This approach depends much on peer review. Peer review offers an outside viewpoint that helps to find flaws or blind spots in the work, whether using official academic channels or casual comments from coworkers or supervisors [87]. Reviewers may comment on the relevancy of literature, the strength of evidence, and the organization of arguments. This approach guarantees consistency with disciplinary standards and provides a system for quality control. It also promotes intellectual humility by motivating authors to be receptive to criticism and changes. Feedback-driven iteration changes first manuscripts into cohesive, convincing, well-substantiated academic works [88].

4.2 Proofreading Tools and Revision Techniques

Revising an academic paper calls for a careful, thorough method. It much surpasses superficial fixes of spelling or punctuation [89]. Effective revision is judging the logical flow of arguments, considering if the assertions are well supported, and ensuring the text keeps a consistent and formal academic tone. Structural consistency is essential: every paragraph should logically relate to the prior and following portions and support a key argument. Equally crucial are the strengthening of transitions and the removal of redundancies. Redundant sentences can cloud meaning and weaken the force of arguments. Clear transitions between paragraphs and parts improve general readability and assist the reader in navigating the argument. Digital tools significantly help throughout modern academic practice's editing and proofreading phases [90]. Writing tools like Microsoft Editor or Grammarly can find grammatical errors, recommend stylistic changes, and increase sentence clarity [91]. Tools for managing citations like Zotero, EndNote, and Mendeley help to keep referencing consistent and avoid citation mistakes [92]. These technologies simplify the writing process and guarantee the efficient handling of technical elements so academics may concentrate on material creation. Though helpful, digital technologies should enhance, not supplant, critical self-revision. Writers must actively assess strength, clarity, and academic worth [93]. Final proofreading should consist of reading the material aloud, verifying formatting consistency, and ensuring accurate citations and cross-references to the bibliography.

4.3 Ethical Aspects of Academic Writing

The activity of academic writing is intertwined with ethical issues. Ethical writing shows integrity, openness, and appreciation of other people's intellectual gifts [94]. Maintaining ethical standards is not simply an intellectual question but also a question of personal and institutional legitimacy. Correct source credit is one of the fundamental ethical values in academic writing. Correct citation lets readers check information, follow up on cited sources, and give credit to the original writers [95]. Whether deliberate or unintentional, failing to cite correctly constitutes plagiarism, a significant violation of academic integrity. Plagiarism erodes confidence in academic communication and may result in disciplinary penalties, harm to reputation, or legal repercussions. Data fabrication or falsification is yet another kind of unethical conduct. Presenting fabricated or altered data distorts the scientific record, misleads other academics, undermines public confidence, and compromises the research. Academic integrity calls for all facts to be gathered, examined, and honestly presented regardless of whether the findings support or refute first ideas [96]. Other ethical issues include ghostwriting and duplicate publication. Ghostwriting is when someone helps a paper that is far under stated authorship. Simultaneously, duplicate publication is publishing in many media or submitting the same material

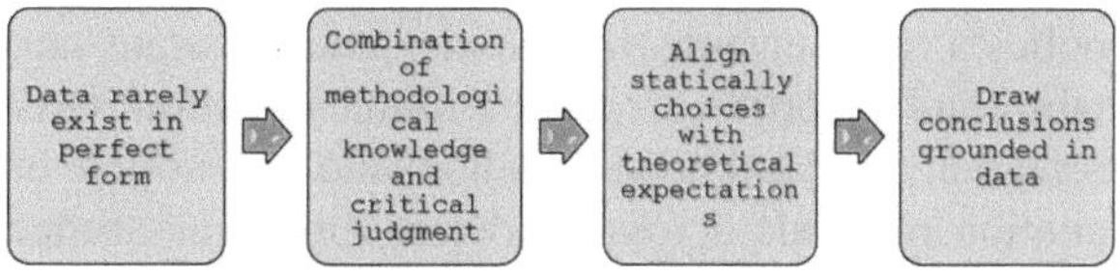

Fig. 9 Workflow of advanced data analysis in quantitative research (*Source* Authors)

to several publications without appropriate disclosure. Both approaches distort the writers' contribution and undermine the integrity of the academic record. Reporting results honestly is also quite important. Writers must reflect their studies' breadth, constraints, and consequences [97]. Exaggerating findings, downplaying negative consequences, or hiding conflicts of interest might lead readers astray and give them a misleading sense of the importance of the work. Ethical academic writing benefits the immediate scholarly community and the larger society, depending on reliable research by fostering openness, justice, and responsibility. Producing high-quality, trustworthy, and influential research requires improving academic writing using iteration and ethical practice. Drafting starts the process; several rounds of rewriting and peer comments follow to progressively shape a clear and logical story [103]. Along with structural improvement, ethical values guarantee that academic writings remain intellectually honest and significantly add to continuous scholarly debate. Combining ethical responsibility with iterative techniques helps researchers preserve academic achievement ideals and support a society of honesty, openness, and critical investigation. The stepwise workflow of advanced data analysis in quantitative research is illustrated in Fig. 9. It then progresses to data preparation and model selection, beginning with identifying data flaws. Model fitting and evaluation continue the process toward outcome interpretation and reporting. Each stage underscores the necessity for critical judgment, methodological rigor, and alignment with theoretical frameworks to ensure accurate, reliable, and meaningful research results.

5 Conclusion

Academic writing is a challenging, reflective, and ethical process fundamental to scholarly research, rather than merely an exercise in language or structure. This chapter has demonstrated that the two interdependent pillars of academic style and academic rigor determine the quality and legitimacy of academic communication. When applied correctly, these components enhance the clarity and persuasiveness of research and uphold the essential principles of the academic community: transparency, responsibility, honesty, and intellectual coherence. Academic style is defined by formality, objectivity, and precision. These characteristics influence academics' rhetorical and linguistic choices in constructing arguments, presenting evidence, and engaging with existing literature. The credibility and accessibility of academic texts rely on the consistent use of discipline-specific terminology, logical organization, and evidence-based reasoning. Adhering to established practices, such as citation policies, also ensures traceability and accurate attribution by minimizing uncertainty and distortion. Cross-disciplinary citation is essential to acknowledge previous research, facilitating verification and promoting ethical responsibility. Different citation methods across academic disciplines reflect various epistemic traditions while serving the common goal of enhancing academic discourse and methodological openness. Academic rigor embodies the intellectual and methodological strength that differentiates academic work. Thoughtful formulation of research questions, transparent application of techniques, and rigorous analysis of findings define rigor. The chapter emphasized that rigor represents a commitment to depth, consistency, and validity, not simply complexity for its own sake. Ensuring research processes are reproducible, logically sound, ethically conducted, and conceptually grounded helps demonstrate this commitment. A conceptual framework supporting rigorous scholarship comprises the three elements of validity, dependability, and depth. Depth highlights the analytical complexity and theoretical importance of academic work, validity ensures studies accurately represent the phenomena being examined, and dependability confirms the consistency of findings across time and context. Throughout the chapter, various strategies for enhancing these elements, such as triangulation, audit trails, pre-testing tools, and methodological alignment, have been discussed. Each approach bolsters academic confidence and facilitates the replication and expansion of research. Furthermore, academic writing incorporating these elements reflects a writer's respect for disciplinary standards and signifies deep engagement with their subject. Critically, academic writing and rigor are evolving practices rather than fixed accomplishments. Scholars must be adaptable, ethically grounded, and critically vigilant as research environments become increasingly multidisciplinary and technologically advanced. Writers are challenged to uphold traditional academic principles while navigating new forms of communication and evaluation, particularly when integrating digital tools, collaborative platforms, and open-access models. Ethical writing, which includes proper crediting, avoiding plagiarism, and managing open data, is increasingly vital in this context. Beyond technical proficiency, effective academic writing demands intellectual reflexivity, the ability to scrutinize one's assumptions, openness to diverse perspectives, and the willingness to adjust arguments in light of new evidence. This self-awareness ensures that academic writing continues to achieve its fundamental objectives: the dissemination of knowledge, the advancement of research, and the maintenance of academic integrity by enriching scholarly contributions. The chapter illustrates that academic writing style and rigor are intertwined reflections of scholarly duty rather than distinct skills. They function as instruments of quality assurance, intellectual engagement, and ethical stewardship, reflecting the standards and aspirations of the academic community. Those who internalize these values and write with clarity, reason precision, cite with integrity, and conduct thorough research are optimally positioned to make significant contributions to their fields and the broader advancement of knowledge. Consequently, academic writing embodies the academic endeavor itself: committed to truth, driven by method, and open to critical examination rather than merely serving as a vehicle for communication.

References

1. Luzón MJ, Pérez-Llantada C (2022) Digital genres in academic knowledge production and communication: perspectives and practices
2. Woods P, Sikes P (2022) Successful writing for qualitative researchers. Routledge
3. Chauhan P (2022) Fundamentals of academic writing: a literature review. J NELTA 27(1–2):161–180
4. Marsh JD, Campion J (2018) Academic integrity and referencing: whose responsibility is it? J Acad Lang Learn 12(1):A213–A226
5. Alabi TA, Olaleye GA (2017) Documentation in research: in-text citations and end-referencing using APA and MLA formatting styles. In: Perspectives on conducting and reporting research in the humanities
6. Wise B, Emerson L, Van Luyn A, Dyson B, Bjork C, Thomas SE (2024) A scholarly dialogue: writing scholarship, authorship, academic integrity and the challenges of AI. High Educ Res Dev 43(3):578–590
7. Hyland K (2004) Disciplinary discourses, Michigan classics ed.: social interactions in academic writing. University of Michigan Press
8. Guelmami N, Ben Ezzeddine L, Hatem G, Trabelsi O, Ben Saad H, Glenn JM, Knechtle B (2024) The ethical compass: establishing ethical guidelines for research practices in sports medicine and exercise science. Int J Sport Stud Health 7(2):31–46
9. Xiong Y, Luo Z (2025) How can individuals develop critical thinking skills to evaluate online information? In: Digital citizenship and building a responsible online presence. IGI Global Scientific Publishing, pp 165–186
10. Ramzan M, Mushtaq A, Ashraf Z (2023) Evacuation of difficulties and challenges for academic writing in ESL learning. University of Chitral J Linguist Liter 7(I):42–49
11. Heston TF (2024) Foundations of scholarly writing. BP International
12. Miao J, Thongprayoon C, Suppadungsuk S, Garcia Valencia OA, Qureshi F, Cheungpasitporn W (2023) Ethical dilemmas in using AI for academic writing and an example framework for peer review in nephrology academia: a narrative review. Clin Pract 14(1):89–105
13. Al Zumor AQ (2021) Stance in advanced academic writing by Saudi EFL postgraduates: a corpus-based study of critique writing. J Lang Lang Teach 9(4):371
14. Digmayer C (2024) Automated economic welfare for everyone? Examining barriers to adopting robo-advisors from the perspective of explainable artificial intelligence. J Interdiscip Econ 36(2):224–245
15. Cooksey RW (2024) Making convincing choices: sharing/communicating research outcomes. In: Unity from diversity: pluralist systemic thinking for social and behavioural research. Springer Nature Singapore, Singapore, pp 421–465
16. Apairach S (2025) Learning to argue and arguing to learn: unpacking the nature of criticality development in English for academic purposes for political science in a Thai university context. Doctoral dissertation, University of Southampton
17. Davenport MJ (2024) The state of law: a legal pandemic. Open J Modern Linguist 14(5):860–906
18. Liu Z, Zhao YG, Ma L, Lin S (2024) Review on high-strength recycled aggregate concrete: mix design, properties, models and structural behaviour. In: Structures, vol 64. Elsevier, p 106598
19. Teng MF, Yue M (2023) Metacognitive writing strategies, critical thinking skills, and academic writing performance: a structural equation modeling approach. Metacogn Learn 18(1):237–260
20. Kumar V, Verma A, Aggarwal SP (2023) Reviewing academic integrity: assessing the influence of corrective measures on adverse attitudes and plagiaristic behavior. J Acad Ethics 21(3):497–518
21. Omrany H, Al-Obaidi KM (2024) Application of digital twin technology for urban heat Island mitigation: review and conceptual framework. Smart Sustain Built Environ
22. Zamiri M, Esmaeili A (2024) Methods and technologies for supporting knowledge sharing within learning communities: a systematic literature review. Admin Sci 14(1):17

23. Piran MJ, Tran NH (2024) Enhancing research methodology and academic publishing: a structured framework for quality and integrity. arXiv preprint arXiv:2412.05683
24. Kowalski CP, Nevedal AL, Finley EP, Young JP, Lewinski AA, Midboe AM, Hamilton AB (2024) Planning for and assessing rigor in rapid qualitative analysis (PARRQA): a consensus-based framework for designing, conducting, and reporting. Implement Sci 19(1):71
25. Wahle JP, Ruas T, Abdalla M, Gipp B, Mohammad SM (2024) Citation amnesia: on the recency bias of NLP and other academic fields. arXiv preprint arXiv: 2402.12046
26. Abdallah MM (2024) Research project in tesol/tefl
27. Flynn BF, Bince RL, Layden S (2024) Critical rhetoric meets community-based participatory research. Commun Stud 75(4):445–463
28. Cole S, Breuer K, Palmer S, Blakeslee B (2022) How history is made: a student's guide to reading, writing, and thinking in the discipline
29. Giampalmi J (2021) Apa style & citations for dummies. John Wiley & Sons
30. Zhang A (2024) Impact of LLMs on academic literature synthesis: influence and oversight in business economics and other disciplines
31. Bell K, Kingori P, Mills D (2024) Scholarly publishing, boundary processes, and the problem of fake peer reviews. Sci Technol Human Values 49(1):78–104
32. Bean JC, Melzer D (2021) Engaging ideas: the professor's guide to integrating writing, critical thinking, and active learning in the classroom. John Wiley & Sons
33. Mulenga R, Shilongo H (2024) Academic integrity in higher education: understanding and addressing plagiarism. Acta Pedagogia Asiana 3(1):30–43
34. Çelik Ö, Razı S (2023) Facilitators and barriers to creating a culture of academic integrity at secondary schools: an exploratory case study. Int J Educ Integr 19(1):4
35. McGuire A (2023) Leveraging ChatGPT for rethinking plagiarism, digital literacy, and the ethics of Co-authorship in higher education: a position paper and comparative critical reflection of composing processes. Irish J Technol Enhanced Learn 7(2):21–31
36. Triggle CR, MacDonald R, Triggle DJ, Grierson D (2022) Requiem for impact factors and high publication charges. Acc Res 29(3):133–164
37. Alnaimat F, Al-Halaseh S, AlSamhori ARF (2024) Evolution of research reporting standards: adapting to the influence of artificial intelligence, statistics software, and writing tools. J Korean Med Sci 39(32)
38. Nguyen-Hoang TA, Hoang NC, Hua PT, Thi MTN, Ta TT, Nguyen T, Nguyen HT (2024) Advancing scholarship management: a blockchain-enhanced platform with privacy-secure identities and AI-driven recommendations. IEEE Access
39. Gulumbe BH, Audu SM, Hashim AM (2024) Balancing AI and academic integrity: what are the positions of scholarly publishers and universities? AI Soc 1–10
40. Tirthali D, Murai Y (2024) Facilitating open online discussions: speech acts inspire and hinder deep conversations. Open Learn J Open Dist e-Learn 39(3):280–296
41. Sain ZH, Serban R, Abdullah NB, Thelma CC (2025) Benefits and drawbacks of leveraging ChatGPT to enhance writing skills in secondary education. At-Tadzkir Islam Educ J 4(1):40–52
42. Guillermo JCL, Garcia YA, Hechavarria EL, Lozano WG, Maguiña WME, Yon GG, Cáceres YS Integrating quantitative and qualitative paradigms to improve transdisciplinary research and educational innovation: interpretation of the scientific method
43. Moreno-Armella L, Brady C (2025) Part 3: mathematical thinking and rigor first steps. Math Enthus 22(3):229–256
44. Elibox W, Khan K (2025) Enhancing examination integrity through constructivist and reflective practices: a case study of the science faculty at a commonwealth university. Qual Assur Educ
45. Almusaed A, Almssad A, Yitmen I (2024) Sustainable built environment and its implications on real estate development: a comprehensive analysis. In: Integrative approaches in urban sustainability-architectural design, technological innovations and social dynamics in global contexts. IntechOpen
46. Haynes E, Marawili M, Mitchell A, Walker R, Katzenellenbogen J, Bessarab D (2022) "Weaving a mat that we can all sit on": qualitative research approaches for productive dialogue in the intercultural space. Int J Environ Res Public Health 19(6):3654

47. Kieser A, Nicolai A, Seidl D (2015) The practical relevance of management research: turning the debate on relevance into a rigorous scientific research program. Acad Manag Ann 9(1):143–233
48. Kastrup B (2014) Why materialism is baloney: how true skeptics know there is no death and fathom answers to life, the universe, and everything. John Hunt Publishing
49. National Academies of Sciences, Medicine, Global Affairs, & Committee on Responsible Science (2017) Fostering integrity in research. National Academies Press
50. Kretser A, Murphy D, Bertuzzi S, Abraham T, Allison DB, Boor KJ, Yada R (2019) Scientific integrity principles and best practices: recommendations from a scientific integrity consortium. Sci Eng Ethics 25:327–355
51. Honig B, Lampel J, Baum JA, Glynn MA, Jing R, Lounsbury M, Van Witteloostuijn A (2018) Reflections on scientific misconduct in management: unfortunate incidents or a normative crisis? Acad Manag Perspect 32(4):412–442
52. Cypress BS (2017) Rigor or reliability and validity in qualitative research: perspectives, strategies, reconceptualization, and recommendations. Dimens Crit Care Nurs 36(4):253–263
53. Yi S, Sintawati W, Zhang Y (2025) Facilitating depth of explanation: utilising reflective feedback in collaborative learning environments. J Comput Assist Learn 41(2):e70010
54. Bruton A, Conway JH, Holgate ST (2000) Reliability: what is it, and how is it measured? Physiotherapy 86(2):94–99
55. Lim WM (2024). What is qualitative research? An overview and guidelines. Australasian Mark J 14413582241264619
56. Amadi A (2023) Integration in a mixed-method case study of construction phenomena: from data to theory. Eng Constr Archit Manag 30(1):210–237
57. Maxwell JA (2017) The validity and reliability of research: a realist perspective. BERA/SAGE Handbook Educ Res 1:116–140
58. Levitt HM, Motulsky SL, Wertz FJ, Morrow SL, Ponterotto JG (2017) Recommendations for designing and reviewing qualitative research in psychology: promoting methodological integrity. Qualit Psychol 4(1):2
59. Lambert LS, Newman DA (2023) Construct development and validation in three practical steps: recommendations for reviewers, editors, and authors. Organ Res Methods 26(4):574–607
60. Ravitch SM, Riggan M (2016) Reason & rigor: how conceptual frameworks guide research. Sage publications
61. Adeoye MA (2024) Mastering the basics: a guide to research methodology for effective writing and publication. Chalim J Teach Learn 4(1):30–41
62. Raunig DL, McShane LM, Pennello G, Gatsonis C, Carson PL, Voyvodic JT, QIBA Technical Performance Working Group2015) Quantitative imaging biomarkers: a review of statistical methods for technical performance assessment. Stat Methods Med Res 24(1):27–67
63. Chilcott J, Tappenden P, Rawdin A, Johnson M, Kaltenthaler E, Paisley S, Shippam A (2010) Avoiding and identifying errors in health technology assessment models. Health Technol Assess 14(25):1–135
64. Cole R (2024) Inter-rater reliability methods in qualitative case study research. Sociol Methods Res 53(4):1944–1975
65. Petroşanu DM, Căruţaşu G, Căruţaşu NL, Pîrjan A (2019) A review of the recent developments in integrating machine learning models with sensor devices in the smart buildings sector with a view to attaining enhanced sensing, energy efficiency, and optimal building management. Energies 12(24):4745
66. Barasa D (2024) Demystifying the discourse: techniques to effective academic writing. J Res Acad Writ 1(1):13–21
67. Zhou H (2025) Exploring the dynamic teaching-learning relationship in interactive learning environments. Interactive Learn Environ 1–31
68. Paudel P (2024) Examining paradigmatic shifts: unveiling the philosophical foundations shaping social research methodologies. J Univ Ruhuna 12(1)
69. Gitonga N (2025) Bridging micro and macro level theories in the social sciences: strategies, challenges, and innovations in theory construction. J Afric Interdiscipl Stud 9(1):88–105

70. Coccia M (2024) Converging artificial intelligence and quantum technologies: accelerated growth effects in technological evolution. Technologies 12(5):66
71. Omodan BI (2024) Research paradigms and their methodological alignment in social sciences: a practical guide for researchers. Taylor & Francis
72. Uzun K (2024) Enhancing written communication skills for academic purposes. In: Teaching English for academic purposes: theory into practice. Springer Nature Switzerland, Cham, pp. 169–190
73. Chere MM (2024) Exploring grammatical cohesion in Grade 10 learners' academic texts at two selected schools in Maseru. Master's thesis, National University of Lesotho
74. Panjaitan AO, Novelyn S, Angreni F (2024) The impact of proper citation on academic integrity and scholarship. Asian J Adv Res Rep 18(10):192–201
75. Kouam AWF (2024) Navigating the publication imperative: a critical reflection on strategies for success as an academic scholar. Exchanges Interdiscipl Res J 12(1):27–40
76. Foderaro A, Gunnarsson Lorentzen D (2024) Traditional, dialogical and complex scholarly communication: towards a renewed trust in science. J Document 80(6):1313–1332
77. Kimera R, Kim YS, Choi H (2024) Advancing AI with integrity: ethical challenges and solutions in neural machine translation. arXiv preprint arXiv:2404.01070
78. Uludag K (2024) Rethinking the publish or do not graduate paradigm: balancing graduation requirements and scientific integrity. In: Engaging higher education teachers and students with transnational leadership. IGI Global, pp 165–177
79. Bai X, Uddenberg S, Labbree BP, Todorov A (2025) Learning too much from too little: false face stereotypes emerge from a few exemplars and persist via insufficient sampling. J Pers Soc Psychol 128(1):61
80. Tracy SJ (2024) Qualitative research methods: collecting evidence, crafting analysis, communicating impact. John Wiley & Sons
81. Brodeur A, Mikola D, Cook N (2024) Mass reproducibility and replicability: a new hope (No. 16912). IZA Discussion Papers
82. George AS, Baskar T, Srikaanth PB (2024) The erosion of cognitive skills in the technological age: how reliance on technology impacts critical thinking, problem-solving, and creativity. Partners Univ Innov Res Publ 2(3):147–163
83. Tusha A, Bulut S, Al-Hendawi M (2024) Promoting a healthy school environment via social-emotional learning in the high school setting: an overview. Adv Med Psychol Public Health 1(3):156–163
84. Compo·X (2024) Enacting writing differently through a collective liminal experience. Cult Organ 1–17
85. Lu Q, Yao Y, Xiao L, Yuan M, Wang J, Zhu X (2024) Can ChatGPT effectively complement teacher assessment of undergraduate students' academic writing? Assess Eval High Educ 49(5):616–633
86. Tseng YC, Lin YH (2024) Enhancing English as a Foreign Language (EFL) learners' writing with ChatGPT: a University-level course design. Electron J e-Learn 22(2):78–97
87. Dedema M, Rosenbaum H (2024) Socio-technical issues in the platform-mediated gig economy: a systematic literature review: an annual review of information science and technology (ARIST) paper. J Am Soc Inf Sci 75(3):344–374
88. Typografia D, Eldridge SA, Harbers F, Banjac S (2024) Media and communication, vol 12 unpacking innovation: media and the locus of change
89. Pasandeh N (2024) Issues in ESL academic writing, Doctoral dissertation, UNSW Sydney
90. Harwood N (ed) (2024) Proofreading and editing in student and research publication contexts: international perspectives. Taylor & Francis
91. Apriani E, Gusmuliana P, Cahyani YD (2025) An analysis Efl students' linguistic problems using grammarly as automated writing evaluation feedback, Doctoral dissertation, Institut Agama Islam Negeri Curup
92. Sabhapandit T (2024) Reference management tool: a boon for managing research references. Research 324

93. Prebel J, Li J (2024) Multifaceted equity: critiquing a first-year writing assessment through curricular, performance, and reliability lenses. J Writ Assess 17(2)
94. Eaton SE (2024) Decolonizing academic integrity: knowledge caretaking as ethical practice. Assess Eval High Educ 49(7):962–977
95. Pechenik JA (2024) A short guide to writing about biology. Waveland Press
96. Newton PM (2024) Design, run, and interpret survey-based research in the fields of academic integrity and misconduct. In: Second handbook of academic integrity. Springer Nature Switzerland, Cham, pp 1689–1706
97. Cui W, Driscoll DL (2025) From telling to crafting: an analysis of peer review feedback and revision processes of emerging scholars writing for publication. Early Child Educ J 1–18

Writing Articles and Theses: Bachelor's, Master's, and Ph.D

Abstract This chapter thoroughly examines the ideas, strategies, and procedures necessary for crafting rigorous and impactful academic articles at the undergraduate, graduate, and doctoral levels. Meeting the fundamental standards and expectations of academic writing provides a solid framework for students and academics aiming to excel in scholarly communication and information dissemination. The chapter methodically analyzes the essential stages of academic writing, from developing research questions to disseminating results, emphasizing the importance of accuracy, coherence, and originality. It outlines the distinct requirements and contributions of academic work across various levels. It shows that bachelor's theses focus on foundational knowledge and application, master's projects emphasize critical evaluation and interdisciplinary methods, and doctoral research prioritizes innovation and theoretical advancement. Concrete examples relevant to each academic level illustrate the development of skills and intellectual rigor necessary to tackle increasingly complex research tasks. The chapter also discusses practical strategies for conducting comprehensive literature reviews, managing citations, and maintaining ethical standards to uphold academic integrity during the writing process. It underscores the iterative nature of academic writing, where drafting, feedback, and editing are vital for refining ideas and producing high-quality work. This chapter equips readers with advanced strategies, creating a valuable foundation for producing scholarly outputs that meet academic standards and significantly contribute to knowledge in their fields.

1 Introduction

Articles and theses at the undergraduate, graduate, and doctoral levels are essential to academic work throughout all stages of higher education. These forms of academic writing are crucial to undergraduate education, postgraduate study, and advanced doctoral research. Inouye and McAlpine [1] note that research on academic writing, particularly related to second- and third-cycle education, has garnered significant interest in higher education over the past twenty years. The focus has been on identity formation through doctoral writing, a key outcome of doctoral education. The

A. Almusaed et al., *Practice of Research Methodology in Civil Engineering and Architecture*, https://doi.org/10.1007/978-3-031-97393-2_24

PhD program aims to develop students into independent scholars capable of making meaningful contributions to their academic disciplines [1]. This includes the creation of academic articles and theses. This chapter presents a systematic framework for producing rigorous and compelling scholarly work at the Bachelor's, Master's, and Ph.D. levels. It achieves this by examining the procedures, norms, and expectations pertinent to academic writing. By critically analyzing these methods, students can acquire the skills necessary to significantly contribute to the academic fields in which they are involved. Writing for academic purposes is more than merely expressing ideas; it entails conducting a methodical investigation, analyzing information, and communicating concepts [2]. At the bachelor's level, the thesis represents an initial foray into the structured research process. This process involves formulating specific research questions, systematically gathering data, thoroughly analyzing the data, and coherently presenting the findings [3]. Developing critical thinking skills, fostering the ability to integrate knowledge from various sources, and adhering to established academic standards are the primary focuses of this foundational stage [4]. These standards encompass correct citation practices, logical reasoning, and proper formatting. Academic writing at the Master's level demands greater intellectual rigor and deeper engagement with the subject matter than at the undergraduate level [5]. Mansour [6] emphasizes that synchronous and asynchronous online discussions and e-facilitation significantly enhance students' and facilitators' understanding of complex topics [6]. The master's thesis, often a fully independent research project, underscores the importance of creativity, methodological clarity, and the ability to explore multi-faceted topics [7]. To complete this academic writing level, students must contextualize their work within applicable theoretical and practical frameworks, emphasizing the articulation of findings within a broader scholarly context. Writing at the doctoral level is regarded as the pinnacle of academic scholarship. It is characterized by the creation of original research that expands the boundaries of knowledge within a specific discipline [8]. The doctoral dissertation or a collection of peer-reviewed publications demonstrates a high level of creativity, theoretical sophistication, and methodological rigor. Writing at this level necessitates a thorough understanding of advanced research techniques, ethical considerations, and appropriate terminology [9]. Furthermore, Ph.D. candidates are expected to actively engage with the academic community by publishing their research in reputable journals, presenting their findings at conferences, and collaborating with professionals across various disciplines [10]. Despite the differing writing requirements across these three levels, several fundamental concepts govern the production of academic articles and theses. The most critical requirements are precision, coherence, and clarity. Whether composing a bachelor's thesis or a doctoral dissertation, the primary goal is to convey complex ideas in a clear and persuasive manner to an academic audience. To accomplish this, information must be meticulously organized, arguments logically constructed, and language precise and straightforward. Furthermore, the ability to build compelling arguments and substantiate claims with solid evidence is crucial to the success of academic writing. Similarly, adherence to ethical standards is equally important as they are essential to preserving the integrity and legitimacy of academic work [11]. These standards include meticulous citation, avoiding plagiarism, and openly and honestly disclosing research methodologies and results. At the PhD level, where the potential implications of research findings can be far-reaching, ethical considerations become even more salient than they would be otherwise. Another important topic discussed in this chapter is the iterative nature of academic writing. To produce high-quality academic work, multiple stages of writing, editing, and refining must occur [12]. This process is greatly enhanced by constructive feedback from supervisors, peers, and reviewers. Feedback helps authors identify areas for improvement, address gaps, and enhance the overall quality of their work. For academic writers, cultivating resilience and a constructive approach to criticism is a vital skill that should be developed throughout their careers. Additionally, the chapter delves into the significant role of literature reviews as a foundational element of academic writing. A literature review frames research identifies existing knowledge gaps, and positions the writer's work within the broader academic discourse [13]. A thoroughly executed literature review is indispensable. This section offers practical guidance for systematic research, critically evaluating sources, and synthesizing material to produce a comprehensive and cohesive literature review. In conclusion, the chapter presents practical strategies that can be applied to navigate the academic writing process efficiently. Key factors to consider include time management, goal setting, and maintaining motivation throughout the research and writing journey. Additionally, students will learn to leverage advanced tools and technologies, such as reference management software, plagiarism detection tools, and collaborative writing platforms, to enhance the quality of their work and streamline their workflow. As students progress through the various stages of higher education, producing academic articles and theses becomes increasingly intricate and sophisticated [14]. Students can handle the obstacles of academic writing with confidence, rigor, and dedication to quality if they have a solid knowledge of the principles and practices described in this chapter. Consequently, they can significantly contribute to the disciplines in which they are academically and professionally engaged. To be successful in academic writing, one must have a comprehensive grasp of research methodology, critical analysis, and excellent communication skills. This process is both structured and is constantly growing. Undergraduate students must exhibit increasing complexity, creativity, and engagement with academic discourse in their writing as they proceed from the undergraduate level to the PhD level. Figure 1 illustrates the progression of academic writing from the undergraduate level up to the doctoral level in a hierarchical fashion. This figure illustrates the evolutionary nature of academic writing. Each level of academic writing builds upon the one that came before it, necessitating higher-order thinking abilities, methodological accuracy, and theoretical depth. Writing at the master's level requires more critical engagement and knowledge synthesis than writing at the undergraduate level. Undergraduate theses serve as an introduction to structured research. When writing at the PhD level, it is necessary to demonstrate originality, sophisticated analytical abilities, and a significant contribution to the academic community. This method emphasizes fundamental ideas, ethical considerations, iterative refining, literature studies, and strategies for strategic writing. Students can improve the clarity, coherence, and impact of their

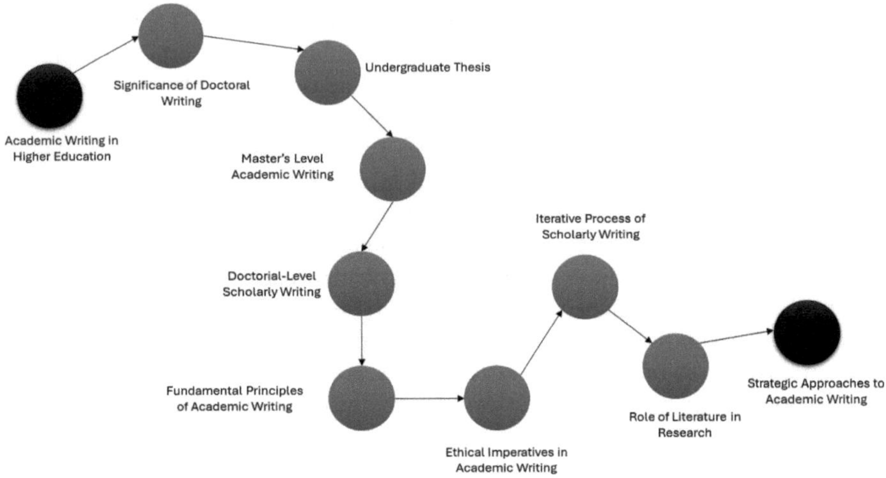

Fig. 1 Stages of academic writing (*Source* Authors)

academic writing while adhering to strict scholarly norms when they have a solid understanding of these components.

2 Comprehending Academic Writing at Various Levels

The progression of scholarly knowledge is reflected in the evolution of academic writing, which displays various traits and requirements at the bachelor's, master's, and doctoral levels [15]. At the undergraduate level, the primary objective is to demonstrate an understanding of core information, critical thinking, and the ability to organize ideas within accepted academic standards frameworks. Rather than producing novel contributions, these publications frequently synthesize previously published literature [16]. At the master's level of academic writing, the approach shifts to more analytical and evaluative [17]. This shift requires students to engage thoroughly with intricate ideas and methodologies. With greater emphasis on critical assessment and scholarly reasoning, it focuses on using advanced concepts to study specific research issues to conduct research [18]. Writing at the doctoral level is considered the pinnacle of academic rigor, as it emphasizes originality and significantly contributes to the field. Extensive literature reviews, complex methodological designs, and the articulation of innovative ideas are all required components of doctoral theses [19]. When ensuring effective communication at any level, adhering to academic traditions and maintaining scholarly accuracy is necessary.

2.1 The Purpose of Academic Writing

Academic writing is crucial for creating, expressing, and disseminating information in both educational and professional fields. [20]. The main objective is to promote intellectual exploration by systematically analyzing, evaluating, and synthesizing current research [21]. By following established disciplinary rules, academic writing guarantees clarity, coherence, and transparency, allowing researchers to express their findings and arguments effectively [22]. Additionally, it promotes critical thinking, rigorous methodological application, and the cultivation of innovative insights. Academic writing, in addition to individual study, advances the communal development of knowledge, promoting scholarly discourse and collaboration while maintaining the integrity and ethical norms of academia [23].

2.1.1 Differences Between Bachelor's, Master's, and Ph.D. Theses/ Articles

There is a significant difference in academic theses and articles and the demands placed on them at the bachelor's, master's, and doctoral degrees. These disparities are especially noticeable in architectural engineering, where they manifest themselves in the breadth of study, the depth of analysis, the creativity, and the contribution to the field.

- Articles and Theses for the Bachelor's Degree

A bachelor's thesis aims to serve as an introduction to academic research and demonstrate that the student can apply theoretical knowledge to real situations. In engineering, the focus is often on finding solutions to issues that are well-defined and clear by using established principles. For example, a bachelor's thesis in architectural engineering may entail designing the structural framework for a residential structure with a mid-rise orientation. This would require doing fundamental load calculations, choosing materials, and ensuring the construction complies with established local building rules. Tools such as basic AutoCAD for drawing and simple finite element software for stress analysis are frequently included in the process [24]. The research aims not to develop new ideas but to demonstrate competence in using established procedures and instruments. The analysis is not complicated, and students are not expected to provide substantial criticism of previously acquired information or develop original concepts. The primary focuses of this course are the knowledge of technical abilities, the comprehension of fundamental concepts, and the adherence to defined academic rules [25]. Textbooks and essential journal articles are frequently used as references rather than cutting-edge research. The structural components of a bachelor's thesis are illustrated methodically in Figure 2, with particular emphasis placed on the thesis's aim, methodological framework, analytical complexity, and academic references. It emphasizes the transition from theoretical knowledge to practical research, ensuring that academic traditions and ethical issues

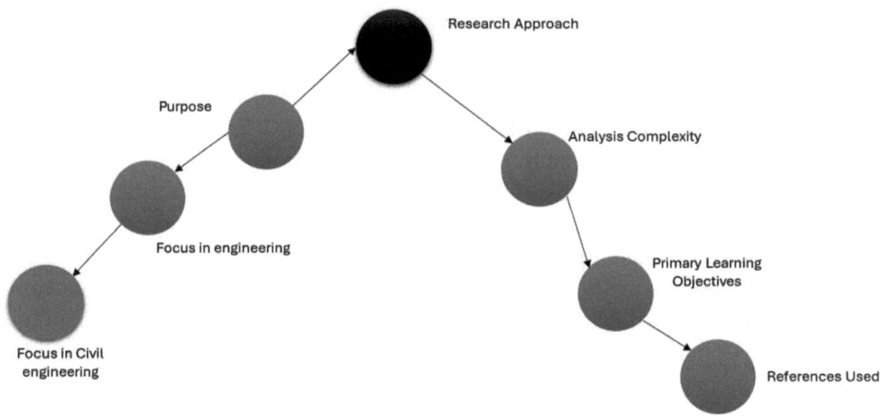

Fig. 2 Academic writing structure in bachelor's theses (*Source* Authors)

are adhered to and fostering the development of necessary technical and analytical abilities in undergraduate research.

- Both Master's Theses and Articles

Master's theses require a better grasp of engineering research's theoretical and practical components [26]. They are built on the foundation of a bachelor's degree in architectural engineering knowledge. These projects frequently deal with more complicated issues, necessitating sophisticated analytical abilities, critical assessment, and the integration of several different fields of study [27]. As an illustration, a master's thesis may concentrate on the energy efficiency of high-performance buildings. This could involve investigating the influence of double-skin facades or sophisticated insulation materials on energy consumption [28]. The technique may use advanced simulation tools like ANSYS or EnergyPlus, and students are expected to analyze various design ideas critically [29]. This contrasts with the research conducted for a bachelor's degree. Integrating multidisciplinary topics in the study area, such as sustainable development, urban planning, or novel building methods, is now possible. A literature analysis of academic papers is also required for master's level research to discover research gaps [30]. Practical advice or solutions for industry-related difficulties are frequently included in the final product of this type of study. The contribution is often an application or modification of previously conducted research, with less focus on creating entirely new information. Originality is vital, but the contribution is typically this [31]. In Figure 3, the key components of master's theses and academic articles are illustrated. These components include emphasizing theoretical and practical knowledge, advanced analytical skills, multidisciplinary integration, and literature evaluation. It shows the progression from undergraduate to postgraduate scholarly contributions by highlighting advanced research methodologies, the application of previous studies, and industry-focused solutions.

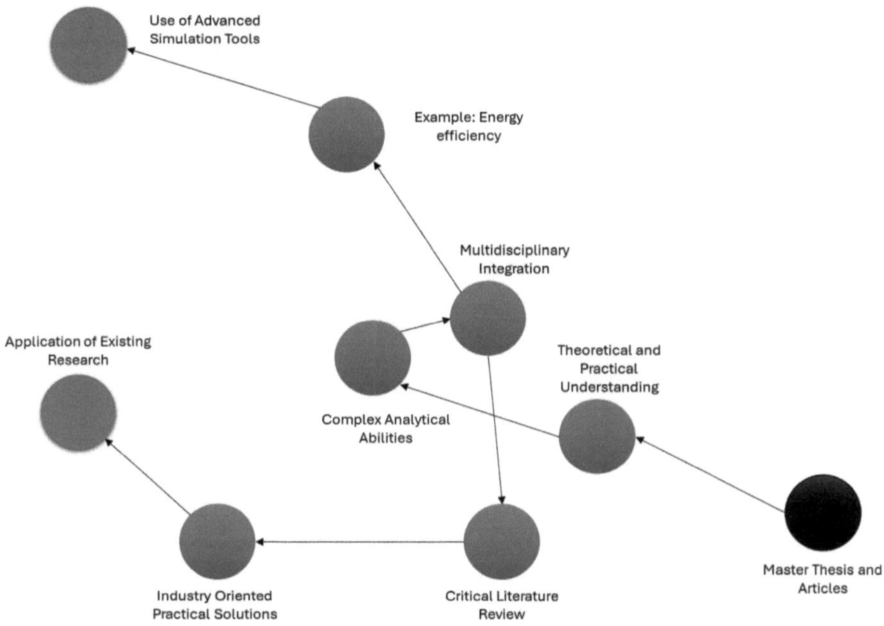

Fig. 3 Structural components of master's theses and articles (*Source* Authors)

- Theses and Articles for the Ph.D.

The Ph.D. research reflects the highest degree of academic inquiry and is characterized by its originality, thoroughness, and significant contributions to the field [32]. When writing a thesis for a doctoral degree in architectural engineering, it is essential to address a research gap and propose original solutions or hypotheses. For example, a Ph.D. thesis may involve the creation of innovative materials, such as bio-based composites, and exploring how these materials can be used in earthquake-resistant constructions [33]. Another approach could focus on developing advanced computer models for predicting the performance of structures under extreme conditions, such as wind or seismic stress [34]. The methodology is complex at this level, often incorporating experimental work, extensive mathematical modeling, and validation through large-scale simulations or laboratory testing. Furthermore, the methodology is frequently integrated with other techniques [35]. Given the expansive nature of a Ph.D. thesis, it is crucial to thoroughly review the existing literature to position the study within the current academic landscape [36]. The work is expected to meet the publishing standards required for peer-reviewed journals, often leading to multiple publications. The contributions are practical and theoretical, providing new frameworks, models, or methodologies that could influence the field for years [37]. Figure 4 illustrates the key components of Ph.D. theses and academic publications, emphasizing the originality of research, refined methods, and contributions independently reviewed by peers. It highlights experimental work, mathematical modeling, and

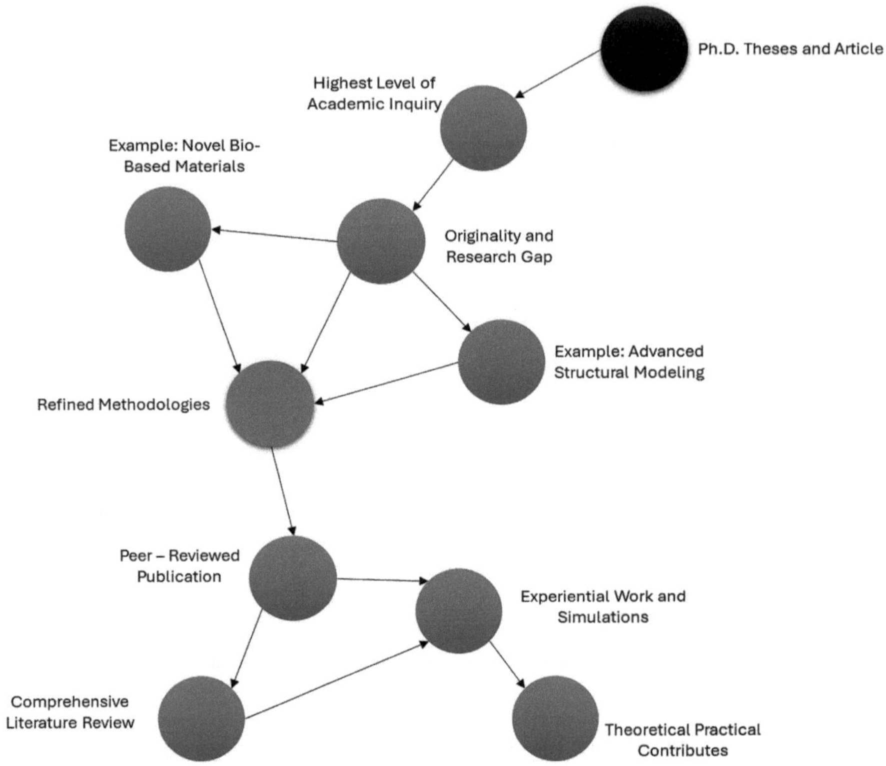

Fig. 4 Structure of Ph.D. theses and articles (*Source* Authors)

literature evaluation, demonstrating how doctoral research advances theoretical and practical knowledge, addresses research gaps, and fosters innovation in specialized academic domains.

- Key Distinctions Between Different Levels

Scope and Complexity: The work associated with a bachelor's degree focuses on clearly defined problems, whereas a master's degree tackles complex challenges through an approach that draws from multiple disciplines [38]. Research undertaken for a doctoral degree seeks to fill knowledge gaps with original contributions. In terms of methodology, projects for a bachelor's degree employ conventional tools and methods; master's degree projects use more sophisticated tools and critical evaluation; and doctoral research entails the development of innovative resources and techniques [7]. Regarding contributions, bachelor's theses demonstrate technical proficiency, master's theses connect theory with practice, and doctoral theses either redefine or expand the overall area of study [39]. Work from a bachelor's degree is primarily aimed at academic assessment. In contrast, master's degree work often emphasizes professional application, while doctoral work contributes to scholarly debate and broadens global knowledge [40]. The progression from grasping fundamental concepts to fostering innovation and making substantial contributions to architectural engineering is reflected at each successive level, highlighting the scholar's growth in knowledge and increasing expectations [41]. In terms of goals, scope, research focus, methods, literature review, analysis, contributions, outputs, and target audience, bachelor's theses and articles, master's theses, and Ph.D. theses and articles are distinguished by the significant differences outlined in Table 1. This evolution illustrates a transition from basic knowledge and practical application at the bachelor's level to critical evaluation at the master's level, culminating in creative, influential research at the doctoral level.

The increasing complexity, breadth, and demands of academic theses and articles at the bachelor's, master's, and doctoral levels, particularly in architectural engineering, are highlighted in Table 2. It illustrates how each phase builds upon the previous one, progressing from basic education to more complex knowledge and unique contributions to the field. The main goals at the bachelor's level are demonstrating technical skills and applying fundamental ideas to straightforward, well-defined issues [48]. For example, students may use standard software like AutoCAD or STAAD to design a small residential building [49]. Only textbooks and introductory materials are included in literature reviews, and the final product usually consists of a project report demonstrating the use of fundamental techniques. Opportunities include gaining technical skills and being prepared for entry-level positions [50]. However, obstacles like a narrow focus and a lack of access to cutting-edge equipment may stifle originality and creativity. The scope of study expands to tackle intricate, multidisciplinary issues at the master's level. Students investigate practical applications, adapt approaches, and critically assess current ideas. For in-depth study, sophisticated technologies like ANSYS or EnergyPlus are utilized [51]. As the literature evaluation grows more thorough, it identifies research gaps in academic publications [52]. The outcomes are comprehensive theses that connect theory and practice with practical suggestions. While maintaining higher educational standards and striking a balance between the breadth and depth of study poses challenges, this stage enhances specialization and problem-solving skills [53]. The emphasis shifts to making noteworthy, unique contributions to discipline at the Ph.D. level. Research addresses significant and wide-ranging issues, often employing state-of-the-art computational and experimental techniques [34]. Developing novel hypotheses or innovations is supported by thorough literature surveys and rigorous assessments [54]. Peer-reviewed articles and a comprehensive thesis are among the results. While there are many opportunities, such as advancing knowledge and impacting global discourse, there are also numerous difficulties, including managing intricate multi-year projects, securing funding, and meeting strict standards for rigor and innovation [55]. This evolution illustrates how scholarly research is transforming and increasingly advancing our understanding. The bar chart shown in Fig. 5 compares the detail provided on various research components across three distinct academic levels: bachelor's, master's, and doctoral. Doctoral students typically offer

Table 1 Comparison of bachelor's, master's, and Ph.D. theses/articles

Aspect	Bachelor's thesis/article	Master's thesis/article	Ph.D. thesis/article
Purpose [42]	Demonstrate foundational knowledge, technical skills, and ability to apply standard methodologies to solve fundamental problems	Tackle complex, multidisciplinary problems with critical thinking, applying advanced theories and methodologies	Contribute significant, original research that advances the field's body of knowledge
Scope [43]	Limited to solving straightforward, well-defined problems (e.g., designing a small residential structure)	Addresses complex, interdisciplinary problems (e.g., improving energy efficiency or sustainability in urban structures)	Explores broad, impactful questions (e.g., developing innovative materials or predictive models for structural behavior)
Research focus [44]	Application of existing principles, with no expectation of innovation	Critical evaluation of existing theories, adaptation of methodologies, and exploration of practical applications	Development of original theories, new methods, or groundbreaking innovations
Methodology [45]	Essential tools and standard practices (e.g., AutoCAD, STAAD.Pro, fundamental finite element analysis)	Advanced tools and software (e.g., ANSYS, EnergyPlus, BIM tools, CFD modeling) integrating various systems	Cutting-edge experimental and computational methods (e.g., machine learning, optimization algorithms, or large-scale experimental setups)
Literature review [16]	Limited to textbooks and a few introductory research papers	Comprehensive review of scholarly articles and conference papers to identify research gaps	Exhaustive review, synthesizing decades of research to identify gaps and position original work
Analysis [46]	Straightforward application of established methods to solve defined problems	In-depth evaluation of strategies, integrating interdisciplinary concepts, and interpreting results critically	Rigorous, multifaceted analysis to develop and validate new models, methods, or systems with high scrutiny
Contribution [47]	Demonstrates competence in applying basic knowledge and tools	Offers practical solutions or recommendations for industry-related challenges, bridging theory and practice	Provides novel insights that advance the field, often influencing future research and practices

more extensive descriptions, particularly in methodology, literature review, and analysis. The graphic illustrates how academic writing becomes increasingly challenging as one progresses through these levels.

Table 2 The evolution of research contributions across academic levels

Academic level	Research focus	Knowledge contribution
Undergraduate [60]	Synthesis and application of existing knowledge to predefined problems	Limited contributions, mainly demonstrating understanding and application of fundamental concepts
Master's [61]	Critical evaluation of current ideas, methods, and practices, often proposing modified solutions	Moderate contributions, refining and adapting knowledge to specific challenges
Ph.D. [62]	Generation of new knowledge through original research, theory development, or innovative approaches	Significant contributions, addressing unresolved questions and advancing disciplinary knowledge

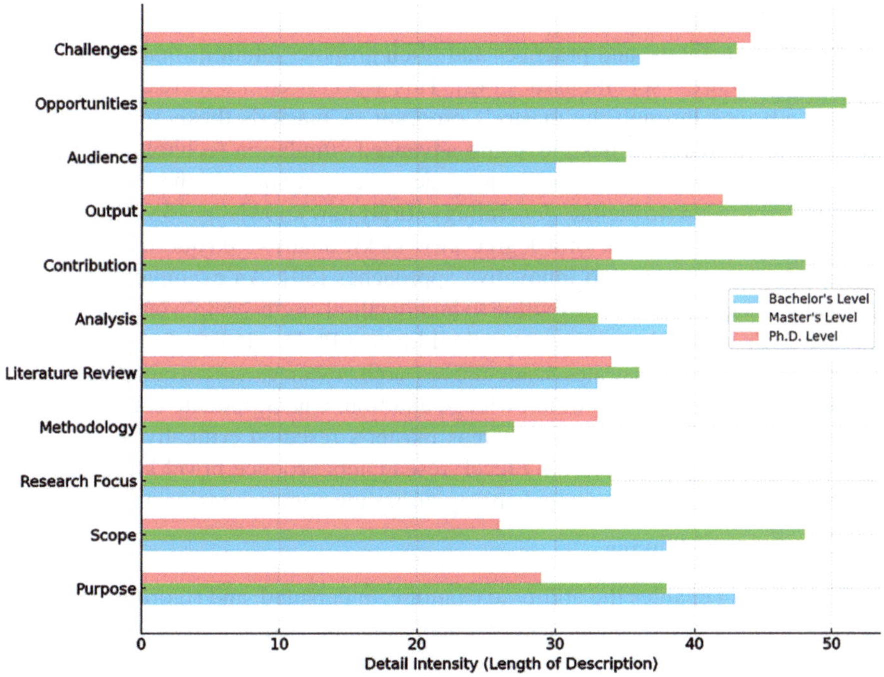

Fig. 5 Detailed analysis of the comparison table: bachelor's, master's, and Ph.D. theses/articles (*Source* Authors)

2.1.2 Critical Thinking, Original Research, and Knowledge Contribution

Three main goals are the foundation for academic research: contributing to the body of knowledge, engaging in critical thinking, and doing original research. These characteristics define the evolution of academic activity from the undergraduate level to the master's and doctorate levels, which form the cornerstone of scholarly research. Accomplishing each aim advances disciplinary knowledge, promotes intellectual growth, and accelerates innovation.

- Impact on the Acquiring of Knowledge

Research conducted in academic institutions is undertaken to contribute to the current body of knowledge by addressing pertinent issues and resolving gaps in understanding [56]. The undergraduate level is characterized by relatively few contributions, the majority of which consist of synthesizing and applying previously acquired information to issues that have been predetermined. In architectural engineering, for example, a bachelor's thesis may concentrate on the design of a straightforward structural framework [57]. This would demonstrate the student's ability to utilize fundamental engineering concepts. Master's level students must critically evaluate current ideas, methods, and practices and often propose solutions that are modified or tailored to address specific issues [42]. This aspect becomes increasingly vital as students advance through the master's program. A master's thesis might examine energy-efficient building designs using advanced simulation technologies to offer meaningful insights to the industry or academic community. At the PhD level, the expectation broadens to include generating entirely new knowledge [58]. The primary research objectives for a doctoral degree are to tackle unanswered questions, establish original theories, or develop innovative approaches [59]. An example of a significant and groundbreaking contribution to architectural engineering would be creating new materials for earthquake-resistant buildings, a Ph.D. candidate in architectural engineering. Table 2 illustrates academic research advances via certain phases, each marked by differing degrees of knowledge addition. Undergraduate courses emphasize applying established ideas, whereas master's research critically assesses and enhances approaches. At the PhD level, researchers produce novel knowledge via original research, substantially enhancing discipline comprehension and creativity.

- Using Critical Thinking

Critical thinking enables the ability to examine, understand, and rigorously evaluate material, making it an essential component of academic research [63]. When students reach college, they utilize critical thinking to comprehend and integrate previously established ideas [64]. Master's level work requires more advanced skills, such as analyzing the strengths and limitations of techniques or comparing competing theoretical frameworks [65]. Critical thinking is vital for doctoral research, as it helps identify knowledge gaps, formulate research questions, and interpret complex data [66]. This may involve assessing the long-term sustainability of new materials or

Table 3 The role of critical thinking in academic research

Academic level	Critical thinking focus	Application in research
Undergraduate [67]	Understanding and integrating established ideas through examination and evaluation	Utilized for comprehension and integration of existing concepts
Master's [68]	Analyzing strengths and limitations of techniques or comparing competing theoretical frameworks	Applied to refine methodologies and evaluate theoretical perspectives
Ph.D. [56]	Identifying knowledge gaps, formulating research questions, and interpreting complex data with intellectual rigor	Essential for developing original theories and assessing foundational assumptions in research

questioning the assumptions underlying conventional structural designs. In architectural engineering, this approach can be applied. Employing such intellectual rigor ensures that the study is based on logic, relevance, and the integrity of the academic community. Table 3 illustrates that Critical thinking is essential at all educational stages, progressing from comprehension of established concepts in undergraduate studies to analysis of methodology in master's programs and identification of research needs in doctorate research. This advancement improves intellectual rigor, guaranteeing logical reasoning, methodological validity, and the integrity of academic research.

- Authentic Investigations

Original research is the defining characteristic of work at the doctorate level and a differentiating component in the progression of academic careers [69]. Research for a bachelor's degree and a master's degree might build upon previously acquired information, whereas research for a doctoral degree aims to innovate. For example, PhD research in architectural engineering may entail the development of adaptable facade systems that can react dynamically to changes in the surrounding environment. This would be a unique addition to both theory and practice. Contributing to knowledge, engaging in critical thinking, and doing original research are the three pillars that support academic inquiry. They ensure the ongoing development of disciplines while addressing modern concerns and encouraging innovation.

2.1.3 A Guide to Selecting a Research Topic: Guidelines and Striking a Balance Both the Possibility and the Importance

Choosing a research subject is a critical phase in academic research. It also significantly impacts the work's quality, relevance, and effectiveness [70]. The process involves carefully evaluating the criteria for subject selection and maintaining a deliberate balance between feasibility and relevance, which varies based on the academic level—whether it's a bachelor's degree, a master's degree, or a doctoral degree.

- Guidelines for Topic Selection

Selecting a research topic must be strategic and methodical, ensuring alignment with contemporary challenges and disciplinary advancements [71]. A well-chosen topic should be relevant, addressing pressing issues or emerging trends within the field. In architectural engineering, for example, research on energy-efficient building designs, sustainable materials, or intelligent construction technologies aligns with critical global concerns such as urbanization and climate change. By adhering to these principles, researchers contribute meaningfully to their discipline while enhancing the practical applicability of their findings. Defined Scope: It is essential to have a clear and focused research topic. In contrast, overly narrow themes may lack adequate depth, while broad issues may result in superficial examination [72]. For example, "The Role of Smart Materials in Sustainable Architecture" is specific and manageable, while "Advancements in All Sustainable Building Materials" would be a broad topic. Consider the availability of relevant data, tools, and expertise when evaluating resource access. For instance, it may be necessary to have access to well-equipped facilities and financial resources to conduct experimental testing on specific topics, such as developing a unique concrete mix. Similarly, problems requiring computational modeling, like maximizing wind loads on high-rise structures, necessitate the right software and technology. Compatibility with the Academic Degree: Both the difficulty level and the expectations for the subject of study must align with the degree being pursued. Research for a bachelor's degree emphasizes foundational skills, while research for a doctoral degree demands uniqueness and innovation. For longer research projects, such as a Ph.D. dissertation, choosing a topic that matches the researcher's interests fosters ongoing engagement and commitment. This is especially relevant when the topic is selected because it aligns with the researcher's interests [73].

- Balancing Feasibility and Significance

It is essential to find a subject that balances relevance (its contribution to the field) and practicality (the ability to be accomplished within constraints) [74]. At different academic levels, this balance varies: understanding and applying fundamental concepts is the primary focus of research at the bachelor's level [75]. The topics must be easy to grasp, clearly stated, and achievable within a limited timeframe [76]. For example, a bachelor's thesis might explore the structural design of a modest residential building using widely accepted codes and software. While these projects contribute minimally to the profession, they help students develop essential technical skills and intellectual knowledge. Research becomes more advanced and multidisciplinary at the master's level, requiring a critical review of established theories or practices. Topics should be substantive yet manageable. For example, analyzing the effectiveness of green roofs in reducing urban heat islands using simulation tools strikes a balance between the feasibility and relevance of the project. Such work enhances practical insights and is achievable with the resources typically available to master's degree students. Originality and technological advancement are paramount at the Ph.D. level. Topics must address gaps in existing knowledge and provide substantial contributions, theoretically or practically [77]. For example, a doctoral candidate might develop novel self-healing concrete material for earthquake situations. Although this topic is significant, it is essential to ensure feasibility by carefully planning experimental activities, resource availability, and effective time management.

- Practical Considerations

Selecting a research topic requires a deliberate and well-informed approach that accounts for several critical factors, regardless of the academic level [78]. One essential consideration is the supervisor's expertise, as their specialized knowledge provides invaluable guidance, ensuring methodological rigor and increasing the likelihood of research success [79]. A well-aligned supervisor helps maintain scholarly standards and disciplinary expectations, strengthening the study's foundation [80]. Furthermore, ethical and sustainability considerations must be prioritized, as research should adhere to established ethical principles and avoid causing harm to human well-being or the environment [81]. This is particularly crucial in fields where ethical implications play a significant role in shaping responsible academic inquiry. Additionally, the chosen topic should align with the researcher's long-term academic and professional objectives, extending its relevance beyond the confines of an educational setting. A topic that contributes to a researcher's intellectual development and career aspirations fosters deeper engagement and enhances the overall impact of the study [82]. By carefully integrating these strategic considerations, researchers ensure their work is academically rigorous and socially and professionally meaningful.

- Final Thoughts

When selecting a topic for study, it is necessary to consider academic goals, personal interests, and the restrictions of practical situations [83]. Whether the topic is centered on basic skills at the bachelor's level, critical analysis at the master's level, or creativity at the doctoral level, the topic must balance feasible and significant. If they adhere to these criteria, researchers can guarantee that their work is practical and attainable and make a meaningful contribution to their profession.

Academic level	Key considerations for topic selection	Practical constraints
Undergraduate [84]	It focuses on understanding and applying fundamental concepts with clearly defined, manageable topics	Topics should be straightforward, achievable within a limited timeframe, and supported by available resources
Master's [85]	Requires a balance between feasibility and significance, involving critical evaluation of theories and methodologies	Topics must be substantive yet realistic, ensuring access to necessary tools, data, and expertise
Ph.D. [86]	Emphasizes originality and innovation, addressing gaps in knowledge with a substantial theoretical or practical contribution	Topics should be ambitious but feasible, requiring careful planning of experimental work, resources, and time management

2.1.4 A Comparative Study at All Academic Levels on the Organization of Academic Theses and Articles

Organizing a thesis or article is crucial for producing cohesive and impactful work in academic writing [87]. The following comprehensively analyzes the fundamental aspects of academic labor and discusses the distinctions in breadth and depth that exist between the bachelor's, master's, and doctorate degrees of education offered.

- Abstract

Typically, the abstract is between 150 and 300 words long and serves as a condensed overview that captures the core of the study investigation [88]. The problems, aims, methods, central results, and conclusions of the research are all included in this section. Regarding a bachelor's thesis, the abstract often provides a concise and straightforward account of the work, emphasizing clarity [89]. In addition to incorporating nuanced results and methodological insights, master's level abstracts demonstrate a deeper level of complexity. On the other hand, abstracts written at the doctoral level emphasize originality, sophisticated methodology, and the more enormous ramifications of the study [23]. These abstracts serve as a doorway to the scholar's contribution to the topic under investigation.

- Introduction

A definition of the problem, an establishment of the context, and an outline of the objectives are all included in the introduction, which sets the stage for the study [90]. The purpose of a bachelor's thesis is often to provide a broad overview of the subject matter and to emphasize the significance of the subject matter, with only a limited amount of critical examination. A more in-depth literature review is incorporated into the introduction at the master's level. This review places the study within theoretical frameworks and addresses specifically identified knowledge gaps [91]. The introductions for doctoral programs go much further, displaying a profound grasp of the area and expressing how the study contributes to advancing academic discourse and resolving complicated challenges [92].

- The Methodology

A description of the study design, techniques, and procedures that were utilized to answer the research question is provided in the methodology section [93]. To demonstrate a fundamental mastery of research processes, bachelor's theses frequently depend on tried-and-true methodologies that are easy and require little to no adaptation. More complexity is reflected in the approaches used at the master's level. These methodologies may utilize a combination of methods or novel modifications to answer nuanced research problems. At the PhD level, the methodology becomes an essential component of the work. It is distinguished by its rigor, creativity, and the possibility of repeatability, and it frequently involves the creation or improvement of more sophisticated approaches [94].

- Results

A dispassionate and impartial presentation of the research findings is provided in the section titled "Results" [95]. This section uses relevant visual aids such as tables, graphs, and charts. Generally speaking, the findings of bachelor's theses are straightforward and closely connected to the initial study issue. Results obtained at the master's level are substantially more involved, frequently incorporating intricate datasets and in-depth investigations [96]. Results obtained at the doctoral level are differentiated by their degree of depth, typically utilizing sophisticated statistical models, longitudinal research, or multidisciplinary techniques that contribute to the area overall [97].

- Discussion

The commentary interprets the findings about the study's aims and places them within the larger context of the already available literature [98]. Discussions at the bachelor's level are often descriptive and center on connecting the findings and the research question. A more analytical approach is used at master's level talks, which include critical evaluation of findings, exploration of theoretical implications, and acknowledgment of limits [99]. At the PhD level, the discussion exhibits a high degree of intellectual engagement, delivering significant insights, addressing theoretical and practical ramifications, and proposing possibilities for future study that emphasize the scholar's competence and inventiveness [100].

- Conclusion

The conclusion provides a synthesis of the research by summarizing the most critical findings, reaffirming the relevance of those findings, and addressing the initial objectives [101]. It is common practice to briefly conclude bachelor's theses to reiterate the primary ideas without requiring significant extrapolation. This is expanded upon in the conclusions of master's level research by exploring the ramifications of the results and making either theoretical or practical recommendations. The findings of a doctoral dissertation are more ambitious than those of a master's thesis [102]. They highlight the work's contribution to the field, identify challenges that have not been resolved, and propose precise study objectives that expand academic and professional knowledge. Differences in Extent and In-Depth Coverage Across Different Academic Levels As one progresses through the bachelor's, master's, and doctorate levels of education, the breadth and depth of each component undergo substantial changes. A bachelor's thesis aims to demonstrate basic knowledge and the capacity to apply recognized approaches to well-defined issues [103]. Most of the writing is descriptive, with just a little critical involvement. It is necessary to have a higher degree of analytical and critical thinking to complete a master's thesis. The thesis should concentrate on filling specific knowledge gaps and thoroughly engaging with theoretical and empirical frameworks. It is common for research to aim to make significant contributions, either in terms of insights or applications, within a particular field of study.

A doctoral-level study is considered the pinnacle of academic research because it is distinguished by its originality, depth, and ability to extend the boundaries of knowledge [104]. Doctoral scholars must tackle complex and original research questions, employ innovative methodologies, and actively contribute to academic discourse [105]. A well-structured research study should demonstrate a deep understanding of the subject, contribute substantially to the field, and reflect intellectual rigor while advancing knowledge within the discipline. Ensuring clarity, coherence, and methodological precision requires aligning the study's structure and depth with the academic level's expectations. This alignment facilitates the development of a compelling and impactful thesis or scholarly article. Table 4 illustrates the evolution of research scope, depth, and outcomes across bachelor's, master's, and doctoral degrees. The academic progression from undergraduate to postgraduate studies reflects increasing complexity, originality, and critical engagement. While bachelor's and master's research primarily establish foundational knowledge, master's and doctoral studies address specific gaps, integrate interdisciplinary perspectives, and contribute to both academic and professional domains.

Table 4 Comparative analysis of academic scope, depth, and outcomes across educational levels

Academic level	Scope	Depth	Key outcome
Bachelor's [106]	Exploration of fundamental principles; application of existing theories and methodologies to clearly defined straightforward problems within a narrow focus	Primarily descriptive and structured, it focuses on demonstrating an understanding of core concepts and applying basic methodologies with limited interpretation of results	Proves ability to comprehend and implement foundational research techniques; establishes competence in addressing simple research questions with accuracy and clarity
Master's [107]	In-depth analysis of specific knowledge gaps; integration of multiple theoretical perspectives; development of nuanced understanding within a specialized area of study	Balances description with critical analysis; employs sophisticated methodologies; interprets results in existing literature to draw meaningful conclusions and implications	Produces insights that address specific gaps in the field; prepare graduates for professional or academic advancement by demonstrating analytical and methodological proficiency
Doctoral [108]	Comprehensive investigation of novel, complex questions; synthesis of interdisciplinary knowledge; substantial contribution to advancing theoretical and empirical frameworks in the field	Demonstrates advanced critical thinking and originality; develops or refines innovative methodologies; engages deeply with theoretical debates and situates findings within a global academic discourse	Delivers groundbreaking, original contributions to the field; positions the scholar as an expert capable of leading academic or applied research initiatives

3 Writing Strategies for Success

The ability to write well for academic purposes is a skill that facilitates the dissemination of research findings and reflects the writer's professionalism and credibility [109]. Writing style and tone, managing sources and citations, and organizing time and drafting are the three components of writing methods discussed in depth in this section. Each component aims to provide authors with practical insights and best practices that will help them tackle the challenges of academic writing with confidence and precision [110]. To accommodate a diverse range of educational requirements, specific examples for bachelor's, master's, and doctoral degrees are provided.

3.1 Writing Style and Tone

3.1.1 Academic Language Conventions

If you want to explain complicated concepts through your writing effectively, academic writing requires a formal tone, accuracy, and clarity [111]. By adhering to linguistic rules, one may ensure that the work is up to the standards expected within the academic community. Among the fundamental principles are:

- To ensure clarity and simplicity, express your thoughts using straightforward language. Unless necessary, avoid jargon and clarify technical terms to make them accessible. A Ph.D. [112]. Dissertation, for instance, employs highly specialized vocabulary to cater to an expert audience, whereas a bachelor's thesis may favor more straightforward explanations to establish fundamental understanding.
- In academic writing, presenting facts and evidence precedes expressing personal beliefs. Avoiding biased terminology and maintaining a neutral perspective is essential [113]. Essays written for a bachelor's degree typically emphasize achieving objectivity through organized arguments, while doctoral writing requires rigorous evidence and an impartial interpretation of advanced research.
- Consistent formatting, language, and citation styles throughout the text are vital for coherence [114]. For example, the presentation of a master's thesis must adhere to the specific style guide of the institution, such as the American Psychological Association or the Modern Language Association.
- The passive voice can highlight research procedures rather than the researcher. While the active voice is often employed to make writing more engaging, the passive voice may be more suitable in specific contexts [115]. For example, "The data were analyzed using statistical software" is appropriate for a bachelor's degree report. In contrast, "WE analyzed the data to investigate trends" might be used in a journal article resulting from research conducted for a doctoral degree.

3.1.2 Tailoring Style for Articles Versus Theses

In academic writing, style and organization are significantly influenced by the purpose of writing and the audience. Although both are scholarly works, the articles have different requirements:

- **Articles**: Academic articles are characterized by their conciseness and focus, aiming to address specific research topics or present findings [116]. Journal-style essays may require a clear and brief exposition of a single idea, such as "The Effects of Urbanization on Local Biodiversity." These essays are typically written by students graduating with a bachelor's degree. A Ph.D. candidate may need to submit an article with a comprehensive methodology, nuanced commentary, and adherence to the specific publication's structure [117]. For example, the essay may be titled "Machine Learning Applications in Predicting Urban Air Quality."
- **Theses**: A thesis is an extensive document demonstrating the writer's expertise and contributing to academic discourse [118]. These are also prevalent in academic writing. Bachelor's theses focus on synthesizing core information, such as "A Comparative Study of Solar Panel Efficiency," while master's theses involve more in-depth research, such as "Integration of Solar Energy into Smart Grids" [119]. Original contributions are essential for doctoral dissertations, as illustrated by the project titled "Advanced Optimization Algorithms for Renewable Energy Systems."

In either scenario, good communication must balance the specifics of the technical detail with the overall ideas being presented.

3.2 *Managing Sources and Citations*

3.2.1 Best Practices for Literature Review and Citation Management

A comprehensive literature review is the most critical component of any academic publication. It provides context and demonstrates the study's relevance [120]. Good source management is essential to guaranteeing completeness and precision.

- Systematic searching involves using academic databases such as PubMed, Scopus, and Google Scholar to locate reliable and relevant sources [121]. Students pursuing a bachelor's degree can start with the readings required for their courses. In contrast, students pursuing a master's or doctoral degree should conduct exhaustive keyword searches and investigate specialist databases.
- The process of taking notes involves summarizing the most essential elements from the sources and documenting how each reference helps your study [91]. For example, a student working on a bachelor's degree may keep a straightforward spreadsheet of sources. Still, a student working toward a doctoral degree would

utilize more sophisticated software such as Zotero, Mendeley, or EndNote to arrange references according to topics or methodology.

- Assessing the quality and relevance of sources is an integral part of critical evaluation. While a bachelor's degree essay can rely on reviewing articles and textbooks, work for a master's or doctoral degree requires reliance on peer-reviewed journal articles, pivotal papers, and original data sources [122].
- According to the theme organization framework, • The literature review should be organized thematically rather than chronologically [123]. For example, a review of climate change for a bachelor's degree may classify sources as "Causes" and "Effects." Still, a dissertation for a doctoral degree might discuss "Quantitative Models," "Case Studies," and "Policy Implications."

3.2.2 Avoiding Plagiarism and Ensuring Academic Integrity

For scholarly work to be considered credible, academic integrity is essential. To avoid plagiarism, it is necessary to pay close attention to the credit of sources and the original contribution:

- Use the relevant citation formats (such as APA, MLA, or Chicago) to credit all sources where you obtained the information. As an illustration, a project at the bachelor's level may mention fundamental publications in the APA format, but a thesis at the master's level might comprise a combination of datasets, policy papers, and articles that have been peer-reviewed.

Skills in paraphrasing involve rephrasing concepts in your own words while maintaining the meaning of the original ideas [124]. Paraphrasing complex theories while preserving their original intent is essential for a doctoral degree candidate's work.

- Plagiarism Detection Tools: Utilize software like Turnitin or Grammarly's plagiarism checker to identify unintended similarities with published content. This is crucial for master's and doctoral students, whose work is often scrutinized.
- Establishing clear agreements about authorship and responsibilities when collaborating on a project to reduce conflicts is essential. For example, a collaborative project for a bachelor's degree might involve sharing writing tasks, but doctoral-level collaborations require a defined order of authorship based on contributions. When authors follow these guidelines, they can uphold academic standards and contribute responsibly to their fields.

3.3 *Managing Resources and Drafting Documents*

3.3.1 Practical Advice for Managing Deadlines and Breaking Down Tasks

Regarding academic writing, time management is an essential component that must be included. It is possible to reduce stress and increase productivity by using effective planning and work separation.

- Divide the writing process into small milestones to ensure your goals are attainable. For example, research, analysis, and drafting for a dissertation for a doctoral degree can take many months [125]. Still, a bachelor's degree project might have weekly deadlines for each component.
- Set priorities for your tasks and concentrate on high-priority portions first [126]. For example, if a student is writing a report that is 5,000 words long, they may begin with the literature review. On the other hand, a student writing a thesis that is 20,000 words long might emphasize the research technique approach more.
- Scheduling tools like Google Calendar, Trello, or Notion can help visualize timelines and track ongoing progress [127]. Ph.D. candidates may find it beneficial to use Gantt charts to coordinate their research activity with their writing deadlines [128].
- Devote Time to Research: Investigating various sources ensures your writing is well-informed and finds no evidence [128]. A student working toward a bachelor's degree would need two weeks of preliminary research, but a project for a master's or doctoral degree might take months of data collection.

3.3.2 Iterative Writing and the Importance of Feedback

The drafting process is an iterative process that involves reviewing and refining ideas to improve clarity and coherence. The importance of feedback cannot be overstated in this cycle:

- During the first draft, you should concentrate on getting your thoughts down on paper without worrying too much about language or style [129]. In the initial draft of their dissertation, a Ph.D. candidate could include comprehensive arguments and interpretations of data, whereas a student working on a bachelor's degree might strive for a fundamental framework.
- Reviewing and revising drafts is essential for enhancing the document's flow, organization, and coherence of arguments. In a master's thesis, the introduction may require revision after completing the discussion section to ensure alignment with the study's findings and overall structure [130].
- Seek Feedback: To receive constructive critique, you should share your manuscripts with your peers, mentors, or writing groups. Students pursuing a bachelor's degree frequently rely on the feedback provided by their instructors,

but applicants for a doctoral degree may seek advice from supervisors, co-authors, and external reviewers.
- Proofreading the final document before submission is essential to ensure it is free of typos, grammatical errors, and formatting inconsistencies [131]. Hemingway Editor and Grammarly are two examples of tools that may be used to help polish the content.
- Writing iteratively improves the document's quality and leads to a more profound comprehension of the topic, raising the work's overall standard.

4 Publishing and Presentation

Academic publishing and presentation are critical components of scholarly work, providing opportunities to share research findings, receive feedback, and build professional networks [132]. This section explores strategies for success in preparing for publication, defending your work, and leveraging opportunities beyond the thesis. These insights are tailored for bachelor's, master's, and Ph.D. students, addressing their needs and goals.

4.1 Preparing for Publication

4.1.1 Turning a Thesis into an Article: Guidelines and Challenges

Converting a thesis into a journal article involves distilling a lengthy, detailed document into a concise, focused manuscript. The process presents both opportunities and challenges:

- Identify Core Contributions: Select your research's most significant findings or unique aspects [133]. For a bachelor's thesis, this might mean emphasizing key results or novel observations, while a Ph.D. dissertation often includes advanced methodologies or groundbreaking conclusions. For example, a bachelor's study on renewable energy adoption could focus on survey results, while a Ph.D. article might detail a new algorithm for optimizing energy systems.
- Adapt Structure and Style: Journal articles require a focused narrative. Remove redundant background information and simplify complex sections [134]. For example, a master's thesis chapter on methodology might be condensed into a single concise paragraph for publication while ensuring that essential details remain intact.
- Comply with Journal Guidelines: Tailor the manuscript to the target journal's submission requirements, including formatting, word count, and citation style [135]. For example, a natural sciences journal may emphasize data visualization, whereas a humanities journal prioritizes detailed arguments.

- Challenges: Adapting a comprehensive thesis into a succinct article often means cutting details. Balancing depth and brevity is particularly challenging for Ph.D. candidates with complex research [136]. Address these challenges by seeking feedback from advisors or experienced colleagues.

4.1.2 Selecting Journals and Understanding Submission Requirements

Choosing the right journal is crucial for maximizing the impact and reach of your work. Follow these steps:

- Identify Scope and Audience: Match your research topic with the journal's focus. For example, a bachelor's research project on renewable energy policy might target a regional journal, whereas a Ph.D. dissertation on advanced algorithms could fit an international engineering journal [137].
- Review Submission Guidelines: Familiarize yourself with the journal's requirements, such as word count, formatting, and citation style. Most journals provide detailed author guidelines on their websites.
- Assess Impact Factor and Reputation: Prioritize reputable journals with high visibility in your field [138]. For example, master's students aspiring for industry recognition might aim for applied research journals.
- Prepare all necessary documents, including a cover letter, abstract, and figures. Ensuring that the manuscript complies with technical requirements and adheres to ethical guidelines, such as declaring conflicts of interest, enhances the integrity and professionalism of the submission [139].

4.2 Defending Your Work

4.2.1 Strategies for Oral Defenses (Master's and Ph.D.)

Oral defenses are a critical milestone in academic journeys, requiring thorough preparation and practical communication skills:

- Understand the Format: Get familiar with the defense format, including time limits and question-and-answer sessions [140]. For a master's thesis, defenses typically focus on the research scope and methodology, while Ph.D. defenses highlight originality and contributions to the field.
- Prepare a Strong Presentation: Use slides to emphasize key points, such as research objectives, methodologies, findings, and implications [141]. Make sure visuals are precise, professional, and engaging. For example, master's students might include diagrams illustrating experimental setups, while Ph.D. candidates may present comparative tables summarizing novel findings.
- Anticipate Questions: Prepare concise responses to potential questions from the committee. Practice explaining complex ideas in straightforward terms to engage diverse audiences [142].

- Confidence and Clarity: Maintain a professional demeanor, speak clearly, and avoid excessive jargon. Rehearse with peers or mentors to improve your delivery.

4.2.2 Presenting Key Findings Effectively

Effective presentation of research findings is essential for engaging your audience and demonstrating the significance of your work:

- Begin with a Hook: Start your presentation with an intriguing fact, question, or anecdote related to your research. [143]. For example, a bachelor's student might open with a real-world example, while a Ph.D. candidate could highlight a knowledge gap that their research addresses.
- Focus on Key Results: Emphasize the most significant findings using visuals such as charts, graphs, or images [144]. For example, a master's student might display a bar graph comparing treatment outcomes, while a Ph.D. candidate might show a heatmap illustrating complex data patterns.
- Connect to Broader Implications: Describe how your research contributes to the field or addresses real-world issues. For instance, illustrate how your findings inform policy, advance technology, or create opportunities for future research.

4.3 Beyond the Thesis

4.3.1 Opportunities for Academic and Professional Advancement

Completing a thesis opens doors to diverse opportunities in academia and industry. Consider the following pathways:

- Further Studies: Bachelor's students can pursue master's degrees to deepen their expertise, while master's graduates can consider Ph.D. programs for advanced specialization.
- Academic Careers: Publish your research and present at conferences to build a scholarly reputation [145]. Ph.D. graduates aiming for faculty positions should focus on developing a strong research portfolio and gaining teaching experience.
- Industry Applications: Tailor your research skills for industry roles. For example, a master's in data science might apply thesis skills to solve real-world business challenges, while a Ph.D. in engineering could lead R&D projects.
- Networking: Attend academic and professional events to connect experts and peers [146]. Engage with alum networks and professional organizations to explore career opportunities.

4.3.2 Building a Portfolio for Future Research or Industry Applications

A well-curated portfolio showcases your skills and accomplishments, demonstrating your potential to employers or academic institutions [147]:

- Include Key Works: Highlight your thesis, published articles, and conference presentations. For bachelor's students, this may include capstone projects or internships. Master's and Ph.D. graduates should emphasize advanced research outputs and awards.
- Showcase Transferable Skills: Emphasize your data analysis, project management, and technical writing skills. For example, a bachelor's portfolio might showcase team projects, while a Ph.D. portfolio could emphasize independent research.
- Leverage Online Platforms: Create a professional website or LinkedIn profile to share your achievements. Include links to publications, project summaries, and testimonials.
- Prepare for Interviews: Use your portfolio to illustrate your expertise during interviews or presentations. Tailor your examples to the specific role or opportunity.

4.3.3 Disseminating Knowledge Beyond Academia

Publishing and presenting research extend its reach beyond academic circles. Consider opportunities like [148]:

- Workshops and Seminars: Present findings in non-academic settings to engage policymakers, industry leaders, and community groups. For instance, a Ph.D. graduate studying environmental policy may attend local government meetings.
- Open Access and Preprints: To enhance the accessibility of your work, consider open-access journals or platforms.
- Collaboration Opportunities: Work with interdisciplinary teams to apply your research in new contexts. For example, a master's degree in healthcare technology could partner with clinicians to implement innovative solutions.

5 Conclusion

Specifically geared for students at the bachelor's, master's, and doctoral levels, this chapter provides a critical consolidation of the fundamental ideas and procedures that are necessary for academic writing. This chapter systematically examines writing practices and expectations and highlights the unifying features of academic writing across disciplines. This chapter also emphasizes the growing complexity and intellectual rigor expected at each higher education degree. At the undergraduate level, academic writing is a gateway into the methodical inquiry and analytical thought processes associated with higher education. The development of core competencies, such as the formulation of research questions, the adherence to accepted techniques, and the presentation of clear arguments, is given a lot of attention. Applying previously established theories and techniques to issues that have been specified is the primary emphasis of bachelor's theses. This helps to cultivate critical thinking and technical expertise. To meet the more advanced academic requirements that are introduced at the master's level, students are required to have a more in-depth engagement with theoretical frameworks and interdisciplinary methods. Critical examination, methodological rigor, and the capacity to incorporate a variety of views are the defining characteristics of writing at this level. It is common for master's theses to focus on practical difficulties, integrating theoretical concepts with their applications in the real world and preparing students for academic and professional pathways. At the PhD level, academic writing is considered the pinnacle of intellectual success. It emphasizes unique contributions that push the frontiers of knowledge. Ph.D. research is distinguished from other types of study by its ability to handle complicated and unsolved research topics, as well as its innovative methodological approaches and theoretical depth. In addition to requiring the greatest analytical precision, inventiveness, and ethical rigor, doctoral dissertations demonstrate an advanced mastery of the subject matter. The repetitive aspect of academic writing continues to be an essential component at all levels, as this chapter has highlighted on several occasions. Drafting, rewriting, and refining procedures are not only procedural stages but fundamental activities necessary to improve academic work's clarity, coherence, and impact. Academic writers' feedback from supervisors, colleagues, and reviewers is essential to this refining process because it helps them become more resilient and adaptable. The consideration of ethical issues is another cornerstone covered in this chapter. The necessity of appropriately attributing sources, maintaining transparency in research methodology, and adhering to norms of academic integrity is emphasized during this process. When sustaining the credibility and validity of scholarly work, these principles are essential, particularly at advanced levels when the research results may impact larger academic and societal settings. In addition, the chapter emphasizes the critical function of literature reviews in constructing research within the context of the larger academic discourse. Literature reviews are an important part of the research process, as they identify knowledge gaps and provide a formal framework for placing new research within current academic discussions. There is an examination of practical ways for doing comprehensive and systematic reviews, critically analyzing sources, and successfully synthesizing material, which highlights the significance of these tactics at all phases of the academic research process. This chapter provides further enrichment by providing valuable advice on the management of time, the allocation of resources, and the prioritizing of tasks. When negotiating the demands of academic writing, these tactics are essential, particularly for large-scale projects such as master's theses and doctoral dissertations. Students can improve the quality of their work and expedite their workflow by employing sophisticated tools for reference management and collaborative writing. This is accomplished by breaking down difficult assignments into more achievable stages. The shift from preparing a thesis to publishing it in academic publications is another topic that receives much attention in this chapter. It discusses ways for translating extensive research into brief and compelling papers, adhering to rules particular to the publication, and effectively distributing discoveries to a wider audience. When it comes to higher academic levels, when publication is an essential means of contributing to the international academic community, this specific feature is very pertinent. In conclusion, the chapter emphasizes the wider consequences of academic writing beyond the parameters of the academic world. Critical thinking, analytical reasoning, and effective communication are some of the talents that may be transferred to a variety of professional settings, and this article illustrates how these skills can be acquired via the process of writing theses and articles. In addition, academic writing is portrayed as a medium through which knowledge may be advanced, intellectual curiosity can be fostered, and societal problems can be addressed. This chapter aims to offer a complete and methodical approach to academic writing at all levels of higher education. Students are provided with the tools and methods necessary to produce rigorous and compelling academic work by addressing the obstacles and possibilities presented at each process step. The concepts that are given in this chapter serve as a crucial framework for academic and professional success. These principles can be utilized for traversing basic research at the undergraduate level, participating in critical analysis at the master's level, or pioneering creative contributions at the doctorate level. Students can convert their writing into a strong medium for intellectual growth and important contributions to their areas if they dedicate themselves to quality, demonstrate a commitment to ethics, and demonstrate accuracy in their writing.

References

1. Inouye K, McAlpine L (2019) Developing academic identity: a literature review on doctoral writing and feedback. Int J Dr Stud 14:1
2. Tracy SJ (2024) Qualitative research methods: collecting evidence, crafting analysis, communicating impact. John Wiley & Sons
3. Al-Dhaqm A, Ikuesan RA, Kebande VR, Razak S, Ghabban FM (2021) Research challenges and opportunities in drone forensics models. Electronics 10(13):1519
4. Abulibdeh A, Zaidan E, Abulibdeh R (2024) Navigating the confluence of artificial intelligence and education for sustainable development in the era of industry 4.0: challenges, opportunities, and ethical dimensions. J Clean Prod 140527
5. Culver KC, Braxton JM, Pascarella ET (2021) What we talk about when we talk about rigor: examining conceptions of academic rigor. J High Educ 92(7):1140–1163
6. Mansour N (2024) Students and facilitators' experiences with synchronous and asynchronous online dialogic discussions and e-facilitation in understanding the Nature of Science. Educ Inform Technol 1–33
7. Sholeh MI, Fathurro'uf M (2024) Evolution of final projects in higher college. ISAR J Multidiscipl Res Stud 2(1)
8. Bhatt I, Samanhudi U (2022) From academic writing to academics writing: transitioning towards literacies for research productivity. Int J Educ Res 111:101917
9. Casal JE, Kessler M (2023) Can linguists distinguish between ChatGPT/AI and human writing? A study of research ethics and academic publishing. Res Meth Appl Linguist 2(3):100068
10. Watts FM, Rodriguez JMG (2023) A review of course-based undergraduate research experiences in chemistry. J Chem Educ 100(9):3261–3275

11. Himcinschi M (2021) Ethics and academic integrity-conceptual aspects. SCIENTIA MORALITAS-Int J Multidiscipl Res 6(2):164–190
12. Kim J, Yu S, Detrick R, Li N (2024) Exploring students' perspectives on generative AI-assisted academic writing. Educ Inform Technol 1–36
13. Chigbu UE, Atiku SO, Du Plessis CC (2023) The science of literature reviews: searching, identifying, selecting, and synthesising. Publications 11(1):2
14. Team G, Anil R, Borgeaud S, Alayrac JB, Yu J, Soricut R, Schalkwyk J, Dai AM, Hauth A, Millican K, Silver D (2023) Gemini: a family of highly capable multimodal models. arXiv preprint arXiv:2312.11805
15. Nerad M, Evans B, Bernstein BL, Evans B, Fyffe J, Halai N, Hall FL, Jensen HS, Marsh H, Ortega S (2014) The continuing evolution of the research doctorate. In: Globalization and its impacts on the quality of PhD education: forces and forms in doctoral education worldwide. pp 5–30
16. Karunarathna I, De Alvis K, Gunasena P, Jayawardana A (2024) Creating value through literature reviews: techniques for identifying research gaps
17. Liang JC, Hwang GJ, Chen MRA, Darmawansah D (2023) Roles and research foci of artificial intelligence in language education: an integrated bibliographic analysis and systematic review approach. Interact Learn Environ 31(7):4270–4296
18. Smith SK, Benbenek MM, Bakker CJ, Bockwoldt D (2022) Scoping review: diagnostic reasoning as a component of clinical reasoning in the US primary care nurse practitioner education. J Adv Nurs 78(12):3869–3896
19. Antons D, Breidbach CF, Joshi AM, Salge TO (2023) Computational literature reviews: method, algorithms, and roadmap. Organ Res Methods 26(1):107–138
20. Sari YI, Utomo DH, Astina IK (2021) The effect of problem based learning on problem solving and scientific writing skills. Int J Instr 14(2):11–26
21. Halkiopoulos C, Gkintoni E (2024) Leveraging AI in e-learning: personalized learning and adaptive assessment through cognitive neuropsychology—a systematic analysis. Electronics 13(18):3762
22. Ghafar ZN (2024) An investigation of the enigmas and accessibility of academic literature, writing, and peer review. Int J High Educ Pedagogies 5(2):1–18
23. Miao J, Thongprayoon C, Suppadungsuk S, Garcia Valencia OA, Qureshi F, Cheungpasitporn W (2023) Ethical dilemmas in using AI for academic writing and an example framework for peer review in nephrology academia: a narrative review. Clin Pract 14(1):89–105
24. Cacciuttolo C, Vásquez Y, Cano D, Valenzuela F (2023) Research thesis for undergraduate engineering programs in the digitalization era: learning strategies and responsible research conduct road to a university education 4.0 paradigm. Sustainability 15(14):11206
25. Shiohira K (2021) Understanding the impact of Artificial Intelligence on skills development. education 2030. UNESCO-UNEVOC International Centre for Technical and Vocational Education and Training
26. Knauss E (2021) Constructive master's thesis work in industry: guidelines for applying design science research. In: 2021 IEEE/ACM 43rd international conference on software engineering: software engineering education and training (ICSE-SEET). IEEE, pp 110–121
27. Sahoo SK, Goswami SS (2023) A comprehensive review of multiple criteria decision-making (MCDM) methods: advancements, applications, and future directions. Decis Making Adv 1(1):25–48
28. Ascione F, Bianco N, Iovane T, Mastellone M, Mauro GM (2021) The evolution of building energy retrofit via double-skin and responsive façades: a review. Sol Energy 224:703–717
29. Kaltenegger J, Frandsen KM, Petrova E (2024) A data management perspective on building material classification: a systematic review. J Build Eng 109172
30. Misra S (2021) A step by step guide for choosing project topics and writing research papers in ICT related disciplines. In: Information and communication technology and applications: third international conference, ICTA 2020, Minna, Nigeria, November 24–27, 2020, Revised Selected Papers 3. Springer International Publishing, pp 727–744

31. Carmen E, Fazey I, Ross H, Bedinger M, Smith FM, Prager K, McClymont K, Morrison D (2022) Building community resilience in a context of climate change: the role of social capital. Ambio 51(6):1371–1387
32. Bhagyamma G (2023) Research as a profound journey of knowledge and understanding. J Res Ethics (JRE) 1(1):34–41
33. Zechmeister C, Gil Pérez M, Dambrosio N, Knippers J, Menges A (2023) Extension of computational co-design methods for modular, prefabricated composite building components using bio-based material systems. Sustainability 15(16):12189
34. Krzywanski J, Sosnowski M, Grabowska K, Zylka A, Lasek L, Kijo-Kleczkowska A (2024) Advanced computational methods for modeling, prediction and optimization—a review. Materials 17(14):3521
35. Lavin A, Krakauer D, Zenil H, Gottschlich J, Mattson T, Brehmer J, Anandkumar A, Choudry S, Rocki K, Baydin AG, Prunkl C (2021) Simulation intelligence: towards a new generation of scientific methods. arXiv preprint arXiv:2112.03235
36. Sun S, Crosthwaite P (2022) "Establish a niche" via negation: a corpus-based study of negation within the Move 2 sections of PhD thesis introductions. Open Linguist 8(1):189–208
37. Sancar R, Atal D, Deryakulu D (2021) A new framework for teachers' professional development. Teach Teach Educ 101:103305
38. Edgell RA, Lee D (2023) Theorizing creative challenges: why are social creativity and reimagined universities necessary for tackling society's problems? J Creat 33(2):100051
39. Ankersborg V, Pogner KH (2022) Conform, transform, resist: the Scandinavian way of master's thesis supervision and its contribution to acquiring research literacy and practice. In: Negotiating the intersections of writing and writing instruction. The WAC Clearinghouse/ University Press of Colorado, pp 195–231
40. Friesen S, Jacobsen M (2021) Collaborative design of professional graduate programs in education. Int J Des Learn 12(1):64–76
41. Onatayo D, Onososen A, Oyediran AO, Oyediran H, Arowoiya V, Onatayo E (2024) Generative AI applications in architecture, engineering, and construction: trends, implications for practice, education & imperatives for upskilling—a review. Architecture 4(4):877–902
42. Shanta S, Wells JG (2022) T/E design based learning: assessing student critical thinking and problem solving abilities. Int J Technol Des Educ 32(1):267–285
43. Bloch T, Borrmann A, Pauwels P (2023) Graph-based learning for automated code checking–exploring the application of graph neural networks for design review. Adv Eng Inform 58:102137
44. Mazzetto S (2024) Interdisciplinary perspectives on agent-based modeling in the architecture, engineering, and construction industry: a comprehensive review. Buildings 14(11):3480
45. Abbasabadi N, Ashayeri M (2024) Digital twin for citywide energy modeling. In: Artificial intelligence in performance-driven design: theories, methods, and tools. p 251
46. Osaba E, Villar-Rodriguez E, Del Ser J, Nebro AJ, Molina D, LaTorre A, Suganthan PN, Coello CAC, Herrera F (2021) A tutorial on the design, experimentation and application of metaheuristic algorithms to real-world optimization problems. Swarm Evol Comput 64:100888
47. Singh AK, Meshram SA, Khandelwal V, Tiwari P, Singh P (2024) Implementation of industry-based perspective in technical education: a qualitative and quantitative analysis. In: Industrial and manufacturing designs: quantitative and qualitative analysis. pp 321–346
48. Akour M, Alenezi M (2022) Higher education future in the era of digital transformation. Educ Sci 12(11):784
49. Ahmed MI, Singh LN, Osamah MM, Afzal M, Mohiuddin DAT Design and analysis of commercial building (C+ G+ 5) under wind load anaysis using staad pro
50. Hirudayaraj M, Baker R, Baker F, Eastman M (2021) Soft skills for entry-level engineers: what employers want. Educ Sci 11(10):641
51. Qi K, Yang J, Gao Z, Xu F (2024) A review of modelica language in building and energy: development, applications, and future prospect. Energy Build 113998

52. Garwe EC (2023) Innovating while maintaining standards: exploring the tension between vertical qualification alignment and innovation in Zimbabwean higher education. Qeios
53. Karunarathna I, De Alvis K, Gunasena P, Dissanayake D (2024) Critical approaches to writing literature reviews: guidelines for success
54. Shahzad M, Qu Y, Rehman SU, Zafar AU (2022) Adoption of green innovation technology to accelerate sustainable development among manufacturing industry. J Innov Knowl 7(4):100231
55. Sillé F, Busquet F, Fitzpatrick S, Herrmann K, Leenhouts-Martin L, Luechtefeld T, Maertens A, Miller GW, Smirnova L, Hartung T (2024) The implementation moonshot project for alternative chemical testing (IMPACT) toward a human exposome project
56. Dwivedi YK, Jeyaraj A, Hughes L, Davies GH, Ahuja M, Albashrawi MA, Al-Busaidi AS, Al-Sharhan S, Al-Sulaiti KI, Altinay L, Amalaya S (2024) "Real impact": challenges and opportunities in bridging the gap between research and practice–Making a difference in industry, policy, and society. Int J Inform Manag 102750
57. Savov A (2021) The use of games and crowdsourcing for the fabrication-aware design of residential buildings
58. Wickert C, Post C, Doh JP, Prescott JE, Prencipe A (2021) Management research that makes a difference: broadening the meaning of impact. J Manage Stud 58(2):297–320
59. GHR G, Aithal PS (2022) How to choose an appropriate research data collection method and method choice among various research data collection methods and method choices during Ph. D. program in India? Int J Manag Technol Soc Sci (IJMTS) 7(2):455–489
60. Varadarajan S, Ladage S (2024) Problem-based learning (PBL): a literature review of theory and practice in undergraduate chemistry laboratories. J Chem Educ 101(8):3027–3038
61. Göttgens I, Oertelt-Prigione S (2021) The application of human-centered design approaches in health research and innovation: a narrative review of current practices. JMIR Mhealth Uhealth 9(12):e28102
62. Golder PN, Dekimpe MG, An JT, van Heerde HJ, Kim DS, Alba JW (2023) Learning from data: an empirics-first approach to relevant knowledge generation. J Mark 87(3):319–336
63. Chukwuere JE (2024) Critical research thinking: a recipe for academic writing success and publications
64. Bean JC, Melzer D (2021) Engaging ideas: the professor's guide to integrating writing, critical thinking, and active learning in the classroom. John Wiley & Sons
65. Hendriarto P, Mursidi A, Kalbuana N, Aini N, Aslan A (2021) Understanding the implications of research skills development framework for indonesian academic outcomes improvement. J Iqra' Kajian Ilmu Pendidikan 6(2):51–60
66. Bellaera L, Weinstein-Jones Y, Ilie S, Baker ST (2021) Critical thinking in practice: the priorities and practices of instructors teaching in higher education. Think Skills Creat 41:100856
67. Capone R (2022) Blended learning and student-centered active learning environment: a case study with STEM undergraduate students. Can J Sci Math Technol Educ 22(1):210–236
68. Abaku EA, Edunjobi TE, Odimarha AC (2024) Theoretical approaches to AI in supply chain optimization: pathways to efficiency and resilience. Int J Sci Technol Res Arch 6(1):092–107
69. Gardner SK, Mendoza P (eds) (2023) On becoming a scholar: Socialization and development in doctoral education. Taylor & Francis
70. Alam A (2022) Impact of university's human resources practices on professors' occupational performance: empirical evidence from India's higher education sector. Inclusive businesses in developing economies: converging people, profit, and corporate citizenship. Springer International Publishing, Cham, pp 107–131
71. Maiya AK, Aithal PS (2023) A review-based research topic identification on how to improve the quality services of higher education institutions in academic, administrative, and research areas. Int J Manag Technol Soc Sci (IJMTS) 8(3):103–153
72. Schiff D, Borenstein J, Biddle J, Laas K (2021) AI ethics in the public, private, and NGO sectors: a review of a global document collection. IEEE Trans Technol Soc 2(1):31–42

73. Plantec Q, Cabanes B, Le Masson P, Weil B (2023) Early-career academic engagement in university–industry collaborative PhDs: research orientation and project performance. Res Policy 52(9):104856
74. Althubaiti A (2023) Sample size determination: a practical guide for health researchers. J Gen Fam Med 24(2):72–78
75. Pascoli DU, Aui A, Frank J, Therasme O, Dixon K, Gustafson R, Wright MM (2022) The US bioeconomy at the intersection of technology, policy, and education. Biofuels, Bioprod Biorefin 16(1):9–26
76. Hancock DR, Algozzine B, Lim JH (2021) Doing case study research: a practical guide for beginning researchers
77. Yu T, Yang W, Xu J, Pan Y (2024) Barriers to industry adoption of AI video generation tools: a study based on the perspectives of video production professionals in China. Appl Sci 14(13):5770
78. Al-Gerafi MA, Goswami SS, Sahoo SK, Kumar R, Simic V, Bacanin N, Lasisi A (2024) Promoting inclusivity in education amid the post-COVID-19 challenges: an interval-valued fuzzy model for pedagogy method selection. Int J Manag Educ 22(3):101018
79. One essential consideration is the supervisor's expertise, as their specialized knowledge provides invaluable guidance, ensuring methodological rigor and increasing the likelihood of research success
80. Baydarova I, Collins HE, Saadi IA (2021) Alignment of doctoral student and supervisor expectations in malaysia. Int J Doctor Stud 16(1)
81. Wilmer H, Meadow AM, Brymer AB, Carroll SR, Ferguson DB, Garba I, Peck DE (2021) Expanded ethical principles for research partnership and transdisciplinary natural resource management science. Environ Manage 68(4):453–467
82. Kilag OK, Tokong C, Enriquez B, Deiparine J, Purisima R, Zamora M (2023) School leaders: the extent of management empowerment and its impact on teacher and school effectiveness. Excellen Int Multi-discipl J Educ (2994–9521) 1(1):127–140
83. Bernacki ML, Greene MJ, Lobczowski NG (2021) A systematic review of research on personalized learning: personalized by whom, to what, how, and for what purpose (s)? Educ Psychol Rev 33(4):1675–1715
84. Uru OB, Sudirman A, Nugroho AD (2021) Exploring cohesions in EFL academic writing: A state of the art on the study of cohesions. Elsya J Eng Lang Stud 3(2):141–149
85. Lapitan LD Jr, Chan ALA, Sabarillo NS, Sumalinog DAG, Diaz JMS (2023) Design, implement, and evaluate an online flipped classroom with a collaborative learning model in an undergraduate chemical engineering course. Educ Chem Eng 43:58–72
86. Khan ZA, Aziz IA, Osman NAB, Ullah I (2023) A review on task scheduling techniques in cloud and fog computing: taxonomy, tools, open issues, challenges, and future directions. IEEE Access 11:143417–143445
87. Rådberg KK, Löfsten H (2024) The entrepreneurial university and development of large-scale research infrastructure: exploring the emerging university function of collaboration and leadership. J Technol Transf 49(1):334–366
88. Shakil H, Farooq A, Kalita J (2024) Abstractive text summarization: state of the art, challenges, and improvements. Neurocomputing 128255
89. Tikhonova, E. V., Kosycheva, M. A., & Mezentseva, D. A. (2024). Ineffective strategies in scientific communication: textual wordiness vs. clarity of thought in thesis Conclusion section. Интеграция образования 28(2):249–265
90. Peters MD, Godfrey C, McInerney P, Khalil H, Larsen P, Marnie C, Munn Z (2022) Best practice guidance and reporting items for the development of scoping review protocols. JBI Evid Synth 20(4):953–968
91. Karunarathna I, De Alvis K, Gunasena P, Jayawardana A (2024) Bridging research gaps: how to write a focused and critical literature review
92. Zhong T, Liu Z, Pan Y, Zhang Y, Zhou Y, Liang S, Liu T (2024) Evaluation of openai o1: opportunities and challenges of agi. arXiv preprint arXiv:2409.18486

93. Taherdoost H (2021) Data collection methods and tools for research; a step-by-step guide to choose data collection technique for academic and business research projects. Int J Acad Res Manag (IJARM) 10(1):10–38
94. Chang WY, Garner M, Basner J, Weinberg B, Owen-Smith J (2022) A linked data mosaic for policy-relevant research on science and innovation: value, transparency, rigor, and community. Harvard Data Sci Rev 4(2)
95. Uttley L, Quintana DS, Montgomery P, Carroll C, Page MJ, Falzon L, Moher D (2023) The problems with systematic reviews: a living systematic review. J Clin Epidemiol 156:30–41
96. O'Connor LM, O'Connor BA, Lim SB, Zeng J, Lo CH (2023) Integrative multi-omics and systems bioinformatics in translational neuroscience: a data mining perspective. J Pharmaceut Anal 13(8):836–850
97. Patil DD, Wagh BD (2023) Research methodologies In: Multidisciplinary subjects, vol 1. Academic Guru Publishing House
98. Karunarathna I, Gunasena P, Hapuarachchi T, Gunathilake S (2024) Data collection fundamentals: a guide to effective research methodologies and ethical practices
99. Wood D, Moss SH (2024) Evaluating the impact of students' generative AI use in educational contexts. J Res Innov Teach Learn 17(2):152–167
100. Polatuly S, Ramankulov S, Usembayeva I, Kazakhbayeva D, Shektibayev N, Kurbanbekov B, Dosymov E (2024) Empowering future physicists: the role of case technology in research competency development. Cogent Educ 11(1):2362547
101. Suntharalingam H (2024) Enhancing digital learning outcomes through the application of artificial intelligence: a comprehensive review. Int J Innov Sci Res Technol 9(4):718–727
102. Siddiqui N, Gorard S (eds) (2022) Making your doctoral research project ambitious: developing large-scale studies with real-world impact. Taylor & Francis
103. Fernández-Cano MI, Marañón AA, Feijoo-Cid M (2021) The bachelor's thesis in nursing: characteristics and students' approach and satisfaction. Nurse Educ Pract 53:103067
104. Elliot DL (2022) A 'doctoral compass': strategic reflection, self-assessment and recalibration for navigating the 'twin'doctoral journey. Stud High Educ 47(8):1652–1665
105. Li X, Li Y (2023) Individualized and innovation-centered general education in a Chinese STEM University. Educ Sci 13(8):846
106. Humayyim MB (2024) Investigating language usage in the language classroom: a comparative analysis of operationalization methods and implications for educational research, Master's thesis, Syracuse University
107. Abuhassna H, Alnawajha S (2023) The transactional distance theory and distance learning contexts: theory integration, research gaps, and future agenda. Educ Sci 13(2):112
108. Dupin N (2023) Interdisciplinarity as a political instrument of governance and its consequences for doctoral training
109. Yu C, Zhao CG (2021) Continuing the dialogue between writing experts and academic librarians: a conceptual model of information-based academic writing in higher education. J Acad Librariansh 47(6):102454
110. Lin LH, Morrison B (2021) Challenges in academic writing: perspectives of engineering faculty and L2 postgraduate research students. Engl Specif Purp 63:59–70
111. Casal JE, Lu X (2021) 'Maybe complicated is a better word': second-language English graduate student responses to syntactic complexity in a genre-based academic writing course. Int J Eng Acad Purposes Res Pract 2021(Spring):95–114
112. Sharma Bagale K (2024) An investigation into usability and universal design of website cookie settings interfaces. Master's thesis, Oslo Metropolitan University
113. Chatfield T (2022) Critical thinking: your guide to effective argument, successful analysis and independent study
114. Saux G, Britt MA, Vibert N, Rouet JF (2021) Building mental models from multiple texts: how readers construct coherence from inconsistent sources. Lang Linguist Compass 15(3):e12409
115. Banks D (2021) Passive voice, first person pronouns and mental process verbs in the physical sciences reseach article. Rev Lingüíst Lenguas Aplicadas 16:37–48

116. Suleimanova OA, Lykova TA (2023) Academic writing details in critical perspective. Train Lang Cult 7(3):94–102
117. Xu X, Liu J, Zhang Y, Zhang H (2024) Development and implementation of personal learning environment-based writing for publication scaffolding platform for Ph. D. Students. Human Soc Sci Commun 11(1):1–12
118. A thesis is an extensive document demonstrating the writer's expertise and contributing to academic discourse
119. Henttonen A, Ahlberg K, Scheja M, Fossum B, Westerbotn M (2023) Students' ways of experiencing writing a bachelor's thesis: a phenomenographic interview study. High Educ Res Dev 42(7):1640–1653
120. Lim WM, Kumar S, Ali F (2022) Advancing knowledge through literature reviews:'what', 'why', and 'how to contribute.' Serv Ind J 42(7–8):481–513
121. Wanyama SB, McQuaid RW, Kittler M (2022) Where you search determines what you find: the effects of bibliographic databases on systematic reviews. Int J Soc Res Methodol 25(3):409–422
122. Agbo S (2024) The scholars' publication sourcebook: preparing academic manuscripts for journals in the humanities and social sciences. Taylor & Francis
123. Karunarathna I, Gunasena P, De Alvis K, Jayawardana A (2024) Structured reviews: organizing, synthesizing, and analyzing scientific literature. ResearchGate
124. Miranda D (2022) The impact of paraphrasing tools on students paraphrasing skills. Doctoral dissertation, UIn Ar-Raniry
125. Single PB, Reis RM (2023) Demystifying dissertation writing: a streamlined process from choice of topic to final text. Routledge
126. Samanci S, Atalay KD, Isin FB (2021) Focusing on the big picture while observing the concerns of both managers and passengers in the post-covid era. J Air Transp Manag 90:101970
127. Weeks A (2023) Planning and managing library projects. In: Essential leadership skills for health sciences information professionals, vol 95
128. Brunet M (2022) Conducting a PhD as a project: sharing insights from my doctoral journey. Int J Manag Proj Bus 15(7):1032–1047
129. Wan Q, Hu S, Zhang Y, Wang P, Wen B, Lu Z (2024) "It felt like having a second mind": investigating human-AI Co-creativity in prewriting with large language models. Proc ACM Hum-Comp Inter 8(CSCW1):1–26
130. Zhou H, Jiang FK (2023) 'The study has clear limitations': presentation of limitations in conclusion sections of PhD dissertations and research articles in applied linguistics. Engl Specif Purp 71:34–47
131. Bouachiba, A. M. I. R. A. (2021). The Importance of Proofreading in Reducing Students' Mistakes in Writing (Master's thesis).
132. Heffernan T (2021) Academic networks and career trajectory: 'There's no career in academia without networks.' High Educ Res Dev 40(5):981–994
133. Bawack RE, Wamba SF, Carillo KDA, Akter S (2022) Artificial intelligence in E-Commerce: a bibliometric study and literature review. Electron Mark 32(1):297–338
134. Ghodratnama S, Beheshti A, Zakershahrak M, Sobhanmanesh F (2021) Intelligent narrative summaries: from indicative to informative summarization. Big Data Res 26:100257
135. Tailor the manuscript to the target journal's submission requirements, including formatting, word count, and citation style
136. Montakhaby Nodeh S (2024) Crafting research narratives that transcend technical Jargon
137. Sharma P, Said Z, Kumar A, Nizetic S, Pandey A, Hoang AT, Tran VD (2022) Recent advances in machine learning research for nanofluid-based heat transfer in renewable energy system. Energy Fuels 36(13):6626–6658
138. Lenarduzzi V, Besker T, Taibi D, Martini A, Fontana FA (2021) A systematic literature review on technical debt prioritization: strategies, processes, factors, and tools. J Syst Softw 171:110827

139. Prasanna B, Kothapalli P, Vasanthan M (2024) The role of quality assurance in clinical trials: safeguarding data integrity and compliance. Cureus 16(8):e67573
140. Martínez-Vázquez RM, Canovas-García F, Milan-García J, De Pablo-Valenciano J (2024) First steps in ai to support university teaching in the subject of economics of labor incentiVES. In: ICERI2024 proceedings. IATED, pp 3643–3649
141. León SP, García-Martínez I (2021) Impact of the provision of PowerPoint slides on learning. Comput Educ 173:104283
142. Meredith D (2021) Explaining research: How to reach key audiences to advance your work. Oxford University Press
143. Youknovsky A, Bowers J (2024) Hook your audience: catch the attention of your audience right from the start. Sell your research: public speaking for scientists. Springer International Publishing, Cham, pp 65–77
144. Andry T, Hurter C, Lambotte F, Fastrez P, Telea A (2021) Interpreting the effect of embellishment on chart visualizations. In: Proceedings of the 2021 CHI conference on human factors in computing systems. pp 1–15
145. Gregor S, Hevner AR (2013) Positioning and presenting design science research for maximum impact. MIS Quart 337–355
146. Harasim LM (1995) Learning networks: a field guide to teaching and learning online. MIT Press
147. Elliott CA (2019) Designing technical communications portfolios: an evaluation of current practices. Doctoral dissertation, University of Wisconsin-Stout
148. Marginson S, Rhoades G (2002) Beyond national states, markets, and systems of higher education: a glonacal agency heuristic. High Educ 43:281–309

Ethical Considerations in Research and Writing

Abstract In an increasingly complex academic environment, this work offers a vital and interesting investigation of how integrity forms the core of scholarly activity. It explores the fundamental ideals of honesty, responsibility, and openness that safeguard the reputation of individual academics and the confidence in academic organizations. It addresses the grave effects of misconduct, from data falsification to unethical writing practices, using accessible knowledge and practical relevance, and the increasing difficulties of artificial intelligence and worldwide cooperation. Emphasizing preventive education, responsible data management, fair authorship, and institutional accountability, readers are presented with practical tools for avoiding ethical traps. Especially pertinent for students, academics, and research leaders, this study provides a clear road map for fostering a responsible research culture based on moral behavior. Whether in science, engineering, or the social sciences, the message is clear: ethical research is good practice and the basis for innovation, credibility, and public trust. Anyone dedicated to preserving the integrity and influence of academic work in the fast-changing research context should read this interesting chapter.

1 Introduction

Ethical issues fundamentally shape the work of research and academic writing. They offer the normative bases on which academic trust, intellectual rigor, and scientific investigation are founded. The ethical aspects of research and writing have grown more complicated as academia operates increasingly in a global, multidisciplinary, and technologically sophisticated context, demanding careful comprehension and vigilant application [1]. Ethical debates determine how academic communities establish legitimacy, acknowledge intellectual achievements, protect vulnerable stakeholders, and create, understand, and share knowledge. The foundation of ethical research is integrity. This concept includes respect for individuals, organizations, knowledge systems; responsibility; honesty; and openness. Research integrity ensures that inquiry processes are conducted with methodological rigor, that data is collected and recorded accurately, and that findings are disseminated in a way

A. Almusaed et al., *Practice of Research Methodology in Civil Engineering and Architecture*, https://doi.org/10.1007/978-3-031-97393-2_25

that does not mislead or distort results [2]. Serious violations of academic standards include breaches of integrity such as fabrication, falsification, and plagiarism, which have far-reaching consequences. These types of misconduct undermine the empirical foundation of knowledge and damage public trust in academic institutions, thereby impeding the cumulative growth of research. In turn, scholarly societies have established official codes of conduct, institutional review processes, and educational systems to protect against such abuses. However, ethical scholarship goes beyond avoiding unethical behavior; it also involves developing values and professional attitudes that guide researchers toward responsible practices [3]. Among these characteristics are intellectual humility, epistemic honesty, and a readiness to critically examine the boundaries of one's output. Researchers must disclose methodological decisions, acknowledge uncertainties, and contextualize their work within broader scholarly discussions. Transparency in methodological reporting and data interpretation represents an ethical commitment and an epistemic necessity. It facilitates informed peer assessment, encourages constructive criticism, and aids reproducibility. From this perspective, ethical research contributes to advancing knowledge and the validity and sustainability of academic inquiry [4]. Ethical research encompasses several facets, with one being responsibility. It operates at multiple levels—individual, collective, institutional, and societal. Individually, academics are expected to adhere to disciplinary standards and conduct research that respects the rights and dignity of all individuals. In collaborative environments, shared ethical understanding and equitable attribution of credit and responsibility are paramount, especially in transdisciplinary or global initiatives [5]. Universities and funding agencies are responsible for establishing robust oversight systems, creating environments conducive to ethical behavior, and responding appropriately to allegations of misconduct at the institutional level. At the societal level, researchers must remain cognizant of their work's significant, broader implications, especially when results influence public policy, clinical practice, or cultural perceptions. The ethical treatment of human subjects underscores the moral obligations associated with empirical research even further [6]. Any ethically sound study involving human participants is based on informed consent, confidentiality, minimizing harm, and respecting participants' autonomy. When working with vulnerable populations, researchers must exercise caution to ensure that participation is voluntary and safeguards are in place to prevent exploitation or coercion. Institutional ethics review boards and research ethics committees are crucial in providing oversight and ensuring that study designs adhere to established ethical standards. The ethical dimensions of academic writing and publication are equally significant. Authorship should be assigned according to clear criteria and reflect genuine intellectual contributions [7]. Ghost authorship and honorary authorship are two practices that distort the attribution of academic work and undermine the integrity of academic production. Proper citation is essential for acknowledging previous work, enabling verification, and fostering open academic discourse. Engagement with predatory publishing outlets, excessive self-citation, and citation manipulation further compromise the integrity of the scholarly record. Consequently, ethical writing should be grounded in fairness, honesty, and intellectual responsibility

Fig. 1 Core ethical dimensions in research and academic writing

principles [8]. The rise of open science initiatives, digital archives, and collaborative research platforms has created ethical opportunities and challenges in recent years. Although these developments promote transparency and enhance access to information, they raise concerns about intellectual property, data privacy, and the equitable distribution of academic credit. Ethical frameworks must be reassessed and redefined to remain relevant as academic institutions adapt to these evolving circumstances. This necessitates ongoing reflection, education, and the establishment of flexible guidelines that support ethical behavior across diverse research contexts [9]. In academic practice, ethical issues in research and writing are not peripheral; they constitute its core. Upholding the integrity of scholarly work, protecting the welfare of research participants, and fostering an academic community grounded in mutual respect and shared intellectual values all depend on ethical principles. The necessity for heightened ethical awareness and robust institutional support is increasingly urgent as the research landscape becomes more complex and interconnected. This chapter aims to elucidate the various dimensions of ethical responsibility in research and writing by providing a comprehensive framework that academics can utilize to navigate the normative aspects of their professional practices [10]. The fundamental ethical values of integrity, accountability, respect, and openness that shape academic research are illustrated in Fig. 1. In an increasingly global and multidisciplinary research landscape, these interconnected ideals ensure methodological integrity, safeguard research participants, and uphold academic legitimacy.

2 Research Integrity and Academic Misconduct

A basic foundation of academic and scientific activity, research integrity includes ideas and beliefs guaranteeing research reliability, openness, and ethical soundness [11]. From developing research questions and creating procedures to collecting data, analyzing, reporting, and publishing findings, this is a normative framework guiding all phases of academic inquiry. In the natural and social sciences, preservation of the validity and trustworthiness of scientific information, as well as the credibility of academic institutions and the larger research ecosystem, depends on research integrity. It promotes a culture of intellectual honesty, responsibility, and mutual respect, creating an environment where knowledge may develop cooperatively and credibly. At the same time, academic dishonesty is a serious breach of fundamental ethical imperatives involving intentional acts that undermine research integrity such as fabrication, falsification, and plagiarism [12]. This behavior can have significant consequences, from misallocating research funds to losing public confidence in science by distorting the scientific record and compromising the reproducibility of results. People and organizations involved in academic dishonesty can suffer tremendous and lasting reputational harm. Sometimes, the effects go beyond personal or corporate reputation and might question whole research projects or public policy efforts based on erroneous or false data. Romenti [13] claims that the increasing integration of artificial intelligence (AI) into company operations at all levels brings creative possibilities and various complicated ethical issues that require thoughtful analysis [13]. In a time of more public scrutiny, worldwide cooperation, and competitive research settings, the effects of neglecting research integrity are especially severe. Therefore, encouraging an ethical research culture calls for the proactive development of ethical awareness, applying unambiguous institutional policies, supporting activities that strengthen integrity at all stages of the research process, and enforcing harsh penalties [14]. Focusing on the typologies and repercussions of academic fraud, this part analyzes critically the fundamental elements of research integrity. It also addresses educational and strategic initiatives to promote a research culture based on accountability, rigor, and respect for the academic community by preventing ethical violations. From developing research questions and methodological design to data collecting and analysis to reporting and publishing findings, this part provides a normative framework guiding all phases of academic inquiry [15]. A visual framework comparing research integrity and academic dishonesty is shown in Fig. 2. It describes fundamental ethical values such as honesty, responsibility, and misconduct categories, including fabrication and plagiarism. The figure underlines the need to promote a culture of academic ethics and responsibility by showing how both routes have significant effects.

Research integrity is crucial in the natural and social sciences to protect scientific knowledge's legitimacy, reliability, and trustworthiness. It also upholds the integrity of academic institutions and the broader research community. According to Kretser et al. [16], a Scientific Integrity Consortium outlined recommended principles and best practices that could support a cross-disciplinary framework to foster consensus on standards of scientific integrity. The new guidance aims to better equip researchers for the challenges of an ever-changing research landscape. Two guiding principles at the center of these recommendations serve as touchstones for ethical and scientifically rigorous conduct [16]. It nurtures a culture of intellectual honesty, responsibility, and mutual respect, fostering conditions for knowledge to develop cooperatively and credibly. Conversely, academic dishonesty seriously violates essential ethical principles, typically involving fabrication, falsification, and plagiarism, collectively referred to as FFP. These actions are deliberate and directly undermine

Fig. 2 Conceptual overview of research integrity and academic misconduct

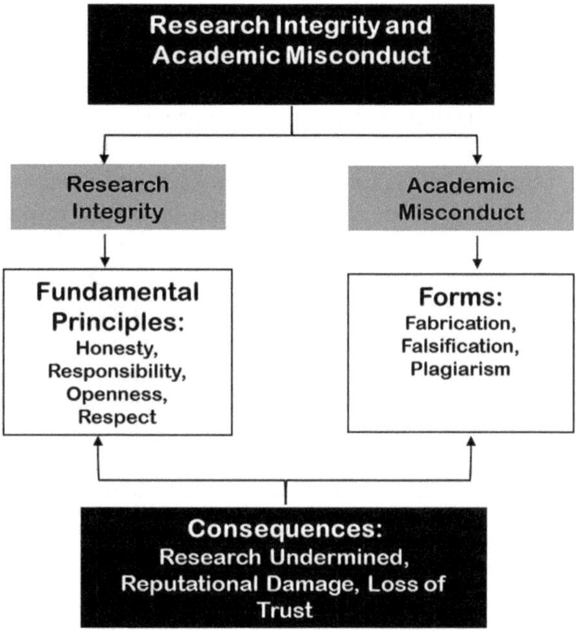

the integrity of the scientific process. Bhaskar [17] argues that publication fraud encompasses broader forms of dishonesty in research design, implementation, and reporting beyond FFP. This includes deviations from accepted protocols, violations of disciplinary standards, and active engagement in or concealment of unethical conduct. Notably, plagiarism is the unauthorized use of another individual's intellectual property without proper attribution. In contrast, research integrity signifies a commitment to vital ethical values such as responsibility, openness, honesty, and respect that ensure academic research's legitimacy and moral obligation [17]. Such misconduct can cause significant harm, ranging from misallocation of research funds to a decline in public confidence in science; it also disturbs the scientific record and compromises the reproducibility of results. The reputational damage experienced by individuals and organizations involved in misconduct can be substantial and enduring, sometimes jeopardizing entire research agendas or policy projects stemming from flawed or fraudulent work. The consequences of neglecting research integrity are particularly evident in an era characterized by increased public scrutiny, global collaboration, and competitive research environments. Therefore, fostering an ethical research culture requires the proactive development of ethical awareness, the implementation of clear institutional policies, and the promotion of activities that support integrity at all stages of the research process, rather than merely imposing punitive measures [18]. This section critically analyzes the fundamental elements of research integrity, focusing primarily on the typologies and ramifications of academic dishonesty. It also addresses educational and strategic initiatives to prevent ethical violations and cultivate a research culture grounded in accountability, rigor, and respect for the academic community. Including definitions, difficulties, and their relevance in scientific research, Table 1 outlines important features of research integrity and academic dishonesty. It emphasizes the requirement of ethical behaviors, unambiguous institutional norms, and appropriate education to preserve public confidence and guarantee the validity of academic work in all fields.

Sustainability, for example, in civil and architectural engineering, is not just a technical goal but also a reflection of ethical responsibility in research and practice. Designing energy-efficient, net-zero buildings and developing environmentally friendly materials like low-carbon concrete call for high research integrity. In these situations, the correctness of lifespan assessments, environmental impact studies, and performance simulations is vital. Any change or distortion of data, whether intentional or accidental, can lead to the application of unsafe or ineffective building methods. Such infractions erode trust in sustainable engineering

Table 1 Overview of research integrity and academic misconduct

Theme	What it means	Main challenges	Why it matters
Research integrity	Being honest, ethical, and transparent in all parts of research	Pressure to publish or succeed can lead to cutting corners	Builds trust in science and ensures valid, reliable results
Importance in science	Makes research believable and supports the reputation of institutions	Weak policies or lack of awareness about ethics	Protects scientific progress and public confidence
Academic misconduct	Doing wrong on purpose like making up data, changing results, or copying	Misconduct can be hidden and hard to prove	Can ruin careers, waste money, and mislead other researchers
Other ethical violations	Breaking rules in how research is planned, shared, or reported	Different fields may have other ethical standards	May damage entire research projects or decisions based on false results
AI and ethics	Using AI tools can help, but also bring new ethical questions	Lack of clear rules for AI use in research	New guidelines are needed to ensure AI is used fairly and responsibly
Institutional role	Universities must set clear rules and support honest research	Some may not take action until problems arise	Strong policies prevent problems and promote a healthy research culture
Education and training	Teaching researchers about ethics and good practices	Not always required or equally offered to all	Helps prevent mistakes and builds shared responsibility for integrity
Common standards	Shared ethical rules across all research fields	Hard to agree across different disciplines	Makes collaboration easier and sets clear expectations for all researchers

solutions, public safety, and resource management. Maintaining ethical standards and supporting long-term social and environmental well-being depend on ensuring openness, reproducibility, and scientific rigor in sustainability-related research.

2.1 Avoiding Misconduct

Fundamentally, academic dishonesty includes behaviors intended to mislead or falsify the research process, conclusions, or intellectual contributions. These actions undermine the reputation of the broader academic community as well as the integrity of individual scholars. Among the most egregious and well-known forms of academic dishonesty are fabrication, falsification, and plagiarism, often referred to collectively as FFP. These behaviors represent fundamental violations of ethical research standards and can significantly impact scientific advancement, institutional trust, and public perception of research [19]. Fabrication is the intentional creation of facts, outcomes, or discoveries that are later presented as though they originated from actual observation, experimentation, or empirical research. This kind of wrongdoing is particularly egregious as it generates an entirely false research record that cannot be independently validated or duplicated, undermining the foundation of scientific trustworthiness and transparency. Resnik [20] asserts that the scientific community upholds rigorous policies against data fabrication and falsification, imposing severe penalties such as loss of research funding, termination of employment, or even legal repercussions. These rules apply to data concerning both observable and unobservable phenomena. Resnik argues that such stringent penalties imply a widespread belief among scientists that research's principal aim is to pursue genuine and accurate hypotheses about actual phenomena, regardless of whether those realities are physically observed. Therefore, the prohibition of fabrication entails procedural adherence and a more profound epistemic commitment to truth-seeking in research [20]. For example, a researcher might create false responses in a sociological survey to bolster an argument or falsify data in a laboratory setting to support a preferred theory. The harm of fabrication extends beyond the apparent effort; once fabricated data enters the academic record, it can mislead other scholars who, unaware of the deception, may rely on these results as a foundation for their research. Entire areas of research risk being built on false or inadequate data, distorting the trajectory of scientific discovery and potentially leading to erroneous findings, misguided policies, or harmful treatment procedures [21]. While distinct from fabrication, falsification is equally harmful to research integrity. It involves intentionally distorting research materials, techniques, or data to misrepresent findings. This can include altering visual presentations such as graphs, images, or charts to exaggerate effects or conceal discrepancies, intentionally modifying statistical analyses to yield significant results, or selectively excluding data points contradicting a researcher's premise. Unlike fabrication, falsification is the deliberate distortion of properly gathered data. This may involve altering numbers, skewing statistical conclusions, or omitting opposing data to misrepresent results. Falsification, as Kennedy [22] notes, undermines the reliability and transparency of research by presenting manipulated data as truthful, thereby deceiving colleagues and jeopardizing the integrity of scientific inquiry investigation. Its often-subtle nature makes falsification particularly insidious; minor modifications in data reporting or analysis may evade detection during peer review, especially in complex or specialized fields [22]. However, the overall impact of such distortions can be substantial, undermining the credibility of published results and jeopardizing the repeatability essential to robust science. Focusing on the unlawful use of another person's intellectual work without appropriate credit, plagiarism represents another significant violation of research ethics. Although many people consider plagiarism to be the exact duplication of material, it extends far beyond this narrow definition. It includes misappropriating data, statistics, or conceptual frameworks; paraphrasing another's ideas or arguments without attribution; using previously published content (including one's own) without citation, often called self-plagiarism; and data theft [23]. Plagiarism is a serious violation of academic integrity, as it involves using someone else's intellectual work without proper acknowledgment. It misleads readers about the true source of ideas and disregards original authorship. Plagiarism severely undermines personal and institutional integrity in academic environments, which rely on originality, intellectual contribution, and ethical conduct, profoundly impacting reputation, funding, and professional advancement.

– *Plagiarism's Impact on Education*

According to Mulenga [24], plagiarism poses a significant threat to the legitimacy of an academic system, affecting students, teachers, and institutions, and therefore has far-reaching consequences for education. This unethical practice deceives individuals and ultimately contributes to widespread plagiarism in higher education, leading to a decline in educational quality, which emphasizes the importance of this essay. Higher education institutions are responsible for nurturing academic knowledge and cultivating moral character in their students. Thus, in pursuing effective and viable solutions to the issue of plagiarism, it is crucial to investigate its root causes, including academic pressure, neglect of ethical principles, and the inability of academic institutions to implement a strong and adaptive plagiarism detection system. It can unjustly advantage those who engage in it over more ethically conscientious colleagues [24]. Moreover, when academic papers contain plagiarism, it erodes the trust of editors, reviewers, and readers, thus fostering a more profound skepticism toward academia and publishing. Beyond the triangle of FFP, other unethical behaviors in writing and research, although perhaps less frequently debated, also bear significant ethical consequences [25]. Ghost authorship is one such practice; it involves individuals who have substantially contributed to the study, such as conducting experiments, performing statistical analysis, or writing key sections of the manuscript, not being acknowledged as co-authors or collaborators [26]. Intentional exclusion or institutional pressure may lead to this omission, compromising transparency in attributing academic efforts. Conversely, gift authorship pertains to including those who have done little or no work on the study as co-authors, often as a courtesy or to gain favor. This practice undermines the meritocratic foundation of authorship and can obscure the true origins of ideas or discoveries. Redundant or duplicate publication, publishing the same or substantially similar content in multiple venues without appropriate disclosure, is another ethically questionable activity [27]. Although studies might be driven to improve the exposure or influence of their work, redundant or duplicate publishing compromises academic integrity by intentionally inflating publication records, distorting citation statistics, and abusing editorial and peer review resources. Nazarovets [28] draws attention to the increasing influence of unethical third-party services, "paper mills," supporting and enabling such wrongdoing. Among their many fraudulent services are paper submission and communication help, the assignment of phony co-authors, citation manipulation, and the false boosting of bibliometric measures. These companies methodically create and distribute phony academic papers using misleading techniques like data and picture forgery, plagiarism, concurrent submissions, and the generation of fake peer review reports. According to Nazarovets, paper mills' output usually lacks genuine experimental data. Inherently deceptive, they seriously endanger the legitimacy of intellectual communication and the accuracy of the academic record [28]. In fields where publication counts influence funding decisions or career advancement, this behavior skews evaluations of scholarly output and undermines the fair distribution of academic accolades. These actions, whether overt or covert, threaten the integrity of academic research. They distort the mechanisms by which information is verified, disseminated, and rewarded. Furthermore, particularly when misconduct occurs in high-stakes or publicly funded research, the consequences can be severe, including retractions, disciplinary measures, loss of credibility, and legal repercussions. Institutions, publications, and professional bodies increasingly emphasize ethical awareness and preventive education to tackle these issues [29]. Ultimately, individual academics bear the responsibility for ethical conduct; they must commit to upholding the standards of their fields and genuinely support the collective endeavor of knowledge generation. Figure 3 illustrates plainly how plagiarism begins and what harm it produces. Beginning with pressures such as excessive expectations and low ethical standards, it leads to issues such declining educational quality and lack of confidence. It also highlights more negative behaviors such duplicate publishing and ghost authoring. Ultimately, it tells us that resolving the issue depends on strong ethics, good policies, and personal integrity.

Students in civil and architectural engineering education, particularly Ph.D studies, frequently face notable academic and institutional constraints that might lead to unethical behavior [30]. This could show up as falsifying sustainability performance data, copying structural design reports, or tampering with technical tests, such as concrete resistance or load capacity. Such infractions compromise academic honesty, scientific evidence, and the safety of human-built surroundings. A mix of thorough ethics education, clear oversight, and enforceable integrity policies is essential to guarantee epistemic reliability, professional responsibility, and ongoing public confidence in the results of the engineering enterprise, given the nature of engineering research, which directly affects vital infrastructure in society. Along with other unethical actions like ghost authorship, gift authorship, and repeating publications, Table 2

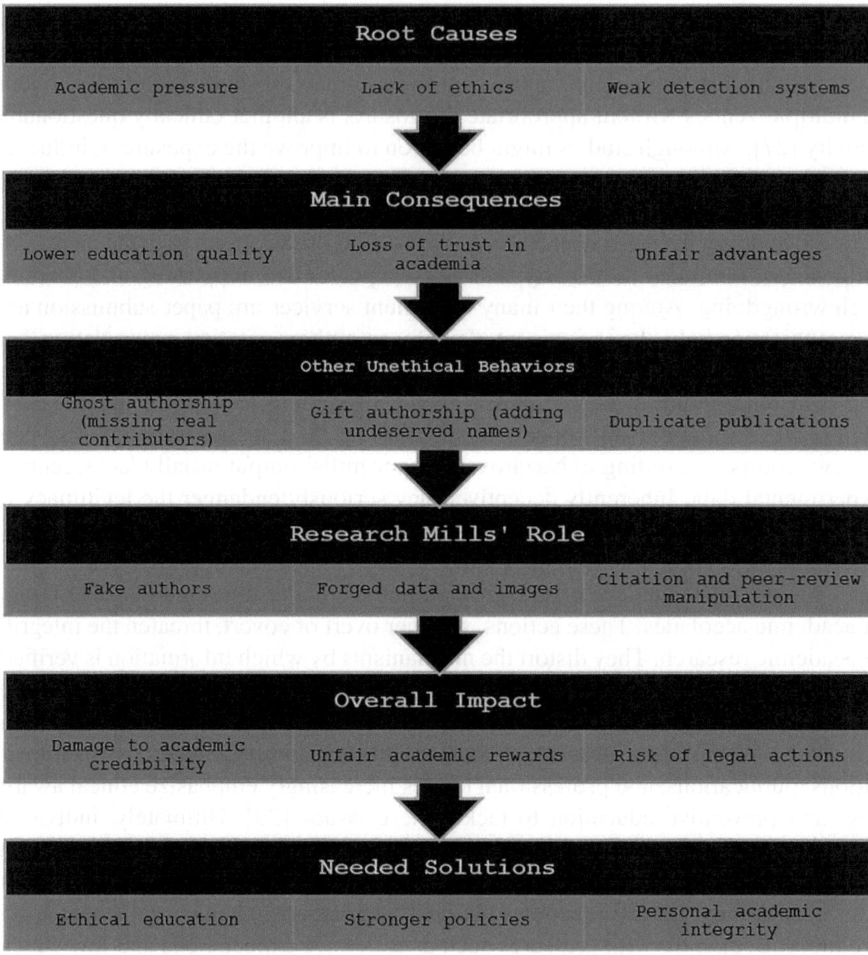

Fig. 3 How plagiarism affects education and research

comprehensively summarizes standard forms of academic dishonesty: fabrication, falsification, and plagiarism. Clearly defined with pertinent ethical consequences and specific illustrations, every category offers a fast yet comprehensive insight into how these actions undermine scientific output's legitimacy, fairness, and integrity.

2.2 *Maintaining Transparency*

Transparency in academic research is a fundamental ethical and epistemological criterion [30]. Researchers are responsible for thoroughly and accurately revealing the methods, data sources, analytical techniques, and interpretive frameworks that

Table 2 Categorization of common forms of academic misconduct and their ethical implications

Type of misconduct	Definition	Examples	Ethical implications
Fabrication	Inventing data or results and presenting them as genuine	Creating fake survey responses or experimental data to support a hypothesis	Produce a false research record, mislead future studies, and damage scientific credibility
Falsification	Manipulating research materials, processes, or data to misrepresent findings	Altering graphs, omitting conflicting data, and adjusting statistical outputs to achieve significance	Undermines transparency, objectivity, and the reproducibility of research
Plagiarism	Using another's work without proper attribution, including ideas, text, or data	Copying content, paraphrasing without citation, and reusing previous work (self-plagiarism)	Violates intellectual property rights, misleads readers, and compromises academic fairness
Ghost authorship	Excluding individuals who significantly contributed to the research	Omitting a lab technician or co-author who helped write significant parts of the manuscript	Obscures actual contribution, diminishes transparency, and breaches collaborative ethics
Gift authorship	Including individuals who made little or no contribution as authors	Adding senior faculty to a paper to gain favor despite no involvement	Distorts academic credit systems and weakens merit-based authorship
Redundant publication	Publishing the same or substantially similar work in multiple venues without disclosure	Submitting a previously published article to a second journal to inflate the publication count	Misleads the scholarly record, wastes editorial resources, and distorts productivity metrics

support their results [31]. Transparency functions as a normative value and a practical need in modern academic research, where reproducibility, critical examination, and cumulative knowledge generation are fundamental to scientific progress. Freese [33] claims that recent advances in encouraging openness and reproducibility in quantitative social science mirror a more general path of methodological evolution predating present worries about research dependability. It encourages methodological rigor, helps confirm assertions, and fosters an atmosphere that promotes academic cooperation and critical engagement [32]. Methodological transparency refers to the systematic, consistent, and open recording of all methodologies utilized during the study process. Rigorous and methodologically sophisticated quantitative research, when conducted openly and reproducibly, is, as Strobl et al. [34] assert, essential for the evolution of theoretical frameworks, the enhancement of professional practice, and the development of evidence-based policy. Conversely, poorly designed studies, reliance on outdated or unclear analytical techniques, and insufficient communication of their methodology and results significantly undermine such research's credibility and practical value [33]. This includes analytical methodologies, data collection methods, instrumentation, inclusion and exclusion criteria, and sample frames. Inadequate reporting of these factors impedes other researchers' ability to assess a study's internal validity, replicate its results under similar circumstances, or critically evaluate the strength of its conclusions. Establishing the validity and replicability of empirical claims in fields increasingly focused on reproducibility and open research relies on comprehensive documentation of methods [34]. The accurate, thorough, and unembellished depiction of scientific evidence is essential, along with methodological disclosure and data transparency. Ethically speaking, researchers must reveal all pertinent information, including negative, ambiguous, or conflicting results, since such openness preserves the integrity of scientific investigation. According to Hammerschmidt et al. [36], deontological ethics emphasizing responsibilities and moral obligations have become the dominant framework in discussions on research ethics. Notably, influential global players' normative criteria and institutional practices help shape this ethical paradigm, thereby highlighting the need for critical thought regarding the cultural and geopolitical aspects of ethical values in research [35]. Such inclusiveness reduces selective reporting and the concealment of results that some may find uncomfortable or conceptually challenging. Furthermore, techniques like data dredging (the post hoc investigation of data to produce unsupported hypotheses) and p-hacking (the manipulation of statistical analyses to achieve significance) pose significant challenges to the integrity of study results. Although occasionally subtle, these methods fundamentally distort the empirical foundation of academic assertions and contribute to disseminating non-reproducible results throughout the literature. Conversely, data openness enhances the epistemic reliability of published work and promotes a more accurate and fair academic record [36]. Equally important is acknowledging study limitations, a fundamental aspect of ethical openness. Every empirical study has limitations, including sample size restrictions, methodological flaws, measurement inaccuracies, and contextual factors [37]. Admitting the shortcomings of one's study is not a sign of failure but rather a demonstration of academic integrity and epistemic humility, traits vital to advancing knowledge. Such openness reflects a researcher's readiness to engage critically with the boundaries of their investigation, fostering trust and intellectual integrity within the academic community. Matheson [39] emphasizes the relevance of epistemic autonomy in this regard, as it is closely related to a constellation of intellectual characteristics, such as curiosity, inquisitiveness, intellectual tenacity, and intellectual boldness. These attributes promote a scholarly attitude in which academics pursue knowledge while recognizing its provisional nature, thus supporting a culture of thoughtful and responsible scientific work. This approach enables future researchers and readers to understand results within their relevant context, evaluate their significance, and identify avenues for further investigation. Conversely, concealing or downplaying methodological flaws distorts the evidential power of research and undermines the standards for responsible scientific discourse [38]. Data misinterpretation or erroneous assumptions are natural and unavoidable parts of the scientific process. Therefore, the validity and advancement of research depend on how the academic community responds to mistakes rather than on their absence. Ethical research requires prompt and transparent correction of errors using well-established tools such as errata, corrigenda, and retractions. While these remain the standard methods for addressing post-publication errors, Teixeira da Silva [40] emphasizes that the academic publishing landscape has evolved to include more complex approaches, such as manuscript versioning, modifications, partial retractions, and "retract and replace" protocols. These changes demonstrate an increasing awareness that maintaining scientific integrity, dependability, and credibility requires institutional procedures and a culture of responsibility. The academic community reinforces its commitment to truth-seeking and public trust by regarding error correction not as a failure but as a necessary part of rigorous research. Although these actions may provoke professional discomfort or reputational risks, they affirm a researcher's dedication to truth and accountability. More broadly, they support the self-correcting nature of science, thus helping to ensure the resilience and reliability of academic endeavors [39]. Table 3 shows why openness is so important in academic research. It reveals the key concepts researchers should emphasize, the usual obstacles that could hinder progress, and pragmatic solutions to address them. Important first steps are being transparent about techniques, acknowledging shortcomings, honestly correcting errors, and freely exchanging data. These methods taken together assist to increase the reliability, credibility, and value of research for the academic community and society.

The recent emergence of open scientific practices has significantly shaped the institutionalization of openness in academic research. These practices, including the public archiving of research materials and protocols, the open sharing of raw datasets and statistical code, and the pre-registration of study designs and hypotheses, have transformed the conventions of academic communication toward greater responsibility and reproducibility. Pre-registration and Registered Reports have arisen as particularly influential drivers among these advancements. These formats aim to mitigate systematic biases such as p-hacking and result switching by clearly distinguishing hypotheses and analytical procedures before data collection begins, as Lakens et al. [41] noted. This allows the scientific community to evaluate the strength of evidence for assertions with greater transparency and epistemic rigor. Ultimately, such policies cultivate confidence and integrity in scientific results and bolster knowledge creation's cumulative and self-correcting nature. These policies strive to reduce questionable research practices, enhance the repeatability of results, and democratize access to knowledge production. Open science projects advocate the ethical imperative of openness by providing practical frameworks that promote collaborative inquiry and interdisciplinary cooperation through external verification and reuse [40]. Openness is a fundamental ethical commitment that emphasizes academic research's legitimacy, reliability, and collective power and is not merely a procedural requirement. Researchers fulfill their obligations to the academic community and society through comprehensive methodological reporting, truthful data presentation, transparent acknowledgment of limitations, and accountable error correction. While principles of openness evolve with methodological and technological advancements in academia, they remain essential to the pursuit of meaningful

Table 3 Ethical foundations and best practices for maintaining transparency in academic research

Theme	Main idea	Challenges	Solutions/best practices
Transparency in research	Researchers must fully disclose methods, data sources, and interpretations	Poor documentation, selective reporting, unclear methods	Full, open reporting of methodologies and results
Methodological rigor	Open, systematic recording of all methodologies ensures validity and reproducibility	Outdated techniques, vague methodological descriptions	Consistent, transparent methodology documentation
Ethical integrity	Ethical research demands revealing all results, including negative or ambiguous findings	Data dredging, p-hacking, concealment of inconvenient results	Honest presentation of all findings and acknowledgment of study limitations
Correction of errors	Prompt correction of mistakes is part of ethical research	Professional discomfort, reputational risks	Use of errata, corrigenda, retractions, and manuscript versioning
Acknowledgment of limitations	Recognizing study flaws demonstrates academic integrity	Concealing weaknesses distorts research credibility	Open discussion of limitations and methodological flaws
Open science practices	Public sharing of data, code, and pre-registered protocols enhances reproducibility	Bias, limited access to research materials	Pre-registration, Registered Reports, open data repositories
Evolution of academic norms	Cultural shift toward openness and reproducibility in research communication	Resistance to change, lack of institutional support	Institutional promotion of open science policies and collaborative frameworks
Ethical commitment to openness	Openness is a core ethical obligation, not just a formal requirement	Procedural focus without ethical grounding	Emphasize openness as fundamental to public trust and scientific legitimacy

and reliable information [41]. For instance, transparency is a core ethical duty in civil and architectural engineering doctoral research. It also enables researchers to provide a comprehensive report when documenting study activities, including the methodology used, outlined analysis, the data collection process, and any limitations. P-hacking, selective reporting, and hiding equivocal findings undermine scientific integrity, hinder reproducibility, and distort evidence-based decision-making. Open science practices—such as pre-registration, public data repositories, and formal error correction—enhance academic accountability. This improves the epistemic

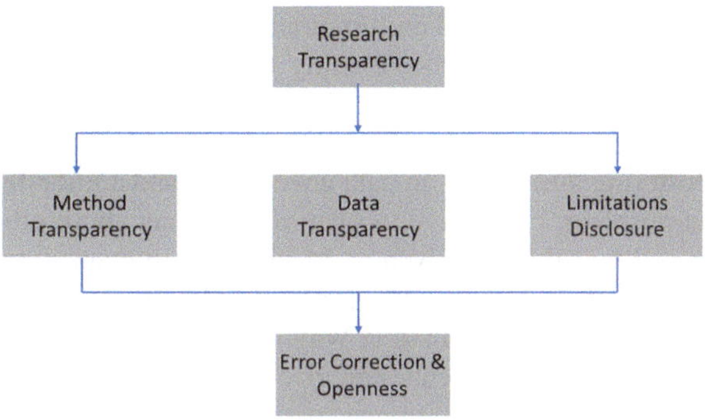

Fig. 4 Key dimensions of research transparency in scholarly inquiry (*Source* Authors)

validity of research findings and ensures that engineering knowledge meets the scientific community's ethical obligations to be verifiable and credible. Research transparency revolves around four interrelated components: methodological transparency, data transparency, disclosures of limitations, and error correction through open science practices, as illustrated in Fig. 4. These factors contribute to establishing validity, reproducibility, and ethical conduct in research and scholarship. Incorporating these practices at every stage of the research process strengthens critical evaluation, nurtures accountability, and reinforces public and academic confidence in the integrity and reliability of scientific knowledge.

2.3 *Ensuring Accountability*

Accountability in research is the shared responsibility of individuals, research groups, and organizations to uphold ethical standards throughout the research process. It is not a solitary issue but a collective commitment across all academic and scientific activity levels. Research is conducted with honesty, transparency, and a sense of responsibility toward the academic community and broader society through accountability. Four interrelated levels—individual, collaborative, institutional, and public—help clarify this concept [42]. Accountability at the personal level begins with each researcher's commitment to ethical conduct, which entails honestly reporting results even when they are unexpected or inconclusive, fairly assigning authorship, and managing data correctly and securely. Researchers should always disclose potential conflicts of interest that could compromise the objectivity of their work and avoid misconduct, such as data fabrication, falsification, or plagiarism. These responsibilities govern all stages of research, from formulating the research topic to collecting and analyzing data, as well as writing and disseminating outcomes [43]. Personal responsibility also includes ongoing self-reflection and a willingness to learn and adapt to ethical challenges. In group settings, accountability expands to encompass team-based obligations [44]. Team composition usually includes individuals with different responsibilities, disciplinary backgrounds, and degrees of accountability in interdisciplinary and multinational research projects. The principal investigator (PI), lead researcher, or coordinating author is responsible for clearly expressing, implementing, and upholding strong ethical standards throughout the research team within such complex systems. This duty includes clearly defining personal responsibilities, fostering agreement on data governance and authorship policies, and establishing an atmosphere conducive to open discussions about ethical issues. Mathur et al. [46] noted that the effectiveness of multidisciplinary research depends on participants' shared ethical awareness. Empirical data from a required Responsible Conduct of Research (RCR) course at Wayne State University reveal notable disciplinary variations in how students perceive the validity and applicability of ethical training, with pronounced differences particularly apparent in the social sciences and humanities. These results underscore the pressing need for morally inclusive and epistemologically sensitive training frameworks that reflect the field's diverse norms, techniques, and epistemic cultures. Such policies are essential to promote rigor, ensure repeatability, and preserve ethical integrity in modern cooperative research environments [45]. In collaborative research, effective leadership means appropriately managing power dynamics, such as the influence of senior researchers on junior colleagues, while fostering a culture that values openness and mutual respect. A foundation of shared responsibility enables fair and productive resolution when issues arise, such as disputes over authorship or ethical decisions [46]. Institutional accountability is embedded in the structures and mechanisms that support ethical research. Universities, research institutions, and funding organizations are responsible for establishing explicit ethical guidelines and implementing systems to enforce them. This includes developing codes of conduct, establishing ethical committees or review panels, and organizing offices focused on research integrity. These organizations must also provide academics with resources and training to help them understand and apply ethical principles in practice [47]. Institutional accountability also encompasses the protection of those who report alleged misconduct. Whistleblowers should be taken seriously, kept confidential, and treated fairly; they should feel safe expressing concerns without fear of retaliation. Institutions must also ensure that investigations into misconduct are transparent, prompt, and based on clear policies. Beyond the academic community, accountability extends to the public, mainly when research influences public policy or is funded by taxpayers. Unethical research practices in these instances can have far-reaching societal consequences, including misallocating public funds, implementing flawed policies based on false information, or eroding trust in Scientific and academic institutions [3] face significant challenges due to unethical behavior by researchers, which may have far-reaching effects that harm not only the reputations of individuals and organizations but also compromise the broader credibility of the scientific endeavor in the eyes of the public. Thus, public responsibility urges research institutions and individual academics to be transparent, ensure the responsible dissemination of results, and clearly articulate the social value of their work. Driven by limited funding, overemphasizing performance metrics, and increased competition, Edwards and Roy [49] argue that current academic and institutional incentive systems have fostered a research environment increasingly susceptible to ethical compromises. These systemic pressures pose a serious threat to public

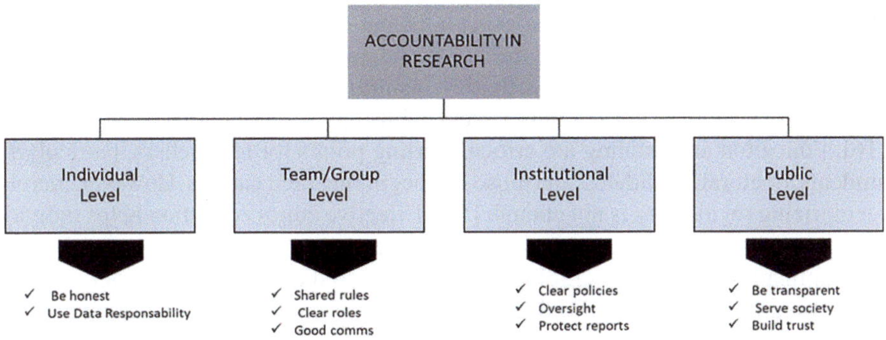

Fig. 5 Core dimensions of accountability in research

trust in science. The authors advocate for comprehensive changes to align incentives with ethical integrity values and reaffirm the role of research as a public benefit to mitigate these risks [48]. Ensuring responsibility in research is not merely about adhering to rules or avoiding misconduct. It is a broader commitment to a culture of integrity, where ethical principles guide decision-making at all levels [49]. By embracing accountability personally, collaboratively, institutionally, and publicly, the research community enhances its credibility, maintains the quality and relevance of its work, and meets its obligations to society. For example, several stages of civil and architectural engineering education, including PhD study, demand responsibility. Accurate data administration, equitable authorship policies, and conflict of interest disclosure are all duties for individual researchers. Principal investigators in team-based studies must promote ethical clarity, fair responsibility, and open communication. Institutions provide significant assistance by developing research integrity procedures, safeguarding whistleblowers, and ensuring a transparent misconduct inquiry. Equally important is public responsibility, as engineering research often guides public safety and infrastructure policies. Ethical lapses can undermine public confidence and jeopardize society's welfare, highlighting the necessity for a culture of integrity across all fields. The four essential degrees of responsibility in research, individual, team/group, institutional, and public, are shown in Fig. 5. Every level has specific duties that together support the integrity of research. The chart emphasizes how ethical behavior is ingrained at several levels, from personal integrity and responsible data management to institutional control and social openness, strengthening trust, cooperation, and credibility throughout the research process.

2.4 Preventive Practices

Preventing research misconduct involves more than simply identifying issues post-hoc. Although reactive measures such as retractions or inquiries are necessary, cultivating a strong, ethical research environment requires proactive thinking [10].

Preventive measures focus not on exceptions but on establishing norms that create a research culture where ethical behavior is standard. These practices should be embedded in our daily academic activities, institutional management, and researcher training to develop systems, mindsets, and habits that foster integrity from the outset [16]. Education and training are critical starting points for researchers, particularly students, doctoral candidates, and those just beginning their careers. However, merely memorizing regulations is not enough [50]. Effective ethics education helps individuals appreciate the importance of these rules, encouraging academics to reflect on the implications of providing reliable knowledge to the world while fostering open discussions about values such as honesty, justice, and responsibility. Case studies and real-life scenarios can be beneficial in this context, enabling academics to envision their responses to ethical dilemmas, such as publishing pressures, ambiguous authorship responsibilities, or data that does not meet expectations [51]. Research integrity and transparency depend on good documentation and data management. Systematic recording of experimental procedures, data collection techniques, analytical processes, and decision-making justifications enables other researchers to interpret, reproduce, and critically assess the work. In addition to allowing reproducibility, such practices also serve as protection; in the event of an inquiry or conflict, a well-maintained audit trail provides accountability and transparency regarding the research process. In cooperative research settings, where coordination and traceability are vital, employing techniques such as regular file labeling, version control systems, and secure data storage enhances organizational efficiency and upholds ethical standards. Trust and confidence in academic outputs, as Eddy et al. [54] contend in relation to healthcare modeling, rely on two main elements: transparency, allowing stakeholders to investigate how a model or process is constructed, and validation, demonstrating how accurately it reflects empirical reality. By analogy, these concepts apply equally in research more generally, where systematic data stewardship and open documentation are prerequisites for responsible and credible knowledge generation [52].

Speaking of teams, internal peer review and mentoring are invaluable. Regularly sharing work with colleagues before submission allows researchers to catch errors, gather feedback, and critically assess ethical considerations. This informal review process can highlight essential questions:

- *Are we reporting all results, even unexpected ones?*
- *Are we giving proper credit to contributors?*
- *Are we fair in how we present our findings?*

Mentoring provides additional support. Seasoned researchers greatly influence how their students or junior colleagues navigate ethical decisions. A mentor who openly discusses research challenges and responsibilities, sharing their mistakes and how they resolved them, fosters a culture of trust and learning. Institutions can facilitate this by ensuring that early-career researchers have access to guidance when needed and providing resources and recognition for effective mentoring. They can also assist in guaranteeing early-career researchers access to support. Moreover, many technological tools support ethical research in the digital era [53]. Data repositories, for example, allow for more open sharing of raw data; plagiarism detection software helps authors identify uncredited material before submitting their work. Version control tools, useful for collaboration and accountability, enable researchers to track changes over time. Electronic lab notebooks also create secure and searchable records of experimental work by automatically logging the date and time of entries. While these tools provide a solid infrastructure for maintaining research integrity, they do not replace the need for ethical awareness. Lastly, we cannot overlook the institutions' role in this. Universities and research centers need to clearly define acceptable ethical behavior and provide policies that are accessible, straightforward, and easy to follow. This includes ensuring protection for whistleblowers and having well-defined procedures for reporting concerns. Leadership also plays a critical role in shaping research culture. Senior faculty and administrators set a positive example by taking ethics seriously, acknowledging mistakes, and encouraging dialogue, conveying that integrity is paramount. Ethics seminars, awareness campaigns, and open discussions about real ethical dilemmas are among the activities that can help keep these issues at the forefront of academic life [54]. Preventive strategies are most effective when integrated into daily practices rather than considered solely in response to challenges. On a personal level, researchers should remain honest and reflective in their work. At the group level, teams must be open and transparent with one another and assist each other in identifying blind spots. Institutional policies and leadership should also foster transparency and accountability. These levels reinforce one another, contributing to an environment where ethical research can thrive. Promoting research integrity is not about achieving excellence but a commitment to truth, justice, education, and the people impacted by our work. Even in times of pressure or uncertainty, preventive measures empower researchers to feel more confident and supported in ethical decisions. They also help maintain the integrity of science as a whole. These proactive strategies will become vital as research grows more complex and interconnected globally. Research misconduct is not only about finding issues post-hoc [55]. Although reactive measures such as retractions or inquiries are required, building a good, ethical research environment calls for more than just reaction; it calls for forward thinking. Preventive measures emphasize not the exception but rather the norm in shaping a research culture where doing the right thing is the norm. Our daily academic work, institutional management, and research training should include these practices. They are about creating systems, mindsets, and habits supporting integrity from the start. Education and training are among the most crucial starting points. Researchers must understand what ethical research entails, especially students, doctoral candidates, and those just beginning their careers. However, merely memorizing laws is insufficient. Effective ethics education helps individuals appreciate the significance of these rules. It encourages scholars to reflect on what it means to provide reliable knowledge and fosters open discussions about values such as honesty, justice, and responsibility. In this context, case studies and real-life situations are instrumental. They allow researchers to envision how they might respond in ethically ambiguous situations, such as publishing pressure, unclear authorship responsibilities, or unexpected data outcomes. Moreover, proper documentation and data management techniques are essential beyond training. Keeping detailed and comprehensive records, including notes from experiments, data collection methods, and decision-making processes, ensures that others can understand, replicate, or verify your work [56]. This documentation also serves as a defensive strategy; if questions about your research emerge, a clear audit trail can elucidate what occurred. Simple ways to promote transparency and accountability include clearly labeling files, maintaining version control, and utilizing secure storage methods. Especially in team environments, these practices enhance research's ethicality, efficiency, and reliability. Speaking of teams, internal peer review and mentoring are invaluable. Regularly sharing their work with colleagues before publication enables researchers to catch errors, receive feedback, and critically assess ethical decisions [57]. This informal review process can raise important questions: Are we reporting all outcomes, including the unexpected ones? Are we appropriately crediting contributors? Are we presenting our findings fairly? Mentoring offers additional support, as experienced researchers greatly influence how their students or junior colleagues navigate ethical choices. A mentor who candidly discusses research challenges and responsibilities, sharing their mistakes and the strategies they employed to address them, fosters a culture of trust and learning. Institutions can facilitate this by ensuring that early-career researchers have access to guidance when needed and providing resources that recognize effective mentoring. They can also aid in guiding early-career researchers. Additionally, various technological tools promote ethical research in the digital era. Data repositories, for instance, facilitate more open sharing of raw data; plagiarism checkers help authors identify uncredited material before submission. Useful for both collaboration and accountability, version control tools enable researchers to track changes over time. Electronic lab notebooks also create secure and searchable records of experimental work by automatically logging entry dates and times. While these tools provide a solid infrastructure for maintaining research transparency and organization, they do not substitute for ethical awareness. Lastly, we cannot overlook institutions' critical role in this process. This includes ensuring protections for individuals who raise concerns and establishing formal Universities and research centers must clearly articulate ethical behaviors and implement policies that are accessible, clear, and easy to follow [58] processes for reporting issues. Leadership also shapes research culture; senior professors and administrators set a strong example by taking ethics seriously, admitting errors, encouraging discussions, and conveying the importance of integrity. Ethics seminars, awareness campaigns, and open dialogues about real ethical dilemmas are among the events that can help keep these issues at the forefront of academic life. Preventive measures are most effective when woven into daily routines instead of being considered only in reaction to problems. Figure 6 illustrates the interaction of many components of a research environment working in concert to promote ethical behavior. Research integrity is at the center, enhanced by sincere collaboration, competent mentoring, useful digital tools, and unambiguous institutional backing. These factors help researchers to openly discuss, learn from one another, and properly manage difficulties. Ethical decision-making becomes a component of daily academic life when these supports are in place.

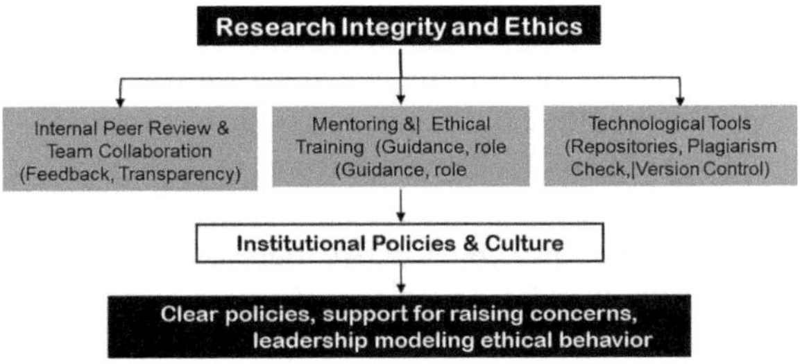

Fig. 6 Building blocks for ethical and collaborative research (*Source* Authors)

At a personal level, researchers should remain truthful and reflective in their work. Teams must be open and honest with one another, assisting each other in recognizing blind spots [59]. Policies and leadership should also cultivate transparency and accountability at the institutional level. These levels reinforce one another and help create an environment where ethical research can flourish. Promoting research integrity is not just about achieving excellence; it is a commitment to truth, justice, education, and the individuals affected by our work. Even amidst pressure or uncertainty, preventive practices empower researchers to feel more confident and supported in making the right decisions. Furthermore, they serve to uphold the integrity of science as a whole. These proactive measures will be increasingly critical as research grows more complex and globally interconnected. Investing in training, mentoring, transparency, and robust institutions helps foster a trustworthy research culture that benefits society [60]. Table 2 comprehensively summarizes fundamental preventive tactics designed to cultivate and uphold research integrity in academic settings. Organized into distinct yet interconnected areas, these policies encompass ethical education and training, systematic documentation and data management, internal peer review procedures, mentoring systems, technological supports, institutional policy frameworks, and ethical leadership. Each entry outlines the essential elements of the practice and its intended role within the broader research ecosystem. By embedding values of openness, responsibility, and accountability into the daily activities of individuals and the operational standards of organizations, these efforts contribute to establishing a proactive, ethically robust research culture [61]. Every entry describes the necessary elements of the practice and its intended role within the larger research ecosystem. By embedding values of openness, responsibility, and accountability into the daily activities of individuals and the operational standards of organizations, these actions help to create a proactive, ethically strong research culture. This approach emphasizes the need to go beyond reactive responses to wrongdoing and stresses the development of preventive standards that support integrity from the beginning of the research process [22]. Offers a thorough summary of fundamental preventive tactics meant to foster and maintain research integrity in academic settings. These policies are organized into separate but related areas, including ethical education and training, systematic documentation and data management, internal peer review procedures, mentoring systems, technological supports, institutional policy frameworks, and moral leadership. Every entry describes the necessary elements of the practice and its intended role inside the more extensive research ecosystem. By embedding values of openness, responsibility, and accountability into the daily activities of people and the operational standards of organizations, these actions help to create a proactive, ethically strong research culture [61]. This approach emphasizes the need to go beyond reactive measures to wrongdoing and stresses the development of preventive standards supporting integrity from the beginning of the research process. Table 4 shows easy yet effective methods to maintain the integrity and dependability of research. It addresses all aspects, from obtaining appropriate training and keeping records solid to sharing work early with coworkers and having effective mentors. It also reveals how much aid certain institutions' clear policies and proper technology use. Leadership creates a culture where doing the correct thing becomes second nature by setting the appropriate example.

Table 4 Preventive practices for promoting research integrity

Preventive area	Description	Purpose and impact
Education and training	Formal instruction in responsible conduct of research (RCR), using case studies and ethical frameworks to promote reflection	Cultivates ethical awareness, critical thinking, and preparedness for real-world dilemmas
Documentation and data management	Keeping accurate, detailed, and time-stamped records; employing clear file labeling, version control, and secure storage	Ensures research transparency, enables reproducibility, and provides protection in case of disputes
Internal peer review	Regular sharing of research with colleagues before submission to identify errors and reflect on ethical choices	Enhances research quality, accountability, and ethical compliance
Mentorship	Guidance from experienced researchers on ethical decision-making and responsible research practices	Builds a culture of trust and models integrity for early-career researchers
Technological tools	Use digital tools such as plagiarism checkers, data repositories, version control systems, and electronic lab notebooks	Supports traceability, collaboration, and compliance with transparency standards
Institutional policies	Clear, accessible codes of conduct; procedures for reporting misconduct; protection for whistleblowers	Provides structure for ethical oversight and reinforces institutional accountability
Leadership and culture	Active leadership promoting ethics through example, awareness campaigns, and open dialogue about real ethical challenges	Embeds integrity in academic culture and fosters an environment where ethical conduct is normalized

3 Informed Consent and Human Subject Research

Studies involving human subjects carry higher ethical liabilities and require a stricter adherence to ethical codes at all stages of the study [62]. Studies involving individuals, in contrast to studies in laboratories or the abstract, confront the fundamental issues of human dignity and privacy, rights, and autonomy [63]. This means that many essential principles must be respected to ensure participants' well-being and preserve the research process's integrity. Such principles include informed consent, voluntary and autonomous choice, confidentiality, and ethical oversight. These components protect participants' rights and encourage ethical research conduct.

3.1 Informed Consent

A fundamental pillar of ethical research, including human participants, is informed consent. It goes beyond procedural formality to reflect a significant ethical dedication to respect for personal autonomy, dignity, and the right to self-determination [64]. For example, in civil engineering and architecture research, where interdisciplinary studies might touch on social, environmental, or health-related aspects, informed consent guarantees that individuals are given sufficient, correct, and understandable information required for making a voluntary and well-considered choice about their involvement. Transparency and proactive disclosure should drive the informed consent process. Researchers have a legal and ethical duty to specify the study's goals, methodological approaches, projected length, and expected legal, physical, psychological, or social consequences. Possible advantages must also be unambiguously stated, whether they accrue directly to individuals or support larger societal or disciplinary progress [65]. The guarantee of voluntariness requires that participants must be explicitly told their involvement is optional and they may withdraw at any point without adverse effects or loss of entitlements, which is a fundamental ethical principle. All available choices in studies on comparative treatments, such as in design intervention or clinical engineering settings, must be shown without coercion or undue influence. Participants' linguistic, cognitive, and educational backgrounds should guide information communication. Often, especially when research crosses multilingual or cultural contexts, avoiding technical jargon, providing materials in accessible language, and using visual or interactive tools are required. Appropriate use of translation and professional interpreter services will help to guarantee understanding. Canonical frameworks like the Belmont Report (1979), which articulates three fundamental values [66]: Respect for Persons, Beneficence, and Justice, ground the theoretical and ethical foundations of informed consent. Respect for people includes acknowledging them as autonomous agents. Researchers have to use more protections in situations where autonomy is reduced, such as among minors, cognitively challenged people, or institutionalized populations, including the use of proxy or legally authorized consent systems following institutional and legal criteria.

Importantly, informed consent should be considered a continuous process instead of a one-time occurrence. Participants in longitudinal studies, multi-phase interventions, or projects using changing methodologies such as iterative urban design trials or adaptive environmental assessments must be regularly informed about protocol changes, emerging risks, or discoveries. Should such changes significantly alter the foundation for ongoing involvement, renewed consent or "re-consent" has to be acquired. Ethically strong consent also means judging participant knowledge rather than getting a signature. Investigators should assess understanding using verification techniques, including participant paraphrasing or focused questioning. If knowledge is insufficient, further clarification should be provided through simplified formats or additional explanatory content. Legal requirements in many areas highlight the need for informed consent. Regulatory systems like the U.S. Common Rule (45 CFR 46), the European Union's General Data Protection Regulation (GDPR), and national ethical review standards set official consent documentation and monitoring standards [67]. However, legal compliance is not enough; the ethical validity of consent is in its substantive quality, and participants have to grasp and sign. Involving vulnerable populations, such as patients, imprisoned people, refugees, or those experiencing socioeconomic struggle, calls for exceptional sensitivity. In such situations, power imbalances, such as when a treating doctor is also the principal investigator, may subtly push people to participate. Researchers must confirm voluntarism, stating that non-participation will not affect service or treatment access. Research digitization has created fresh complexity. Digital consent systems used in remote sensing studies, online surveys, or mobile apps may not have the personal immediacy that helps clarify questions. Figure 7 details the elements that provide informed consent significance in studies including humans. It indicates that academics must explicitly describe the study, guarantee voluntary involvement, and verify that participants grasp what they are consenting to. These three easy actions help to create confidence, safeguard participants, and make research more respectful and responsible by informing, enabling free choice, and verifying understanding.

Digital platforms, therefore, have to be user-oriented, interactive, and accessible, with embedded explanations and responsive elements to help participants understand. Ethically acceptable in emergency research or studies involving incapacitated people, such as disaster response simulations or biomedical facility interventions, deferred consent or consent via legally authorized representatives may be ethically acceptable, subject to Institutional Review Board (IRB) or equivalent ethics body approval. Participants have to be told retrospectively at the earliest possible time, and their right to leave has to be respected [68].

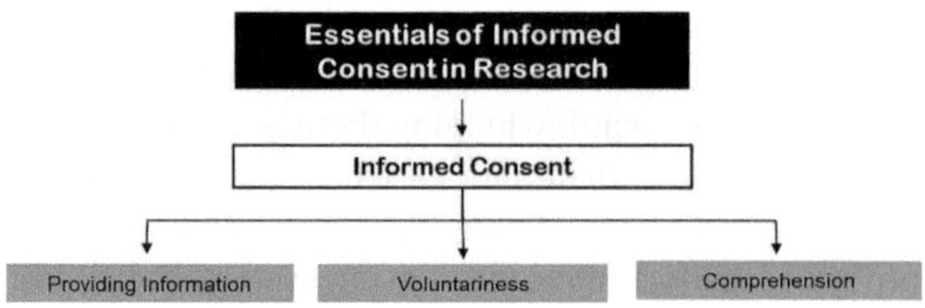

Fig. 7 Core components of informed consent in research (*Source* Authors)

3.2 Voluntariness and Autonomy

In civil engineering and architectural research, the ethical values of voluntarism and autonomy form a foundation of responsible research behavior, including human subjects [69]. These values are closely related to the concept of informed consent, which guarantees that involvement is the outcome of a free and intentional decision. While voluntariness emphasizes the need for those decisions to be made free of coercion, manipulation, or undue influence, autonomy shows the acknowledgment of people as self-governing agents, able to make rational decisions. Determining voluntarism in research methodology requires strict protections against obvious and subtle coercion. Coercion might appear openly, for example, using the threat of lost services or opportunities, or more subtly, using too strong incentives that could compromise independent judgment. This is especially relevant in civil and architectural fieldwork, where participants might include construction workers, students, tenants, or municipal employees, who may feel participation is required because of institutional hierarchies or affiliations. Researchers must proactively develop a setting where non-participation is a neutral and acceptable choice. Engaging with vulnerable populations makes the ethical need to respect autonomy clearer. These consist of children, seniors, people with cognitive disabilities, low-income communities, and those under institutional control. In such settings, more ethical procedures are needed. For example, in studies including children, such as those on safety education on construction sites or exposure to build environment risks, researchers must get informed consent from a legal guardian and the child's agreement, modified to their degree of maturity. A legally authorized representative has to be involved for those without decision-making capacity to guarantee ethical legitimacy. Figure 8 illustrates the interaction between voluntariness and autonomy in fostering ethical research. Voluntariness is the absence of coercion for anyone to participate; autonomy is the liberty of individuals to decide for themselves. They work together to foster a secure, courteous atmosphere where involvement is based on free will. Working with disadvantaged populations or environments where power relations could affect choices makes this more crucial.

Voluntariness is also greatly challenged by professional and academic hierarchies. The power difference in supervisory relationships, such as between a senior engineer, an intern, a faculty member, and a student, may prevent real consent [70]. Researchers have to follow policies that separate participation from any felt duty, so explicitly stating that academic assessment, employment status, or access to institutional resources would not be affected by participation decisions. Research environments, including construction sites, housing projects, or public space assessments, need careful management of the temporal and environmental circumstances under which consent is sought. Consent acquired during occupational stress or environmental risk, such as during active site inspections, may not satisfy free and informed decision-making criteria [71]. Ethical research design has to prioritize environments that encourage cognitive clarity and voluntary involvement. Incentives of money or goods meant to motivate participation should be measured to prevent excessive

Fig. 8 Voluntariness and autonomy in ethical research participation

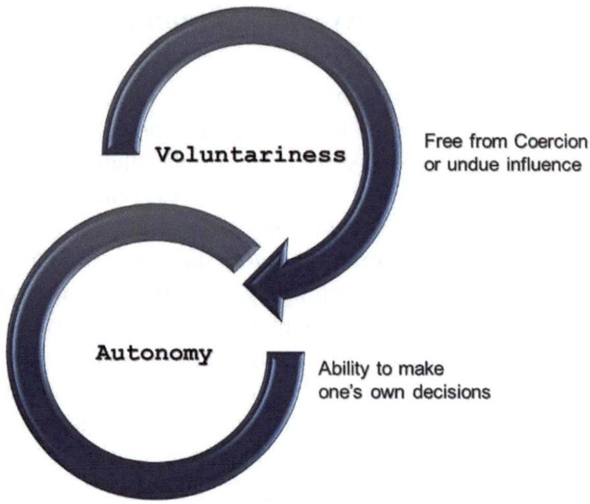

influence. Although reasonable pay for time and effort is acceptable, massive incentives, especially in socioeconomically disadvantaged populations, risk undermining voluntariness. Compensation systems should be driven by fairness and proportionality ideas. Voluntarism and autonomy must be read in cross-cultural or international studies, typical in urban development and comparative infrastructure studies within culturally varied ethical frameworks [72]. Though many societies value group decision-making, individual consent is a moral and legal requirement. Researchers have to walk these culturally sensitive yet ethically rigorous waters.

3.3 Confidentiality

Confidentiality, including human participants, is a basic ethical and legal duty in civil engineering and architectural studies. It relates to the duty of researchers to protect PII and sensitive information gathered throughout the research process [73]. Under the assumption that this information will be handled with discretion, security, and integrity, participants trust researchers with private data, ranging from personal identification and health issues to occupational roles and community ties. Any violation of this confidence could cause reputational damage, psychological suffering, or legal repercussions for participants and undermine the study's validity. All research lifecycle stages must be kept confidential, from first data gathering to last distribution and long-term storage. Maintaining confidentiality depends on the use of de-identification or anonymization techniques. Anonymization is the irreversible removal of all identifiable features from data sets, preventing future re-linkage to individual subjects. Pseudonymization, on the other hand, lets you assign coded identifiers and keeps the decryption key stored apart under tight access controls. Longitudinal studies in civil engineering, such as repeated evaluations of housing quality, construction site safety behavior, or community infrastructure feedback, where follow-up involvement with participants is required, particularly benefit from this. Data protection policies have to fit internationally accepted legal systems. The General Data Protection Regulation (GDPR) sets rigorous data collection, processing, and storage criteria in the European setting. Data minimization, purpose restriction, responsibility, and explicit consent are essential ideas [74]. Similar obligations govern health-related research in the United States as those in the Health Insurance Portability and Accountability Act (HIPAA). Civil engineering and architecture researchers who manage data relating to health, disability, or environmental exposure have to follow these legal requirements to prevent ethical breaches and institutional sanctions [75]. Participants must be thoroughly informed about how their data will be gathered, handled, examined, shared, and stored to guarantee ethical compliance. This covers the kinds of recorded data (e.g., biometric, spatial, demographic), the planned storage duration, any data sharing agreements (e.g., with engineering consultants or urban planning authorities), and the techniques used to maintain confidentiality. Especially in areas where GDPR applies, participants must be informed of their rights to access, correct, or request the deletion of their data [76]. Furthermore, the distribution of research results calls for strict confidentiality. Researchers should avoid any direct or indirect identifiers that might let people be identified when presenting or publishing data, whether in technical reports, academic journals, or design proposals. This is particularly important in small-scale studies or community-based initiatives where the sample population is localized or demographically unique.

3.4 Ethical Oversight

Ethical supervision is a vital protection in civil engineering and architecture, two fields where research increasingly interacts with human participants in constructed environments, guaranteeing all concerned stakeholders' rights, dignity, and safety. Usually, ethical supervision is provided using official channels like Institutional Review Boards (IRBs) or Research Ethics Committees (RECs), which are responsible for evaluating the ethical and methodological rigor of suggested projects [77]. These autonomous organizations guarantee the study fits the set ethical values and pertinent legal and institutional frameworks by acting as objective assessors. Comprehensive and multi-dimensional, the ethical review process Key research elements, including the study's rationale and design, participant recruitment techniques, informed consent procedures, data confidentiality strategies, and a thorough examination of the risk–benefit balance, are evaluated under this framework. Ethics committees particularly examine whether the research approach reduces possible risks, be they physical, psychological, social, or legal, and whether these risks are justifiable given the expected contributions to knowledge or practice of the study [78]. In civil and architectural research could mean evaluating exposure risks during site visits, socio-environmental impact studies, or post-occupancy assessments in residential areas.

Such evaluations' ethical basis is anchored in globally acknowledged frameworks, including the Declaration of Helsinki (World Medical Association, 2013) and the International Ethical Guidelines for Health-Related Research Involving Humans published by CIOMS (2016) [79]. These documents emphasize fundamental ethical values, respect for people, beneficence, non-maleficence, and justice, which must be ingrained throughout the research lifecycle from planning and implementation to the distribution of results. In built environment research, these ideas sometimes overlap with public interest issues and long-term societal effects. Ethical supervision helps preserve research integrity rather than being a procedural formality. Ethics committees help to prevent bias, institutional conflicts of interest, and methodological errors that could compromise participant welfare or data validity, utilizing independent and expert examination. Constructive comments throughout the review process can result in methodological improvement, more precise consent records, or better community involvement strategies, thereby raising the study's general quality and ethical standing. In many countries, ethical oversight is not only a best practice but also a legal obligation. For example, the Common Rule of the United States (45 CFR 46) calls for IRB clearance for all human subject research funded by the federal government [80]. Research activities involving personal data or direct human interaction in the European Union are subject to ethical review and compliance with the General Data Protection Regulation (GDPR) [74]. Institutional research policies may call for official permission before starting studies, particularly when the study involves vulnerable populations, invasive data collecting, or fieldwork in public or sensitive areas. Not getting appropriate ethical approval could have significant consequences, including the retraction of published work, loss of research funding, legal responsibilities, and harm to the researcher and the host institution. More importantly, it might cause direct or indirect damage to participants and decline public confidence in scientific work.

3.5 Ongoing Ethical Considerations

Ethical responsibility in civil engineering and architectural studies involving human participants is not confined to acquiring initial approval from institutional review bodies, such as an Institutional Review Board (IRB) [81]. Instead, it is a continuous ethical obligation covering the whole path of the research lifecycle from recruitment and data collecting to analytical interpretation to the public distribution of results. Particularly in studies embedded in complicated, inhabited, or vulnerable built settings, this continuous dedication is essential to maintain the values of respect for people, beneficence, and justice. The systematic tracking of participant welfare across the study is a core element of continuous ethical practice. Researchers have an ethical and professional responsibility to find and handle unanticipated adverse events that might happen during the execution of the study. These could be physical hazards, such as injuries suffered on active construction sites, psychological distress caused by protracted interviews or surveys, or social damage resulting from the unintentional release of sensitive personal or community-level information. Longitudinal or multi-phase studies, including post-occupancy evaluations or environmental health assessments, demand ongoing risk–benefit profile reassessment. Immediate protocol change and re-ethical clearance are required should new risks arise that materially change the ethical calculus on which original IRB or ethics committee approval was granted [82]. Equally important is the availability of clearly defined, private channels whereby individuals can express complaints or concerns. These systems must be customized to fit people's settings and accessibility requirements in the built environment sectors, where participants might include residents, public sector workers, construction workers, or clients of municipal services. Researchers have to ensure such systems let people voice worries free of retaliation or social disadvantage. Dealing with issues transparently, promptly, and with cultural sensitivity strengthens the research process's validity and responsibility. Responsible reporting and distribution of research results also fall under ethical stewardship, extending ethical stewardship [83]. Especially when those results pertain directly to their living conditions, occupational safety, or environmental exposures, participants who offer their experiences, viewpoints, or contextual knowledge deserve to be informed of significant research findings. Dissemination formats such as lay summaries, infographics, or community forums should be chosen with respect for clarity, accessibility, and relevance, to promote mutual respect and reciprocal knowledge exchange. Researchers must preserve strong safeguards for participant confidentiality and anonymity when sharing findings. Contextual variables such as neighborhood, profession, or building typology—particularly in small or demographically unusual populations—can allow deductive disclosure even in anonymized datasets. Researchers are responsible for carefully evaluating disclosure risks and applying suitable data suppression or aggregation strategies. Moreover, when publishing results on vulnerable populations, infrastructure shortcomings, or environmental hazards, researchers must be mindful of the possibility of stigmatization [84]. The ethical terrain gets more complicated in community-engaged approaches, action research, and participatory design projects. These methods call for reconfiguring the conventional researcher-subject relationship toward one of cooperative epistemology. In such systems, people are co-creators of knowledge; ethical best practices call for group analysis of results, co-authorship possibilities, and participatory distribution techniques that fit community values and information requirements.

4 Responsible Authorship, Citation, and Publication Practices

The ethical communication of academic work is fundamental to academic integrity and professional behavior in civil engineering and architectural studies [85]. Responsible authorship goes beyond simply avoiding plagiarism; it also includes correct attribution of contributions, thorough source citation, complete openness about possible conflicts of interest, and rigorous adherence to fair publication criteria. Considering the built environment disciplines' applied and socially relevant character, ethical writing practices are essential to guarantee credibility, relevance, and lasting value of research contributions.

4.1 Authorship

Consistent with international academic codes such those set by the International Committee of Medical Journal Editors (ICMJE) and the Committee on Publication Ethics (COPE), authorship in civil engineering and architectural research must be granted only to those who have significantly, obviously intellectual contributed to the idea, development, execution, and presentation of the work [86]. Such contributions might include, but are not limited to, formulating research hypotheses based on engineering issues or architectural theory, creating computational models or analytical techniques, performing experimental tests or simulation studies, interpreting results within regulatory or design frameworks, and writing or significantly revising the manuscript. Research in the built environment is collaborative and multidisciplinary, so early and continuous discussion on authorship duties is essential [87]. Projects usually include people from several fields: structural engineers, architects, urban planners, environmental scientists, and construction managers, each adding unique knowledge. Establishing authorship criteria at the project's beginning, officially agreed upon by all contributors and modified as the research develops, is thus essential. This strategy avoids arguments about inclusion or order of authorship and promotes mutual respect. Every writer must be able to publicly support at least one significant aspect of the study and sign off on the last version of the paper. Besides intellectual contribution, authors are responsible for the published work's integrity and correctness. Particularly in civil engineering and architecture, where technical data, performance analysis, and spatial modeling have real-world safety and sustainability consequences, this shared responsibility strengthens the credibility of scholarly literature. By contrast, those who help with logistical coordination, financial management, routine data collecting, graphic rendering, or other tasks may not be eligible for authorship but should still be recognized suitably. Including those who helped in the acknowledgments honors their contribution and preserves the clarity of intellectual responsibility [88]. Examples include undergraduate assistants, administrative support staff, laboratory technicians, or field survey teams. Across fields, unethical authorship techniques still raise questions. "Guest authorship" refers to people being named without significant contribution; it usually happens to exploit reputational benefit. "Ghost authorship" refers to actual contributors being left out; it can result from hierarchical pressure or business interests. Both methods taint the academic record, exaggerate qualifications, and undermine the integrity of the scientific process. Maintaining openness to authorship is vital in civil engineering and architectural research, where teamwork and industry involvement are prevalent, to guarantee professional responsibility and public confidence.

4.2 Accurate Citation

Precise civil engineering and architecture citation techniques are fundamental to academic integrity and cumulative knowledge growth. Correct citation has two essential purposes: first, to recognize the intellectual work and original ideas of other academics; second, to allow readers, reviewers, and practitioners to track, verify, and interact with the sources of knowledge. Transparent and correct citing is vital in fields that often include complicated methodologies, empirical data, and technical criteria, such as studying structural behavior, applying environmental assessment frameworks, or creating building simulation models [89]. Researchers have an ethical and professional responsibility to cite only those materials they have read in full and critically assessed. This obligation guarantees that the cited work is used in relevant and meaningful settings and strengthens the validity of the citations. Academic fraud is misrepresenting sources, citing a paper supporting a claim that the original work does not endorse. Similarly, quoting literature depending only on its abstract, title, or secondary mention in another article compromises the credibility of academic writing [90]. Although sometimes required, self-citation should be used sparingly and with relevance. Excessive or calculated self-citation to boost bibliometric indicators, such as the h-index or citation counts, distorts the objectivity of the scholarly record and distorts impressions of research impact. Ethical self-citation should be restricted to instances when prior work is directly relevant to the present study, such as the continuation of a research project, application of a previously created method, or expansion of theoretical frameworks. Overreliance on secondary citations, where a researcher cites a source discussed in another work without reading the original, is another cause for worry. Unless the primary source is unreachable, this practice should be shunned. Secondary citations should be marked and shown to preserve academic integrity. Ensuring the reproducibility and traceability of research also depends on precise and consistent referencing. Precision in citations in civil and architectural studies, especially those involving case-based investigations, site-specific analyses, or regionally adapted construction standards, enables others to replicate techniques, validate findings, and assess regional relevance [91]. This is particularly relevant in applied research areas like seismic design, sustainable building materials, or urban resilience, where changing standards and localized datasets are vital for contextual interpretation. Moreover, citation methods have epistemological consequences. They determine the visibility of knowledge systems, guide whose voices are amplified, and impact which theoretical and methodological frameworks rise to prominence. Equitable citation practices in globalized fields like architecture and urban studies call for intentionally diversifying referenced sources, including work from underrepresented areas, minority scholars, and non-English-language publications. Tools for managing citations like EndNote, Zotero, or Mendeley can help researchers maintain accuracy and consistency in their references. However, technological tools do not replace critical source interaction. Researchers should routinely check the citation style rules of target journals or academic institutions to guarantee compliance with set formatting standards, including APA, IEEE, or Chicago [92].

4.3 Conflict of Interest Disclosure

Maintaining scholarly objectivity and protecting public confidence in civil engineering and architectural research depend on openness in revealing conflicts of interest (COIs) [93]. Given these sectors' applied, multidisciplinary, and sometimes commercially involved character, academics might often work with private sector companies, government agencies, or professional associations. Although such partnerships can improve the relevance and applicability of studies, they also create possible conflicts of objectivity that need to be carefully controlled by complete disclosure. A conflict of interest occurs when a researcher's financial, professional, or personal ties could affect, or seem to affect, the design, execution, interpretation, or reporting of research results [94]. In the built environment fields, COIs could play advisory roles in urban planning authorities, ownership of pertinent patents, equity in engineering consultancies, funding from manufacturers of building materials, or consultancy contracts with construction companies. Though they are undisclosed and might affect the impression of neutrality, these connections can be concerning since they do not naturally compromise the integrity of the study. Academic papers, funding organizations, and research universities are demanding more and more that writers provide official COI disclosures at the time of grant proposal or manuscript submission. Usually covering all sources of financial support, including research grants, fellowships, travel allowances, and in-kind contributions like equipment or software, these declarations require researchers' full non-financial interests that could affect their findings, such as leadership roles in standard-setting committees, editorial board memberships, or institutional ties. Conflict of interest disclosure's primary goal is not to exclude academics from obtaining funding or publishing, but to offer openness so editors, reviewers, and readers may justly evaluate the work's objectivity. These disclosures must be proactive and thorough. Whether intentional or unintentional, omissions can cause reputational harm, retractions, or disqualification from future funding and publishing prospects. Maintaining a conflict-of-interest log for ongoing projects, addressing possible conflicts during project planning stages, and asking institutional ethics committees or legal advisors for advice when in doubt are all best practices in handling COIs [95]. In team-based projects, particularly those involving interdisciplinary or cross-sectoral cooperation, it is recommended to do a group examination of possible conflicts and record personal disclosures. Training on COI policies should be included in graduate courses and ongoing professional development in educational environments. Ethical literacy on COIs is crucial for civil engineering and architectural professionals who sometimes straddle the line between academia and practice as they negotiate dual obligations and preserve

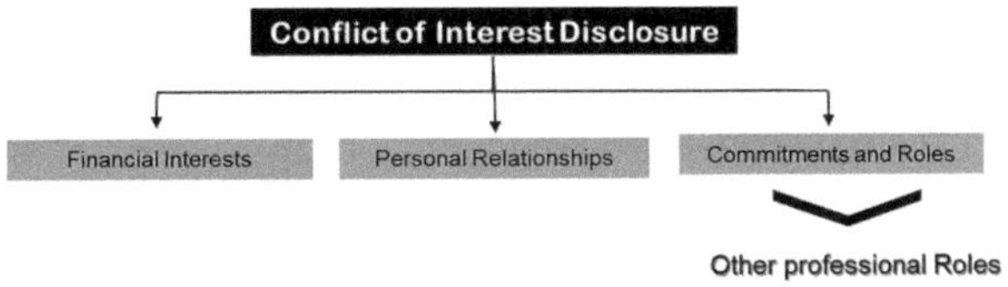

Fig. 9 Key areas of conflict of interest disclosure in research

research integrity and professional standards. In the end, complete and honest disclosure of conflicts of interest supports the credibility of academic work and enhances the ethical framework that promotes responsible research in the built environment. Researchers guarantee their responsibility to the academic community and the stakeholders, clients, and communities impacted by their work by promising open and consistent disclosure policies [96]. Three key areas researchers must be honest about, financial ties, personal relationships, and professional roles, are depicted in Fig. 9. Being honest about these links helps people to see where prejudice could arise and fosters confidence in the study. It's not about avoiding alliances; it's about clarity so that everyone can justly evaluate the job. Maintaining responsible and open research depends on simple, honest disclosure.

4.4 *Fair Publication Practices*

In civil engineering and architecture, where the spread of technical knowledge must show originality, openness, and responsibility, fair publication practices are a fundamental component of research ethics [97]. Given that these fields frequently guide choices with far-reaching consequences for public safety, resource management, and environmental impact, the integrity of the academic publication process is paramount. Duplicate submission, the practice of sending the same manuscript to more than one journal concurrently, is among the most flagrant breaches of fair publication standards. This compromises the peer review process, squanders the time and resources of editors and reviewers, and could lead to the same results being published several times in various places, distorting the academic record. Likewise, salami slicing, dividing a single coherent study into several minimally publishable papers, is frowned upon. Although justified by size and depth, publishing various facets of a major project is academically dishonest; artificial segmentation to boost publication numbers is not. Another major ethical violation is manipulating the peer review process [98]. Practices like suggesting prejudiced reviewers, taking advantage of editorial oversight holes, or participating in reciprocal review agreements (where researchers agree to evaluate one another's work favorably) undermine the objectivity and credibility of peer assessment. Maintaining the integrity of peer review is non-negotiable in civil and architectural engineering, where validating methodologies, material specifications, design standards, and simulation models depend on it. Authors should also be careful in their choice of publishing channels. The growing danger to academic integrity is posed by the proliferation of predatory publications exploiting the open-access model for monetary gain and often lacking thorough peer review. These journals usually offer fast publication, charge steep prices, and show poor editorial standards. Publishing in such venues could lead to low visibility, reputational damage, and exclusion from citation databases. Researchers should prioritize journals linked to reputable academic publishers, professional organizations, or research institutions, and those listed in acknowledged databases, including Scopus, Web of Science, or DOAJ [99]. The duty to amend the academic record is another aspect of ethical publishing. Authors are meant to work with editors to publish corrections, clarifications, or, in extreme cases, full retractions when post-publication errors, be they methodological, analytical, or interpretive, are found. Timely and open correction of the academic record not only preserves the credibility of the authors but also guarantees the dependability of cumulative scientific knowledge. Furthermore, openness about funding sources and contributions increases the responsibility of the publication process. Many journals now call for author contribution statements and financial support disclosures, which enable readers to evaluate possible biases and grasp the cooperative nature of the study [100]. Ethical research should also promote the publication of replication studies and unfavorable findings. Reporting what does not work is as vital as stressing success in civil engineering and architecture, where theoretical models and design assumptions must be validated in the real world. Ignoring such results because of publication bias could result in ongoing mistakes, wasteful use of resources, and erroneous design or policy suggestions.

5 Conclusion

In research and writing, ethical issues are not a secondary concern; they form the core. Maintaining integrity becomes more urgent as the research sector becomes more global, multidisciplinary, and technologically sophisticated. Ethics is about building a culture of honesty, openness, and responsibility that enhances individual work and the larger academic community, not only avoiding errors or wrongdoing. This chapter has examined how ethical research calls for action on several levels: personal, collaborative, institutional, and societal. It calls for personal honesty and methodological diligence from individual researchers, team fairness and openness, institutional integrity supported by unambiguous policies and leadership, and responsible communication to preserve society's confidence in science. Ethical research is keeping the clarity and accuracy of the academic record, properly acknowledging contributions, careful data management, and respect for human participants. Particularly crucial are preventive plans. Ethical awareness has to be developed early and constantly strengthened using education, mentoring, internal evaluation, and sound institutional processes. Waiting to respond to wrongdoing after it happens is insufficient; universities, research institutions, and people must be proactive in fostering a culture where doing the right thing is not just anticipated but ingrained into daily research activities. The core of this initiative is openness. Transparency about techniques, findings, constraints, and mistakes increases the reproducibility of science and fosters public confidence. It enables academics to discuss their failures, doubts, and triumphs. Transparency thus promotes real progress instead of brittle appearances. Although digital tools such as version control systems, data repositories, and plagiarism checkers help to preserve openness, they cannot replace the intrinsic dedication to honesty that every researcher must bear. Responsibility is also quite important. Every researcher has a duty not just to their direct colleagues but also to the larger public, who rely on reliable studies to make choices that influence society. Institutions should penalize mistakes and praise and set an example for ethical behavior. Research groups also have to foster a sense of shared accountability by helping one another in ethical decision-making and learning from errors with humility instead of dread. Equally important is responsible writing and publishing. Authorship must show genuine intellectual contributions; citations must be honest and accurate; conflicts of interest must be revealed freely. Ethical writing techniques guarantee the academic record is an authentic and dependable reflection of group intellectual activity, not warped by self-interest or shortcuts. Ethical research is about principles, not alone laws. These values empower and provide significance to research: integrity, transparency, fairness, and respect. Maintaining a strong ethical basis will be crucial for guaranteeing that research stays a force for good, advancing knowledge, enhancing lives, and preserving the public trust as the difficulties confronting researchers get more complicated. Investing in ethical education, robust mentoring, transparent processes, and institutional leadership dedicated to integrity can help us create a research culture that promotes ethical excellence rather than just responding to lapses. Maintaining these obligations is a professional responsibility and a significant contribution to the future of knowledge and society itself.

References

1. Miao J, Thongprayoon C, Suppadungsuk S, Garcia Valencia OA, Qureshi F, Cheungpasitporn W (2023) Ethical dilemmas in using AI for academic writing and an example framework for peer review in nephrology academia: a narrative review. Clin Pract 14(1):89–105
2. Adeoye MA (2024) Mastering the basics: a guide to research methodology for effective writing and publication. Chalim J Teach Learn 4(1):30–41
3. Steneck NH (1994) Research universities and scientific misconduct: history, policies, and the future. J High Educ 65(3):310–330
4. Moravcsik A (2020) Transparency in qualitative research. SAGE Publications Limited, London
5. Urbanovič J, Tauginienė L (2013) Institutional responsibility vs individual responsibility: ethical issues in the management of research performance. Procedia Soc Behav Sci 81:72–78
6. Molewijk B, Stiggelbout AM, Otten W, Dupuis HM, Kievit J (2004) Scientific contribution. Empirical data and moral theory. A plea for integrated empirical ethics. Med Health Care Philos 7:55–69
7. Osborne JW, Holland A (2019) What is authorship, and what should it be? a survey of prominent guidelines for determining authorship in scientific publications. Pract Assess Res Eval 14(1):15

8. Pruschak G, Hopp C (2022) And the credit goes to…-Ghost and honorary authorship among social scientists. PLoS ONE 17(5):e0267312
9. Ul Hassan M, Murtaza A, Rashid K (2025) Redefining higher education institutions (HEIs) in the era of globalisation and global crises: a proposal for future sustainability. Eur J Educ 60(1):e12822
10. National Academies of Sciences, Medicine, Global Affairs and Committee on Responsible Science (2017) Fostering integrity in research. National Academies Press
11. Kocaman Kalkan K, Kekeçoğlu M (2024) Ethical principles and rules in experimental animal studies: a comprehensive review. Duzce Med J 26
12. Ramesh MR (2024) Ethical publication and dissemination of research: principles and practices. In: Ethical frameworks in special education: a guide for researchers. p 29
13. Romenti S, Colleoni E, Murtarelli G, Moroni C (2025) Artificial intelligence ethical violations and corporate response strategies. In: Responsibility in strategic communication: truth or trap?. Emerald Publishing Limited, pp 147–164
14. Skiba R (2024) Leading and influencing ethical practice. After Midnight Publishing
15. Ravitch SM, Riggan M (2016) Reason & rigor: how conceptual frameworks guide research. Sage publications
16. Kretser A, Murphy D, Bertuzzi S, Abraham T, Allison DB, Boor KJ, Dwyer J, Grantham A, Harris LJ, Hollander R, Jacobs-Young C (2019) Scientific integrity principles and best practices: recommendations from a scientific integrity consortium. Sci Eng Ethics 25:327–355
17. Bhaskar R, Ola M (2024) Publication misconduct, plagiarism, and research integrity. In: Principles of research methodology and ethics in pharmaceutical sciences. CRC Press, pp 271–310
18. Roje R, Reyes Elizondo A, Kaltenbrunner W, Buljan I, Marušić A (2023) Factors influencing the promotion and implementation of research integrity in research performing and research funding organizations: a scoping review. Account Res 30(8):633–671
19. Youssef PEIM (2023) Dealing with research misconduct in Egyptian public academic institutions a dynamic performance governance approach to enhance collaboration
20. Resnik DB (2014) Data fabrication and falsification and empiricist philosophy of science. Sci Eng Ethics 20:423–431
21. Chevassus-au-Louis N (2019) Fraud in the lab: the high stakes of scientific research. Harvard University Press
22. Kennedy JE (2024) Addressing researcher fraud: retrospective, real-time, and preventive strategies–including legal points and data management that prevents fraud. Front Res Metr Anal 9:1397649
23. Posner RA (2009) The little book of plagiarism. Pantheon
24. Mulenga R, Shilongo H (2024) Academic integrity in higher education: understanding and addressing plagiarism. Acta Pedagogia Asiana 3(1):30–43
25. Yeo-The NSL, Tang BL (2024) On "intent" in research misconduct. Account Res 1–19
26. Teixeira da Silva JA, Dobránszki J (2016) Multiple authorship in scientific manuscripts: ethical challenges, ghost and guest/gift authorship, and the cultural/disciplinary perspective. Sci Eng Ethics 22:1457–1472
27. Yeo M, Lewis M (2019) Co-authoring in action: practice, problems and possibilities. Iran J Lang Teach Res 7(3):109–123
28. Nazarovets S (2024) Dealing with research paper mills, tortured phrases, and data fabrication and falsification in scientific papers. In: Scientific publishing ecosystem: an author-editor-reviewer axis. Springer Nature Singapore, Singapore, pp 233–254
29. Shapiro JP, Stefkovich JA (2016) Ethical leadership and decision making in education: applying theoretical perspectives to complex dilemmas. Routledge
30. Bielefeldt AR, Polmear M, Canney N, Swan C, Knight D (2018) Ethics education of undergraduate and graduate students in environmental engineering and related disciplines. Environ Eng Sci 35(7):684–695
31. Pratt MG, Kaplan S, Whittington R (2020) Editorial essay: the tumult over transparency: decoupling transparency from replication in establishing trustworthy qualitative research. Adm Sci Q 65(1):1–19

32. Tracy SJ (2024) Qualitative research methods: collecting evidence, crafting analysis, communicating impact. John Wiley & Sons
33. Freese J, Rauf T, Voelkel JG (2022) Advances in transparency and reproducibility in the social sciences. Soc Sci Res 107:102770
34. Strobl A, Fernández-Mesa A, Miroshnychenko I, Öztura P, Korzynski P (2025) Publishing quantitative research in EMJ: some editorial guidelines and recommendations. Eur Manag J
35. Hou Z, Li X, Yang L, Liu T, Lv H, Sun Q (2025) Prediction models for sarcopenia risk in dialysis patients: a systematic review and critical appraisal. Aging Clin Exp Res 37(1):18
36. Hammerschmidt T, Hafner A, Stolz K, Passlack N, Posegga O, Gerholz K-H (2025) A review of how different views on ethics shape perceptions of morality and responsibility within AI transformation. Inf Syst Front 1–87
37. Martin BR (2025) Ethics and integrity in publishing. How Get Publ Best Manag JS 28–50
38. Twietmeyer G (2025).The reproducibility crisis and kinesiology: confronting the human & statistical limitations. Int J Kinesiol High Educ 1–16
39. Matheson J (2024) Epistemic autonomy and intellectual humility: mutually supporting virtues. Soc Epistemol 38(3):318–330
40. Teixeira da Silva JA (2022) A synthesis of the formats for correcting erroneous and fraudulent academic literature, and associated challenges. J Gen Philos Sci 53(4):583–599
41. Lakens D, Mesquida C, Rasti S, Ditroilo M (2024) The benefits of preregistration and registered reports. Evid-Based Toxicol 2(1):2376046
42. Leonelli S (2023) Philosophy of open science. Cambridge University Press
43. Levinson MP (2010) Accountability to research participants: unresolved dilemmas and unravelling ethics. Ethnogr Educ 5(2):193–207
44. Lal KK, Sharma B (2023) Research integrity & ethics: scientific misconduct. In: National seminar on academic integrity and research ethics. National seminar on academic integrity and research ethics
45. Moles K, Robinson L, Merritt J, Warner KD (2024) Strengthening ethical commitments to the common good: team-based learning in the Ignatian pedagogical paradigm. Relig & Educ 51(1–2):85–100
46. Mathur A, Lean SF, Maun C, Walker N, Cano A, Wood ME (2019) Research ethics in inter-and multi-disciplinary teams: differences in disciplinary interpretations. PLoS ONE 14(11):e0225837
47. Menkel-Meadow C (2018) Ethics in alternative dispute resolution: new issues, no answers from the adversary conception of lawyers' responsibilities. In: Mediation. Routledge, pp 429–476
48. Azibaraniyar O (2023) An overview of financial mismanagement and financial institutions in Nigeria. FUO J Manag Sci 7(Special):153–161
49. Edwards MA, Roy S (2017) Academic research in the 21st century: maintaining scientific integrity in a climate of perverse incentives and hypercompetition. Environ Eng Sci 34(1):51–61
50. Ostler E, Williams T, Schultz J (2025) Leading with data, leading with ethics: a practical ethics guide in statistics and data analysis for educational leaders. Sch LeadShip Rev 20(1):3
51. Smith NJ (2008) Achieving your professional doctorate. McGraw-Hill Education (UK)
52. Eddy DM, Hollingworth W, Caro JJ, Tsevat J, McDonald KM, Wong JB (2012) Model transparency and validation: a report of the ISPOR-SMDM modeling good research practices task force–7. Med Decis Making 32(5):733–743
53. Borges do Nascimento IJ, Abdulazeem H, Vasanthan LT, Martinez EZ, Zucoloto ML, Østengaard L, Novillo-Ortiz D (2023) Barriers and facilitators to utilizing digital health technologies by healthcare professionals. NPJ digital medicine 6(1):161
54. Chukwu E, Adu-Baah A, Niaz M, Nwagwu U, Chukwu MU (2023) Navigating ethical supply chains: the intersection of diplomatic management and theological ethics. Int J Multidiscip Sci Arts 2(3):127–139
55. De Peuter S, Conix S (2023) Fostering a research integrity culture: actionable advice for institutions. Sci Public Policy 50(1):133–145

56. Mutai N (2024) Ethical decision-making in data analysis: navigating challenges and ensuring integrity. Ethics Stat: Oppor Chall 183
57. Drolet MJ, Rose-Derouin E, Leblanc JC, Ruest M, Williams-Jones B (2023) Ethical issues in research: perceptions of researchers, research ethics board members and research ethics experts. J Acad Ethics 21(2):269–292
58. Dwivedi YK, Kshetri N, Hughes L, Slade EL, Jeyaraj A, Kar AK, Wright R (2023) Opinion paper: "So what if ChatGPT wrote it?" multidisciplinary perspectives on opportunities, challenges and implications of generative conversational AI for research, practice and policy. Int J Inf Manage 71:102642
59. Hagendorff T (2022) Blind spots in AI ethics. AI Ethics 2(4):851–867
60. Kayyali M (2024) Promoting research and innovation through accreditation. In: Quality assurance and accreditation in higher education: issues, models, and best practices. Springer Nature Switzerland, Cham, pp 399–426
61. Onyekwelu NP, Ezeafulukwe C, Owolabi OR, Asuzu OF, Bello BG, Onyekwelu SC (2024) Ethics and corporate social responsibility in HR: a comprehensive review of policies and practices. Int J Sci Res Arch 11(1):1294–1303
62. Walker RL, MacKay D, Waltz M, Lyerly AD, Fisher JA (2022) Ethical criteria for improved human subject protections in phase I healthy volunteer trials. Ethics & Hum Res 44(5):2–21
63. Aggarwal N, Kare PK, Datta SK (2021) Ethics in laboratory medicine: an overview of considerations for ethical issues. Bioeth Med Soc
64. Mondal MP, Bhutia MNN, Nagesh M, Prasad MB, Khati MA, Saha MS, Singh MS (2025) Navigating the ethical spectrum: principles and practices in research. OrangeBooks Publication
65. Bombardini M, Trebbi F, Zhang MB (2024) Measuring the costs and benefits of regulation. Annu Rev Econ 17
66. Tariq MU (2025) The Belmont report: guiding ethical principles in human research. In: IRB, human research protections, and data ethics for researchers. IGI Global Scientific Publishing, pp 245–268
67. Mistretta S (2021) Amending federal regulations to counteract language barriers in the informed consent process
68. Rubinstein I, Margulies P (2022) Risk and rights in transatlantic data transfers: EU privacy law, US surveillance, and the search for common ground. Conn L Rev 54:391
69. Al Kakoun N (2023) Priming and mining the civil engineering mindset: how personal values and perfectionism shape societal engagement and consideration in design (Doctoral dissertation, Swansea University)
70. Bahnson M, Satterfield D, Wyer M, Kirn A (2022) Interacting with ruling relations: engineering graduate student experiences of discrimination. Stud Eng Educ 3(1)
71. Godskesen T, Björk J, Juth N (2023) Challenges regarding informed consent in recruitment to clinical research: a qualitative study of clinical research nurses' experiences. Trials 24(1):801
72. Toner J, Martins JT (2022) Institutional isomorphism in collaborative, cross-cultural, project-based development work: an inquiry into the knowledge sharing behaviour of volunteers. J Knowl Manag 26(7):1763–1788
73. Dash B, Sharma P, Ali A (2022) Federated learning for privacy-preserving: a review of PII data analysis in Fintech. Int J Softw Eng & Appl (IJSEA) 13(4)
74. Labadie C, Legner C (2023) Building data management capabilities to address data protection regulations: learnings from EU-GDPR. J Inf Technol 38(1):16–44
75. Isibor E (2024) Regulation of healthcare data security: legal obligations in a digital age. SSRN 4957244
76. Caride CG (2021) Respecting the GDPR in online interviewed focus groups. J Audiov Transl 4(2):42–61
77. Mehta P, Zimba O, Gasparyan AY, Seiil B, Yessirkepov M (2023) Ethics committees: structure, roles, and issues. J Korean Med Sci 38(25)
78. Mathews B, MacMillan HL, Meinck F, Finkelhor D, Haslam D, Tonmyr L, Walsh K (2022) The ethics of child maltreatment surveys in relation to participant distress: implications of social science evidence, ethical guidelines, and law. Child Abuse Negl 123:105424

79. Rägo L, Zweygarth M (2023) CIOMS research guidelines: considering the needs of developing countries. In: Ethical innovation for global health: pandemic, democracy and ethics in research. Springer Nature Singapore, Singapore, pp 175–194
80. Marciano MA, Hall A, Mozayani A, Cromp T, Maynard HP III (2023) Enhancing research and collaboration in forensic science: a primer on human subjects' research protection. Forensic Sci Int: Synerg 7:100443
81. McNair L (2022) Ethical and regulatory oversight of clinical research: the role of the institutional review board. Exp Biol Med 247(7):561–566
82. Bukoski J, Erbes C (2025) Incorporating ethical reasoning in calculus I. PRIMUS 1–16
83. Capizzo L, Luisi M (2024) Responsibility, sustainability, or environmental justice? strategic communication and evolving expectations for stewardship and citizenship. Public Relat Inq 13(2):161–187
84. Gabbidon K, Chenneville T (2021) Strategies to minimize further stigmatization of communities experiencing stigma: a guide for qualitative researchers. Stigma Health 6(1):32
85. Chance S, Lawlor R, Direito I, Mitchell J (2021) Above and beyond: ethics and responsibility in civil engineering. Australas J Eng Educ 26(1):93–116
86. Kambhampati SB, Maini L (2023) Authorship in scientific manuscripts. Indian J Orthop 57(6):783–788
87. Pérez Vega C, Zielinska-Dabkowska KM, Hölker F (2021) Urban lighting research transdisciplinary framework—a collaborative process with lighting professionals. Int J Environ Res Public Health 18(2):624
88. Cabaniss ER, Davis SN, Parrotta KL (2025) Self-care rhetoric and institutional culpability: theorizing the academy and intellectual labor. J Fam Theory Rev 17(1):60–72
89. Pan Y, Zhu M, Lv Y, Yang Y, Liang Y, Yin R, Yuan X (2023) Building energy simulation and its application for building performance optimization: a review of methods, tools, and case studies. Adv Appl Energy 10:100135
90. Jarrah AM, Wardat Y, Fidalgo P (2023) Using ChatGPT in academic writing is (not) a form of plagiarism: what does the literature say. Online J Commun Media Technol 13(4):e202346
91. Heinz M, Turek ME, Schaefli B, Keiser A, Holzkämper A (2024) Can adaptations of crop and soil management prevent yield losses during water scarcity?–a modelling study. EGUsphere 2024:1–33
92. Piran MJ, Tran NH (2024) Enhancing research methodology and academic publishing: a structured framework for quality and integrity. arXiv:2412.05683
93. Thirumoorthy T (2023) Conflicts of interest in medicine: understanding the concepts to preserve the integrity of professional judgement and promote trust in the profession. Singapore Med J 64(2):121–126
94. Graham SS, Karnes MS, Jensen JT, Sharma N, Barbour JB, Majdik ZP, Rousseau JF (2022) Evidence for stratified conflicts of interest policies in research contexts: a methodological review. BMJ Open 12(9):e063501
95. Traversy G, Barnieh L, Akl EA, Allan GM, Brouwers M, Ganache I, Tonelli M (2021) Managing conflicts of interest in the development of health guidelines. CMAJ 193(2):E49–E54
96. Li W, Yigitcanlar T, Nili A, Browne W (2023) Tech giants' responsible innovation and technology strategy: an international policy review. Smart Cities 6(6):3454–3492
97. Weber-Lewerenz B (2021) Corporate digital responsibility (CDR) in construction engineering—ethical guidelines for the application of digital transformation and artificial intelligence (AI) in user practice. SN Appl Sci 3:1–25
98. Bell K, Kingori P, Mills D (2024) Scholarly publishing, boundary processes, and the problem of fake peer reviews. Sci, Technol, & Hum Valu
99. Maddi A, Maisonobe M, Boukacem-Zeghmouri C (2025) Geographical and disciplinary coverage of open access journals: OpenAlex, Scopus, and WoS. PLoS ONE 20(4):e0320347
100. Razlan NM, Samsuddin SF, Abrizah A (2024) Embracing transparency and openness in scholarly publishing: Insights from the Malaysian social sciences researchers. Malays J Libr Inf Sci 29(3):19–45

Dissemination of Research Findings

Abstract The effective publication of research findings is crucial for transforming scholarly work into useful insights that influence policy, practice, and innovation. Chapter 26 explores the theoretical underpinnings and practical methodologies crucial for effective research dissemination. It presents frameworks like Knowledge Translation (KT), Diffusion of Innovation, and stakeholder-driven Co-Creation, illustrating how these concepts connect evidence to implementation. The chapter highlights systematic approaches for customizing knowledge for various audiences, such as policymakers, practitioners, and the public. Case examples demonstrate effective public health, technology, and education distribution, showcasing the transformational capacity of strategic involvement and communication. Furthermore, it examines quantitative and qualitative criteria for measuring diffusion impact, providing instruments for analyzing success in practical implementations. Ethical concerns are included throughout, guaranteeing inclusion, precision, and cultural sensitivity in the communication process. This chapter provides researchers with useful insights and illustrative examples to enhance their work's societal and academic effects. This chapter is essential for scientists seeking to guarantee that their discoveries have an impact beyond academia and foster significant change across several fields.

1 Introduction

1.1 Introduction to the Theoretical Foundations of Research Dissemination

The dissemination of research findings is a key element in the scientific process. It facilitates the transmission of knowledge created via methodical inquiry to audiences relevant to the sought understanding [1]. By following this procedure, the research output will significantly contribute to formulating policies, future investigations, and practice. Research dissemination is influenced by principles that increase the efficacy of information sharing and encourage its acceptance. These principles are grounded in sound theoretical frameworks. Knowledge Translation Frameworks, Diffusion of

A. Almusaed et al., *Practice of Research Methodology in Civil Engineering and Architecture*, https://doi.org/10.1007/978-3-031-97393-2_26

Innovation Theory, Stakeholder Engagement, and Co-Creation are the three primary theoretical domains discussed in this section [2]. These domains provide the foundation for effective dissemination. One way to bridge the gap between evidence and application is by utilizing knowledge translation frameworks, which are systematic procedures. These frameworks emphasize contextualizing results to cater to a wide range of audiences, which helps to build a practical grasp of how research outputs may be effectively applied. Everett Rogers' Diffusion of Innovation Theory provides a complete model that explains how new ideas, habits, or technology spread throughout a population. Rogers established this theory [3]. The characteristics of the invention, the transmission routes, the passage of time, and the social structure in which diffusion occurs are some essential elements this theory highlights as impacting the acceptance of innovations. Finally, the Stakeholder Engagement and Co-Creation approach emphasizes the active participation of various stakeholders in the distribution technique. This strategy guarantees that research results are not just accessible but also connect with the goals and priorities of end-users by creating collaborative partnerships. This technique also ensures that the research outcome is accessible. An increase in the relevance and sustainability of research applications is achieved through stakeholder-driven co-creation, which also encourages shared ownership of the consequences of knowledge [4]. These theoretical underpinnings, when taken as a whole, offer a solid basis for the development and execution of dissemination strategies that are designed to maximize the impact and usability of research findings. For example, in architectural engineering research, communicating results on energy-efficient building materials entails converting experimental data into actionable guidelines for architects and engineers. Employing knowledge translation frameworks guarantees that this information is both accessible and actionable. Engagement with enterprise stakeholders promotes co-creation, increasing acceptance rates, and advancing sustainable methods in practical implementations [2].

1.2 Knowledge Translation (KT) Framework

Knowledge translation (KT) is a dynamic, cyclical process that improves results across many settings by creating, disseminating, and using knowledge [5]. The primary objective of KT is to guarantee that insights generated by systematic research are accessible, applicable, and pertinent to essential stakeholders, including policymakers, practitioners, and the public. KT reconciles the chasm between knowledge creation and actionable measures by addressing the enduring disparity between research outcomes and their practical implementation. Knowledge transfer transcends mere information dissemination; it requires knowledge customization to meet unique requirements and situations to enhance its acceptance and efficacy. This methodology converts research into practical ideas, guaranteeing its significance transcends academics. Through KT, information catalyzes transformation, empowering stakeholders to make educated decisions and effectively produce significant enhancements [6]. By promoting cooperation between researchers and end-users, KT establishes a framework conducive to flourishing evidence-based practices, yielding quantifiable and durable results. For example, in architectural engineering research, KT has been successfully utilized to execute sustainable building designs. Research on energy-efficient materials and passive cooling systems is converted into recommendations and decision-making instruments for architects and builders. These technologies assist experts in incorporating creative ideas into construction projects, decreasing energy consumption, and enhancing environmental performance. These instances illustrate how KT connects theoretical research with practical execution, guaranteeing tangible results [7].

1.2.1 Defining Knowledge Translation (KT)

Knowledge Translation (KT) tackles the essential issue of transforming intricate research outcomes into understandable and practical formats for non-academic audiences [8]. This method is necessary for closing the "know-do" gap between information acquired via systematic investigation and its practical application in policies or practices. The continuation of this gap frequently stems from the complex and specialized characteristics of research outputs, which may not successfully engage practitioners, policymakers, or other end-users. KT prioritizes the adaptation and contextualization of knowledge to address the distinct requirements of various audiences rather than just distributing information [9]. KT customizes results to correspond with the specific conditions of their intended application, promoting significant and enduring alterations in practice or policy. This method guarantees that research surpasses academic limitations, allowing stakeholders to interact with facts in manners that guide decision-making and stimulate enhancements in practical settings.

1.2.2 Core Principles of Knowledge Translation

Knowledge Translation (KT) fundamentally represents a complex and methodical methodology that integrates efficient tactics for knowledge dissemination with strong mechanisms to promote its acceptance and implementation. KT's success is attributed to its capacity to involve end-users throughout the process, cultivating collective ownership over the outcomes [10]. This participative method increases the probability of practical implementation and guarantees that the generated information aligns with its intended recipients' unique requirements and goals. A defining characteristic of KT is its dependence on organized models and frameworks to facilitate the conversion of research findings into practical implementations. These models offer a systematic approach, allowing academics and practitioners to connect knowledge development with practical applications. Among these frameworks, the most notable are:

- The Integrated Information Translation (IKT) Method: This approach highlights the active cooperation between academics and stakeholders, guaranteeing that information is co-created and directly relevant to real-world issues [11].
- The PARIHS Framework (Promoting Action on Research Implementation in Health Services): This paradigm emphasizes the interaction of evidence quality, contextual elements, and facilitation measures to promote effective implementation [12].
- The Knowledge-to-Action (KTA) Framework, developed by the Canadian Institutes of Health Research: The KTA framework outlines a cyclical process that combines knowledge development with actionable tasks, facilitating ongoing learning and enhancement [13].

1.2.3 The Knowledge-To-Action Framework

The Knowledge-to-Action (KTA) Framework, created by the Canadian Institutes of Health Research (CIHR), serves as a fundamental framework in knowledge translation [14]. This structure divides the process into two interrelated cycles, each vitally connecting knowledge development with practical implementation.

- The Cycle of Knowledge Creation: This initial phase focuses on the methodical production, synthesis, and adaptation of information to improve its applicability and relevance for target audiences. Essential tasks in this cycle encompass identifying credible evidence, synthesizing research findings, and adapting outputs into accessible formats, like policy briefings, toolkits, or practical suggestions [15]. These measures are intended to guarantee that knowledge applies and is significant to many stakeholders.
- The Cycle of Action: The second phase emphasizes the practical implementation of information in real-world contexts [16]. It includes tasks like recognizing obstacles to implementation, customizing treatments to context-specific requirements, and methodically assessing the results of these efforts. This phase's iterative and cyclical characteristics promote ongoing refining and learning, guaranteeing that knowledge application adapts to new problems and possibilities.

The KTA Framework integrates these two cycles, offering a resilient and flexible method for turning research into effective practices and policies, promoting sustainable enhancements across several areas.

1.2.4 The PARIHS Framework

The Promoting Action on Research Implementation in Health Services (PARIHS) Framework offers a thorough methodology for executing evidence-based practices in intricate organizational settings [17]. This paradigm highlights the dynamic interaction among three essential components:

- For knowledge to facilitate successful action, it must be believable, pertinent, and substantial. Robust evidence underpins effective implementation.
- The context in which information is utilized is crucial, involving elements such as corporate culture, leadership, and the organization's preparedness for change. An accommodating and flexible environment enables implementation [18].
- Facilitation: The methods and personnel that aid in applying knowledge are equally essential. Facilitation entails empowering and encouraging stakeholders to embrace and maintain new practices, assuring their effective integration into current systems.

The PARIHS Framework integrates these components to provide a systematic and comprehensive approach for translating evidence into practice, allowing businesses to tackle obstacles and attain lasting enhancements [19].

1.2.5 Integrated Knowledge Translation (IKT)

The Integrated Knowledge Translation (IKT) methodology underscores the active participation of stakeholders across the whole research process [20]. IKT improves research findings' relevance, application, and impact by including end-users as co-creators of knowledge. This participatory technique cultivates trust, connects research with stakeholder interests, and promotes collaborative partnerships that facilitate significant and enduring change.

Framework for Efficient Knowledge Dissemination

The framework for effective knowledge translation is a systematic and iterative process to convert research findings into practical applications. Essential elements comprise:

- Stakeholder Engagement: Actively incorporating stakeholders from the beginning guarantees that the information produced corresponds with their requirements and goals. This collaborative method fosters trust and collective response and promotes the implementation of outcomes [21].
- Contextualization of Knowledge: Adapting research findings to circumstances is essential. This involves transforming complex data into understandable representations, such as decision aids, suggestions, or training materials tailored to the target audience's needs.
- Facilitating Adoption: It is essential to identify and surmount implementation hurdles. Strategies may include implementing training programs, distributing necessary resources, or enhancing organizational readiness for change [22].
- Continuous Evaluation: Regular implementation process assessment ensures ongoing learning and improvement. Feedback loops enhance technique optimization and elevate total outcomes [23].
- Enduring Influence: To ensure sustained effectiveness of knowledge translation, organizations must augment capabilities, cultivate a culture of continuous improvement, and integrate KT methodologies into routine practices [24].

Integrated Knowledge Translation (IKT) embodies a dynamic and collaborative methodology for connecting research with practice. IKT guarantees that research has significant, lasting outcomes by emphasizing stakeholder interaction, contextualizing results, and facilitating adoption [25]. Frameworks like KTA and PARIHS offer structured direction for executing knowledge translation, highlighting the significance of collaboration, contextual comprehension, and facilitation in attaining enduring advancement and advantages across many domains. Table 1 compares the Knowledge-to-Action (KTA) and PARIHS frameworks, underscoring their distinct objectives and utilization. The KTA Framework prioritizes a cyclical process that amalgamates knowledge generation and practical execution, rendering it applicable across various fields such as health, education, and engineering. Conversely, the PARIHS Framework concentrates on implementing evidence-based practices within intricate organizational environments, mainly health services. Fundamental elements of KTA encompass knowledge and action cycles, whereas PARIHS accentuates evidence, context, and facilitation. KTA's methodical, iterative approach promotes adaptability, stakeholder involvement, and enduring enhancements [26]. PARIHS underscores the dynamic interactions among its components, empowering stakeholders and addressing organizational preparedness for transformation. Both frameworks connect research and practice, yet they diverge in focus and scope of application.

1.3 Diffusion of Innovation Theory

Everett Rogers's Diffusion of Innovation Theory provides a significant framework for comprehending the dissemination of new ideas, technologies, or behaviors within a social system throughout time [35]. This theory is especially important for research dissemination since it elucidates the elements that affect the acceptance and adaptation of innovations. This theory facilitates successful communication and the integration of new concepts into practice by examining the stages of dissemination, classifying adopters, and acknowledging the roles of important influencers. Rogers' diffusion model delineates five essential stages in disseminating innovations within a social system [36]. The initial step is Knowledge, wherein awareness of the invention is established among prospective adopters. This is succeeded by persuasion when individuals develop attitudes toward the invention shaped by variables such as perceived advantages and societal norms. The third stage, Decision, entails embracing or rejecting the innovation based on the facts and attitudes developed in the preceding stages [37]. The subsequent phase is the Implementation stage, whereby innovation is executed, necessitating practical tools, resources, and modifications to established procedures. The Confirmation stage reinforces the decision by validating favorable outcomes and feedback.

Rogers classifies adopters into five categories according to their readiness and pace in adopting innovation [38]. The Innovators are the initial adopters of novel concepts, frequently motivated by curiosity and a propensity for risk-taking. The

Table 1 Differences between KTA and PARIHS frameworks

Aspect	Knowledge-to-action (KTA) framework	PARIHS framework
Purpose [27]	It focuses on a cyclical process that connects knowledge creation with its application in practice	Focus on implementing evidence-based practices in complex organizational settings
Core components [28]	– Knowledge creation cycle: involves generating, synthesizing, and tailoring knowledge – Action cycle: deals with implementing knowledge, addressing barriers, and assessing outcomes	– Evidence: ensures knowledge is credible, relevant, and significant – Context: highlights the importance of organizational culture, leadership, and readiness – Facilitation: supports the process with resources and personnel
Approach [29]	Provides a structured and iterative model that integrates research findings into actionable practices or policies	Emphasizes the dynamic interplay of evidence, context, and facilitation for effective implementation
Stakeholder engagement [30]	Actively involves stakeholders in both the knowledge creation and action cycles	Focuses on enabling and empowering stakeholders through facilitation for sustained implementation
Focus areas [31]	Broader applicability across multiple domains such as health, education, and engineering	Primarily designed for health services and similar settings requiring evidence-based practice
Implementation strategy [32]	Uses systematic feedback loops to refine and optimize implementation processes continuously	Encourages adapting to the specific organizational environment and focusing on cultural readiness
Flexibility [33]	Offers a more general framework adaptable to various disciplines	Primarily tailored to health service contexts but can be adapted with effort to other fields
Outcome goals [34]	Achieves sustainable improvements through a blend of ongoing knowledge creation and application	Attains effective integration of evidence-based practices by addressing barriers and promoting readiness

Early Adopters thereafter emerge, serving as pivotal role models and influencers within their communities, facilitating the legitimization of the invention. The Early Majority embraces innovations after recognizing their advantages and obtaining validation from the experiences of early adopters. The Late Majority is often more suspicious, needing further confirmation before adopting change. The Laggards are the most resistive, typically embracing ideas only if they become widespread or essential. Key individuals and groups assume distinct roles in the diffusion process, profoundly impacting the rate and scope of adoption. Early adopters function as trendsetters whose behaviors engender trust and emulation [39]. Opinion leaders, often esteemed individuals within a community, offer endorsements that authenticate the invention and promote broader adoption. Simultaneously, Change Agents effectively promote adoption by providing tools, training, and continuous support to guarantee a seamless transfer to the new practice. The distribution of research and innovations must be carefully coordinated with the phases of the diffusion process to optimize efficacy. In the Knowledge phase, focused outreach and explicit communication are crucial for enhancing awareness. During the Persuasion phase, narratives, case studies, and demonstrable success effectively cultivate trust and engagement [40]. During the Decision and Implementation phases, offering practical resources, training, and continuous assistance guarantees the smooth integration of the innovation into reality. Ultimately, in the Confirmation phase, ongoing assessment and feedback systems facilitate the reinforcement of favorable results and consolidate the integration of innovation [41]. Utilizing the Diffusion of Innovation Theory, companies and researchers may comprehend the dynamics of change, allowing them to formulate customized strategies that promote the acceptance of novel ideas and practices within their target populations.

For example, The Diffusion of Innovation Theory is especially pertinent in architectural engineering research, notably in facilitating the adoption of energy-efficient building designs. Like pioneering architects, early adopters may adopt passive design ideas and incorporate them into their projects [42]. Opinion leaders may authenticate these designs through case studies, promoting the widespread adoption of sustainable methods throughout the sector. Figure 1 depicts the phases of the diffusion of innovation model along with the corresponding adopter types. The upper portion delineates the decision-making process into five stages: Knowledge, Persuasion, Decision, Implementation, and Confirmation. These phases assist individuals or organizations in the adoption of innovations. The lower half categorizes the population into five adopter groups according to their propensity to embrace innovations: Innovators, Early Adopters, Early Majority, Late Majority, and Laggards. Moreover, distinct influencer positions such as Trendsetters, Opinion Leaders, and Change Agents are highly influential among early adopters. This paradigm is extensively utilized in marketing, technology, and sociology to comprehend the dissemination of innovations inside a community.

1.4 Stakeholder Engagement and Co-Creation

Stakeholder engagement is a cornerstone of effective research dissemination, ensuring that findings are relevant and actionable within their intended contexts [43]. By integrating stakeholders into the research process, dissemination strategies become more tailored and impactful, addressing real-world needs. Co-creation, which involves stakeholders in designing and implementing these strategies, further

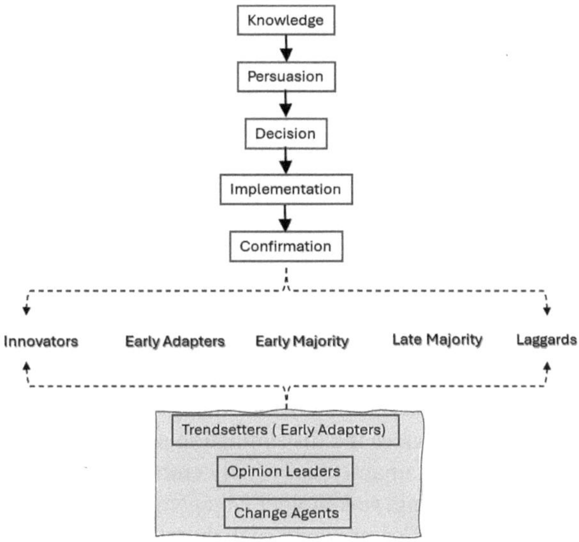

Fig. 1 Stages of adoption and diffusion of innovations (*Source* Authors)

strengthens the process by fostering shared ownership and commitment, ultimately enhancing the application and sustainability of research outcomes. The theoretical rationale for involving stakeholders is their crucial role in shaping the environments where research findings are applied [44]. Stakeholders—such as practitioners, policymakers, community members, and industry representatives—possess valuable insights and practical knowledge that can align research outputs with societal and organizational priorities. Engaging stakeholders early in the research process ensures that their needs, values, and expectations are incorporated, making the findings more relevant and actionable. The Community-Based Participatory Research (CBPR) model exemplifies this approach, emphasizing equitable partnerships between researchers and stakeholders [45]. This framework advocates shared decision-making, collaborative problem-solving, and mutual respect, which are vital for generating knowledge that resonates across diverse audiences and contexts. Effective stakeholder engagement begins with stakeholder mapping and prioritization. Stakeholder mapping involves identifying individuals or groups affected by or can influence the research outcomes. These stakeholders are then categorized based on their level of interest and influence. High-priority stakeholders are those who stand to benefit directly from research or whose support is essential for successful implementation. This approach helps researchers allocate their efforts strategically, focusing on stakeholders who can drive the adoption and impact of their findings. Researchers can create stronger connections and partnerships that enhance the dissemination process by prioritizing engagement with influential stakeholders. Co-creating dissemination strategies with stakeholders offers several benefits that significantly improve the relevance and effectiveness of research applications [46]. One key benefit is relevance. Researchers can develop dissemination strategies tailored to specific contexts' unique needs, challenges, and opportunities by collaborating with stakeholders. For example, in architectural engineering research, stakeholders such as developers and contractors can contribute insights to make sustainable building practices more practical and appealing. Another critical benefit of co-creation is buy-in. When stakeholders are actively involved in developing dissemination strategies, they are more likely to take ownership of the research findings and commit to their implementation. This shared ownership builds trust and fosters a sense of partnership, increasing the likelihood of successful application. Finally, co-creation often leads to innovation [2]. Collaborative approaches bring together diverse perspectives, enabling creative and practical solutions that may not emerge from traditional dissemination methods. Researchers and stakeholders' combined expertise and experiences can uncover new pathways for applying research findings to maximize impact. Researchers can bridge the gap between knowledge generation and practical application by adopting co-creation practices and prioritizing stakeholder engagement. This collaborative approach enhances the dissemination process and ensures that research findings are impactful, sustainable, and widely embraced by those they aim to serve. In architectural engineering, stakeholder engagement can enhance the adoption of green infrastructure [47]. For example, collaborating with urban planners, local government, and community members during research on sustainable stormwater management ensures practical application. Co-creation, such as designing rain gardens with input from these stakeholders, increases relevance, builds support, and fosters innovative, site-specific solutions.

2 Methods for Efficient Exchanging of Information and Knowledge

A deliberate and strategic strategy is required to communicate research findings to a wide range of people effectively [1]. Researchers must consider their target groups' specific requirements, expectations, and preferences when picking relevant instruments and methodologies. This is necessary to ensure that their work has the most possible impact. The utilization of various tactics in four significant areas is required to accomplish this goal. Before everything else, academic distribution routes continue to be essential. Engaging in activities like publishing in journals subject to peer review, presenting at conferences, and participating in educational seminars are all ways to communicate with scholarly audiences. Second, policy-oriented distribution is centered on converting research findings into formats that are easily accessible to policymakers [48]. These formats include policy briefs, white papers, or customized reports that provide information that may be used to inform decision-making processes. Third, engaging with the community and the public is essential to connect with non-intellectual audiences. This can be accomplished through participatory workshops, public lectures, or community events, which create an environment conducive to conversation and guarantee that the study is accessible and pertinent. In

Fig. 2 Key strategies for effective research dissemination (*Source* Authors)

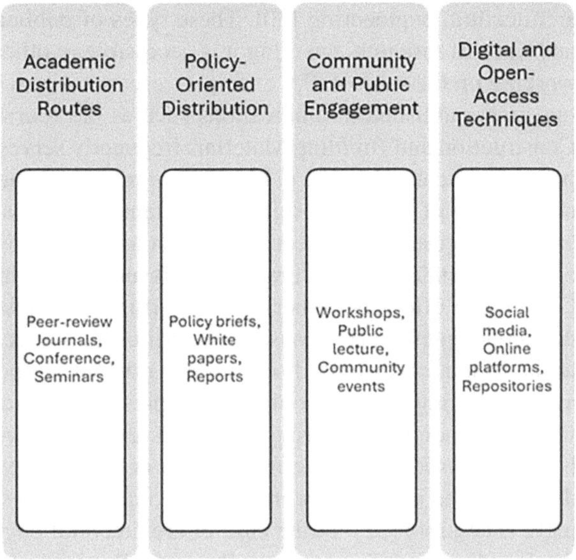

conclusion, digital and open-access techniques are critical roles that play a significant part in expanding reach and accessibility. Researchers can publish their findings worldwide via internet platforms, social media, and open-access repositories, which creates additional avenues for collaboration and citizen involvement. Researchers may increase their results' visibility, relevance, and impact by deliberately combining these strategies [49]. This ensures that their findings connect with a wide range of audiences and significantly contribute to knowledge and practice. The following four key tactics are illustrated in Fig. 2, which highlights how to disseminate research to various audiences effectively. Among these are the following: Academic Distribution, which makes use of scholarly journals and conferences; Policy-Oriented Distribution, which emphasizes formats such as policy briefs to influence decision-makers; Community Engagement, which emphasizes public interaction through workshops and lectures; and Digital and Open Access methods, which make use of online platforms to improve accessibility and collaboration. When it comes to ensuring that research findings are compelling, widely understood, and adaptable to real-world situations, each strategy plays a distinct role in facilitating knowledge sharing across various sectors and stakeholders.

2.1 Multiple Methods of Academic Dissemination

When exchanging knowledge among scholarly societies, transmitting academic information is essential. Publishing articles in journals that are peer-reviewed and have a high impact is a crucial component of academic dissemination in the field of architectural engineering [50]. These types of publications authenticate the quality and rigor of research, ensuring it is accessible to other professionals and academics working on the subject. For example, research on environmentally friendly building materials published in publications such as the Journal of Building Engineering or Construction and Building Materials frequently serves as academic and professional practice benchmarks. The publication of research in academic and professional journals is one of the most significant channels for dissemination; another essential channel is research presentations at conferences [51]. The International Conference on Sustainable Building Materials and the Architectural Engineering Institute Conference offer researchers the opportunity to present their results, interact with their colleagues, and get constructive criticism. Interactions like these frequently lead to the formation of multidisciplinary partnerships and the opening of new research avenues. Using a dynamic conference presentation, researchers can interact directly with their audience, ensuring their work resonates beyond the written word. This can be accomplished by delivering a poster or through a vocal presentation. In addition, databases and repositories are critical components in the process of making research more accessible [52]. In the case of architectural engineering studies, for example, the Building and Construction Research Repository methodically preserves them, ensuring they are accessible for an extended period. The research findings can reach a wider audience and have a more significant influence thanks on these platforms, encouraging more widespread participation and boosting innovation in architectural engineering [53]. The many approaches to academic distribution are discussed in Table 2, emphasizing the significance of these approaches and practical examples and obstacles. Publication in journals subject to peer review, presentation at conferences, utilization of dynamic presentation formats, and utilization of databases, repositories, and open-access platforms are some of the ways that fall under this category. Each approach plays a vital role in ensuring that the study's findings are appropriately communicated to various audiences. Although these methods expand the scope of research and its effect, they have drawbacks, including accessibility, prices, and long-term viability. Researchers can maximize their distribution methods by solving these problems, which will, in turn, stimulate greater cooperation, creativity, and engagement within their respective academic and professional networks among their peers.

2.2 *Policy-Oriented Circulation of Information*

It is necessary to employ distribution tactics that are both targeted and strategic to bridge the gap between research and policymaking [60]. Policy briefs and white papers are essential for converting complicated research findings into insights that decision-makers can implement. The essential findings are condensed into formats that are easy to understand and are presented in these publications. Additionally, specific recommendations for policies or practices backed by data are included. The

Table 2 Key methods, examples, and challenges in academic dissemination

Method	Description	Examples	Challenges
Peer-reviewed journals [54]	High-impact journals authenticate research rigor and ensure accessibility to academics and professionals	Research on sustainable materials published in the journal of building engineering or construction and building materials	Lengthy peer-review process and limited accessibility due to paywalls
Conference presentations [55]	Conferences provide platforms for presenting research, networking, receiving feedback, and fostering collaborations	The international conference on sustainable building materials and architectural engineering institute conference	High costs for travel and participation; limited audience beyond attendees
Dynamic presentations [56]	Allows direct audience interaction through vocal or poster presentations, enhancing engagement and comprehension	Poster presentations at architectural engineering institute conference	Time constraints and limited reach compared to digital dissemination
Databases and repositories [57]	Systematic archiving ensures the long-term accessibility of research findings	The building and construction research repository preserves architectural engineering studies for extended periods	Maintain repositories and ensure data standards across studies
Interdisciplinary partnerships [58]	Interactions at conferences and collaborations across disciplines open new research avenues	Multidisciplinary research partnerships emerging from conferences like the international conference on sustainable building materials	Differences in terminologies and methodologies across disciplines
Open-access platforms [59]	Platforms expand the reach and impact of research findings, fostering broader participation and innovation	Use of open-access platforms for disseminating innovative architectural engineering solutions	Sustainability of funding for open-access publishing and quality control

most effective policy briefs address contemporary concerns and connect with policymakers' goals and informational requirements [61]. This ensures that the policy briefs are relevant. A further strengthening of involvement may be achieved using interactive methods such as workshops, forums, or discussions with public authorities. These events provide researchers with the opportunity to directly convey their results to stakeholders involved in the creation and implementation of policies. This technique encourages communication and cooperation, guaranteeing that decision-making processes are informed by research that is based on evidence [62]. The concrete influence of research on policy decisions is demonstrated by case studies showcasing successful policy translation. These case studies give significant insights. These examples motivate scholars and provide them with specific instructions on achieving comparable results in their study. For example, research in architectural engineering has impacted urban planning by creating new standards for constructing more energy-efficient buildings. In response to these results, which were effectively communicated to legislators, improved construction rules that promote sustainability have been implemented. Engaging in strategic communication and active cooperation is vital to bridge the gap between research and policy. This will ensure the development of evidence-based policies to bring about meaningful change [63]. Table 3 summarizes critical aspects of bridging research and policymaking through strategic communication and collaboration. It highlights the importance of tools like policy briefs, interactive methods, and case studies in converting complex research findings into actionable insights. Effective implementation ensures that decisions are evidence-based and address contemporary issues. However, challenges like simplifying complex data, maintaining stakeholder engagement, and overcoming legislative resistance can hinder progress. An example from architectural engineering illustrates the practical impact of research, where findings on energy-efficient buildings influenced sustainable urban planning policies. By addressing these challenges and emphasizing targeted strategies, researchers can strengthen the link between academia and policy, ensuring meaningful and informed change in practice.

2.3 Engagement with the Community and the Public

Adjusting one's communication style and approach is necessary to successfully engage with communities and the public [70]. Simulating complex language research is essential to guarantee accessibility and comprehension. Creative tools such as blogs, films, and infographics are handy for converting research findings into more easily understood formats by lay audiences. Through these platforms, researchers can effectively communicate their messages, expanding the scope of their study beyond the conventional confines of academic institutions. Research dissemination can have a more significant impact if done in collaboration with community groups [71]. Researchers can plan outreach activities that cater to the community's particular requirements and objectives by forming partnerships with local organizations. In addition to fostering trust, these partnerships guarantee that the research findings are applicable and practical within the community context. Social media platforms have become increasingly common and influential instruments for public involvement. Twitter, LinkedIn, and Facebook offer academics direct avenues to communicate their findings with audiences worldwide [72]. Using these platforms, researchers can participate in real-time debates, provide updates, and encourage the broader dissemination of their discoveries. It is possible to increase the exposure

Table 3 Insights and challenges in bridging research and policy

Key aspects	Examples	Outcome	Challenges
Targeted and strategic distribution tactics [64]	Focused strategies to connect research findings with policymaking	Enhances the relevance of research to policy challenges	Difficulty in tailoring strategies to diverse policymaker needs
Policy briefs and white papers [65]	Condense complex research into actionable insights, addressing policymakers' needs and priorities	Ensuring findings are clear, concise, and aligned with current priorities	Simplifying complex findings without losing essential details
Interactive methods [66]	Engagement through workshops, forums, and discussions to convey research findings to stakeholders	Promotes communication and collaboration, improving policy implementation	Ensuring active participation and interest from stakeholders
Case studies [67]	Showcase successful research-to-policy translation, offering practical insights and inspiration	Provides guidelines for researchers to replicate successful outcomes	Identifying and documenting relevant successful examples
Example from architectural engineering [68]	Impact on urban planning by creating energy-efficient building standards and influencing updated construction rules	Promote sustainability through informed urban planning policies	Overcoming legislative resistance to implementing new standards
Strategic communication and cooperation [69]	Essential for bridging research and policy, ensuring evidence-based decisions and meaningful changes	Facilitate evidence-based policymaking, ensuring impactful change	Maintaining consistent and effective communication channels

and effect of research by taking a strategic approach to social media. This strategy includes the utilization of specific hashtags, the tagging of key stakeholders, and the creation of material that is engaging. For example, digital twins research can engage communities by creating interactive visualizations demonstrating how virtual models mirror real-world systems like urban infrastructure or energy grids. These tools make complex data relatable and accessible. Researchers can host workshops or public forums by partnering with local groups, allowing community members to interact with and contribute to the research, fostering deeper involvement and practical application [73]. Figure 3 depicts "Engagement with the Community and the Public," highlighting five interrelated tactics. The central subject is situated in the core, extending to essential components: Modify Communication Style, underscoring the necessity of customizing language for varied audiences; Innovative Tools,

Fig. 3 Strategies for effective engagement with communities and the public in research dissemination (*Source* Authors)

including blogs and infographics, that elucidate intricate research results; Cooperative Partnerships, accentuating collaborative endeavors with local organizations to guarantee pertinence and credibility; Social Media, leveraging platforms such as Twitter and Facebook to broaden outreach; and Interactive Visualizations, providing accessible representations of data. These tactics combine to enhance research impact and foster more community engagement.

2.4 *Methodologies that are Digital and Open-Access*

The advent of the digital era has brought about a revolution in the distribution of knowledge, which has brought about several benefits for audiences and scholars [74]. Publishing research findings in an open-access format has become more popular, allowing unfettered and unlimited access. Through the implementation of this strategy, information is democratized and made available to individuals and institutions regardless of their financial circumstances [75]. Preprint repositories further support open access, allowing researchers to share their work before it is subjected to formal peer review. Examples of such repositories are bioRxiv and SSRN. Webinars, podcasts, and other interactive online tools are innovative digital formats that enable engaging communication with audiences worldwide. Webinars mix visual and aural aspects to produce an entertaining presentation, whereas podcasts offer a more personal platform for in-depth conversations and stories [76]. Audiences can dynamically study research findings using interactive tools such as online simulations and data visualizations. This results in better comprehension and increased engagement. Alternative metrics, also known as altmetrics, are helpful when it comes to monitoring the impact of research that has been distributed. Alternative metrics, in contrast to standard metrics focused on citations, can measure various activities, such as mentions on social media, downloads, and blog referrals [77]. Utilizing these metrics offers researchers significant insights into how the findings of their study are being debated and exploited in digital spaces. This enables researchers to adjust their dissemination tactics to have a higher effect. Knowledge sharing, efficient and effective, is necessary to maximize the impact of research and ensure that it is relevant to a wide range of audiences. Researchers can extend the scope of their study and increase its effects by utilizing various methods, including academic dissemination channels, policy-oriented tactics, community participation, and digital technologies. Every strategy has its own set of benefits. It is essential to customize them so that they are in line with the objectives of the dissemination effort and the requirements of the audience that is being targeted [78]. Strategic dissemination ensures that research findings contribute significantly to advancing knowledge and resolving real-world difficulties. This is accomplished by fostering increased cooperation, creativity, and social impact during dissemination. By way of illustration, researchers may utilize open-access platforms to publish sustainable construction techniques in the context of Construction 5.0 research. This makes the methodologies available to many people. The influence of artificial intelligence-driven building technology may be demonstrated through interactive digital tools such as online simulations [79]. Webinars connect worldwide players, creating an environment conducive to innovation and cooperation in building techniques, while social media platforms enhance the distribution of information. Figure 4 illustrates the implementation of digital and open-access methodologies in research dissemination. The central theme connects to five key components: Open-Access Platforms, enabling democratized knowledge sharing regardless of financial constraints; Preprint Repositories, such as bioRxiv and SSRN, for early research distribution; Interactive Tools, including simulations and data visualizations, enhancing engagement and comprehension; Digital Formats, like webinars and podcasts, offering dynamic and accessible communication; and Alternative Metrics, tracking research impact through diverse indicators such as social media mentions and downloads. Together, these strategies expand the reach and influence of research, fostering innovation and global collaboration.

3 Assessing the Effects of the Dissemination of Research Resources

Analyzing the consequences of distributing research resources is essential to knowing the effectiveness and results of the various dissemination techniques [80]. Researchers and institutions might find effective practices by systematically examining these efforts to discover beneficial methods and address possible limits in the

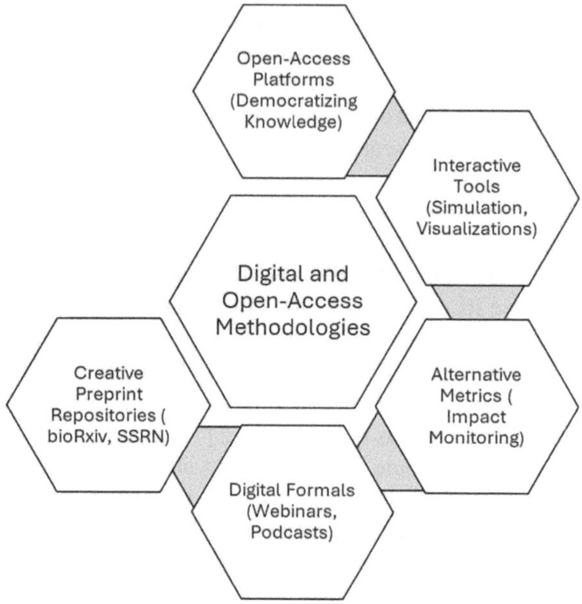

Fig. 4 Framework for digital and open-access methodologies in research dissemination (*Source* Authors)

transfer of information. A thorough evaluation framework must incorporate quantitative and qualitative research approaches to provide a detailed examination of the effects of dissemination [81]. Quantitative methods may involve the utilization of metrics such as citation analysis, download counts, and data derived from surveys to evaluate the extent of reach and usage. Using qualitative methodologies, such as interviews, focus groups, and case studies, one may gain a more profound understanding of the involvement of stakeholders and the contextual elements that influence the acceptance of knowledge that has been conveyed. These approaches, when combined, offer a solid basis that may be used to evaluate the extent to which dissemination initiatives can reach their target audience and their ability to impact practice, policy, or more research. As part of this research, we will also investigate the many difficulties inherent in quantifying the effects of diffusion. Variability in access to stakeholders, varying interpretations of study findings, and the lengthy time frames that are sometimes necessary to see real change contribute to these challenges [82]. This section intends to shed light on effective tactics and develop procedures for enhancing research dissemination efforts' social and academic value by combining various approaches and addressing the problems presented. For example, in passive building research, assessing the dissemination of findings related to energy-efficient designs could involve measuring citations in architectural journals and gathering feedback from practitioners about how these insights are applied in construction projects. Challenges such as variability in stakeholder access, differing interpretations of findings, and the lengthy timelines required for measurable change highlight the complexities of evaluation [83]. This section aims to shed light on effective tactics and develop procedures to enhance research dissemination efforts' societal and academic value by integrating diverse approaches and addressing these challenges.

3.1 Metrics Related to Quantitative Impact

The utilization, recognition, and appreciation of research by various audiences may be better understood through quantitative metrics, which give significant insights. They assist in determining the extent to which a study has an effect inside academic circles and the extent to which it is relevant to the public. Using citation analysis and alternative metrics (altmetrics) are two key strategies utilized in evaluating the effect of research [84]. Academic recognition is the primary emphasis of citation analysis, which involves keeping track of the number of times a study is mentioned in scholarly literature. On the other hand, alternative metrics quantify involvement outside of the academic world by analyzing online interactions such as mentions on social media, downloads, and media attention [85]. Taken as a whole, these measures provide an all-encompassing perspective on the influence of research.

3.1.1 Citation Analysis

Citation analysis is a well-established method for evaluating research influence within academic communities [86]. It employs citation counts, h-index scores, and journal impact factors to measure how frequently a study is referenced in other scholarly publications. For example, a study with high citations often signifies its findings have been pivotal in advancing knowledge, shaping methodologies, or guiding subsequent studies within a particular field. Such metrics serve as indicators of the work's importance and its long-term contribution to academic discourse. Despite its utility, citation analysis has notable limitations. It focuses exclusively on academic recognition and overlooks research's broader societal or practical impacts [87]. For example, studies that inform policy changes, inspire industry innovation, or raise public awareness may not be immediately reflected in citation metrics, limiting the method's comprehensiveness. Additionally, the accumulation of citations is a gradual process. It can take years for a study to gain widespread acknowledgment and reach a critical citation threshold. This time lag renders citation analysis less effective for assessing the immediate influence of newly published research. As a result, while citation analysis remains a valuable tool, it should be complemented by other metrics to capture a more holistic view of research impact [88].

3.1.2 Altmetrics

Altmetrics offers a contemporary method for evaluating the impact of research by monitoring the level of public interest and interaction on the internet [89]. A few

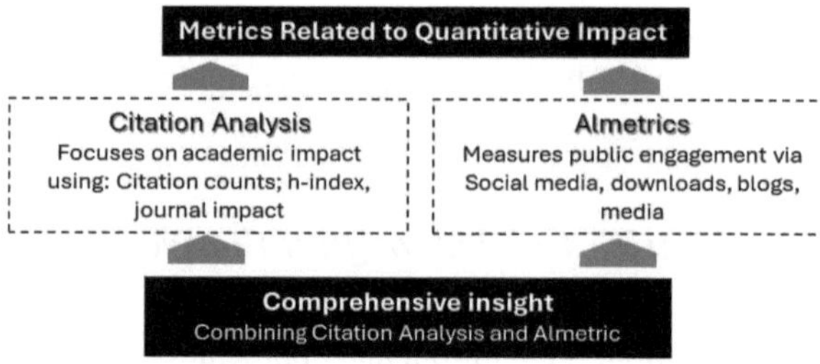

Fig. 5 Quantitative impact metrics for research evaluation (*Source* Authors)

examples of these indicators are mentions on social media, downloads of articles, references to blogs, and publicity throughout the media. Real-time data on how research is discussed outside of academic circles can be obtained through tools such as Altmetric.com. When comprehending research's impact on society, altmetrics are helpful [90]. These studies shed insight on how the general public, policymakers, and organizations engage with academic results, providing a more comprehensive and immediate perspective on the influence of research. For example, a study on sustainable material passports might gain attention through LinkedIn posts from construction professionals, blogs on green architecture, or media articles about zero-waste building practices. Tools like Altmetric.com can reveal how this research is discussed and shared, emphasizing its relevance to policymakers, industry leaders, and the public [91]. Altmetrics thus offer a broader and more immediate view of the research's societal impact. Altmetrics and Citation Analysis are two important methods used to measure the quantitative effect of research, and Fig. 5 demonstrates both methods. An evaluation of academic recognition is carried out by tracking citations in scholarly work. This is accomplished through several techniques, including citation counts, h-index scores, and journal impact factors. It reflects the effect that a study has inside academic circles, but it does not have any immediacy or societal context. Altermetrics, on the other hand, is a method of measuring public involvement by examining online interactions. These interactions include mentions on social networking platforms, downloads, blogs, and media coverage. These measures offer a more comprehensive and up-to-date perspective on how research resonates beyond academia. Researchers obtain a more comprehensive understanding of the impact of their work by combining the two methodologies, which consider both the academic contributions and the social significance of their findings.

3.2 Methods of Evaluation that are Qualitative

Although quantitative approaches provide helpful information on the quantifiable results of research dissemination, they frequently fail to capture its more complex and wide-ranging effects [92]. Because of this constraint, qualitative methods emphasizing narrative evidence, user experiences, and contextual knowledge are required.

These approaches highlight the situational and human aspects of research dissemination, offering a more thorough and nuanced framework for evaluation. Interviews, focus groups, case studies, and content analysis are frequently used in qualitative evaluation. Interviewing research end users, for example, might provide their opinions on how the results affected their business procedures or choices [93]. While case studies offer a detailed examination of examples of effective distribution, focus groups bring together a variety of stakeholders and facilitate conversations that emphasize shared experiences and collective perceptions. Recurring themes and attitudes that are more difficult to measure can be found through content analysis of reports, social media conversations, and user feedback. The capacity of qualitative approaches to evaluate the contextual suitability of dissemination initiatives is another of their advantages [94]. They can determine, for instance, if distribution tactics meet sector-specific, organizational, or cultural requirements. These approaches guarantee that research dissemination is assessed in a manner that accurately represents its actual societal and professional implications by gathering the opinions of many stakeholders [95]. By emphasizing the depth of knowledge, the breadth of experiences, and the wider ramifications of research dissemination, qualitative techniques enhance quantitative methods and offer a comprehensive framework for assessment.

3.2.1 Gathering People's Opinions

Although quantitative approaches can produce accurate measurements and statistical insights, they frequently fail to consider the complex and far-reaching effects caused by the dissemination of research. The usefulness of qualitative techniques, which concentrate on obtaining narrative evidence and user viewpoints, is brought into the foreground by this constraint. The qualitative approaches provide a more in-depth comprehension of how research is received, understood, and used in real-world situations. This is because qualitative methods prioritize the voices and experiences of affected people. Researchers can collect comprehensive ideas and input from various stakeholders through qualitative approaches such as interviews, focus groups, and open-ended surveys [96]. These techniques provide psychological and situational insights and illuminate unintended repercussions and obstacles that stand in the way of adoption. Considering this perspective, the qualitative methods offer a more all-encompassing analysis of the broader social and practical ramifications of study dissemination. In architectural engineering, interviews with architects and builders on sustainable construction practices can highlight practical obstacles and regional adjustments that are not obvious in quantitative statistics. Such insights may bridge the gap between theory and practice, assisting in refining guidelines to enable successful implementation across various scenarios.

3.2.2 Gathering User Feedback

Qualitative evaluation emphasizes the contributions of end-users, such as policymakers, practitioners, and community members [97]. This input is extremely helpful in gaining important insights into the perceived value of study findings and their practical implementation. Essential instruments for the collection of this input include:

- In-depth, one-on-one conversations that investigate participants' viewpoints and experiences regarding the research are referred to as interviews [98].
- In the context of research, focus groups are group talks that are facilitated to collect a variety of perspectives and stimulate discourse on the relevance and usability of research findings.
- Surveys are structured questionnaires that aim to obtain insights from a wider population. They frequently include qualitative and quantitative components [99].

3.2.3 A Narrative of the Results

The documentation of real-world instances that indicate how research has changed policies, practices, or behaviors is another component of qualitative assessment [100]. Case studies and thorough reports are among the instruments utilized in presenting these tales, which provide tangible proof of the impact of research. To demonstrate the practical implications of their work, researchers can, for instance, examine how their results have resulted in changes to organizational policies or other organizational practices. Qualitative approaches thoroughly grasp societal relevance and wider repercussions of study dissemination [101]. This is accomplished via the incorporation of user input and narrative evidence. Figure 6 depicts the Qualitative Evaluation Methods, emphasizing the objective of documenting extensive research impacts through User Feedback and Narrative Evidence. User feedback encompasses the collection of insights from end-users using techniques such as interviews, focus groups, and surveys. Interviews offer in-depth individual insights, focus groups gather varied opinions in collective environments, and surveys facilitate extensive population engagement and evaluation. Narrative Evidence enhances this by providing empirical illustrations of the research effect [102]. This encompasses case studies that illustrate results and reports that exhibit alterations in policies or practices prompted by the survey. These elements collectively provide a comprehensive evaluation that amalgamates user perspectives and empirical narratives to comprehend the social and practical implications of research thoroughly. This methodology guarantees that the influence of research transcends academic measures to encompass its practical relevance.

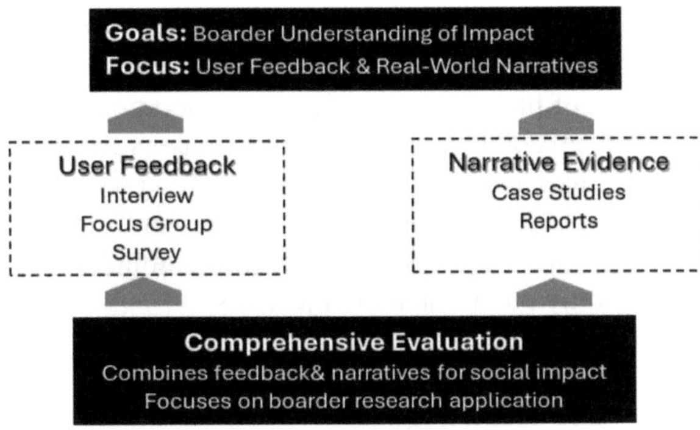

Fig. 6 Qualitative methods for evaluating broader research impact (*Source* Authors)

3.3 Case Studies of the Impact that was Measured

Case studies are essential for analyzing how research findings, when communicated through targeted and strategic distribution strategies, may be successfully converted into concrete, real-world effects [103]. They provide thorough reports of individual events, offering a realistic framework for assessing the effectiveness of research communication initiatives. Moreover, case studies enhance comprehension of how research findings might lead to significant social, environmental, or technical progress. They assess the degree to which research-based treatments may induce significant change and emphasize the contextual aspects that facilitate or obstruct these processes [104]. This renders them essential for recognizing excellent practices, enhancing dissemination strategies, and guaranteeing that research endeavors fulfill their intended goals. Consequently, case studies serve as a critical link between theoretical knowledge and practical, significant results, highlighting the necessity of customized tactics to enhance the value of academic research.

3.3.1 The Health of the Public

Efficient distribution has consistently been fundamental to public health, significantly contributing to resolving global health issues [105]. This was particularly apparent during the COVID-19 pandemic when the prompt distribution of epidemiological data greatly impacted vaccination plans and public health measures. The primary variables contributing to these results included the swift sharing of data, the wise utilization of digital platforms, and the proactive involvement of stakeholders. These strategies accelerated decision-making and enhanced public confidence in scientific research, eventually promoting broad compliance with preventative measures [106]. The swift dissemination of epidemiological data allowed governments and health organizations to efficiently customize responses to emerging patterns. Digital platforms functioned as essential channels for real-time information distribution to various audiences, including policymakers, healthcare professionals, and the public. Simultaneously, stakeholder participation guaranteed that communication initiatives were pertinent and actionable, fulfilling the urgent needs of communities while tackling complaints and disinformation. Furthermore, the pandemic highlighted the necessity of promptly and efficiently incorporating research findings into practice. Researchers collaborated with policymakers to guarantee their findings were scientifically valid and practical for real-world application [107]. The synchronization of research, policy, and community requirements improved the efficacy of public health initiatives and developed a cooperative atmosphere in which science and society could jointly pursue shared objectives.

3.3.2 The Technology

The distribution of research has been a driving force behind breakthroughs in artificial intelligence (AI) in the technology industry, notably in sectors such as autonomous cars and smart city infrastructure [108]. Stakeholder participation has increased through interactive approaches such as webinars, open-access databases, and collaborative workshops. These tactics have significantly accelerated the incorporation of artificial intelligence (AI) developments into practical applications, so emphasizing the significance of distribution practices that are both accessible and inclusive in the process of accelerating technological change.

3.3.3 Education and Learning

In education, the distribution of research has significantly impacted the creation of curricula and instruction methodology [109]. Educator workshops and interactive guides are two examples of targeted initiatives that have resulted in widespread adoption of the results about digital learning tools. To ensure the successful deployment of these tools stimulates pedagogical innovation and increases learning outcomes, the distribution process needs to be aligned with the unique needs of educators. As an illustration, the distribution of research has also contributed to the advancement of passive building design. In this context, the insights gained from energy-efficient construction processes have influenced architectural practices [110]. For example, Research on thermal insulation and natural ventilation has been extensively disseminated through open-access journals, design guidelines, and industry forums. Because of these efforts, architects and engineers have included environmentally friendly techniques, which have reduced energy consumption and promoted environmental resilience in the building industry. In Fig. 7, we illustrate the influence that the transmission of knowledge has had on three critical areas: education, technology, and public health. Specifically, it covers accomplishments in each area, such as improvements in public health programs, developments in artificial intelligence, innovations in education, and an educational approach. Under each category, the figure explains the operational procedures responsible for achieving these results. For instance, during the COVID-19 pandemic, successes in public health were driven

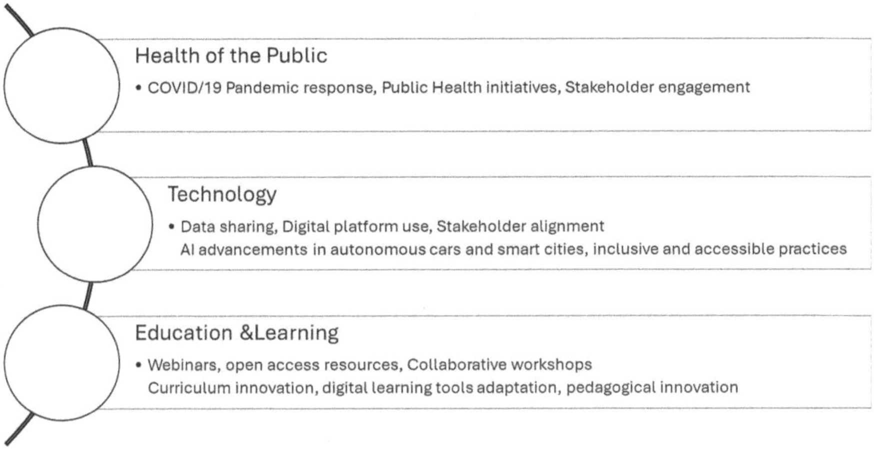

Fig. 7 Case studies of impact and working processes in research dissemination (*Source* Authors)

by prompt data exchange, exploitation of digital platforms, and participation of stakeholders. Additionally, the incorporation of artificial intelligence into practical applications was sped up by using open-access materials and collaborative workshops. The following diagram illustrates how strategic procedures and focused distribution efforts allow research discoveries to be translated into real-world applications that significantly influence various industries.

3.4 *Obstacles to Evaluating the Impact of Research Dissemination*

Evaluating the impact of research dissemination involves several key issues that need to be solved to guarantee a relevant and comprehensive assessment [111]. These problems include the temporal gap between distribution and quantifiable consequences, which can sometimes span years. This can be seen in sectors such as climate policy or sustainable construction practices, where changes in behavior or policy take time to manifest. However, these challenges are not without their challenges. Another challenge is the intricacy of attribution, which is especially problematic when research is disseminated across various venues, such as conference proceedings, journals, and social media. Determining which channels or stakeholders contributed the most to the impact can be challenging, particularly when engaging in collaboration or multidisciplinary initiatives [112]. A thorough assessment incorporating quantitative measures such as citation analysis with qualitative insights derived from user feedback and case studies is necessary to overcome these obstacles. This evaluation should be designed to reflect the full breadth of the impact of distribution.

3.4.1 Temporal Lag

When analyzing the dissemination of research, one of the most significant challenges is the time lag between the dissemination efforts and the consequences that may be observed [113]. This is especially true in public policy, sustainable development, and scientific innovation, where research effects can take years or even decades to become apparent. Because of this delay, it is difficult to determine the real success of distribution tactics within the short period for which the assessment is conducted. For example, it may take many years for research on climate change policy to impact legislative frameworks or the behavior of the general population. Similarly, research that promotes sustainable building methods, such as renovating buildings to be more energy efficient or using low-carbon construction materials, frequently necessitates extended industry adaptation, engagement of stakeholders, and regulatory alignment before meaningful outcomes can be seen. Extensive timeframes can conceal the causal relationships between distribution attempts and the impact that ultimately occurs. Because they cannot consider delayed impacts, short-term assessments are especially prone to underestimate the broader effects of research [113]. The relevance of longitudinal study and continuous monitoring is brought to light by this problem, which demonstrates the need for precise effect evaluation. Researchers monitoring their work overtime can better recognize patterns, pinpoint when impacts become apparent and adjust their distribution techniques accordingly. When the temporal leg is addressed, it enables complete knowledge of the effect of research. It provides information that may inspire future attempts to expand the study's reach and contribute more to society.

3.4.2 Attribution Complexity

Determining the specific factors contributing to research success is another critical challenge. Research is often disseminated through multiple channels—academic journals, conferences, blogs, social media, and more—making it difficult to attribute impact to a specific source [114]. This difficulty is especially pronounced in collaborative initiatives involving numerous stakeholders, where individual contributions may be indistinct or overshadowed by collective efforts. Understanding the relative effectiveness of different dissemination channels requires nuanced, multi-faceted analysis.

3.4.3 Comprehensive Evaluation

Researchers must adopt a holistic approach to evaluating the impact of dissemination to address these challenges. Quantitative measures, such as citation analysis, usage statistics, and altmetrics, provide valuable metrics for assessing reach and influence [84]. Complementing these with qualitative methods—such as user feedback, narrative evidence, and case studies—yields a richer, more nuanced understanding of how research is utilized and perceived. Public health, technology, and education examples highlight successful approaches to overcoming these challenges. They demonstrate that robust evaluation frameworks enhance dissemination strategies and ensure that research contributes meaningfully to societal progress. Overcoming obstacles like temporal delays and attribution difficulties is vital to achieving impactful research dissemination that advances knowledge and fosters positive change [115].

4 Guidelines and Examples for Writing an Effective Results Section

The "Results" part of a thesis or dissertation is an essential component that is intended to convey the findings of your study in an unambiguous, rational, and objective manner [116]. The primary emphasis is the data obtained and analyzed throughout the research; no subjective judgments or interpretations are contained. Thanks to this separation, readers can analyze the data independently before examining the analysis presented in the discussion section. To avoid using subjective phrases such as "appears" or "implies," the findings should be succinct, targeted, and connected to the study questions or hypotheses. Any discoveries not linked to the main topic might be included in an appendix. This part is often written in the past tense to reflect work that has been finished. The essential findings are presented in a well-organized results section, followed by the specifics discussed. This hierarchical arrangement allows the reader to understand the broad themes better before digging into the specifics [117]. An example of this would be research on customer behavior that presents general purchasing trends before analyzing patterns specific to a particular group. Depending on the study, the findings may center on statistical analysis (quantitative) or theme findings (qualitative). Investigating hypotheses using quantitative research is possible by employing statistical tests such as t-tests or regression analysis. It is common practice to present descriptive statistics, such as averages and standard deviations, and inferential statistics, such as p-values and confidence intervals, using visual aids such as graphs and tables to improve clarity. In addition to being immediately relevant, visual aids must be correctly identified and attributed within the text.

It is common practice to report qualitative research findings in a theme format. Each theme draws attention to a significant discovery backed up with proof such as quotations or snippets from the data [100]. Details about the participants, such as their age and gender, serve to contextualize the findings. Appendices are where additional qualitative data can be given to maintain the main text's conciseness. The discussion part is responsible for interpreting the data, considering consequences, and investigating constraints, whereas the results section must maintain objectivity and provide findings without interpretation. Researchers can generate a cohesive and powerful results section that makes a meaningful contribution to their area if they adhere to these principles, which include striking a balance between having sufficient material and being brief, making good use of images, and conforming to style standards [118]. To present their findings in a manner that is both clear and precise, researchers can use Table 4, which offers an organized and complete approach. Some key principles and checklists are outlined in the table for the purpose of organizing and disseminating results while keeping the rigor of scientific research. It emphasizes the significance of objectivity, logical structure, and connection with the study's aims. Among the most important recommendations are the presentation of the findings in a hierarchical fashion, the utilization of suitable graphics, and the utilization of the past tense to guarantee stability. The problems column discusses frequent issues experienced while creating the results section. These challenges include balancing conciseness and comprehensiveness, minimizing repetition, and keeping clear boundaries between the results and their interpretation. The problems presented highlight the areas that require special attention to guarantee coherence and accuracy. The advantages of adhering to certain best practices are highlighted in the opportunity column. Researchers can improve the readability, trustworthiness, and effect of their findings by adhering to the framework that is advised. This methodical approach encourages a more substantial level of involvement with the target audience. It increases the value of the study to the academic subject in which it is being conducted. In general, Table 4 is an invaluable resource for researchers, particularly when it comes to ensuring that the section on the results is methodologically sound, logically presented, and academically rigorous.

5 Conclusion

The dissemination of research findings is an essential component in the process of bridging the gap between academic discovery and practical application. This process involves the transformation of theoretical insights into information that can be put into practice, which in turn influences policy, changes practices, and facilitates innovation. This chapter has provided a detailed analysis of the theoretical frameworks, procedures, and issues that relate to effectively conveying the research results. The chapter has also emphasized the significance of strategic dissemination in contributing to the achievement of societal and academic impact. The theoretical underpinnings, including Knowledge Translation (KT), Diffusion of Innovation, and Stakeholder Engagement and Co-Creation, offer comprehensive models that may be utilized to understand how research results can be effectively translated, shared, and implemented. The necessity for individualized ways of distribution is brought to light by these frameworks. This ensures that research is accessible but also relevant and practical for a wide range of audiences. KT frameworks, for example, emphasize adapting results to unique circumstances. Still, the Diffusion of Innovation theory emphasizes the necessity of engaging multiple adopter types to enable wider adoption and utilization of new ideas. Several fields, including public health, technology, and education, were investigated to examine the feasibility of various diffusion tactics. The purpose of these case studies was to demonstrate how utilizing targeted distribution strategies, the participation of stakeholders, and implementing novel communication methods

Table 4 Comprehensive framework for structuring and presenting research results

Section	Checklist/guidelines	Challenges	Opportunities
Introduction to results section [119]	– State findings objectively, free from subjective terms like "appears" or "implies." – Use past tense consistently – Avoid interpretation of data in this section	Avoiding subjective interpretations while maintaining readability and clarity in presenting findings	Establishing a strong foundation for the credibility and reliability of the research
The framework of the results section [120]	– Organize findings logically, starting with general trends and narrowing to specifics – Ensure data presentation follows a logical sequence aligned with research questions	Ensuring logical progression while avoiding redundancy in presenting findings	Enhancing reader understanding by guiding them through the findings in a structured and logical manner
Logical organization [54]	– Present general trends first, followed by specific, detailed findings – Use subheadings for clarity and structure – Ensure results align with hypotheses or research questions	Balancing detailed analysis with an overview to prevent overwhelming the reader	A straightforward narrative engages the reader and facilitates comprehension of complex findings
Comparing qualitative and quantitative results [121]	– Tailor results presentation based on the research type – Highlight statistical results (quantitative) or thematic patterns (qualitative) – Avoid mixing findings from different methods unless integrated analysis is required	Effectively integrating qualitative themes with quantitative data for comprehensive understanding	Demonstrating the complementary strengths of both qualitative and quantitative approaches

(continued)

Table 4 (continued)

Section	Checklist/guidelines	Challenges	Opportunities
Reporting quantitative research [122]	– Restate hypotheses or research questions before presenting findings – Summarize analysis methods (e.g., t-tests, regressions) – Include descriptive (e.g., means) and inferential (e.g., p-values) statistics	Presenting complex statistical data in a concise and accessible format for all audiences	Providing precise, evidence-based support for hypotheses and research questions
Statistical analysis [123]	– Report statistical outcomes, including effect sizes and confidence intervals – Ensure results are replicable and avoid unnecessary technical jargon – Include significance levels for inferential tests	Avoiding technical jargon while accurately reporting statistical significance and reliability	Offering robust and replicable insights that strengthen the study's contributions to the field
Visual aids [124]	– Use tables for precise data and graphs for trends – Include clear titles, axis labels, and legends – Reference visuals in the text for context	Designing visuals that enhance understanding without duplicating information in the text	Creating engaging and visually appealing representations that aid in data comprehension
Reporting qualitative research [100]	– Organize findings thematically, supported by direct evidence – Include participant details for context (e.g., age, gender, occupation) – Use quotations to highlight key insights	Ensuring representative and unbiased interpretation of qualitative data while maintaining reader engagement	Highlighting the richness of human experiences and perspectives to provide deeper insights
Organization of themes [125]	– Group findings into clear, logical themes – Begin with broad themes, narrowing down to specific insights – Use subheadings to delineate themes	Avoiding thematic overlaps and maintaining coherence across themes	Demonstrating the interconnectedness of themes to build a cohesive narrative

(continued)

Table 4 (continued)

Section	Checklist/guidelines	Challenges	Opportunities
Evidence for themes [126]	– Provide supporting evidence (e.g., quotes or excerpts) for each theme – Ensure evidence is representative and relevant to the research questions – Contextualize data within the study's scope	Selecting illustrative quotes or examples without overloading the text with excessive evidence	Offering vivid, real-world insights that enhance the authenticity and depth of the findings
Results, discussion, and conclusion [116]	– Separate results from the discussion and conclusion – Avoid interpretative statements in the results section – Ensure logical flow between sections	Maintaining clear boundaries between reporting findings and interpreting their implications	Facilitating a seamless transition from reporting to interpreting and contextualizing findings
Challenges in writing results [127]	– Balance detail with conciseness – Prioritize results that directly address research questions – Use visuals efficiently and avoid redundancy	Striking the right balance between brevity and comprehensiveness while effectively using visuals	Refining the communication of findings to cater to both expert and non-expert audiences
Recommendations [128]	– Adhere to a logical structure, using past tense – Exclude subjective interpretations – Follow style guide requirements (e.g., APA, MLA)	Following style guide requirements consistently while maintaining a logical and concise flow	Aligning with established standards to ensure professional and academic integrity
Closing remarks [129]	– Maintain objectivity and clarity – Use visuals to enhance, not repeat, textual content – Distinctly separate results, discussion, and conclusion	Avoiding overlaps between results, discussion, and conclusion to ensure clarity and readability	Enhancing the clarity of the research's contributions and ensuring its relevance to the academic community

may double the reach and effect of research. In public health, for instance, the quick transmission of epidemiological data during the COVID-19 pandemic highlighted the crucial role that timely communication matched with stakeholders' interests plays in driving effective responses. Similarly, developments in artificial intelligence (AI) and sustainable educational methods have shown that strategic research distribution has the capacity to bring about a transformational change.

One of the recurrent ideas that is discussed throughout the chapter is the need to ensure that distribution efforts are in line with the requirements and objectives of end-users. Because of this alignment, the findings of the research are not only disseminated but also utilized efficiently. The participation of stakeholders and the co-creation of content emerged as crucial components in accomplishing this objective, which aimed to cultivate trust, shared ownership, and the continued implementation of research findings. Researchers can improve the relevance, acceptability, and durability of their work by facilitating stakeholders' participation in designing and implementing dissemination strategies. Evaluating the impact of the dissemination of research is a task that is equally necessary and difficult. Several factors, including temporal delays and attribution complexity, can impede the appropriate evaluation of results. However, a comprehensive knowledge of the efficacy of distribution may be obtained by applying a combination of qualitative approaches and quantitative measurements, such as narrative evidence and user feedback. Examples of quantitative metrics include citation analysis and altmetrics. Not only does this dual approach consider the immediate and physical benefits of research, but it also considers the larger and more long-term implications that research has for society.

Despite the apparent advantages, several challenges still stand in the way of the efficient distribution of research evidence. The time lag between the distribution efforts and the observable consequences, particularly in domains such as climate policy and sustainable construction practices, underscores the necessity of patience and long-term monitoring. An additional factor that makes the review process more difficult is the complexity of attribution, which results from the utilization of different dissemination channels. To effectively address these problems, comprehensive assessment frameworks are required. These frameworks must incorporate a variety of approaches and take into consideration the multidimensional character of research impact. It has become more apparent that digital and open-access approaches are significant instruments that may be utilized to broaden the scope of the study. Knowledge is democratized using open-access platforms, preprint repositories, webinars, podcasts, and interactive visualizations. This ensures that individuals from all over the world have access to the findings of research initiatives. Altmetrics provide vital insights into how research is debated and disseminated in digital environments. These metrics should be considered in addition to standard citation metrics. These technologies can improve visibility and encourage cooperation and invention across a wide range of specializations and geographical locations. Additionally, the chapter emphasized the significant part that community and public participation play in disseminating research. Researchers can make their work accessible to audiences who are not specialists by explaining difficult results and using creative technologies such as infographics, blogs, and interactive visualizations. Researchers may further improve the relevance and application of their findings by forming collaborative collaborations with community groups. This helps to ensure that the study addresses the needs and goals of the actual world. Ultimately, the chapter emphasizes the significance of taking a strategic, inclusive, and evaluative approach to disseminating research. Researchers can increase the effect of their work by combining theoretical ideas with practical strategies and overcoming the inherent obstacles present in scientific research. This chapter presents a route for effectively turning academic results into meaningful, real-world consequences, supporting advancement across various areas, and contributing to improving society. The tactics and frameworks that are covered in this chapter give a roadmap.

References

1. National Academies of Sciences, Division of Behavioral, Social Sciences, Committee on the Science of Science Communication and A Research Agenda (2017) Communicating science effectively: A research agenda
2. Loureiro SMC, Romero J, Bilro RG (2020) Stakeholder engagement in co-creation processes for innovation: a systematic literature review and case study. J Bus Res 119:388–409
3. Bandura ALBERT (2006) On integrating social cognitive and social diffusion theories 1. In: Communication of innovations: a journey with Ev Rogers. Sage Publications India Pvt Ltd, pp 111–135
4. Cole NL, Reichmann S, Ross-Hellauer T (2023) Toward equitable open research: stakeholder co-created recommendations for research institutions, funders and researchers. Royal Society Open Science 10(2):221460
5. Olson A, Oudshoorn A (2020) Knowledge translation: a concept analysis. Nurs Forum 55(2):157–164
6. Zhang L, Banihashemi S, Zhu L, Molavi H, Odacioglu E, Shan M (2024) A scientometric analysis of knowledge transfer partnerships in digital transformation. J Open Innov: Technol, Mark, Complex 100325
7. Koedinger KR, Corbett AT, Perfetti C (2012) The Knowledge-learning-instruction framework: bridging the science-practice chasm to enhance robust student learning. Cogn Sci 36(5):757–798
8. Pearce S (2023) Exploring the role of knowledge synthesis and knowledge translation methodologies in one health. Doctoral dissertation, University of Guelph
9. Provvidenza C, Kingsnorth S, Dawson J, Ledoux AA, Zemek R, Reed N (2022) Applying knowledge translation frameworks, approaches and principles to co-create a dissemination plan: optimizing the spread of a pediatric concussion guideline. J Concussion 6:20597002221116240
10. Langley J, Wolstenholme D, Cooke J (2018) 'Collective making'as knowledge moustralian-nthe contribution of participatory design in the co-creation of knowledge in healthcare. BMC Health Serv Res 18:1–10
11. Ferreira C (2024) Empower collaboration by developing KAMK living labs to connect students and companies
12. Stetler CB, Damschroder LJ, Helfrich CD, Hagedorn HJ (2011) A guide for applying a revised version of the PARIHS framework for implementation. Implement Sci 6:1–10
13. Damba FU, Mtshali NG, Chimbari MJ (2023) Development of a conceptual framework to guide uptake of PhD-generated knowledge by policymakers. Open Public Health J 16(1)
14. Campbell B (2010) Applying knowledge to generate action: a community-based knowledge translation framework. J Contin Educ Heal Prof 30(1):65–71

15. Leask CF, Sandlund M, Skelton DA, Altenburg TM, Cardon G, Chinapaw MJ, De Bourdeaudhuij I, Verloigne M, Chastin SF (2019) GrandStand, safe step and teenage girls on the move research groups. Framework, principles and recommendations for utustraliaarticipatory methodologies in the co-creation and evaluation of public health interventions. Res Involv Engag 5:1–16
16. Bamberger M, Mabry L (2019) RealWorld evaluation: working under budget, time, data, and political constraints. Sage publications
17. Roberts N (2021) Implementing patient: reported outcomes in a medical oncology setting. Doctoral dissertation, Queensland University of Technology
18. Clark CE, Cavanaugh NC, Brown CV, Sambamurthy V (1997) Building change-readiness capabilities in the IS organization: insights from the Bell Atlantic experience. MIS Q 425–455
19. Curtis K, Fry M, Shaban RZ, Considine J (2017) Translating research findings to clinical nursing practice. J Clin Nurs 26(5–6):862–872
20. Gagliardi AR, Berta W, Kothari A, Boyko J, Urquhart R (2015) Integrated knowledge translation (IKT) in health care: a scoping review. Implement Sci 11:1–12
21. Mussehl ML, Horne AC, Webb JA, Poff NL (2022) Purposeful stakeholder engagement for improved environmental flow outcomes. Front Environ Sci 9:749864
22. Ayu R, Abdullah NA, Sulaiman WSW, Selamat MNB (2023) The analysis of organizational changes using structural equation modelling with mediating readiness to change in higher education. JOIV: Int J Inform Vis 7(4):2387–2393
23. Tatineni S, Allam K (2024) AI-driven continuous feedback mechanisms in DevOps for proactive performance optimization and user experience enhancement in software development. J AI Healthc Med 4(1):114–151
24. Mallidou AA, Atherton P, Chan L, Frisch N, Glegg S, Scarrow G (2018) Core knowledge translation competencies: a scoping review. BMC Health Serv Res 18:1–15
25. Berrone P, Rousseau HE, Ricart JE, Brito E, Giuliodori A (2023) How can research contribute to the implementation of sustainable development goals? an interpretive review of SDG literature in management. Int J Manag Rev 25(2):318–339
26. da Silva Miranda É (2021) Knowledge translation in public health, a case study in Manguinhos, Brazil
27. Dryden-Palmer KD, Parshuram CS, Berta WB (2020) Context, complexity and process in the implementation of evidence-based innovation: a realist informed review. BMC Health Serv Res 20:1–15
28. Graham ID, Tetroe JM (2010) The knowledge to action framework. Model Fram Implement Evid-Based Pract: Link Evid Action 207:222
29. Luke DA, Powell BJ, Paniagua-Avila A (2024) Bridges and mechanisms: integrating systems science thinking into implementation research. Annu Rev Public Health 45
30. Pollock A, Campbell P, Struthers C, Synnot A, Nunn J, Hill S, Goodare H, Morris J, Watts C, Morley R (2019) Development of the ACTIVE framework to describe stakeholder involvement in systematic reviews. J Health Serv Res Policy 24(4):245–255
31. Van Hoof J, Rutten PG, Struck C, Huisman ER, Kort HS (2015) The integrated and evidence-based design of healthcare environments. Arch Eng Des Manag 11(4):243–263
32. Viterouli M, Belias D, Koustelios A, Tsigilis N, Papademetriou C (2024) Time for change: designing tailored training initiatives for organizational transformation. In: Organizational behavior and human resource management for complex work environments. IGI Global, pp 267–307
33. Iflaifel M, Lim RH, Ryan K, Crowley C (2020) Resilient health care: a systematic review of conceptualisations, study methods and factors that develop resilience. BMC Health Serv Res 20:1–21
34. Saunders H (2016) Nurses' readiness for evidence-based practice: implementing the paradigm shift of transforming evidence for clinical practice. Doctoral dissertation, Itä-Suomen yliopisto
35. Curtis M (2020) Toward understanding secondary teachers' decisions to adopt geospatial technologies: an examination of Everett Rogers' diffusion of innovation framework. J Geogr 119(5):147–158

36. Miller RL (2015) Rogers' innovation diffusion theory (1962, 1995). In Information seeking behavior and technology adoption: theories and trends. IGI Global, pp 261–274
37. Talke K, Heidenreich S (2014) How to overcome pro-change bias: incorporating passive and active innovation resistance in innovation decision models. J Prod Innov Manag 31(5):894–907
38. Gomes R, Osman SS (2019) Managing organizational adoption of IoT: revisiting rogers' 'iffusion of innovation theory
39. Brdesee H (2013) Exploring factors impacting e-commerce adoption in tourism industry in Saudi Arabia. Doctoral dissertation, RMIT University
40. Rodgers S, Stemmle J (2020) Are "well-told" stories of cancer worn out? insights on persuasion characteristics used in cancer narrative PSAs. J Curr Issues & Res Advert 41(3):257–283
41. Rai A, Brown P, Tang X (2009) Organizational assimilation of electronic procurement innovations. J Manag Inf Syst 26(1):257–296
42. Mlecnik E (2013) Innovation development for highly energy-efficient housing: opportunities and challenges related to the adoption of passive houses
43. Ishimwe C, Sanja M (2024) Stakeholder engagement and successful implementation of social economic development project: a case of Jyambere project. J Entrep & Proj Manag 8(4):27–49
44. Patriotta G, Gond JP, Schultz F (2011) Maintaining legitimacy: controversies, orders of worth, and public justifications. J Manage Stud 48(8):1804–1836
45. Oetzel JG, Wallerstein N, Duran B, Sanchez-Youngman S, Nguyen T, Woo K, Wang J, Schulz A, Keawe 'aimoku Kaholokula J, Israel B, Alegria M (2018) Impact of participatory health research: a test of the community-based participatory research conceptual model. Biomed Res Int 2018(1):7281405
46. Pérez Jolles M, Willging CE, Stadnick NA, Crable EL, Lengnick-Hall R, Hawkins J, Aarons GA (2022) Understanding implementation research collaborations from a co-creation lens: recommendations for a path forward. Front Health Serv 2:942658
47. Zuniga-Teran AA, Staddon C, de Vito L, Gerlak AK, Ward S, Schoeman Y, Hart A, Booth G (2020) Challenges of mainstreaming green infrastructure in built environment professions. J Environ Planning Manage 63(4):710–732
48. Bogenschneider K, Corbett T (2011) Evidence-based policymaking: insights from policy-minded researchers and research-minded policymakers. Routledge
49. Näkki P, Bäck A, Ropponen T, Kronqvist J, Hintikka KA, Harju A, Pöyhtäri R, Kola P (2011) Social media for citizen participation, vol 755. Report on the Somus project, VTT publications
50. Manca S (2018) ResearchGate and academia. edu as networked socio-technical systems for scholarly communication: a literature review. Res Learn Technol 26
51. Marín-González E, Malmusi D, Camprubí L, Borrell C (2017) The role of dissemination as a fundamental part of a research project: lessons learned from SOPHIE. Int J Health Serv 47(2):258–276
52. Shen W, Hao Q, Mak H, Neelamkavil J, Xie H, Dickinson J, Thomas R, Pardasani A, Xue H (2010) Systems integration and collaboration in architecture, engineering, construction, and facilities management: a review. Adv Eng Inform 24(2):196–207
53. Brem A, Bilgram V, Marchuk A (2019) How crowdfunding platforms change the nature of user innovation–from problem-solving to entrepreneurship. Technol Forecast Soc Chang 144:348–360
54. Piran MJ, Tran NH (2024) Enhancing research methodology and academic publishing: a structured framework for quality and integrity. arXiv:2412.05683
55. Kuo CL (2023) Conferencing and presentation strategies for young academics in the global landscape. Arch Otorhinolaryngol-Head & Neck Surg (AOHNS)
56. Valeiras-Jurado J (2017) A multimodal approach to persuasion in oral presentations: the case of conference presentations, research dissemination talks and product pitches. Doctoral dissertation, Ghent University
57. Marsh RM (2015) The role of institutional repositories in developing the communication of scholarly research. OCLC Syst & Serv: Int Digit Libr Perspect 31(4):163–195

58. Schummer J (2004) Multidisciplinarity, interdisciplinarity, and patterns of research collaboration in nanoscience and nanotechnology. Scientometrics 59:425–465
59. Vicente-Saez R, Gustafsson R, Van den Brande L (2020) The dawn of an open exploration era: emergent principles and practices of open science and innovation of university research teams in a digital world. Technol Forecast Soc Chang 156:120037
60. Lavis JN (2006) Research, public policymaking, and knowledge-translation processes: Canadian efforts to build bridges. J Contin Educ Heal Prof 26(1):37–45
61. Davis G, Althaus C (2020) The auustralianolicy handbook: a practical guide to the policy making process. Routledge
62. Baba VV, HakemZadeh F (2012) Toward a theory of evidence based decision making. Manag Decis 50(5):832–867
63. Mallonee S, Fowler C, Istre GR (2006) Bridging the gap between research and practice: a continuing challenge. Inj Prev 12(6):357–359
64. Grimshaw JM, Eccles MP, Lavis JN, Hill SJ, Squires JE (2012) Knowledge translation of research findings. Implement Sci 7:1–17
65. Arnautu D, Dagenais C (2021) Use and effectiveness of policy briefs as a knowledge transfer tool: a scoping review. HumIties Soc Sci Commun 8(1):1–14
66. Deverka PA, Lavallee DC, Desai PJ, Esmail LC, Ramsey SD, Veenstra DL, Tunis SR (2012) Stakeholder participation in comparative effectiveness research: defining a framework for effective engagement. J Comp Eff Res 1(2):181–194
67. Nguyen NHD (2023) Translation and/in development: promoting more effective policy Interventions in Vietnam. Doctoral dissertation, Dublin City University
68. Pacheco R, Ordóñez J, Martínez G (2012) Energy efficient design of building: a review. Renew Sustain Energy Rev 16(6):3559–3573
69. Wentworth L, Mazzeo C, Connolly F (2017) Research practice partnerships: a strategy for promoting evidence-based decision-making in education. Educ Res 59(2):241–255
70. Innes JE, Booher DE (2004) Reframing public participation: strategies for the 21st century. Plan Theory Pract 5(4):419–436
71. Macpherson H, Davies C, Hart A, Eryigit-Madzwamuse S, Rathbone A, Gagnon E, Buttery L, Dennis S (2017) Collaborative community research dissemination and networking: experiences and challenges. GatewS: Int J Community Res Engag 10:298–312
72. Thapliyal K, Thapliyal M, Thapliyal D (2024) Social media and health communication: A review of advantages, challenges, and best practices. Emerg Technol Health Lit Med Pract 364–384
73. Isiaka AO, Soliu A, Aremu BA, Bamidele BA, Saba-Jibril S, Ibitoye AR (2024) The evolving role of libraries in the fourth industrial revolution: navigating digital transformation. Library Philosophy and Practice, pp 1–26
74. Onunka O, Onunka T, Fawole AA, Adeleke IJ, Daraojimba C (2023) Library and information services in the digital age: opportunities and challenges. Acta Inform Malays 7(1):113–121
75. Yue T, Au D, Au CC, Iu KY (2023) Democratizing financial knowledge with ChatGPT by OpenAI: unleashing the power of technology. SSRN 4346152
76. Das MK (2024) Teacher education with ICT tools and techniques. Academic Guru Publishing House
77. Olive R, Townsend S, Phillips MG (2023) 'Not everything that can be counted counts, and not everything that counts can be counted': searching for the value of metrics and altmetrics in sociology of sport journals. Int Rev Sociol Sport 58(3):431–454
78. Gómez-Carmona O, Buján-Carballal D, Casado-Mansilla D, López-de-Ipiña D, Cano-Benito J, Cimmino A, Poveda-Villalón M, García-Castro R, Almela-Miralles J, Apostolidis D, Drosou A (2023) Mind the gap: the AURORAL ecosystem for the digital transformation of smart communities and rural areas. Technol Soc 74:102304
79. Yitmen I, Almusaed A, Alizadehsalehi S (2024) Facilitating construction 5.0 for smart, sustainable and resilient buildings: opportunities and challenges for implementation. Smart Sustain Built Environ

80. Vinsensia D, Amri S, Sihotang J, Sihotang HT (2024) New method for identification and response to infectious disease patterns based on comprehensive health service data. MATRIK: J Manaj, Tek Inform Dan Rekayasa Komput 23(3):583–592
81. Rositch AF, Unger-Saldaña K, DeBoer RJ, Ng'ang'a A, Weiner BJ (2020) The role of dissemination and implementation science in global breast cancer control programs: Frameworks, methods, and examples. Cancer 126:2394–2404
82. Petersson L, Larsson I, Nygren JM, Nilsen P, Neher M, Reed JE, Tyskbo D, Svedberg P (2022) Challenges to implementing artificial intelligence in healthcare: a qualitative interview study with healthcare leaders in Sweden. BMC Health Serv Res 22(1):850
83. Skivington K, Matthews L, Simpson SA, Craig P, Baird J, Blazeby JM, Boyd KA, Craig N, French DP, McIntosh E, Petticrew M (2021) Framework for the development and evaluation of complex interventions: gap analysis, workshop and consultation-informed update. Health Technol Assess (Winch, Engl) 25(57):1
84. Kavic MS, Satava RM (2021) Scientific literature and evaluation metrics: impact factor, usage metrics, and altmetrics. JSLS: J Soc Laparosc & Robot Surg 25(3)
85. Del Rocío Bonilla M, Perea E, del Olmo JL, Corrons A (2020) Insights into user engagement on social media. Case study of a higher education institution. J Mark High Educ 30(1):145–160
86. Passas I (2024) Bibliometric analysis: the main steps. Encyclopedia 4(2)
87. Anderson MH, Lemken RK (2023) Citation context analysis as a method for conducting rigorous and impactful literature reviews. Organ Res Methods 26(1):77–106
88. Ye G, Hudders L, De Jans S, De Veirman M (2021) The value of influencer marketing for business: a bibliometric analysis and managerial implications. J Advert 50(2):160–178
89. Wang X, Feng X (2024) Research on the relationships between discourse leading indicators and citations: perspectives from altmetrics indicators of international multidisciplinary academic journals. Library Hi Tech 42(4):1165–1190
90. Gholampour S, Lim WM, Lund BD, Noruzi A, Elahi A, Saboury AA, Nawaz R, Gholampour B (2024) Does social media contribute to research impact? an altmetric study of highly-cited marketing research. Total Qual Manag Bus Excell 35(13–14):1671–1701
91. Williams K (2022) What counts: making sense of metrics of research value. Sci Public Policy 49(3):518–531
92. Cresswell K, Anderson S, Montgomery C, Weir CJ, Atter M, Williams R (2023) Evaluation of digitalisation in healthcare and the quantification of the "unmeasurable." J Gen Intern Med 38(16):3610–3615
93. Thelwall M, Nevill T (2021) Is research with qualitative data more prevalent and impactful now? interviews, case studies, focus groups and ethnographies. Libr Inf Sci Res 43(2):101094
94. McNamara KE, Clissold R, Westoby R, Piggott-McKellar AE, Kumar R, Clarke T, Namoumou F, Areki F, Joseph E, Warrick O, Nunn PD (2020) An assessment of community-based adaptation initiatives in the Pacific Islands. Nat Clim Chang 10(7):628–639
95. Smit JP, Hessels LK (2021) The production of scientific and societal value in research evaluation: a review of societal impact assessment methods. Res Eval 30(3):323–335
96. Maurer M, Mangrum R, Hilliard-Boone T, Amolegbe A, Carman KL, Forsythe L, Mosbacher R, Lesch JK, Woodward K (2022) Understanding the influence and impact of stakeholder engagement in patient-centered outcomes research: a qualitative study. J Gen Intern Med 37(Suppl 1):6–13
97. Shah A (2011) User involvement in medical device development: an empirical study (Doctoral dissertation. Brunel University, School of Information Systems, Computing and Mathematics
98. Osborne N, Grant-Smith D (2021) In-depth interviewing. In: Methods in urban analysis. Springer Singapore, Singapore, pp 105–125
99. Nardi PM (2018) Doing survey research: a guide to quantitative methods. Routledge
100. Tracy SJ (2024) Qualitative research methods: collecting evidence, crafting analysis, communicating impact. John Wiley & Sons
101. May T, Perry B (2022) Social research: issues, methods and process. McGraw-Hill Education (UK)

102. Good C (2010) Persuasive effect of narrative and statistical evidence combinations. Doctoral dissertation, Kansas State University
103. Bergmann M, Schäpke N, Marg O, Stelzer F, Lang DJ, Bossert M, Gantert M, Häußler E, Marquardt E, Piontek FM, Potthast T (2021) Transdisciplinary sustainability research in real-world labs: success factors and methods for change. Sustain Sci 16:541–564
104. Davies HT, Nutley S (2008) Learning more about how research-based knowledge gets used. Unpublished Working Paper. William T. Grant Foundation
105. Chan M (2018) Ten years in public health 2007–2017: report by dr margaret chan director-general world health organization. World Health Organization
106. Lindell MK, Perry RW (2012) The protective action decision model: theoretical modifications and additional evidence. Risk Anal: Int J 32(4):616–632
107. Cook NL, Lauer MS (2021) Biomedical research COVID-19 impact assessment: lessons learned and compelling needs. NAM perspectives
108. Allam Z, Allam Z (2021) Big data, artificial intelligence and the rise of autonomous smart cities. In: The rise of autonomous smart cities: technology, economic performance and climate resilience. pp 7–30
109. Reeves TC, Herrington J, Oliver R (2005) Design research: a socially responsible approach to instructional technology research in higher education. J Comput High Educ 16:96–115
110. Umoh AA, Adefemi A, Ibewe KI, Etukudoh EA, Ilojianya VI, Nwokediegwu ZQS (2024) Green architecture and energy efficiency: a review of innovative design and construction techniques. Eng Sci & Technol J 5(1):185–200
111. Tabak RG, Khoong EC, Chambers DA, Brownson RC (2012) Bridging research and practice: models for dissemination and implementation research. Am J Prev Med 43(3):337–350
112. Podestá GP, Natenzon CE, Hidalgo C, Toranzo FR (2013) Interdisciplinary production of knowledge with participation of stakeholders: a case study of a collaborative project on climate variability, human decisions and agricultural ecosystems in the Argentine Pampas. Environ Sci Policy 26:40–48
113. Brownson RC, Colditz GA, Proctor EK (eds) (2018) Dissemination and implementation research in health: translating science to practice. Oxford University Press
114. Powell DA, Jacob CJ, Chapman BJ (2012) Using blogs and new media in academic practice: potential roles in research, teaching, learning, and extension. Innov High Educ 37:271–282
115. Morrison-Smith S, Ruiz J (2020) Challenges and barriers in virtual teams: a literature review. SN Appl Sci 2(6):1–33
116. Turabian KL (2018) A manual for writers of research papers, theses, and dissertations: Chicago style for students and researchers. University of Chicago Press
117. Leong FT, Austin JT (eds) (2023) The psychology research handbook: a guide for graduate students and research assistants. SAGE publications
118. Lidwell W, Holden K, Butler J (2010) Universal principles of design revised and updated: 125 ways to enhance usability, influence perception, increase appeal, make better design decisions, and teach through design. Rockport Pub
119. Pennings P, Kleinnijenhuis J, Keman H (2005) Doing research in political science: an introduction to comparative methods and statistics
120. Ravitch SM, Riggan M (2016) Reason & rigor: how conceptual frameworks guide research. Sage Publications
121. Curry LA, Nembhard IM, Bradley EH (2009) Qualitative and mixed methods provide unique contributions to outcomes research. Circulation 119(10):1442–1452
122. Alem DD (2020) An overview of data analysis and interpretations in research. Int J Acad Res Educ Rev 8(1):1–27
123. Pek J, Flora DB (2018) Reporting effect sizes in original psychological research: a discussion and tutorial. Psychol Methods 23(2):208
124. Yau N (2024) Visualize this: the flowing data guide to design, visualization, and statistics. John Wiley & Sons
125. Dougherty D (2017) Grounded theory research methods. The Blackwell companion to organizations. pp 849–866

126. Bandara W, Furtmueller E, Gorbacheva E, Miskon S, Beekhuyzen J (2015) Achieving rigor in literature reviews: insights from qualitative data analysis and tool-support. Commun Assoc Inf Syst 37(1):8
127. Wallwork A (2024) English for academic research: writing exercises. Springer Nature
128. Sternberg RJ (2003) The psychologist's companion: a guide to scientific writing for students and researchers. Cambridge University Press
129. Bazerman C, Prior P (eds) (2003) What writing does and how it does it: an introduction to analyzing texts and textual practices. Routledge

MIX
Papier aus verantwortungsvollen Quellen
Paper from responsible sources
FSC® C105338

If you have any concerns about our products,
you can contact us on
ProductSafety@springernature.com

In case Publisher is established outside the EU,
the EU authorized representative is:
Springer Nature Customer Service Center GmbH
Europaplatz 3, 69115 Heidelberg, Germany

Printed by Libri Plureos GmbH
in Hamburg, Germany